C0-AVJ-396

**Prentice Hall Advanced Reference Series**

*Engineering*

ANTOGNETTI AND MILUTINOVIC, EDS. *Neural Networks: Concepts, Applications, and Implementations, Volumes I, II, III, and IV*
DENNO *Power System Design and Applications for Alternative Energy Sources*
ESOGBUE, ED. *Dynamic Programming for Optimal Water Resources Systems Analysis*
FERRY, AKERS, AND GREENEICH *Ultra Large Scale Integrated Microelectronics*
GNADT AND LAWLER, EDS. *Automating Electric Utility Distribution Systems: The Athens Automation and Control Experiment*
HALL *Biosensors*
HAYKIN, ED. *Advances in Spectrum Analysis and Array Processing, Volumes I and II*
HAYKIN, ED. *Selected Topics in Signal Processing*
HENSEL *Inverse Theory and Applications for Engineers*
JOHNSON *Lectures on Adaptive Parameter Estimation*
MILUTINOVIC, ED. *Microprocessor Design for GaAs Technology*
QUACKENBUSH, BARNWELL III, AND CLEMENTS *Objective Measures of Speech Quality*
ROFFEL, VERMEER, AND CHIN *Simulation and Implementation of Self-Tuning Controllers*
SASTRY AND BODSON *Adaptive Control: Stability, Convergence, and Robustness*
SWAMINATHAN AND MACRANDER *Materials Aspects of GaAs and InP Based Structures*
TOMBS *Biotechnology in the Food Industry*

# Materials Aspects of GaAs and InP Based Structures

V. Swaminathan

*AT&T Bell Laboratories*
*Breinigsville, Pennsylvania*

A. T. Macrander

*Argonne National Laboratory*
*Argonne, Illinois*

Prentice Hall, Englewood Cliffs, New Jersey 07632

*Library of Congress Cataloging-in-Publication Data*

Swaminathan, V.
Materials aspects of GaAs and InP based structures / by
V. Swaminathan, A. T. Macrander.
p. cm.
Includes bibliographical references and index.
ISBN 0-13-346826-7
1. Semiconductors—Materials. 2. Gallium arsenide semiconductors.
3. Epitaxy. I. Macrander, A. T. II. Title.
TK7871.85.S937 1991 90-41035
621.381'52—dc20 CIP

Cover design: *Karen A. Stephens*
Manufacturing buyer: *Kelly Behr* and *Susan Brunke*

**Prentice Hall Advanced Reference Series**

Published by Prentice-Hall, Inc.
A Simon & Schuster Company
Englewood Cliffs, New Jersey 07632

Printed in the United States of America
10 9 8 7 6 5 4 3 2 1

ISBN 0-13-346826-7

PRENTICE-HALL INTERNATIONAL (UK) LIMITED, *London*
PRENTICE-HALL OF AUSTRALIA PTY. LIMITED, *Sydney*
PRENTICE-HALL CANADA INC., *Toronto*
PRENTICE-HALL HISPANOAMERICANA, S.A., *Mexico*
PRENTICE-HALL OF INDIA PRIVATE LIMITED, *New Delhi*
PRENTICE-HALL OF JAPAN, INC., *Tokyo*
SIMON & SCHUSTER ASIA PTE. LTD., *Singapore*
EDITORA PRENTICE-HALL DO BRASIL, LTDA., *Rio de Janeiro*

# CONTENTS

# PREFACE

Of all the III-V compounds GaAs and InP occupy the center stage today. These compounds and the related III-V alloys are important because they are used to make a variety of photonic and electronic devices. Current research and development in these compounds are focussed on areas of improved materials growth, development of suitable process technologies, fabrication of novel devices using artificially layered structures and better understanding of degradation mechanisms in electronic and photonic devices for improved device reliability. Significant progress has been made in these areas but further work is needed on many issues. At present the research direction is determined by the devices and often good devices are linked with good "quality" material. But the quality metric for the material varies and depends on the device. For a p-i-n photodiode, it is important to tailor growth parameters to achieve a low background doping level. On the other hand in a laser, minimization of defect density in the active layer is important. While the basic material quality is determined by the crystal growth conditions it is profoundly influenced by the process conditions and also the conditions under which the device is operated. A good example for the latter case is the phenomenon of recombination enhanced motion of defects in lasers. In making a good device it is important to understand the material issues that are related to device fabrication and device operation and to achieve synergies between material preparation, processing and device functions. Once suitable quality metrics are identified, necessary material characterization techniques must be employed to select wafers for further processing. These selection procedures serve two purposes; one, they provide crystal growers valuable feedback and second, they weed out poor wafers from further processing and thus save time and money. To develop these strategies an understanding of the materials aspects of the semiconductors is crucial.

The purpose of writing this book is to provide a comprehensive treatment of the materials aspects of GaAs, InP and some related alloys used in the fabrication of photonic and electronic devices. The existing books that deal primarily with these devices give only an incomplete treatment of the materials issues. For example, discussions on point defects and dislocations in III-V compounds are lacking, for the most part, from comparable texts. We hope that this book covering exclusively the materials aspects complements the books written with a device flavor. We have mainly concentrated on GaAs, InP and the alloys AlGaAs, GaInAs and GaInAsP. We have omitted the Sb based compounds or the alloys like AlInAs, GaAsP. But the basic concepts which we have covered should apply equally well to these material systems.

The book is intended for researchers in the field of III-V compounds, in particular, GaAs and InP. It should serve as a reference book and can also provide the basis for a graduate level course on materials aspects of semiconductors. We have given a large number of references useful for further study.

In selecting topics we have been guided by our own research experience. Accordingly we have omitted diffusion and ion implantation except in the contexts of out-diffusion of impurities and defects and implantation damage gettering. We have a limited discussion of superlattices and quantum wells. We have concentrated on topics such as material characterization techniques, defects, impurities, and effects of defects on devices. We felt that these have not been adequately covered in existing books.

The book is organized as follows. Chapter 1 discusses bonding and the crystal lattice and band structure of the binary compounds and their alloys. Physical properties of these materials are summarized in several tables. The physical basis for obtaining the relevant parameters in the alloys is given. Chapter 2 covers various crystal growth techniques. These are bulk crystal growth, liquid phase epitaxy, hydride vapor phase epitaxy, chloride vapor phase epitaxy, metal organic chemical vapor deposition, molecular beam epitaxy(MBE), gas source MBE and chemical beam epitaxy. Chapters 3 through 5 cover the structural, electrical and optical characterization techniques. Chapter 6 deals with defects and impurities. It covers shallow impurities, deep impurities, theories of defects, and dislocations. Chapter 7 discusses the effects of defects on device performance. It covers surface and interface states as appropriate for metal-semiconductor and dielectric-semiconductor junctions. Thermal conversion and out-diffusion of defects and impurities and their effects on devices are also discussed. Defect gettering is included next. Finally, the role defects in the performance and reliability of optoelectronic and electronic devices is treated.

We have benefited a great deal over the years by our interaction with a number of our colleagues at AT&T Bell Laboratories who have provided insights on many of the topics. We acknowledge all their help. We are thankful to S.J.Pearton and E.M.Monberg for their critical reading of the whole book and for their suggestions. We thank U.K.Chakrabarti, M.Hybertsen and D.P.Wilt for their comments on Chapter 1; W.D.Johnston,Jr., M.A.Digiuseppe, R.F.Karlicek and D.P.Wilt on Chapter 2; W.Lowe on Chapter 3; G.Higashi on Chapter 4; A.Jayaraman, P.Parayanthal and H.Temkin on Chapter 5; G.A.Baraff and E.A.Fitzgerald,Jr. on Chapter 6 and N.Chand and C.Zipfel on Chapter 7. We are grateful to the many authors whose work we have freely drawn upon to illustrate many points. We thank the many publishing houses who have granted permission to use copyright material. Lisa Hailey and Gerry Moore have done a wonderful job of typing the manuscript and the numerous iterations. Special thanks are due to Doreen Micheletti for diligently preparing the final camera ready copy. Danuta Sowinska-Kahn did an excellent job in coordinating all the art work. The support of AT&T Bell Laboratories management for this project is greatfully acknowledged. Finally, we thank our wives and children for their patient endurance of our preoccupation for nearly two years. Without their support and encouragement we could not have succeeded in this endeavor.

**V.Swaminathan**
**A.T.Macrander**

# CHAPTER 1

# INTRODUCTION

## 1.1 EMERGENCE OF GaAs AND InP

The statement "GaAs is the material of the future" has often elicited the humorous response from the antagonists, and sometimes from the protagonists as well, that GaAs will remain the material of the future! What is implied is that GaAs will not replace Si. Perhaps this may be true. The first paper on the semiconducting properties of III-V compounds by Welker in 1952 (H. Welker, Z. Naturforsch 79, 744 (1952)) suggested that GaAs might be a useful semiconductor. Since that time, GaAs has come a long way. At the present time, among the III-V semiconductors, GaAs and InP have taken the position of being the most technologically important compounds. They have indeed become the materials of the present in their own right without having to gain that position by replacing Si.

The research that took place in the 60s since the first paper in 1952 was mainly devoted to the physical properties of the III-V compounds. In the decades that followed, research emphasis was shifted to technology and applications. In the late 70s and the early 80s, phenomenal progress was made in the preparation and processing of GaAs and InP and they emerged as viable technological materials. In both these areas, these materials have not yet reached a stage comparable to Si. Nevertheless, the interest in them for many device applications has not diminished. If the number of papers published is a measure of growth of the field, then it indicates a rapidly growing field. In the 1960s the number of papers published per year on GaAs was 200 which increased to about 1200 in the 70s and to 3000 in the first eight years of the 80s. In the same period the number of papers per year on InP grew from a mere 15 to 115 to 600.

The technological importance of GaAs and InP comes about because of their use in both electronic and photonic devices. In this latter aspect they have an advantage over Si, which being an indirect band gap semiconductor, is not an efficient photonic material. Besides, they are attractive for high-speed devices since the electron saturation velocity in GaAs is 1.5 times that of Si. The low-field electron mobility in GaAs or InP is also higher than that in Si. The advantages of high speed and low power dissipation of GaAs devices appeal to the communications industry. When compared to Si, III-V compounds have higher radiation hardness and are capable of operating over a wider temperature range from -196°C to 300°C. Because of this, GaAs devices are of special interest for space and military applications.

The major use of GaAs at present is in the area of microwave devices, high speed digital integrated circuits, and as substrates for epitaxial layer growth to fabricate photonic and electronic devices. For InP, the use is almost exclusively as substrates for growing lattice matched epitaxial films of alloy semiconductors such as GaInAs and GaInAsP. These ternary and quarternary compounds are the materials of choice for making light sources and detectors for the present day long haul fiber optic communication systems. In the remainder of this chapter we describe the bonding, crystal structure, band structure and physical properties of GaAs and InP, and some of the alloy semiconductors closely related to them.

## 1.2 BONDING AND CRYSTAL STRUCTURE

Elemental semiconductors such as Si, Ge crystallize in the diamond cubic structure. The space lattice of the diamond cubic structure is the face centered cubic (fcc) lattice. The diamond structure consists of two interpenetrating fcc lattices, one lattice shifted by $a/4$ [111] relative to the other fcc lattice, $a$ being the length of the fcc cube edge. Since each fcc lattice has 4 atoms per unit cell, the diamond lattice contains 8 atoms. In the unit cell, the atoms are arranged such that for each atom there are four equally distant atoms arranged at the corners of a regular tetrahedron. The tetrahedral arrangement of atoms is the hallmark of directional covalent bonding of semiconductors having 8 valence electrons per pair of atoms in $sp^3$ hybridized orbitals.

III-V compound semiconductors such as GaAs and InP crystallize in the cubic zinc blende (sphalerite) structure which is closely related to the diamond cubic structure. In this structure, the two fcc lattices of the diamond structure are occupied by two different atoms, for example, Ga and As, respectively. In the tetrahedral arrangement of atoms in the sphalerite structure, each group III atom is bonded to four group V atoms in the tetrahedron and vice versa. The cubic unit cell of a sphalerite structure is shown in Fig. 1.1(a). If the coordinates of the group III atoms are $000,\ 0\frac{1}{2}\frac{1}{2},\ \frac{1}{2}0\frac{1}{2},\ \frac{1}{2}\frac{1}{2}0$, that of the group V atoms are $\frac{1}{4}\frac{1}{4}\frac{1}{4},\ \frac{1}{4}\frac{3}{4}\frac{3}{4},\ \frac{3}{4}\frac{1}{4}\frac{3}{4},\ \frac{3}{4},\frac{3}{4},\frac{1}{4}$. There are four molecules of the compound AB per unit cell. Unlike the diamond structure, the sphalerite structure does not have inversion symmetry. If we consider the arrangement of atoms along the body diagonal, the order AB..AB..AB, where dots represent vacant sites, is not invariant under inversion.

The stacking of {111} planes in the cubic semiconductors follows the ABCABC stacking of the close packed {111} planes in the fcc lattice. However, the diamond cubic or the sphalerite structure is relatively empty in that the fraction of the total volume filled by spheres is only 0.34 compared to 0.74 of a close packed structure. In the sphalerite structure, because of the presence of two types of atoms, there is a polarity in the stacking of the {111} planes, that is,

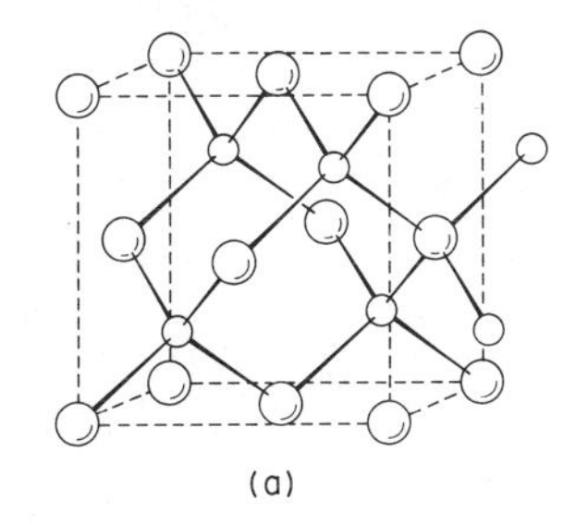


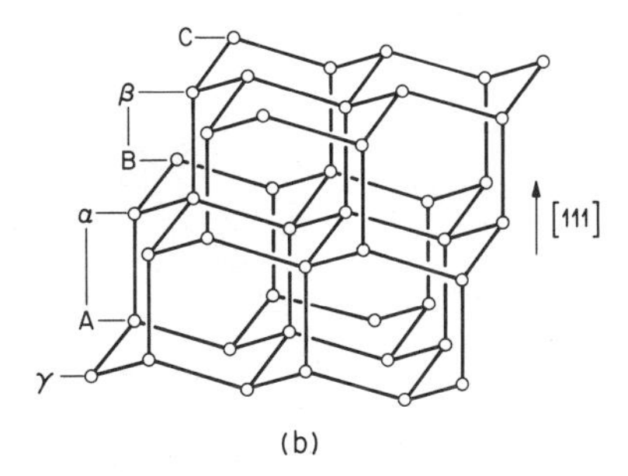


**Figure 1.1** (a) The sphalerite structure (b) stacking of (111) planes in the sphalerite structure. ABC ... is the stacking order in one fcc lattice and αβγ is the stacking order in the other.

[111] ≠ [111]. In the sphalerite lattice, each A atom has four nearest neighbor B atoms and has 12 next-nearest neighbor A atoms and so on. Table 1.1 shows the neighbors within a distance of 2.5 bond lengths in the sphalerite lattice [1]. The shortest circuit (closed loop, or *n*-ring consisting of *n* adjacent bonds, *n* connected sites) in the sphalerite structure is a 6-atom ring as shown in Fig. 1.1(b). The configuration of the 6-atom rings is referred to as the "*chair*" configuration.

### 1.2.1 Fractional Ionic Character in Bonding in III-V Semiconductors

Although the bonding in the sphalerite structure consists of $sp^3$ hybridized orbitals, unlike in diamond cubic semiconductors, there is some charge transfer between the two types of atoms giving rise to a partial ionic character to the bonding. If $f_i$ and $f_h$ denote the fractions,

**Table 1.1**

Number of Nearest Neighbors Within a Distance of 2.5 Bond Lengths in the Sphalerite Structure [1]

| Number of bond steps from the initial atom A | Distance from the initial atom A in units of bond length | Type of neighboring atom | Number of neighbors of the specified type of atom |
|---|---|---|---|
| 1 | 1.0 | B | 4 |
| 2 | $(8/3)^{1/2}$ = 1.633 | A | 12 |
| 3 | $(11/3)^{1/2}$ = 1.915 | B | 12 |
| 4 | $(16/3)^{1/2}$ = 2.309 | A | 6 |
| 3 | $(19/3)^{1/2}$ = 2.517 | B | 12 |

**Table 1.2**

Fractional Ionic Character of Bonds and Cohesive Energy of III-V Semiconductors [2]

| Crystal | Fractional ionic Character, $f_i$ | Cohesive Energy* kcal/mole |
|---|---|---|
| BAs | 0.002 | |
| BP | 0.006 | |
| AlSb | 0.250 | |
| BN | 0.256 | -293.9 |
| GaSb | 0.261 | -118.3 |
| AlAs | 0.274 | -157.3 |
| AlP | 0.307 | -169.5 |
| GaAs | 0.310 | -135.4 |
| InSb | 0.321 | -108.5 |
| GaP | 0.327 | -152.4 |
| InAs | 0.357 | -125.6 |
| InP | 0.421 | -132.9 |
| AlN | 0.449 | |
| GaN | 0.500 | |

*Gibbs free energy of sublimation into neutral atoms at STP.

respectively, of ionic or heteropolar character and covalent or homopolar character in the bond, then the sum $f_i + f_h$ equals unity. In elemental semiconductors $f_h = 1$ and $f_i = 0$. A semiempirical theory of fractional ionic or covalent character of the bond has been developed by Phillips [2] which involves estimating the ionic contribution to the energy gap of the AB semiconductor. Table 1.2 lists the $f_i$ values for III-V compounds as obtained by Phillips. Also listed in Table 1.2 are the cohesive energies, denoted by the Gibbs free energies of atomization at standard temperature and pressure (STP), of the respective compounds. It can be seen that, in general, the cohesive energy decreases with increasing ionicity.

### 1.2.2 Brillouin Zones

The electronic and vibronic states in crystals are best described in reciprocal-space or k-space. The smallest unit cell in reciprocal space is called the first Brillouin zone. While a/2 [110], a/2 [101], a/2 [011] are the primitive translation vectors of the fcc lattice, $2\pi/a$ $(11\bar{1})$, $2\pi/a$ $[1,\bar{1},1]$ and $2\pi/a$ $[\bar{1}11]$ are the primitive translation vectors in the reciprocal lattice. Incidentally, the latter are the primitive translation vectors of a body centered cubic (bcc) lattice, so that the bcc lattice is the reciprocal lattice of the fcc lattice. The volume of the primitive cell in the reciprocal fcc lattice is $4\,(2\pi/a)^3$.

The first Brillouin zone for the cubic semiconductors is the truncated octahedron shown in Fig. 1.2. The principal symmetry points and lines in k-space are labeled using the standard nomenclature in Fig. 1.2 and their Cartesian coordinates are given in Table 1.3. All points

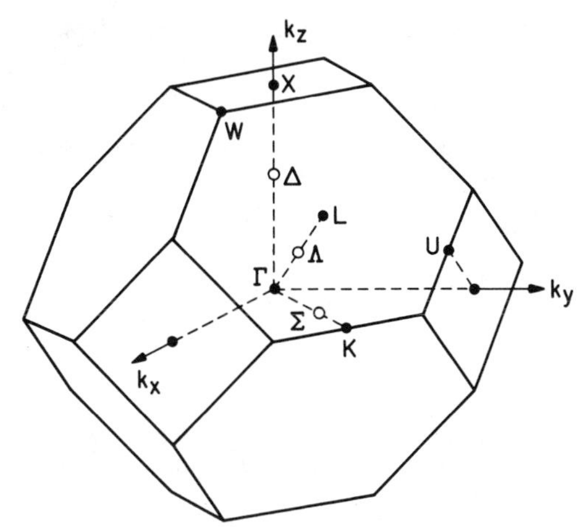


**Figure 1.2** The first Brillouin Zone for the cubic semiconductor showing the symmetry points and axes [1].

obtainable from one another by cubic symmetry operations are given the same name. For example, there are six symmetry equivalent Δ points (2x, 0, 0), (0, 2x, 0), (0, 0, 2x), (–2x, 0, 0), (0, –2x, 0), (0, 0, –2x).

## 1.3 ENERGY BAND STRUCTURE

The energy band theory of semiconductors is based on two major simplifying assumptions. First, the atoms in the solid are assumed to be stationary and form a lattice with perfect translational symmetry. Second, one electron at a time is considered and the influence of all electrons is

**Table 1.3**

Cartesian Coordinates of Symmetry Points and Lines in the fcc Brillouin Zone [2]

| Label | Coordinates |
|---|---|
| Γ | (0, 0, 0) |
| X | (2, 0, 0) |
| L | (1, 1, 1) |
| K = U | $(\frac{3}{2}, \frac{3}{2}, 0)$ |
| W | (2, 1, 0) |
| Δ | (2x, 0, 0) |
| Λ | (x, x, x) |
| Σ | $(\frac{3}{2}x, \frac{3}{2}x, 0)$ |

x denotes a number between 0 and 1 and the coordinates are given in units of $(\pi/a)$ where a is the cubic lattice constant.

represented by an average over their wave functions. The functional dependence of energy on wave vector k for the various bands, $E_n(k)$ is defined by the Schrodinger equation of the corresponding one electron problem, namely

$$H\psi(r) = E\psi(r)$$

$$H = p^2 / 2m_0 + V(r) \tag{1.1}$$

$$\frac{p^2}{2m_0} = -\frac{\hbar^2}{2m_0}\left[\frac{\partial^2}{\partial x^2} + \frac{\partial^2}{\partial y^2} + \frac{\partial^2}{\partial z^2}\right]$$

where E is the energy eigenvalue, V(r) is the potential energy, $p^2 / 2m_0$ is the kinetic energy and $m_0$ is the free electron mass. The wave function $\psi(r)$ of the electrons in the crystal lattice is expressed by the well known Bloch's theorem

$$\psi(r) = \exp(ik\cdot r)U_k(r) \tag{1.2}$$

where the function $U_k(r)$ has the period of the crystal lattice such that $U_k(r) = U_k(r+T)$ where T is any vector of the Bravais lattice. The nearly free-electron approximation is a good starting point for discussing the energy band theory. It explains the origin of the band gap and that of the effective mass $m^*$ which is defined as the reciprocal of the curvature of E versus k diagram. Once the concept of effective mass is introduced, the band model becomes the Sommerfeld free electron model with Eq. (1.1) modified to

$$-\frac{\hbar^2}{2m^*}\left[\frac{\partial^2}{\partial x^2} + \frac{\partial^2}{\partial y^2} + \frac{\partial^2}{\partial z^2}\right]\psi(r) = E\psi(r) \tag{1.3}$$

Equation (1.3) differs from the free electron Schrodinger equation only in the use of $m^*$ instead of $m_0$, the value of $m^*$ taking into account the periodic potential in Eq. (1.1).

### 1.3.1 The k·p Method

The k·p method is a method to calculate the band structure in the vicinity of most important symmetry points in great detail [3]. This theory forms the foundation for the discussion of the many experimental results. This theory will be described in detail.

If we substitute the Bloch functions given by Eq. (1.2) in Eq. (1.1) we obtain

$$\left[\frac{p^2}{2m_0} + \frac{\hbar}{m_0}k\cdot p + \frac{\hbar^2 k^2}{2m_0} + V(r)\right]U_k(r) = E(k)U_k(r) \tag{1.4}$$

where p is the momentum operator $-i\hbar\nabla$. The complete set of eigen functions represented by Eq. (1.4) is called the k·p representation. The spin-orbit interaction that plays an important role in many semiconductors can also be included in Eq. (1.4) which can be represented in a simpler form in terms of the various Hamiltonians

$$(H_0 + H_1 + H_2 + H_3)\, U_k = E_k\, U_k \tag{1.5}$$

where $H_0 = \frac{p^2}{2m_0} + V(r)$ the Hamiltonian of the unperturbed problem, $H_1 = \frac{\hbar}{m_0}\, k{\cdot}p$ (for small k) is spin-independent perturbation,

$$H_2 = \frac{\hbar^2}{4m_0^2c^2}(\nabla V \times p) . \sigma \text{ and } H_3 = \frac{\hbar^2}{4m_0^2c^2}(\nabla V \times k) . \sigma$$

are spin dependent perturbations and $\sigma$ is the Pauli spin operators. $H_3$ is usually very small compared to $H_2$ since the main contribution to the spin-orbit interaction comes from the core region where $\nabla V$ and p are very large. $H_3$ is frequently neglected for this reason. Relativistic corrections to the Hamiltonian which do not involve the spin are also important. Such corrections are implicitly included in the empirically adjusted atomic data that often serve as the starting point for band structure calculations [3].

The $\Gamma$ point ($k = 0, 0, 0$) which is the point of highest symmetry in the Brillouin zone is the most suitable point to use as the basis for a $k{\cdot}p$ representation to cover the entire zone [3]. Semiconductors with tetrahedral coordination have valence band maxima at the $\Gamma$ point and many have conduction band minima also at the $\Gamma$ point (e.g., GaAs, InP). The valence band maxima is derived from p like orbitals, $p_x, p_y, p_z$ which remain degenerate under the tetrahedral group of the zinc blende lattice. The representation of the valence band is $\Gamma_{15}$. When electron spin is included there are altogether six states. The six states are further split by spin-orbit interaction into a four fold degenerate $J = 3/2$ and a two fold degenerate $J = 1/2$ states, the $J = 1/2$ band being the lowest band. The value of the spin-orbit splitting for III-V semiconductors is given in Table 1.4 [2]. The $J = 3/2$ bands are called the light hole and heavy hole band. These two bands can be labeled by the azimuthal quantum number of J relative to the k axis, $m_J = 3/2$ or $1/2$.

**Table 1.4**

Spin-orbit Splittings [2]

| Compound | $\Delta_0$ (eV) |
|---|---|
| GaP | 0.127 |
| GaAs | 0.34 |
| GaSb | 0.80 |
| InP | 0.11 |
| InAs | 0.38 |
| InSb | 0.82 |

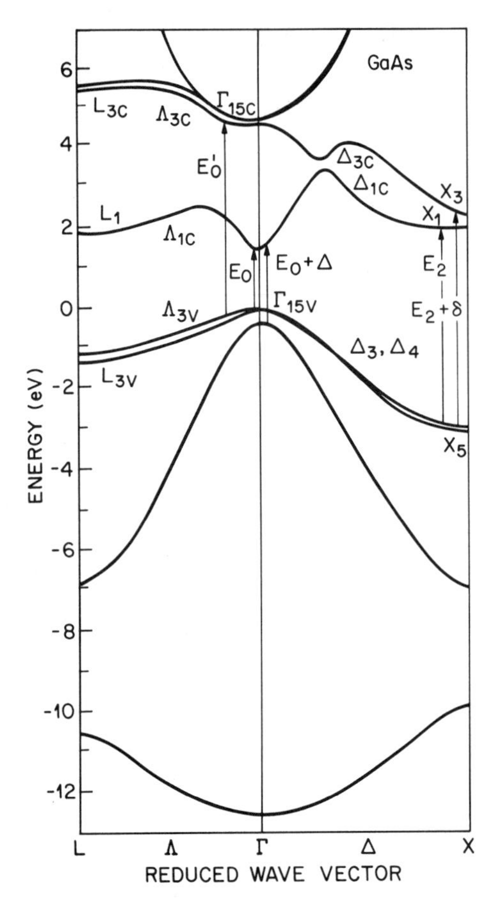


**Figure 1.3** Calculated energy band structure of GaAs in two principal directions in the Brillouin Zone [5]. The various critical point energies are indicated.

The conduction band is derived from s-like orbitals and denoted by the representation $\Gamma_1$. It is very much like the mirror image of the split-off spin orbit valence band. Starting with the two s-functions (with opposite spins) and the six p-functions as the zero-order base functions for the expansion of the wave function in Eq. (1.5), an $8 \times 8$ secular determinant is obtained. On virtue of the symmetry of the base functions, the solution of the secular determinant finally yields four eigen values at $k = 0$ corresponding to the bottom of the conduction band edge, the top of the heavy and light hole bands and the spin-orbit band. The influence of the higher and lower lying bands is then included by a second-order perturbation calculation in $H_1$ and $H_2$ to finally yield the E versus $k$ relation.

The nonparabolic corrections to the conduction band are generally small in semiconductors like GaAs with large energy gap and hence E versus $k$ relations are approximated to only $k^4$ terms. Accordingly, the conduction band energy for GaAs is given by [4]

$$E(k) = \frac{\hbar^2 k^2}{2m^*} - \left[1 - \frac{m^*}{m_0}\right]^2 \left[\frac{\hbar^2 k^2}{2m^*}\right]^2 \left\{\frac{3E_g + 4\Delta + 2\Delta^2 / E_g}{(E_g + \Delta)(3E_g + 2\Delta)}\right\} \tag{1.6}$$

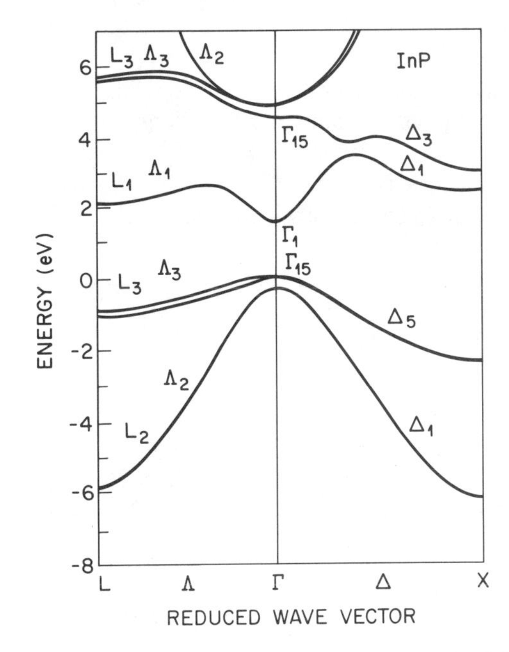


**Figure 1.4** Calculated energy bandstructure of InP in two principal directions in the Brillouin Zone [5].

where $\Delta$ is the spin-orbit splitting, $E_g$ is the energy gap, $m^*$ is the effective mass of the electron. The zero of the energy is taken at the bottom of the $\Gamma_1$ conduction band. $m^*$ is given by

$$\frac{1}{m^*} = \frac{1}{m} + \frac{2P^2}{3\hbar^2}\left[\frac{2}{E_g} + \frac{1}{E_g + \Delta}\right] \tag{1.7}$$

where P is the matrix element connecting the conduction band and the light hole and spin orbit split valence bands.

The calculated energy band structure of GaAs is shown in Fig. 1.3 [5]. The lowest conduction band minimum $\Gamma_1$ occurs at $k=0$ with higher minima occurring in the <100> ($\Delta$) and <111> ($\Lambda$) directions. The L and X minima are, respectively, 0.31 eV and 0.52 eV above the $\Gamma_1$ minimum [5]. The valence band consists of heavy hole, light hole bands degenerate at $k=0$ and a spin orbit split band at 0.34 eV lower energy. The various critical point energies are also indicated in Fig. 1.3. These critical points satisfy the condition $\nabla_k E(k) = 0$. They can be readily seen in modulated reflectivity spectrum and their location and nature can be identified from calculated energy bands. The electronic band structure of InP is shown in Fig. 1.4 [5].

## 1.4 ENERGY BAND STRUCTURE OF ALLOYS

III-V alloy semiconductors have assumed an important role in the semiconductor industry because of their utility in many electronic and photonic devices such as heterostructure transistors, light emitting diodes, and lasers. Let us consider two III-V compounds AC and BC which form a cation alloy $A_xB_{1-x}C$ (e.g. $Al_xGa_{1-x}As$). Similarly, compounds AC and AD form an anion alloy $AC_xD_{1-x}$ (e.g. $GaAs_xP_{1-x}$). Mixed compounds of this type are called pseudo

**Table 1.5**

Comparison of theory and experiment for the bowing parameter pertaining to the minimum direct energy gap [6].
$c_i$ and $c_e$ are the intrinsic (virtual crystal) and extrinsic (disorder) bowing parameters.
$c_{cal} = c_i + c_e$

| Alloy | $c_i$ (ev) | $c_e$ (eV) | $c_{cal}$ (eV) | $c_{exp}$ (eV) |
|---|---|---|---|---|
| GaAs-P | 0.21 | 0.09 | 0.30 | 0.21 |
| InAs-P | 0.15 | 0.08 | 0.23 | 0.20, 0.26 |
| Ga-InSb | 0.12 | 0.24 | 0.36 | 0.43 |
| Ga-InAs | 0.28 | 0.29 | 0.57 | 0.33, 0.56 |
| Ga-AlAs | 0.0 | 0.03 | 0.03 | ~0.2 |
| Ga-InP | 0.39 | 0.31 | 0.70 | 0.88 |

binary alloys. If the two compounds do not share a cation or an anion, one gets a quaternary alloy, for example, $Ga_xIn_{1-x}As_yP_{1-y}$.

It has been found experimentally, that in many semiconductor alloys the direct energy gap ($\Gamma_{15}^v - \Gamma_1^c$) varies nonlinearly with composition. For an $A_xB_{1-x}C$ compound the energy gap can be expressed in the form [6]

$$E_0(x) = a + bx + cx^2 \tag{1.8}$$

where c is called the bowing parameter and is four times the deviation of $E_0(x)$ from linearity at $x = 0.5$. The other two parameters a and b are determined by the values of $E_0$ observed in the pure binary compounds. The bowing parameter has been calculated using the dielectric method of band structure calculation [7, 8]. The calculations were made in the virtual crystal approximation where the parameters of the model and the lattice constant are assumed to vary linearly with alloy composition. The total bowing parameter has been taken to be the sum of two terms, an intrinsic bowing found in the virtual crystal approximation and an extrinsic bowing as a result of the aperiodic crystal potential caused by disorder in the alloys. The latter contribution is essentially a short range (within one unit cell) effect. The calculated values of these two bowing parameters and the experimental value pertaining to the minimum direct gap are given for some III-V compounds in Table 1.5 [6]. Note that the bowing parameters for the compounds listed in Table 1.5 are all positive which means that the energy bands bow downward.

### 1.4.1 Material Parameters of Alloy Semiconductors

In deriving many physical parameters of the alloys, an interpolation scheme is generally adopted using the values of the related binary compounds. For a ternary compound $A_xB_{1-x}C$, the parameter P is derived from

$$P(A_xB_{1-x}C) = xP_{AC} + (1-x)P_{BC} \,. \tag{1.9}$$

Material parameters such as lattice constant vary linearly with composition as per Eq. (1.9). As noted in the last section, some parameters such as the energy gap deviate from linearity and in that case Eq. (1.8) is more appropriate.

The parameter of a quaternary compound such as $A_xB_{1-x}C_yD_{1-y}$ can be obtained from the respective values of the four binaries, AC, AD, BC and BD as per

$$P(A_xB_{1-x}C_yD_{1-y}) = xy\,P_{AC} + x(1-y)P_{AD} + (1-x)y\,P_{BC} + (1-x)(1-y)P_{BD}\,. \tag{1.10}$$

If the parameters for the ternary alloy $A_xB_{1-x}C$, $A_xB_{1-x}D$, $AC_yD_{1-y}$ and $BC_yD_{1-y}$ are available then Eq. (1.10) can be modified to give [9]

$$P(A_xB_{1-x}C_yD_{1-y}) = \frac{x(1-x)[(1-y)P_{ABD} + y\,P_{ABC}] + y(1-y)\,[x\,P_{ACD} + (1-x)\,P_{BCD}]}{x(1-x) + y(1-y)} \tag{1.11}$$

The ternary parameters $P_{ABD}$ and so on, may include a quadratic dependence given by Eq. (1.8). This interpolation equation reduces to the average of the four ternary parameters at x=y=0.5.

For determining the band gap of the quaternary alloy Moon et al. [10] have used the following equation

$$E_g(A_xB_{1-x}C_yD_{1-y}) = xE_{ACD} + (1-x)\,E_{BCD} - \Delta \tag{1.12}$$

where $\Delta$ is the bowing parameter term for the quaternary alloy and is given by

$$\Delta = x(1-x)[(1-y)c_{ABD} + y\,c_{ABC}] + y(1-y)\,[x\,c_{ACD} + (1-x)\,c_{BCD}]\,. \tag{1.13}$$

where $c_{ABD}$ and so on, are the ternary bowing parameters of Eq. (1.8). For $x = y = 0.5$, $\Delta$ is $1/8^{th}$ of the sum of the four ternary bowing parameters which is twice as large as the quaternary bowing parameter given by Eq. (1.11). A comparison of the two interpolation schemes given by Eqs. (1.11) and (1.12) with measured values of the band gap in $Ga_xIn_{1-x}As_yP_{1-y}$ alloys showed that both schemes give comparable errors [9]. However, for calculation of the band gap, Eq. (1.12) which is the quaternary version of Eq. (1.8) is preferred since it has some theoretical basis. Using the band gaps and lattice constants of the binary compounds Glisson et al., [9] have derived the lowest energy band gap and lattice constant of several quaternary alloys of the III-V semiconductors at room temperature. For calculating the band gap they used Eq. (1.12). The unknown bowing parameters of the ternary semiconductors were obtained by extrapolation of the data shown in Table 1.5. When the c values shown in Table 1.5 were plotted against the difference in the lattice constants of the two end point binary compounds, a definite trend toward larger c with larger lattice constant difference can be observed. Glisson et al. [9] obtained the quaternary lattice constants from the binary values using Eq. (1.11). The lattice constants of the ternaries are assumed to follow the linear relation given by Eq. (1.9). The energy band gap and lattice constant contours of several III-V quaternary alloys as derived by Glisson et al., are reproduced in Figs. 1.5 to 1.9.

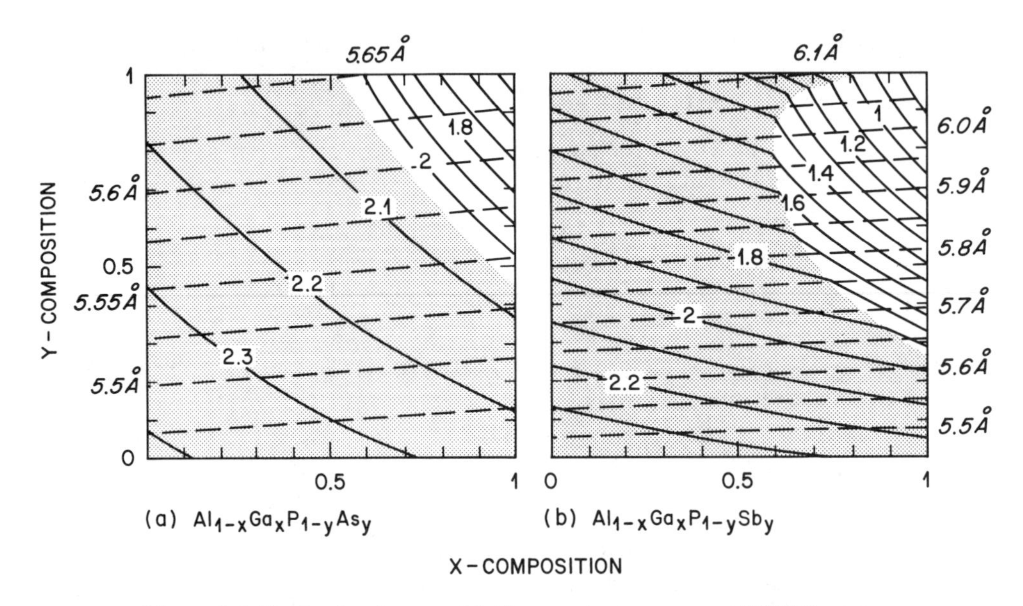


**Figure 1.5** Energy band gap and lattice constant contours at 300K for (a) $Al_{1-x}Ga_xP_{1-y}As_y$ and (b) $Al_{1-x}Ga_xP_{1-y}Sb_y$ alloys. The solid curves are the energy gap contours and the dashed curves are lattice constant contours. The shaded region shows the compositional range over which the material has an indirect band gap [9].

**Figure 1.6** Energy band gap and lattice constant contours at 300K for (a) $Al_{1-x}Ga_xAs_{1-y}Sb_y$ and (b) $Al_{1-x}In_xP_{1-y}As_y$ alloys. The solid curves are the energy gap contours and the dashed curves are lattice constant contours. The shaded region shows the compositional range over which the material has an indirect band gap [9].

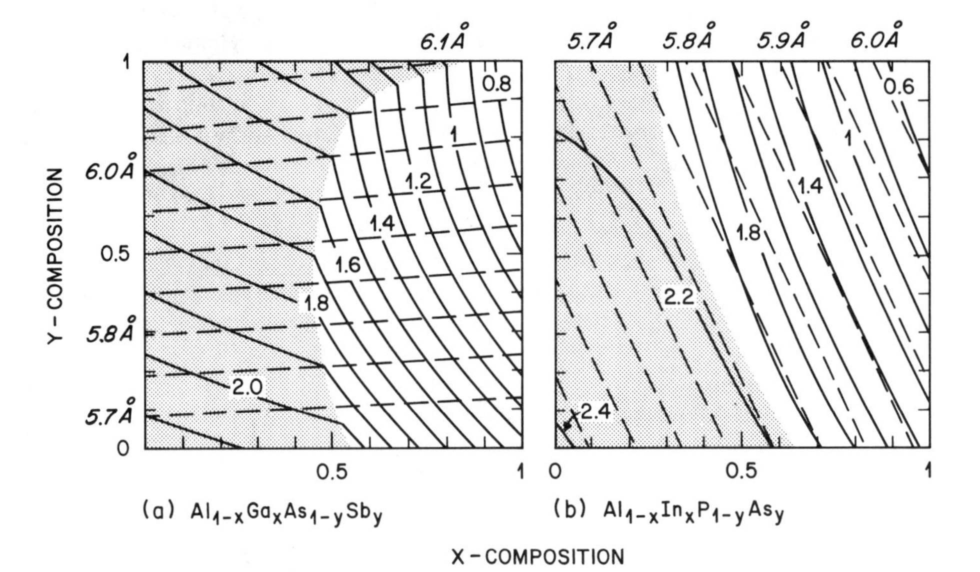

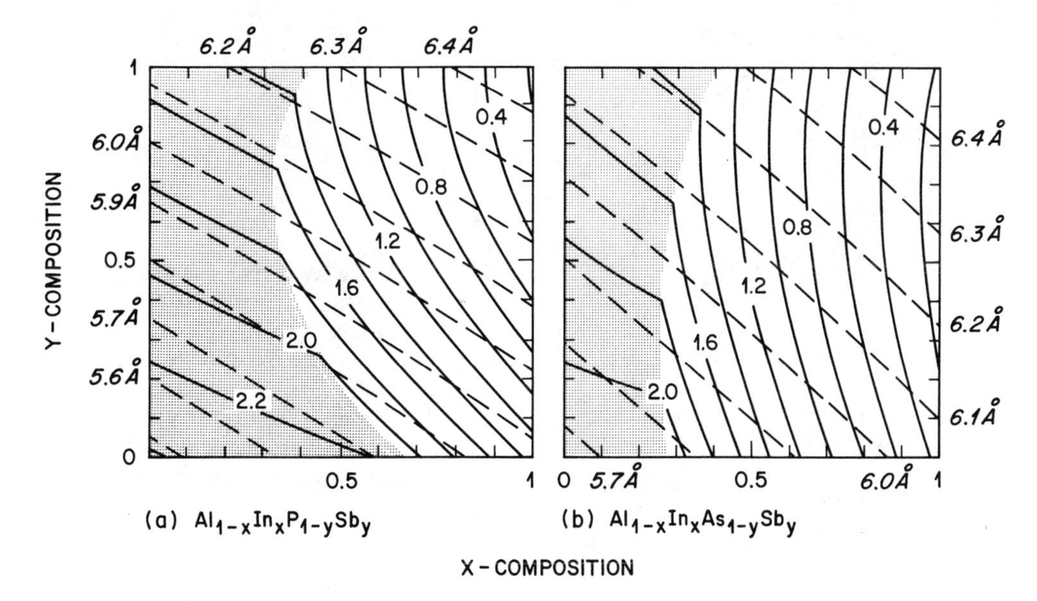


**Figure 1.7** Energy band gap and lattice constant contours at 300K for (a) $Al_{1-x}In_xP_{1-y}Sb_y$ and (b) $Al_{1-x}In_xAs_{1-y}Sb_y$ alloys. The solid curves are the energy gap contours and the dashed curves are lattice constant contours. The shaded region shows the compositional range over which the material has an indirect band gap [9].

**Figure 1.8** Energy band gap and lattice constant contours at 300K for (a) $Ga_{1-x}In_xAs_yP_{1-y}$ and (b) $Ga_{1-x}In_xP_{1-y}Sb_y$ alloys. The solid curves are the energy gap contours and the dashed curves are lattice constant contours. The shaded region shows the compositional range over which the material has an indirect band gap [9].

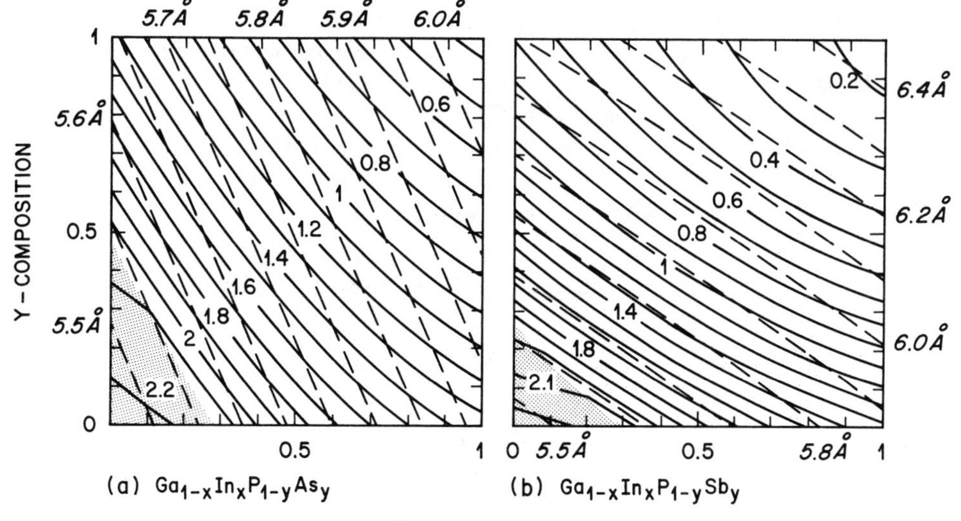

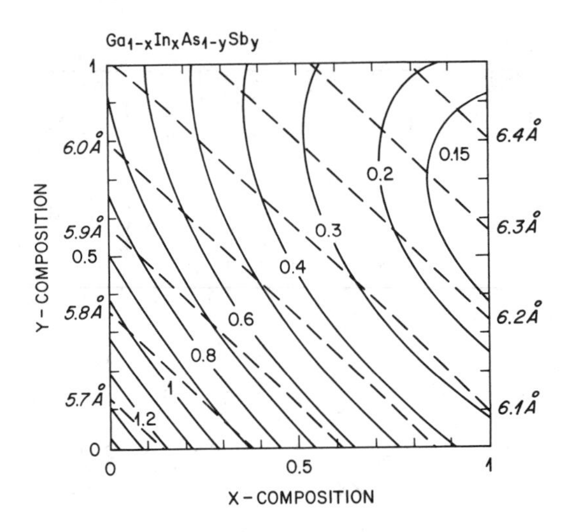


**Figure 1.9** Energy band gap and lattice constant contours at 300K for $Ga_{1-x}In_xAs_{1-y}Sb_y$ alloys. The solid curves are the energy gap contours and the dashed curves are lattice constant contours. The shaded region shows the compositional range over which the material has an indirect band gap [9].

## 1.5 TABLES OF MATERIAL PROPERTIES

### 1.5.1 Binary Compounds

In this section we have summarized the material properties of the important III-V binary compounds. The values for these properties unless otherwise noted are taken from the Landolt-Bornstein series [11, 12]. Table 1.6 lists the lattice constant, density, melting point, Debye temperature, coefficient of thermal expansion, and thermal conductivity. For the last two parameters the values at or close to 300K are given.

**Band structure**

The parameters pertinent to the electronic band structure are summarized in Tables 1.7 and 1.8. The direct and indirect energy gaps, their temperature dependences, and the pressure dependence of the minimum energy gap are given in Table 1.7. For AlP, AlAs, AlSb and GaP the conduction band minimum is located at the $\Delta$ axes near the Brillouin zone boundary. For AlAs the k value of the conduction band minimum is 0.087 $k_{max}$ from the X point where $k_{max} = 2\pi/a$, a being the lattice constant. For GaP the corresponding value is 0.043 $k_{max}$ from the X point. The valence band structure of all the compounds listed in the tables is very similar. The valence band maximum is located at $k=0$ and it is characterized by a four-fold degenerate heavy hole and light-hole bands and a two-fold degenerate spin-orbit band separated by the energy $\Delta_0$ given in Table 1.4.

The temperature dependence of the band gap is generally expressed by [13]

$$E_g(T) = E_g(0) - \frac{\alpha T^2}{(T+\beta)} \qquad (1.14)$$

where $E_g(0)$ is the energy gap at 0K, $\alpha$ and $\beta$ are constants with $\beta$ having a value close to the Debye temperature. Near room temperature and above, $E_g(T)$ varies linearly with temperature.

**Table 1.6**

Lattice Constant, Density, Melting Point, Debye Temperature, Coefficient of Thermal Expansion, and Thermal Conductivity

| Compound | Lattice Constant (Å) | Density (gm cm$^{-3}$) | Melting Point (K) | Debye Temperature[a] (K) | Coefficient of Thermal Expansion[b] $10^{-6}$/°C | Thermal[a,b] Conductivity (W cm$^{-1}$ K$^{-1}$) |
|---|---|---|---|---|---|---|
| AlP | 5.467 | 2.40 | 2823 | 588 | 4.5 | 0.9 |
| AlAs | 5.660 | 3.70 | 2013 | 417 | 4.9 | 0.8 |
| AlSb | 6.136 | 4.26 | 1338 | 292 | 4.0 | 0.57 |
| GaP | 5.4512 | 4.138 | 1740 | 456 | 4.5 | 0.77 |
| GaAs | 5.6532 | 5.3161 | 1513 | 344 | 6.86 | 0.46 |
| GaSb | 6.0959 | 5.6137 | 985 | 266 | 7.75 | 0.39 |
| InP | 5.8687 | 4.81 | 1335 | 321 | 4.75 | 0.68 |
| InAs | 6.0583 | 5.667 | 1215 | 249 | 4.52 | 0.273 |
| InSb | 6.4794 | 5.7747 | 800 | 203 | 5.37 | 0.166 |

(a) M. G. Holland *Semiconductors and Semimetals, Vol. 2:3*, (ed. R. K. Willardson and A. C. Beer). New York; Academic, 1966, p. 3.

(b) Values near or at 300K.

**Table 1.7**

Energy Band Structure

| Parameter | | AlP | AlAs | AlSb | GaP | GaAs | GaSb | InP | InAs | InSb |
|---|---|---|---|---|---|---|---|---|---|---|
| Direct Energy gap (eV) | $\Gamma_{15}-\Gamma_1$ | 3.62 (77K) | 3.14 | 2.22 | 2.78 | 1.424 | 0.70 | 1.34 | 0.356 | 0.180 |
| Indirect Energy gap (eV) | $\Gamma_{15}-X_1$ | 2.45 | 2.14[a] | 1.63[a] | 2.268[a] | 1.804 | 1.25 (10K) | 2.74 | | |
| Temperature Dependence of direct and indirect energy gap ($\times 10^{-4}$ eV $K^{-1}$) | $dE_d / dT$ | | –5.2 | | –4.5 | –3.9 | –3.7 | –2.9 | –3.5 | –2.8 |
| | $dE_{ind} / dT$ | –3.6 | –4.0 | –3.5 | –5.2 | –2.4[(b)] | | –3.7[(b)] | | |
| Pressure Dependence of minimum energy gap ($\times 10^{-6}$eV $^{-1}$) | dE/dP | | | –1.5 | –1.6 | 12.0 | 14.7 | 8.8 | 10.6 | 15.9 |
| Indirect Energy gap (eV) | $\Gamma_{15}-L_1$ | | | | | 1.81 (110K) | 0.81 | 1.74 | | |

(a) Minimum situated at the Δ axes near the boundary: k = (0.903, 0, 0) for AlAs; k = (0.95, 0, 0) for GaP.

(b) Temperature dependence of the energy separation between Γ and X minima.

**Table 1.8**

Critical Point Energies (eV)

| Critical Points | AlP | AlAs | AlSb | GaP | GaAs | GaSb | InP | InAs | InSb |
|---|---|---|---|---|---|---|---|---|---|
| $E_0'$ $(\Gamma_{15V}-\Gamma_{15C})$ | | 4.34 | 3.7(77K) | 4.8 (80K) | 4.49 (4.2K) | 3.2 (80K) | 4.8 (77K) | 4.5 (77K) | 3.16 |
| $E_1$ $(\Lambda_{3V}-\Lambda_{1C})$ | | | 2.78 | 3.7 (80K) | 2.90 | | 3.15 | 2.5 | 1.88 |
| $E_1+\Delta$ $(\Lambda_{4V}-\Lambda_{1C})$ | | | 3.18 | 3.9 (80K) | 3.17 | | 3.30 | 2.75 | 2.38 |
| $E_2$ $(X_{5V}-X_{1C})$ | | 4.54 | | | | | 5.04 (at X?) | 4.72 | 4.08 |
| $E_2+\delta$ $(X_{5V}-X_{3C})$ | | 4.89 | | | | | 5.6 | 5.3 | 4.6 (110K) |

**Table 1.9**

Temperature dependence of the minimum energy gap. Parameters $\alpha$ and $\beta$ as defined in Eq. (1.14) are obtained from Ref. 14.

| Compound | $E_g$ (0) (eV) | $\alpha$ $\times 10^{-4}$ eV $K^{-1}$ | $\beta$ |
|---|---|---|---|
| AlP | 2.52 | 3.18 | 588 |
| AlAs | 2.239 | 6.0 | 408 |
| AlSb | 1.687 | 4.97 | 213 |
| GaP | 2.338 | 5.771 | 372 |
| GaAs | 1.519 | 5.405 | 204 |
| GaSb | 0.810 | 3.78 | 94 |
| InP | 1.421 | 3.63 | 162 |
| InAs | 0.420 | 2.50 | 75 |
| InSb | 0.236 | 2.99 | 140 |

The values of the linear temperature coefficient of the energy gap are given in Table 1.7. Table 1.9 gives the $\alpha$ and $\beta$ values for obtaining $E_g$ below room temperature [14].

Table 1.8 gives the critical point energies of the different binary III-V compounds. These critical point energies are indicated in Fig. 1.3. The temperature dependence of the critical point energies also follows Eq. (1.14) with different $\alpha$ and $\beta$ values.

**Effective mass**

Table 1.10 lists the effective masses at conduction and valence band edges. Semiconductors which have conduction band minimum at $\Gamma = 0$ are characterized by one mass only. Those which have minimum near the X point are characterized by longitudinal and transverse effective masses, $m_l$ and $m_t$. Because of lack of inversion symmetry, the degenerate bands at the X point are split into a higher lying $X_3$ band and a lower lying $X_1$ band. As a result, these semiconductors also exhibit what is known as a camel's back structure in the conduction band. The presence of the camel's back structure introduces extreme non parabolicity and as a result, considerable scatter exists in the experimental values of the effective masses [15]. For the indirect band gap semiconductors the density of states effective mass is given by $\nu^{2/3}$ $(m_l m_t^2)^{1/3}$ where $\nu$ is the number of equivalent conduction band minima. When the minimum occurs at the $X$-point $\nu = 3$. The valence band is characterized by three effective masses corresponding to the heavy hole, light hole, and spin-orbit bands. The density of states effective mass for holes is given by $(m_{hh}^{3/2} + m_{lh}^{3/2})^{2/3}$.

The conduction band effective mass is also a function of doping. When the doping level increases, the bottom of the conduction band becomes filled and the electron gas becomes degenerate. Since the non parabolicity of the band becomes significant as the Fermi level moves

**Table 1.10**

Effective Mass Parameters

| | Conduction Band (units of $m_o$) | | | Valence Band (units of $m_o$) | | | |
|---|---|---|---|---|---|---|---|
| Compound | $m_l$ | $m_t$ | $m_e$ | $m_{hh}$ | $m_{lh}$ | $m_{hso}$ | $m_h^{(b)}$ |
| AlP | | | | 0.63 | 0.20 | 0.29 | 0.70 |
| AlAs | 1.5 | 0.19 | 0.79[a] | 0.76 | 0.15 | 0.24 | 0.80 |
| AlSb | 1.64 | 0.23 | 0.92[a] | 0.94 | 0.14 | 0.29 | 0.98 |
| GaP | 7.25 | 0.313 | 1.86[c] | 0.54 | 0.16 | 0.24 | |
| GaAs | | | 0.067[d] | 0.49 | 0.08 | 0.15 | |
| GaSb | | | 0.044 | 0.34 | 0.044 | 0.13 | |
| InP | | | 0.075[d] | 0.56 | 0.12 | 0.12 | |
| InAs | | | 0.024 | 0.37 | 0.025 | 0.14 | |
| InSb | | | 0.014 | 0.39 | 0.016 | 0.47 | |

(a) For three equivalent minima, $m_e$ is given by $3^{2/3}(m_l m_t^2)^{1/3}$

(b) $m_h = (m_{hh}^{3/2} + m_{lh}^{3/2})^{2/3}$

(c) Because of the presence of a camel's back conduction band structure and extremely high non parabolicity, there is considerable scatter in the effective mass values reported in the literature.

(d) The concentration dependence of the effective mass is shown, respectively, in Figs. (1.10) and (1.11) for GaAs and InP.

up in the band, the electron mass increases with doping level for carrier concentrations larger than $10^{18}$ cm$^{-3}$. Figures 1.10 and 1.11 show the dependence of electron effective mass on carrier density for GaAs [16] and InP [17], respectively.

**Optical properties**

Table 1.11 lists the optical properties-refractive index near the band gap wavelength, its temperature dependence, the static and high frequency dielectric constants and the frequencies of the zone center LO and TO phonons. The optical absorption of solids can be described either by the complex index of refraction $n+ik$ or by the complex dielectric function $\epsilon + i\epsilon_2$, where n, k, $\epsilon_1$ and $\epsilon_2$ are all real functions of the frequency $\omega$.

$$\epsilon_1 = n^2 - k^2, \quad \epsilon_2 = 2nk \tag{1.15}$$

The functions $\epsilon_1(\omega)$ and $\epsilon_2(\omega)$ are connected by the Kramers-Kronig relations. The absorption coefficient $\alpha$ is related to k by

$$\alpha = \frac{2\omega k}{c} \tag{1.16}$$

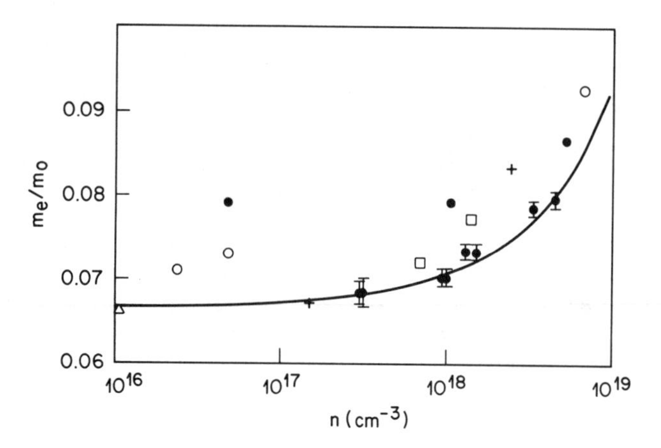


**Figure 1.10** The electron effective mass versus carrier density in GaAs. The data points represent experimental results of several authors. The solid line is a calculated curve [16].

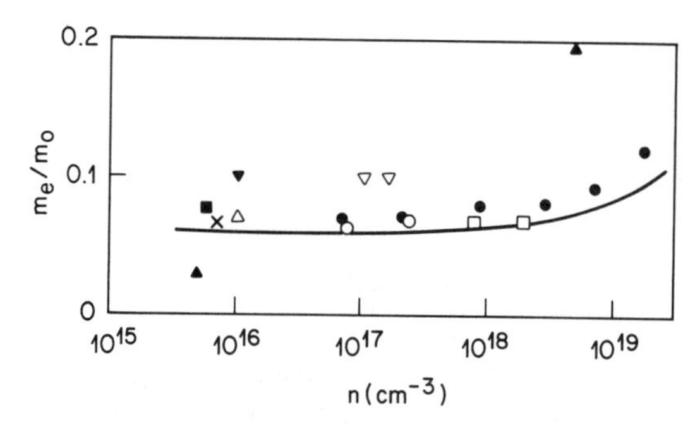


**Figure 1.11** The electron effective mass versus carrier concentration in InP. The data points represent experimental results of several authors. The solid line is a calculated curve [17].

where k the imaginary part of the refractive index is the extinction coefficient. One quantity which is measured in experiments is reflectance. For normal incidence it is given by

$$R = \frac{(n-1)^2 + k^2}{(n+1)^2 + k^2} \tag{1.17}$$

when $k = 0$, that is, in the transparent range, then

$$R = \frac{(n-1)^2}{(n+1)^2} \tag{1.18}$$

**Table 1.11**

Optical Properties

| Compound | Refractive index n near $E_g$ | $\frac{1}{n}\frac{dn}{dT}$ ($K^{-1}$) | Dielectric constant[a] $\epsilon(0)$ | $\epsilon(\infty)$ | Wavenumber of Raman Phonon ($cm^{-1}$) LO | TO |
|---|---|---|---|---|---|---|
| AlP | 3.03 | $3.5\times10^{-5}$ | 9.8 | 7.54 | | |
| AlAs | 3.18 | $4.6\times10^{-5}$ | 10.06 | 8.16 | 404.1 | 360.9 |
| AlSb | 3.4 | $3.5\times10^{-5}$[b] | 12.04 | 10.24 | 339.6 | 318.8 |
| GaP | 3.45 | $2.5\times10^{-5}$[b] | 11.1 | 9.08 | 403.0 | 367.3 |
| GaAs | 3.65 | $4.5\times10^{-5}$ | 12.91 | 10.9 | 291.9 | 268.6 |
| GaSb | 3.82 | $8.2\times10^{-5}$ | 15.69 | 14.44 | 233.0 | 224.0 |
| InP | 3.41 | $2.7\times10^{-5}$ | 12.61 | 9.61 | 345.0 | 303.7 |
| InAs | 3.52 | $6.5\times10^{-5}$[b] | 15.15 | 12.25 | 238.6 | 217.3 |
| InSb | 4.00 | $1.2\times10^{-4}$ | 17.7 | 15.68 | 190.8 | 179.8 |

(a) In units of $\epsilon_0$, the permittivity of free space $8.85\times10^{-14}$ F $cm^{-1}$.

(b) D. K. Ghosh, L. K. Samanta, and G. C. Bhar, *Infrared Phys.* *26*;111, 1986.

In the limit $\omega \to 0$, $\epsilon(0)$ the static dielectric constant is given by the relation [2]

$$\epsilon(o) = 1 + \left(\frac{\hbar\omega_p}{E_h}\right)^2 A \tag{1.19}$$

where A is a number close to unity, $\omega_p$ is the free electron plasma frequency and $E_h$ is the homopolar energy gap. The free electron plasma frequency is given by

$$\omega_p^2 = \frac{4\pi ne^2}{m} \tag{1.20}$$

where m is the free electron mass and n = 8 electrons per diatomic unit cell volume.

The *static* and *high* frequency dielectric constants are related by

$$\epsilon(0) = \epsilon(\infty) + \frac{4\pi Ne_T^2}{\omega_{TO}^2 M} \tag{1.21}$$

where M is the reduced mass, $e_T$ the effective charge, $\omega_{TO}$ the TO phonon frequency, and N the number of ion pairs. Note from Table 1.11 that both $\epsilon(0)$ and $\epsilon(\infty)$ show an inverse dependence on $E_g$, decreasing with increasing $E_g$.

In the study of transport and optical properties of polar semiconductors, the coupling between the electron and LO phonons becomes important. The electron - LO phonon interaction is given by the well known Frohlich coupling constant [18]:

$$\alpha_F = \frac{e^2}{4\pi\epsilon_0} \left[ \frac{m^*}{2\hbar^3 \omega_{LO}} \right]^{1/2} \left[ \frac{1}{\epsilon(\infty)} - \frac{1}{\epsilon(0)} \right] \tag{1.22}$$

where $\omega_{LO}$ is the LO phonon frequency. Equation (1.22) shows that $\alpha_F$ depends on the ionic polarizability of the crystal which is related to $\epsilon(\infty)$ and $\epsilon(0)$.

Values of k and n over a wide range of photon energies for many III-V semiconductors has been compiled by Seraphin and Bennett [19].

**Elastic properties**

The elastic compliances $S_{ij}$ in units of $10^{-12}$ cm$^2$ dyne$^{-1}$ are listed in Table 1.12. For a cubic crystal, there are three independent elastic constants $S_{11}$, $S_{12}$, and $S_{44}$. The elastic compliance relates strain to stress

$$\epsilon_i = \sum_j S_{ij} \sigma_j \tag{1.23}$$

**Table 1.12**

Elastic Compliances in Units of $10^{-12}$ cm$^2$ dyne$^{-1}$

| Compound | $S_{11}$ | $S_{12}$ | $S_{44}$ |
|---|---|---|---|
| AlP | 1.090 | −0.350 | 1.630 |
| AlAs | 1.070 | −0.320 | 1.840 |
| AlSb | 1.696 | −0.562 | 2.453 |
| GaP | 0.973 | −0.298 | 1.419 |
| GaAs | 1.176 | −0.365 | 1.684 |
| GaSb | 1.582 | −0.495 | 2.314 |
| InP | 1.650 | −0.594 | 2.170 |
| InAs | 1.945 | −0.685 | 2.525 |
| InSb | 2.443 | −0.863 | 3.311 |

NOTE: The elastic stiffnesses $C_{ij}$ is related to $S_{ij}$:
$C_{11}=(S_{11}+S_{12})/S$;
$C_{12}=-S_{12}/S$;
$C_{44}=1/S_{44}$;
where $S = (S_{11}-S_{12})(S_{11}+2S_{12})$.

where the strain $\epsilon_i$ and stress $\sigma_j$ are the components of 2nd rank tensors and $S_{ij}$s are components of a 4th rank tensor. Equation (1.23) can be expressed in its covariant form

$$\sigma_i = \sum_j C_{ij} \epsilon_j \tag{1.24}$$

where $C_{ij}$s are the elastic stiffness constants. The compliance tensor is the reciprocal of the stiffness tensor. For a cubic crystal $C_{ij}$s are related to $S_{ij}$s according to [20]

$$C_{11} = \frac{S_{11} + S_{12}}{S}$$

$$C_{12} = -\frac{S_{12}}{S} \tag{1.25}$$

$$C_{44} = \frac{1}{S_{44}}$$

where

$$S = (S_{11} - S_{12})(S_{11} + 2S_{12})$$

From the point of practical utility, the elastic constant of interest is the Young's modulus E. Another elastic property of interest is the Poisson's ratio $\nu$. In the cubic crystals both E and $\nu$ are anisotropic. The expressions for E and $\nu$ for an arbitrary crystallographic direction in a cubic crystal are given by Brantley [21]

$$\frac{1}{E} = S_{11} - 2(S_{11} - S_{12} - \frac{1}{2} S_{44})(l_1^2 l_2^2 + l_2^2 l_3^2 + l_1^2 l_3^2)$$

$$\nu = -\frac{S_{12} + (S_{11} - S_{12} - \frac{1}{2} S_{44})(l_1^2 m_1^2 + l_2^2 m_2^2 + l_3^2 m_3^2)}{S_{11} - 2(S_{11} - S_{12} - \frac{1}{2} S_{44})(l_1^2 l_2^2 + l_2^2 l_3^2 + l_1^2 l_3^2)} \tag{1.26}$$

where $l$ is the longitudinal stress axis and m is an orthogonal direction to $l$ and the $l_s$ and $m_s$ are the direction cosines for $l$ and m with respect to the cube axes. For the <100> axes

$$\frac{1}{E} = S_{11}$$

$$\nu = -\frac{S_{12}}{S_{11}} \tag{1.27}$$

From Eq. (1.26) it follows that the factor $2(S_{11} - S_{12})/S_{44}$ is a measure of deviation from isotropy. For the biaxial plane stress conditions associated with thin films, often encountered in the case of heterostructures, longitudinal stresses and strains parallel to the film substrate are related by $E/(1-\nu)$. The expressions for this composite elastic constant are easily obtained from Eq. (1.26).

The bulk modulus for zinc blende type crystals is given by

$$B = \frac{1}{3(S_{11}+2S_{12})} \tag{1.28}$$

For <100> axes $B = E/3(1-2\nu)$. Knowing the density $\rho$ and $C_{ij}$ one can obtain the sound velocity v using the relation

$$v = (C_{ij}/\rho)^{1/2} \tag{1.29}$$

**Deformation potential**

Application of an external stress has a profound effect on the band structure. Hydrostatic stress causes a shift of the energy states. Uniaxial stress lowers the symmetry of the lattice and splits some degenerate states. In cubic semiconductors, the 4-fold degeneracy of the valence band is lifted resulting in a $J = 3/2$ $m_j = 3/2$ and $J = 3/2$ $m_j = 1/2$ heavy hole and light hole bands. The shift of $E_g$ with hydrostatic stress and the splitting of the valence band $\delta E_v$ are given by the deformation potential constants. The values of these constants are listed in Table 1.13. Variation of direct energy gap with hydrostatic pressure p is described by constant a

$$a = -\frac{1}{3}(C_{11}+2C_{12})\frac{dE_g}{dp} \tag{1.30}$$

**Table 1.13**

Deformation Potential Constants

| Compound | a (eV) | | b (eV) | d (eV) |
|---|---|---|---|---|
| | Direct gap | Indirect gap | | |
| AlP | | | | |
| AlAs | | | | |
| AlSb | −5.9 | 2.2 | −1.35 | −4.3 |
| GaP | −9.6 | | −1.65 | −4.5 |
| GaAs | −9.77 | | −1.70 | −4.55 |
| GaSb | −8.28 | | −2.0 | −4.7 |
| InP | −6.35 | | −2.0 | −5.0 |
| InAs | −6.0 | | −1.8 | −3.6 |
| InSb | −7.7 | | −2.0 | −4.9 |

The splitting of the valence band under [100] stress is described by the deformation potential constant b and the splitting under [111] stress by constant d, according to the relation

$$\delta E_v = b(S_{11} - S_{12})\sigma$$

$$\delta E_v = \left[\frac{d}{2\sqrt{3}}\right] S_{44}\, \sigma \tag{1.31}$$

where σ is the applied uniaxial stress.

### 1.5.2 Alloy Semiconductors

Figure 1.12 shows the lattice parameter versus energy gap at room temperature for various III-V compounds and their alloys [22]. The composition of the lattice matched ternary alloys and the corresponding substrates on which they are grown are indicated in Fig. 1.12. The shaded regions I and II represent the quaternary alloys $Ga_xIn_{1-x}As_yP_{1-y}$ and $Ga_xIn_{1-x}As_ySb_{1-y}$, respectively.

**Figure 1.12** Lattice parameter versus energy gap at room temperature for various III-V compounds and their alloys [22].

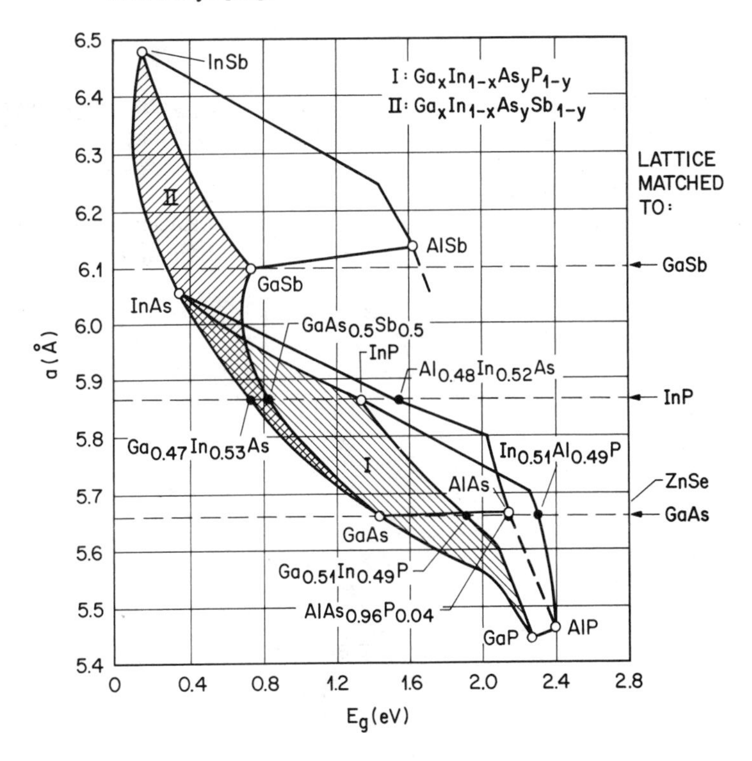

The former quaternary alloy can be grown lattice matched on GaAs or InP substrates. When GaAs is used as the substrate, the lattice matched GaInAsP alloy covers the energy range 1.42-1.91 eV, more or less the same energy range as that covered by the ternary alloy AlGaAs. The GaInAsP alloy grown on InP substrate spans the energy range 0.75-1.35 eV between the band gap of $Ga_{0.47}In_{0.53}As$ (0.75 eV) and that of InP (1.35 eV). To obtain most physical parameters of alloy semiconductors linear variation with composition may not be generally applicable. However, when specific experimental data are unavailable, a linear interpolation may be used using the formulas given in Sec. 1.4.

### Lattice constant

The lattice constant of ternary III-V alloys generally varies linearly with composition (Vegard's law). This is illustrated in Fig. 1.13 for the $Al_xGa_{1-x}As$ alloy [23]. Vegard's law can also be assumed for the quaternary alloys. The $Ga_xIn_{1-x}As_yP_{1-y}$ alloys grown lattice matched on InP substrates can be considered part of a pseudo binary system between $Ga_{0.47}In_{0.53}As$ and InP. The lattice matching relation between x and y can be expressed as

$$x \cong 0.47y \quad (0 < y < 1.0)$$

or using Eq. (1.10) more rigidly as

$$x = \frac{0.1896\,y}{0.4175 - 0.012\,y} \quad (0 < y < 1) \tag{1.32}$$

### Band gap

A parameter which is of interest often is the variation of $E_g$ with composition. Besides the variation of the minimum energy gap, the variation of the energy separation between the different

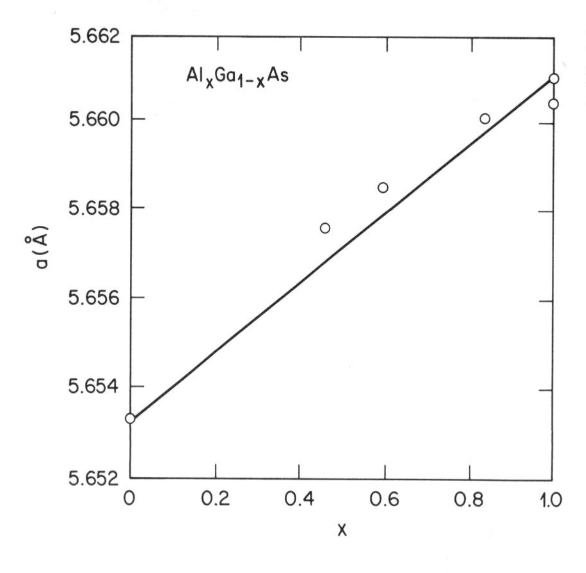


**Figure 1.13** Room temperature lattice parameter versus composition for $Al_xGa_{1-x}As$ [23].

conduction band minima are also of interest. The composition dependence of the energy gap is often represented by Eq. (1.8). Table 1.14 gives $E_g$ versus composition for some important ternary alloy semiconductors [24 - 26]. Figure 1.14 shows the Γ,X and L energy gaps versus composition for $Al_xGa_{1-x}As$, the most intensively studied III-V ternary alloy [27]. The calculated compositional dependence of the Γ,X and L energy gaps for another well studied ternary alloy $Ga_xIn_{1-x}As$ is shown in Fig. 1.15 [28]. The variation of the Γ gap at 2K in a narrow range of composition for unstrained $Ga_xIn_{1-x}As$ layers is shown in Fig. 1.16 [29]. Away from the lattice matched composition considerable shift in $E_g$ from the calculated value can occur caused by built-in elastic strain [30 - 32].

### $Ga_xIn_{1-x}As_yP_{1-y}$ on InP substrate

GaInAsP lattice matched to InP has a direct gap over the entire range of alloy compositions. By fitting photoluminescence and electroreflectance data at room temperature to Eq. (1.8) the following expression for $E_g$ has been obtained [33]

$$E_g(y) = 1.35 - 0.775\, y + 0.149\, y^2 \tag{1.33}$$

**Table 1.14**

Compositional Dependence of the Energy Gap in III-V Ternary Alloy Semiconductors at 300K

| Alloy | Direct Energy Gap $E_\Gamma$ | Indirect Energy Gap $E_X$ | Indirect Energy Gap $E_L$ |
|---|---|---|---|
| $Al_xIn_{1-x}P$ | $1.34+2.23x$ | $2.24+0.18x$ | |
| $Al_xGa_{1-x}As$ | $1.424+1.247x$ $(x<0.45)$<br>$1.424+1.087x+0.438x^2$ | $1.905+0.10x+0.16x^2$ | $1.705+0.695x$ |
| $Al_xIn_{1-x}As$ | $0.36+2.35x+0.24x^2$ | $1.8+0.4x$ | |
| $Al_xGa_{1-x}Sb$ | $0.73+1.10x+0.47x^2$ | $1.05+0.56x$ | |
| $Al_xIn_{1-x}Sb$ | $0.172+1.621x+0.43x^2$ | | |
| $Ga_xIn_{1-x}P$ | $1.34+0.511x+0.6043x^2$<br>$(0.49<x<0.55$, VPE layer) | | |
| $Ga_xIn_{1-x}As$ | $0.356+0.7x+0.4x^2$ | | |
| $Ga_xIn_{1-x}Sb$ | $0.172+0.165x+0.413x^2$ | | |
| $GaP_xAs_{1-x}$ | $1.424+1.172x+0.186x^2$ | | |
| $GaAs_xSb_{1-x}$ | $0.73-0.5x+1.2x^2$ | | |
| $InP_xAs_{1-x}$ | $0.356+0.675x+0.32x^2$ | | |
| $InAs_xSb_{1-x}$ | $0.18-0.41x+0.58x^2$ | | |

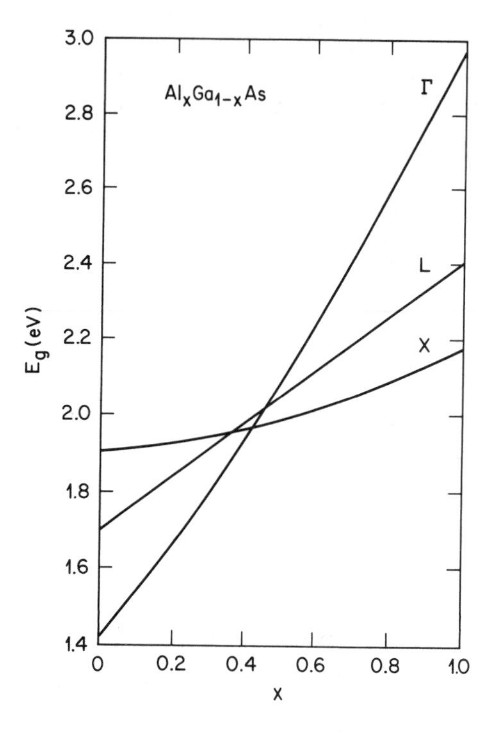


**Figure 1.14** Position of the Γ, X and L conduction band minima relative to the top of the valence band versus composition in $Al_xGa_{1-x}As$ at room temperature [27].

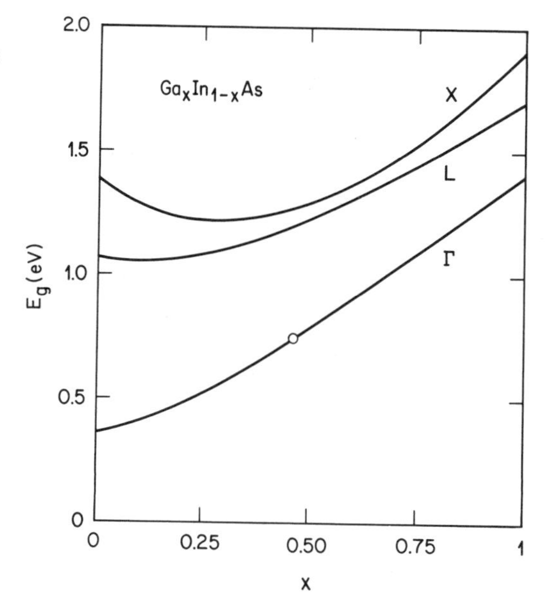


**Figure 1.15** Calculated compositional dependence of the Γ, X and L energy gaps in room temperature in $Ga_xIn_{1-x}As$ [28].

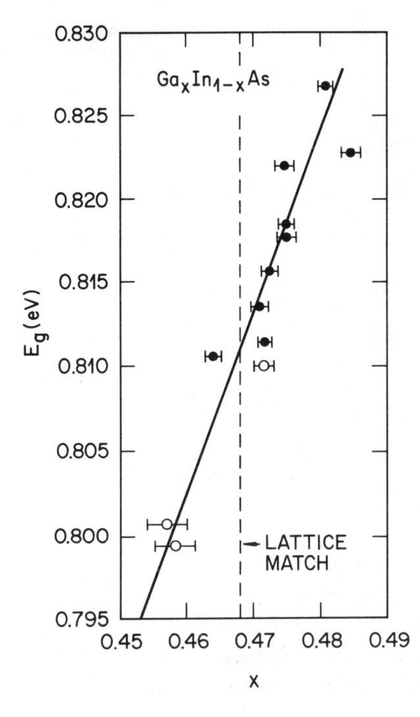


**Figure 1.16** Energy gap at 2K versus composition for $Ga_xIn_{1-x}As$ near the lattice matched compositions on InP substrates. Full circles: LPE layers, open circles: VPE layers. Solid line is given by $E_g = 0.4105 + 0.6337x + 0.475\, x^2$ [29].

For the quaternary alloy, the bowing parameter c in Eq. (1.8) is the sum of the contribution from each sub lattice

$$E_o(x,y) = a + by + c_{III}\, x(1-x) + c_V\, y(1-y) \tag{1.34}$$

For the lattice matched compositions, the only contribution to bowing is the extrinsic bowing caused by alloy disorder since the intrinsic bowing parameter is, in principle, zero. Pearsall [33] calculated the disorder bowing parameter according to the prescription in Ref. 6 and obtained the following expression

$$c\,(y) = 0.219\, y - 0.149\, y^2 \tag{1.35}$$

Substituting Eq. (1.35) in Eq. (1.8) then gives Eq. (1.33). Figure 1.17 shows the experimental and calculated $E_g(y)$ and spin-orbit splitting $\Delta(y)$ values.

For the general case of arbitrary x and y, Kuphal [34] has given the following expression for $E_g$

$$E_g(x,y) = 1.35+0.668x-1.068y+0.758x^2+0.078y^2-0.069xy-0.322x^2y+0.03xy^2 \tag{1.36}$$

This general expression for $E_g$ closely reproduces Eq. (1.33) for lattice matched alloy. The composition parameters x and y can be obtained from $E_g$ measurement (e.g., photoluminescence)

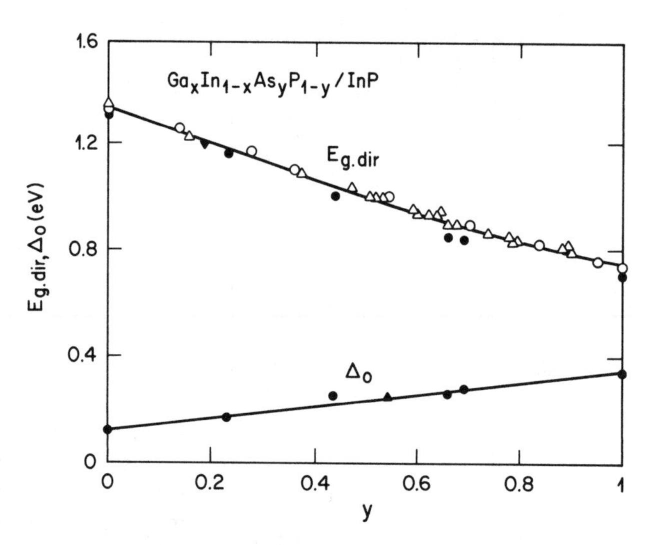


**Figure 1.17** Direct energy gap $Eg,_{dir}$ and spin orbit splitting $\Delta_0$ versus composition in $Ga_xIn_{1-x}As_yP_{1-y}$ / InP at room temperature. The data points represent several experimental sources and the solid lines are calculated [33].

and lattice mismatch Δa/a measurement (e.g., X-ray rocking curve) using Eq. (1.36) and Vegard's law given by

$$a(x,y) = 0.1896y - 0.4175x + 0.0124xy + 5.8687 = a_{InP}\,[1 + \left(\frac{\Delta a}{a}\right)_0] \tag{1.37}$$

$(\Delta a / a)_0$ in Eq. (1.37) is the relaxed mismatch which is related to the measured $(\Delta a / a)_\perp$ for the case of thin epitaxial layers [35]

$$(\Delta a / a)_\perp = \frac{1+\nu}{(1-\nu)}\,(\Delta a / a)_0 \tag{1.38}$$

which for {100} crystals according to Eq. (1.27) reduces to

$$\left(\frac{\Delta a}{a}\right)_{100} = \frac{S_{11} - S_{12}}{S_{11} + S_{12}} \left(\frac{\Delta a}{a}\right)_0 \tag{1.39}$$

Using the $S_{ij}$ values in Table 1.12 we obtain

$$\left(\frac{\Delta a}{a}\right)_{100} = 2.125\,(\Delta a / a)_0 \tag{1.40}$$

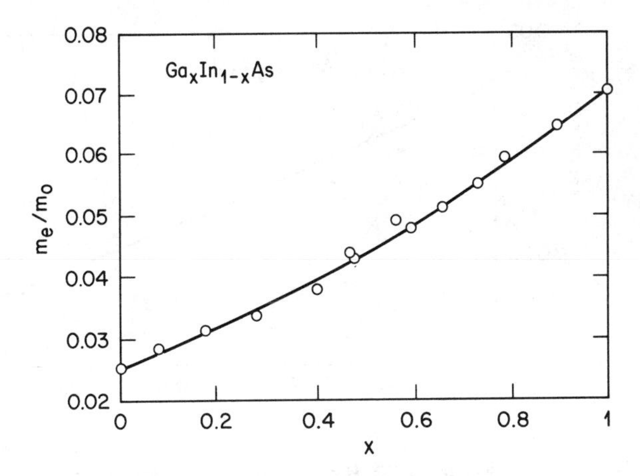


**Figure 1.18** Electron effective mass versus composition in $Ga_xIn_{1-x}As$ [37].

**Effective mass in ternary and quaternary alloys**

According to the k·p approximation, the effective mass of the carrier is related to the square of the matrix element $P^2$ via Eq. (1.7). It has been found in many ternary alloys that the effective mass calculated under the assumption that $P^2$ varied linearly between the values for the binary compounds, does not agree well with the experimental values. The origin of this discrepancy has been attributed to disorder induced mixing of the conduction and valence bands at the Γ point [36]. The effect of this mixing is to reduce $P^2$ and thus to increase the effective mass above that which would be obtained by the k·p calculation. Figures 1.18 and 1.19 show the electron effective mass for $Ga_xIn_{1-x}As$ [37] and $Al_xIn_{1-x}As$ [38], respectively. Figures 1.20 and 1.21 show the electron effective mass and light hole effective mass for lattice matched $Ga_xIn_{1-x}As_yP_{1-y}$, respectively [33]. The heavy hole and split-off band masses can be obtained from the values of the binary compounds using Eq. (1.10).

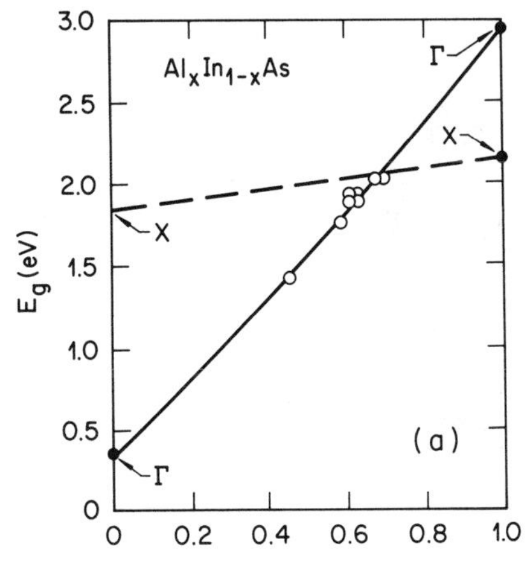


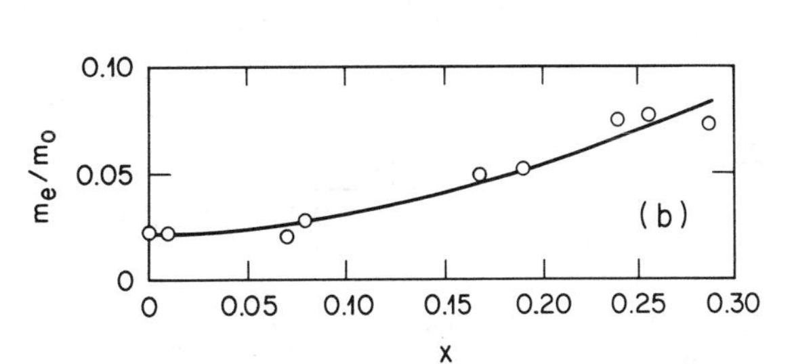


**Figure 1.19** Γ and X energy gaps and electron effective mass in $Al_xIn_{1-x}As$ [38].

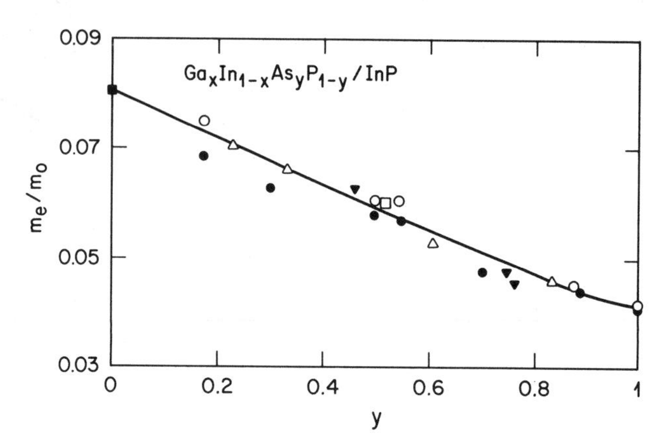


**Figure 1.20** Electron effective mass versus composition in $Ga_xIn_{1-x}As_yP_{1-y}$ / InP [33].

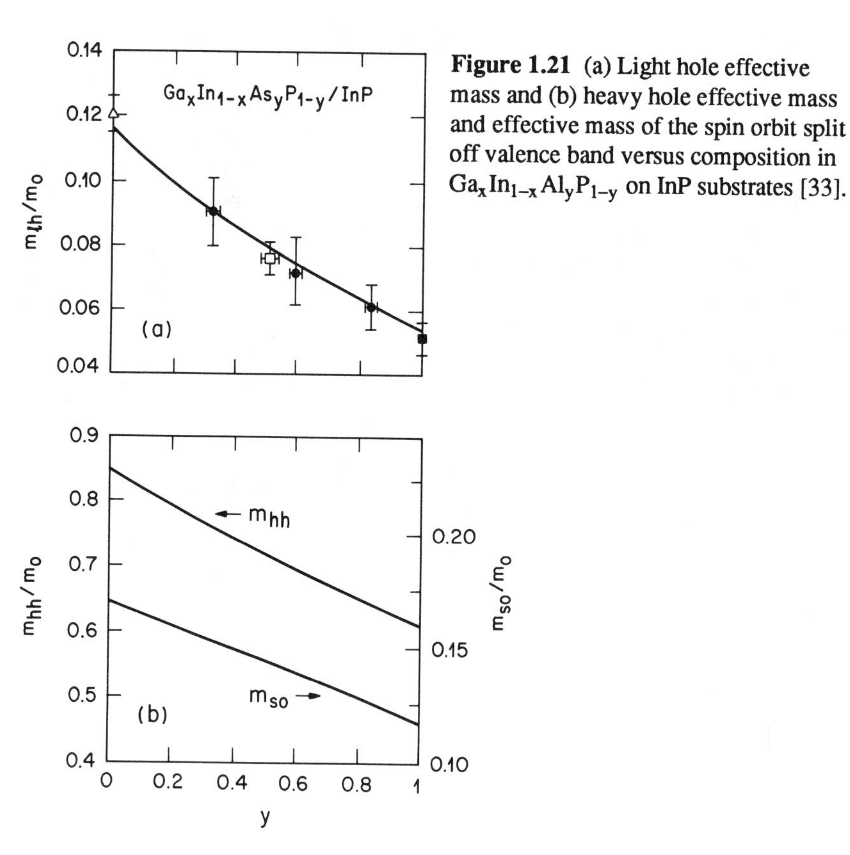


**Figure 1.21** (a) Light hole effective mass and (b) heavy hole effective mass and effective mass of the spin orbit split off valence band versus composition in $Ga_xIn_{1-x}Al_yP_{1-y}$ on InP substrates [33].

**Thermal resistivity**

Thermal resistivity of semiconductors is an important parameter in the design of power dissipating devices such as semiconductor lasers. It is also important in calculating the figure of merit of thermoelectric devices (e.g., Peltier devices). The thermal resistivity of ternary and quaternary alloys follows a quadratic relation the same as $E_g$. For the ternary alloy $A_xB_{1-x}C$ it can be written as

$$W(x) = x\,W_{AC} + (1-x)\,W_{BC} + c_{A-B}x(1-x) \tag{1.41}$$

where $W_{AC}$ and $W_{BC}$ are the thermal resistivities of the binary compounds. The bowing parameter $c_{AB}$ for several ternary alloys has been obtained by Adachi [39] and are given in Table 1.15. Note that $c_{As-P}$ values of InAsP and GaAsP alloys are considerably smaller than $c_{In-Ga}$ of InGaAs and that $c_{Ga-Al} \approx c_{As-P}$. The mean atomic weight of In-Ga is about two times heavier than that of As-P which is nearly the same as that of Al-Ga. The anharmonic contribution to thermal resistivity resulting from mass difference between the atoms is supposed to be responsible for the differences and similarities in the c values.

The thermal resistivity of a quaternary alloy $A_xB_{1-x}C_yD_{1-y}$ can be expressed in the same way as Eq. (1.41) but with two bowing parameters $c_{A-B}$ and $c_{C-D}$. The thermal resistivity of $Ga_xIn_{1-x}As_yP_{1-y}$ lattice matched to InP calculated using $c_{In-Ga} = 72\ W^{-1}$ deg cm and $c_{As-P} \simeq 25\ W^{-1}$ deg cm is shown in Fig. 1.22 [39]. The thermal resistivity increases with increasing As content and reaches a maximum value of 24 $W^{-1}$ deg cm at $y \sim 0.75$.

**Thermal expansion coefficient**

In double heterostructures consisting of compositionally different layers, differences in the thermal expansion coefficient of the layers can generate elastic stresses during cooling from the growth temperature to room temperature or during thermal cycling of heterostructure devices. In

**Table 1.15**

The alloy disorder bowing parameter $c_{A-B}$ denoting the deviation from linearity of the thermal resistivity of ternary alloys.
The bowing parameter is obtained by fitting experimental data to Eq. (1.41) [39].

| Ternary alloy | $C_{A-B}$ ($W^{-1}$ deg cm) |
|---|---|
| InGaAs | 72 |
| InAsP | 30 |
| GaAsP | 20 |
| AlGaAs | 30 |
| InGaP | 72 $^{est}$ |
| InGaSb | 72 $^{est}$ |

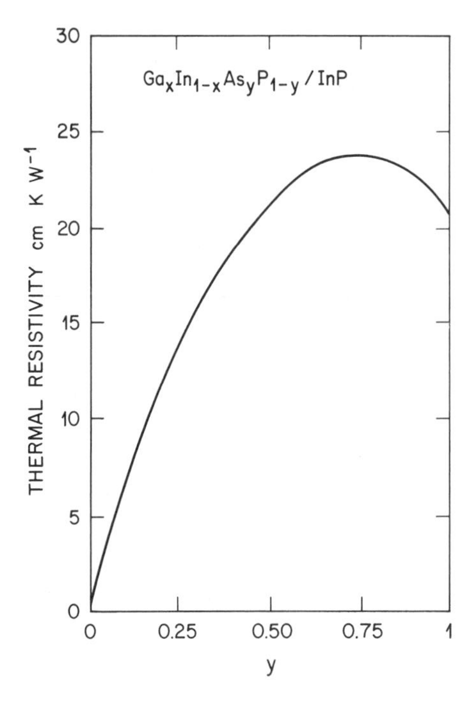


**Figure 1.22** Thermal resistivity versus composition calculated for $Ga_xIn_{1-x}As_yP_{1-y}$ / InP [39].

Fig. 1.23, the thermal expansion coefficient of $Ga_xIn_{1-x}As_yP_{1-y}$ alloy lattice matched to InP obtained from linear interpolation is shown together with experimental data at selected y values [40]. Pietsch and Marlow [41] found that the thermal expansion coefficient of the relaxed lattice (see Eq. 1.40) is about 30 percent smaller than the values shown in Fig. 1.23.

### Refractive index

One of the important factors in the design of heterostructure lasers and optoelectronic devices is the refractive index, n. Often knowledge of n as a function of photon energy near and below the band gap is required. Several models have been proposed to obtain n as a function of photon energy and composition in ternary and quaternary alloys. In many models the starting point for calculating n in alloys is the semi empirical single effective oscillator model proposed by Wemple and DiDomenico [42] to analyze refractive index dispersion in many covalent, ionic, and amorphous materials. According to this model, n is given at a photon energy E

$$n^2 - 1 = E_d E_0 / (E_0^2 - E^2) \tag{1.42}$$

where $E_0$ is the single oscillator energy and $E_d$ is the dispersion energy which is a measure of the strength of interband optical transitions. The single oscillator parameters $E_0$ and $E_d$ of the binaries can be interpolated according to Eq. (1.9) or Eq. (1.10) to yield n of ternary or quaternary alloys. However, description of n near the band edge, the region of most interest, using this model appears to be inadequate even for binary compounds. Afromowitz [43] proposed a modified single oscillator model to explain dispersion of n near the band gap. In this model $n^2$ is expressed as

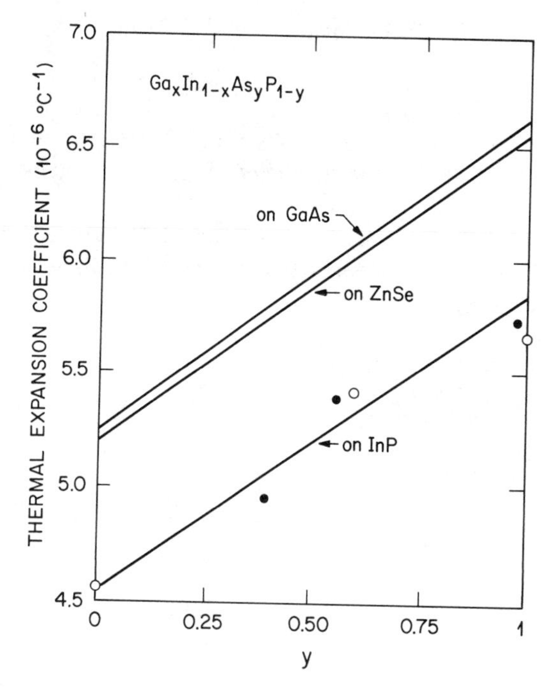


**Figure 1.23** Thermal expansion coefficient of $Ga_xIn_{1-x}Al_yP_{1-y}$ lattice matched to InP, ZnSe and GaAs as a function of composition [40].

$$n^2 - 1 = \frac{E_d}{E_0} + \frac{E_d E^2}{E_0^3} + \frac{\eta E^4}{\pi} \ln\left[\frac{2E_0^2 - E_g^2 - E^2}{E_g^2 - E^2}\right] \tag{1.43}$$

where

$$\eta = \frac{\pi E_d}{2E_0^3(E_0^2 - E_g^2)} \tag{1.44}$$

The index describing the group velocity, $\bar{n}$ which is related to the longitudinal mode spacing in lasers, is given by

$$\bar{n} = n + E\, dn / dE \tag{1.45}$$

and can be obtained from Eq. (1.43).

Buus and Adams [44] used Eq. (1.43) to calculate n as a function of E and y in $Ga_xIn_{1-x}As_yP_{1-y}$ lattice matched to InP. The values of $E_0$ and $E_d$ were obtained from their respective values in the binary compounds using Eq. (1.10) and are given by

$$\begin{aligned} E_0(y) &= 3.391 - 1.652y + 0.863y^2 \\ E_d(y) &= 28.91 - 9.278y + 5.626y^2 \end{aligned} \tag{1.46}$$

Equation (1.46) together with Eq. (1.33) for $E_g(y)$ can be substituted in Eqs. (1.43) and (1.44) to obtain n as a function of E and y.

Burkhard et al. [45] have measured n for $Ga_xIn_{1-x}As_yP_{1-y}$ lattice matched to InP as a function of y and E for $E > E_g$. They fitted their experimental results using the interpolation given by Eq. (1.10) and using the quantity $(\epsilon - 1)/(\epsilon + 2)$ as the parameter. Then n was obtained taking $n^2 \approx \epsilon$.

Adachi [46] proposed a generalized model of dielectric constants of semiconductors based on models of the intraband transitions. According to this model the real part of the dielectric constant $\epsilon$ is expressed as a function of photon energy E as

$$\epsilon(E) = A_0\{f(x) + \frac{1}{2}\left[\frac{E_g}{E_g + \Delta_0}\right]^{3/2} f(x_0)\} + B_0 \tag{1.47}$$

where $A_0$ and $B_0$ are constants obtained by fitting experimental data to Eq. (1.47), and

$$f(x) = x^{-2}[2 - (1+x)^{½} - (1-x)^{½}]$$

$$x = E/E_g \tag{1.48}$$

$$x_0 = \frac{E}{E_g + \Delta_0}$$

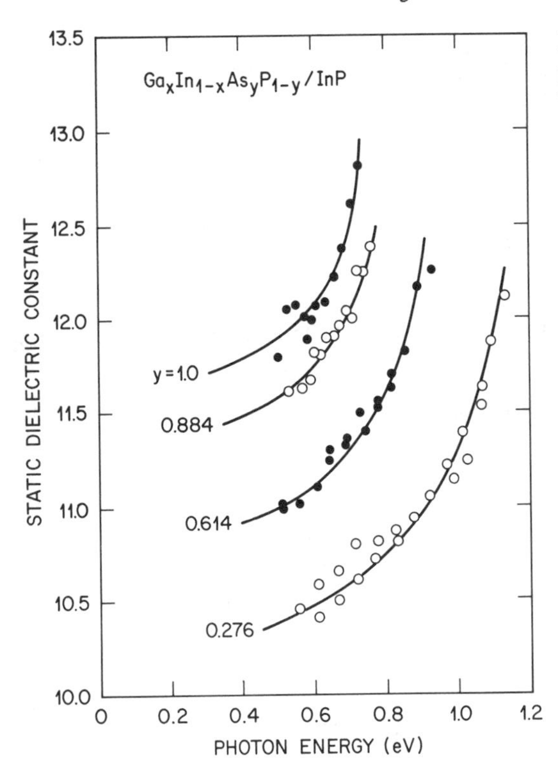


**Figure 1.24** Dielectric constant versus photon energy for $Ga_xIn_{1-x}As_yP_{1-y}$ / InP at various compositions [47].

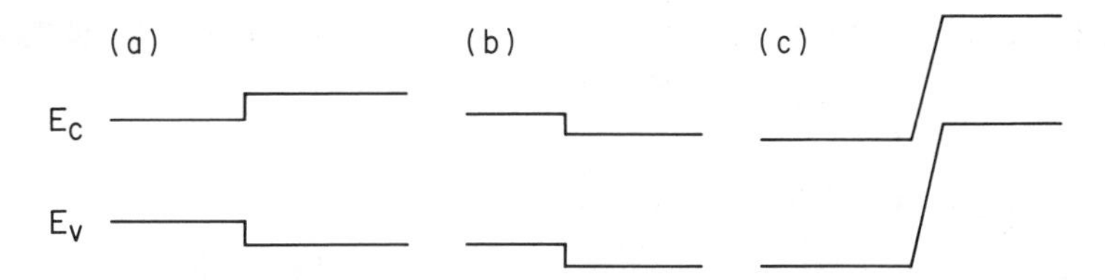


**Figure 1.25** Schematic band gap diagrams showing the different heterojunction band lineup: (a) Straddling lineup (b) Staggered lineup and (c) broken-gap lineup.

Figure 1.24 shows the experimental values of $\epsilon(\approx n^2)$ for $Ga_xIn_{1-x}As_yP_{1-y}$ as a function of wavelength [47] for several compositions. The dispersion curves calculated based on Eqs. (1.47) and (1.48) are also shown in Fig. 1.24. The agreement with experimental curves is quite good over most of the wavelength range.

**Heterojunctions**

Heterostructures have attracted much interest because of scientific and technological interest [48-51]. The primary reason for the consideration of heterostructures for semiconductor device applications is the abrupt change in the energy band structure at the heterointerface. This leads to discontinuities, or offsets, in the conduction and valence band edges. Such offsets act as potential steps in addition to pure electrostatic potential variation that may be present and as such can be used to control the flow and distribution of electrons and holes in device structures.

Heterojunctions can be classified by their band lineup. The types of band offsets that occur at abrupt semiconductor heterojunctions are illustrated in Fig. 1.25. In the straddling band lineup shown in Fig. 1.25a, the band offsets in both the conduction and valence bands act as potential barriers and keep electrons and holes in the smaller band gap semiconductors. Heterostructures that have straddling lineups are called type I heterostructures. Notable examples of these are the GaAs-$Al_xGa_{1-x}As$ and the InP-GaInAsP heterojunctions. Some heterojunctions have staggered (Fig. 1.25b) or broken-gap (Fig. 1.25c) band lineups. These are called type II heterostructures. The heterojunction InAs/GaSb is of the broken-gap type in which the top of valence band of GaSb is higher than the bottom of conduction band of InAs. In this situation, the electrons at the top region of valence band of GaSb can freely enter the conduction band of InAs when the two semiconductors are brought into contact with each other. Photoluminescence measurement of the valence band offset in $Al_{0.3}Ga_{0.7}As$-AlAs and $Al_{0.3}Ga_{0.7}As$-$Al_{0.7}Ga_{0.3}As$ heterostructures shows that the band alignments for these two systems change from type I to type II as the lowest conduction band state becomes the X band in AlAs or $Al_{0.7}Ga_{0.3}As$ [52].

Apart from the classification of heterostructures based on band lineups, a further important distinction to be made between them is whether they are lattice matched or lattice mismatched structures (see Fig. 1.12). The availability of good quality substrates dictates to a large extent as to which of the lattice matched heterostructures would be of use for device structures. At present, technologically important III-V heterostructures are limited to those which are grown on GaAs and InP substrates.

The parameter of utmost importance in heterostructures for device applications is the band edge offset. There have been many experimental and theoretical determination of band offsets over the last 20 years. The band edge offsets in the conduction and valence bands are related to the difference in the band gap of the two components of the heterojunction as

$$\Delta E_V + \Delta E_C = \Delta E_g \tag{1.49}$$

Thus if one of the offsets is calculated the other is easily obtained since $\Delta E_g$ is reasonably well known. The calculation of $\Delta E_V$, for example, requires knowledge of the absolute energies of the valence band edges of the semiconductors. Further, when the two semiconductors are brought into contact with each other, the charge redistribution in the interface region and the associated electronic interface dipole effects may be significant. Given these two facts, $\Delta E_V$ may be written as

$$\Delta E_V = (E^A_{V_0} - E^B_{V_0}) - \delta E_V \tag{1.50}$$

where $E^A_{V_0}$ and $E^B_{V_0}$ are the energies with respect to the vacuum level of the valence band maximum of the two semiconductors A and B, respectively, and $\delta E_V$ is the energy shift in the band edges at the interface caused by the formation of the interface dipole. Apart from charge redistribution at the interface, the interface dipole effect may also be affected by the presence of localized interface defects, particularly in mismatched heterostructures [53].

**Theories of band offsets**

Anderson [54] has suggested the electron affinity rule (EAR) to calculate the band offset according to which $\Delta E_C$ is equal to the difference in electron affinities of A and B as shown schematically in Fig. 1.26.

$$\Delta E_C = \chi_A - \chi_B \tag{1.51}$$

$\Delta E_V$ is obtained from Eq. (1.49) neglecting contribution of $\delta E_V$.

In the Harrison atomic orbital (HAO) model [55] $\Delta E_V$ is estimated as the difference in the ionization energies of A and B which are calculated using linear combination of atomic orbitals (LCAO) model. From Fig. 1.26

$$\Delta E_V = \phi_B - \phi_A = E^B_{V_0} - E^A_{V_0} \tag{1.52}$$

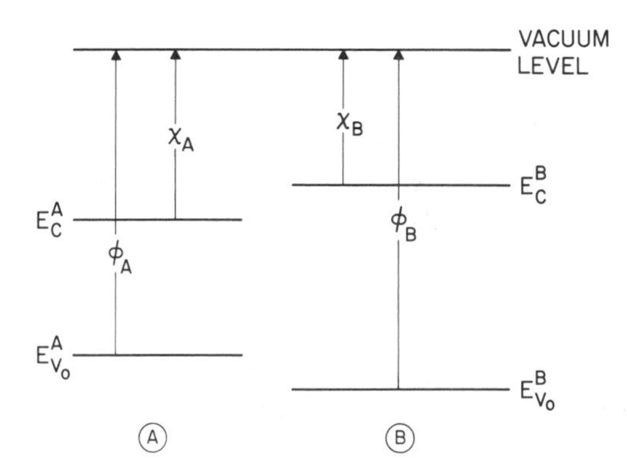


**Figure 1.26** Schematic band diagram showing the differences in electron affinity and ionization potential of semiconductors A and B.

Once again $\delta E_V$ is neglected in determining $\Delta E_V$.

Frensley-Kroemer pseudopotential theory [56] preceeds the HAO model. It is similar in principle to HAO but differs in calculational methodology.

Tersoff has proposed the "midgap lineup" model to treat the interface dipole effects on band offsets [57]. In this model, a midgap level is assigned to each semiconductor. When two semiconductors form a heterojunction, their midgap levels are lined up. Good agreement has been obtained between offsets calculated by this model and determined experimentally for many III-V semiconductors on Ge and Si.

Ruan and Ching [58] recently proposed an effective dipole theory for band lineups. They consider the penetration of the effective-mass electrons from the material with higher valence band top into the other which results in an effective charge transfer across the heterointerface and the formation of an effective dipole. The offsets predicted by this model differ from experimental values by ~0.1 eV on average for 30 heterojunctions. Table 1.16 lists the valence band offsets for some important GaAs, InP based heterostructures [58].

**Table 1.16**

Calculated and experimental valence band offsets (eV) for GaAs, InP based heterostructures as compiled in Ref. [58].

| Heterostructure | Experiment | EAR | HAO | Tersoff's model | Effective dipole model[a] |
|---|---|---|---|---|---|
| Ge/InP | 0.64 | 0.90 | 0.64 | 0.58 | 0.67 |
| Ge/GaAs | 0.25-0.65 | 0.70 | 0.41 | 0.52 | 0.51 |
| Si/InP | 0.57 | 0.58 | 0.24 | 0.40 | 0.36 |
| Si/GaAs | 0.05 | 0.38 | 0.03 | 0.34 | 0.22 |
| GaAs/AlAs | 0.19-0.50 | 0.52 | 0.04 | 0.35 | 0.36 |
| GaAs/InAs | 0.17 | 0.19 | 0.32 | 0.20 | 0.16 |
| GaAs/ZnSe | 0.96-1.10 | 1.31 | 1.05 | – | 0.95 |
| InP/CdS | 1.63 | 1.40 | 1.36 | – | 0.97 |
| InSb/InP | – | 0.87 | | | 0.71 |
| InSb/GaAs | – | 0.67 | | | 0.53 |
| InAs/InP | | 0.39 | | | 0.29 |
| GaAs/InP | – | 0.20 | | | 0.13 |
| $Ga_{0.47}In_{0.53}As$ / InP | $0.346 \pm 0.01$[b]<br>0.39[c] | | | | |

(a) Ref. [58].

(b) D. V. Lang, *Heterojunction Band Discontinuities,* eds. F. Capasso and G. Margaritondo, Amsterdam; North-Holland, 1987, p. 377.

(c) S. R. Forrest, *Heterojunction Band Discontinuities,* eds. F. Capasso and G. Margaritondo, Amsterdam; North-Holland, 1987 p. 311.

## REFERENCES


1. R. Zallen *Handbook on Semiconductors, Vol. 1,* ed. W. Paul. Amsterdam; North Holland, 1982, Chap. 1.
2. J. C. Phillips, *Bonds and Bands in Semiconductors.* New York; Academic, 1973.
3. E. O. Kane, *Handbook on Semiconductors, Vol. 1,* ed. W. Paul, Amsterdam; North Holland, 1982, Chap. 4A.
4. Q. H. F. Vrehen, *J. Phys. Chem. Solids 29:* 129, 1968.
5. M. L. Cohen and J. R. Chelikowsky *Electronic Structure and Optical Properties of Semiconductors.* Berlin; Springer-Verlag, 1988.
6. J. A. Van Vechten and T. K. Bergstresser, *Phys. Rev.* B 1:3351, 1970.
7. J. C. Phillips and J. A. Van Vechten, *Phys. Rev. Lett.* 22:705, 1969.
8. J. A. Van Vechten, *Phys. Rev.* 187; 1007, 1969.
9. T. H. Glisson, J. R. Hauser, M. A. Littlejohn and C. K. Williams, *J. Electron. Mat.* 7:1, 1978.
10. R. L. Moon, G. A. Antypas and L. W. James, *J. Electron. Mater.* 3:635, 1974.
11. Landolt-Bornstein, "Numerical Data and Functional Relationships", in *Science and Technology,* Vol. 17: Semiconductors subvol. a, (eds. O. Madelung, M. Schulz and H. Weiss. New York; Springer, 1982.
12. Landolt-Bornstein, "Numerical Data and Functional Relationships" in, *Science and Technology,* Vol. 17: Semiconductors subvol. d, (eds. O. Madelung, M. Schulz and H. Weiss) New York; Springer, 1984.
13. Y. P. Varshni, *Physica* , 34, 149 1967.
14. H. C. Casey, Jr. and M. B. Panish, *Heterostructure Lasers Part B: Materials and Operating Characteristics.* New York; Academic, 1978, p. 9.
15. D. Bimberg, M. S. Skolnick, and L. M. Sander, *Phys. Rev.* B 19; 2231, 1979.
16. A. Raymond, J. L. Robert and C. Bernard, *J. Phys.* C 12; 2289, 1979.
17. F. P. Kesamanly, D. N. Nasledov, A. Ya. Nashelskii, and V. A. Skripkin, *Sov. Phys. Semicond.* 2; 1221, 1969.
18. J. T. DeVreese, "Polarons" in *Ionic Crystals and Polar Semiconductors.* Amsterdam; North Holland, 1972.
19. B. O. Seraphin and H. E. Bennett, in *Semiconductors and Semimetals,* eds. R. K. Willardson and A. C. Beer. Academic, New York; 1967, p. 499.
20. J. F. Nye, *Physical Properties of Crystals.* Oxford; Clarendon Press, 1957.

21. W. A. Brantley, *J. Appl. Phys.* 44:534, 1973.

22. Landolt-Bornstein, "Numerical Data and Functional Relationships" in *Science and Technology, Vol. 22,* Semiconductors, Subvol. a, (eds. O. Madelung and M. Schulz), Springer, New York; 1987.

23. S. Adachi, *J. Appl. Phys. 58*:R1 1985.

24. Ref. 14, p. 14.

25. M. Neuberger, *Handbook of Electronic Materials, Vol. 7 III-V Ternary Semiconducting Compounds - Data Tables*. New York; IFI/Plenum, 1972.

26. A. Onton *Festkorper Probleme XIII, Advances in Solid State Phys.*, ed. H. J. Queisser. Oxford; Pergamon, 1973, p. 59.

27. A. K. Saxena, *Phys. Stat. Solidi, 105b*:777, 1981.

28. W. Porod and D. K. Ferry, *Phys. Rev. 27B*:2587, 1983.

29. K. H. Goetz, D. Bimberg, H. Jurgensen, J. Selders, A. V. Solomonov, G. F. Glinskii, and M. Razhegi, *J. Appl. Phys. 54*:4543, 1983.

30. H. Asai and K. Oe, *J. Appl. Phys. 54*:2052, 1983.

31. C. P. Kuo, S. K. Vong, R. M. Cohen and G. B. Stringfellow, *J. Appl. Phys. 57*:5428, 1985.

32. A. T. Macrander and V. Swaminathan, *J. Electrochem. Soc. 134*:1247, 1987.

33. T. P. Pearsall, *GaInAsP Alloy Semiconductors, (ed. T. P. Pearsall). New York; 1982, p. 295.*

34. *E. Kuphal, J. Cryst. Growth 67*:441, 1984.

35. J. Hornstra and W. J. Bartels, *J. Cryst. Growth, 44*:513, 1978.

36. O. Berolo, J. C. Woolley and J. A. Van Vechten, *Phys. Rev. B8*:3794, 1973.

37. M. B. Thomas and J. C. Woolley, *Can. J. Phys. 49*;2052, 1971.

38. E. E. Matyas and A. G. Karoza, *Phys. Stat. Solidi 111b*;K45, 1982.

39. S. Adachi, *J. Appl. Phys. 54*;1844, 1983.

40. R. Bisaro, P. Mevenda and T. P. Pearsall, *Appl. Phys. Lett. 34*;100, 1979.

41. U. Pietsch and D. Marlow, *Phys. Stat. Solidi 93a*;143, 1986.

42. S. H. Wemple and M. DiDomenico, Jr., *Phys. Rev. B3*;1338, 1971.

43. M. A. Afromowitz, *Solid State Commun. 15*;59, 1974.

44. J. Buus and M. J. Adams, Proc. IEE Solid State Elect. Dev., *3*, 189, 1979.

45. H. Burkhard, H. W. Dinges, and E. Kuphal, *J. Appl. Phys. 53*;655, 1982.

46. S. Adachi, *J. Appl. Phys. 53*;5863, 1982.

47. P. Chandra, L. A. Coldren and K. E. Strege, *Electron. Lett. 17*;6, 1981.

48. A. G. Milnes and D. L. Feucht, *Heterostructures and Metal Semiconductor Junctions.* New York; Academic, 1982.

49. B. L. Sharma and R. K. Purohit, *Semiconductor Heterojunctions.* Oxford, Pergamon; 1974.

50. H. C. Casey and M. B. Panish, *Heterostructure Lasers.* New York; Academic, 1978, Parts A and B.

51. H. Kroemer, in *VLSI Electronics Microstructure Science,* Vol. 10, "Surface and Interface Effects in VLSI" eds. N. G. Einspruch and R. S. Bauer) New York; Academic, 1985, Chap 4.

52. P. Dawson, B. A. Wilson, C. W. Tu, and R. C. Miller, *Appl. Phys. Lett. 48*;541, 1986.

53. V. Tejedor and F. Flores, *J. Phys. C1*;L19, 1978.

54. R. L. Anderson, *Solid State Electron. 5*;341, 1962.

55. W. A. Harrison, *J. Vac. Sci. Technol. 14*;, 1016, 1977.

56. W. R. Frensley and H. Kröemer, *Phys. Rev. B16*;2642, 1977.

57. J. Tersoff, *Phys. Rev. Lett. 52*;465, 1984; *Phys. Rev. B32*;3968, 1985.

58. Y. C. Ruan and W. Y. Ching, *J. Appl. Phys. 262*;2885, 1987.

# CHAPTER 2

# CRYSTAL GROWTH

The significant growth of GaAs and InP based semiconductor devices in the last decade or so has been to a large extent a result of the progress made in the preparation of bulk crystals and in the various epitaxial techniques for growing thin layers. Improvements in bulk crystal growth concerning uniformity, reproducibility, and reduction in defect density have produced successful results. In epitaxial growth techniques, molecular beam epitaxy (MBE) and vapor phase epitaxy (VPE) including metal organic chemical vapor deposition (MOCVD) have ceased to be essentially laboratory experiments and are now considered as viable manufacturing technologies. Hybrid epitaxial techniques such as gas source MBE [1] which utilized group V hydride sources and metal organic MBE (MOMBE) [2, 3] where organometallic group III sources are used, have also been successfully applied for the growth of many GaAs, InP based heterostructures. The liquid phase epitaxial (LPE) technique in spite of the stiff competition from MBE, VPE, and MOCVD, still remains popular owing to its simplicity, economy and versatility. Even now it remains as the workhorse of manufacturing GaInAsP based heterostructures. In this chapter we review the bulk crystal growth techniques and the epitaxial techniques-LPE, MBE, VPE, MOCVD, and hybrid MBE.

## 2.1 GROWTH OF BULK CRYSTALS

Whether devices are made directly on the substrates, as in GaAs field effect transistors (FET) made by ion implantation into semi-insulating substrates, or they are made from epitaxial structures grown on substrates, as in heterojunction devices such as heterojunction bipolar transistors, heterojunction lasers, and so on, the quality of substrates is an important aspect of device design. Significant improvements have been made in the areas of bulk crystal growth dealing with uniformity, reproducibility, reducing dislocation density, thermal stability, diameter control, and impurity and dopant control. Single crystals of GaAs and InP have been grown from the melt by one of several techniques which include horizontal and vertical Bridgman, gradient freeze, Czochralski, liquid encapsulated Czochralski (LEC), liquid-encapsulated Kyropoulus (LEK), float-zone, horizontal and vertical zone melting and magnetic LEC. They have also been grown from the solution by the solid diffusion method (InP) [4] or by the travelling heater method (GaAs) [5]. The most widely used techniques are the horizontal Bridgman, gradient freeze, and LEC which usually employ *in situ* synthesis of the compounds from the elements.

### 2.1.1 Phase Equilibria

The temperature-composition phase diagram describing the equilibrium of solid with a liquid of binary III-V compounds can be schematically represented as shown in Fig. 2.1. The solid at the congruent melting point (the melting point at which the solid and liquid have the same composition) does not necessarily have the ideal composition of 50 percent A and 50 percent B atoms. There is a range of composition around 50 percent which defines the extent of the solid phase. The extent of the solid phase field is called the existence region. Figure 2.2 shows the calculated solidus curve of GaAs [6]. Note that the congruent melting point occurs slightly to the Ga rich side of the stoichiometric composition. For compositions in the existence region different from the 50 percent value, the solid contains nonstoichiometric crystal defects such as vacancies, interstitials, and antisite defects (see Chap. 6). Since deviation from stoichiometry can have a

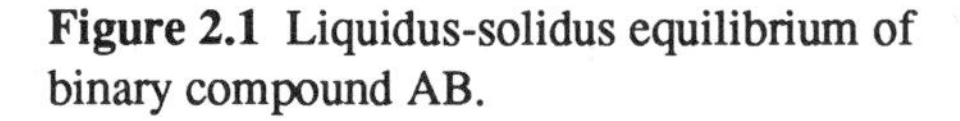

**Figure 2.1** Liquidus-solidus equilibrium of binary compound AB.

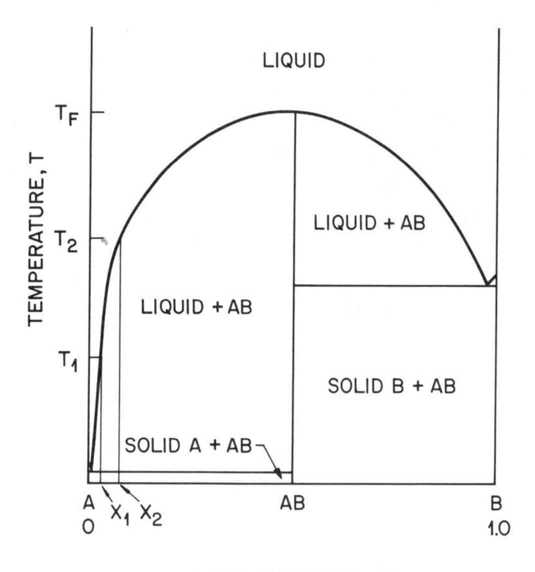

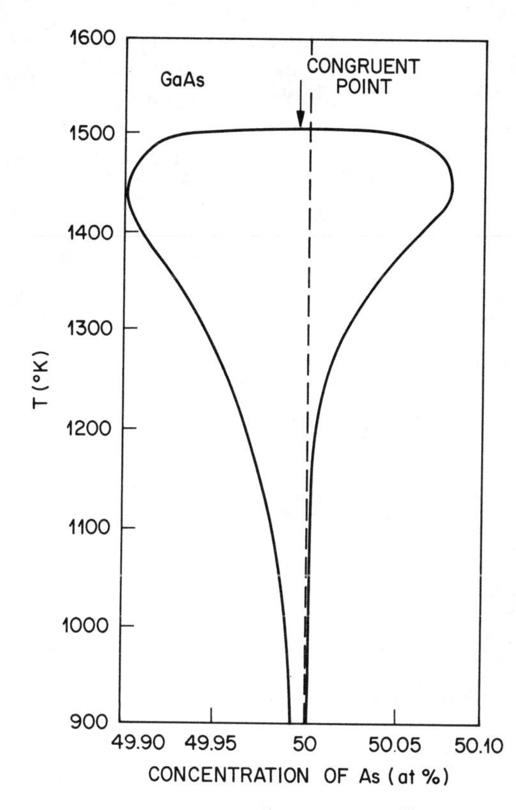


**Figure 2.2** Calculated temperature versus As concentration solidus curve in GaAs [6]. The area inside the curve is the existence region of the solid. At the congruent melting point the solid is in equilibrium with a melt of the same composition. The congruent melting point lies slightly on the Ga rich side.

profound influence on the electrical properties [7, 8] one desires the crystal to be as close to the stoichiometric composition as possible in order that the density of these native defects is very low. This can be achieved only by controlling the stoichiometry of the melt during crystal growth which is by no means an easy task.

The III-V compounds generally dissociate into the elements at temperatures near the melting point. The vapor pressure at the melting point is determined by the vapor pressure of the more volatile group V element rather than the group III element whose vapor pressure is lower by several orders of magnitude. Depending on the temperature, the group V element in the vapor phase is monoatomic or forms molecules of 2 or 4 atoms. Figure 2.3 shows the partial pressures of As, $As_2$, $As_4$, and Ga in equilibrium with the Ga-As liquidus and the solid as a function of reciprocal temperature [9]. Figure 2.4 shows the partial pressures of $P_2$ and $P_4$ in equilibrium with the liquidus in the In-P system as a function of reciprocal temperature [10]. Near the melting point, the vapor pressure of As in equilibrium with GaAs (melting point 1511°K) is ~1 atm while the vapor pressure of P in equilibrium with InP (melting point 1335°K) is ~27 atm. These high pressures dictate that sufficient precautions should be taken to prevent As or P depletion from the melt during crystal growth. The higher pressure in the case of InP compared to GaAs further necessitates that the growth furnaces should be suitably designed to withstand high pressures.

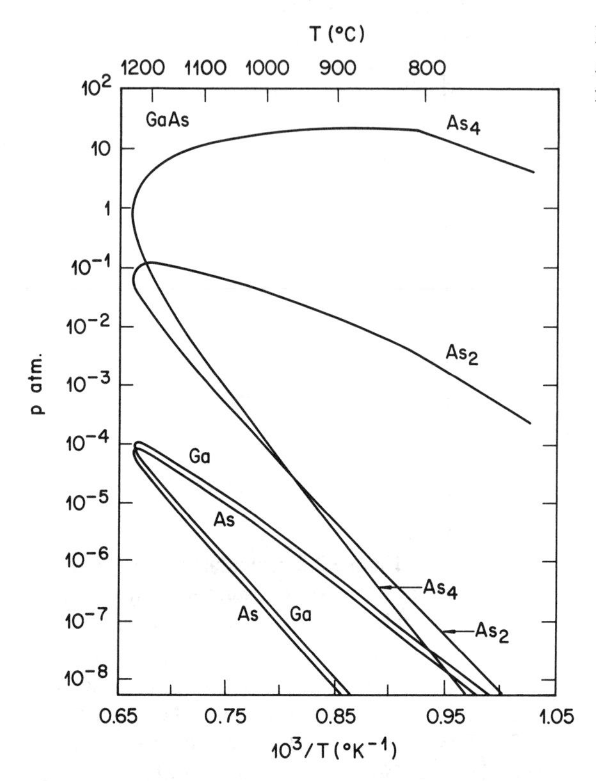


**Figure 2.3** Partial pressures p of $As_4$, $As_2$, As and Ga along the Ga-As liquidus versus reciprocal temperature [9].

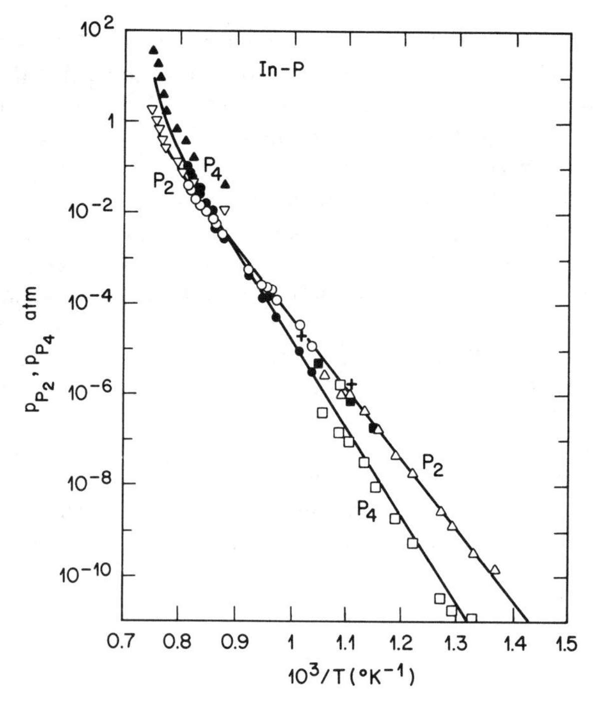


**Figure 2.4** Partial pressures p of $p_2$ and $p_4$ along the In-P liquidus versus reciprocal temperature [10].

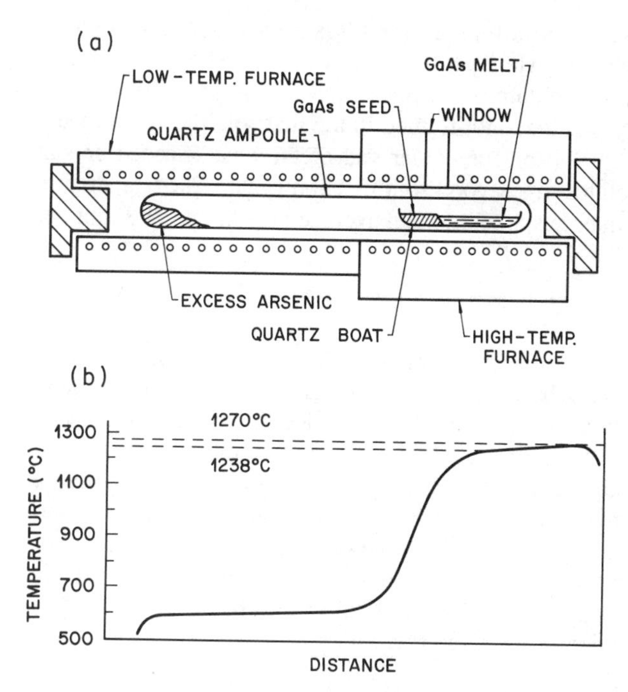


**Figure 2.5** Horizontal gradient freeze furnace (a) and a typical temperature profile (b) for growth of GaAs [11].

### 2.1.2 Crystal Growth Techniques

#### Horizontal gradient freeze method

The gradient freeze technique is a static technique where the melt is gradually solidified by the movement of a temperature gradient along the melt. Figure 2.5 shows schematically the horizontal gradient furnace used for growing GaAs [11]. In this technique a sealed quartz tube contains high purity (6-9 s pure) Ga in a quartz boat at one end and excess high-purity As (6-9 s pure) at the other end. A porous quartz plug is placed between the two sections which acts as a radiation shield and diffusion barrier. The quartz tube is placed in a two-zone furnace which has a temperature profile as shown in Fig. 2.5. The arsenic is kept at the lower temperature of the two zone furnace. The temperature at this zone is kept around 620°C to yield approximately 1 atm of arsenic vapor pressure. A heat pipe may be used in this section to establish a flat temperature profile and maintain a constant arsenic vapor pressure [12]. The quartz boat containing Ga and the seed is located in the higher temperature zone of the furnace with the Ga metal above the melting point of GaAs. The arsenic vapor reacts with gallium to form the GaAs melt. Contact with the seed crystal is made by slight tilting of the quartz tube and the temperature is adjusted to melt a small part of the seed. The hot zone temperature is then decreased at a controlled rate to achieve single crystal growth while maintaining the temperature of the cooler end at ~620°C. The crystal has to be cooled at a slow and even rate to reduce thermal stress induced dislocation generation. Also to achieve good crystal growth, a planar to slightly convex solid-liquid interface is required. This may be achieved by providing additional heating elements below the boat together with a radiation shield above the melt [12].

The horizontal gradient freeze technique has also been used for InP [13]. Indium kept in a boat is placed at one end of a quartz ampoule and the other end contains red phosphorus in excess of the amount required for producing stoichiometric InP. The red P is slowly heated to ~540°C to maintain the P pressure at 27.5 atm. The boat with In is heated to 1075°C. After the synthesis of InP, which takes about ~12 hours, the temperature of the hotter zone is cooled to 1000°C at a constant rate of 15°C per day. The polycrystalline ingot formed in the first solidification process is remelted until only a few crystallites are left on the colder end of the boat. Seeded crystal growth is accomplished from these crystallites in a second solidification cycle. The temperature gradient in the melt is typically 2°C per cm corresponding to an average growth rate of 3 mm per hour.

**Horizontal Bridgman method**

In the horizontal Bridgman growth of GaAs, the arrangement of the charge and the excess As for providing the As pressure inside the quartz boat are the same in Fig. 2.5. The charge formed separately or compounded *in situ* is kept in a quartz or pyrolytic BN (PBN) crucible with a seed crystal. The entire furnace is moved along the quartz tube such that the solidification of the melt starting from the seed crystal is achieved as the ampoule moves through a temperature gradient from the hotter to cooler section of the furnace [14, 15]. The temperature profile is kept constant. It is also important to avoid vibrations that would otherwise cause imperfections. The use of multizone furnace with a sodium heat pipe would help to achieve precise control of As pressure [14]. Thermal uniformity in the hot zone of ±0.5°C vertically and better than ±0.1°C horizontally has been achieved in such multizone furnaces [14].

In the horizontal growth techniques, the shape of the crystal is constrained by the walls of the quartz boat. The crystals are typically D-shaped and are seeded in the <111> direction. To obtain (100) oriented circular slices from these crystals requires sawing with excessive material loss. On the other hand, the stringent diameter control required in Czochralski technique (to be described later) is avoided. In general, significantly lower thermal gradients and hence reduced convective flows in the melt and extremely low thermal stresses can be realized in the horizontal technique compared to the Czochralski technique. As a result, convection induced impurity striations are considerably reduced in crystals grown by the horizontal technique. Further, owing to the reduced thermal stresses and precise control of the As over pressure, dislocation densities tend to be much less in these crystals. Typical dislocation densities in GaAs crystals grown by the horizontal Bridgman or the horizontal gradient freeze technique are 3000 to 5000 $cm^{-2}$ in 5 to 7.5 cm wide D-shaped crystals.

**Vertical growth techniques**

In vertical Bridgman growth, the quartz or PBN crucible contains the seed in a well at the bottom and polycrystalline GaAs above it. The arrangement of the crucible in the vertical furnace and the temperature profile are shown schematically in Fig. 2.6. The seed crystal has to be accurately positioned in the crucible and the temperature carefully adjusted to avoid dissolution of it. For this reason, *in situ* synthesis is extremely difficult to incorporate in this technique. For growth to occur, the initial charge and a portion of the seed is melted and the crucible is lowered slowly into the bottom section of the furnace. Instead of moving the crucible, the furnace can be moved relative to the crucible.

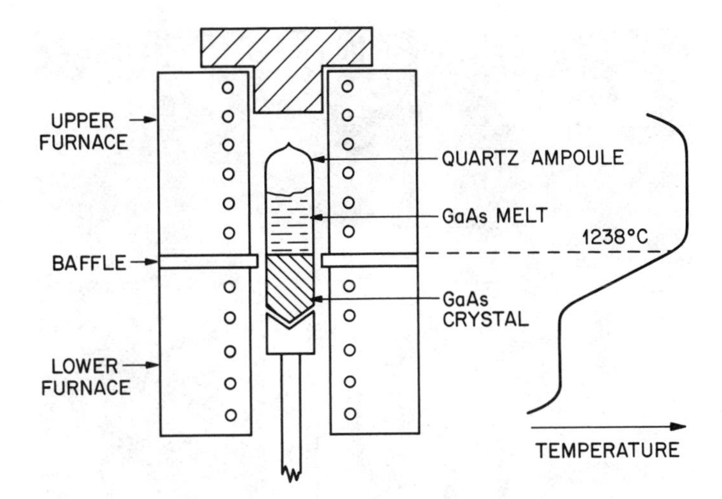


**Figure 2.6** Vertical Bridgman furnace and a typical temperature profile for growth of GaAs [11].

In vertical gradient freeze (VGF) technique, the crucible and the furnace are kept stationary and the growth is achieved by slowly cooling the melt in an appropriate temperature gradient. One of the principal advantages of the VGF technique is the much reduced axial and radial temperature gradients with concomittant advantages of reduced convective flow and thermal stresses. These benefits translate into low dislocation densities $\leq 100$ to $3000\ cm^{-2}$ in crystals that are 5 to 7.5 cm in diameter.

Reproducible growth of large, low dislocation density crystals has been achieved by a vertical gradient freeze method designed specifically to grow GaP, GaAs, and InP crystals [15]. A cross sectional view of the equipment used is shown in Fig. 2.7. A growth vessel containing elemental phosphorus (or arsenic) and polycrystalline InP (or GaAs) is placed in a vertical two-zone heater assembly, which is enclosed in a water cooled pressure vessel. The reservoir containing the red phosphorus is heated in the lower zone to provide the P pressure. The upper zone heats a bottom seeded crucible that contains the InP charge. This zone controls solidification and heat transfer during crystal growth. The desired axial and radial temperature gradients within the upper zone are achieved by optimizing the insulation, heat shields and the spatial distribution of heat input. The growth vessel is also rotated to average out thermal asymmetries in the furnace. Further, the growth vessel can be translated in the heater to adjust the crucible's position in the temperature profile.

Stainless steel lined pressure vessel is used for cleanliness, safety, and operating convenience. The vessel generally operates at 7 MPa but is designed for 17 MPa operation. Pyrolytic boron nitride is used as the crucible. The entire furnace is kept in an individual room with appropriate exhausting. Also a comprehensive fail-safe system shuts down the furnace in case of serious

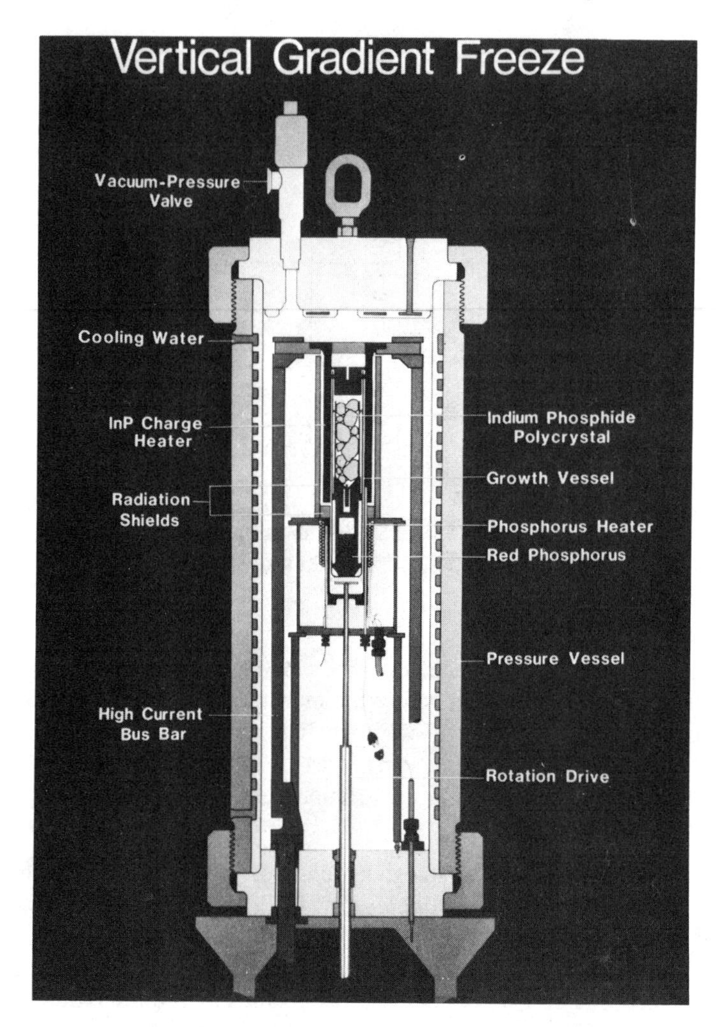


**Figure 2.7** Vertical gradient freeze furnace for growth of InP [15].

problems. Figure 2.8 shows a 750 g <111> seeded 50 mm diameter InP crystal (Fig. 2.8a) and a 1200 g <100> seeded 50 mm diameter GaAs crystal (Fig. 2.8b) grown by this technique. As a result of the low-thermal gradients during growth, the dislocation densities in these crystals are typically $\lesssim 1000\ \text{cm}^{-2}$.

**Crystal growth by zone melting**

In zone melting, a fraction of a polycrystalline ingot is molten and this zone is passed through the ingot by relative displacement of the heater and the crystal. By passing the molten zone several times, high-purity crystals can be obtained provided the distribution coefficient of the impurities is less than 1. Vertical float zone technique has been used to grow high-purity GaAs crystals [16, 17]. A vertical zone melting technique which employs PBN crucibles and $B_2O_3$

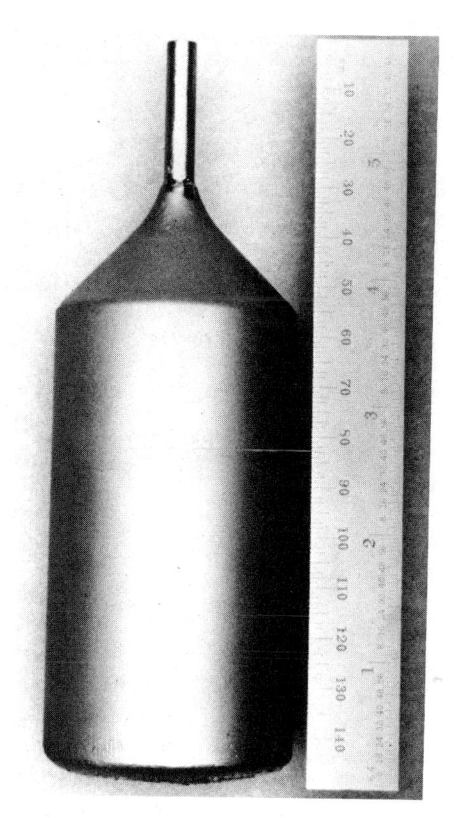

**Figure 2.8** Photomicrograph of a 750 g <111> seeded 50 mm diameter InP crystal (a) and a 1200 g <100> seeded 50 mm diameter GaAs (b) crystal grown by the vertical gradient freeze furnace [15].

encapsulant has been used to grow 38 mm diameter semi-insulating GaAs crystals with dislocation densities below 3000 $cm^{-2}$ [18]. The main difficulty with float zone growth of GaAs is the instability of the molten zone because of variation in the surface tension caused by temperature gradient in the zone and corresponding variation in the melt composition (see the phase diagram in Fig. 2.1).

The growth of single crystals of GaAs has also been achieved by horizontal zone melting [19, 20]. A heat pipe is used to maintain a constant arsenic partial pressure. GaAs crystals of 55 mm diameter with dislocation densities of 6000-9000 $cm^{-2}$ have been obtained by this technique [20].

**Crystal growth by pulling from a crucible**

One of the most important techniques for the growth of large round single crystals of GaAs and InP is the Czochralski pulling method [21]. Here, a seeded single crystal is withdrawn from the melt and is rotated to maintain thermal geometry and cylindrical geometry. The basic components of the technique are illustrated in Fig. 2.9 [11]. The seed is dipped into the melt whose temperature is lowered until a small amount of crystalline material is solidified. The seed

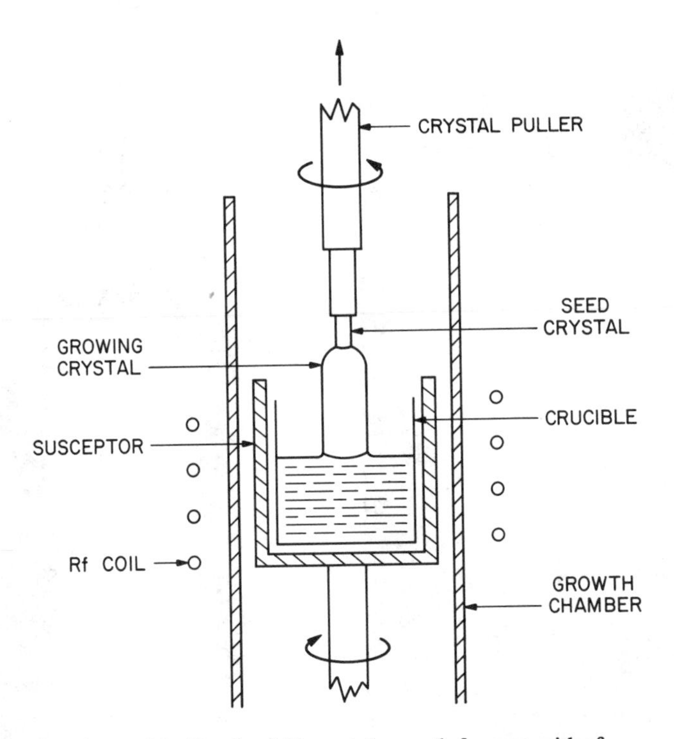


**Figure 2.9** Czochralski crystal growth furnace with rf induction heating [11].

is then withdrawn from the melt at a rate of 1 to 10 mm per hour. The melt temperature is lowered slowly and the diameter of the crystal increases. Once the desired diameter is reached, the lowering of the temperature is stopped. Growth at a constant diameter is maintained till the desired length is grown.

One of the problems in the pulling of volatile compounds like GaAs and InP is the decomposition of the melt. The hot-wall technique is helpful to prevent the condensation of the volatile group V element. In the Gremmelmaier method, the whole pulling system is enclosed in a sealed quartz ampoule and magnetic coupling is used for rotation and withdrawl of the crystal [22].

An alternative method to prevent loss of group V element from the melt is the encapsulation technique where the melt is covered by a thin layer of molten $B_2O_3$ [23]. A counter pressure of an inert gas which is higher than the partial pressure of the group V element is maintained on top of the $B_2O_3$. This technique is called the liquid encapsulated Czochralski (LEC) technique. Figure 2.10 shows the LEC furnace used for growing of InP crystals [24]. LEC furnaces that can be used up to pressures of 100 bar are available commercially. During LEC growth automatic control of diameter by continuous monitoring of the weight of the growing crystal is also usually incorporated.

The material to be used as an encapsulant should have low vapor pressure, low viscosity, density lower than that of the melt, and it should not mix with the melt or react with the melt and

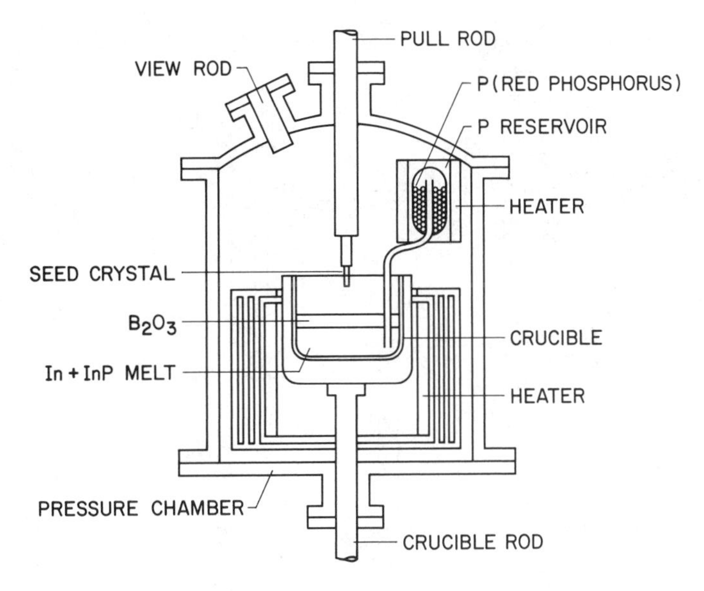


**Figure 2.10** In situ liquid encapsulated Czochralski method for growing InP [24].

the crucible, and should melt before significant decomposition of the III-V compound occurs. $B_2O_3$ satisfies these requirements and is the commonly used encapsulant. It is less dense than the GaAs or InP melts. The density of $B_2O_3$ is 1.5 g $cm^{-3}$ while that of GaAs and InP melts are, respectively, 5.71 and 5.1 g $cm^{-3}$. It softens and begins to flow at 450°C and has a vapor pressure of only 0.1 mm at the melting of GaAs. It reacts to some extent with silica or alumina crucibles. It is immiscible with the melts of III-V compounds. The residual water content in $B_2O_3$ plays an important role in its reaction with the melt, especially Si-doped melts [25]. The water in the $B_2O_3$ causes oxidation of the melt and subsequent incorporation of the oxide in the $B_2O_3$. Also large amounts of water will lead to severe bubbling and loss of the volatile group V element [25]. In the case of InP, there is a strong correlation between etch pit clusters in the grown crystal and water content in $B_2O_3$ [26]. The water from $B_2O_3$ is removed by vacuum baking it at ~ 1000°C [25]. Large crystals of GaAs [11, 27-29] and InP [24, 30-34] have been grown by the LEC technique using $B_2O_3$ as the encapsulant.

The LEC growth can be accomplished in two ways. One in which the growth consists of two steps: preparing of the compound separately and subsequent crystal growth, and the other in which the compound is synthesized *in situ* prior to growth in the same system.

In the low-pressure LEC growth of GaAs, arsenic contained in a quartz ampoule is heated and the vapor is transported into $B_2O_3$-covered molten gallium to form the compound [35]. The compounding *in situ* has also been achieved in the low pressure system by using an arsenic cell-injection system [36]. The cell is used to bubble arsenic into hot gallium to form the compound and is subsequently removed from the system. In another variation, the arsenic cell is mounted within the pulling chamber and can be used to inject arsenic into the melt continuously or periodically. Precise control of the stoichiometry of the crystal is possible by proper design of the injection nozzle.

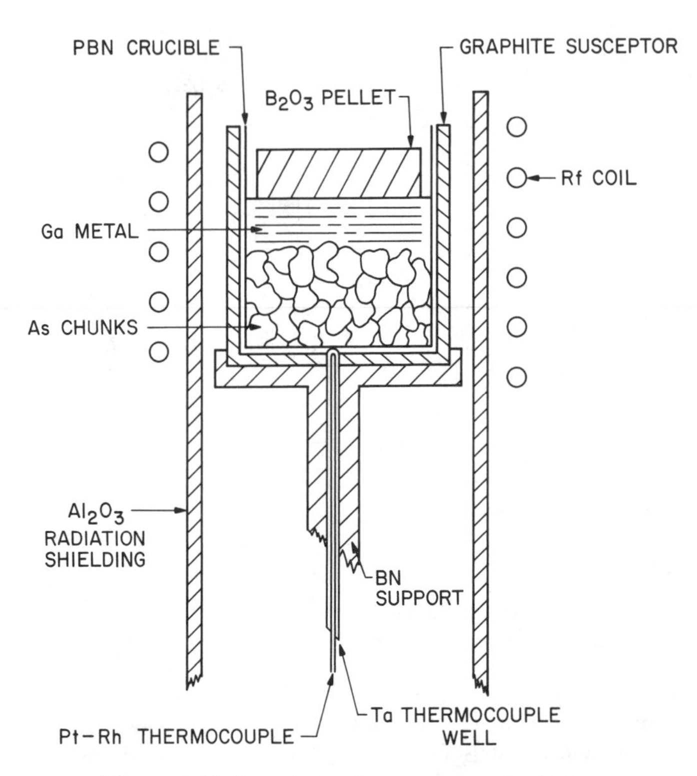


**Figure 2.11** Experimental arrangement for high pressure in situ synthesis and growth of GaAs [11].

The high-pressure (HP) LEC system was initially developed for the growth of GaP and InP. It was subsequently applied to *in situ* direct synthesis and growth of GaAs [29, 37]. The direct synthesis of GaAs occurs exothermically at about 700°C under a high purity nitrogen or argon pressure of 60 atm. The starting materials for the synthesis are stoichiometric quantities of high purity Ga and As. A dehydrated pellet of $B_2O_3$ is placed on top of the charge which when molten, encapsulates the charge. High-purity pyrolytic boron nitride (PBN) crucibles are used to grow consistently pure materials. Although the initial cost of the PBN crucibles is high, the crucible can be cleaned and reused about 15 times. In fact, the use of PBN crucibles has become a standard in the growth of undoped semi-insulating GaAs crystals. Figure 2.11 shows the experimental arrangement for HP LEC growth of GaAs [11]. After the compound is synthesized, the temperature is increased above the melting point of GaAs and the pressure is reduced to 2 to 4 atm. Single crystal growth then proceeds as in the conventional LEC technique. A major advantage of the HP LEC technique is the ability to grow <100> crystals with high yield.

In the growth of InP single crystals by HP LEC the compounding is done *ex situ.* The polycrystalline material then serves as the charge for LEC growth. As in the growth of GaAs, here also PBN crucibles are used and $B_2O_3$ as the encapsulant. Efforts have also been made towards *in situ* synthesis of InP prior to growth in order to improve the purity of the crystals [38,

39]. However, the optimum temperature profiles for synthesis and growth are different enough that compromises have to be made in the final yield of crystals. Further, improvements in purity are offset by increased twinning.

**Magnetic LEC growth**

In conventional Czochralski and LEC growth, thermal convection currents that are present affect crystal homogeneity and interface shape. The use of a magnetic field can decrease thermal convection. This is brought about by the "magnetic viscosity" effect by which the viscosity of a heated conductive liquid moving through a magnetic field increases thereby reducing thermal convection [40]. Temperature fluctuations in GaAs melts have been found to be reduced from ±18°C to 0.1°C at a magnetic field of 1.3k Gauss [41]. Application of the magnetic field of 2000 Oe also reduces growth striations [42]. Application of either transverse or longitudinal magnetic field produces beneficial effects in reducing dislocation density and growth striations. Terashima et al., [41] found that the $EL_2$ concentration decreased and conductivity increased with the application of the magnetic field. This result suggests that perhaps the native point defect concentration is reduced in magnetic LEC growth. If this is so, it would be difficult to reproducibly grow semi-insulating GaAs crystals since with low $EL_2$ concentration, proper compensation of carbon acceptors may not be readily achieved.

The application of a magnetic field during LEC growth of InP has also been reported. At a magnetic field of 1000 Oe, growth striations are absent in the crystals [34].

**Diameter control**

The LEC crystal growth technique produces cylindrical boules from which circular wafers suitable for IC fabrication can be cut. The desire to get round wafers has led to significant efforts in the area of diameter control of the growing crystal. The diameter of the crystal can be controlled either by continuously monitoring the crystal weight, or by the coracle process [29]. In the first technique the pull rod is equipped with a high-sensitivity load cell to continuously monitor the weight of the growing crystal. An increase or decrease in the differential weight signal indicates increase or decrease in the diameter. Using this signal and by visual monitoring of the growth interfaces, adjustments are made to the heater temperature and the pull rate to keep the crystal diameter constant. Automatic computer-controlled LEC growth employing closed-loop control using crystal weight has made possible diameter control with ±1 to 2 percent [43]. A complete analysis of the closed-loop control of diameter including effects of the $B_2O_3$ liquid encapsulant and the capillary forces can be found in Ref. [44]. The variations in radius are small in the large temperature gradient conventional LEC process (100-200°C $cm^{-1}$) where automatic diameter control can be easily achieved. On the other hand, in the low-temperature gradient, LEC shape instabilities are too severe for current diameter control techniques.

In the coracle method of diameter control, a coracle which is a $Si_3N_4$ die with a round hole in the center floats on top of the GaAs melt. As the crystal is pulled through, the die good diameter control is achieved. The coracle technique is well suited for growths of large diameter <111> crystals and diameter control to within ±2 percent is possible. However, for growing <100> crystals, the technique has been found to be unsuitable because of the tendency of the crystals to twin at the early stages of growth [29].

### 2.1.3 Crystalline Imperfections

Crystalline defects which are introduced into the crystal during crystal growth are point defects, dislocations, stacking faults, twins, inclusions, precipitates, and microdefects. Point defects are native point defects such as vacancies, interstitials and antisite defects, and foreign atoms on substitutional or interstitial sites. A more detailed description of these defects is given in Chap. 6.

#### Point defects

Melt stoichiometry controls the concentration of point defects and hence the electrical properties of the crystal. Figure 2.12 illustrates the dependence of resistivity and free carrier concentration of LEC undoped GaAs on arsenic atom fraction in the melt [7]. The crystal is p-type below 0.475 As atm fraction and semi-insulating above it. The defect responsible for the semi-insulating character of undoped GaAs is the deep donor referred to as $EL_2$ [45] (see Sec. 7.6

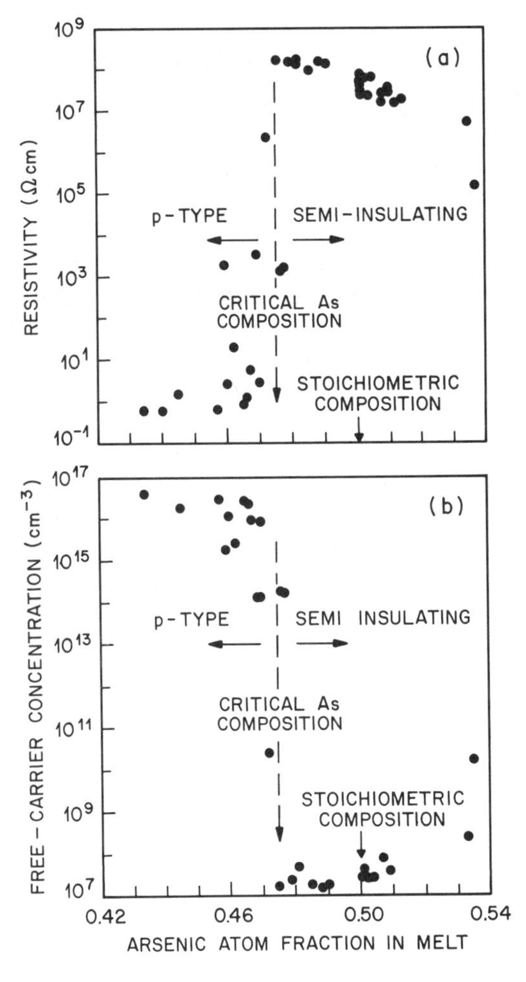


**Figure 2.12** Variation of electrical resistivity (a) and carrier concentration (b) of LEC GaAs with melt stoichiometry [7].

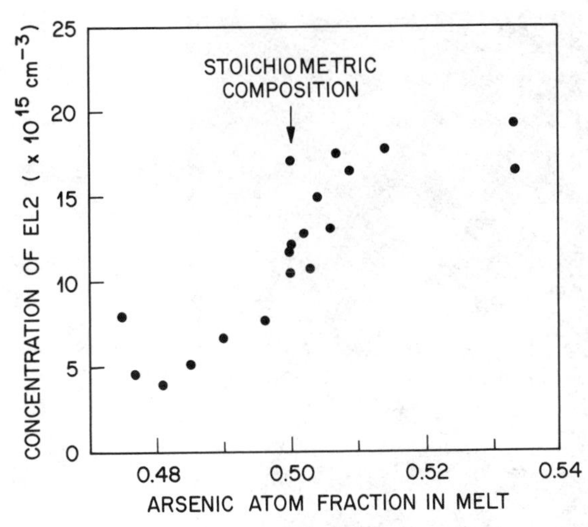


**Figure 2.13** Concentration of $EL_2$ ($cm^{-3}$) determined by optical absorption as a function of melt stoichiometry [7].

for further discussion). The concentration of $EL_2$ depends on the melt stoichiometry as shown in Fig. 2.13 [7], increasing from about $5 \times 10^{15}$ $cm^{-3}$ to $1.7 \times 10^{16}$ $cm^{-3}$ as the As atom fraction increased from 0.48 to 0.51. Since the compensation of shallow acceptors (mostly carbon acceptors) by the deep $EL_2$ donor produces semi-insulating material, the dependence of resistivity on As atom fraction in Fig. 2.12 is easily explained.

In InP bulk crystals presence of melt stoichiometry related point defects and point defect complexes has been inferred from low-temperature photoluminescence study [46]. Based on the changes in the intensity of the emission bands seen in LEC crystals and polycrystalline InP obtained by direct synthesis [47] after post growth annealing treatments in excess P pressure conditions, specific assignments of the bands to native point defects were made. For example, an emission band at 1.2 eV has been assigned to In vacancy related complexes. Similarly, a band at 0.99 eV has been assigned to P vacancy related defects.

**Dislocations**

Dislocation generation during growth of bulk crystals can occur by one of three mechanisms:

1. nonuniform heat flow during solidification and the ensuing thermal stresses causing plastic deformation
2. condensation of excess point defects present near the growth temperature to form prismatic dislocation loops
3. propagation of dislocations from defective seed crystal or punching out of dislocations from foreign particles and inclusions.

These mechanisms and methods for dislocation reduction are outlined in detail in Sec. 6.7.4.

For growing crystals of smaller diameters (≤ 15-20 mm) by LEC, thermal stresses are low and dislocation free crystals can be easily grown. This is generally achieved by the use of Dash type seeding [48]. In this method of seeding, dislocations in the seed are removed by growing a thin neck before increasing the diameter to form the crystal cone. Once successful seeding is

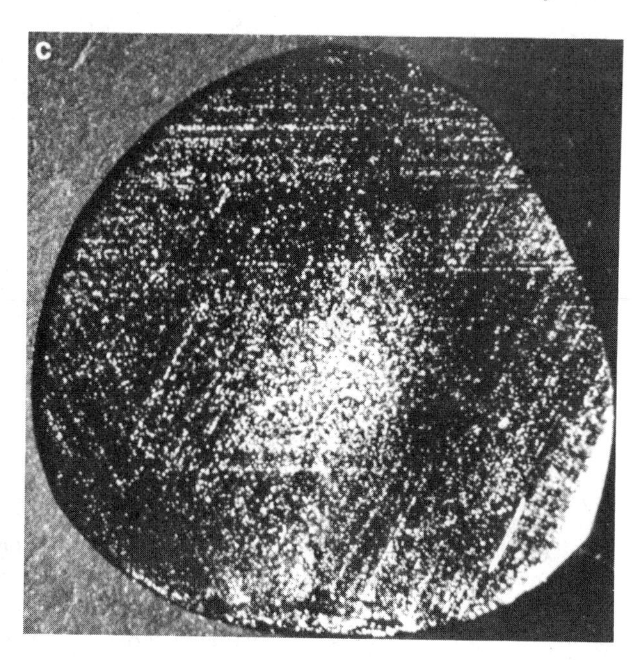


**Figure 2.14** Photomicrograph of the dislocation distribution in (1111) wafers of LEC InP showing a sixfold symmetry [49].

achieved, then controlling melt stoichiometry, temperature gradients at the growth interface and angle of the crystal cone as it emerges from the encapsulant would yield low dislocation crystals. In large diameter (>20 mm) LEC crystals, dislocation generation is primarily caused by thermal stresses because of large radial and axial temperature gradients.

In large diameter LEC crystals, dislocation distribution across the crystal diameter is not homogeneous. The radial distribution is crystallographic orientation dependent. For example, the arrangement of dislocations in {111} wafers shows a six-fold symmetry as shown in Fig. 2.14 for a LEC grown InP crystal [49]. The observed pattern can be explained on the basis of the thermal stress model for dislocation generation [49]. The dislocation distribution exhibits a U-shaped pattern, low at the center and maximum at the circumference. In contrast, the dislocation distribution in {100} wafer is always W-shaped as shown in the photomicrograph of a KOH etched (100) GaAs wafer obtained from the seed end of a LEC crystal shown in Fig. 2.15 [50]. The dislocation density is maximum near the center and edges. Once again, the distribution for {100} wafers is in accord with the thermal stress model [50] (see Sec. 6.7.4 for details). Figure 2.16 shows the etch pit density as a function of radial distance along the <100> direction for a 50 mm diameter and a 75 mm diameter LEC GaAs (100) wafers [29]. The W-shaped distribution is observed for both wafers. Distribution along the <110> direction also exhibits the W-shaped distribution but the dislocation density is lower than that along the <100> direction. Figure 2.17 shows the W-shaped distribution for a {100} LEC InP crystal [34].

In the thermal stress model [51], the dislocation density is assumed to be proportional to the

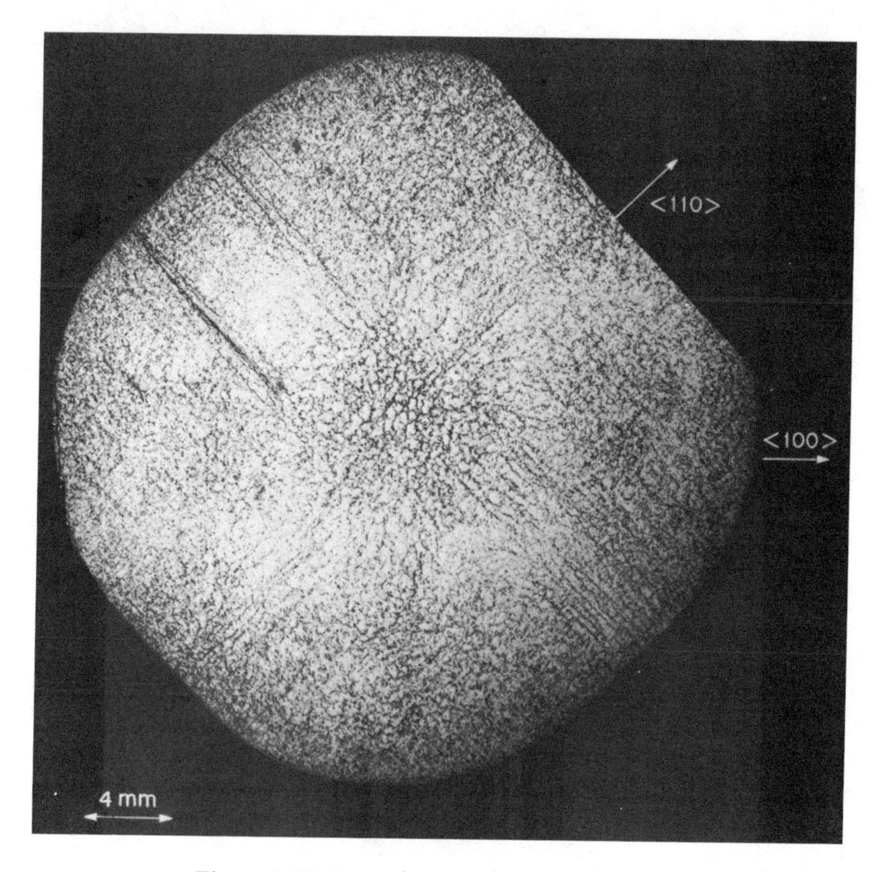


**Figure 2.15** Photomicrograph of a KOH etched (100) GaAs wafer from the seed end of a LEC crystal showing the W-shaped dislocation distribution [50].

total resolved shear stresses, calculated in terms of the twelve {111} <110> slip systems in the diamond cubic or sphalerite lattice, minus the critical resolved shear stress for the on set of plastic deformation. The excess shear stress is maximum near the edge and center of the crystal, the edge stress being higher than the center stress. The stress is minimum at about one half of the radius. Further, in moving from the <110> to the <100> direction on a {100} wafer the stress increases. Both the observed W-shaped pattern of the dislocation distribution and the relatively higher dislocation density along the <100> direction compared to the <110> direction are in excellent agreement with the calculated thermal stress distribution indicating that dislocation generation in large diameter (>20 mm) LEC crystals is primarily a result of the thermal stress induced plastic deformation.

The clearly defined radial distribution of dislocation in the seed end of the crystal becomes more diffuse as the tail end is approached. The dislocation distribution is dense and no definite symmetry is observed. This is believed to be caused by the increasing dislocation multiplication

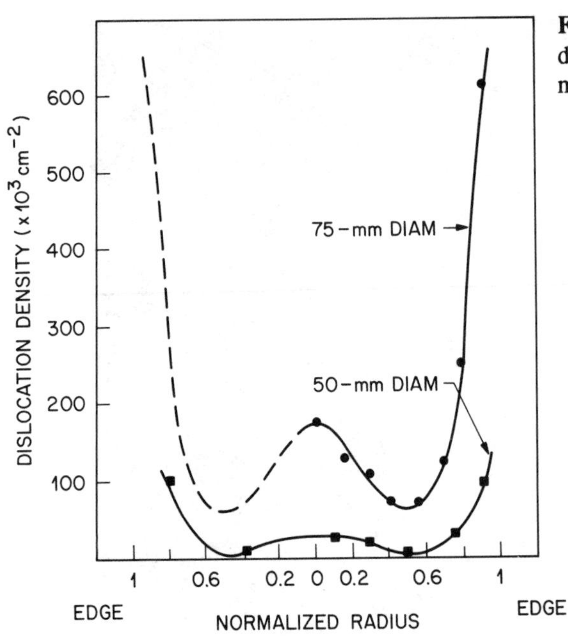


**Figure 2.16** Radial dislocation density distribution along <100> direction in 50 and 75 mm diameter LEC (100) GaAs slices [29].

**Figure 2.17** Radial dislocation density distribution along <100> direction for a (100) LEC InP crystal showing the W-shaped distribution [34].

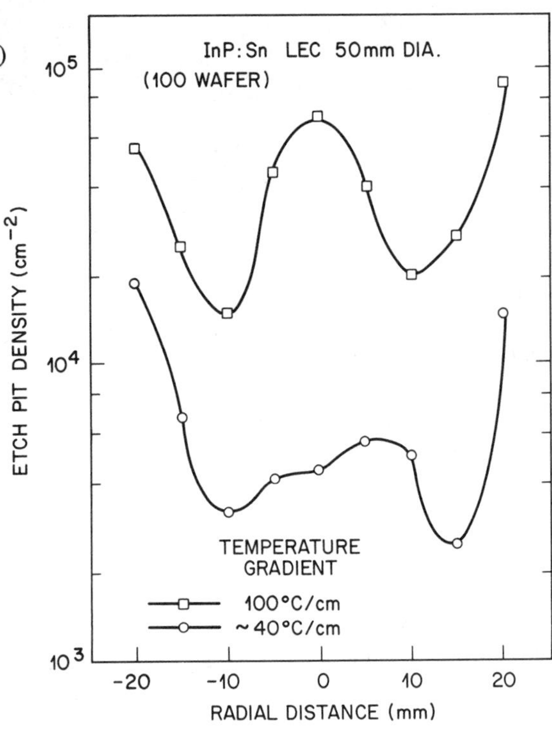

along the length of the crystal by dislocation climb and other mechanisms and because of dislocation-dislocation interactions. The tail end distribution can not be explained on the basis of this thermal stress model described which does not consider these dynamic aspects of dislocation multiplication.

**Parameters affecting dislocation density**

The parameters which affect dislocation generation during LEC growth are

1. seed quality and necking
2. cone angle
3. ambient pressure
4. melt stoichiometry
5. crystal diameter
6. boric oxide thickness

As stated earlier, the impact of seed necking is less for large diameter crystals even though necking helps to reduce dislocation density independent of the seed quality [27]. Since the convective heat transfer coefficient at the crystal-ambient interface is proportional to the square root of the ambient pressure [52], decreasing the pressure reduces the heat transfer coefficient and thus the dislocation density [51]. However, reducing the ambient pressure leads to thermal decomposition of the surface of the crystal as it is pulled from the encapsulant. Since the volatile group V element escapes excess group III element is left on the surface. Droplets of Ga in the case of GaAs or In in the case of InP are thermally migrated through the crystal giving rise to inclusions and polycrystalline material [27]. Also the excess metal runs down to the meniscus and causes twinning of the crystals.

The cone angle has only minimal effect on dislocation density [27]. The dependence of dislocation density on cone angle is nil for angles greater than 25 degrees. For shallow angles the radial gradient gives rise to high densities.

In horizontal Bridgman crystal growth melt stoichiometry has a significant effect on dislocation density [14] owing to the fact that excess point defects condensing to prismatic dislocation loops is a major source of dislocations. Kirkpatrick et al., [27] have reported that in LEC GaAs growth controlling the As atom fraction in the melt between 0.505 and 0.535 may be advantageous for minimizing dislocation density. Tomizawa et al., [53] have obtained 5 cm diameter GaAs crystals with dislocation densities as low as 2000 $cm^{-2}$ by Czochralski growth (without the encapsulant) under controlled As pressure. The observed dislocation density is lower by an order of magnitude than that in conventional LEC grown crystals. These authors attributed the reduction in dislocation density to improved melt composition by As vapor pressure control, reduced thermal stress caused by low-temperature gradient just above the melt and the diameter control. The crystals grown by this technique were found to be p-type and have a resistivity of $10^4$ Ω cm owing to high C acceptors. Reducing the C level would help to achieve close compensation by $EL_2$ deep donors and higher resistivity.

Jordan and Parsey [54] have considered in detail the interrelationship between GaAs crystal diameter and dislocation density. They find that at high-temperature gradients, ($T_f - T_a = 200K$

where $T_f$ is the melting point and $T_a$ is the temperature of the ambient around the boule) dislocation density increases with increasing diameter and levels off at 4 cm diameter. The core region of the larger diameter crystal exhibits a lower dislocation density. Further, the larger diameter crystals have greater dislocation free regions than the smaller ones. Thus it is possible to obtain processable size dislocation free wafers by increasing the diameter and subsequent coring [55].

Kirkpartick et al., [27] investigated the effect of varying the thickness of the $B_2O_3$ layer on dislocation density. As the $B_2O_3$ thickness increased from 8 mm to 17 mm, dislocation density in the seed end of the crystal decreased from $(6\pm3)\times10^4$ cm$^{-2}$ to $1.5\times10^4$ cm$^{-2}$. A similar effect was observed in the tail end as well but the magnitude of the decrease was much smaller. The reduction in dislocation density with increasing thickness of $B_2O_3$ is thought to arise because of reduction in the radial temperature gradients near the crystal-melt interface as a result of more effective thermal isolation between the crystal and the Ar ambient.

By far the most important parameter which influences dislocation generation in large diameter LEC crystals is the axial and radial temperature gradients existing at the melt-$B_2O_3$ interface and across the $B_2O_3$ layer itself. Reducing these gradients reduces the thermal stresses and hence the dislocation density. For example, thermal stress calculations show that a 10-fold decrease in ambient temperature gradient ($T_f - T_a$ is reduced by a factor of 10) results in a ~ 13-fold reduction in dislocation density at the periphery. However, modifications required in the conventional LEC growth furnace to reduce temperature gradients are by no means simple.

The advantage of low thermal gradients in reducing dislocation density in LEC crystals has been demonstrated in the case of both GaAs [50, 56-58] and InP [34, 59-62]. Reducing the temperature gradient can be accomplished by increasing the thickness of the $B_2O_3$ encapsulant and/or introducing radiation shields and after-heaters (to increase $T_a$) in the growth chamber [51]. By using a heat shield, Elliot et al., [57] were able to reduce the gradient to 6°C/cm compared to a typical value of 100-200°C/cm, and achieved average dislocation densities of 5000 cm$^{-2}$ over 75 percent of the area in 5 cm diameter undoped GaAs crystal. In a RF heated LEC puller to grow 5 cm diameter undoped GaAs, Von Neida et al., [50, 56] reduced the gradient to 20°C/cm by increasing the thickness of the $B_2O_3$. This resulted in a dislocation density of 5000 cm$^{-2}$ at the center and 8000 cm$^{-2}$ near the periphery of seed end wafers.

Similar benefits of reducing thermal gradient on dislocation density have been realized in InP LEC growth also. In growing <111> InP crystals, Muller et al., [62] increased the $B_2O_3$ thickness from 7 to 30 mm which decreased the gradient from 80 to 20°C/cm. For a 7.5 cm diameter crystal, the average dislocation density was reduced from $10^5$ to $10^4$ cm$^{-2}$. Shinoyama et al., [60, 61] also adopted the method of increasing $B_2O_3$ thickness to reduce temperature gradients and achieved a nearly proportional reduction in dislocation density. Katagiri et al., [59] employed a heat shield above the heater and the crucible and succeeded in reducing the temperature gradient near the melt-crystal interface from 100°C/cm to 40°C/cm. As a result, the dislocation density was reduced from $8\times10^4$ cm$^{-2}$ to $3\times10^4$ cm$^{-2}$. A similar approach was also adopted by Morioka et al. [34] in their growth of 5 cm diameter Sn-doped InP crystals. Under the reduced temperature gradient not only the dislocation density reduces but its radial distribution becomes more uniform [34, 59]. This is illustrated by the radial dislocation distribution in a InP wafer grown under reduced gradient shown in Fig. 2.17. The W-shaped profile is less evident in this case.

Low gradient LEC growth is, however, not without problems. One of them is the diameter instability associated with the lack of confining radial gradient present in standard LEC growth. Further the lower temperature gradient makes the surface temperature of the crystal as it emerges from the encapsulant higher. This leads to decomposition of the surface giving rise excess Ga or In droplets. These droplets can cause polycrystalline growth if they migrate through the crystal or reach the solid-liquid interface [27, 34]. In the growth of InP, twinning becomes a problem under low temperature gradient.

To achieve diameter control and suppress thermal decomposition under low-gradient LEC growth several schemes have been tried. To reduce surface damage, the pulled crystal is kept in $B_2O_3$ at all times. This technique is called the fully encapsulated Czochralski (FEC) method [63]. The application of vertical magnetic field has been found to improve diameter control by suppression of laminar thermal convection in the melt [64]. Using vertical magnetic field in the FEC growth, Kohda et al., [65] grew 5 cm diameter completely dislocation free semi-insulating GaAs under a temperature gradient of 30-50°C/cm. By using an X-ray imaging scheme to monitor crystal diameter, Ozawa et al., [66] were able to achieve diameter control under a low temperature gradient GaAs LEC growth which also had an As ambient to minimize surface decomposition. In the absence of these modifications, low-gradient LEC growth is a difficult technique for routine applications. For achieving low-dislocation density crystals under low thermal gradients, vertical or horizontal gradient freeze techniques appear more promising.

Another method to achieve low dislocation crystals is to use dopants which harden the semiconductor, that is, increase the critical resolved shear stress (CRSS). In general, n-type dopants such as Si, Te in GaAs, S, Ge in InP increase the CRSS. In the case of InP, high concentrations of Zn, a p-type dopant, also hardens the lattice. Although the microscopic mechanisms underlying the hardening of III-V semiconductors by solute atoms are not clearly identified (for more detailed discussion, see Sec. 6.7.3) it is well established that impurity hardening reduces dislocation density. The concentration of these mentioned impurities needed for significant dislocation reduction is typically very high, $>3-5\times10^{18}$ $cm^{-3}$. Increasing the doping concentration decreases the dislocation density as illustrated in Fig. 2.18 for GaAs containing S, Te or Zn, and InP containing S or Zn [67].

Doping with high concentration of electrically active impurities would not, however, be practical for growing semi-insulating crystals. Low dislocation density conductive wafers are suitable for use as substrates required for such applications as heterostructure lasers, LEDs, and photodetectors. Even in these optical devices, if light is allowed to pass through the substrates, optical absorption as a result of high doping can be a problem. In addition, impurities from the substrates can diffuse into epitaxial layers grown on them. For these reasons, it is highly desirable to grow low dislocation density substrates by decreasing the doping level and by using electrically inactive impurities. In this regard, isovalent impurities such as In, Ga, and Sb have been remarkably effective in reducing dislocation density [68]. Particularly, addition of $10^{19}-10^{20}$ $cm^{-3}$ In has produced 7.5 cm diameter GaAs wafers with substantial areas (up to 70% of the diameter) having less than 400 $cm^{-2}$ dislocations [69].

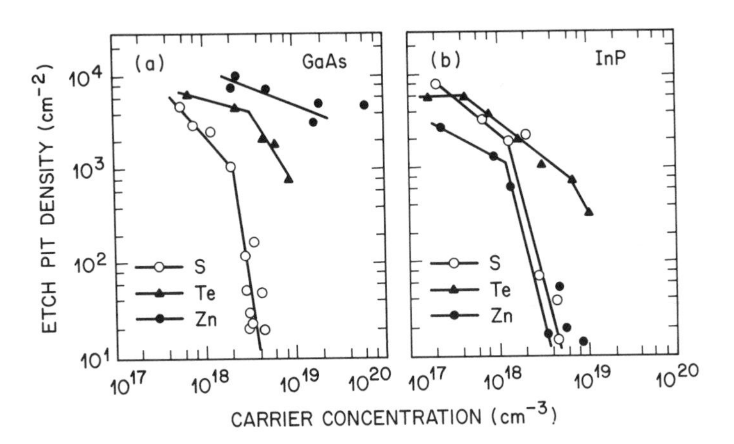


**Figure 2.18** Etch pit density versus carrier concentration in (a) GaAs crystals and (b) InP crystals [67].

In the case of InP growth, isovalent impurities such as Sb, Ga, and As have been used [60, 61, 68, 70, 71]. Of these, Ga is preferred since it has a distribution coefficient $k > 1$ ($k \approx 2.5 - 4.0$) [72] which permits lower concentration of impurities, in the melt and minimize constitutional supercooling problems. Essentially dislocation free crystals of 25 to 30 mm diameter have been obtained with $\sim 10^{19}$ $cm^{-3}$ Ga doping.

Since the distribution coefficient of Ga > 1, its concentration is lower at the tail end of the crystal which has higher dislocation density. To achieve uniform dislocation reduction along the entire crystal double doping method, with one impurity having $k > 1$ and another having $k < 1$, has been adopted. In the case of InP, Ga and Sb ($k < 1$) [61] or Ga and As ($k < 1$) [34] have been used as co-dopants.

Doping the crystals for reducing dislocation density even with isovalent impurities has its problems. First of all, impurity hardening whether achieved by electrically active dopants or isovalent impurities, is effective in reducing dislocations particularly in small diameter crystals. When the thermal gradients are high dislocations near the edges of the crystals are relatively insensitive to the degree of hardening [54]. Impurity hardening combined with smaller thermal gradients render these regions dislocation free. At high concentration of impurities, the lattice parameter of the crystal changes which causes lattice mismatch when 2 to 3 μm epitaxial layers are grown.

When GaAs substrates contained $\geq 6 \times 10^{19}$ $cm^{-3}$ In atoms, the lattice constant increases towards that of InAs and causes lattice mismatch when GaAs layers are grown. The degree of mismatch measured by double crystal X-ray rocking curves show that the difference in the diffraction angle between the GaAs:In substrate and 2 to 3 μm thick GaAs epitaxial layer increases from 12 to 44 arc sec as the In doping of the substrate varies from zero to $6.2 \times 10^{19}$ $cm^{-3}$ [73]. For 2 to 3 μm thick layers In concentration of $3.7 \times 10^{19}$ $cm^{-3}$ introduced no misfit dislocations. For In $> 6.2 \times 10^{19}$ $cm^{-3}$ many misfit dislocations were generated independent of epitaxial layer thickness.

The change in the lattice parameter of the substrate by isovalent doping is also encountered in

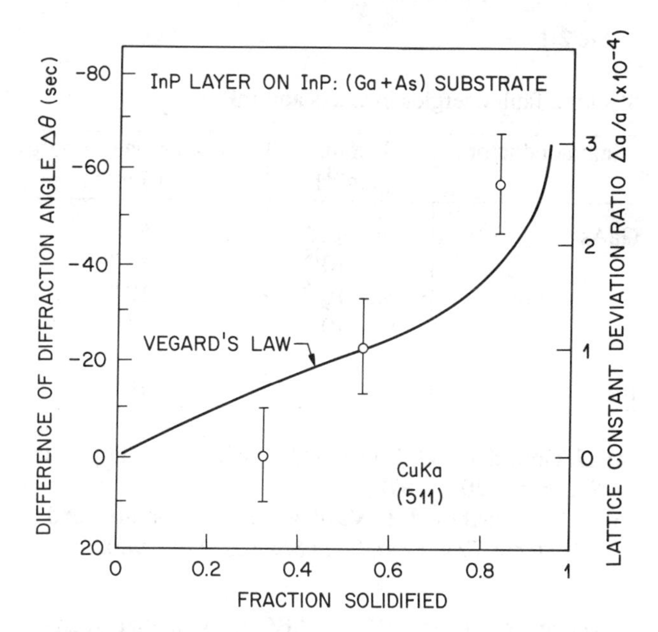


**Figure 2.19** Difference in the lattice constants for an InP substrate doped with Ga and As and the InP layer grown on it as a function of fraction of melt solidified [34].

InP co-doped with Ga and As. Figure 2.19 shows the difference in the diffraction angles (or lattice constants) for an InP substrate doped with Ga and As to a level of $5–10\times10^{19}$ cm$^{-3}$ and an InP layer grown on it, as a function of the fraction of the melt solidified. Ga decreases the lattice constant towards GaP and As increases it toward InAs. The experimental data agrees fairly well with the curve calculated using Vegard's law and assuming that the crystal is a GaInAsP quaternary alloy [34]. Since As with $k<1$ segregates near the tail end of the ingot, a maximum deviation of lattice constant, $\Delta a/a \sim 4\times10^{-4}$, is observed in this region. The possibility of misfit dislocations is greatest when substrates from the tail end of this ingot are used for epitaxial growth.

Because of lattice mismatch problems, it is desirable to achieve dislocation reduction at lower concentrations of isovalent impurities. This is possible in gradient freeze techniques which are inherently low temperature gradient processes. Young et al., [74] obtained 5 cm diameter (D-shaped) GaAs crystals with dislocation densities ~300 cm$^{-2}$ using only $2\times10^{19}$ cm$^{-3}$ In in a horizontal gradient freeze technique with a 5°C/cm gradient. Inoue et al., [73] obtained similar results in a three temperature zone horizontal Bridgman method.

### Stacking faults and twins

Stacking faults and twins are two dimensional defects observed frequently in III-V semiconductor crystals. A stacking fault is a flaw in the stacking sequence of atomic planes in the crystal lattice. It occurs in the sphalerite structure in the same way as it does in fcc lattices. If the $A\alpha B\beta C\gamma$ stacking sequence of {111} planes shown in Fig. 1.1b is written simply as ABCABC,

**Table 2.1**

Stacking fault energies in GaAs and InP

| Semiconductor | Doping ($cm^{-3}$) | Stacking Fault Energy ($mJ\ m^{-2}$) |
|---|---|---|
| GaAs | undoped | 55[a] |
| | Te ~ $10^{18}$ | 23[b] |
| | Sn ~ $10^{16}$ | 10[b] |
| | Sn ~ $10^{17}$ | 6[b] |
| InP | | 18[a] |

(a) H. Gottschalk, G. Patzer, and H. Alexander, Phys. Stat. Solidi *a45*;207, (1978).
(b) V.M. Astakhov, L.F. Vasileva, Yu.G. Siodorov, and S.I. Stenin, *Sov. Phys. Solid State 22*;279, 1980.

then the insertion of a B plane results in ABC · B · ABC stacking that is, an extrinsic stacking fault and the removal of a B plane results in A · C ABC that is, an intrinsic fault. Unlike in the fcc lattice, in the sphalerite structure, stacking faults involve insertion or removal of a pair of planes instead of a single plane. The stacking fault energies of GaAs and InP are given in Table 2.1. The stacking fault energy depends on doping. It decreases with increasing doping an effect which is responsible for the formation of stacking faults in dislocation free heavily doped GaAs [67].

A twin represents two regions in the crystal with one region a mirror reflection of the other across a lattice plane that is common to both regions. The reflection plane is called a twin plane. A volume of the crystal bounded in between parallel twin planes, which are separated only by a few lattice planes, is called a twin lamella. Since the space lattice for the cubic semiconductors is fcc, the twin plane is {111} as in fcc crystals. In the ABCABC stacking sequence of {111} planes, the mirror symmetry and hence the twin corresponds to a stacking sequence

where the line denotes the twin plane. The formation of twin is inherently connected with that of the stacking fault [75]. For example, let us consider an intrinsic stacking fault denoted by

| |
ABC BCABC · · ·
| |

where plane A is missing. The removal of this plane can be regarded as equivalent to two twinning operations, with C and B as the twin planes, separated by one atomic layer. Similarly, an extrinsic stacking fault

$$\mathrm{A\,B\,C\,|\,B\,|\,A\,B\,C} \cdots$$

is equivalent to two twin planes separated by two atomic layers. Because of the connection between twinning and stacking faults, the tendency for twinning is large when the stacking fault energy is low. The lower stacking fault energy of InP would therefore suggest that growing twin-free InP crystals is more difficult compared to GaAs.

Twinning during growth can occur as a result of (a) nucleation at the crucible wall, (b) deviation from melt stoichiometry, (c) excessive thermal stresses due to variations in crystal diameter, (d) instabilities in the shape of the crystal growth front associated with the emergence of the crystal through the boric oxide layer, (e) thermal decomposition following growth, and (f) excessive facet formation.

In the growth of large diameter GaAs crystals, the formation of twins is significantly reduced when crystals are pulled from As-rich melts [27]. The use of As-rich melts has also helped to reduce twins in Bridgman [76] and modified Gremmelmaier [77] growth experiments. The twins are mostly nucleated in the growth of the shoulder part of the crystal and the initiation of a twin plane can be coincident with the As or Ga facet. A gradual increase of the crystal diameter therefore helps to reduce twinning in the early stages of growth [47, 78].

Removing the moisture from $B_2O_3$ helps to reduce twinning [37, 79]. The effect of using "dry" $B_2O_3$ on twinning in the growth of <100> GaAs crystals is illustrated in Fig. 2.20 [29].

**Figure 2.20** Effect of [OH] content of boric oxide on twinning in large diameter <100> GaAs crystals pulled from (a) fused silica and (b) PBN crucibles [29].

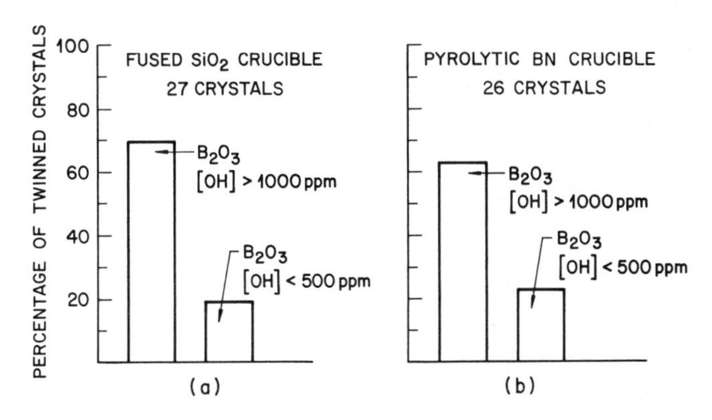

For growths from fused $SiO_2$ or PBN crucibles, the incidence of twinning within the first 75 percent of growth is substantially reduced when vacuum baked $B_2O_3$ (< 500 ppm wt [OH]) is used.

Thermal decomposition of the crystal has been associated with twinning in InP crystal growth [78]. When the crystal emerges from the $B_2O_3$ layer, thermal decomposition of the crystal surface occurs as a reuslt of preferential evaporation of P. Thermal decomposition leaves small cavities which reach the growth interface and cause twins. A thin protective skin of $B_2O_3$ on the crystal withdrawn from the $B_2O_3$ layer suppresses thermal decomposition. The formation of the protective skin which depends on the viscosity of the $B_2O_3$ occurs when the surface temperature of the layer is less than 550°C. However, keeping the outer surface of $B_2O_3$ at low temperature increases the thermal gradient near the InP melt surface and in turn the thermal stresses.

**Inclusions, precipitates, and dislocation clusters**

When the doping concentrations exceed their solubility limit, precipitates are formed. For example, Zn in InP at a doping level exceeding $1 \times 10^{18}$ $cm^{-3}$ forms precipitates [80, 81]. Boron contamination and precipitation have also been identified in LEC InP [82]. Formation of precipitates or the incorporation of inclusions during the growth process also gives rise to dislocation clusters [26, 83]. These clusters are believed to be the initial stage of the formation of the "grappe" defect [84]. Both the dislocation cluster and the "grappe" in LEC InP crystals are shown in Figs. 2.21 and 2.22, respectively [85]. The grappe consists of a central core region

**Figure 2.21** Photomicrograph of a LEC InP crystal showing dislocation cluster [85].

**Figure 2.22** Photomicrograph of a LEC InP crystal showing the "grappe" defect [85].

surrounded by dense dislocation etch pits. The core of the clusters and the "grappe" is believed to be impurity precipitates or inclusion of the matrix elements, most likely In [26].

**Growth striations**

The incorporation of impurities into the growing crystal is characterized by a distribution coefficient, k, defined as $k = c_s / c_l$, where $c_s$ and $c_l$ denote concentrations of the impurity in the solid ($c_s$) which is in equilibrium with the melt ($c_l$). The concentration in the crystal obeys the normal freezing law

$$c = c_o k (1 - g)^{k-1} \tag{2.1}$$

where $c_o$ is the initial concentration in the melt and g is the fraction of the melt solidified. For an impurity with $k < 1$ its concentration will increase along the length of the crystal as it solidifies. Equation (2.1) is the basis of purifying crystals by zone refining [86].

The k used in Eq. (2.1) corresponds to an effective distribution coefficient which depends on the growth rate, stirring condition, crystallographic orientation of the growing interface and on the stoichiometry. Tables 2.2 and 2.3 list the effective distribution coefficients for various impurities in GaAs [87, 88] and InP [89], respectively. For some of the impurities, the effective free electron distribution coefficient is given assuming that $c_s$ is equal to the free electron concentration at room temperature.

**Table 2.2**

Effective distribution coefficients of impurities in GaAs

| Impurity | Effective Distribution Coefficients | |
|---|---|---|
| | LEC growth | Non-LEC growth |
| Be | | 3 |
| Cr | $1.03 \times 10^{-3}$ | $5.7 \times 10^{-4}$ |
| Ge | $2.8 \times 10^{-3}$ a) | $1.0 \times 10^{-2}$ |
| Si | $1.85 \times 10^{-2}$ a) | $1.4 \times 10^{-1}$ |
| Sn | $5.2 \times 10^{-3}$ a) | $8.0 \times 10^{-2}$ |
| Se | $5.0 \times 10^{-2}$ a) | $3.0 \times 10^{-1}$ |
| Te | $6.8 \times 10^{-2}$ a)b) | $5.9 \times 10^{-2}$ b) |
| | $3.0 \times 10^{-2}$ d) | $2.8 \times 10^{-2}$ c) |
| Fe | | $1.0 \times 10^{-3}$ |
| Zn | | $4.0 \times 10^{-1}$ |
| In | 0.10 e) | |
| Mn | | 0.02 |
| Co | | $4.0 \times 10^{-4}$ |
| Ni | | $4.0 \times 10^{-3}$ |
| O | | 0.3 |
| Sb | | $1.6 \times 10^{-2}$ |
| Pb | | $<1.0 \times 10^{-3}$ |

(a) Effective free electron distribution coefficient
(b) [111] As
(c) $[\bar{1}\bar{1}\bar{1}]$ Ga
(d) [100]
(e) Ref. [69]

If the distribution coefficient $k \neq 1$, then impurity striations occur. Striations are periodic variations of the concentration of an impurity in the crystal. Rotation of the seed-crucible gives rise to symmetric or rotational striations. Nonrotational striations are mainly caused by thermal convection flows especially in large volume melts. The convective flows cause severe temperature fluctuations. Measurements of the temperature fluctuations $\Delta T$ in a 150 mm diameter $B_2O_3$ encapsulated 3 kg GaAs melt in a high pressure (20 atm Ar) commercial puller indicated that $\Delta T > 3°C$ at the $B_2O_3$-GaAs interface and $\Delta T \sim 9°C$ at 1 cm below the surface of the melt [29]. Thermal asymmetries at the crystal-melt interface in turn give rise to fluctuations in the microscopic growth rate causing impurity striations.

Figure 2.23 is a Nomarski optical photomicrograph showing the growth striations in a (100) InP wafer cut from a (111) oriented LEC crystal [85]. The polished wafer was etched in a solution of $1HNO_3$:3HBr to reveal the striations [90]. The wide parallel strips are twin plates.

**Table 2.3**

Effective distribution coefficient of impurities in InP

| Impurity | Effective Distribution Coefficient | |
|---|---|---|
| | LEC growth | Non-LEC growth |
| Co | $4\times10^{-5}$ | |
| Cr | $(1...6)\times10^{-4}$ | |
| Fe | $1.6\times10^{-3}$ | |
| Mn | 0.4 | |
| Ge | $2.4\times10^{-2}$ | $5\times10^{-3}$ (LPE)[a] |
| S | 0.5 | |
| Se | | 4.2 (LPE)[a] |
| Sn | $2.2\times10^{-2}$ | $2.0\times10^{-3}$ (LPE)[a] |
| Si | 0.55[b]<br>0.15[c]<br>>1.0[d] | 30 (LPE)[a] |
| Te | | 0.4 (LPE)[a] |
| Sb | $2.6\times10^{-2}$ | |

(a) E. Kuphal, *J. Cryst. Growth 54*;117, 1981.
(b) effective free electron distribution coefficient.
(c) LEC in $SiO_2$ crucible
(d) LEC in PBN crucible

These plates are parallel to (111) planes which intersect the (100) plane along [011] directions. These closely spaced twin plates are probably deformation twins caused by thermal stresses.

Growth striations can also be observed by transmission cathodoluminescence (TCL) [91] and by x-ray topography. The variation in the dopant concentration across the growth striation causes change in the luminescence efficiency. By observing the spatial variation of luminescence efficiency, one can identify impurity striations. Figure 2.24 is a TCL micrograph showing the growth striations in S-doped (100) InP [85]. Note the irregular spacing between the striations which denote nonrotational striations.

The change in lattice constant of the crystal caused by impurity doping gives rise to strain contrast in x-ray topography which is another method to study growth striations. Figure 2.25 shows transmission x-ray topographs from S-doped InP [85]. In Fig. 2.25(a), the diffraction condition is favorable to reveal the striations. In Fig. 2.25(b) the diffraction vector is parallel to the striations such that the strain contrast vanishes indicating that the displacement field associated with the dopant fluctuation is perpendicular to the melt-solid interface.

Several methods of characterization are used to characterize the various defects described. These are x-ray topography, transmission electron microscopy, electron beam induced current

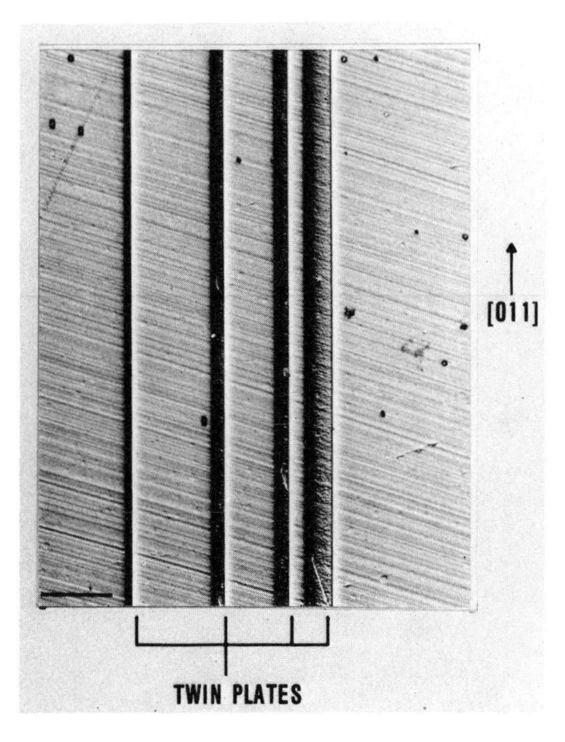


**Figure 2.23** Nomarski optical photomicrograph showing growth striations in a (100) LEC InP wafer [85]. The wide parallel strips are twin plates.

**Figure 2.24** Transmission cathodoluminescence micrograph showing growth striations in S-doped (100) InP [85]. The irregular spacing between the striations denote non-rotational striations.

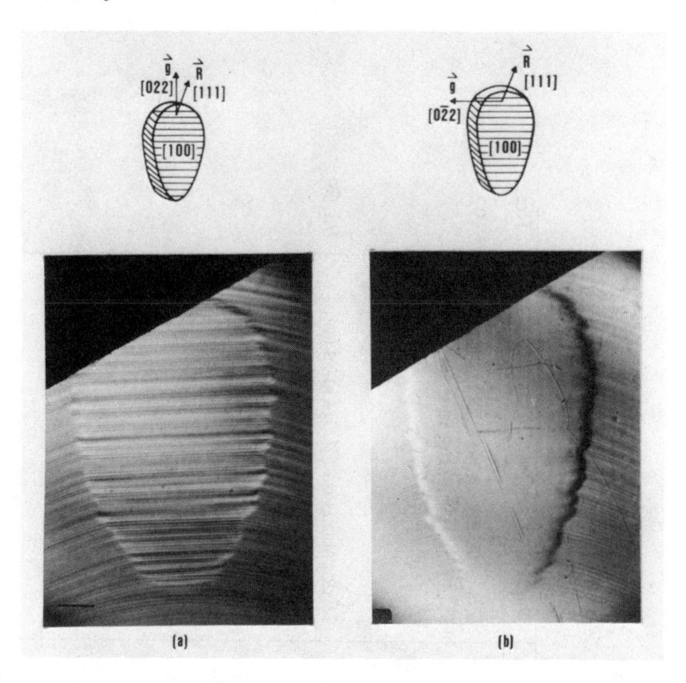


**Figure 2.25** Transmission x-ray topographs from S-doped InP: (a) the diffraction condition is favorable to show the striations (b) the diffraction vector is parallel to striations such that the strain contrast vanishes [85].

(EBIC) and cathodoluminescence, and optical microscopy. Of these, optical microscopy is a frequently used method for the detection of defects when the defects are revealed by special etching techniques. The etchants used for this purpose to reveal defects in GaAs and InP are given in Table 2.4. The substrates are polished before the etchants are used.

**Impurity contamination**

The growth of high purity crystals mandates high purity starting materials. Besides the starting materials, the other sources of contamination are the materials which come into contact with the crystal during growth. For example, the use of quartz crucibles would lead to Si incorporating in the crystal. Replacing quartz crucibles with pyrolytic BN crucibles would minimize Si contamination but increases B contamination. Another impurity which is readily incorporated from impure quartz is Cu [103]. The use of $B_2O_3$ as encapsulant leads to B contamination of the crystals [82, 104-106].

Although several analytical methods have been employed for detecting impurities, atomic absorption spectroscopy (AAS), spark source mass spectrometry (SSMS), secondary ion mass spectrometry (SIMS), and local mode infrared (LVM) absorption are the ones frequently used [107]. Table 2.5 shows the results of SIMS analysis of LEC semi-insulating GaAs crystals pulled from high-purity pyrolytic BN crucibles [29]. The residual Si concentrations in crystals grown in quartz crucibles are typically a factor of 10 higher than the values shown in Table 2.5. Note the

**Table 2.4**

Etchants used for revealing defects in GaAs and InP

| Compound | Etchant | Recipe | Defect | Ref. |
|---|---|---|---|---|
| GaAs | AB-etch | 10 ml $H_2O$, 40 mg $AgNO_3$, 5g $CrO_3$ and 5 ml HF etching 10 min at 65°C | Dislocations on (100), (110) and (111) Ga. Stacking faults on (100), by deletion of $CrO_3$ also on (111) As | 92 |
| | KOH-etch | Molten KOH at 330°C | Dislocations on (100) GaAs and stacking faults | 93, 94 |
| InP | $HNO_3$: HBr | Volume ratio 1:3. The solution should be freshly prepared. Before etching, the polished wafers are dipped in concentrated $HNO_3$ for 1 min and rinsed clean in deionized water. Etching times at room temperature are 20-30 sec for the (100) surface and 10 sec for the (111) P surface | Dislocations on (100) and (111) P surfaces | 90 |

**Table 2.4 (cont'd)**

Etchants used for revealing defects in GaAs and InP

| Compound | Etchant | Recipe | Defect | Ref. |
|---|---|---|---|---|
| | $H_3PO_4$ (85%) :HBr (47%) (Huber etch) | Volume ratio 2:1. Etching time of 2 min at 24°C | Dislocations on (111) P (100) and (110) | 95 |
| | $HCl:HNO_3:Br$ | Volume ratio 20:10:0.25. Wait 4 min after mixing. Etching for 0.1 to 2.0 min. | Dislocations on (100) and (111) surfaces | 96 |
| InP | HBr/HF<br>$HBr/CH_3COOH$ | | Dislocations on (100) and (111) InP | 97 |
| GaAs | $H_2SO_4:H_2O_2$ (30%) $:H_2O$ | 3:1:1 mixture<br>Etching under illumination | Striations, precipitates (111) and (100) GaAs | 98 |
| | $HF:H_2O_2:H_2O$ | 1:1:10, under illumination | Dislocations (100) plane | 99 |
| | $H_3PO_4$ (85%)$:H_2O_2$ | 10:1, 1:1, 1:10<br>illumination 3 to 15 min | Dislocations and microdefects (100) plane | 100 |
| | AB etch | Under illumination | Striations, dislocations | 101 |
| InP | $H_3PO_4$ (85%) $:H_2O_2$ (30%) | 1:1, 50W lamp,<br>10-20 min | Dislocations and microdefects | 102 |

**Table 2.5**

High Sensitivity Secondary Ion Mass Spectroscopy Analysis of LEC Semi-Insulating GaAs Crystals Pulled from High-Purity Pyrolytic Boron Nitride Crucibles [29]

| Crystal[a] | C[b] | O[b] | Si | S | Se | Te | Cr | Mg | Mn | B | Doped | |
|---|---|---|---|---|---|---|---|---|---|---|---|---|
| | | | | | | | | | | | Yes | No |
| BN-1 (s) | 2e15 | 1.5e16 | <8e14 | 5.7e14 | 3e13 | 1e12 | <4e14 | 1e15 | 7e14 | 2.3e15 | | √ |
| BN-2 (s) | 1e16 | 6e16 | 6e14 | 2e15 | 1e15 | 7e13 | 5.6e14 | 4e14 | 8e14 | 1.7e17 | | √ |
| BN-3 (s) | 3e15 | 1.3e16 | 9.3e14 | 4.2e14 | 3e14 | <5e12 | 4.3e14 | 2e15 | 1e15 | 5.4e17 | | √ |
| BN-4 (s) | - | - | 2e15 | 8e14 | 2e14 | 1e14 | 1e16 | 2.5e15 | 2e15 | 1.5e17 | Cr | |
| BN-6 (s) | 2.9e15 | 9.8e15 | 5.4e14 | 5e14 | 9.4e13 | <5e12 | 5.4e14 | 2.5e15 | 1.7e15 | 6.9e17 | | √ |
| BN-10 (s) | 2e15 | 1e16 | 8.6e14 | 1.5e15 | 7e14 | 7e12 | <4e14 | 3.6e15 | 1.2e15 | 2e18 | | √ |
| BN-11 A (s) | 4.3e15 | 1.8e16 | 6.4e14 | 1e15 | 8e13 | 6e12 | 6.4e15 | 3.2e15 | 1.4e15 | 1.9e17 | Cr | |
| BN-11 B (s) | 3e15 | 1.3e16 | 7.6e14 | 8.7e14 | 1e14 | 8e12 | 6.3e15 | 4.4e15 | 1.6e15 | 1.9e17 | Cr | |
| BN-12 (s) | 2e16 | 7e16 | 1e15 | 2e15 | 5e14 | 6e13 | 3.2e15 | 3e14 | 8e14 | 1.8e17 | Cr | |
| BN-13 (s) | - | - | <3e14 | 4e15 | 2e14 | 1e13 | 2e15 | 5e14 | 7e14 | 7e16 | Cr | |
| BN-13 (t) | - | - | <3e14 | 3e15 | 5e14 | 1e13 | 4e15 | 8e14 | 7e14 | 2e17 | Cr | |
| BN-14 (s) | - | - | 4e14 | 8e14 | 9e14 | 8e12 | 6e14 | 1e15 | 8e14 | 3e17 | | √ |
| BN-14 (t) | - | - | 7e14 | 1e15 | 1e15 | 1e13 | 1e15 | 9e14 | 1e15 | 7e17 | | √ |
| BN-15 (s) | - | - | <3e14 | 1e15 | 1e14 | 9e12 | 3e15 | 7e14 | 8e14 | 7e16 | Cr | |
| BN-15 (t) | - | - | <3e14 | 2e15 | 5e14 | 2e13 | 5e15 | 1e15 | 8e14 | 2e17 | Cr | |
| BN-18 (s) | - | - | 4e14 | <3e14 | 7e13 | 1e14 | 3e15 | <3e14 | <7e14 | 6e16 | Cr | |
| BN-20 (s) | - | - | 3e14 | 6e14 | 8e13 | 6e13 | 2e15 | 9e13 | <7e14 | 1e17 | Cr | |
| BN-21 (s) | - | - | 2e15 | <3e14 | 3e13 | 7e13 | 3e15 | - | - | 9e16 | Cr | |
| BN-22 (s) | - | - | 9e14 | 6e14 | 1e14 | 6e13 | 1e15 | 5e13 | <7e14 | 2e16 | Cr | |
| BN-23 (s) | 9e15 | 5e16 | 1e15 | 7e14 | 3e14 | 3e13 | <2e15 | <3e14 | 1e15 | 1e17 | | √ |
| BN-28 (s) | 1e16 | 6e16 | 1e15 | 1e15 | 3e14 | 3e13 | 4e14 | <3e14 | 8e14 | 6e16 | | √ |
| | | | | | | | | | | | | |
| Detection Limit | - | - | 3e14 | 3e14 | 5e12 | 5e12 | 4e14 | 3e14 | 7e14 | 8e12 | | |

(a) (s) seed end sample and (t) Tang-end sample.
(b) Detection limits for C and O not well defined.

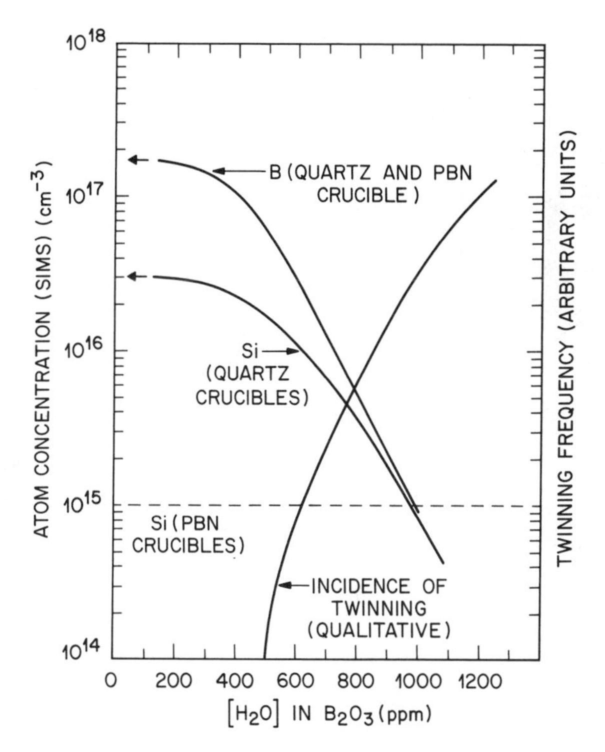


**Figure 2.26** Dependence of Si and B concentrations and twinning in LEC GaAs on water content of $B_2O_3$ encapsulant [108].

high concentrations of B in the crystals shown in Table 2.5. The carbon concentration determined by infrared LVM technique is found to be lowest in LEC GaAs crystals grown from quartz crucibles ranging from $<2\times10^{15}$ cm$^{-3}$ (detection limit) to $9\times10^{15}$ cm$^{-3}$ [108]. In contrast, crystals grown from PBN crucibles contain C at levels $2\times10^{15}-1.5\times10^{16}$ cm$^{-3}$.

The B concentration generally varies from $10^{15}$ to $10^{17}$ cm$^{-3}$ and it mainly comes from the $B_2O_3$ encapsulant. It is observed that the incorporation of both B and Si is affected by the water content of $B_2O_3$ [29, 108]. Both decrease as the water content of $B_2O_3$ increases as shown in Fig. 2.26 [108]. Note also the dependence of twinning on the water content of $B_2O_3$. If quartz crucibles are used, use of wet $B_2O_3$ reduces the Si level and semi-insulating GaAs can be obtained. On the other hand, the crystals are likely to contain twins. If, however, PBN crucibles are used, Si contamination is eliminated and dry $B_2O_3$ can be used to grow twin free semi-insulating material. Since B is an isovalent impurity in GaAs, no electrical activity associated with B has been found. However, a deep acceptor level (~0.07 eV above the valence band) found in low-resistivity p-type GaAs crystals has been found to depend on B concentration. This level has been associated with a complex involving B and a non-stoichiometric defect caused by excess Ga [29].

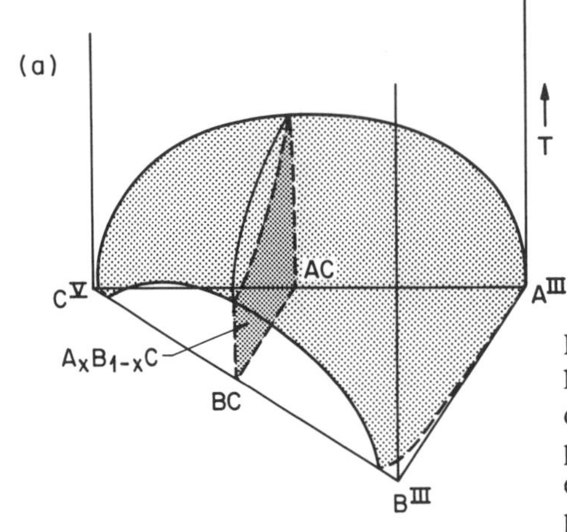


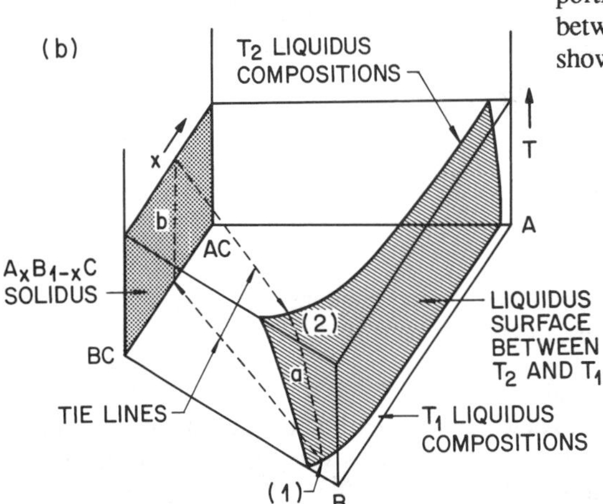


**Figure 2.27** a) Schematic representation of the liquidus and solidus surfaces in the ternary phase diagram of alloy $A^{III}B^{III}C^{III}$ b) Three dimensional phase diagram for the liquidus and solidus in equilibrium with the solid $A_xB_{1-x}C$. Only the portion of the system in which $X_A + X_B > X_C$ between two closely lying temperatures $T_2$ and $T_1$ is shown in [109].

## 2.2 LIQUID PHASE EPITAXY

Liquid phase epitaxy (LPE) is the simplest and most widely used crystal growth technique for growing epitaxial layers of many III-V compounds. Basically, LPE involves the growth of an epitaxial layer on a single crystal substrate from a solution saturated or supersaturated with the material to be grown. The substrate has similar crystal structure and lattice constant to the growing layer to allow the continuation of a coherent crystal structure. The growth solution is rich in one of the major components of the solid and dilute in all others (e.g., growth of GaAs from a Ga-rich solution saturated with As).

The thermodynamic basis of LPE can be illustrated by the binary phase diagram in Fig. 2.1. A liquid solution of composition $X_2$ at temperature $T_2$ is cooled to temperature $T_1$ resulting in the deposition onto a substrate of an amount of solid AB equivalent to the loss of ($X_2$−$X_1$) atom fraction of B and an equal amount of A. The growth of an alloy $A_xB_{1-x}C$ where A and B can be group III elements and C is a group V element can be understood by means of a ternary phase diagram illustrated in Fig. 2.27a. Figure 2.27b shows the group III rich region between two close isotherms. The compositions of liquid A-B-C that can be in existence with the solid $A_xB_{1-x}C$ at a given temperature are completely represented by the liquidus curve and the corresponding tie lines. In growing $A_xB_{1-x}C$, as the liquid solution at 2 in Fig. 2.27b is cooled to 1 from $T_2$ to $T_1$ its composition follows curve a. A solid richer in A than the liquid will deposit epitaxially and its composition follows curve b [109].

### 2.2.1 Crystal Growth Apparatus

The LPE growth apparatus simply allows a growth solution of the desired composition to be placed in contact with the substrate for a certain time under controlled temperature conditions. The solution supersaturation necessary for deposition is achieved by reducing the temperature utilizing the fixed relationship between temperature and solubility as predicted by the phase diagram. Besides thermodynamic considerations there are other factors such as diffusion of constituents in the solution, nucleation and the mechanism of growth at the surface and convection as the result of temperature and compositional gradients that affect the LPE process. Some of these considerations are system and procedure dependent and as such the details of the apparatus and procedures used by different workers vary considerably. In general, each growth procedure is evolved in a semiempirical way depending upon the particular furnace design, boat design, and time-temperature considerations.

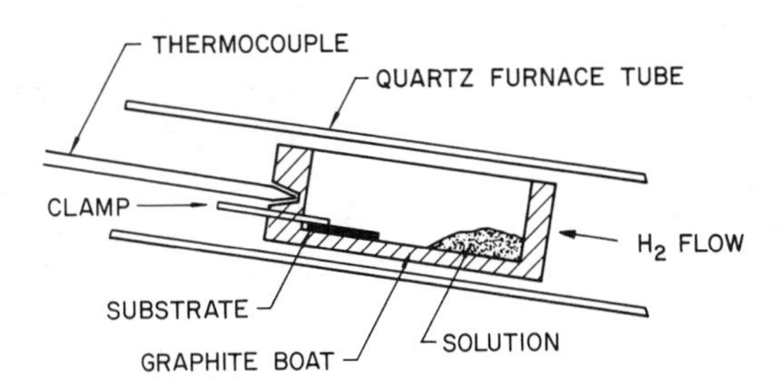


**Figure 2.28** Schematic representation of the tipping furnace [110].

There are three principal LPE growth techniques: tipping, dipping, and sliding. In the tipping technique the solution is brought into contact with the substrate by tipping the furnace [110]. The dipping technique uses a vertical furnace in which the substrate is immersed in the solution. The sliding technique is used for growing multilayers and employs a multibin system in which layers are sequentially grown by bringing the substrate into contact with different solutions.

The original tipping furnace is shown in Fig. 2.28 [110]. The substrate is held tightly at the upper end of a graphite boat and the growth solution is placed at the other end. The solution is brought into contact with the substrate by tipping the solution. The furnace is then slowly cooled and an epitaxial layer is grown on the substrate. The solution remains in contact with the substrate for a defined temperature interval and growth is terminated by tipping the furnace back to its original position. The solution remaining on the film surface is removed by wiping and dissolving in a suitable solvent.

The dipping technique [111] uses a vertical furnace as shown in Fig. 2.29. The solution is contained in a graphite or alumina crucible at the lower end of the furnace and the substrate fixed

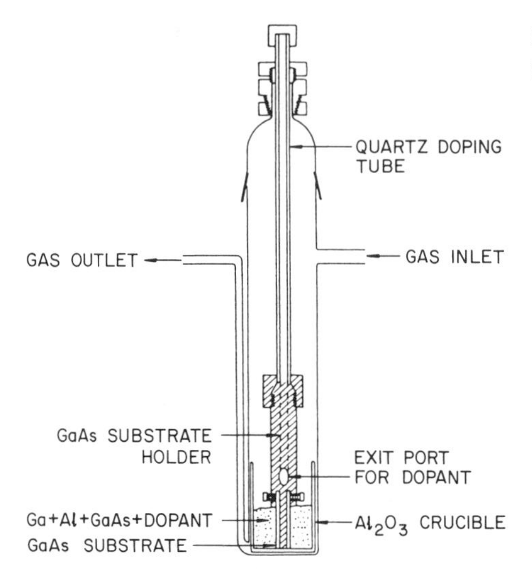


**Figure 2.29** Growth apparatus for LPE growth using the dipping technique [111].

in a movable holder is initially positioned above the solution. At the desired temperature growth is initiated by immersing the substrate in the solution and it is terminated by withdrawal of the substrate from the solution.

The apparatus used for the tipping and dipping techniques is very simple and easy to operate. High quality GaAs layers have been grown by both techniques. However, growth of multiple layers by these techniques would require considerably more complex apparatus. The third LPE technique, the sliding technique, uses a multibin boat [112, 113] to grow multiple epitaxial layers. This is illustrated in Fig. 2.30 [113]. The principal components of this apparatus are a massive split-graphite barrel with a graphite slider, a fused silica growth tube to provide a protective atmosphere and a horizontal resistance furnace. The graphite barrel has the desired number (usually six) solution chambers depending on the number of layers to be grown, and the slider has two slots for the precursor seed substrate and the growth substrate. The function of the precursor seed is to saturate the growth solution before it is brought into contact with the growth substrate. The substrates are brought into contact with the solutions by motion of the barrel over the slider. This operation can easily be automated. The fused silica tube is usually within a heat pipe thermal liner [114] in the furnace to ensure uniform temperature. Alternatively, a multizone furnace without a heat pipe can also be used. Growth is usually carried out in an atmosphere of $H_2$. An approximate temperature-time profile and the relative substrate and solution positions are also shown in Fig. 2.30.

A vertical LPE furnace can also be employed for multilayer growth in which a graphite boat is rotated to transport seeds from one solution to another solution. With the seed at the hottest region, vertical temperature gradients are obtained by regulation of the different zones of a multizone vertical furnace while maintaining radial temperature uniformity. Thompson and Kirkby [115] used a vertical LPE furnace to grow GaAs-AlGaAs heterostructures.

Another variation of the horizontal boat technique that has been used by several workers [112, 116-120] is to use very thin (~1 mm per gm of solution) solutions. This procedure is found to be useful to grow very thin layers with excellent uniformity and for suppressing convection and edge

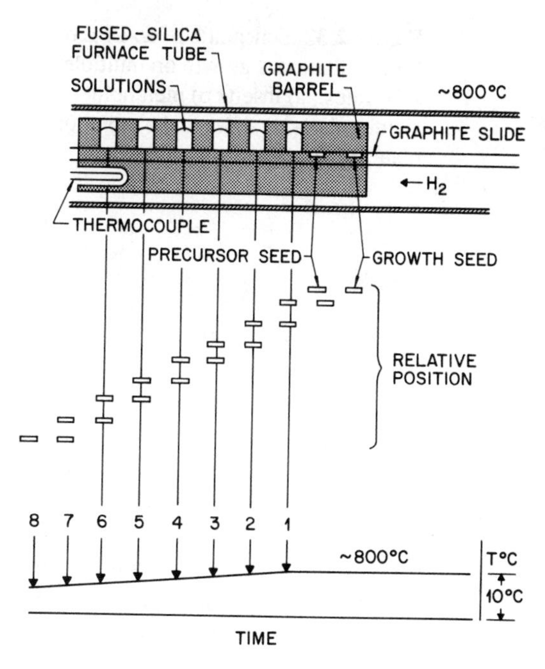


**Figure 2.30** Multibin LPE growth apparatus. The relative positions of the precursor and growth substrates and solutions during growth and the temperature profile are shown [113].

growth. The confined [116] and the baffled [120] melt geometries employed for reducing the solution volume are illustrated in Fig. 2.31.

Horikoshi [121] used a "wiping less" LPE method to grow multilayer AlGaAs heterostructure. In this technique multilayers are grown in such a manner that the solution used

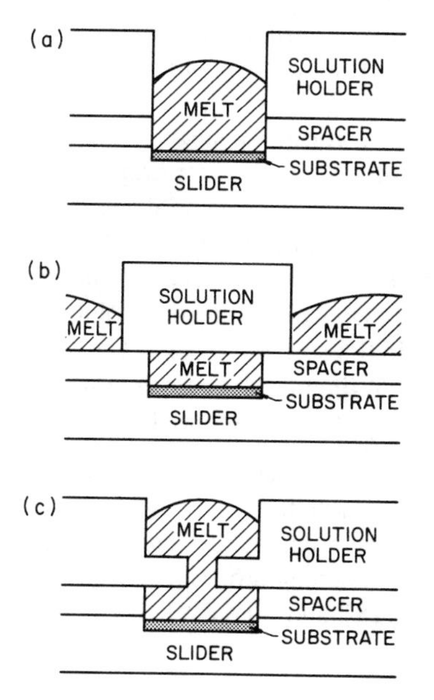


**Figure 2.31** Schematic diagrams of the a) conventional, b) confined and c) baffled solution holder [117].

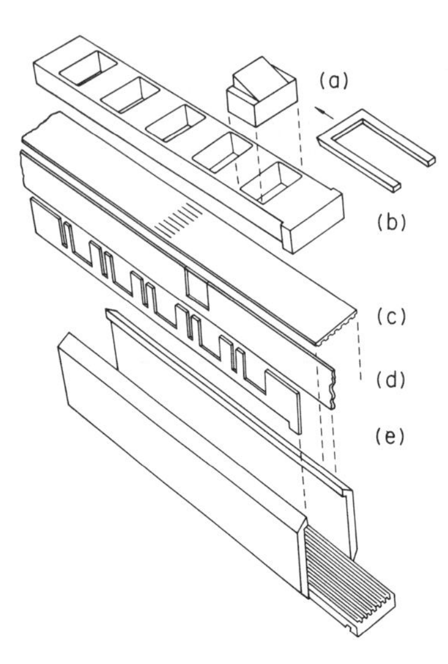


**Figure 2.32** Schematic of an LPE boat for simultaneous growth on multiple substrates. a) inserts b) melt holder c) shutter plates d) slider and e) solution plate [124].

for the growth of one layer is pushed out by the solution for the subsequent layer growth. In this way scratches on epitaxial layer surface brought about by the mechanical sliding process were eliminated. Further, contamination of interfaces between epitaxial layers by impurities such as oxygen was thought to be reduced.

The conventional multibin sliding LPE technique illustrated in Fig. 2.30 permits epitaxy on one substrate only. To overcome this disadvantage and improve throughput, multisubstrate LPE has been developed [122 - 124]. A schematic of one such multisubstrate LPE used for the growth of lattice matched GaInAs on InP is illustrated in Fig. 2.32 [124]. The substrates are stacked vertically side-by-side in respective cavities of the substrate holders and they are slid across a number of aliquots of each solution which are arranged one after another. Each substrate holder can contain two substrates sandwiching each aliquot allowing simultaneous LPE growth on two substrates from a thin aliquot as temperature is lowered. The solution chambers located in the upper section are separated from the lower vertically sandwiched stack of solution plates and sliders by a shutter. The shutter consists of a set of slots for dropping the solutions into a number of aliquots in the solution plate. The shutter allows sequential selection and dropping of the LPE solutions at desired intervals of time. The bottom of the boat and the shutter plate are corrugated as shown to improve the alignment of the sandwich stack of sliders and solution plates. To improve wiping and to prevent solution carry-over from one solution to another, small wiping walls are provided between the aliquots in the solution plates. The solution plates also contained reservoirs with a volatile source to prevent thermal decomposition of the substrate (e.g., InP-Sn solution to provide an over pressure of P in order to suppress the thermal dissociation of InP).

Instead of relying on the cooling of the furnace to reduce the temperature to achieve LPE growth, another method has been adopted which uses Peltier cooling at the substrate-solution interface produced by passing electric current across the interface while the temperature of the growth system is maintained constant. This method of LPE called electroepitaxy has been used to grow GaAs [125, 126], AlGaAs [127, 128], and InP [129].

### 2.2.2 Thermodynamic Principles of LPE Growth

The thermodynamic basis of LPE growth is the phase diagram that provides information on the composition of the solution and the solid in equilibrium as a function of temperature. The feature that compound AB (in Fig. 2.1) can be in equilibrium with liquid solutions that are dilute in B is the basis for depositing AB from a liquid that primarily contains A. Several authors have calculated the phase diagrams for binary [130, 131], ternary [132, 133], and quaternary [134, 135] compounds using simple solution models and experimental data on temperature and entropy of fusion of binary compounds.

Let us consider the binary compound AB. The equilibrium between solid AB and liquid solution of A and B at temperature T can be described by the reaction

$$A(\ell) + B(\ell) \rightleftarrows AB(s) \tag{2.2}$$

At thermodynamic equilibrium, the Gibbs free energy change is zero which gives the relation

$$\mu_A(T) + \mu_B(T) - \mu_{AB}(T) = 0 \tag{2.3}$$

where $\mu_s$ are the chemical potentials, given by

$$\mu_i = \mu_i^o + RT \ln a_i = \mu_i^o + RT \ln \gamma_i X_i \tag{2.4}$$

where $a_i$, $X_i$, and $\gamma_i$ are, respectively, the activity, mole fraction and activity coefficient of component i [136]. The chemical potential of the pure component is denoted by $\mu_i^o$. For the pure solid AB the activity $a_{AB}$ is unity giving

$$\mu_{AB} = \mu_{AB}^o \tag{2.5}$$

Using Eq. (2.4), Eq. (2.3) can be written as

$$\mu_A^o(T) + RT \ln[\gamma_A(T)\, X_A(T)] + \mu_B^o(T) + RT \ln[\gamma_B(T)\, X_B(T)] - \mu_{AB}^o(T) = 0 \tag{2.6}$$

The chemical potential change between temperatures $T_2$ and $T_1$ with $T_2 > T_1$ can be obtained from Eq. (2.6)

$$\mu_A^o(T_2) - \mu_A^o(T_1) + \mu_B^o(T_2) - \mu_B^o(T_1) - \mu_{AB}^o(T_2) + \mu_{AB}^o(T_1)$$

$$\equiv \Delta\mu^o = -RT_2 \ln[\gamma_A(T_2)\, \gamma_B(T_2)\, X_A(T_2)\, X_B(T_2)] + RT_1 \ln[\gamma_A(T_1)\, \gamma_B(T_1)\, X_A(T_1)\, X_B(T_1)] \tag{2.7}$$

For a pure component $\mu_i$ can be written between $T_2$ and $T_1$ as [136]

$$\mu_i^o(T_2) - \mu_i^o(T_1) = -\int_{T_1}^{T_2} S_i^o(T)\, dT \tag{2.8}$$

where $S_i^o$ is the entropy of pure component i. For a reversible process [136]

$$dS = \frac{C_P}{T}\, dT \tag{2.9}$$

so that for a pure component in which there is no phase transformation between $T_2$ and T

$$S_i^o(T) = S_i^o(T_2) + \int_{T_2}^{T} \frac{C_{P,i}^o}{T}\, dT \tag{2.10}$$

where $C_{P,i}^o$ is the heat capacity at constant pressure for pure component i. Combining Eqs. (2.8) and (2.10) and evaluating the left hand side of Eq. (2.7) one gets [137]

$$\Delta\mu^o = \Delta S^o(T_2)\,(T_1 - T_2) - \int_{T_1}^{T_2}\int_{T_2}^{T} \frac{\Delta C_P^o}{T'}\, dT'dT \tag{2.11}$$

where $\Delta S^o(T_2) = S_A^o(T_2) + S_B^o(T_2) - S_{AB}^o(T_2)$ and $\Delta C_P^o = C_{P,A}^o + C_{P,B}^o - C_{P,AB}^o$. By setting $T_2$ equal to $T_F$ the melting point of AB and noting that $X_A = X_B = 0.5$ at $T_F$, the right hand side of Eq. (2.7) can be written as

$$RT_1 \ln[\gamma_A(T_1)\,\gamma_B(T_1)\,X_A(T_1)\,X_B(T_1)] - RT_F \ln[\gamma_A(T_F)\,\gamma_B(T_F)/4]$$

$$= -\Delta S^o(T_F)\,(T_F - T_1) - \int_{T_1}^{T_F}\int_{T_F}^{T} \frac{\Delta C_P^o}{T'}\, dT'dT \tag{2.12}$$

Equation (2.12) is the liquidus equation. To calculate the liquidus using this equation requires the activity coefficients, the heat capacities and the entropy change. The $\Delta C_P^o$ term is usually small compared to other terms and may be neglected in Eq. (2.12) [138].

The activity coefficients are represented by a quantity called the interaction parameter [139-141]. The simple solution model and variations of it have been used successfully to estimate this parameter. The liquid and solid solutions are considered simple solutions for which the excess free energy of mixing $G^E_m$ can be expressed in terms of a composition independent interaction parameter $\alpha$ such that [142]

$$G^E_m = \alpha X(1-X) \tag{2.13}$$

The mole fraction of one of the constituents is X. The enthalpy and excess entropy of mixing are given respectively by

$$\begin{aligned} H_m &= (\alpha - T\frac{\partial \alpha}{\partial T})\, X(1-X) \\ S^E_m &= -\frac{\partial \alpha}{\partial T}\, X(1-X) \end{aligned} \tag{2.14}$$

For many III-V binary systems it has been found that $\alpha(T)$ has a linear temperature dependence which may be expressed as [131]

$$\alpha(T) = a - bT \tag{2.15}$$

where a and b are constants. Strictly regular solutions and athermal solutions are limiting special cases of simple solutions. For the former $\alpha = a$ and for the latter $\alpha = -bT$. Simple solutions for which Eq. (2.15) applies have also been designated ''quasi-regular'' solutions but this terminology is, however, not preferred because of the usual association of the term ''regular'' with random mixing [132]. Equation (2.15) used in conjunction with the entropy of fusion equation for the liquidus [130] is found to satisfactorily describe the experimental liquidus data. The use of the relation in Eq. (2.13) and noting that the excess chemical potential of component A, $\mu^E_A$ is simply the nonideal part of $\mu_A$ given by

$$\mu^E_A = RT \ln \gamma_A \tag{2.16}$$

and further noting that [142]

$$\mu^E_A = G^E - X_B \frac{\partial G^E}{\partial X_B} \tag{2.17}$$

would give the following relations for simple binary solutions [132]

$$RT \ln \gamma_A = \alpha(T)\, X_B^2 = \alpha(T)(1-X_A)^2 \tag{2.18}$$

$$RT \ln \gamma_B = \alpha(T)\, X_A^2 = \alpha(T)(1-X_B)^2 \tag{2.19}$$

Substituting Eqs. (2.18) and (2.19) in the liquidus equation (Eq. 2.12) gives for arbitrary temperature $T = T_1$

$$RT \ln [X_B(1-X_B)] + \Delta S^o(T_F)(T_F-T) + RT_F \ln 4$$
$$+ \alpha(T)[2X_B^2-2X_B + 1] - 0.5\alpha(T_F) = 0 \tag{2.20}$$

$\Delta S^o(T_F)$ is the difference in entropy between the solid AB and its components in their pure liquid state. With further algebra and thermodynamic equalities, Eq. (2.20) reduces to

$$0.5[RT \ln 4\ X_B(1-X_B) + \Delta S^F\ (T_F-T)]$$
$$= -\ \alpha(T)(0.5-X_B)^2 \tag{2.21}$$

where $\Delta S^F$ is the entropy of fusion.

The experimental liquidus data for GaAs and InP are given in Fig. 2.33 [143]. The interaction parameter $\alpha(T)$ determined from the data points in Fig. 2.33 at each X and T using Eq. (2.21) is shown in Fig. 2.34. The values of $\Delta S^F$ used in these calculations are shown in Table 2.6 for GaAs, InP as well as other binary III-V compounds [144, 145].

Ilegems and Pearson [146] extended the treatment applied to binary compounds to several of the ternary III-V systems. In this treatment the solid solution of composition $A_{1-x}B_xC$ is treated

**Figure 2.33** Liquidus composition versus reciprocal temperature for Ga-As and In-P binary systems [143].

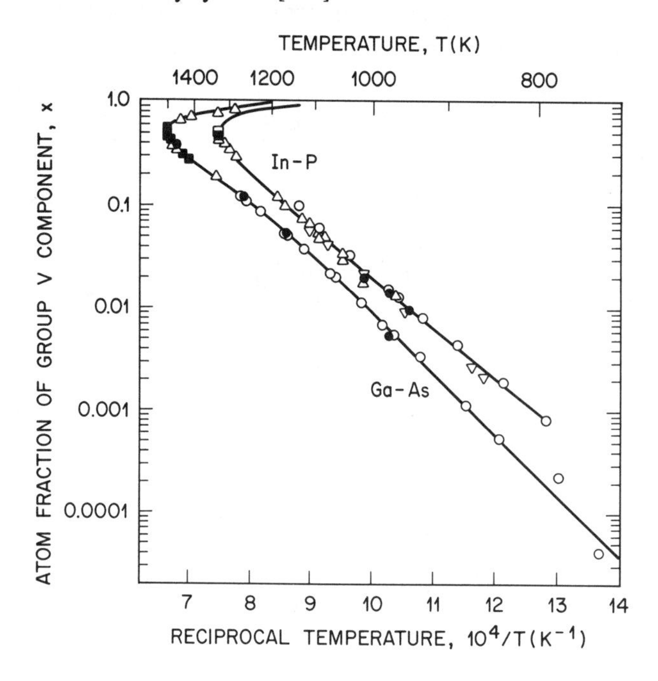

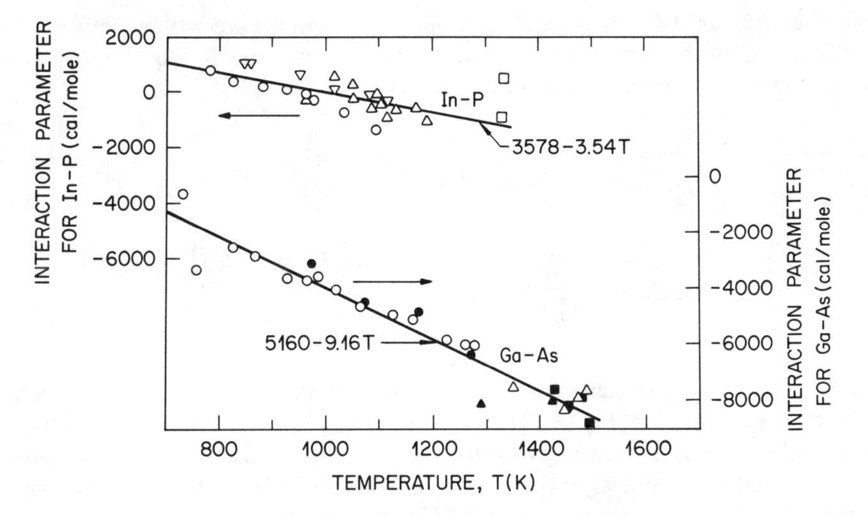


**Figure 2.34** Liquidus interaction parameter α(T) for Ga-As and In-P obtained from the liquidus shown in Fig. 2.33 [143].

**Table 2.6**

Temperature of fusion, entropy of fusion and interaction parameter for binary III-V compounds

| Compound | Temperature of fusion $T^F$ (°K) | Entropy of fusion $\Delta s^F$ (cal/mol °K) | Interaction Parameter α (T) (cal/mole) | Ref. |
|---|---|---|---|---|
| AlP | 2803 | 15.0 | 1750-2.0T | 144 |
| AlAs | 2043 | 15.6 | –6390-5.5T | 144 |
| AlSb | 1333 | 14.74 | 12300-10T | 144 |
| GaP | 1740 | 17.3 | 2120-4.45T | 144 |
| | | | 2870-4.0T* | 145 |
| GaAs | 1511 | 16.64 | 5160-9.16T | 144 |
| | | | 4530-8.93T* | 145 |
| GaSb | 983 | 15.8 | 4700-6.0T | 144 |
| InP | 1335 | 15.2 | 3578-3.54T | 144 |
| | | | 5055-5.0T* | 145 |
| InAs | 1215 | 14.52 | 3860-10.0T | 144 |
| | | | 3740-10.1T* | 145 |
| InSb | 798 | 14.32 | 3400-12.0T | 144 |

* In the temperature range 580 - 670°C

as a mixture of AC and BC and separate equilibrium conditions are written between AC in the solid and A and C in the ternary liquid, and BC in the solid and B and C in the liquid. The following expressions are then derived for the ternary system $A_{1-x}B_xC$

$$\gamma^s_{AC}(1-x) = \frac{4\gamma^\ell_A \gamma^\ell_C}{\gamma^{sl}_A \gamma^{sl}_C} X^\ell_A X^\ell_C \exp\left[\frac{\Delta S^F_{AC}}{RT}(T^F_{AC}-T)\right] \tag{2.22}$$

$$\gamma^s_{BC}x = \frac{4\gamma^\ell_B \gamma^\ell_C}{\gamma^{sl}_B \gamma^{sl}_C} X^\ell_B X^\ell_C \exp\left[\frac{\Delta S^F_{BC}}{RT}(T^F_{BC}-T)\right] \tag{2.23}$$

The superscripts s, l, and sl stand for solid, liquid, and stoichiometric liquid (liquid A–C in Eq. (2.22) and liquid B–C in Eq. (2.23)). By setting $\gamma^s_{AC}(1-x)=1$ in Eq. (2.22) it reduces to Eq. (2.12) (without the $\Delta C^\circ_P$ term), giving the liquidus curve $X^\ell_A(T)$. The pseudobinary phase diagram is obtained by setting $X_C = 0.5$ in Eqs. (2.22) and (2.23). In the pseudobinary limit, the liquid can be regarded as a mixture of AC and BC molecules or as a mixture of A and B atoms on quasicrystalline lattice sites [132].

The ternary liquidus and solidus surfaces can be calculated via Eqs. (2.22) and (2.23) provided the $\gamma_s$ in the liquid and solid solutions can be satisfactorily estimated. The simple solution model with the interaction parameter given by Eq. (2.15) is in general a good description for the liquid phase. The general relation between $\gamma_S$, X, and binary interaction parameters $\alpha_{ij}$ for multicomponent simple solutions at temperature T has been given by Jordan [147] as

$$RT \ln \gamma_i = \sum_{\substack{j=1\\ i\neq j}}^{m} \alpha_{ij} X_j^2 + \sum_{\substack{k=1\\ k<j,\, j\neq i,\, i\neq k}}^{m} \sum_{j=1}^{m} X_k X_j (\alpha_{ij}+\alpha_{ik}-\alpha_{kj}) \tag{2.24}$$

For the ternary liquid Al-Ga-As Eq. (2.24) gives

$$RT \ln \gamma_{As} = \alpha_{Al\text{-}As} X^2_{Al} + \alpha_{Ga\text{-}As} X^2_{Ga} + (\alpha_{Al\text{-}As}+\alpha_{Ga\text{-}As}-\alpha_{Al\text{-}Ga}) X_{Al} X_{Ga} \tag{2.25}$$

$$RT \ln \gamma_{Ga} = \alpha_{Ga\text{-}As} X^2_{As} + \alpha_{Ga\text{-}Al} X^2_{Al} + (\alpha_{Ga\text{-}As}+\alpha_{Ga\text{-}Al}-\alpha_{Al\text{-}As}) X_{As} X_{Al} \tag{2.26}$$

$$RT \ln \gamma_{Al} = \alpha_{Al\text{-}As} X^2_{As} + \alpha_{Al\text{-}Ga} X^2_{Ga} + (\alpha_{Al\text{-}As}+\alpha_{Ga\text{-}Al}-\alpha_{Ga\text{-}As}) X_{Ga} X_{As} \tag{2.27}$$

The interaction parameters $\alpha_{Al\text{-}As}$, $\alpha_{Ga\text{-}As}$, and $\alpha_{Al\text{-}Ga}$ are for the subscripted component binary pairs. Equations (2.25 to 2.27) are used to eliminate the liquid activity coefficients in Eqs. (2.22) and (2.23). Note that the liquidus compositions are related by mass conservation relation

$$X_{Al} + X_{Ga} + X_{AB} = 1 \tag{2.28}$$

For determining the activity coefficients in the solid ternary solution (left side of Eqs. (2.22) and (2.23)) two approaches have been used. One is to use a regular or simple solution treatment [132] which gives expressions similar to those for the liquid $\gamma_s$ and to derive interaction parameters fitting experimental data. Another is a theoretical approach to predict the interaction parameters.

In the simple solution model, Eqs. (2.8 to 2.19) give for the $Al_xGa_{1-x}As$ ternary system

$$RT \ln \gamma_{AlAs} = \alpha_{AlAs\text{-}GaAs}(1-x)^2 \tag{2.29}$$

$$RT \ln \gamma_{GaAs} = \alpha_{AlAs\text{-}GaAs}x^2 \tag{2.30}$$

Equations (2.29 and 2.30) are then used for determining the left side of Eqs. (2.22 and 2.23).

To calculate the ternary phase diagram Eqs. (2.25 to 2.30) can be solved simultaneously, the input data being two independent liquid compositions, liquid and solid interaction parameters, two entropies of fusion, two temperatures of fusion, and one solid composition. For the liquid interaction parameters the values obtained from the analysis of the binary liquidus by means of Eq. (2.21) are used when available. For the solid interaction parameter the value which gives the best fit to the experimental pseudobinary if available, is generally used. When binary liquid or pseudobinary data are not available, curve fitting approach has been used with available ternary data.

The liquidus and the solidus calculated following the above procedure for $Al_xGa_{1-x}As$ are shown in Figs. 2.35 and 2.36, respectively [148]. The parameters used in the calculation are

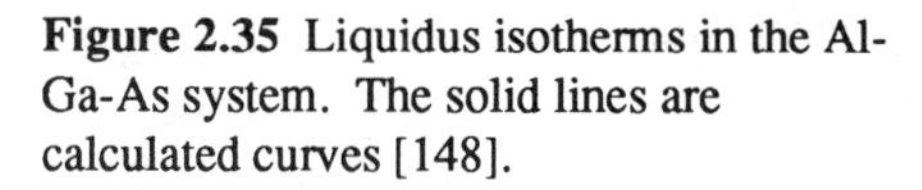

**Figure 2.35** Liquidus isotherms in the Al-Ga-As system. The solid lines are calculated curves [148].

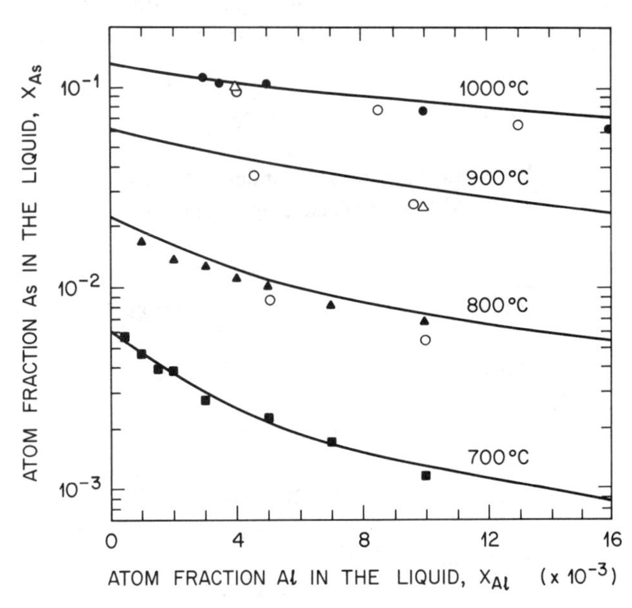

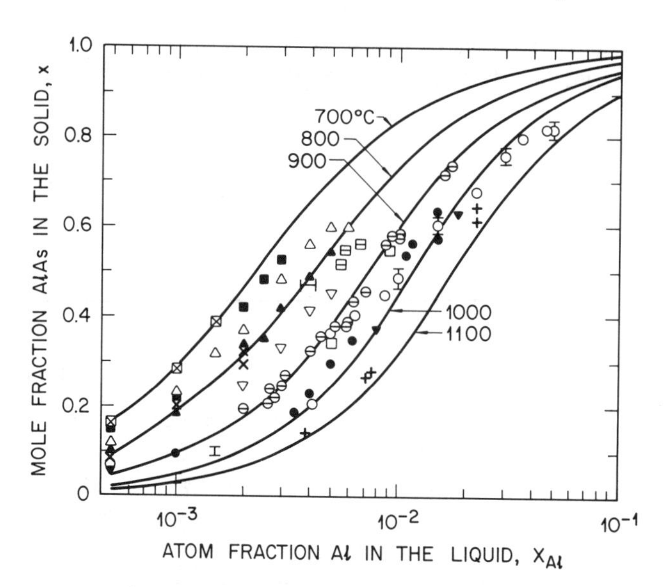


**Figure 2.36** Solidus isotherms as a function of the liquidus composition. The solid lines are calculated curves [148].

listed in Table 2.7 [149]. The liquidus compositions for lattice matched growth of $Ga_{0.47}In_{0.53}As$ on (100) InP as a function of temperature between 570 and 660°C are shown in Fig. 2.37 [150]. The curves calculated according to the simple solution model using the parameters given in Table 2.8 are also shown.

In the theoretical approach, solid solution interaction parameters are obtained from the calculated heats of mixing in a strictly regular solution approximation, (i.e., the distribution of

**Table 2.7**

Interaction parameters for calculation of the Al-Ga-As phase diagram [149]

| | | |
|---|---|---|
| $\alpha_{Ga\text{-}As}$ | = | 5160 – 9.16T cal/mole |
| $\alpha_{Al\text{-}As}$ | = | –6390 – 5.5T cal/mole |
| $\alpha_{Al\text{-}Ga}$ | = | 104 cal/mole |
| $\alpha_{GaAs\text{-}AlAs}$ | = | 400 cal/mole at 973°K |
| $\alpha_{GaAs\text{-}AlAs}$ | = | –3892 + 4T cal/mole from 1073 to 1273°K |

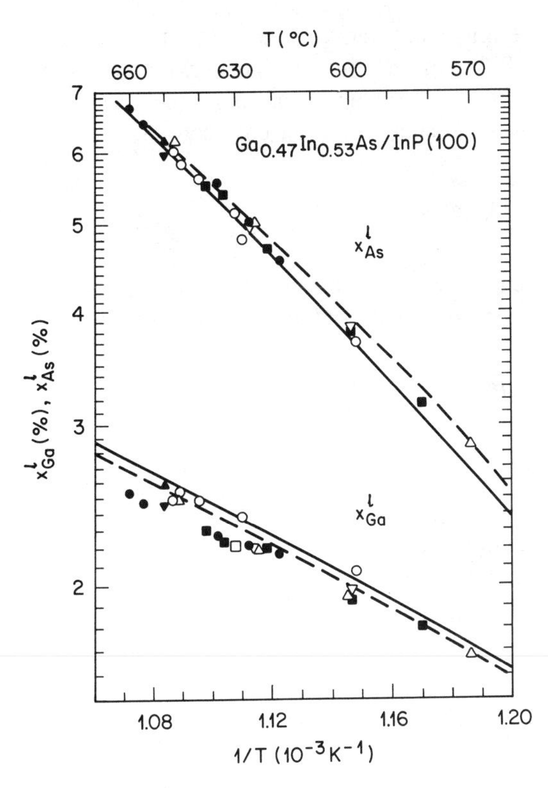


**Figure 2.37** Liquidus composition versus reciprocal temperature for lattice matched growth of $Ga_xIn_{1-x}As$ on InP. The solid and dashed lines are calculated curves using different interaction parameters (see text). The data points represent work of several authors [150].

constituents on the lattice is random). In this approximation, the enthalpy of mixing in a binary solution is given by

$$H^m = \alpha_s\, x(1-x) \tag{2.31}$$

**Table 2.8**

Interaction parameters for calculation of the $Ga_xIn_{1-x}As$ phase diagram

| | | | |
|---|---|---|---|
| $\alpha_{Ga\text{-}As}$ | = | 5160 – 9.16T cal/mole | |
| $\alpha_{In\text{-}As}$ | = | 3860 – 10.0T cal/mole | |
| $\alpha_{In\text{-}Ga}$ | = | 2050 cal/mole | (Ref. 151) |
| | = | 2120 cal/mole | (Ref. 150) |
| $\alpha_{InAs\text{-}GaAs}$ | = | 2650 cal/mole | (Ref.151) |
| | = | (7400 – 5.0T) cal/mole | (Ref. 150) |

The model of semiconductor bonding by Phillips and Van Vechten [152 - 155] is used to predict $H^m$ in solid alloys and hence $\alpha_s$ [156 - 160]. In the Phillips-Van Vechten model the average energy gap of a semiconductor is separated into homopolar and ionic components. The homopolar energy gap $E_h$ is taken to be proportional to the lattice constant

$$E_h \propto a_o^{-2.5} \tag{2.32}$$

Thus neglecting ionicity of bonding the enthalpy of atomization $H^{at}$ which is a measure of the bonding energy (which is linearly related to the band gap) and may be written as [156, 157]

$$H^{at} = K\, a_o^{-2.5} \tag{2.33}$$

$H^m$ may then calculated using the relation

$$H^m = (1-x)H_A^{at} + x\, H_B^{at} - H_{alloy}^{at} \tag{2.34}$$

Taking $\alpha_s = 4\, H^m$ at $x = 0.5$ yields

$$\alpha_s = 4K\left\{\frac{1}{2}\left[a_A^{-2.5}+a_B^{-2.5}\right] - \left[\frac{a_A + a_B}{2}\right]^{-2.5}\right\} \tag{2.35}$$

Expanding Eq. (2.35) in power series and keeping only first and second order terms in $\Delta a = a_A - a_B$ gives

$$\alpha_s \simeq 4.375\ K\, \frac{(\Delta a)^2}{(\bar{a})^{4.5}} \tag{2.36}$$

where $\bar{a} = (a_A + a_B)/2$. The description of $\alpha_s$ in terms of $\Delta a$ is known as the delta lattice parameter (DLP) model [158].

The value of K is obtained by fitting Eq. (2.32) to experimental data for III-V systems giving a value of $1.26\times10^7$ cal/mole Å$^{2.5}$. This agrees with the value of $1.15\times10^7$ cal/mole Å$^{2.5}$ obtained by least-square fit of Eq. (2.35) to available experimental values of III-V pseudo-binary system. It should be noted that in their analysis of pseudobinary data Brebrick and Panlener [159] have proposed a sixth power dependence of $\alpha_s$ with $\frac{\Delta a}{\bar{a}}$ as well as a linear temperature dependence for $\alpha_s$. Stringfellow [158] has used the single solution with $\alpha = a - bT$ for the liquid phase and the DLP model for the solid phase to calculate quaternary III-V phase diagrams.

In extending the thermodynamic treatment to quaternary solid solutions, two types of solid solutions must be considered, one in which mixing is restricted to one of the sublattices (e.g., $(A_x^{III}B_{1-x}^{III})_yC_{1-y}^{III}D^V$ and another in which mixing occurs on both sublattices (e.g.,

$A_x^{III}B_{1-x}^{III}C_y^{V}D_{1-y}^{V}$). In the former case the mixture may be regarded as a ternary mixture of AD, BD CD. In the latter case, the quaternary solid solution may be considered as regular mixtures of ternary compounds [134]. The solid liquid equilibrium in the Al-Ga-P-As system was calculated in this way. Jordan and Ilegems [135] have given a rigorous thermodynamic analysis of solid-liquid equilibria in quaternary systems by treating the quaternary solid as a regular mixture of binary compounds, subject to the restrictions imposed by the crystalline structure of the solid. According to this model the solid-liquid equilibrium for the $Ga_xIn_{1-x}As_{1-y}P_y$ system is described by the following equations:

$$\Delta H_{GaAs}^F - T\Delta S_{GaAs}^F + RT \ln 4\, X_{Ga}^{\ell} X_{As}^{\ell} = M_{GaAs}^{\ell} + RT \ln a_{GaAs}$$

$$\Delta H_{InAs}^F - T\Delta S_{InAs}^F + RT \ln 4\, X_{In}^{\ell} X_{As}^{\ell} = M_{InAs}^{\ell} + RT \ln a_{InAs}$$

$$\Delta H_{GaP}^F - T\Delta S_{GaP}^F + RT \ln 4\, X_{Ga}^{\ell} X_{P}^{\ell} = M_{GaP}^{\ell} + RT \ln a_{GaP}$$

$$\Delta H_{InP}^F - T\Delta S_{InP}^F + RT \ln 4\, X_{In}^{\ell} X_{P}^{\ell} = M_{InP}^{\ell} + RT \ln a_{InP} \qquad (2.37)$$

where

$$M_{ij}^{\ell} = \alpha_{ij}^{\ell}[0.5 - X_i^{\ell}(1-X_j^{\ell}) - X_j^{\ell}(1-X_i^{\ell})]$$
$$+ (\alpha_{ik}^{\ell} X_k^{\ell} + \alpha_{im}^{\ell} X_m^{\ell})(2X_i^{\ell} - 1)$$
$$+ (\alpha_{jk}^{\ell} X_k^{\ell} + \alpha_{im}^{\ell} X_m^{\ell})(2X_j^{\ell} - 1) + 2\alpha_{km}^{\ell} X_k^{\ell} X_m^{\ell} \qquad (2.38)$$

and

$$i, j, k, m = Ga, In, As, P; \quad i \neq j \neq k \neq m$$

If the quaternary solid solution is treated as a regular solid solution then

$$RT \ln a_{GaAs} = RT \ln x(1-y) + \alpha_{Ga\text{-}In}^{s}(1-x)^2 + \alpha_{As\text{-}P}^{s} y^2$$
$$+ \alpha_c (1-x)y$$

$$RT \ln a_{GaP} = RT \ln xy + \alpha_{Ga\text{-}In}^{s}(1-x)^2 + \alpha_{As\text{-}P}^{s}(1-y)^2$$
$$- \alpha_c (1-x)(1-y)$$

$$RT \ln a_{InP} = RT \ln (1-x)y + \alpha_{Ga\text{-}In}^{s} x^2 + \alpha_{As\text{-}P}^{s}(1-y)^2$$
$$+ \alpha_c (1-y)x$$

$$RT \ln a_{InAs} = RT \ln (1-x)(1-y) + \alpha_{Ga\text{-}In}^{s} x^2$$
$$+ \alpha_{As\text{-}P}^{s} y^2 - \alpha_c xy \qquad (2.39)$$

where

$$\alpha_c = \Delta H^F_{GaAs} - T\Delta S^F_{GaAs} + \Delta H^F_{InP} - T\Delta S^F_{InP}$$
$$- \Delta H^F_{GaP} + T\Delta S^F_{GaP} - \Delta H^F_{InAs} + T\Delta S^F_{InAs}$$
$$+ \frac{1}{2}(\alpha^{\ell}_{GaP} + \alpha^{\ell}_{InAs} - \alpha^{\ell}_{InP} - \alpha^{\ell}_{GaAs}) \tag{2.40}$$

and

$$\alpha^s_{Ga\text{-}In} = \frac{1}{2}(\alpha^s_{GaAs\text{-}InAs} + \alpha^s_{GaP\text{-}InP})$$

$$\alpha^s_{As\text{-}P} = \frac{1}{2}(\alpha^s_{GaAs\text{-}GaP} + \alpha^s_{InAs\text{-}InP}) \tag{2.41}$$

The set of Eqs. (2.37 to 2.41) gives complete expressions for calculating the Ga-In-As-P phase diagram.

Figures 2.38 and 2.39 show the liquidus isotherms at, respectively, 650 and 600°C for the Ga-In-As-P system at various $X^{\ell}_{As}$ [160]. These isotherms were determined by the seed dissolution technique [161]. The corresponding solidus curves at these two temperatures are shown in Figs. 2.40 to 2.43 [160]. The epitaxial layers were grown on (111)B InP substrates

**Figure 2.38** Liquidus isotherm at 650°C in the Ga-In-As-P system. The data points are taken from several sources. $X^{\ell}_{As} = 0.0$ (∇, □); 0.0145(■); 0.030 (Δ); 0.40 (Δ); 0.050(●), 0.055(O) [160].

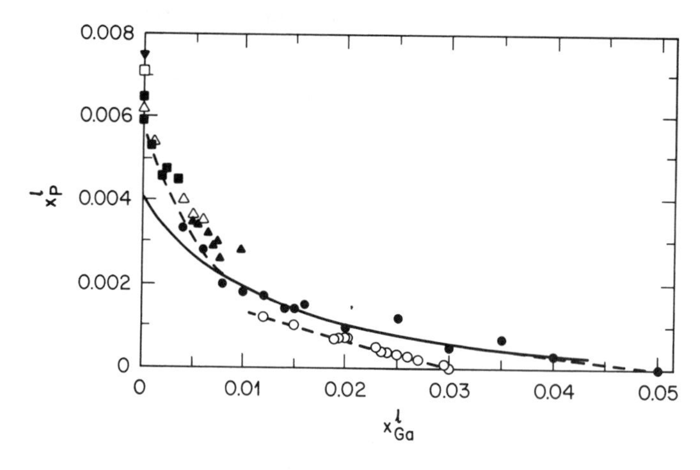

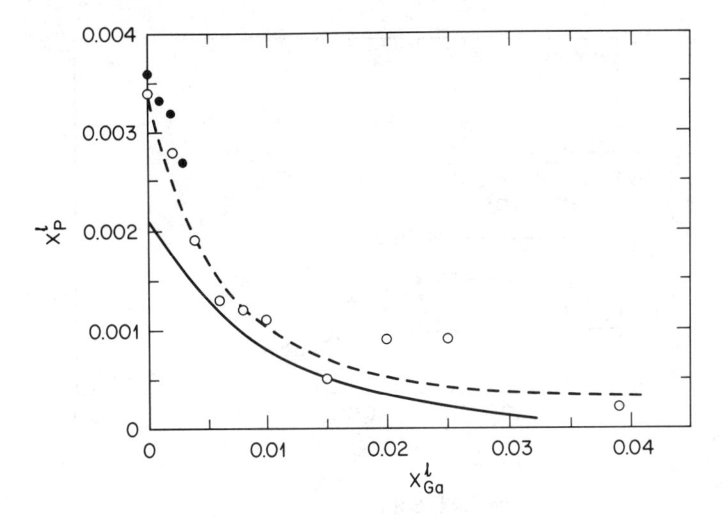


**Figure 2.39** Liquidus isotherm at 600°C in the Ga-In-As-P system. $X_{As}^{\ell} = 0.0$(●); 0.03(O) [160].

**Figure 2.40** Solid composition (x) in $Ga_xIn_{1-x}As_{1-y}P_y$ alloys versus liquidus composition for growth at 650°C. The data points are taken from several sources. $X_{As}^{\ell} = 0.030$(Δ); 0.040(□), 0.050(●); 0.055(O) The solid lines are calculated curves for $X_{As}^{\ell} = 0.03$ and 0.05 [160].

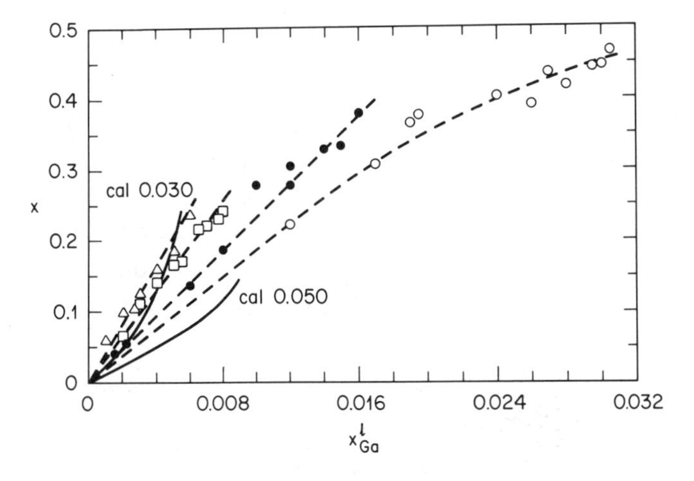

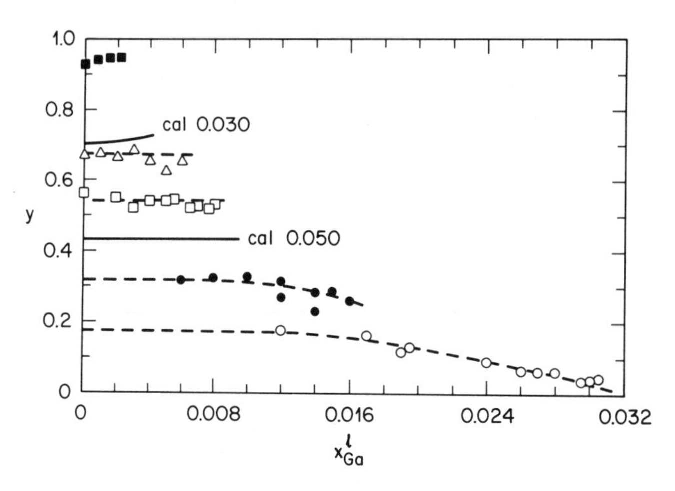


**Figure 2.41** Solid composition (y) in $Ga_xIn_{1-x}As_{1-y}P_y$ alloys versus liquidus composition for growth at 650°C. The data points are taken from several sources. $X^\ell_{As}$ = 0.030(Δ); 0.040(□); 0.050(●); 0.055(O). The solid lines are calculated curves for $X^\ell_{As}$ = 0.03 and 0.05 [160].

**Figure 2.42** Solid composition (x) in $Ga_xIn_{1-x}As_{1-y}P_y$ alloys versus liquidus composition for growth at 600°C at several $X^\ell_{As}$ [160].

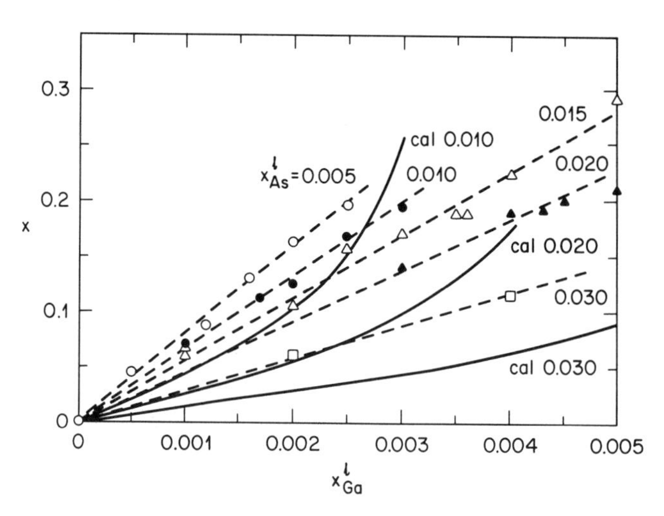

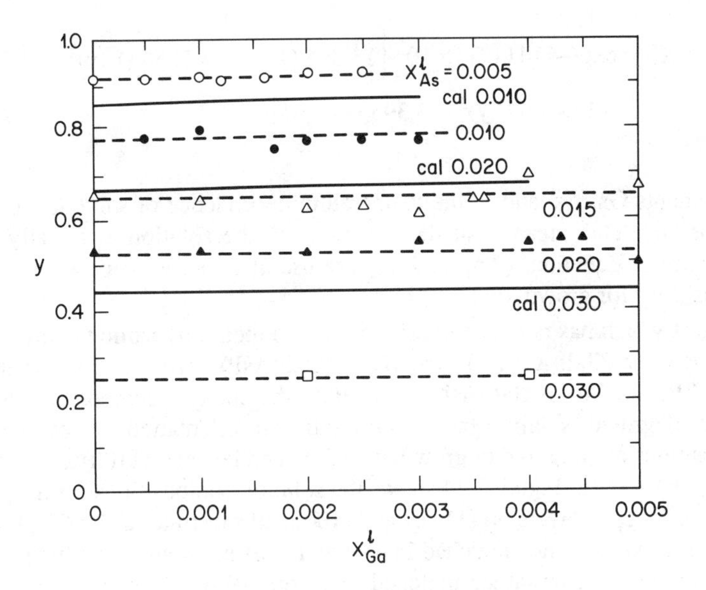


**Figure 2.43** Solid composition (y) in $Ga_xIn_{1-x}As_{1-y}P_y$ alloys versus liquidus composition for growth at 600°C at several $X^{\ell}_{As}$ [160].

under equilibrium conditions. The solid curves in Figs. 2.38 to 2.43 represent the calculated curves using the interaction parameters given in Ref. [132] and [144]. It can be seen that the agreement between the calculations and experiments is very poor owing mainly to the uncertainties in the various interactions parameters. Perea and Fonstad [162] made an empirical correction to the expressions in Eq. (2.39) and obtained better agreement between calculations and the experimental data. Including the elastic strain energy contribution to the regular solution model has also been shown to improve the agreement with experiments [163].

Kuphal [150] has derived empirical expressions relating the solution and solid compositions in the temperature range 570 to 660°C for growth of $Ga_xIn_{1-x}As_yP_{1-y}$ on (100) InP substrates from an analysis of several hundred published data. The following expressions are given:

$$X^{\ell}_{As} = \exp(-7181/T)\Big[3.8451\times10^{4}\, X^{\ell}_{Ga} - 5.6805\times10^{6}\left(X^{\ell}_{Ga}\right)^{2} + 5.0985\times10^{8}\left(X^{\ell}_{Ga}\right)^{3} - 2.6191\times10^{10}\left(X^{\ell}_{Ga}\right)^{4} + 7.0231\times10^{11}\left(X^{\ell}_{Ga}\right)^{5} - 7.6075\times10^{12}\left(X^{\ell}_{Ga}\right)^{6}\Big] \quad (2.42)$$

$$X^{\ell}_{Ga} = \exp(-3584/T)\Big[0.70694\,x + 3.4624\,x^{2} - 8.7492\,x^{3} + 36.554\,x^{4} - 32.878\,x^{5}\Big] \quad (2.43)$$

$$X_P^\ell = \exp(-11411/T) \times 10^2 \Big[13.305\,(1-y) - 4.7256\,(1-y)^2 + 12.417(1-y)^3 - 3.3953\,(1-y)^4\Big] \tag{2.44}$$

Of the three elements Ga, As, and P, the temperature dependence of solubility is strongest for P and weakest for Ga. This means that the P content in the solution essentially determines the liquidus temperature. Equations (2.42 to 2.44) are useful to derive good starting values for the solution compositions for growth on (100) InP.

Another effect which has not been considered in the thermodynamic treatment of solid-liquid equilibrium in ternary (GaInAs) and quaternary (GaInAsP) solid solutions is that of substrate orientation. In LPE growth of GaInAsP and GaInAs the epilayer composition depends not only on composition of growth solution but also on substrate orientation. In other words, different solution compositions are required to grow lattice matched layers on (100) and (111) oriented InP substrates [151, 164 - 167]. Figure 2.44 shows the solution compositions for the growth of lattice matched $Ga_xIn_{1-x}As_{1-y}P_y$ layers on (111)B and (100) InP substrates at 650°C [160]. For a given $X_{As}^\ell$, the solution to grow lattice matched layer on (111)B substrate has a higher $X_{Ga}^\ell$ and lower $X_P^\ell$ than the solution to grow lattice matched layer on (100) substrate. Figure 2.45 shows the composition of the solution for growing lattice matched $In_xGa_{1-x}As$ (x=0.53) on (100), (111)A, and (111)B InP substrates as a function of growth temperature [167]. The solution contains decreasing (increasing) amount of Ga(As) in going from (111)B to (111)A to (100) substrates. It is generally believed that interface kinetics or surface free energy considerations may be important in determining the orientation dependence.

**Figure 2.44** Liquidus compositions for growth of lattice matched $Ga_xIn_{1-x}As_{1-y}P_y$ on (100) and (111)B InP substrates at 650°C. The data points represent experiments of several authors [160].

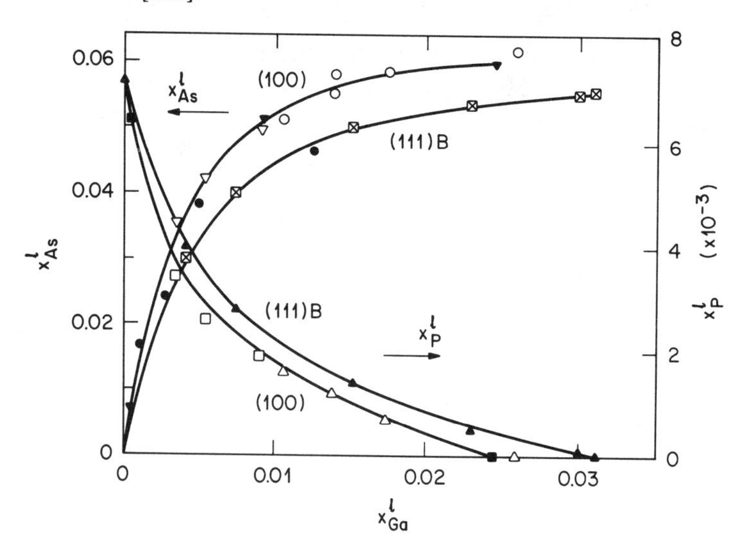

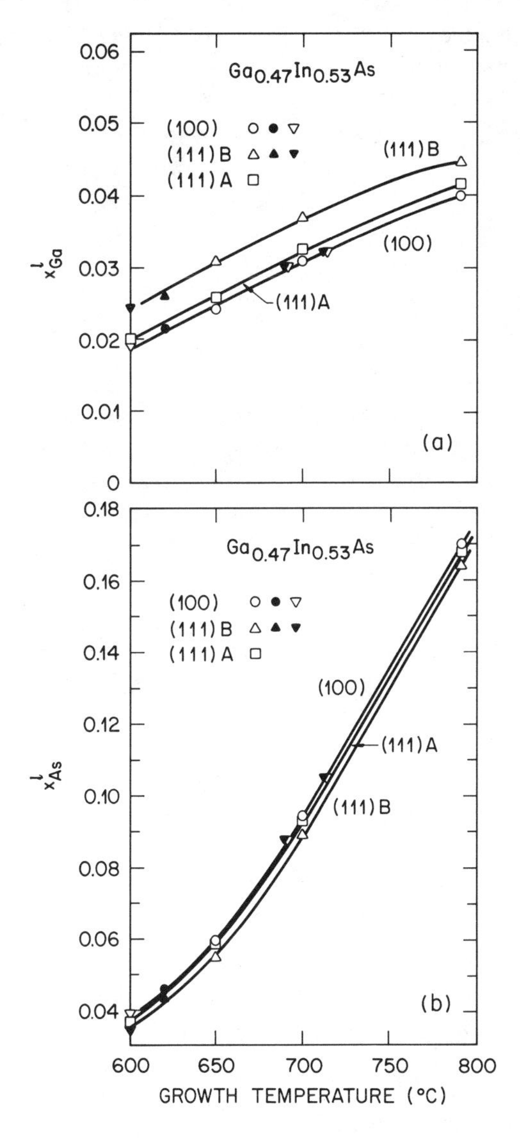


**Figure 2.45** Compositions of Ga, $X^{\ell}_{Ga}$ (top) and As, $X^{\ell}_{As}$ (bottom) for growth of lattice matched $Ga_xIn_{1-x}As(x = 0.47)$ on (100), (111)A and (111)B InP substrates as a function of the growth temperature. The data points represent work of several authors [167]. Reprinted with permission by The Electrochemical Society, Inc.

**Solution compositions**

The phase diagram calculated using the thermodynamic guidelines outlined serves as a first approximation for determining the solution compositions. For example, if a $Al_{0.3}Ga_{0.7}As$ layer has to be grown on a GaAs substrate at 800°C, then the solution compositions can be determined from Figs. 2.35 and 2.36. For a 30 percent Al composition in the solid, the solution should contain $X_{Al} = 2\times10^{-3}$, $X_{As} = 1.7\times10^{-2}$, and $X_{Ga} = 1 - X_{Al} - X_{As}$. If dopants are to be added to the solution, depending on the distribution coefficient of the dopant, the Al composition in the solution needs to be modified. For an impurity with high distribution coefficient (e.g., Te) only small amounts of the impurity are added to the liquid and no change in solution composition is required. On the other hand, for impurities with low distribution coefficient (e.g., Sn) the liquid composition must be modified considering the quaternary solution containing Al, Ga, As, and the impurity [168]. Once the solution composition is determined, the weight of each element i needed in the solution is obtained from the definition

$$X_i = \frac{N_i}{\sum_{i=1}^{n} N_i} \tag{2.45}$$

where $N_i$ is the number of moles (weight of the element/molecular weight) of each element.

### 2.2.3 Crystal Growth Kinetics

Besides equilibrium considerations LPE growth is also greatly influenced by the kinetic effects. Effects such as arrival of solute atoms at the growing interface, convection as the result of compositional and temperature gradients, thermodynamic instability between the solution and the substrate when a new layer is grown as in the case of heterojunction growth, supersaturation of solution during growth, and nucleation and the mechanism of growth at the surface all play a role in determining LPE growth rates.

The LPE growth rate has been analyzed theoretically by several authors using a diffusion controlled model. In this model, growth is assumed to be limited by the diffusion of the solute to the growth interface. Several other assumptions are also made [169]. These assumptions are: (a) the absence of convection transport is assumed, (b) the growth solution is assumed to be isothermal, (c) it is assumed that thermodynamic equilibrium prevails at the growth interface so that the solute concentration in the solution at the interface is given by the liquidus of the phase diagram, and (d) the removal of solute from the solution is assumed to occur only by deposition on the substrate and not by precipitation within the solution or at the boundaries of the solution. With these assumptions, the growth rate and layer thickness can be calculated by solving a one-dimensional diffusion equation with appropriate boundary conditions.

#### Steady-state growth

This method has been used to grow thick layers of uniform composition [170, 171]. A source crystal is immersed in the solution which is separated from the substrate by a distance w. A fixed temperature difference is established between the source and the substrate with the latter being at a lower temperature. The furnace temperature is not changed during growth. Since solubility decreases with temperature, a solute concentration gradient exists from the source to the substrate resulting in the transport of the solute from the source to the substrate. At steady-state the rate of dissolution of the source equals the rate of growth on the substrate. The growth rate is obtained by solving the diffusion equation to be

$$r = \frac{D}{w} \ln \left[ \frac{C_o - C_s}{C_w - C_s} \right] \tag{2.46}$$

where $C_s$ is the solute concentration in the solid, $C_w$ and $C_o$ are the solute concentrations that are in equilibrium with the solution respectively, at the substrate and source temperatures.

**Transient growth**

Epilayers grown by the steady-state growth technique suffer from thickness variations owing to convection in the solution and is not suitable for growing thin (~ 0.1 μm to a few microns) epilayers used in most devices. For growing thin layers, the transient LPE growth technique is adopted. There are four different transient techniques designated as equilibrium or uniform-cooling, step cooling, supercooling, and two-phase-solution cooling techniques.

In the uniform-cooling technique, the substrate is brought into contact with the equilibrated solution at temperature $T_2$ and both the substrate and the solution are then slowly cooled at a constant cooling rate to $T_1$ (as illustrated for the binary phase diagram shown in Fig. 2.1). When source and seed substrates are used, the solution is kept in contact with the source wafer and after the solution is equilibrated, the source is replaced by the substrate and cooling begins. The thickness of the epilayer under this growth condition and for semi-infinite solutions is given by [172]

$$d = \frac{4}{3}\left[\frac{R}{C_s m}\right]\left[\frac{D}{\pi}\right]^{1/2} t^{3/2} \tag{2.47}$$

where d is the thickness of the crystal, R is the cooling rate, $C_s$ is the solute concentration in the solid, m is the slope of the liquidus curve, t is the growth time and D is the diffusivity.

In the step cooling technique, the solution is saturated at some temperature $T_2$, separated from the saturating source, cooled to $T_1$, brought into contact with the substrate at $T_1$ and kept at this temperature until growth is terminated. The temperature difference $T_2 - T_1 = \Delta T$ is kept small such that spontaneous precipitation in the solution does not occur. For this growth the layer thickness is given by [173].

$$d = 2\,\frac{\Delta T}{C_s m}\left[\frac{D}{\pi}\right]^{1/2} t^{1/2} \tag{2.48}$$

The supercooling technique [173] is a combination of the step-cooling technique and the uniform-cooling technique. The solution and the substrate are cooled at a constant rate to temperature $T_1$ and brought into contact. After this, step-cooling is continued at the same rate until growth is terminated. The layer thickness for the super-cooling technique is given by the sum of Eqs. (2.47) and (2.48) as

$$d = (2/C_s m)\,(D/\pi)^{1/2}\left[\Delta T\, t^{1/2} + \frac{2}{3} R t^{3/2}\right] \tag{2.49}$$

The thickness of GaAs layers grown on GaAs substrates from GaAs solutions at 800°C by the three growth techniques is compared in Fig. 2.46 with the predicted values as per Eqs. (2.47 to 2.49) [172]. In all cases, the diffusivity As in the liquid $D_{As}$ was found to be in the range $4 - 5.2 \times 10^{-5}$ $cm^2/sec$. The agreement between experiment and theory is poor only for long growth times (greater than 20 min) because of precipitation of GaAs in the solution and the

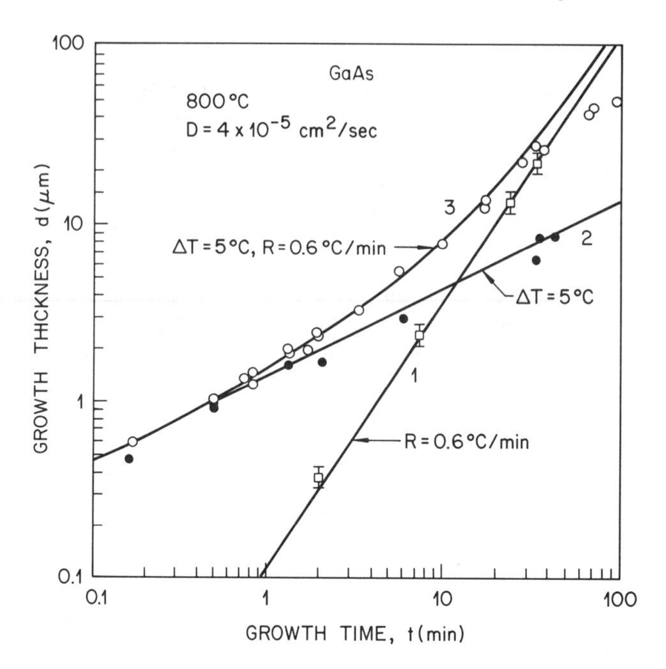


**Figure 2.46** Epitaxial layer thickness d as a function of growth time t for GaAs layers grown at 800°C by uniform cooling (curve 1), step cooling (curve 2) and supercooling (curve 3) techniques. The solid lines are calculated curves using Eqs. (2.47)-(2.49) [172].

invalidity of the semi-infinite solution assumption for long times. Similar considerations apply for the growth of InP from InP solutions [174]. The diffusivity of P that gives best fit to the thickness data at 630°C is obtained to be $4.8 \times 10^{-5}$ cm$^2$/sec.

It can be seen from Fig. 2.46 that with the step-cooling technique it may be difficult to grow very thin layers (~0.1 μm) such as the active layers in heterostructure lasers. The uniform-cooling technique is preferred for the thin layer growth. On the other hand, the step-cooling technique provides layers of uniform thickness and also of improved surface morphology [172, 175].

Figure 2.47 shows the layer thickness of $Ga_xIn_{1-x}As_{1-y}P_y$ grown at 650°C from a finite solution by the uniform-cooling method [176]. Better agreement with theory is obtained when variable m and $C_s$ are used instead of constant m and $C_s$.

The two-phase-solution cooling technique consists of lowering the solution temperature far below $T_2$ for spontaneous precipitation to occur in the solution [172, 177]. The solution and substrate are brought into contact and cooling continues at the same rate till growth is terminated. The technique can be regarded as a variation of the uniform-cooling technique except that the layer thickness is reduced by simultaneous deposition on the precipitates. In a modification of the

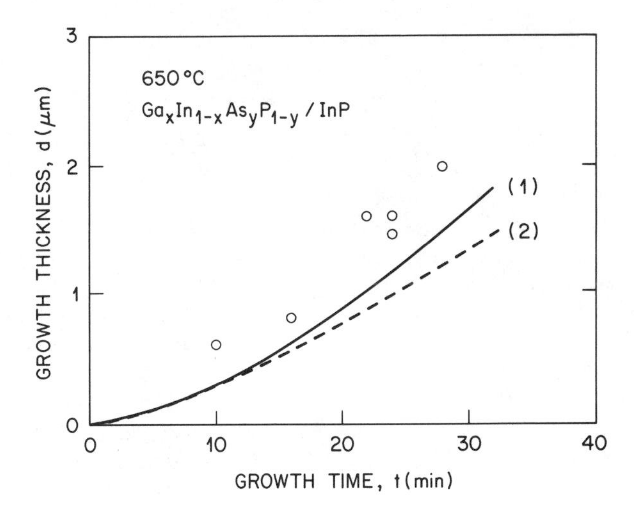


**Figure 2.47** Epitaxial layer thickness d as a function of growth time t for $Ga_xIn_{1-x}As_{1-y}P_y$ layers grown at 650°C from a finite solution of 3mm thickness by uniform cooling. The solid and dashed lines are calculated using variable m and $C_s$ and constant m and $C_s$ in Eq. (2.47) [160].

two-phase-solution technique an excess of source crystal more than that can be dissolved is added to the growth solution. This is accomplished by placing a piece of substrate on top of the solution. The growth of thin layers can be achieved by this technique. However, there can be some supersaturation in the two-phase solution at the time growth begins because of diffusion limited deposition on the individual precipitates and this can give rise to high growth rates [172].

### 2.2.4 Crystal Growth of Ternary and Quaternary Alloys

#### Lattice mismatch

Unlike in homoepitaxy (e.g., GaAs on GaAs), in heteroepitaxy (e.g., GaAs on Ge), and in LPE growth from solutions with more than two components besides dopants (e.g., growth of ternary and quaternary alloys) there are two requirements for successful growth. These are lattice constant matching between the substrate and the epilayer and chemical compatibility between growth solution and substrate.

Successful LPE growth can be achieved if the lattice mismatch is less than 1 percent which means that there is only a limited number of combinations of substrates and epilayers. A comparison of the lattice constants of binary III-V semiconductors shows that the lattice matching condition is readily satisfied only for GaP-AlP, GaAs-AlAs, and GaSb-AlSb pairs. In the case of their ternary alloys (e.g., $Al_xGa_{1-x}As$) the mismatch is further reduced between the alloy and either binary compound since the lattice constant of the alloy is intermediate between the two compounds.

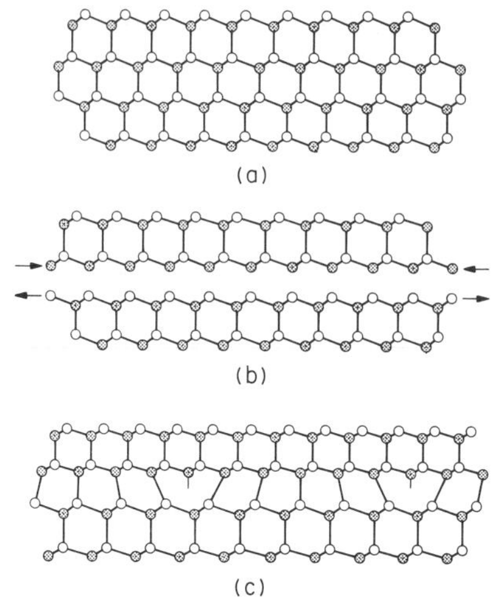


**Figure 2.48** Misfit dislocations in sphalerite lattice seen in the (011) plane: a) perfect crystal; b) crystal cut along the (111) plane and expanded on one side and contracted on the other; c) the two materials bonded together showing edge dislocation with dangling bond [180].

In the case of LPE growth of ternary alloy in which the compound and the alloy have only one common component (e.g., GaInAs on InP, GaInP on GaAs), the alloy has the same lattice constant as the substrate only at a particular composition. For example, $Ga_xIn_{1-x}As$ is lattice matched to InP for $x = 0.47$. For growing ternary alloys when lattice mismatch exceeds 1 percent, several layers of intermediate comparison are grown such that the mismatch between successive layers is kept low [178]. Alternatively, a single layer with continuous grading in comparison can be grown to accommodate the mismatch [179].

When there is a mismatch in the lattice constant between the substrate and the epilayer, it can be accommodated by elastic strain. However, if the strain is relatively large, the strain energy stored in the crystal is reduced by the formation of dislocations at the interface. These dislocations are referred to as misfit dislocations. Only edge dislocations or edge components of mixed dislocations accommodate the strain. Figure 2.48 shows misfit dislocations in a (111) heterojunction in a cubic zinc blende structure seen in the $(01\bar{1})$ plane [180]. The misfit dislocations are usually of the 60 degree type with {111} slide planes.

For a given strain $\Delta a/a$, where $\Delta a$ is the difference in the lattice parameters of the epilayer and the substrate, the generation of misfit dislocations is dependent on the epitaxial layer thickness. The critical layer thickness can be determined from the simple analysis of Matthews [181]. In this analysis, the lattice strain $\Delta a/a_o$ is assumed to be accommodated by an elastic strain in the film $\varepsilon_s$ and by formation of misfit dislocations. Thus

$$\frac{\Delta a}{a_o} = \varepsilon_s + \varepsilon_d \tag{2.50}$$

where $\varepsilon_d$ is the strain relieved by the formation of misfit dislocations. Since two perpendicular sets of parallel edge dislocations are formed to accommodate the two dimensional misfit, the number of dislocations per unit length is $2\varepsilon_d/b$, where b is the Burgers vector of the dislocations. The total strain energy caused by these misfit dislocations is

$$E_d = \frac{1}{2}\left\{\frac{Gb^2}{2\pi(1-\nu)}\ln\frac{R}{b}\right\}\frac{2\varepsilon_d}{b} \quad (2.51)$$

where G is the shear modulus assuming that the film and the substrate have nearly equal values of shear modulus, $\nu$ is the Poisson ratio, and R is the extent of the dislocation strain field. For a plane strain condition, the elastic energy in an elastically isotropic film of thickness d is

$$E_s = \frac{2G(1+\nu)}{(1-\nu)}\, d\, \varepsilon_s^2 \quad (2.52)$$

Substituting for $\varepsilon_d$ from Eq. (2.50) and minimizing the total energy $E_d + E_s$ with respect to $\varepsilon_s$ the following is obtained:

$$\varepsilon_s^* = \frac{b}{8\pi d(1+\nu)}\ln\left[\frac{R}{b}\right] \quad (2.53)$$

The value of R can be taken to be equal to the film thickness if the spacing between the parallel misfit dislocations is greater than half the film thicknesses. If $\varepsilon_s^* > \Delta a/a_o$, the layer remains elastically strained and no misfit dislocations are formed. If $\varepsilon_s^* < \Delta a/a_o$, misfit dislocations will form to relieve the strain $(\Delta a/a_o - \varepsilon_s)$. By setting $\varepsilon_s^* = \Delta a/a_o$ in Eq. (2.53), Matthews [181] obtained the critical layer thickness for dislocation formation

$$d_c = \frac{b}{8\pi\,(1+\nu)\,(\Delta a/a_o)}\ln\left[\frac{d_c}{b}\right] \quad (2.54)$$

In determining the maximum film thickness before misfit dislocations can be formed, it should be recognized that the lattice mismatch has two contributions. The first is the mismatch at the growth temperature and the second is the change of mismatch with temperature as the crystal is cooled to room temperature. A good match in the thermal expansion coefficients will reduce the stress during the cool down period.

Figure 2.49 shows the relationship between epitaxial layer thickness and lattice mismatch for the growth of dislocation free $Ga_xIn_{1-x}As$ layers on InP substrates [182]. It can be seen that misfit dislocation free thick (>4μm) ternary layers cannot be obtained by exact lattice matching at room temperature and that thick layers without misfit dislocations can be grown from 650°C only when $\Delta a/a_o$ is between $-6.5\times10^{-4}$ and $-9\times10^{-4}$. Since the thermal expansion coefficients of InP and $Ga_{0.47}In_{0.53}As$ are $4.56\times10^{-6}$ $C^{-1}$ and $5.66\times10^{-6}$ $C^{-1}$, perfect lattice matching at room temperature would imply a positive mismatch of $\sim6\times10^{-4}$. Conversely, perfect lattice match at a growth temperature would give a negative mismatch of $\sim6\times10^{-4}$ at room temperature.

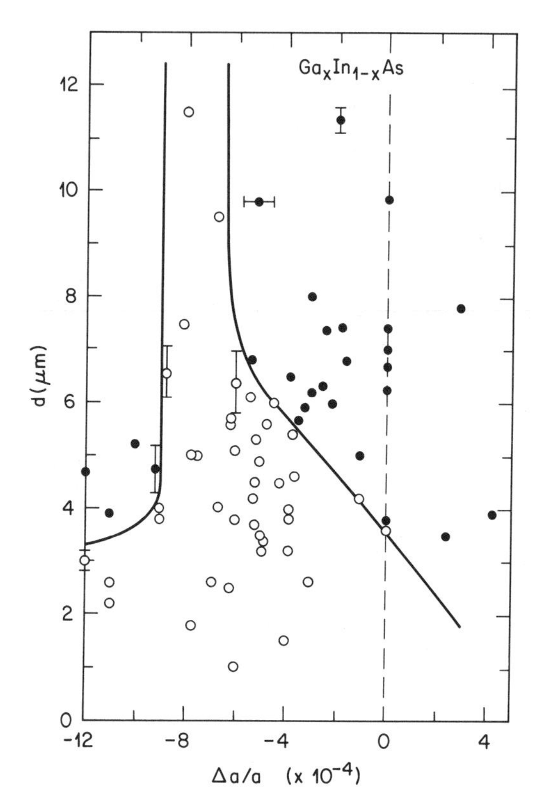


**Figure 2.49** Layer thickness d versus lattice mismatch Δa / a for $Ga_xIn_{1-x}As$ layers grown on (100) InP at 650°C. (O) misfit dislocation free (●) misfit dislocations exist [182].

Therefore, misfit dislocations are easily generated in a layer which is lattice matched at room temperature, but not in a layer which is lattice matched at the growth temperature. A similar observation has been made with respect to the growth of GaInAsP on InP also [183].

Besides misfit dislocations, there are also the so called inclined dislocations which penetrate through the grown structure. These inclined dislocations may have both screw and edge components [184]. Inclined dislocations may propagate into the active region of the device and be detrimental to device operating life. By a judicious choice of one of several epitaxial growth conditions the deleterious effect of dislocations can be reduced: a) the growth of thin layers with thickness smaller than the critical thickness for formation of misfit dislocations. b) selecting closely lattice matched system such as the $Al_xGa_{1-x}As$/GaAs heterostructure. Even in this system, improvements in lattice matching at room temperature can be made by the addition of phosphorus [185]. The use of quaternary system may allow greater flexibility in lattice matching. An example would be the growth of $Ga_xIn_{1-x}As_yP_{1-y}$ on InP, where lattice matching can be achieved over a range of compositions between InP and $Ga_{0.43}In_{0.57}As$. c) by compositional grading in order to change the substrate lattice constant to that of the desired solid solution composition. This can be in a continuous manner (gradual grading) [186] or in small abrupt steps (step grading) [187]. In this manner dislocations are distributed over a large volume so that the average density in the final layer is reduced.

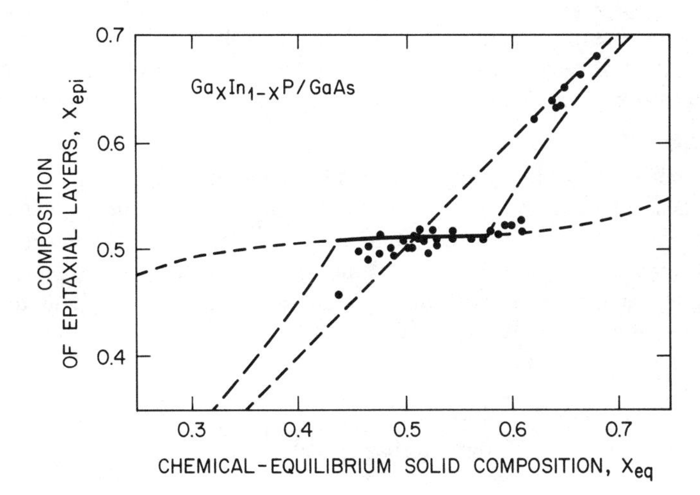


**Figure 2.50** Experimental solid composition versus equilibrium solid composition given by the phase diagram of $Ga_xIn_{1-x}P$ layers grown on GaAs substrates. The solid line is calculated taking into account the contribution of the lattice mismatch strain energy to the total free energy of the system [188].

### Lattice pulling

In the growth of ternary and quaternary solid solutions the tendency of the lattice matched composition to deposit in preference to the nonlattice matched composition predicted by the equilibrium phase diagram is known as "lattice pulling." This is illustrated in Fig. 2.50 which shows the measured solid composition versus that predicted by the phase diagram for $Ga_xIn_{1-x}P$ epilayers grown on GaAs [188]. For a wide range of predicted composition values the measured epilayer composition shows no variation and is close to the value of lattice matched composition. Because of the increase in the total free energy caused by the strain energy associated with the mismatch, the solid composition is stabilized near the lattice matched composition [188]. This means that good quality lattice matched GaInP layers can be grown on GaAs even if the solution composition is not accurately controlled. It is not clear whether the lattice pulling effect occurs in the growth of GaInAs and GaInAsP layers on InP. Results have been reported both for [150, 189 - 191] and against [166, 192, 193] lattice pulling effect.

### Miscibility gap

Miscibility gaps have been reported in several III-V pseudobinary alloys [194] and in quaternary alloys such as AlGaAsSb [195], GaInAsSb [196], and GaInAsP [197, 198]. In fact, thermodynamic calculations predict the existence of miscibility gaps in nearly all quaternary alloys [199 - 203]. The existence of miscibility gaps prevents equilibrium growth of a range of solid compositions when the critical temperature above which no immiscibility exists, is higher than the growth temperature. Within the miscibility gap the alloys are unstable and undergo phase separation. Even if the critical temperature is below growth temperature but above room temperature, processing at intermediate temperatures would cause spinodal decomposition and clustering.

Instability criterion in a binary solution is related to the second derivative of Gibbs free energy with respect to composition x, that is the curvature of the G versus x curve. If $\partial^2G/dx^2 > 0$ (concave curvature) then the solution is stable or metastable as any small change in composition will tend to increase G. On the other hand if $\partial^2G/\partial x^2 < 0$ (convex curvature) the solution is unstable and any composition fluctuation will decrease G and the solution will undergo phase separation. For a quaternary alloy $A_{1-x}B_xC_{1-y}D_y$ the unstable region is defined by the condition

$$\frac{d^2G}{dx^2}\frac{d^2G}{dy^2} - \left[\frac{d^2G}{dxdy}\right]^2 < 0 \tag{2.55}$$

with $d^2G/dx^2 > 0$. A schematic view of the free energy surface when immiscibility occurs is shown in Fig. 2.51 [203]. The spinodal curve represents the boundary between the metastable and unstable regions. From the spinodal curve, which is calculated by equating Eq. (2.55) to zero, the extent of the miscibility gap is predicted. Figure 2.52 shows the spinodal curves for the GaInAsP system calculated by a strictly regular solution model for the solid solution [203]. It can be seen that the critical temperature for the growth of lattice matched GaInAsP on InP without phase separation is ~ 800°C.

Although the thermodynamic calculations predict the existence of miscibility gap, lattice matched GaInAsP layers on InP covering the full wavelength range can be grown by LPE at

**Figure 2.51** Free energy G(x,y) as a function of the solid solution composition x and y for a quaternary alloy $A_xB_{1-x}C_{1-y}D_y$. Regions I, II and III are stable, metastable and unstable regions, respectively. The binodal curve is the boundary between regions I and II and the spinodal curve is the boundary between regions II and III [203].

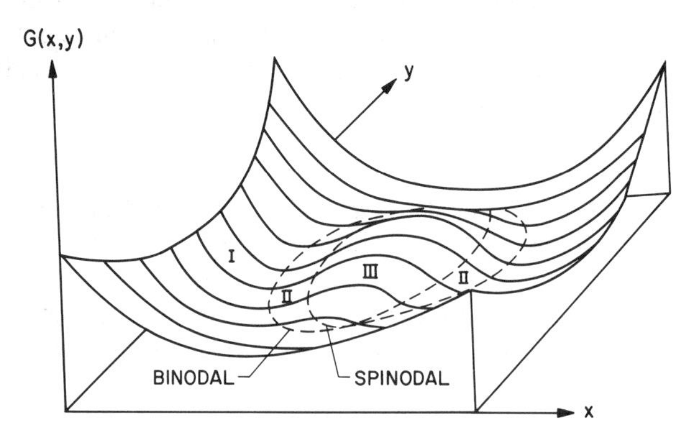

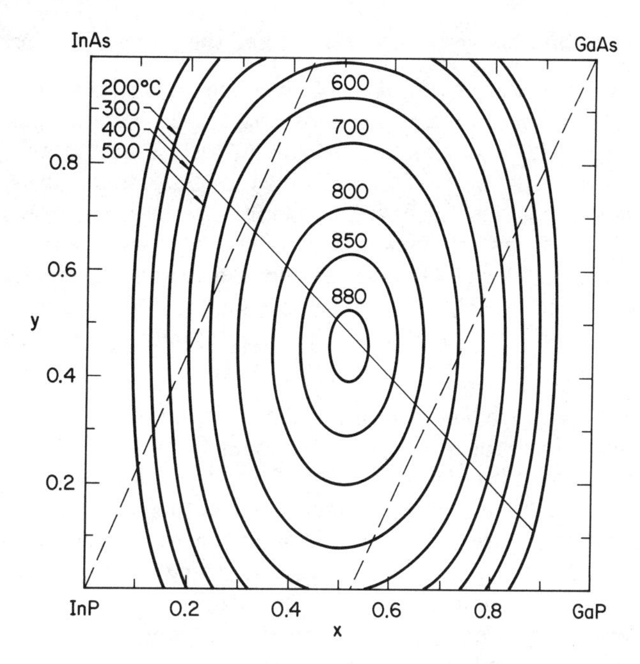


**Figure 2.52** Spinodal curve for $Ga_xIn_{1-x}As_yP_{1-y}$ (solid lines). The phase separation direction is denoted by the straight line. The dashed lines represent compositions for lattice matched growth on InP and GaAs [203].

~600°C without phase separation. The absence of phase separation is attributed to the lattice mismatch strain that stabilizes the one phase solid [201, 202].

**Composition of alloy LPE layers and layer thickness**

In growing binary compounds, the composition of the growth solution does not cause a significant change in the solid composition since the binary solid is nearly stoichiometric. For a ternary or quaternary layer, however, the composition depends on the solution composition. The distribution coefficients of the elements which relate their concentrations in the solid to that in the solution differ from each other and often differ from unity. Consequently, the alloy composition can vary significantly with layer thickness. For example, in the case of LPE growth of $Al_xCa_{1-x}As$ the distribution coefficient of Al is of the order of 100 while that of Ga is less than 1 (see Fig. 2.36). This means that Al is depleted from the solution causing x to decrease with increasing layer thickness. The change of composition with thickness depends on the growth rate, the initial Al concentration in the solution and the growth temperature [111]. For an initial solid composition of x = 0.40 and growth at 850°C with a cooling rate of 0.5°C per minute, the decrease in x was found to be about 0.5 at percent per micrometer of growth [204]. Numerical methods have been used to calculate the layer composition as a function of thickness for diffusion controlled growth [204, 205].

**Thermodynamic instability between the substrate and the growth solution**

Another feature of the LPE growth of alloy layers on binary substrates is that equilibrium cannot exist between the substrate and the growth solution. For example, a GaAs substrate cannot be in equilibrium with a solution that contains Al besides Ga and As. What this means is that the solution and the substrate will react to dissolve the GaAs so that the solution will contain less Al than the starting solution. Substrate instability has been noted in several other III-V systems: Ga-In-P and Ga-In-As solutions in contact with GaAs, [206] In-Ga-As-P solutions in contact with InP [207 - 209] and Ga-As-P solutions in contact with GaAs and GaP. The dissolution of the substrate, ''melt back,'' can be quite limited if it produces local supersaturation of the solution that results in the rapid deposition of a thin alloy layer which prevents further dissolution [210].

The amount of substrate melt back depends on the solubility of the substrate component in the solution so as to achieve local supersaturation. For example, because of the low solubility of Al in Ga-As solutions, less melt back occurs when a GaAs layer is grown on a AlGaAs substrate than in the opposite case. Similarly, less melt back occurs for growth of lattice matched $Ga_{0.47}In_{0.53}As$ on InP than vice versa [169]. For this reason when double heterostructures containing $Ga_{0.47}In_{0.53}As$ or $Ga_xIn_{1-x}As_yP_{1-y}$ (corresponding to band gaps <0.9 eV) layers are grown on InP substrates, an ''antimelt back'' $Ga_xIn_{1-x}As_yP_{1-y}$ (usually corresponding to a band gap of 0.95 eV) layer is grown on top of them before growing a binary layer. Otherwise dissolution of the quaternary or ternary layer produces a ragged and graded heterojunction [211].

**Solution carry-over**

One of the problems encountered with the sliding boat LPE growth employed for growing multilayered structure is the carry-over of one solution into the next [212]. This problem manifests itself in several ways. Mahajan et al., [213] showed that in the GaInAsP layers carry-over manifests itself in the form of dissolution pits and holes. Anomalous photoluminescence bands observed in AlGaAs/GaAs/AsGaAs double heterostructure have been attributed to solution carry-over [214].

Solution carry-over can cause variations in layer thickness, composition, interface properties, and dopant profiles. These inhomogeneities can adversely affect the performance of devices made from LPE layers. For example, carry-over causes variations in lasing wave length [215, 216], increases broad area threshold current densities [216] and increases temperature dependence of threshold current [217] in double heterostructure AlGaAs/GaAs [215, 217] and GaInAsP/InP [216] layers. Since the amount of solution carry-over increases with increasing clearance between the substrate and the solution, [212, 216] minimizing the slider clearance can minimize carry-over. But minimizing the slider clearance can give rise to more scratching of the layers, so that a compromise solution must be reached. Besomi et al., [216] studied in detail carry-over during LPE of InP and GaInAsP as a function of solution height, slider clearance, and push speed. They found that below a critical clearance of 50 to 75 μm carry-over is minimized and nearly independent of clearance. Further, carry-over increased with height of solution suggesting that it should be significantly reduced by the use of confined melts [116].

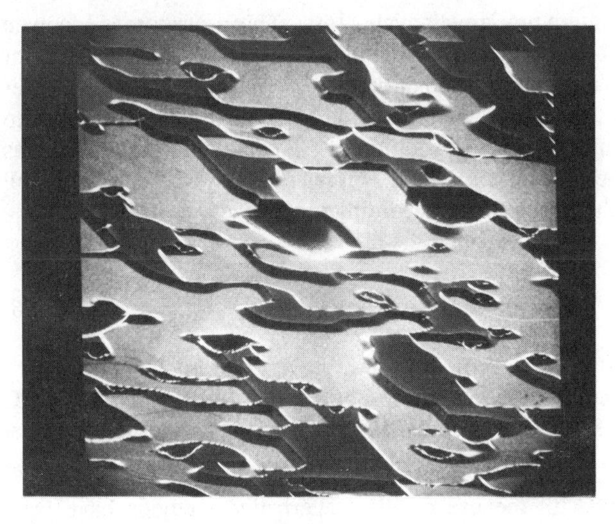


**Figure 2.53** Photomicrograph of an InP layer grown on an InP substrate showing island growth [220].

### 2.2.5 Surface Morphology

One of the limitations of LPE growth is the difficulty in achieving flat and featureless surface morphology of the epitaxial layers. There are three distinct types of surface features associated with LPE layers. These are (a) island growth, (b) terracing, and (c) meniscus lines.

Island growth is a surface feature arising as a result of inadequate nucleation. The island is faceted and the orientation of the facets depends on the orientation of the growth surface. Island formation occurs when the crystal surface is oxidized [218, 219]. Therefore, minimizing contamination by air or water vapor would lead to better growth. Figure 2.53 shows island growth in LPE InP/InP [220].

Contamination of the substrate and/or the growth solutions can also give rise to surface imperfections. The substrate should be carefully polished using mechanical and chemomechanical means and should be thoroughly degreased with solvents before loaded into the growth system. Foreign particles on the substrate surface can generate stacking faults [221, 222]. To minimize such contamination, the growth system should be kept clean and leaktight. The solvent used for the growth solution should be baked at high temperature before each growth.

Thermal decomposition of the substrate caused by preferential evaporation of the volatile group V component (e.g., As in GaAs, P in InP) before it is placed in contact with the growth solution can cause thermal etch pits containing droplets of group III metal. These etch pits and the metal droplets lead to the formation of pinholes in the LPE layers [169].

Obviously, the thermal decomposition of the substrate can be reduced by lowering the temperature of growth and pre-growth equilibration. Alternatively, the substrate can be covered by the graphite slider during high temperature pre-growth cycle [169]. This procedure is adequate for GaAs even at the typical pre-growth equilibration temperatures of 700 to 850°C but not for InP which undergoes more severe thermal decomposition during the pre-heat cycle at much lower temperatures 600 to 650°C. In this case, special provisions should be made in the LPE reactor to minimize P loss. One way to achieve this is the direct addition of a P containing species such as $PH_3$ to the hydrogen introduced into the furnace [223]. For normal LPE growth conditions, InP substrate decomposition in a $H_2$ atmosphere takes place according to the reaction [224]

$$\mathrm{InP\ (s)} + \frac{3}{2}\,\mathrm{H_2\ (g)} \gtrless \mathrm{In}(\ell) + \mathrm{PH_3(g)} \tag{2.56}$$

Therefore, addition of $PH_3$ to the $H_2$ ambient forces the reaction to the left. At about 625°C addition of $PH_3$ at a concentration of $3.5\times10^{-5}$ is found to be sufficient to suppress thermal erosion [223].

Because of safety issues associated with the use of $PH_3$, alternative methods have been developed for substrate protection such as the use of a large InP cover wafer [225] or a Sn-InP solution containing chamber [226]. The P vapor pressure above a Sn-InP solution is much greater than the dissociation pressure of InP thereby preventing thermal decomposition. But the P overpressure produced by the Sn-InP solution may be too high, affecting the composition of adjacent growth solutions. If this is the case, a Sn-In-InP solution may be used to reduce the P pressure but still maintaining it above the dissociation pressure.

In place of or in addition to the surface protection schemes, the use of melt back and/or buffer layers has also been reported. In the melt-back technique an undersaturated solution is brought into contact with the substrate so as to dissolve it to a depth sufficient to remove thermal damage. It is then immediately replaced by the growth solution before thermal erosion of the etched surface can occur [227, 228]. The melt-back technique can also be used in conjunction with growth of a buffer layer of the same composition as the substrate before growing the desired layers. Typically, buffer layers of a few micrometers thickness are sufficient to give smooth epi layers. The use of buffer layers also gives rise to improved photoluminescence efficiency of overlaying layers [229].

**Terraces**

The surface features that are most frequently observed on III-V LPE layers grown on substrates with orientations close to low-index planes are terraces [230 - 232]. Usually, the substrates used in LPE are nominally of (100) or (111) orientations with misorientations of ~ 1°. Such misoriented substrates consist of steps of submicroscopic dimensions. During LPE, a step-bunching process occurs which increases the average height h and distance $\lambda$ of steps at the growth interface (Fig. 2.54) [233]. As a result of the increase in h and $\lambda$, the steps become readily visible at the surfaces of thick LPE layers. These steps are usually referred to as terraces. Figure 2.55 shows the terraces observed on a p-InP layer grown on a n-type InP substrate of nominally (100) orientation [220].

It has been suggested that the terrace formation is related to surface reconstruction [231]. Under the same growth conditions terraces are formed on (100), (111)A and (111)B surfaces

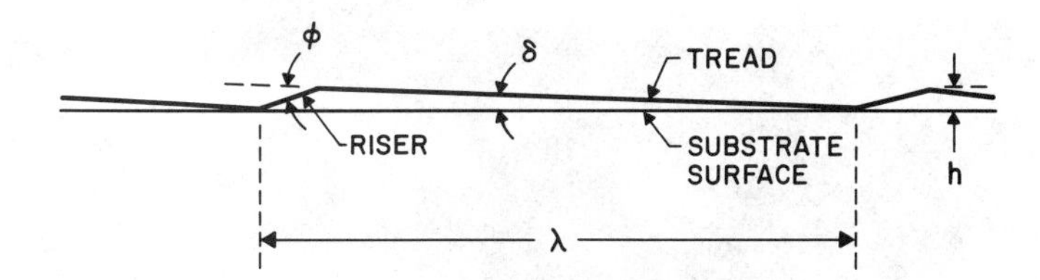


**Figure 2.54** Schematic representation of surface terrace on low indexed substrate.

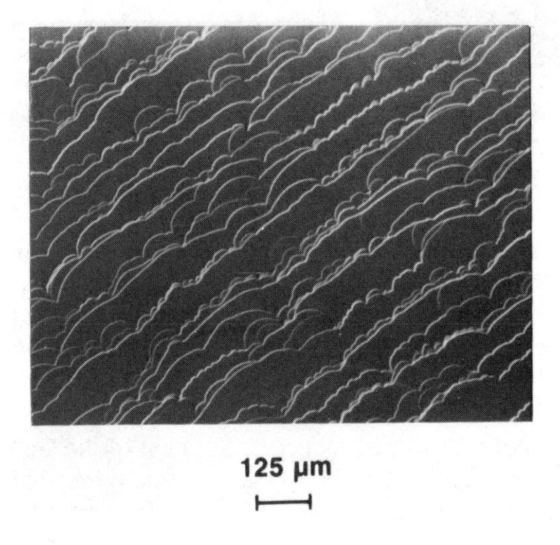


**Figure 2.55** Photomicrograph of a p-type InP layer grown on n-type InP substrate showing surface terraces [220].

which exhibit surface reconstruction but not on (110), (211)A, (211)B and (511) orientations which do not show reconstruction. The surface reconstruction model further suggested that there is a critical misorientation of the substrate which would prevent terrace formation, that is, when $\delta = 0$. For growth of InP (100) and (111) substrates the critical misorientation angle is reported to be 2.6° and 3.3°, respectively [234]. In GaAs a critical misorientation from the (100) surface by 0.8° suppressed terrace formation [231]. The misorientations in both cases are towards the <110> direction. Elimination of terraces has also been reported for growth on accurately oriented substrates with misorientation less than 0.1° [230, 235].

**Meniscus lines**

Meniscus lines are formed as the result of motion of the trailing edges of the liquid as it moved across the crystal during the sliding procedure [236]. The solution leaves the surface in a series of discontinuous movements and at each pause in the wipe off a meniscus line forms. They have been observed in GaAs, AlGaAs, InP, and GaInAsP growth. They also occur in a multilayer structure. The meniscus lines are easily recognized by Nomarski microscopy for they are roughly parallel lines spaced about $100 \pm 50$ μm apart. They can also be decorated by contaminant particles which may originate from the solution or the graphite boat. This is seen in Fig. 2.56 which shows the meniscus lines observed in LPE growth of GaInAs/InP [220].

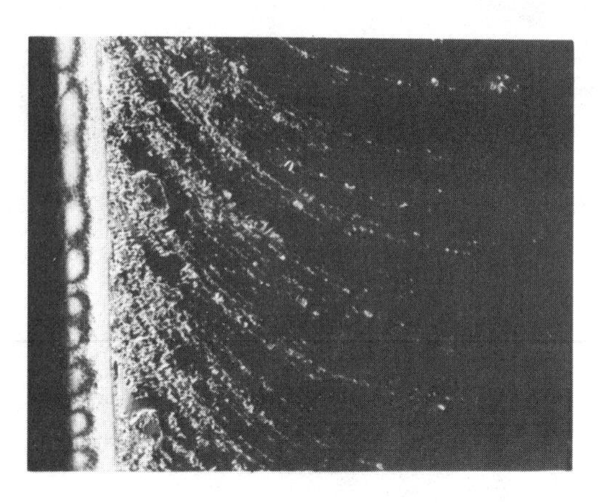


**Figure 2.56** Photomicrograph of a $Ga_xIn_{1-x}As$ (x ~0.47) layer grown on an InP substrate showing the meniscus lines. The meniscus lines are decorated by contaminant particles [220].

**Rake lines**

In the growth of AlGaAs/GaAs/AlGaAs double heterostructure, the absence or near absence of the thin GaAs active layer is called "rake lines" [237 - 239]. The cause of poor device yield and performance of AlGaAs/GaAs semiconductor lasers has frequently been assigned to rake lines. These pathological defects are easily observed in examination of the wafers by photoluminescence or photocurrent techniques [238, 239]. Rake lines are roughly parallel to meniscus lines and in some instances the meniscus line patterns are virtually identical to the rake line contours. The rake lines are supposed to be caused by poor nucleation of the GaAs active layer on n-type AlGaAs layer [239]. The particular n-type dopant used was also found to have an effect on rake lines formation. Te enhanced rake line formation while a nonvolatile dopant like Sn greatly reduced it [239]. Rake lines formation to the same severity as in AlGaAs/GaAs/AlGaAs system has not been observed in the InP/GaInAsP/InP double heterostructure.

## 2.2.6 Crystal Growth of High Purity Epitaxial Layers

The purity of epitaxial layers is an important consideration in many devices such as avalanche photodiodes, p-i-n photodiodes and microwave devices. Further, for characterization of epitaxial layers, high-purity layers would be advantageous especially if one is interested in identifying impurities and or contaminants by photoluminescence exciton spectroscopy and other equivalent techniques. Generally, baking of the growth solutions has been adopted to obtain high-purity layers. A preliminary baking of the graphite boat at high temperature (~ 1600°C) has also been used to reduce impurity levels [240]. Growth solutions are baked in a pure $H_2$ atmosphere for long periods of time. This prolonged baking removes impurities by possible reactions with $H_2$. One such impurity is S which is one of the major background donor impurity [241]. It can be removed as $H_2S$ gas by reaction with $H_2$. The effect of baking on room temperature carrier orientation in LPE InP and GaInAs layers is shown in Fig. 2.57 [242]. The carrier concentration decreases with increasing baking time for both materials but the effect in InP is not as marked as in GaInAs.

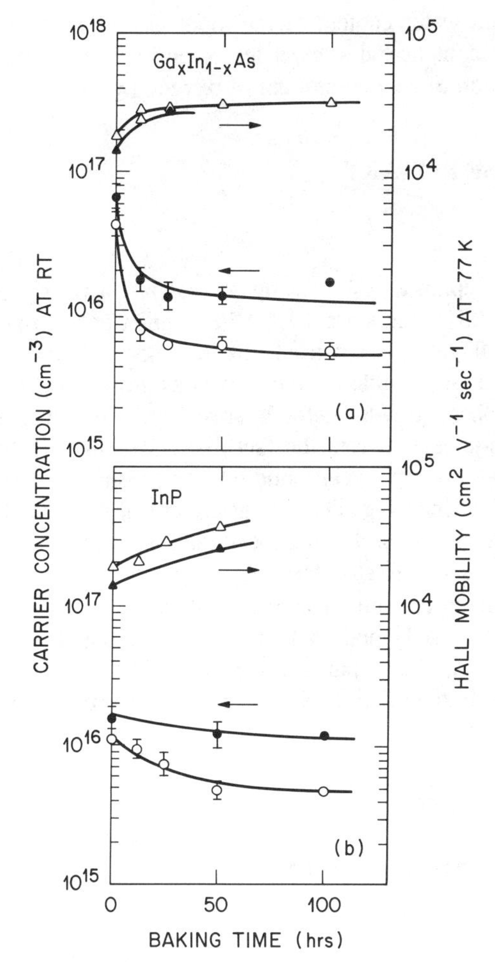


**Figure 2.57** Carrier concentration at room temperature and Hall mobility at 77K of (a) $Ga_xIn_{1-x}As$ layers and (b) InP layers grown on InP from solutions baked at 670°C. (O, Δ) (100) InP; (●, ▲) (111)B InP [242].

The purity of the starting material is also an important consideration for growing high purity layers. The group III element which is the major constituent in the solution should be of highest purity (usually of 99.999 percent purity).

Hydrogen plays also an important role in determining the contamination of solutions with Si. Several chemical reactions involving Si occur [243] of which the predominant one is [244]

$$SiO_2 \text{ (Vitreous)} + 2\, H_2(g) \rightleftarrows Si(\ell \text{ in Ga or In}) + 2\, H_2O(g) \tag{2.57}$$

Since the activities of $SiO_2$ and $H_2$ are close to unity, the equilibrium constant for this reaction is

$$K(T) = [Si]\, [H_2O]^2 \tag{2.58}$$

Equation (2.58) indicates that the equilibrium Si concentration in the solution (and hence in the layer) is a function of temperature and water content of the input gas only. To achieve high-purity layers low in Si, a higher water content and a lower temperature are required [245 - 247]. Some authors have also used the addition of a small amount of oxygen ($\lesssim 0.1$ ppm) to obtain low Si concentrations [248, 249].

## 2.3 TRICHLORIDE VAPOR PHASE EPITAXY

### 2.3.1 Introduction

Since the mid 1960s vapor phase epitaxy (VPE) using an $AsCl_3$-Ga-$H_2$ system has been successfully used to grow GaAs [250, 251], and since 1970 VPE using a $PCl_3$-In-$H_2$ system has been used to grow InP [252]. Crystal growth is carried out in a hot walled reactor and the chemical reactions are reversible making it possible to etch as well as grow. Inherent advantages of trichloride VPE over other vapor phase growth methods arise from the fact that $AsCl_3$ and $PCl_3$ can be distilled into very pure liquids and from the fact that $AsH_3$ and $PH_3$ are not used. The first fact leads to crystals having a very low-background doping density. A low background doping is essential in several key III-V devices (e.g., GaAs field-effect-transistors and GaInAs/InP photodetectors). The fact that toxic group V hydrides are not needed leads to a cost advantage in manufacturing because of the savings of safety related costs.

The basic chemical reactions involved are most easily described with the help of Fig. 2.58 which is due to J. Long et al. [253]. $AsCl_3$ and $PCl_3$ are both volatile liquids with vapor pressures similar to that of water (e.g., the vapor pressure of $PCl_3$ is 100 Torr at room temperature). $H_2$ gas is bubbled through these liquids, and reactions of the type illustrated by the reaction

$$4PCl_3 + 6H_2 \rightarrow P_4 + 12HCl \tag{2.59}$$

**Figure 2.58** Schematic trichloride VPE reactor [253].

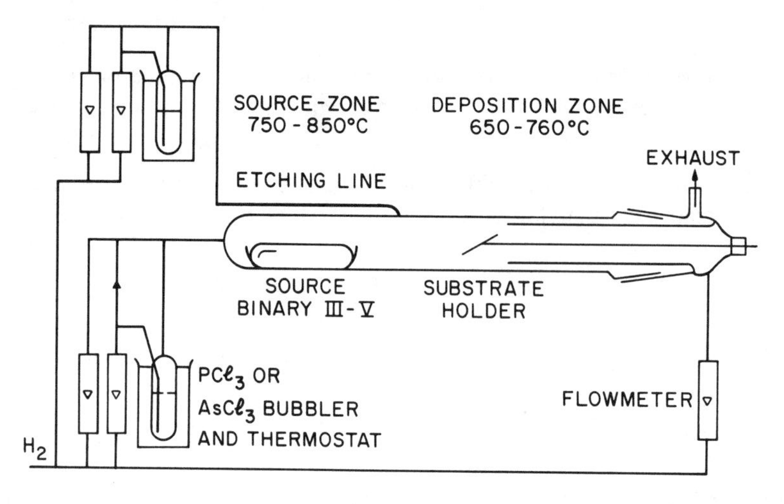

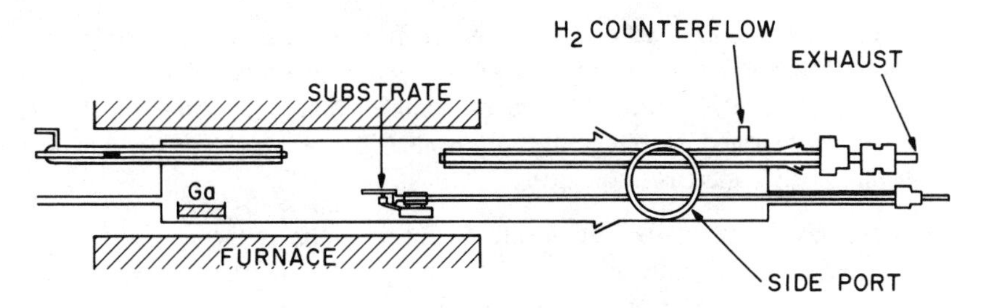


**Figure 2.59** A trichloride VPE reactor which incorporates a counterflow to prevent wall deposits [261].

take place. When the $P_4$ and HCl reaction products are fed to a liquid In source boat, InP forms in the boat, and on saturation the liquid In is covered by a crust of solid InP.

The main disadvantage related to trichloride VPE is that it is difficult to grow III-V alloys by that means. For example, it is not possible to grow AlGaAs since the growth temperature of AlAs, 1100°C, differs too much from that of GaAs, 750°C [254]. Another disadvantage is that both GaAs [255] and InP [256] substrates must be misoriented by 3-6 degrees away from (100) to attain specular epitaxial layer surfaces. Growth facets occur otherwise. Laser device processing is more complex and time consuming as a result. Lastly, the inherent instability [257] of two-phase sources makes compositional control in alloys very difficult [258] and leads to poor reproducibility and yield for the binary InP [256]. However, the use of solid sources in recent years [258, 259] has ameliorated this disadvantage.

### 2.3.2 Crystal Growth of GaAs and InP

For the growth of GaAs the following reaction takes place both above the source boat and above the substrate:

$$4\mathrm{GaAs} + 4\mathrm{HCl} \underset{T_{low}}{\overset{T_{high}}{\rightleftarrows}} 4\mathrm{GaCl} + \mathrm{As}_4 + 2\mathrm{H}_2 \tag{2.60}$$

In the source region, which is at temperature $T_{high}$, the GaAs crust on top of the liquid Ga in the source boat is reacted with HCl (which results from the reaction of $AsCl_3$ and $H_2$) and GaCl and $As_4$ are formed. These are transported to the substrate which is at temperature $T_{low}$ where the reverse reaction happens. The net result is the transferral of GaAs from the crusted source to the substrate.

Several additional features have become standard for trichloride reactors because they solve problems associated with this growth technique.

Wall deposits [260] have plagued crystal growers because such deposits uncontrollably modify the partial pressures of reactants in the growth zone. A counterflow such as shown in Fig. 2.59 (ref. [261] and references therein) reduces wall deposits downstream from the substrate by forcing the exhaust gases out of the growth chamber through a tube which extends into the reactor from the rear (i.e., the exhaust end). $H_2$ fed in from the rear flows in the reactor chamber counter to the main flow and keeps the inside surfaces clean. The tube which extends in from the rear is lined with a removable ''dump'' tube on which effluent deposits form. This tube is designed to be easily removed for cleaning. With such an arrangement the inside walls of a reactor do not need to be cleaned as frequently.

Since a primary advantage of the trichloride VPE technique is the ability to grow material having a high purity, achieving an understanding of the chemistry of impurity incorporation constituted a significant advance. It was recognized by J. V. DiLorenzo [262] that for the growth of high purity GaAs a flow of $AsCl_3$ in $H_2$ carrier gas which bypasses the source is crucial. Such a bypass flow is shown schematically in Fig. 2.58 and is labeled as an etching line since it can also be used to etch. As discussed earlier, HCl is a reaction product when $AsCl_3$ and $H_2$ are mixed, and, consequently, the bypass flow increases the concentration of HCl in the growth zone. As a result the incorporation of Si impurities in GaAs is suppressed. It has been proposed that the Si impurities come from chlorosilanes which form from the reaction of HCl with the reactor walls according to the reaction [255],

$$SiO_2 + nHCl + (4-n)H_2 \rightarrow SiH_{4-n}Cl_n + 2H_2O \tag{2.61}$$

We note, however, that the presence of chlorosilanes has never been demonstrated. The chlorosilanes in turn react with $H_2$ according to the reaction [255],

$$SiH_{(4-n)}Cl_n + (n-2)H_2 \rightleftarrows nHCl + Si \tag{2.62}$$

The Si impurities are incorporated predominantly on As sites in GaAs and form donors. By introducing additional HCl directly into the growth chamber through the bypass, crystal growers can shift the chemical balance in Eq. (2.62) (or of some other reaction which produces both HCl and Si) through the law of mass action and thereby reduce the amount of Si incorporated in the growing layer. This effect has come to be known as the ''mole fraction effect,'' and the efficacy of this procedure for the growth of high purity InP is demonstrated in Fig. 2.60 [256].

In models for vapor phase growth a so called ''stagnant'' layer or ''boundary'' layer is often invoked. Since gas velocity must be zero at the surface of a growing layer, a layer-like region above the substrate in which the gas velocity is small is expected. Calculated velocity profiles of the flow along a reactor axis are shown in Fig. 2.61 [255]. Diffusion of nutrients through the boundary layer and thus the concentration gradient across it is thought to control the growth rate at high temperatures [255]. When the growth rate is mass transport limited in this way, the thickness profile of the boundary layer and the rate of nutrient depletion along the flow direction will determine the thickness uniformity of the grown layer. The boundary layer thickness can be altered by varying the tilt angle of the substrate as shown in Fig. 2.61.

At low temperatures, the growth rate of GaAs is temperature dependent and is kinetically limited by a relatively slow surface process such as adsorption, desorption, or surface reaction [263]. The growth rate of GaAs is also a function of the $AsCl_3$ mole fraction [255]. By optimizing both the growth temperature and the $AsCl_3$ mole fraction, Komeno et al., [264] have shown that GaAs layers with good thickness uniformity can be grown. Their results are shown in

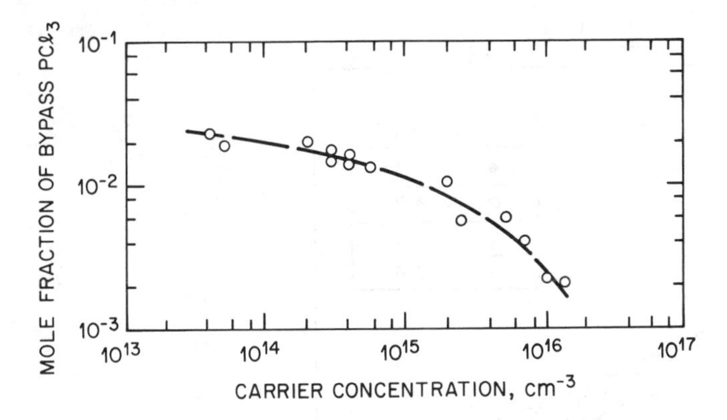


**Figure 2.60** Data showing the relationship between the carrier concentration of InP and the mole fraction of bypass $PCl_3$. This relationship is known as the "mole fraction effect" [256].

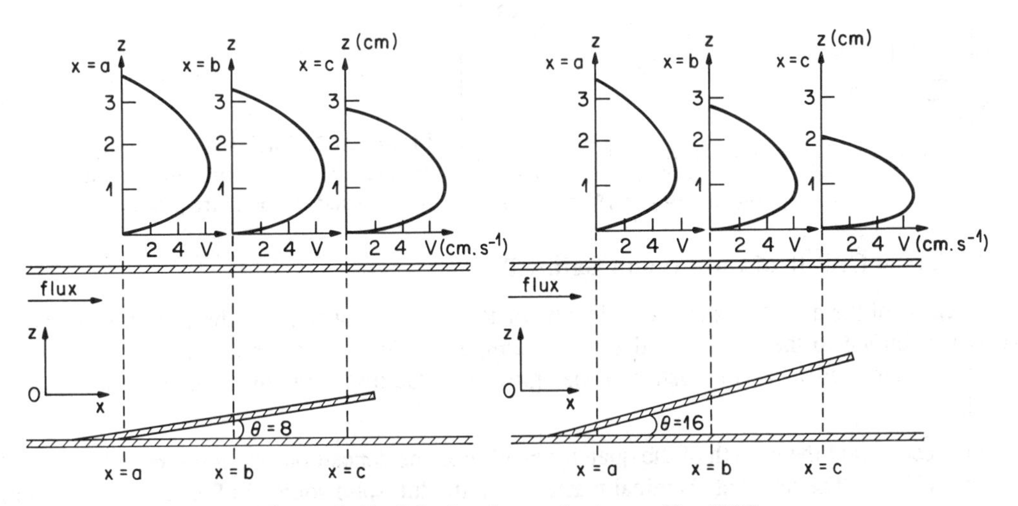


**Figure 2.61** Calculated gas velocity profiles [255].

Fig. 2.62 for wafers of up to 40 $cm^2$ in area. Here uniformity is expressed as [2(max-min)/mean]×100(%). Komeno et al., [264] also report a thickness uniformity of 0.6 percent for a GaAs wafer having a diameter of 50 mm. The mean thickness in this case was 8.91 μm.

These results demonstrate that very tight control over the $AsCl_3$ mole fraction and over the growth temperature are required for good uniformity. For this reason Cox et al., [261] have instead employed a rotating substrate holder to improve the uniformity of GaAs grown by trichloride VPE.

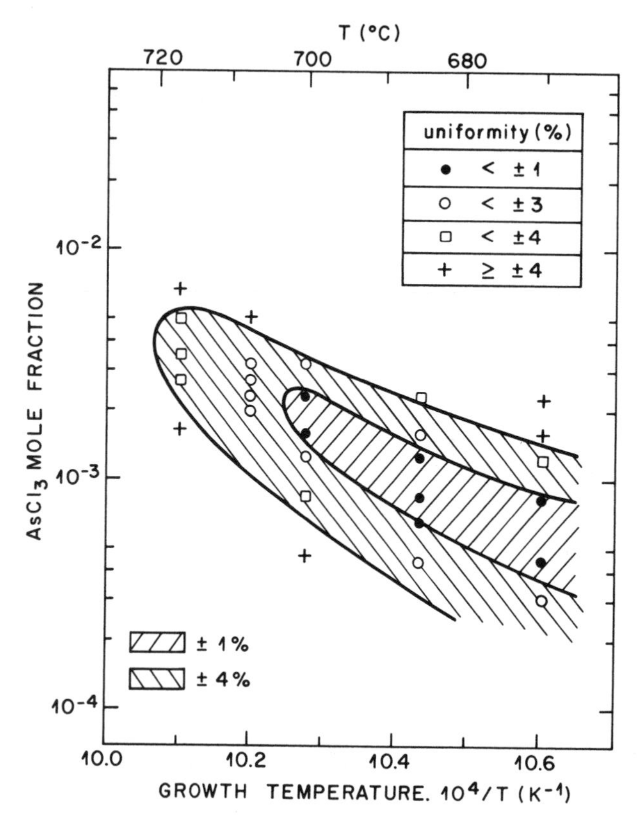


**Figure 2.62** Thickness uniformity of GaAs epitaxial layers as a function of $AsCl_3$ mole fraction and temperature [264].

### 2.3.3 Crystal Growth of GaInAs and GaInAsP

Because of the difficulties connected with crusted liquid metal sources, the emphasis in recent years has shifted to the use of solid sources [258, 259, 265]. A schematic diagram of a solid source reactor used to accomplish epitaxial growth of GaInAsP on InP is shown in Fig. 2.63 [259].

One can regard the growth of the quaternary alloy as the deposition of a mixture of GaAs, InP and InAs [265]. The relevant chemical reactions for the InP solid source in Fig. 2.63 are reaction Eq. (2.59) listed for the reaction of $PCl_3$ with $H_2$ to form the products $P_4$ and HCl and the following reaction,

$$P_4 + 4HCl + 4InP \underset{T_{low}}{\overset{T_{high}}{\rightleftarrows}} 4InCl + 2P_4 + 2H_2 \tag{2.63}$$

The InP solid source essentially replaces the InP crust of a two-phase source. Just as for growth using a two-phase source, crystal growth on a substrate is achieved by keeping the substrate at a lower temperature than the source so that reaction Eq. (2.63) is driven in the reverse direction. Again we note that net result may be viewed as the transferral of InP from the source to the substrate. Similarly, when $AsCl_3$ in $H_2$ is fed to the GaAs and InAs sources these two binaries

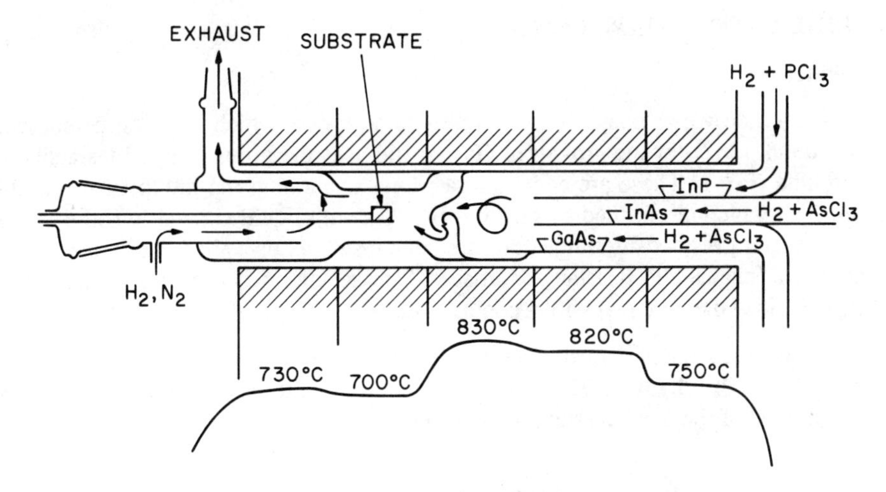


**Figure 2.63** Schematic solid source reactor [259].

are effectively transferred to the substrate. However, because of different incorporation efficiencies for P and As the partial pressures must be carefully adjusted for the growth of lattice matched GaInAsP/InP. Vohl [259] has reported solid source growth of alloy compositions with band gaps in the range 0.90 -1.10 eV ($\lambda = 1.38–1.13\,\mu m$). Lattice mismatches for these alloys fell within the range –0.16% to +0.33%. Cox et al., [258] have used a similar arrangement to grow GaInAs using solid InP and solid GaAs sources in a reactor designed to provide a flow of InCl as well. The InCl flow was achieved by first cracking $AsCl_3$ to form HCl and $As_4$ and then condensing the $As_4$ to remove it. The high purity HCl achieved in this way is then passed over liquid In to form InCl. GaInAs with a carrier concentration as low as $1 \times 10^{14}$ $cm^{-3}$ was reported by Cox et al., [258]. This same scheme has been modified recently [265] with the addition of an InAs solid source. A schematic diagram of the modified reactor is shown in Fig. 2.64. Quaternary GaInAsP alloys spanning the entire range of compositions which are lattice matched to InP can be grown with this scheme [266].

**Figure 2.64** Schematic solid source reactor for the growth of GaInAsP over the entire range of compositions which are lattice matched to InP [265].

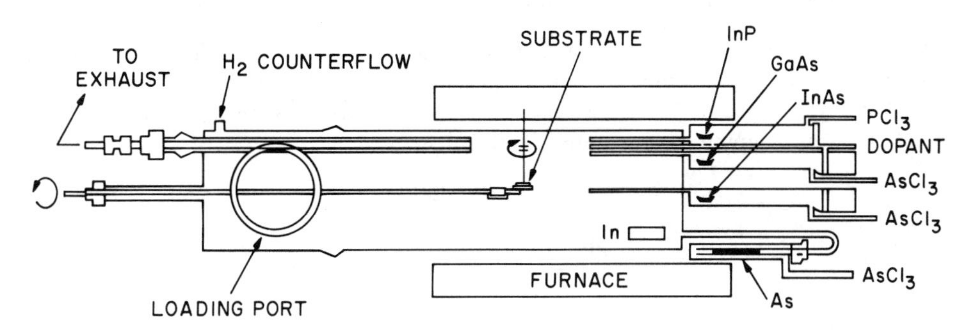

## 2.4 HYDRIDE VAPOR PHASE EPITAXY

### 2.4.1 Introduction

The term "hydride vapor phase epitaxy" refers to a process which was first demonstrated by Tietjen and Amick [267] for the growth of GaAsP and uses the group V hydrides arsine ($AsH_3$) and/or phosphine ($PH_3$). These are added to gas flows containing the group III metal chlorides GaCl and/or InCl which are formed by reacting HCl gas with liquid Ga and In sources. Both the source reactions and the deposition reactions are done inside a hot-walled quartz reactor which is surrounded by a 'clamshell' resistance furnace. The reactions occur under near-equilibrium conditions, and, since they are reversible, controlled in-situ etching is possible.

Since one has independent control over each group III and group V species (which is not the case for trichloride VPE), hydride VPE is generally considered to be a more flexible [265] growth technique, and most of the VPE crystal growth of GaInAsP has been done by the hydride method [268].

Advantages of the hydride VPE method are that epitaxial growth on large wafers and on multiple wafer simultaneously is possible. Smooth surfaces are attainable (unlike in the case of LPE), and all $Ga_xIn_{1-x}As_yP_{1-y}$ alloy compositions which are lattice matched to InP can be grown. Furthermore, the process is well suited for commercial use since several growth runs can be performed in an eight hour day.

The use of $AsH_3$ and $PH_3$ are a disadvantage since these are supplied in compressed gas cylinders and are highly toxic. The reaction chemistry in the hydride process is complicated and is difficult to model, and in general automation is needed to make it a commercially viable process [265, 268]. The performance of hydride VPE reactor is usually dependent on the history of the reactor. Two reasons for this have been clearly identified. First, wall deposits often occur in the growth zone and can significantly alter the partial pressures of the important reactants [269], and, second, there are memory effects associated with the sources [270]. The purity of starting materials is generally not as good as for trichloride VPE. Compressed gas cylinders of HCl are usually used to supply HCl to a reactor. Since HCl is highly corrosive unless the environment is anhydrous and since the cylinders must be made of high-pressure steel, the HCl gas is a likely source of impurities. In addition, impurities in the hydride gas cylinders can be the dominant impurities introduced during growth [271]. Background carrier concentrations for hydride VPE are generally above $3\times10^{15}$ $cm^{-3}$ [265] for these reasons. Finally, we note that Al and Sb compounds are difficult to grow by the hydride VPE technique [268].

### 2.4.2 Basic Chemistry and Crystal Growth

A basic reactor design due to Olsen and co-workers [268] is shown in Fig. 2.65. The reactor has a source zone which contains the liquid metals, a mixing zone in which the hydrides and metal chlorides are mixed and a deposition zone where crystal growth takes place. These are normally held at the following temperatures [268]:

$T_{source}$ : 800-850°C
$T_{mix}$ : 825-900°C
$T_{grow}$ : 650-725°C

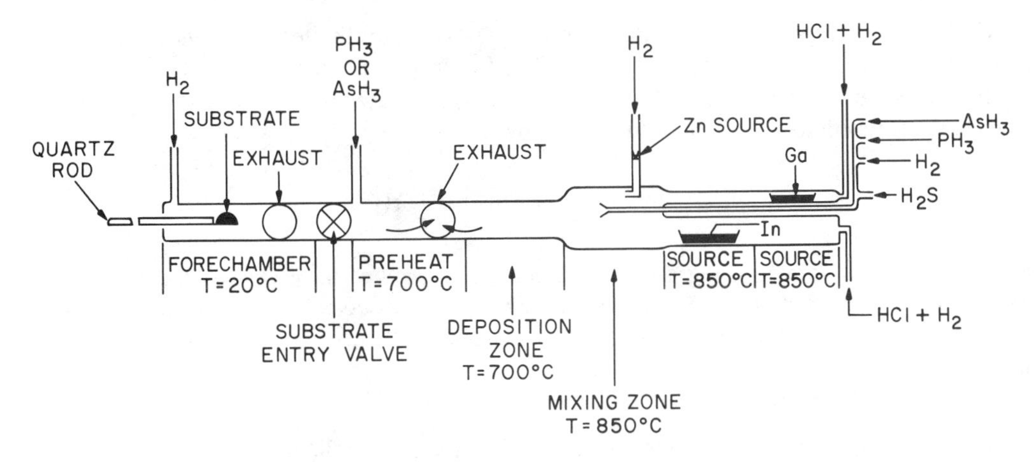


**Figure 2.65** Schematic hydride VPE reactor [268].

Horizontal reactors have historically been more popular than vertical reactors even though substrate rotation for improved uniformity is more easily implemented using a vertical design. A vertical design was used by Zinkiewicz [272].

In the horizontal geometry with the substrate held in the horizontal plane (or at a small tilt with respect to the horizontal plane), the gas flow is often modeled using a ''stagnant'' layer or boundary layer model. (See the discussion in Sec. 2.3 concerning boundary layers). The thickness uniformity of this boundary layer (which is influenced by the flow rate) can influence the thickness uniformity of the grown layer if diffusion across the boundary layer is a significantly rate limiting step.

Just as for the other VPE techniques (chloride-VPE and MOCVD), two regimes of crystal growth are usually identified. For binary compounds these are distinguished by the temperature dependence of the growth rate. For the one case, the growth rate is not a function of temperature (or is only weakly temperature dependent [268]). In this case crystal growth can be limited by the rate of mass transport of the chemical species toward the growth zone or by the rate of mass transport of exhausted reaction products. Growth rates are also temperature independent if mass transfer by diffusion across the boundary layer is the rate limiting step. The other case is characterized by a strongly temperature dependent growth rate. A temperature dependence is found if crystal growth is limited by the rate of a chemical reaction occuring at the surface of the growing layer or by the rate of adsorption or desorption. In the temperature dependent case crystal growth is often referred to as being kinetically limited. Mass transport limited growth usually has only a weak substrate orientation dependence and a high flow rate dependence, whereas kinetically limited growth has a strong substrate orientation dependence and a low flow rate dependence. Whether or not there is a substrate orientation dependence is used to determine whether the growth of GaInAsP alloys is mass transport or kinetically limited. The fact that the binary constituents of GaInAsP have different free energies of formation prevents one from using a temperature dependence to make this determination. For example, Kanbe et al., [273] found that growth of GaInAs on InP at 700°C was 9.2 μm/hr for the (111)A face but only 1.2 μm/hr for the (111)B face. This implies that the growth was kinetically limited. The Ga fraction of the ternary alloy was 0.42 for the (111)A face and 0.36 for the (111)B face.

In the source regions a flow of HCl mixed with $H_2$ is reacted with liquid metal held in boats. The chemical reaction in the case of In is given by

$$In_{(\ell)} + HCl \rightleftarrows InCl + \frac{1}{2} H_2 \tag{2.64}$$

Similarly, GaCl is formed using a liquid Ga boat. For the growth of GaInAsP, the ratio of the HCl over In flow to that of the HCl over Ga flow must be considerably greater than one (this ratio is typically in the range 5 - 15) since InCl is considerably less reactive than GaCl [274]. Once this is taken into account a first order estimate of the metal stoichiometry of GaInAsP can be obtained from the GaCl/InCl mole fraction ratio [265, 274 - 275].

It is quite important that the source reaction be driven to completion. The effect of not doing so on the lattice mismatch of $Ga_{0.26}In_{0.74}As_{0.55}P_{0.45}$ has been calculated from thermodynamics and is shown in Fig. 2.66 [276]. We see from this figure that if the source reactions are less than 99 percent complete, then the lattice mismatch will be bigger (i.e. more negative) than $-1 \times 10^{-3}$. Complete conversion of HCl to metal chloride is more difficult to achieve for In than for Ga, and

**Figure 2.66** Alloy stoichiometry of $Ga_xIn_{1-x}As_yP_{1-y}$ as a function of source reaction completeness. Lattice mismatches are also shown [276]. Reprinted with permission by The Electrochemical Society, Inc.

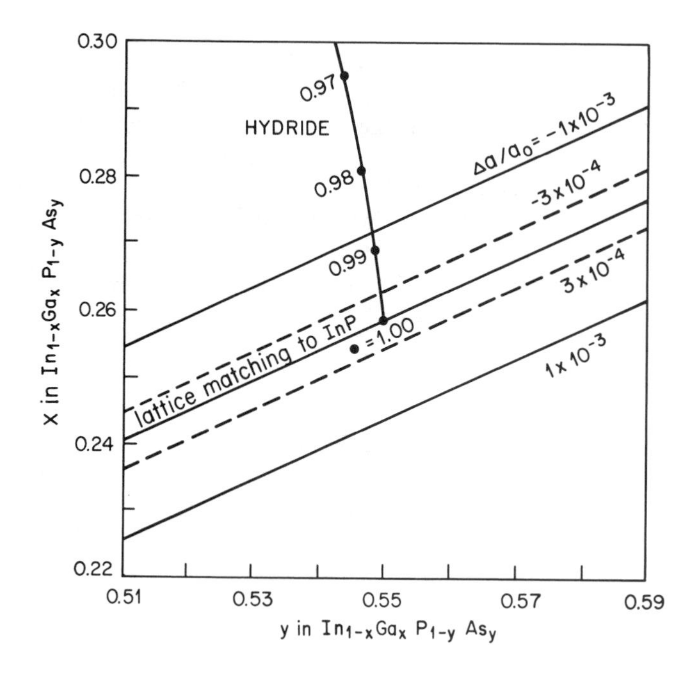

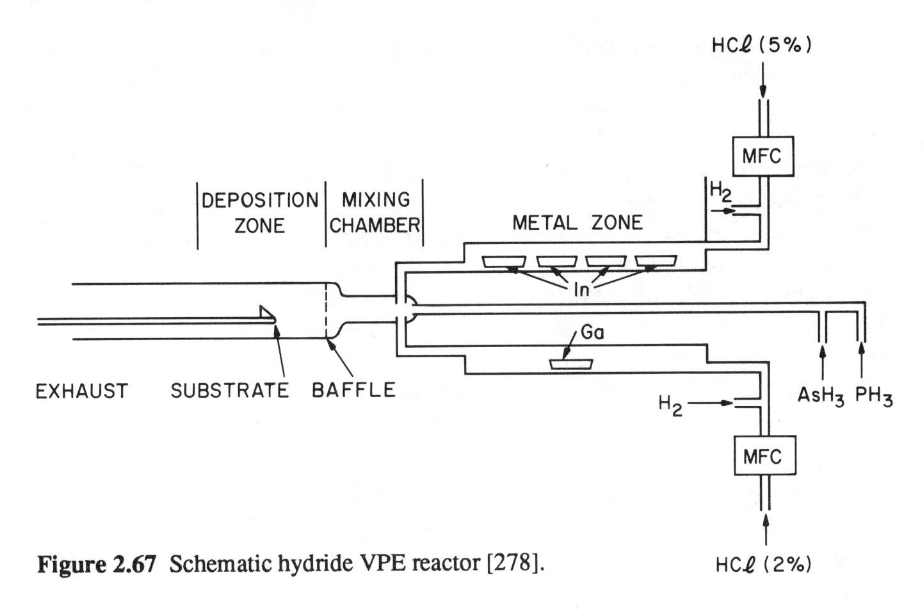


**Figure 2.67** Schematic hydride VPE reactor [278].

a larger In boat is usually required [277]. Johnston and Strege [278] solved this difficulty by using several In boats in succession as shown in Fig. 2.67.

Good mixing of metal chlorides with the hydrides is also crucial for high quality epitaxial layers, and several workers have included baffles [279, 280] or nozzles [265] to promote mixing.

In the growth zone the following set of reactions are thought to predominate [268]

$$2\,\mathrm{GaCl} + \frac{1}{2}\,\mathrm{As_4} + \mathrm{H_2} \rightleftarrows 2\,\mathrm{GaAs}\ (\mathrm{alloy}) + 2\,\mathrm{HCl} \tag{2.65}$$

$$2\,\mathrm{InCl} + \frac{1}{2}\,\mathrm{As_4} + \mathrm{H_2} \rightleftarrows 2\,\mathrm{InAs}\ (\mathrm{alloy}) + 2\,\mathrm{HCl} \tag{2.66}$$

$$2\,\mathrm{GaCl} + \frac{1}{2}\,\mathrm{P_4} + \mathrm{H_2} \rightleftarrows 2\,\mathrm{GaP}\ (\mathrm{alloy}) + 2\,\mathrm{HCl} \tag{2.67}$$

$$2\,\mathrm{InCl} + \frac{1}{2}\,\mathrm{P_4} + \mathrm{H_2} \rightleftarrows 2\,\mathrm{InP}\ (\mathrm{alloy}) + 2\,\mathrm{HCl} \tag{2.68}$$

$$\mathrm{As_4} \rightleftarrows 2\,\mathrm{As_2} \tag{2.69}$$

$$\mathrm{P_4} \rightleftarrows 2\,\mathrm{P_2} \tag{2.70}$$

$$\mathrm{AsH_3} \rightleftarrows \frac{1}{2}\,\mathrm{As_2} + \frac{3}{2}\,\mathrm{H_2} \tag{2.71}$$

$$\mathrm{PH_3} \rightleftarrows \frac{1}{2}\,\mathrm{P_2} + \frac{3}{2}\,\mathrm{H_2} \tag{2.72}$$

Other compounds and reactions involving them can also be included [276] at the expense of additional complexity. These are the higher chlorides such as $GaCl_3$ and $InCl_3$ and compounds of the form $As_nP_m$ (n+m=4). Although the higher chlorides are not present at significant concentrations under typical conditions the $As_nP_m$ are [281].

Since there are many important chemical reactions, modeling of the crystal growth is complex and difficult. Nevertheless, it has been done by several persons who have made the assumption that chemical equilibrium thermodynamics can be applied to calculate GaInAsP stoichiometry from input flows and from thermodynamic data [282 - 284]. The calculation is simplified if $P_2$ and $As_2$ are neglected [282] and results obtained in that case agreed reasonably with the growth results of Sugiyama et al., [279]. Agreement with these data and those of Olsen and Zamerowski [268] were also obtained by Koukitu and Seki [284] who did not neglect $P_2$ and $As_2$. A comparison of these calculations with the data of Olsen and Zamerowski is shown in Fig. 2.68. The comparison reveals that the calculation is within 10 mol percent of the experimental data [268].

Sensitivity of the stoichiometry and lattice mismatch of GaInAsP to input flows and to temperature have been calculated by Yoshida and Watanabe [276]. These persons find that increasing separately flows of HCl over Ga, HCl over In, $AsH_3$, and $PH_3$ by 3 percent produces the stoichiometry changes indicated by the length of the arrows in Fig. 2.69. The effect of increasing the growth temperature by 3 degrees is also shown. This figure shows that greater

**Figure 2.68** $Ga_xIn_{1-x}As_yP_{1-y}$ alloy stoichiometries calculated for a set of hydride VPE growth conditions (triangles) and the actually obtained values (circles) [268].

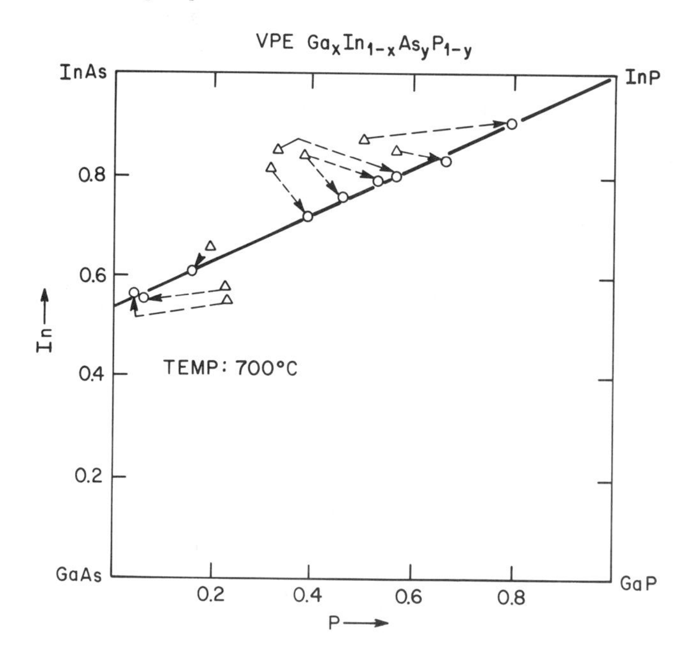

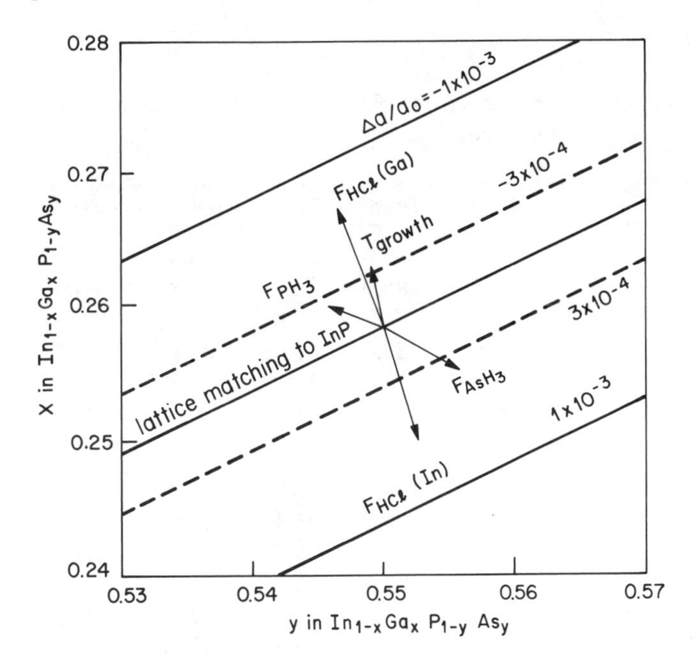


**Figure 2.69** Sensitivity of $Ga_xIn_{1-x}As_yP_{1-y}$ alloy stoichiometry to gas flow and temperature perturbations [276].

control must be achieved over the metal chloride flows than over the hydride flows in order to stay within a given mismatch limit. A similar conclusion was reached by calculating the percent variation required to produce a vector of length 0.004 on the x-y plane (where x is the Ga fraction in $Ga_xIn_{1-x}As_yP_{1-y}$ and y is the As fraction). These variations are given in Table 2.9 [276].

A significant point to be noted about the growth of GaInAsP by hydride VPE is that the

**Table 2.9**

Required accuracy of growth parameters for a length variation of 0.004 on the x-y plane where x and y are the Ga and As mole fractions of the alloy $Ga_xIn_{1-x}As_yP_{1-y}$ (from Yoshida and Watanabe [276]).

| | |
|---|---|
| growth temperature | 2.6°C |
| HCl over Ga | 1.2% |
| HCl over In | 1.3% |
| $AsH_3$ | 1.7% |
| $PH_3$ | 2.4% |

**Table 2.10**

Comparison of partial pressures (p) to obtain alloys with lattice mismatch ($\Delta a/a$) and band gap wavelength ($\lambda_g$).

| | $p_{GaCl}$ | $p_{InCl}$ | $p_{AsH_3}$ | $p_{PH_3}$ | $\Delta a/a$ (%) | $\lambda_g$ ($\mu m$) |
|---|---|---|---|---|---|---|
| ($\times 10^4$ atm) | 2.0 | 86.0 | 3.6 | 46.8 | +0.04 | 1.04 |
| | 3.3 | 86.0 | 5.2 | 46.8 | +0.09 | 1.11 |
| RCA | 2.6 | 84.2 | 8.0 | 22.8 | <0.03 | 1.23 |
| ($H_2 = 5000$ $cm^2$ $min^{-1}$, | 2.6 | 83.4 | 8.4 | 22.8 | <0.03 | 1.28 |
| $T_g = 700°C$) | 3.6 | 83.4 | 17.0 | 22.8 | <0.03 | 1.33 |
| | 4.4 | 46.0 | 55.0 | 6.0 | +0.03 | 1.55 |
| | 4.8 | 46.0 | 55.0 | 0 | +0.07 | 1.66 |
| | 5.5 | 176.4 | 40.0 | 142.0 | <0.03 | 1.17 |
| | 2.7 | 85.5 | 36.4 | 64.5 | <0.03 | 1.28 |
| | 2.7 | 88.5 | 28.2 | 62.7 | <0.03 | 1.30 |
| Varian | 5.5 | 103.6 | 40.9 | 68.2 | <0.03 | 1.45 |
| ($H_2 = 1100$ $cm^3$ $min^{-1}$, | 2.7 | 55.4 | 34.5 | 22.7 | <0.03 | 1.48 |
| $T_g = 688°C$) | 5.5 | 75.5 | 45.5 | 33.6 | <0.03 | 1.53 |
| | 5.5 | 58.6 | 63.6 | 0 | <0.03 | 1.67 |
| | 2.8 | 40.9 | 45.5 | 0 | <0.03 | 1.67 |

growth conditions for a given stoichiometry are not unique. The results listed in Table 2.10 [268] reveal that lattice matched growth over the entire range of alloy compositions which are lattice matched to InP can be achieved at two very different $H_2$ carrier gas flows. Data obtained at two different laboratories, RCA and Varian, are included in this table [268].

Predicting alloy composition from vapor flows is also complicated by the formation of deposits on the walls of the reactor. Such deposits have been widely observed and reported [284, 285], and modern reactors are usually designed to permit easy cleaning of these deposits by injecting HCl directly into the growth zone.

As pointed out by Olsen [268] one cannot simply extrapolate relationships between alloy compositions and gas flows from one reactor to another. Differences in reactor design such as the growth tube diameter, carrier gas flow rates, and substrate positioning can have a great effect even if partial pressures and growth temperatures are the same. This is because kinetic effects can dominate the thermodynamics of crystal growth. Olsen furthermore points out that proper gas flows will always have to be determined empirically. For this reason x-ray diffraction to determine the mismatch and photoluminescence to determine the bandgap should be considered essential capabilities for the successful implementation of hydride VPE [286].

### 2.4.3 Surface Preservation and Reactor Design for Abrupt Interfaces

Deterioration of the surface of a grown layer caused by thermal decomposition is of great concern during VPE. For example, at temperatures above 365°C InP vaporizes incongruently [287] and a pitted surface which is rich in In is formed [288, 289]. For successful multilayer

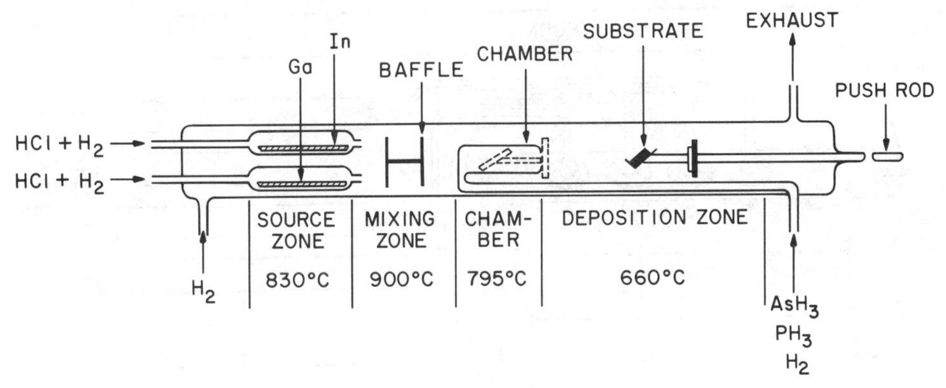


**Figure 2.70** Schematic hydride VPE reactor [280].

growth by VPE care must be taken to preserve the surface not only during the substrate warm-up period but also during the time that gas flows are changed and during the wafer cool-down period.

Good surface quality can be preserved if the sample is held in the proper hydride ambient during these critical times, and this can be accomplished in a preheat zone such as shown in Fig. 2.65. Two quite innovative techniques to accomplish the same thing have also been reported. In the first [269], a quartz 'slider boat' is used to house the wafer in the growth zone. The wafer is protected until growth is initiated by sliding open the boat. In the second, an internal preservation chamber is included in the reactor. This design is schematically shown in Fig. 2.70 [280].

An alternative to holding the sample in a special reactor compartment while the gas flows are adjusted for the growth of a new layer is to have the necessary flows already established. This can be accomplished in a multibarrel reactor. The basic concept is schematically illustrated in Fig. 2.71 which is due to Olsen [268]. An important added benefit is that abrupt transitions between layers are more readily achieved since gas flushing times have been eliminated. Double-barrel reactors have been used by several groups [290, 291, 292, 293] and abrupt doping transitions [290] and 150 Å thick quantum wells have been demonstrated [293]. A four-barrel reactor has also been reported [294], but extending the multibarrel concept still further quickly leads to an impractically large reactor diameter.

A more recent innovation due to Strege and Johnston [265] is shown schematically in Fig. 2.72. In this reactor design, wafers are transferred from one station to another as is done during LPE and during levitation VPE [295]. Pre-established gas flows exist at each station and are sprayed over the wafer from "shower-head" nozzles which contain a quart frit.

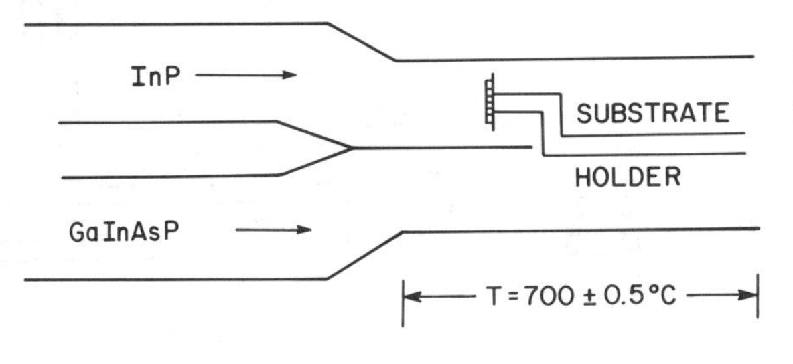


**Figure 2.71** Schematic design for the multi-barrel concept [268].

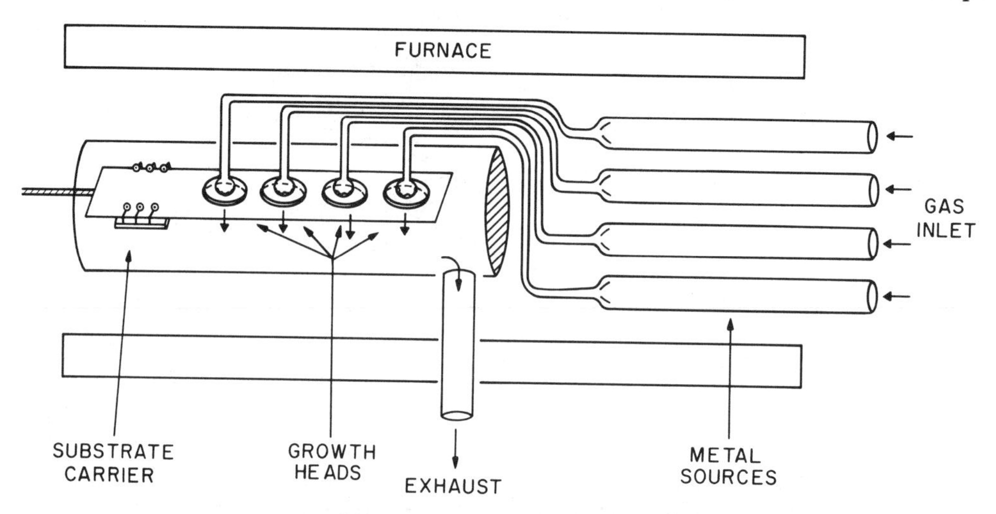


**Figure 2.72** Schematic hydride VPE reactor [265].

We hasten to point out, however, that high quality epitaxial structures suitable for opto-electronic devices can be grown in a single barrel reactor [296, 297]. Using ultraviolet light absorption spectroscopy [298] Karlicek and co-workers not only diagnosed reactor source region designs sufficiently well to ensure that source reactions approach thermodynamic equilibrium [270], but also diagnosed $PH_3$ pyrolysis in the growth region [299]. $PH_3$ pyrolysis is more sluggish than $AsH_3$ pyrolysis. Insufficient $PH_3$ pyrolysis leads to morphological defects often referred to as "hillocks" and results in an increased growth rate [299]. The increased understanding of chemical reactions taking place inside a hydride-VPE reactor as outlined have resulted in design improvements sufficient to make a single barrel reactor viable for production. A schematic of such a reactor is shown in Fig. 2.73 [265]. Noteworthy features of this reactor

**Figure 2.73** Schematic hydride VPE reactor [265].

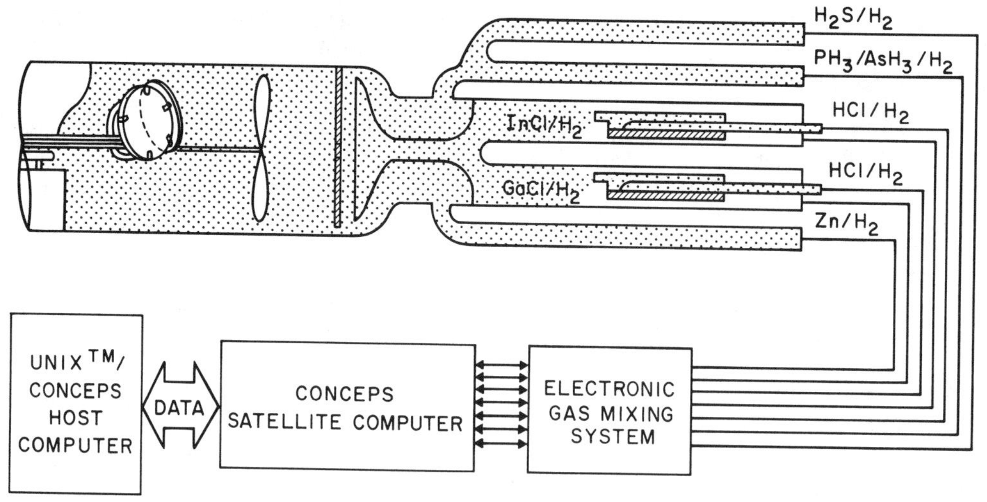

**Table 2.11**

Simple Alkyls for MOCVD*

| Compound | Acronym | Melting Point (°C) | Boiling Point at 760 torr (°C) | Vapor Pressure (torr) |
|---|---|---|---|---|
| Trimethylgallium | TMGa | –15.8 | 55.7 | 64.5 at 0°C |
| Triethylgallium | TEGa | –82.3 | 143 | 18 at 48°C |
| Trimethylaluminum | TMAl | 15.4 | 126 | 8.4 at 20°C |
| Trimethylindium | TMIn | 88.4 | 134 | 1.7 at 20°C |
| Triethylindium | TEIn | –32 | 184 | 3 at 53°C |

* From reference 254.

are that it is designed for growth on two 2 inch diameter wafers at the same time and that a special baffle configuration is used to promote turbulence and to enhance mixing.

## 2.5 METAL-ORGANIC CHEMICAL VAPOR DEPOSITION (MOCVD)

### 2.5.1 Introduction

MOCVD is a very promising crystal growth technique which uses metal alkyls and hydrides as source materials in a cold walled reactor. A brief list of metal alkyls is given in Table 2.11 along with their melting points, boiling points, and vapor pressures. They are all volatile liquids at room temperature with the notable exception of trimethylindium (TMIn) which is a waxy clear solid at room temperature. In general, the lighter methyl compounds have a higher vapor pressure than do the ethyl compounds. This facilitates their transport to the substrate, and in most instances they have been preferred. Substrates are placed on a susceptor which is heated either by RF induction or with infrared irradiation. The technique was pioneered by Manasevit in 1968 who demonstrated that TMGa when mixed with arsine, $AsH_3$, and pyrolized at temperatures between 600°C and 700°C in an $H_2$ atmosphere would grow thin single crystals of GaAs on GaAs, sapphire, and several other substrates [300]. Several other names are sometimes used to refer to this process, including "organometallic CVD" (OMCVD) and "organometallic VPE" (OMVPE).

A schematic diagram of a vertical atmospheric-pressure MOCVD reactor arrangement due to Dupuis [301] for the growth of $Al_xGa_{1-x}As$ is shown in Fig. 2.74. The main elements present in all MOCVD growth systems are illustrated in this figure. An open tube reactor containing a conducting graphite susceptor is heated by RF induction (in this case). Gas flows containing TMGa and TMAl are fed to the growth tube by bubbling $H_2$ through bubblers containing liquid TMGa and TMAl. These are mixed with $AsH_3$ and pyrolized above the substrate. (Note that since $AsH_3$ is highly toxic, possibilities involving organometalic compounds of As have been recently explored [302]). The net reaction for the growth of $Al_xGa_{1-x}As$ is given by

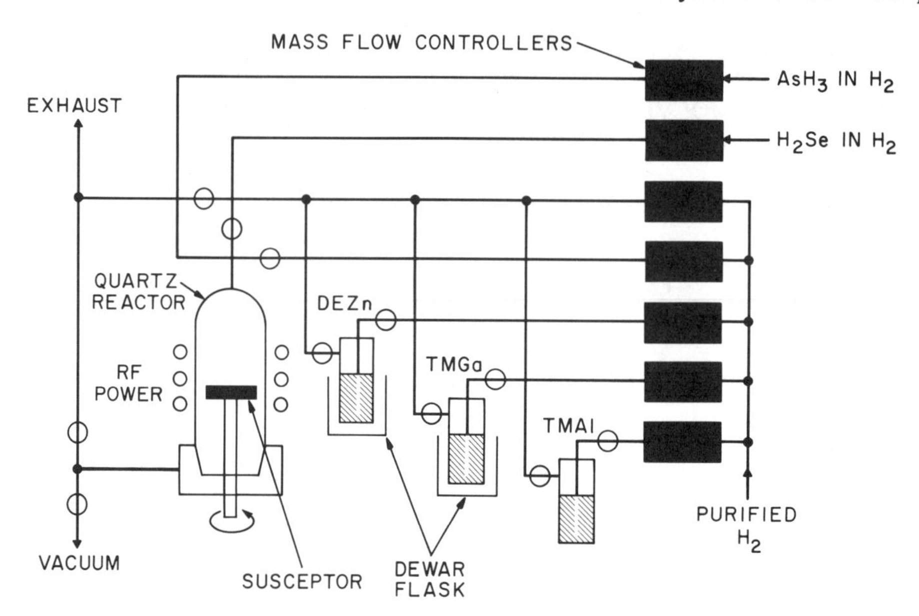


**Figure 2.74** Schematic atmospheric-pressure MOCVD arrangement for the growth of $Al_xGa_{1-x}As$ [301].

$$(1-x)[(CH_3)_3Ga] + x[(CH_3)_3Al] + AsH_3 \xrightarrow[\sim 700^\circ C]{H_2} Al_xGa_{1-x}As + 3CH_4 \qquad (2.73)$$

The alloy composition is directly determined [302] by the relative initial partial pressures of the TMGa and TMAl. This simple relationship between the gas phase and grown layer stoichiometries occurs generally for ternary alloys containing a mixture of group III species (e.g., $Al_xGa_{1-x}As$ and $In_xGa_{1-x}As$) [303] but not for alloys containing a mixture of group V species. This is caused by either a variable pyrolysis rate as in the case of $PH_3$ pyrolysis for the growth of $InAs_{1-x}P_x$ or to thermodynamic constraints on the distribution coefficient as in the case of $InAs_{1-x}Sb_x$ [303]. Dopant flows of diethyl zinc (DEZn) and of the hydride $H_2Se$ are also shown in Fig. 2.74.

A number of advantages for the commercial use of MOCVD compared to other vapor phase growth techniques can be pointed out. First, since the reactions are irreversible, very abrupt transitions in composition of epitaxial structures are possible [254]. Note that this may also be viewed as a disadvantage since in-situ etching is not possible. This implies that surface preparation and cleaning are critical steps for high-quality crystal growth by MOCVD [265]. Autodoping as a result of a significant ambient vapor pressure of a species containing a substrate dopant can occur in some vapor phase growth techniques. No evidence of autodoping during MOCVD has been reported, and abrupt changes in doping from one layer to the next have been achieved. Another advantage is that crystal growth can be performed at relatively low temperatures which minimizes the effects of interdiffusion. The growth of Al containing III-V alloys by chloride VPE is difficult because of the high reactivity of Al metal and AlCl vapor with fused quartz. Although high performance AlAs/GaAs solar cells have been demonstrated using chloride transport VPE in an all-alumina reactor [304], the growth of $Al_xGa_{1-x}As$ by chloride transport is not possible. This is so because the growth of GaAs for $AsCl_3$ and GaCl occurs around 750°C whereas growth of AlAs takes place around 1100°C. However, with MOCVD the growth temperatures of these two binaries are sufficiently similar that the growth of $Al_xGa_{1-x}As$ is relatively straightforward. An advantage of MOCVD over conventional molecular beam epitaxy (MBE) is that phosphorus poses no special problems. The high-vapor pressure of both $P_4$ and $P_2$ above elemental P makes MBE difficult.

### 2.5.2 Reactor Design

Three basic reactor designs have been popular. These are the vertical design shown schematically in Fig. 2.74, the high-capacity barrel reactor design which is shown schematically in Fig. 2.75 [305], and a horizontal design shown schematically in Fig. 2.76 [265]. The geometry of the reaction chamber is quite important because the reaction chamber is the slowest to flush. The abruptness of interfaces can be severely degraded because of dead spaces inside the growth chamber which can trap and slowly release reactant gases. Reactors are made from fused silica because of its chemical inertness, optical transparency (which is important for IR heated susceptors), and low electrical conductivity (which is important for RF induction heated susceptors). The susceptor is usually made of graphite and is usually coated with SiC to make it chemically inert.

The vertical geometry was used in the pioneering work of Manasevit [300], and in this design the inlet gas flow is perpendicular to the substrate. Generally, substrate rotation is used to enhance uniformity. If a flow deflector is used to spread out the gas flow, uniform films over moderately large areas (~ 10-15 $cm^2$) have been reported [305]. Only a few flow analyses have been made, however, since the flow upstream of the susceptor is susceptible to convective changes [305].

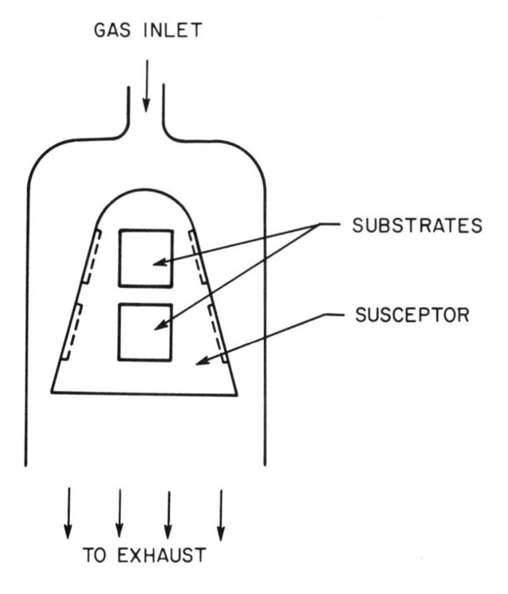


**Figure 2.75** Schematic barrel reactor design for MOCVD growth on several substrates simultaneously [305].

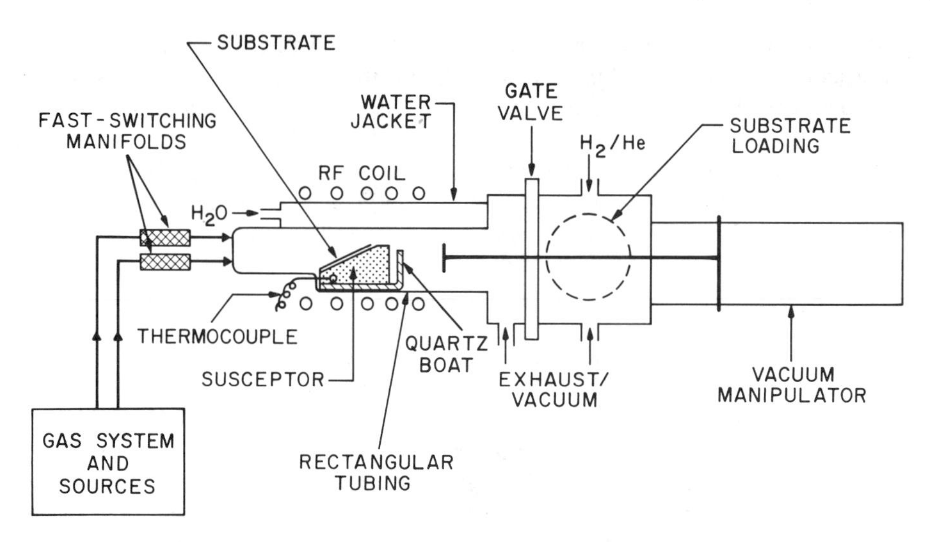


**Figure 2.76** Schematic atmospheric-pressure reactor for the growth of $Ga_xIn_{1-x}As_yP_{1-y}$ [265].

To achieve a more laminar gas flow a horizontal geometry is often preferred. Such a flow is more readily modeled theoretically and is roughly parallel to the substrate. Because there are reactant depletion effects, the susceptor in horizontal systems is often inclined at several degrees from the horizontal. The angle of inclination is usually in the range 15 to 30degrees. An inclined susceptor may not compensate completely for reactant depletion, however [306]. To improve the uniformity, substrate rotation has been successfully implemented. For the growth of GaAs on 2 inch diameter GaAs substrates a multiwafer planetary rotation has been employed to achieve a layer having a thickness of 675 Å with a variation of ±2.0 percent (14 Å) [307]. Substrate rotation of a single 2 inch diameter InP substrate has also been used to achieve excellent uniformity for GaInAsP in an atmospheric pressure MOCVD reactor [308]. The standard deviation of the wavelength of band gap photoluminescence ($\lambda = 1.3\ \mu m$) was found to be 3 nm. The accompanying lattice mismatch had a standard deviation in the range 0.0060 to 0.0090 percent.

A third type of reactor is shown schematically in Fig. 2.75, and is similar to the vertical design in that the gas inlet is positioned at the top. However, the gas flow is not perpendicular to the surface but is approximately parallel to the multiple susceptors as for the horizontal design. However, the flow patterns are expected to be more complicated [305].

### 2.5.3 Models for Crystal Growth

Our discussion will be limited to models applying to horizontal reactors. In that case there is evidence for the existence of a so called "stagnant" layer [309] or "boundary" layer [303] as illustrated schematically in Fig. 2.77. This figure applies to a low-pressure MOCVD (LPMOCVD) system and is due to Hirtz et al., [309] (LPMOCVD will be discussed). Crystal growth is believed to occur in the following way [309]:

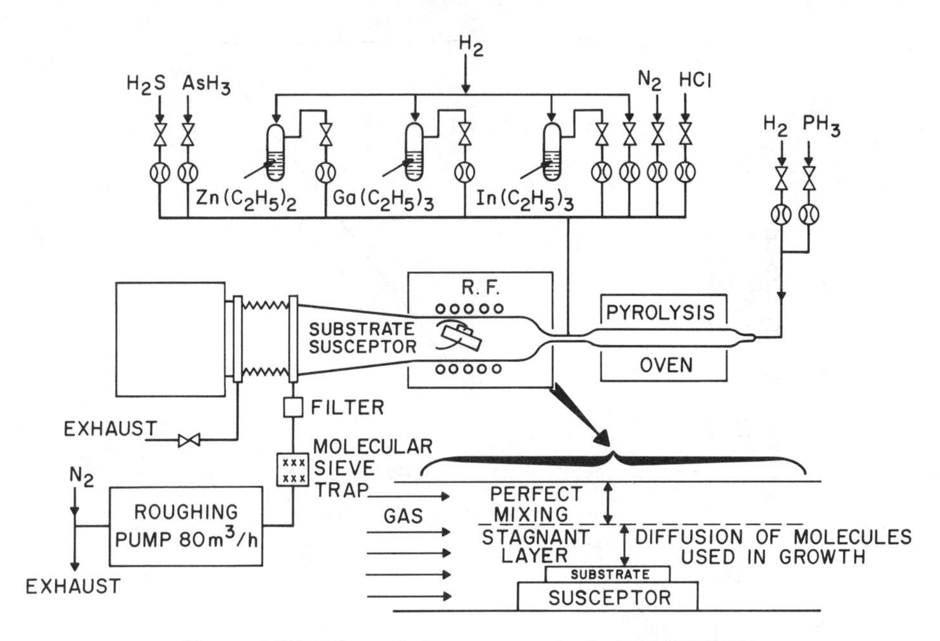


**Figure 2.77** Schematic low-pressure horizontal MOCVD reactor [309].

(a) gas molecules diffuse through the stagnant layer to the substrate,

(b) at the hot surface the metal alkyls and hydrides decompose,

(c) the decomposition products move over the surface until they find available lattice sites where they are incorporated in the lattice.

The identity of the decomposition products has not yet been established, nor is the identity of the species that diffuses over the surface. Some workers believe the decomposition products to be group III and group V elemental species [309] whereas others believe that (for the growth of GaAs) it is most probably the case that "$GaCH_3$ and AsH are the primary reactant species adsorbed on the growing surface" [306]. In any case, only if step (a), namely diffusion across the boundary layer, is rate limiting, then the growth rate is independent of temperature. If one of the subsequent two steps, (b) or (c), is rate limiting, then the growth rate is strongly temperature dependent because the reaction rates depend exponentially on temperature [309].

Evidence that diffusion across the boundary layer is rate limiting for the growth of GaAs can be found from the work of several groups. As shown in Fig. 2.78 (from Stringfellow [303]) a temperature independent growth rate has been reported over the temperature range ~550°C to ~770°C by Manasevit and Simpson [310], Gottschalch et al., [311], Leys and Veenvliet [312], and by Krautle et al., [313]. Ghandi and Bhat, however, report a very weak temperature dependence for the growth rate in this temperature range (1.5-2.2 kcal/mole) [254]. GaAs is usually grown using large V/III ratios in which case the growth rate is independent of the flow

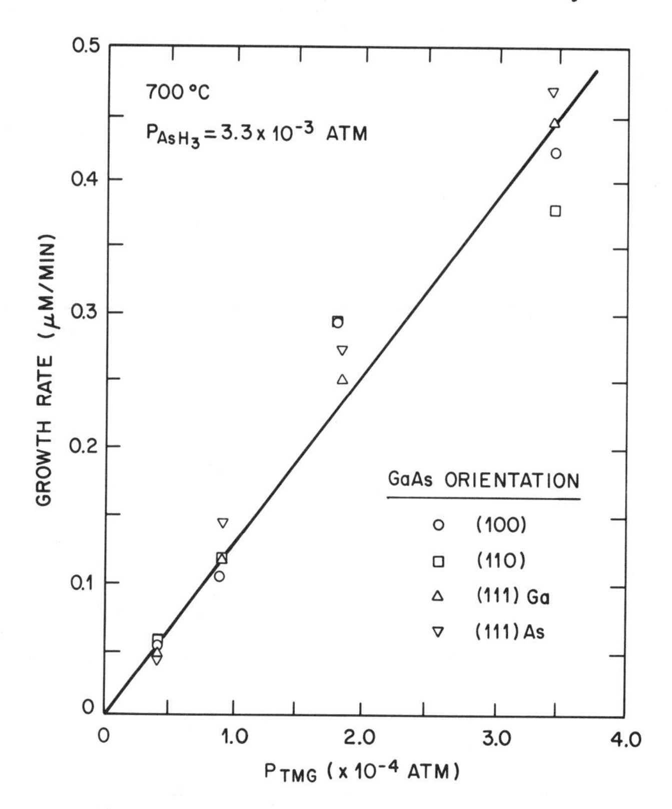


**Figure 2.78** Growth rate as a function of the partial pressure of trimethylgallium. The growth is found to be independent of the crystallographic orientation of the substrate [303].

rate of $AsH_3$ and depends linearly on the flow rate of TMGa as shown in Fig. 2.78. This is expected if the diffusion of TMGa across the boundary layer is rate limiting. Furthermore, if surface reactions are not rate limiting, then the surface orientation should not affect the growth rate. This is also corroborated by the data shown in Fig. 2.78.

### 2.5.4 Indium Depletion, Parasitic Reactions, Low-Pressure MOCVD, and Crystal Growth of InP, GaInAs, and GaInAsP

MOCVD crystal growth involving TEIn or TMIn is susceptible to adduct formation upstream of the susceptor. For the growth of GaAs or AlGaAs parasitic gas phase reactions are not a problem, but for the growth of InP, GaInAs, and GaInAsP such reactions can cause In depletion and must be carefully avoided. The alkyls are acidic and the hydrides are basic, and a mixture of the two will form a Lewis acid-Lewis base compound. This is not a problem if the bond strength is small (5-6 kcal/mole [254]) and the compound is volatile since it will pyrolyze above the hot susceptor. If, however, an involatile adduct is formed (e.g., TMIn-$AsH_3$ or TEIn-$AsH_3$), then the gas flow above the susceptor is depleted, and a black polymer is usually observed to deposit on the walls of the reactor.

Solutions to these problems have been found. Gas phase reactions can be reduced or eliminated either by keeping reactants separated until just ahead of the growth region (See Ref. [305] and references therein) or by reducing the chemical driving force for compound formation. The first approach suffers compared to the second when uniformity is considered because the reactants cannot be as well mixed. The second approach has led to LPMOCVD. In a low pressure system adduct formation is inhibited for several reasons. First, since reactant pressures are reduced, the reaction rate is slowed [254]; second, the reaction time is reduced since [254] higher gas velocities are usually employed; and third, the elimination of recirculation caused by convection [314] also reduces reaction time.

For LPMOCVD the growth rate is generally not temperature dependent which implies that diffusion through the stagnant layer is rate limiting [309]. Also, since gas speeds are usually higher for LPMOCVD, new gas compositions are rapidly established permitting more abrupt changes in composition. Furthermore, higher gas speeds are also thought to produce a more uniform stagnant layer thickness over a large area leading to a good uniformity of thickness and composition of epitaxial layers [309].

LPMOCVD is usually done at a V/III ratio considerably greater than one, and in that case the growth rate of InP is found to be proportional to the TEIn flow rate, independent of the phosphorus partial pressure, and approximately independent of temperature in the range 500°C to 650°C [309]. These results are analogous to those discussed for the growth of GaAs. For the LPMOVD growth of GaInAs lattice matched to InP using TEGa and TEIn, the growth rate was found to vary as the sum of the partial pressures of TEGa and TEIn and did not depend on the arsenic partial pressure [309]. Similarly, for the LPMOCVD growth of GaInAsP lattice matched to InP ($\lambda = 1.3\,\mu m$), the growth rate was found to vary as the sum of the partial pressures of TEGa and TEIn and did not depend on either the arsenic or the phosphorus partial pressures [309]. The $PH_3/AsH_3$ ratio must exceed 20 because of the inefficient incorporation of P [309]. Finally, we note that a critical value for the sum of the arsenic and phosphorus partial pressures has been observed [309]. For growth at higher partial pressures, the composition of the growth becomes strongly temperature dependent [309].

### 2.5.5 Crystal Growth of Semi-Insulating Fe-Doped InP

Epitaxial crystal growth of semi-insulating (SI) InP doped with Fe is readily accomplished by MOCVD [315]. Resistivities as high as $2 \times 10^8$ $\Omega$-cm for epilayers grown on heavily S doped InP substrates are readily achieved [316] using ferrocene, $Fe(CH_5H_5)_2$, a solid which is stable to 450°C and has a vapor pressure of $4.7 \times 10^{-4}$ Torr at 0°C [317]. Ferrocene is not the only compound suitable for Fe doping during MOCVD; less volatile Fe-carbonyl derivatives have also been successfully used [318] to dope InP. Epitaxial InP:Fe is quite important for current blocking in semiconductor lasers [319, 320] and may be used for device isolation for field-effect transistors [321].

Note that SI epitaxial layers of InP have not been achieved by LPE or by conventional hydride VPE which uses $H_2$ as a carrier gas. These crystal growth techniques suffer because of the low-vapor pressure of metallic Fe and the high-HCl concentration needed to transport $FeCl_2$. High resistivity InP with resistivities of $10^8 \Omega$-cm has very recently been achieved by hydride VPE using $N_2$ as a carrier gas [322]. Resistivities as high as $10^8 \Omega$-cm have also recently been achieved for InP:Fe grown with trichloride VPE [323].

Semi-insulating epitaxial InP was first achieved in the atmospheric pressure reactor shown in Fig. 2.76 [315]. Growth of device quality InP was achieved using the Lewis-acid Lewis-base adduct TMIn-TMP [315]. Note that by using this adduct as a source material crystal growers have at their disposal another means of avoiding In depletion and the formation of nonvolatile polymers which was discussed.

### 2.5.6 MOCVD Summary

In summary, we quote from a recent review by W. D. Johnston, Jr. et al., [265]:

> The surface quality and interfacial abruptness of MOCVD grown structures can be excellent, and although this process is less well developed than the older hydride and trichloride processes, it is viewed today as holding the most promise for the future. In the low pressure variation it offers the versatility of the hydride system with purity levels approaching those obtained with trichloride CVD. Although the reason is not fully understood, the impurities residual in arsine appear to be less efficiently incorporated in the growing layer during low pressure MOCVD growth than in atmospheric pressure MOCVD or hydride growth.

## 2.6 MOLECULAR BEAM EPITAXY

Molecular beam epitaxy (MBE) is basically a technique of vacuum evaporation, one of the oldest and easiest techniques for depositing solid films. Although vacuum evaporation was used as early as in the 1950s for preparing semiconductors, epitaxial growth conditions were not realized until improvements occurred in ultrahigh vacuum technology, in the design and the control of the sources and substrate cleaning procedures. MBE has now become a versatile technique for growing epitaxial thin films of semiconductors and metals. The advances that have occurred in MBE over the last several years have made possible the design of novel electronic and photonic devices by "band-gap engineering" using artificially layered materials grown by this technique [324]. In this section the principles and application of MBE to the growth of III-V compounds particularly GaAs and InP and their alloys are reviewed.

MBE is a process of depositing epitaxial films from molecular or atomic beams on a heated substrate under ultrahigh vacuum (UHV) conditions. The beams are thermally generated in Knudsen-type effusion cells which contain the constituent elements or compounds of the desired epitaxial film. The temperature of the cells is accurately controlled to give the thermal beams of appropriate intensity. The thermal beams escaping from orifices in the cells travel in rectilinear paths to the substrate where they condense and grow under kinetically controlled growth conditions. As early as in 1958, MBE of III-V compounds was described by Gunther [325] but single crystalline GaAs films were first reported only in 1968 [326, 327]. Arthur's [327 - 329] seminal work on the kinetic behavior of the Ga and As species incident on a GaAs surface was the impetus for the rapid development and innovation in the field of MBE. Since that work, the

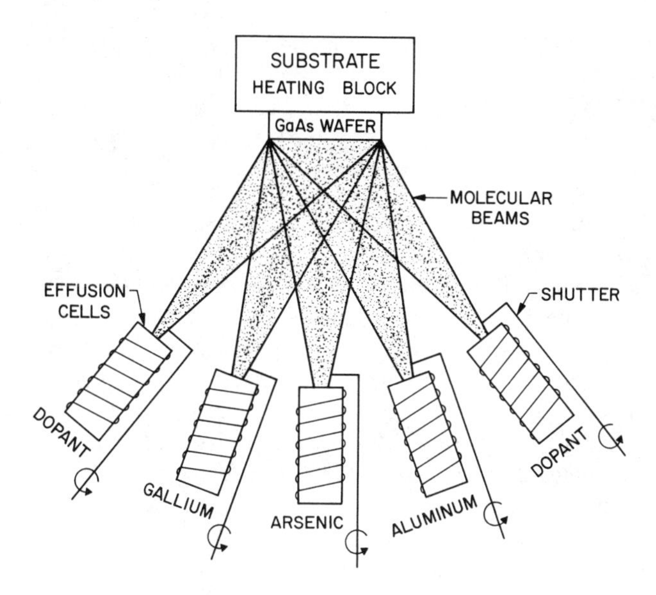


**Figure 2.79** Schematic of the basic evaporation process for MBE of intentionally doped GaAs and $Al_xGa_{1-x}As$ [335].

evolution of MBE has occurred with progress in the growth of device quality GaAs and AlGaAs films [330 - 332], of semiconductor superlattices [333 - 335], and in the understanding of growth kinetics [336, 337].

### 2.6.1 Effusion Cells

The basic MBE process is schematically illustrated in Fig. 2.79 for growing intentionally doped GaAs or $Al_xGa_{1-x}As$ film on a heated GaAs substrate which is usually of (100) orientation. The molecular beams are generated in Knudsen type effusion cells whose temperatures are controlled to an accuracy of $\pm 1°C$ in a UHV system. Externally controlled mechanical shutters in front of each effusion cell facilitate rapid changing of the beam species so that the composition and/or doping of the growing film can be abruptly altered. For an ideal Knudsen cell, assuming that the cell aperture is less than the mean free path of vapor molecules within the cell, the flux density J (molecules $cm^{-2}$ $sec^{-1}$) is given by

$$J = \frac{A}{\pi L^2} \frac{p}{(2\pi mkT)^{1/2}} \cos\theta \tag{2.74}$$

where p is the equilibrium vapor pressure within the cell, T is the temperature, m is the mass of the effusing species, k is the Boltzman constant, A is the area of the orifice, L is the distance between the cell and the substrate, and $\theta$ the angle between the beam and the normal to the substrate. As an illustrative example of Eq. (2.74), let us consider the growth of GaAs from the compound itself. If the effusion cell filled with GaAs is heated to a temperature of 900°C, the volatile As escapes leaving behind a Ga-rich surface. Therefore, the pressures of Ga and $As_2$ ($p_{As_2} > p_{As_4}$) are the ones in equilibrium with the Ga rich side of the liquidus (in Fig. 2.3). At 900°C, $p_{Ga}$ is $5.5 \times 10^{-7}$ atm and $p_{As_2}$ is $1 \cdot 1 \times 10^{-5}$ atm. For a typical geometry of the effusion cell with $A \approx 0.1$ cm$^2$ and $L = 5$ cm, Eq. (2.74) gives $J_{Ga} \approx 6.6 \times 10^{13}$ cm$^{-2}$ sec$^{-1}$ and $J_{As_2} \approx 1.3 \times 10^{15}$ cm$^{-2}$ sec$^{-1}$.

Equation (2.74) serves only as a guideline since in practice the cell aperture is enlarged to enhance the growth rate, resulting in Langmuir-type effusion cells [334]. The beam fluxes emerging from these nonequilibrium effusion cells are generally determined experimentally in most cases using movable nude ionization gauge placed in the substrate position. The cells should be made from nonreactive, refractory materials which can withstand high temperatures and they themselves should not contribute to the molecular beams. Pyrolytic Boron Nitride (PBN) or high-purity graphite are used as the cell materials. The cells consist of an inner crucible and an outside tube which is wound with Ta or Mo wires for resistive heating. A chemically stable W-Re thermocouple facilitates precise control of the cell temperature which is very essential for achieving constant growth rates since small temperature fluctuations of the order $\pm 1$°C can result in $\pm 2$ to 4 percent fluctuations in molecular beam intensity (Eq. (2.74)) which in turn can cause fluctuations in growth rate of the same order. The various cells are all placed and angled in such a way so that their beams converge on the substrate for epitaxial growth. Individual shutters provided for each cell and the cell temperature can be computer controlled to achieve high reproducibility and little human intervention. The cells are individually surrounded by a liquid nitrogen shroud to prevent cross-heating and cross-contamination. For group V elements, a high-temperature cracker with internal buffer is incorporated at the exit end of the effusion cell. This cracker dissociates the tetramers to dimers.

### 2.6.2 Ultra-High-Vacuum Crystal Growth Conditions

Figure 2.80 shows a schematic cross section of a typical MBE system with the effusion cells and the cryopanels [338]. The system is prepumped by standard mechanical and sorption pumps. A combination of ion pump and titanium sublimation pump with cryopanel is used to achieve UHV pressures in the chamber. When large amounts of highly volatile elements such as phosphorous (e.g., in the deposition of GaInAsP) are used in the system the use of diffusion, turbomolecular or closed loop helium cryopumps is desirable. Once the entire system is baked, a base pressure of $\sim 10^{-10}$ torr is easily attained. Additional liquid nitrogen cryopanels around the substrate helps to condense OH containing species such that their partial pressures are kept below $10^{-14}$ torr in the vicinity of the growing crystal. During deposition, the chamber pressure rises above $10^{-9}$ torr because of scattered beam species but UHV condition is always maintained with respect to impurity species.

The substrate holder is usually made out of Mo and the substrate is mounted onto it using In [338]. This procedure ensures good temperature uniformity and is advantageous for mounting substrates of irregular shapes. With the availability of large diameter GaAs substrates In free mounting techniques have also been developed [339]. Resistance heating from behind the Mo block is employed to heat the substrate to the growth temperature. The substrate holder can accommodate 7.5 cm diameter wafers and it can be rotated during deposition at speeds of 0.1 to

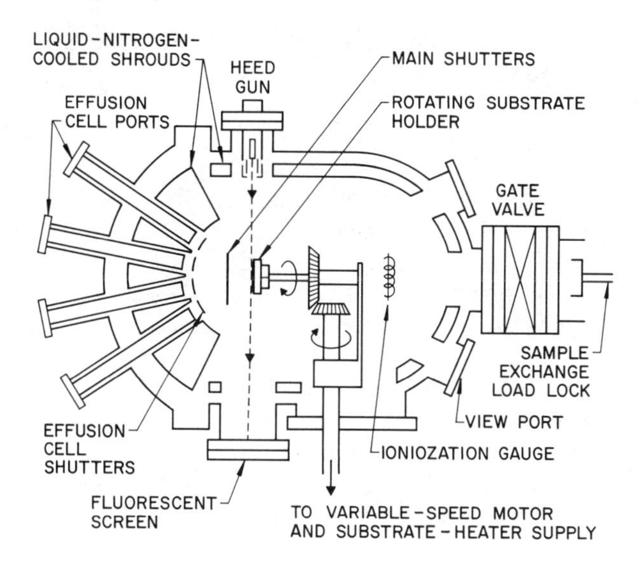


**Figure 2.80** Schematic of a typical MBE system viewed from the top. The rotating sample holder has a variable speed from 0.1 to 5 rpm [338].

120 rpm to achieve uniform epitaxial layers. A sample exchange load-lock system can be used to decrease the time interval between successive growth runs and at the same time to maintain reproducible growth conditions and minimize contamination of the effusion cells by reducing frequent opening of the growth chamber to atmospheric pressure. In the growth of III-V compounds, since the volatile group V sources deplete very rapidly, charge interlocks have been used to recharge group V effusion cells without venting the chamber to atmosphere.

### 2.6.3 Surface Analysis

An unique feature of MBE growth is that the UHV growth conditions allows incorporation of surface analytical tools in the growth chamber for insitu analysis of the growing surface to monitor and control the growth conditions. The typical surface analytical instruments used are Auger electron spectroscopy (AES) and reflection high-energy electron diffraction (RHEED). A quadrupole mass analyser can also be used to detect the residual gas composition in the chamber as well as to detect the molecular beams coming from the effusion cells. The mass analyser should be placed in line of sight from the effusion cells and in close proximity to the substrate. A ion-sputtering gun may also be used to clean the substrate prior to deposition.

The AES technique is mainly used to characterize the initial surface composition of the substrate and to ensure its cleanliness. The RHEED patterns obtained at glancing incidence provides information on surface reconstruction, microstructures and surface smoothness. The use of RHEED has advanced significantly the understanding of the growth mechanisms in MBE. Although difficult in practice to operate during growth, low-energy electron diffraction (LEED) has also been used in MBE kinetic studies [340, 341]. In addition to the instruments mounted directly on the growth system, a separate UHV analysis chamber can also be connected to the growth chamber. The as-grown samples can be transported to the analysis chamber via a gate valve without exposing the sample to atmosphere [335].

### 2.6.4 Substrate Preparation

The preparation of the substrate is an important step for achieving high-quality epitaxial layers. In the MBE growth of III-V semiconductors using GaAs or InP substrates, generally (100) oriented ($\pm 0.1°$) $\sim 250\,\mu m$ thick substrates are used. The substrates are chemo-mechanically polished to obtain mirror-like surface finish. They are thoroughly degreased using organic solvents such as chloroform, trichloroethylene, acetone and methanol. The degreasing may be accomplished in several steps. After the last degreasing step they are rinsed in deionized water. In the case of GaAs substrate, it is then etched in a stagnant solution of $H_2SO_4{:}H_2O_2{:}H_2O$. The composition and temperature of the solution and time of etching vary from laboratory to laboratory. One such variation is etching in a $4H_2SO_4{:}1H_2O_2{:}1H_2O$ solution at 60°C for 10 to 20 minutes [332]. After etching, the wafer is flooded with deionized water to stop the etchant, rinsed in water, and blown dry with filtered nitrogen gas just before soldering onto a preheated Mo block with In under dust-free conditions. It is then immediately loaded into the MBE system.

In the case of InP substrate, it is etched in agitated dilute ($\leq 0.3\%$) Br-methanol solution for 10 to 15 minutes. It is rinsed in methanol and deionized water without exposing the substrate surface and blown dry with filtered nitrogen. It is then soldered onto the Mo block and loaded into the MBE system.

The substrates prepared in this manner when checked by AES indicates the presence of oxygen and a slight amount of carbon on the surface. Heating of the substrate at about 500°C for a few minutes removes the oxygen. This is illustrated in Fig. 2.81 for a GaAs substrate heated to 540°C which is below its congruent evaporation temperature [342]. The Auger spectrum in Fig. 2.81b taken after heating at 540°C shows hardly any trace of oxygen.

Since the congruent evaporation temperature for InP is $\sim 360°C$ [287], substrate cleaning at elevated temperatures requires additional P or As beam directed toward the substrate to prevent decomposition. Figure 2.82 shows the Auger spectra for an InP substrate before and after heating to 500°C under an As molecular beam [343]. The indium oxide formed during the oxide passivation process gets transformed into arsenic oxide which is easily desorbed at $\sim 500°C$.

The removal of carbon from the surface is much more difficult than oxygen. It requires the use of ion-bombardment. However, an annealing treatment is required to remove the surface damage. Excessive C contamination causes facetted and twinned growth. It is found that if the surface is passivated with a protective oxide during etching, it does not readily adsorb carbon containing gases.

### 2.6.5 Crystal Growth Process

MBE growth is a kinetically controlled process which involves (a) adsorption of the constituent atoms or molecules, (b) surface migration and dissociation of the adsorbed molecules, and (c) incorporation of the atoms to the substrate resulting in nucleation and growth. Most of the basic studies of MBE growth has been made on GaAs but the results are applicable to other III-V semiconductors as well.

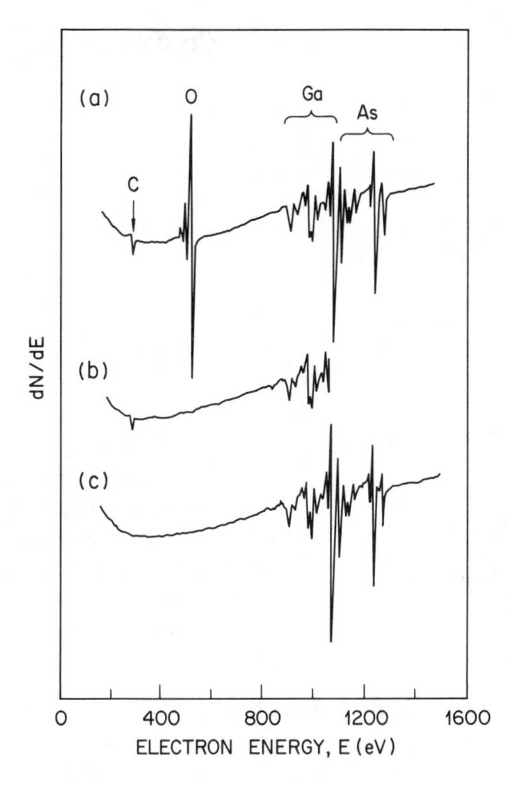


**Figure 2.81** Auger spectra taken with primary electron energy of 4 keV of the surface of (100) GaAs substrate: (a) after chemical etching (b) after heating at 530°C for 1 h and (c) after $Ar^+$ sputtering at $5 \times 10^{-5}$ Torr for 5 min. [334].

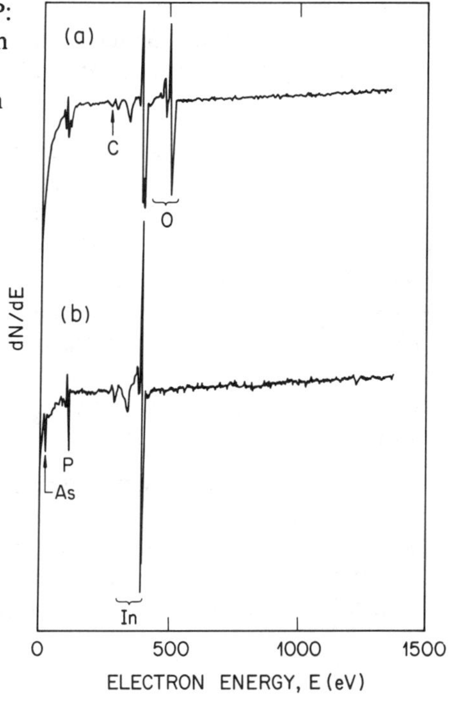


**Figure 2.82** Auger spectra of (100) InP: (a) etched in $Br_2$-methanol and rinsed in deionized water, (b) after heating to 500°C under an arsenic molecular beam [343].

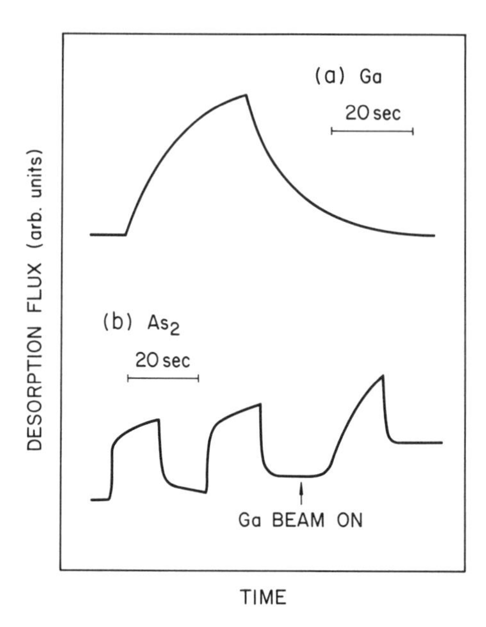


**Figure 2.83** Desorbed fluxes from pulsed beams incident on (111) GaAs: (a) Ga-beam on GaAs at 885K indicating a surface lifetime of 105 (b) $As_2$-beam on GaAs at room temperature, demonstrating its total reflection unless a prior Ga-coverage exists [328,329].

Using modulated incident beams, Arthur investigated the interactions of Ga and $As_2$ beams with (111) GaAs [328]. He observed that Ga has a sticking coefficient of unity below 750K. At higher temperatures, it is desorbed following a first-order reaction with a surface life time of ~ 10s at 885K as shown in Fig. 2.83. On the other hand, As has a negligible sticking coefficient unless Ga atoms are present on the surface. This is illustrated in Fig. 2.83b. Once a prior coverage of Ga is produced either from a Ga flux or from the GaAs substrate itself as a result of preferential loss of volatile As at temperatures above 775K, the incident As atoms react with Ga atoms to form GaAs.

Arthur's observations suggested that the growth of GaAs is kinetically controlled by the adsorption of $As_2$ while the growth rate is determined by the arrival rate of the Ga flux. For example, if GaAs is deposited from the compound itself at 900°C and if the area of effusion cell orifice is 0.1 $cm^2$ and the distance between it and the substrate is 5 cm then (Eq. (2.74)) $J_{Ga} = 6.6 \times 10^{13}$ $cm^{-2}$ $sec^{-1}$ and $J_{As_2} = 1.3 \times 10^{15}$ $cm^{-2}$ $sec^{-1}$. Under these conditions the growth rate would be $J_{Ga} / N \sim 0.3$ Å $sec^{-1}$ where N is the number of GaAs molecules per unit volume $2.2 \times 10^{22}$ $cm^{-3}$. For a 10-fold increase in the orifice area the growth is 3 Å $sec^{-1}$ (1 μm $hr^{-1}$) or about a monolayer per second. As a result of the slow growth rate, operation of the shutters in front of the effusion cells at fractions of a second would result in abrupt interfaces between compositionally different materials.

Further insights about the surface reactions were obtained from RHEED measurements. It was observed that the surface structure show two distinct patterns depending on the incident Ga and $As_2$ fluxes and the substrate temperature. These two characteristic surfaces are termed Ga-stabilized or As-stabilized and these are correlated with the loss or gain of arsenic to the extent of about one-half monolayer from the surface. Figure 2.84 shows the As-stabilizes and Ga-

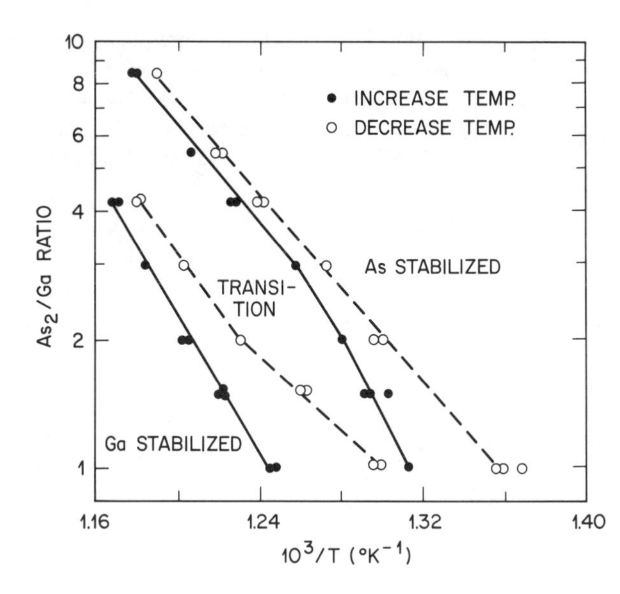


**Figure 2.84** Regions of As- and Ga-stablized surface structures on (100) GaAs in terms of flux ratio and substrate temperature. Mysteresis is observed depending on the increase or decrease of temperature [344].

stabilized regions for different $As_2$/Ga flux ratios and substrate temperatures [344]. Hysteresis is observed depending on the increase or decrease of temperature.

Using similar modulated beam techniques and more refined transform analysis, Foxon and Joyce [345, 346] and Joyce and Foxon [347] have proposed the presently accepted detailed growth model for GaAs from Ga and $As_2$ molecular beams. Figure 2.85 illustrates the model proposed by these authors. According to this model, $As_2$ molecules, generated from heated GaAs source in an effusion cell or by high-temperature dissociation of $As_4$ evaporated from elemental As source, are first adsorbed into a mobile weakly bound precursor state. Dissociative chemisorption of $As_2$ occurs when the migrating $As_2$ molecules encounter single Ga sites. In the absence of Ga adatoms, though $As_2$ has a measurable surface lifetime, no permanent condensation occurs. The sticking coefficient of $As_2$, $S_{As_2}$, is thus proportional to the Ga flux $J_{Ga}$ (a first order process) as shown in Fig. 2.86 [345, 346] for a substrate temperature of 600K. $S_{As_2}$ approaches unity asymptotically and stoichiometric GaAs is formed for $J_{As_2}/J_{Ga} \gtrsim 1$. At substrate temperatures below 600K, a pairwise association of adsorbed $As_2$ molecules and the subsequent desorption of $As_4$ molecules becomes dominant.

When $As_4$ molecular beams, generated from an elemental As source, are involved, they behave similarly to the $As_2$ molecules in that they are adsorbed to form a mobile precursor state. Pairs of $As_4$ molecules react on adjacent Ga sites ($2^{nd}$ order reaction). In the temperature range 450 to 600K, the sticking coefficient of $As_4$, $S_{As_4}$ is proportional to $J_{Ga}$ at low values but most importantly saturates at 0.5 as $J_{Ga}$ is increased as shown in Fig. 2.86. For low As/Ga flux ratios ($S \approx 0.5$), when the $As_4$ surface concentration is small compared to the number of Ga sites, the growth rate limiting step is the encounter and reaction probability between $As_4$ molecules. When $J_{As} > J_{Ga}$ ($S_{As_4} < 0.5$), the probability of $As_4$ molecules finding other $As_4$ molecules on adjacent sites increases. Growth occurs by adsorption and desorption of $As_4$ via a bimolecular interaction

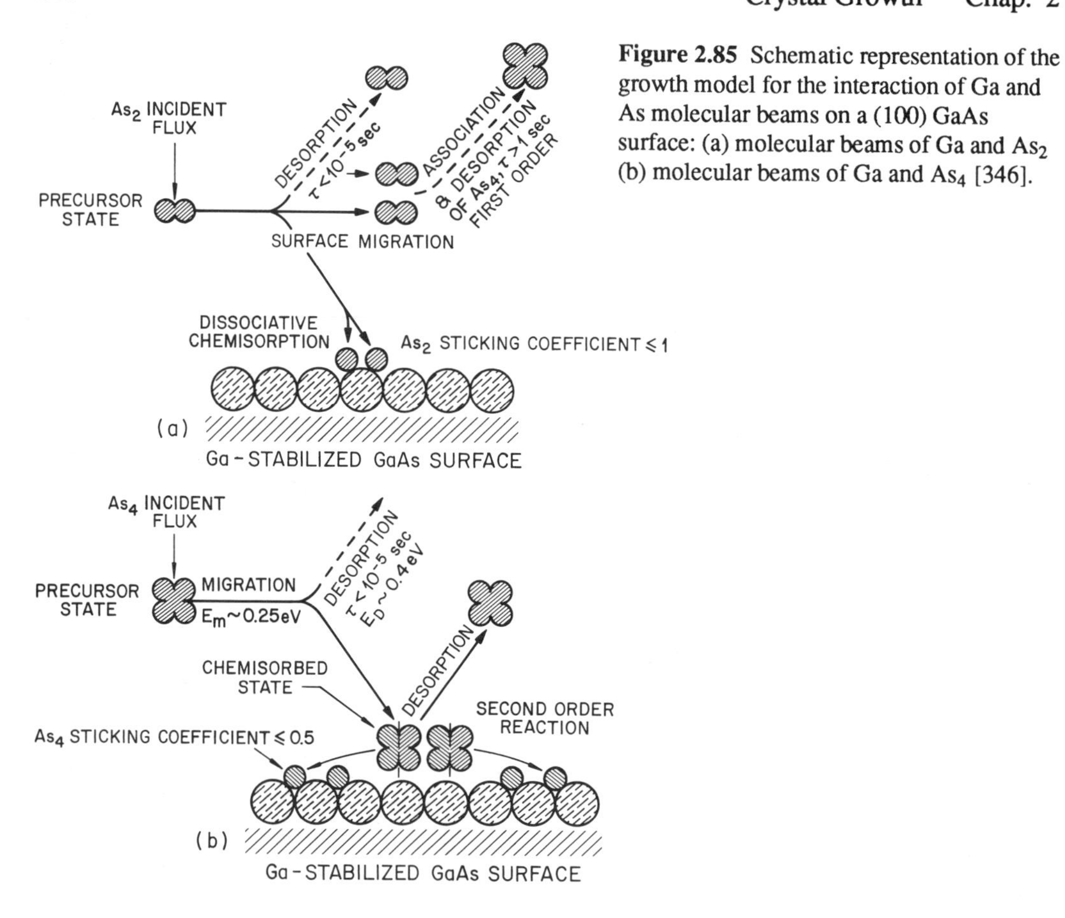


**Figure 2.85** Schematic representation of the growth model for the interaction of Ga and As molecular beams on a (100) GaAs surface: (a) molecular beams of Ga and $As_2$ (b) molecular beams of Ga and $As_4$ [346].

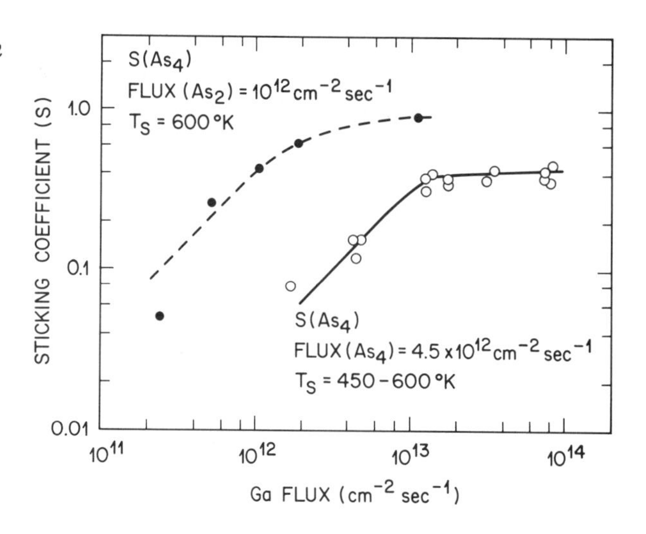


**Figure 2.86** Sticking coefficients of $As_2$ and $As_4$ on (100) GaAs versus Ga-beam fluxes [345,346].

resulting in one As atom sticking for each Ga atom. The different growth process for $As_2$ and $As_4$ may be expected to influence the native defect concentrations in the films grown under otherwise identical conditions.

As the substrate temperature is increased, $As_2$ is lost by desorption (irrespective of the form of the incident flux) from the growing GaAs surface. This results in an increase in Ga adatom population leading to Ga-stabilized surface structure. An As-stabilized structure is achieved only when the supplied As flux exceeds the rate of loss of $As_2$ by thermal desorption. Thus the flux ratios of $J_{As_4}/J_{Ga}$ or $J_{As_2}/J_{Ga}$ required to generate a specific surface structure depend on the substrate temperature.

From the point of view of practical film growth by MBE, these kinetic studies show that the growth rate is governed by the Ga flux as long as a sufficient As beam either as $As_2$ or as $As_4$ is supplied to form the stoichiometric GaAs. The growth rate is given by SJ/N, where S is the sticking coefficient of Ga and N is the number of GaAs molecules per unit volume. S is close to unity for typical MBE growth temperatures under As stabilized growth conditions. At high temperatures, Ga desorption occurs and S becomes less 1. This affects the growth rate as illustrated in Fig. 2.87 [348]. At temperatures above 640°C, the growth rate decreases for the same beam flux because of decreased $S_{Ga}$.

Detailed kinetic studies such as those made in GaAs have not been made for other III-V compound semiconductors. Nevertheless, the basic processes can be expected to be similar for other compounds as well as ternary III-III-V alloys such as $Al_xGa_{1-x}As$, $Ga_xIn_{1-x}As$. Since the

**Figure 2.87** Growth rates normalized to low temperature (≤620°C) values as a function of substrate temperature: (O) AlAs; (Δ) GaAs with Al (represents the Ga fraction of the growth rate); (□) GaAs without Al [348].

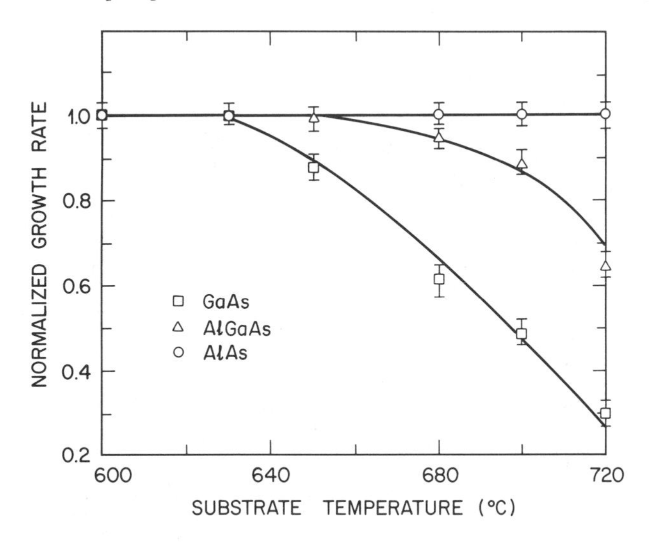

sticking coefficients of the group III elements are unity, the alloy composition in the film is simply determined by the relative fluxes reaching the surface. For example, the Al fraction in $Al_xGa_{1-x}As$ can be determined by the relation [332]

$$x = \frac{R(Al_xGa_{1-x}As) - R(GaAs)}{R(Al_xGa_{1-x}As)} \tag{2.75}$$

where R is the growth rate.

Equation (2.75) is not, however, valid as the substrate temperature is increased because of the different thermal stability of the binary end members of the alloy. One effect is caused by the different rate of loss of group V dimers. For example, in the case of growth of $Ga_xIn_{1-x}As$, above 300°C the rate of loss of As from InAs is much higher than from GaAs [349, 350]. For growth temperatures less than the congruent evaporation temperature of the less stable of the binary compound, good compositional control can be achieved by supplying excess group V species and adjusting the flux densities of the group III beams. At higher growth temperatures, however, preferential evaporation of the more volatile group III element occurs (e.g., Ga from $Al_xGa_{1-x}As$, In from $Ga_xIn_{1-x}As$). Under these circumstances the film composition is not only determined by the added flux ratios but also by the differences in the desorption rates. To a first approximation, the loss of the group III element can be estimated from the vapor pressure data since the vapor pressure of the element over the compounds is similar to that over the element itself. The loss rate of the group III elements at typical growth rates of 1 monolayer per second estimated from the vapor pressure data is shown in Table 2.12 [350]. In the growth of $Ga_xIn_{1-x}As$ above 550°C significant loss of In occurs and the surface is enriched in Ga. In the case of $Al_xGa_{1-x}As$ grown above 650°C loss of Ga occurs.

The situation is more complicated for the growth III-V-V alloys such as $GaAs_yP_{1-y}$, and $InAs_yP_{1-y}$ by MBE. The lower vapor pressure group V element is incorporated preferentially at

**Table 2.12**

Approximate rate of loss of group III elements at monolayer per sec growth rate [355]

| Temperature °C | In | Ga | Al |
|---|---|---|---|
| 550 | 0.03 | – | – |
| 600 | 0.3 | – | – |
| 650 | 1.4 | 0.06 | – |
| 700 | 8 | 0.4 | – |
| 750 | 30 | 2 | 0.05 |

low temperatures. The sticking coefficient decreases in the order Sb, As, P and tetramers have smaller sticking coefficient than dimers. This provides a simple method of controlling the composition of the alloy. For example, for the growth of $GaAs_yP_{1-y}$ from Ga plus $P_4$ and $As_4$ (or $P_2$ and $As_2$), the As flux should be small compared to the Ga flux in order to provide a higher proportion of Ga sites for reaction with $P_4$. At higher growth temperatures, the loss of the group V element makes compositional control difficult. The group V element which has a higher vapor pressure over the alloy will be desorbed preferentially leading to a film deficient in that element [350]. For growing quaternary alloys, for example, GaInAsP, the situation is complex, since for InAsP the vapor pressure of P is higher than that of As at any temperature while for GaAsP the opposite situation holds. Depending on the In to Ga ratio increasing the substrate temperature could increase or decrease the As content of the layers. These considerations show that in the growth of alloys, the control of the substrate temperature and the judicious choice of the flux ratios are of critical importance for obtaining specific composition.

### 2.6.6 Reflection High-Energy Electron Diffraction

Reflection high energy electron diffraction (RHEED) is perhaps the most useful surface analytical equipment in the MBE growth chamber for *in situ* studies of the surface crystallography and kinetics. RHEED is now routinely used in MBE systems to monitor and control the growth.

In RHEED, a collimated monoenergetic electron beam is directed towards the surface at a grazing angle of about 1 degree and orthogonal to the molecular beam paths as shown in Fig. 2.80. The primary electron energy lies in the range 5-40 keV. Since the energy component perpendicular to the substrate is of the order of ~ 100 eV, the penetration depth of the incident electron beam is limited to only the first few atomic layers. As a result, a smooth crystal surface acts as a two-dimensional grating and diffracts the incident electron beam. A fluorescent screen placed diametrically opposite the electron gun records the diffraction pattern. In this configuration the growing surface can be continuously monitored.

Bragg's law relates the wavelength of the incident electron beam $\lambda$ and the diffraction angle $\theta$ according to

$$\lambda = 2\,d\sin\theta \qquad (2.76)$$

where d is the interplanar distance between the diffracting lanes. For a cubic crystal

$$d = \frac{a}{\sqrt{h^2 + k^2 + l^2}} \qquad (2.77)$$

where a is the lattice parameter and h, k, and l are the Miller indices corresponding to the (hkl) plane. The diffraction pattern recorded on the fluorescent screen consists of spots and the spacing between them (D/2) is related to d as

$$d = \frac{2\lambda L}{D} \qquad (2.78)$$

where L is the distance between the substrate and the fluorescent screen. Therefore from the diffraction pattern one can deduce the interplanar spacings.

RHEED is used to study thermal decomposition of oxides prior to growth, and to study specific surface reconstruction that can be correlated to the surface stoichiometry [330, 333]. The temporal RHEED intensity oscillations when growth is initiated are used to study growth dynamics and the formation of heterointerfaces [351 - 354].

The removal of surface oxides and determining the right surface condition for commencing the epitaxial growth can be conveniently studied using RHEED. With the removal of the surface oxide the RHEED changes from diffuse reflection, to amorphous ring patterns and finally spot patterns as the oxide is almost desorbed. This is illustrated in the RHEED pattern and the corresponding photomicrographs of Pt-C replicas of the same surface shown in Fig. 2.88a [355]. The spot pattern similar to the transmission diffraction pattern results caused by penetration of the electron beam through surface asperities.

The spot RHEED pattern changes to a streak pattern after deposition of 150 Å of GaAs (Fig. 2.88b). The RHEED pattern showing a series of streaks perpendicular to the surface is consistent with a two dimensional diffraction. This streak pattern is usually taken as an indication of surface smoothness on an atomic scale. With further deposition, the surface becomes flat and featureless and the corresponding RHEED pattern is uniformly streaked (Fig. 2.88c).

The observation via RHEED patterns of surface smoothing process suggests that the mechanism of growth is via a two dimensional step propagation. Initial growth will be largest at surface steps whose density is highest at irregularities with the effect of reducing the surface roughness.

In Figs. 2.88b and 2.88c, it can be seen that additional streaks appear at half-way positions between the elongated bulk spots as the surface is smoothed out. They represent diffraction from a rearrangement of surface atoms into an ordered array in order to lower the surface free energy. Since surface reconstruction lowers the symmetry of the crystal, extra diffraction lines are seen in the RHEED pattern. A large number of such reconstructed surface structures have been reported under different experimental conditions [330, 333]. The most relevant structures for (100) GaAs are the As-stabilized $(2\times 8)$ or $(2\times 4)$ and the Ga-stabilized $(8\times 2)$ or $(4\times 2)$ patterns. A schematic representation of the $(2\times 8)$ and $(2\times 4)$ reconstructed surface structures in reciprocal space and real shape is shown in Fig. 2.89 [332]. The RHEED patterns of the As-stabilized structure produced under excess As/Ga flux ratios for three azimuthal directions are shown in Fig. 2.90 [334]. The Ga stabilized surface shows the same patterns as in Fig. 2.90 but with the two [110] azimuths interchanged. That is, the two structures can be transformed from one to the other by a 90 degree rotation in real space. Note the 1/4 order streaks at $[1\bar{1}0]$ and 1/2 order streaks at [110] and [100] patterns in Fig. 2.90. They indicate reconstructed surface structures with 4 and 2 times the lattice spacings of the bulk material.

Depending on the incident As/Ga flux ratios and substrate temperature, revisible transitions between the two principal structures on (100) GaAs are possible. Several intermediate structures, for example, $(3\times 1)$, $(1\times 6)$, $(4\times 6)$, $(3\times 6)$ and combinations can be observed within very narrow range of growth conditions. Using the flash desorption technique Arthur [340] has showed that

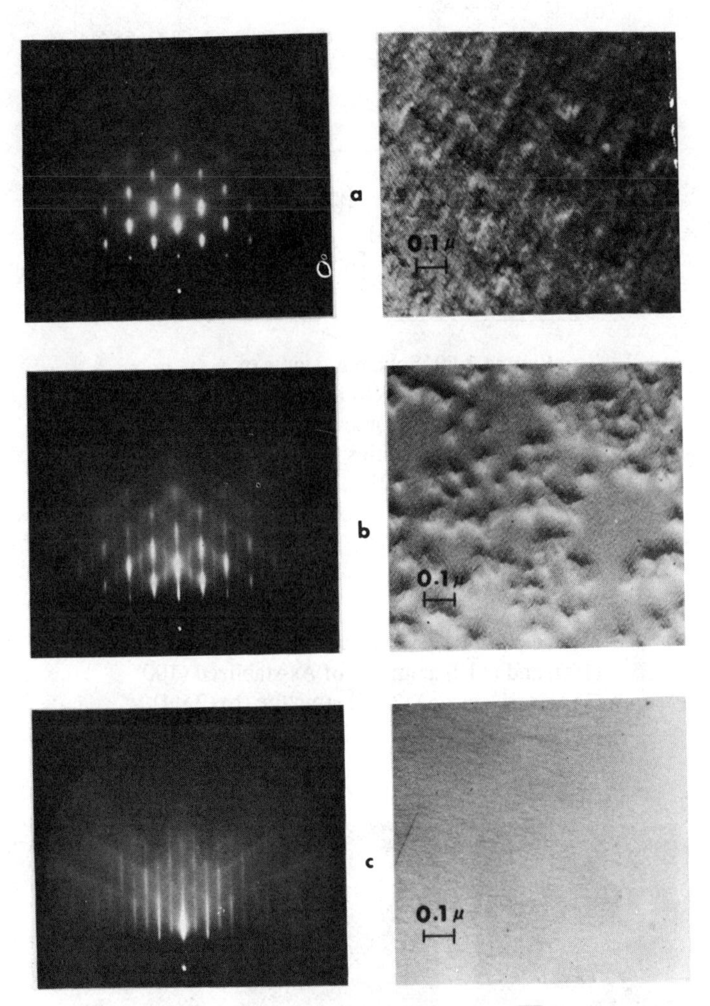


**Figure 2.88** RHEED pattern (40 keV, $[\overline{1}\overline{1}0]$ azimuth) and the corresponding electron micrographs of Pt-C replica of the same surface: (a) $Br_2$-methanol polished GaAs heated in vacuum to 580°C; (b) 150Å layer of GaAs deposited on surface of (a); (c) 1μm GaAs deposited on surface of (a) [355].

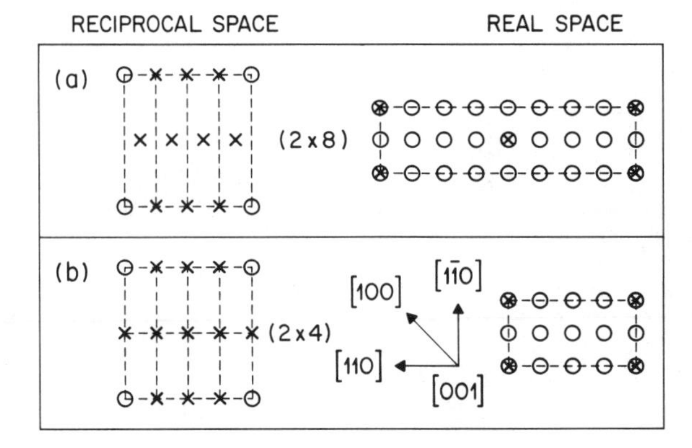


**Figure 2.89** Schematic view of reconstructed (2×8) and (2×4) surface structures in reciprocal and real space (O) "bulk" periodicities; (X) Superlattice periodicities [332].

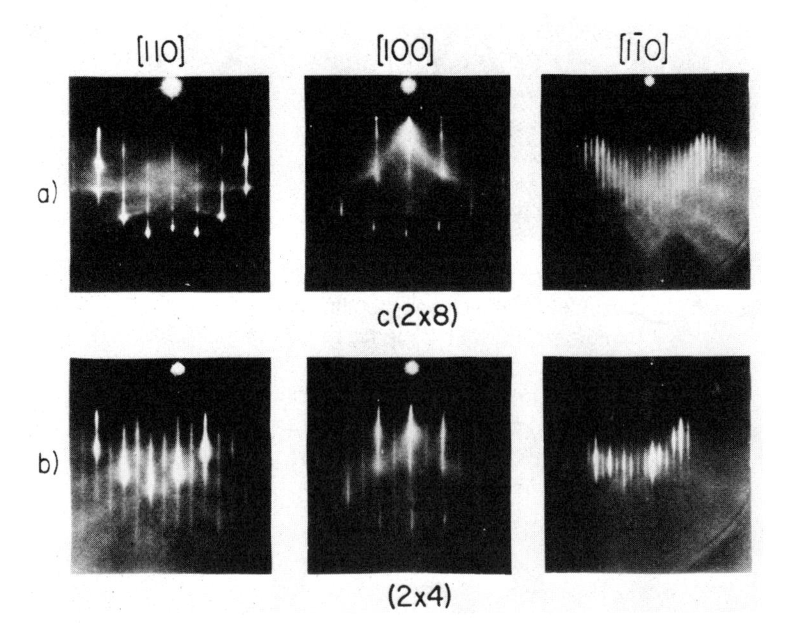


**Figure 2.90** RHEED patterns at 20 keV and [110], [100] and [1̄10] azimuths of As-stablized (100) GaAs surfaces: (a) (2×8) structure (b) (2×4) structure [334].

the fractional coverage of As is smaller than 0.1 for the Ga-stabilized surface and between 0.5 and 0.6 for the As-stabilized surface. This means that surface structure changes with the gain or loss of ~ 0.5 monolayer of As atoms.

The existence of various structures depends not only on the substrate temperature and the ratio of the component fluxes but also on the absolute magnitude of the fluxes themselves [356]. For example, the transition region shown in Fig. 2.84 is widened as the substrate temperature is increased and the Ga flux is decreased. From the point of view of practical GaAs growth the As-stabilized structure is preferred. Extremely smooth (100) surfaces of good quality can be achieved under this condition. On the other hand, even a small increase (~ 3%) in the Ga flux more than above that required to produce the Ga stabilized structure leads to a dull surface caused by the formation of Ga droplets.

Another application of RHEED in MBE growth is to study the dynamic conditions of film growth. This is accomplished by studying the temporal changes in the diffracted intensity (intensity oscillations) that occur during growth [352, 353]. The period of the RHEED intensity oscillations is explicitly related to the growth rate and as such is of practical value in determining the beam fluxes and in controlling the layer thickness to within one monolayer accuracy. The average intensity of the specular beam is a measure of the surface smoothness, the higher the intensity the smoother the surface for a chosen azimuth and glancing angle incidence. For these reasons, even though dynamical nature of electron diffraction [357,358] can complicate the interpretation of the intensity behavior, RHEED intensity oscillations have turned out to be a powerful tool for *in situ* real time study of MBE growth dynamics.

The observation of RHEED intensity oscillations has been seen for GaAs, (Al,Ga)As, (Ga,In)As, Ge and Si [359]. The intensity oscillation during growth of GaAs is illustrated in Fig. 2.91. The period of the oscillation corresponds to the growth of precisely one monolayer, which in the case of GaAs is equal to a thickness of a/2 in the growth direction. For elemental semiconductors the thickness equivalent is a/4. The amplitude of the oscillations is damped but usually several periods can be seen. Similar oscillatory effects have been reported in Auger line

**Figure 2.91** RHEED intensity oscillation of the specularly reflected beam during MBE growth of GaAs; the period of oscillation corresponds precisely to a monolayer, $a_o/2$ for GaAs [332].

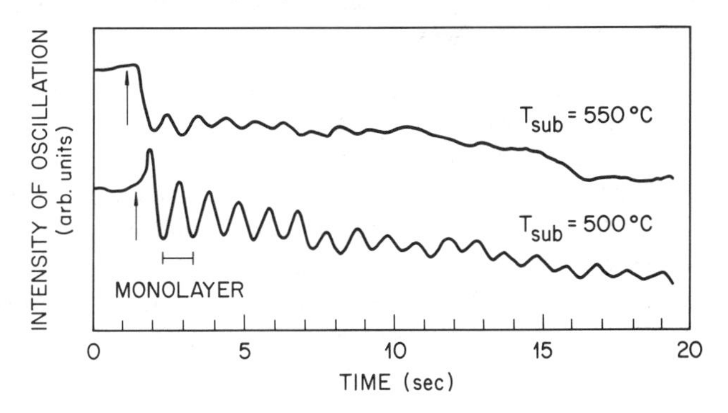

shapes [360] and LEED patterns [361]. In all cases the oscillatory behavior has been ascribed to two-dimensional layer-by-layer growth.

The de Broglie wavelength of electrons used in RHEED is typically <0.1 Å while the monolayer step height in the <100> direction is a/2 (for GaAs a/2 ~ 2.83 Å) so that the scatterer is larger than the wavelength of the electrons. If it is assumed that two-dimensional layer-by-layer growth starts on an atomically flat surface, the formation of monolayer steps reduces the intensity of the specular beam and increases that of the diffuse scattered beam. If growth is regular, the intensity of the specular beam reaches a minimum and returns to maximum only on completion of the layer. The process repeats with each new layer giving rise to oscillatory behavior. The two dimensional layer-by-layer growth continues until the mean terrace width becomes less than group III adatom migration length at which point two dimensional step propagation becomes the dominant growth mode. The damping of the RHEED oscillations is taken to imply that steady state terrace width distribution has reached.

It has been recognized that in the interpretation of RHEED oscillations, effects of multiple scattering should be included. The dynamical nature of the diffraction is revealed in rocking curves of the specular intensity. For a fixed substrate temperature the specular beam intensity shows a complicated relation with the incident angle suggesting dynamical nature of the diffraction [358, 359, 362]. The dynamical diffraction effect has important implications if a direct correlation is to be obtained with specific points in the layer growth sequence, especially when interrupted growth is involved [362].

The RHEED intensity oscillations also provide real time information about compositional effects and growth modes during heterojunction formation when there is no interruption of growth. As an example, in the formation of the $Al_xGa_{1-x}As$/GaAs interface, the sequence of layer growth plays an important role in determining interface roughness. Joyce et al., [362] note that two cases can be distinguished: (a) a steady state terrace width distribution has not reached in the layer preceeding the interface when the interface is reached, that is, the RHEED oscillations are still present; (b) the RHEED oscillation is completely damped and steady state growth is achieved when the interface is reached. Figure 2.92 illustrates the two cases [362]. In Fig. 2.92(a) the layers are thin such that RHEED oscillations are present across the interfaces between GaAs/$Al_{0.3}Ga_{0.7}As$ as well as $Al_{0.3}Ga_{0.7}As$/GaAs. The arrows indicate where the Al flux is switched on or off. The RHEED behavior is insensitive to both the change in composition and the stage in layer growth where it occurred. However, there is a change of period when the flux changes. In case (a) the specular beam intensity can only provide information on growth rates and alloy composition across interfaces.

Case (b) is illustrated in Fig. 2.92(b). It can be observed that the oscillations restart at the GaAs/(Al,Ga)As interface but not at the (Al,Ga)As/GaAs interface. This is entirely consistent with the relative surface migration lengths of the group III adatoms. Under typical MBE growth conditions on (100) surfaces, the migration lengths of Al and Ga are estimated to be 35 Å and 200 Å, respectively [363]. Al has a shorter migration length than Ga, that is, shorter steady state terrace width, so that the initial growth mode of (Al,Ga)As on GaAs will be two-dimensional layer-by-layer growth giving rise to RHEED oscillations. When GaAs is grown on (Al,Ga)As, the terrace width is shorter than the Ga migration length so that step propagation has become the

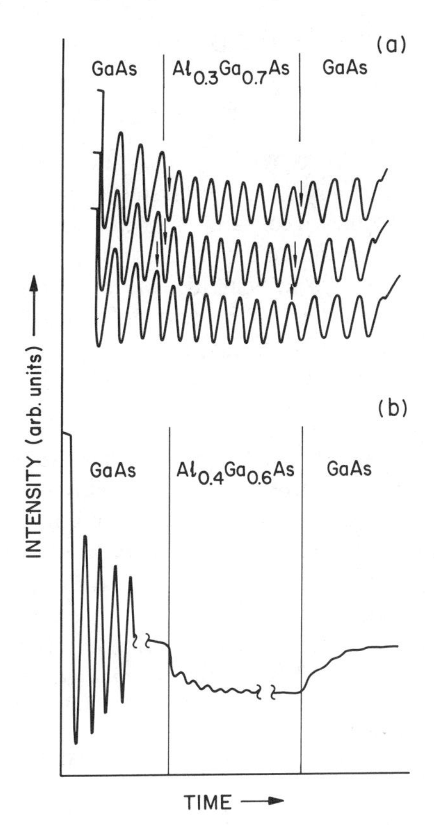


**Figure 2.92** RHEED intensity oscillation across GaAs/$Al_xGa_{1-x}As$/GaAs interfaces during MBE growth on (100) GaAs substrate. [110] azimuth; specular spot 00 rod; angle of incidence 1°: (a) interfaces formed before oscillations have damped. Arrows indicate switching (on or off) of Al flux. $x = 0.3$, growth temperature 580°C (b) interfaces formed when oscillations have almost fully damped. $x = 0.4$, growth temperature 600°C [362].

growth mode and no intensity oscillations. Therefore, in case (b) the migration length of the relevant group III atom in relation to the mean terrace width in the previous layer determines the growth mode of each successive layers.

The differences in the group III adatom diffusion lengths have also consequences on the nature of the interface. A GaAs layer should be covered by smooth terraces of 200 Å average length between monolayer steps. The terraces will be only 35 Å apart in (Al,Ga)As. This implies that the GaAs/(Al,Ga)As interface is much smoother on an atomic scale than the (Al,Ga)As/GaAs interface. This hypothesis is confirmed by high resolution transmission electron microscope lattice images which show that the heterointerface is abrupt to within one atomic layer only when the ternary alloy is grown on the binary compound but not for the inverse growth sequence [364]. In minimizing the interface roughness the "growth interruption" technique at each interface appears to be successful. Growth interruption allows the small terraces to relax into larger terraces via surface diffusion of the adatoms. This reduces the step intensity and thus increases the RHEED specular beam intensity which can be used for real time monitoring. Multiple scattering effects should, however, be considered for quantitative analysis [362]. The time of closing both the Al and Ga shutters (while the As shutter is kept open) depends on the growth conditions. It varies from a few seconds to several minutes [365, 366].

## 2.7 GAS SOURCE MBE AND METAL-ORGANIC MBE

### 2.7.1 Introduction

Although conventional molecular beam epitaxy (MBE) which uses solid sources has been successful for the crystal growth of GaAs, and AlGaAs the crystal growth of III-V compounds which contain phosphorus is extremely difficult by conventional MBE. Because elemental solid phosphorus consists of allotropic forms with different vapor pressures ($P_4$ and $P_2$), a beam flux sufficiently well controlled to be useful for MBE is difficult to achieve. In 1980 Panish described a method for the decomposition of $AsH_3$ and $PH_3$ for MBE of GaAs and InP [1]. The MBE process wherein $AsH_3$ and $PH_3$ are cracked to yield $As_2$ and $P_2$ beams has come to be known as gas source molecular beam epitaxy (GSMBE), and it has been very successful for the crystal growth of thin layers of GaInAsP on InP [367].

The GSMBE process employs elemental Ga and In sources, and the beam profiles from conventional MBE sources set limits on wafer size and on the compositional and thickness uniformity of epitaxial layers grown on large area substrates. As a result, epitaxial growth by MBE and GSMBE is usually done with substrate rotation. Attendant scale-up constraints are relaxed if not only the group V but also the group III species are delivered to the substrate using gas sources. This is conveniently accomplished using metal-organic group III species in addition to hydride group V sources as demonstrated by W. T. Tsang [368, 369]. Dubbed "chemical beam epitaxy" by Tsang, this process will be referred as metal-organic molecular beam epitaxy (MOMBE).

GSMBE and MOMBE can be viewed as evolutions from hydride vapor phase epitaxy (VPE), metal-organic chemical vapor deposition (MOCVD), and MBE. Gas pressures during crystal growth differ among these techniques and can be used to distinguish them as shown in Fig. 2.93. The pressure regimes for GSMBE and MOMBE are determined by the criterion that beam epitaxy occurs. If the occurrence of molecular flow is used to define beam epitaxy, then a pressure of about $10^{-3}$ torr is the maximum possible [370]. The condition that the mean free path be greater than the source to substrate distance also leads to a maximum pressure of $10^{-3}$ torr [370]. When there is beam epitaxy a boundary layer near the substrate is absent and the geometry of the surrounding growth chamber is not influential on the vapor flows. This qualitative difference clearly distinguishes GSMBE and MOMBE on the one hand and MOCVD (including low pressure MOCVD) and VPE on the other hand. There are also significant similarities between the gas source beam techniques and the viscous flow techniques. A quite notable similarity is the common use of mass flow controllers to control the gas flows. As a result, considerations of interface abruptness must share a certain similarity with similar considerations for the viscous flow techniques.

### 2.7.2 Gas Source MBE

A basic growth chamber configuration for GSMBE is shown schematically in Fig. 2.94 [367]. The innovative and novel feature of this MBE method is the $AsH_3$ and $PH_3$ thermal cracker. It has been demonstrated that these hydrides are negligibly cracked at the substrate surface and that a means of decomposing them to produce molecular beams is needed [367]. Two basic cracker designs have been studied by Panish and co-workers [371]. The source most extensively studied

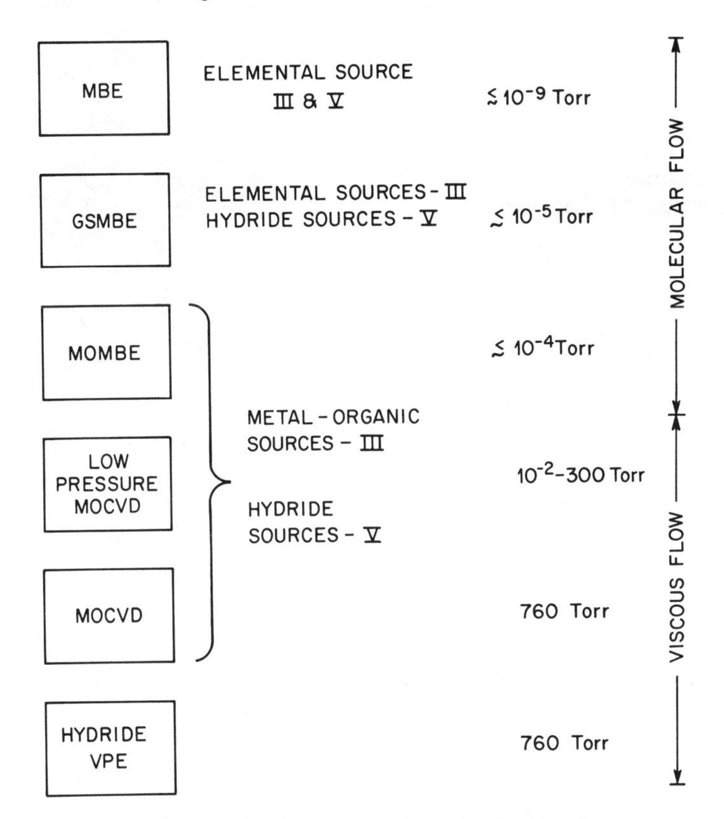


**Figure 2.93** Pressure regimes for MBE and vapor phase epitaxial growth techniques. Basic GSMBE source configuration [367].

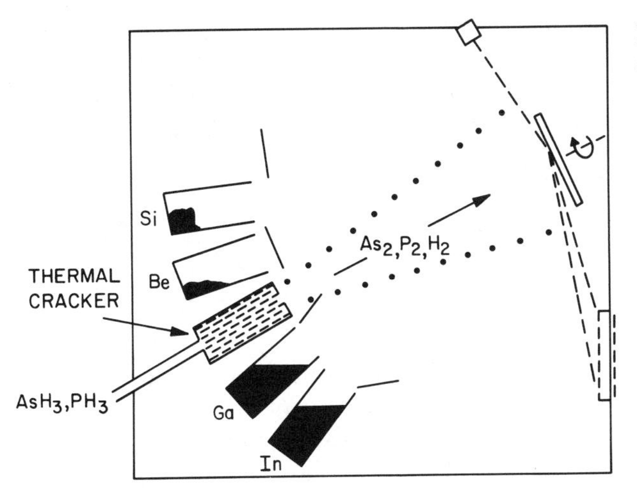


**Figure 2.94** Basic GSMBE source configuration [367].

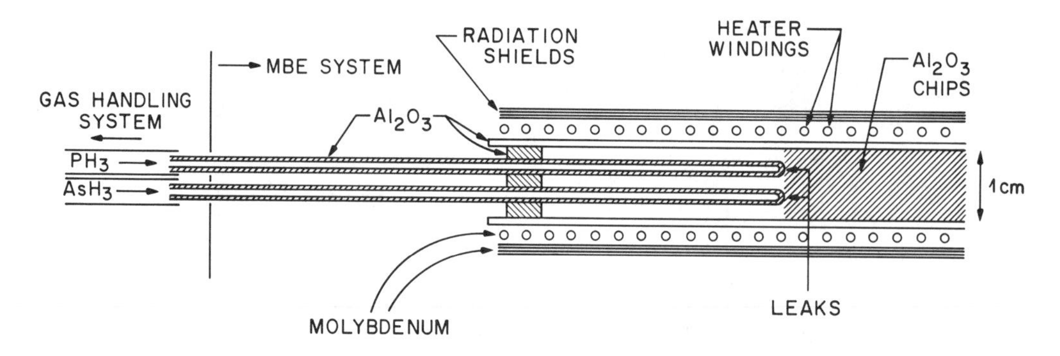


**Figure 2.95** GSMBE source [371].

and used by them is shown in Fig. 2.95. It thermally decomposes 100 percent $AsH_3$ and $PH_3$ to produce $As_2$, $P_2$, and $H_2$. The first of the decompositions reactions is

$$4\,MH_3 \rightarrow M_4 + 6\,H_2$$

where M represents As or P, and occurs in the alumina inlet tubes at pressures between 400 and 1500 torr (0.5 - 2 atm). The $M_4$ and $H_2$ reaction products diffuse through a hot leak into a low pressure region where the tetramers crack to dimers according to the reaction.

$$M_4 \rightarrow 2\,M_2\,.$$

The inlet decomposition tubes have a leaky seal at the delivery end. The leak rate is designed to permit useful growth rates, but it is made small enough to stay within the pumping capacity of the MBE system [1, 372]. The heater region of the gas source is usually held at temperatures between 900 and 1000°C, and at those temperatures the equilibrium species in the alumina tubes are $As_4$ and $P_4$. But dimers and not tetramers are preferred for MBE since nearly every dimer which impinges on the surface of the growing layer is incorporated. For tetramers this is apparently not so [367]. At the much lower pressures after the leaky seal, the equilibrium is shifted to a preponderance of dimers as shown in Fig. 2.96 [367]. The resulting flux of dimers at the substrate as a function of hydride pressure is shown in Fig. 2.97 [371].

Virtually complete decomposition of both $AsH_3$ and $PH_4$ has been achieved with such a gas source [372]. This was evidenced by a negligible build up of arsine and phosphine on the cryopanels. Cracking efficiencies of at least 99.9 percent are a practical necessity in a MBE vacuum system for safety reasons [367].

In the gas source there is a high temperature leak of tetramers from ~ 1 atm into high vacuum, and a transition occurs from viscous hydrodynamic flow to molecular flow. The leak is essentially a jet which should cause a shock wave on the low-pressure side. Ideally the shock wave should result in a thermal velocity distribution having a cosine distribution [371]. The ideal distribution as well as the observed one are shown in Fig. 2.98 for a 10 cm distance away from

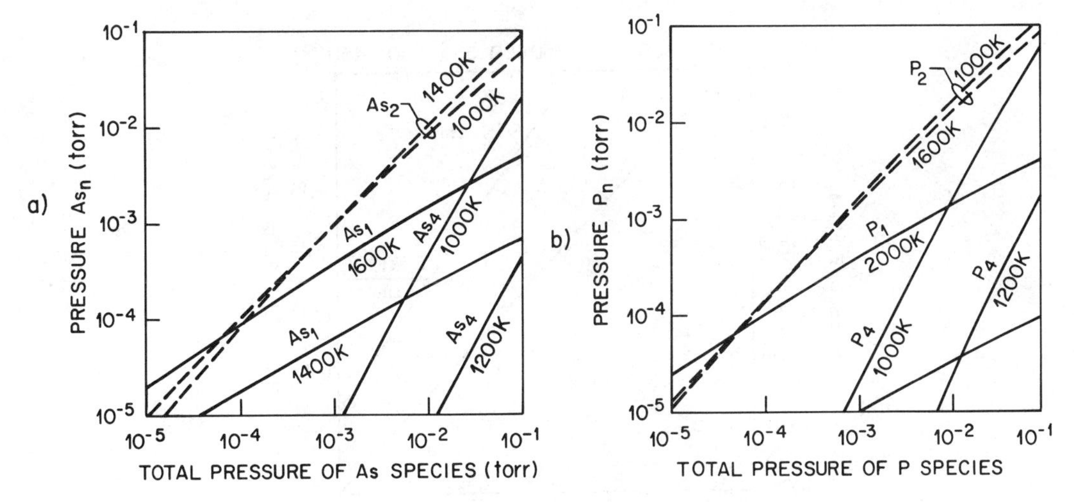


**Figure 2.96** Partial pressures for the source shown in Fig. 2.95 at equilibrium for a) As species and b) P species [367].

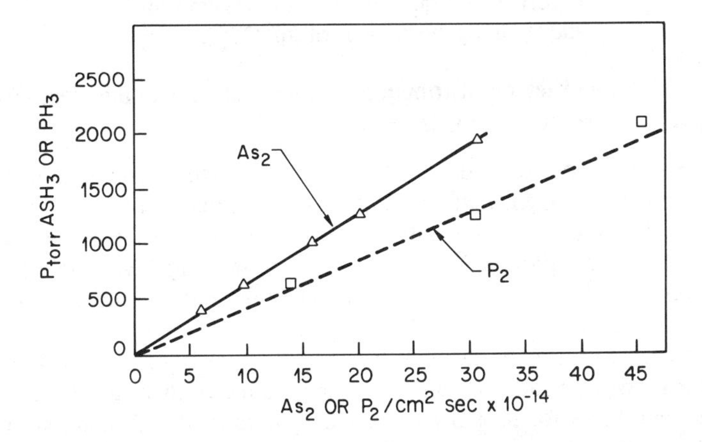


**Figure 2.97** Dimer flux at the substrate for GSMBE [371].

**Figure 2.98** Flux distribution during GSMBE [372].

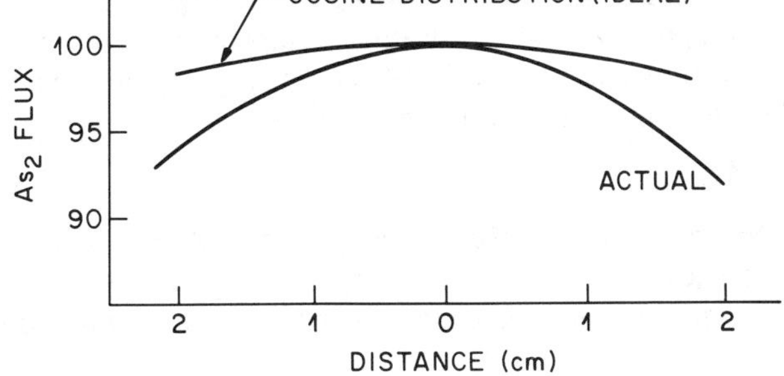

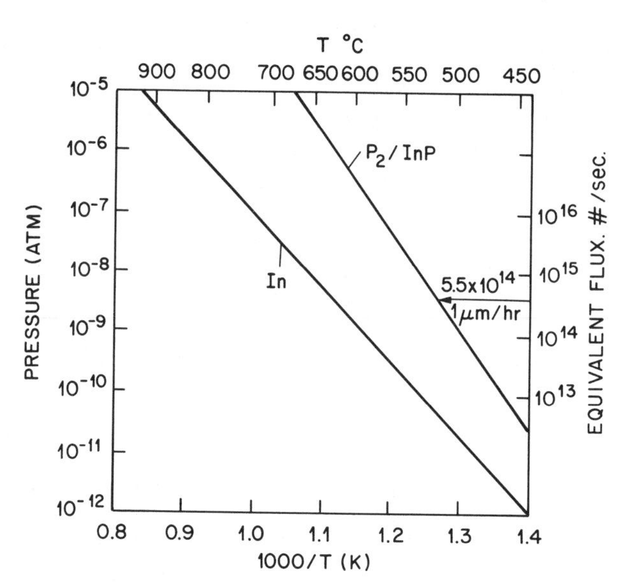


**Figure 2.99** Equilibrium vapor pressures for In and $P_2$ along the liquidus of InP [370].

the gas source [371]. The observed distribution is more peaked than the ideal distribution which most likely is because the source has too large an orifice [371].

During GSMBE, hydride decomposition generates $H_2$ which results in a much higher system pressure than is present during conventional MBE. The fact that so much $H_2$ is generated places an upper limit on the substrate temperature [370]. This occurs because equilibrium vapor pressures of group III and group V species over the growing layer are needed to prevent decomposition. For example, for the growth of InP at 1 μm/hr an impingement rate of $5.5 \times 10^{14}$ In atoms/cm$^2$-sec is typically needed [370]. If a factor of 2 safety margin in the $P_2$ flux is automatically included to make certain that we are on the safe side of the InP decomposition reactions, then for growth we must use a $P_2$ flux equal to the In flux. In Fig. 2.99 are shown the equilibrium vapor pressures for In and $P_2$ along the liquidus of InP along with the equivalent fluxes [370]. The temperature at which the $P_2$ flux is equal to $5.5 \times 10^{14}$ cm$^{-2}$-sec$^{-1}$ is 520°C, and at higher temperatures the flux required is higher still. In other words, for temperatures above 520°C with the In flux set at $5.5 \times 10^{14}$ cm$^{-2}$-sec$^{-1}$ the $P_2$ flux necessary to prevent InP decomposition is higher than that required for growth. By increasing the $P_2$ flux at higher temperatures appropriately GSMBE crystal growers generate exponentially increased amounts of $H_2$. This means that the upper limit of the substrate temperature is reached when the pumping speed of the vacuum system is reached [370]. We note that pumping speeds greater than 2000 l per second are usually sufficient for surface preservation.

### 2.7.3 Metal-Organic MBE

The growth chamber configuration used by Tsang for MOMBE is shown in Fig. 2.100 [373]. Not only the group V species but also the group III species are delivered from a gas source during MOMBE. A blend of metal-organic compounds is fed to a single gas source. An example of

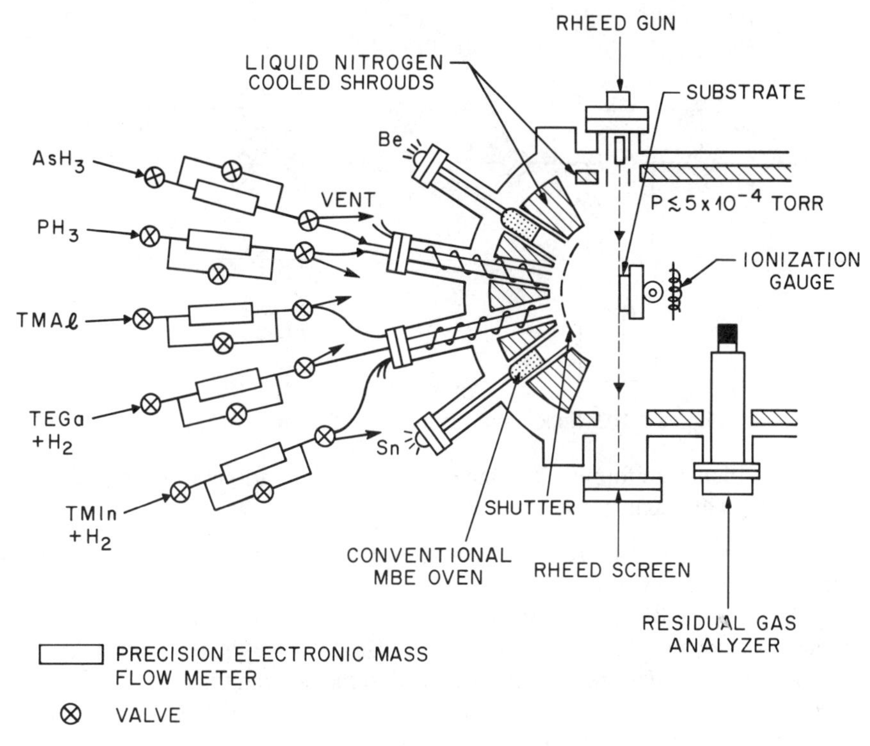


**Figure 2.100** Schematic MOMBE source configuration [373].

such a source is shown schematically in Fig. 2.101 [373]. The source is maintained at only ~50°C, and the metal organics are not decomposed until they are pyrolysed on the heated substrate. The temperature of the source need only be high enough to prevent condensation of the metal organics. These would otherwise condense on the walls because cryopanels cooled by liquid nitrogen are positioned nearby. The quartz source has a narrow inlet tube which ends in a shock chamber covered with an interchangeable baffle. The configuration of holes in the baffle can be tailored to achieve a high degree of thickness uniformity in an epitaxial layer. Just as in the case of the hydride gas a source discussed earlier the shock chamber helps to thermalize the flux. The inlet tube is made narrow to reduce dead volume. A high degree of mixing leading to

**Figure 2.101** A schematic MOMBE source [373].

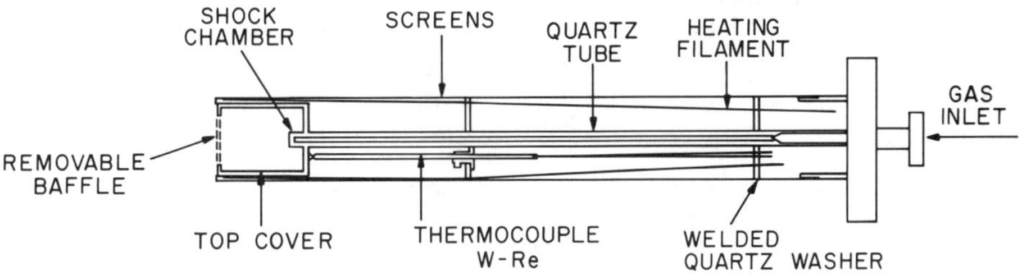

SHOCK CHAMBER
SCREENS
QUARTZ TUBE
HEATING FILAMENT
GAS INLET
REMOVABLE BAFFLE
TOP COVER
THERMOCOUPLE W-Re
WELDED QUARTZ WASHER

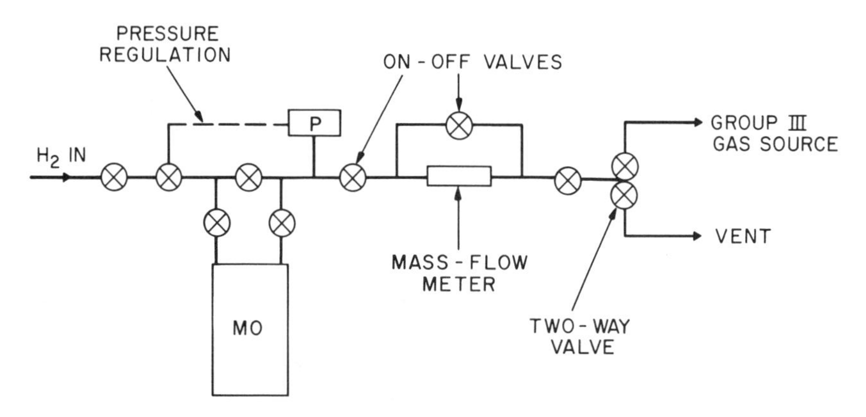


**Figure 2.102** Schematic gas manifold for MOMBE [373].

good alloy compositional uniformity is achieved because the various metal organics are mixed and delivered through the same gas source [373].

All the commonly used group III metal-organics except trimethyl indium are liquids at room temperature with vapor pressures below ~150 torr (see Sec. 2.5 on MOCVD) and are conveniently fed to an MBE gas source with $H_2$ carrier gas. The quantity of $H_2$ carrier gas introduced into the vacuum system is small compared to the quantity generated from the decomposition of $PH_3$ and does not significantly add to the pumping load [373]. A gas-handling system for the group III gas source is shown schematically in Fig. 2.102. A metal organic compound is kept in a cylinder immersed in a thermal bath which holds the temperature stable to better than ±0.05°C [373]. In this way the partial pressure of the metal organic inside the cylinder is kept constant [373]. Experimentally it is found that only when a newly filled cylinder is used or when a cylinder is almost empty does the group III flux vary.

An important difference between the gas manifold arrangement shown in Fig. 2.102 and that used in MOCVD is the location of the mass flow controller. Even for low-pressure MOCVD, the mass flow controller must be placed ahead of a metal-organic bubbler because the apertures inside it can clog. This is especially important in the case of TMIn. However, for MOMBE the pressure inside the mass flow controller is less than ~0.1 torr and metal organics will not easily accumulate. Since the mass flow controller can be used to directly control the metal organic flow in MOMBE, more direct control over the group III concentration as well as a faster flow change response is achieved [373].

Since the metal organics are not pyrolysed before they impinge on the heated substrate, the chemical reactions at the growing surface in MOMBE are expected to be considerably more complex than in the case of GSMBE, and since there is no boundary layer, the chemistry is different than in the case of MOCVD [370]. Insight concerning the chemistry has come from observations on the crystal growth of $Ga_xIn_{1-x}As_{1-y}P_y$ and on the growth rate of GaAs as a function of temperature.

In Fig. 2.103 is shown the P mole fraction of quaternary layers grown lattice matched to an

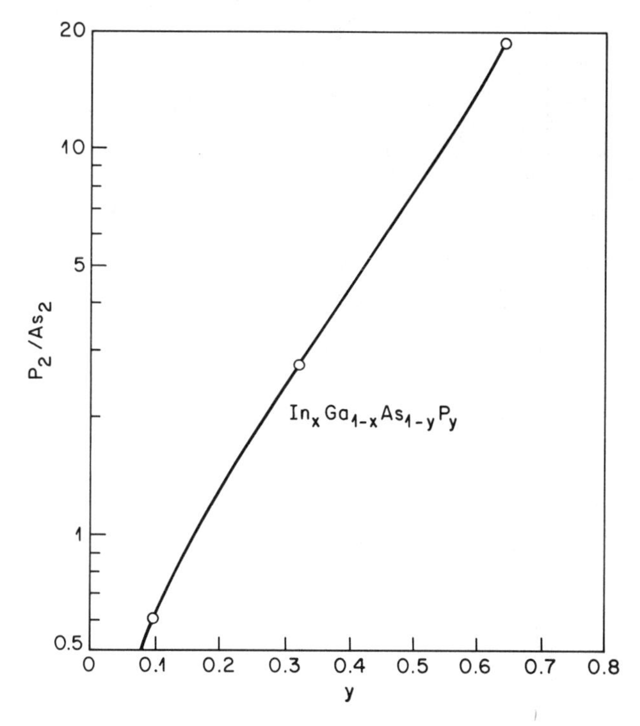


**Figure 2.103** Dimer ratio for the growth of $Ga_xIn_{1-x}As_{1-y}P_y$ lattice matched to InP for both GSMBE and MOMBE [370].

InP substrate at various As to P beam flux ratios during both MOMBE and GSMBE. The results do not depend on which growth technique was used [370]. Furthermore, the stoichiometry of $Ga_xIn_{1-x}As$ is not influenced by the $As_2$/III ratio. These two observations suggest that the presence of group V species does not affect the decomposition of the metal-organic compounds during MOMBE [370].

Precise growth rate studies have been made by Robertson et al., [374]. They used RHEED oscillations to obtain the dependence of the growth rate on temperature and on triethylgallium (TEGa) flux. Their results for GaAs are shown in Fig. 2.104. A model of the surface pyrolysis of TEGa was developed by Robertson et al., [374] which qualitatively explains their results. The model assumes that TEGa, diethylgallium (DEGa), and monoethyl gallium as well as ethyl radicals are all adsorbed on the surface of a growing layer. Elimination of hydrogen from adsorbed ethyl groups to yield ethylene was also included in the model. The growth rate limiting step is assumed to be breaking of the second ethyl-gallium bond. The reasonable fits shown in Fig. 2.104 were obtained [374]. In particular, the model reproduces the fall-off in growth rate at higher temperatures which, according to the model, occurs because DEGa radicals are desorbed. Furthermore, a nonlinear dependence of the growth rate on TEGa flux below 500°C which changes to a linear dependence above 500°C was observed. This was also reproduced in the

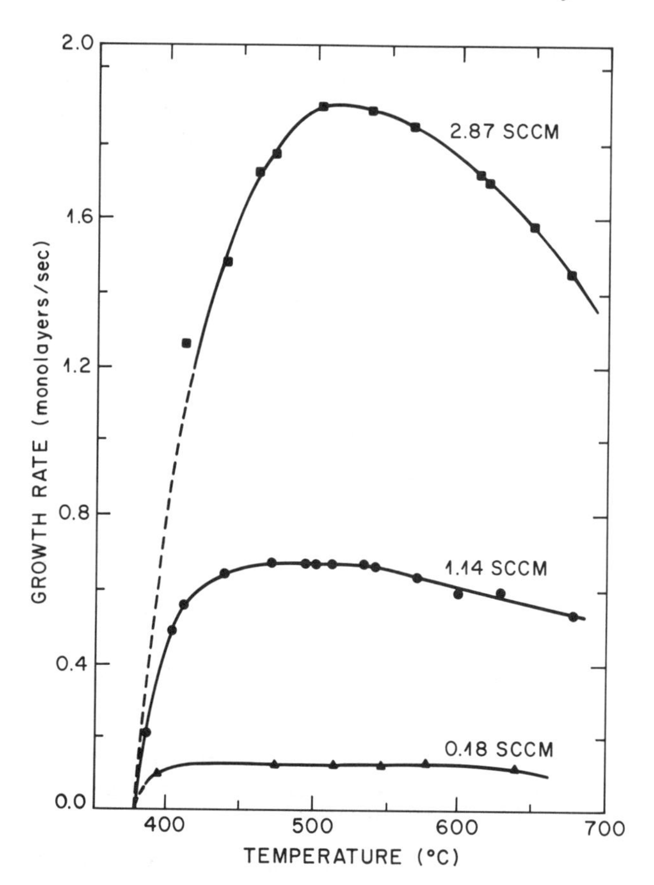


**Figure 2.104** Temperature dependence of growth rate during MOMBE [374].

model. According to the Robertson model this is caused by the second order recombination of DEGa with an ethyl radical followed by desorption of TEGa. We see that even in this simplified [370, 373] model a complex set of surface reactions must be considered for MOMBE.

Finally, note that in the original MOMBE studies of the growth of GaAs with TMGa, the epitaxial layers were found to be heavily p-type with hole concentrations of $10^{18}$–$10^{20}$ $cm^{-3}$ [375,376]. The layers were p-type because of carbon autodoping which can be greatly reduced if TEGa, is used in place of TMGa [377, 378]. Increased C incorporation with TMGa is expected because there is no alternate reaction path such as H elimination to form ethylene [370].

To conclude, note that these epitaxial crystal growth techniques are still young and, quoting Panish, that MBE using gas sources "must still be considered to be in the early stage of development in spite of its success with the GaInAsP/InP system" [370].

## REFERENCES


1. M.B. Panish, J. Electrochem. Soc. *127*;2729, 1980.
2. E. Veuhoff, W. Pletschen, P. Balk, and H. Lüth, *J. Cryst. Growth, 55*;30, 1981.
3. W.T. Tsang, A.N. Dayem, T.H. Chiu, J.E. Cunningham, E.F. Schubert, J.A. Ditzenburger, J. Shah, J.L. Zyskind, and N. Tabatabaie, *Appl. Phys. Lett., 49*;170, 1986.
4. A. J. Marshall, and K. Gillessen, *J. Cryst. Growth, 44*;651, 1978.
5. V.F.S. Yip and W.R. Wilcox, *J. Cryst. Growth, 36*;29, 1976.
6. D.T.J. Hurle, *J. Phys. Chem. Solids, 40*;613, 1979.
7. D.E. Holmes, R.T. Chen, K.R. Elliot, and C.G. Kirkpatrick, *Appl. Phys. Lett., 40*;46, 1982.
8. H.M. Hobgood, L.B. Ta, A. Rohatgi, G.W. Eldridge, and R.N. Thomas, in *Semi-Insulating III-V Materials,* (eds. S. Makram-Ebeid and B. Tuck) Nantwich, England; Shiva Publishing Ltd., p. 28, 1982.
9. C.D. Thurmond, *J. Phys. Chem. Solids, 26*;785, 1965.
10. M.B. Panish and J.R. Arthur, *J. Chem. Thermodynamics, 2*;299, 1970.
11. T.R. AuCoin and R.O. Savage, in *Gallium Arsenide Technology,* (ed. D.K. Ferry) Indiana; Howard W. Sams & Co., p. 47, 1985.
12. P.D. Greene, *J. Cryst. Growth, 50*;612 1980.
13. K.J. Bachman and E. Buehler, *J. Electron. Mater.;3*, 279, 1974.
14. J.M. Parsey, Jr., Y. Nanishi, J. Lagowsky and H.C. Gatos, *J. Electrochem. Soc. 128*; 936, 1981; *129*; 388, 1982.
15. W.A. Gault, E.M. Monberg, and J.E. Clemens, *J. Cryst. Growth, 74*;491, 1986.
16. J.M. Whelan and G.H. Wheatley, *J. Phys. Chem. Solids 6*;169, 1958.
17. L.R. Weisberg, F.D. Rosi, and P.G. Herkart in *Properties of Elemental and Compound Semiconductors,* ed. H.C. Gatos, New York, Interscience; p. 25, 1960.
18. E.M. Swiggard, *J. Cryst. Growth 94*;556, 1989.
19. J.L. Richards, *J. Appl. Phys. 31*;600, 1960.
20. N. Yamenidijian and B.A. Lombos, *J. Cryst. Growth 56*;163, 1982.
21. J. Czochralski, *Z. Physik. Chem. 92*;219, 1917.
22. R. Gremmelmaier, Z. Naturforsch, 11a; 511, 1956.
23. E.P.A. Metz, R.C. Miller and J. Mazelsky, *J. Appl. Phys. 33*;2016, 1962.

24. K.J. Bachmann, E. Buehler, J.L. Shay and A.R. Strand, *J. Electron. Mater., 4*;389, 1975.

25. M.E. Weiner, D.T. Lassota and B. Schwartz, *J. Electrochem. Soc., 118*;301, 1971.

26. B. Cockayne, G.T. Brown and W.R. MacEwan, *J. Cryst. Growth 51*;461, 1981.

27. C.G. Kirkpatrick, R.T. Chen, D.E. Holmes and K.R. Elliot, in *Gallium Arsenide Materials, Devices and Circuits*, (eds. M.J. Howes and D.V. Morgan), New York; Wiley, p. 39, 1985.

28. R.K. Willardson in *Semi-insulating III-V Materials*, (eds. D.C. Look and J.S. Blakemore) Nantwich, England, Shiva Publishing Ltd., p. 96, 1984.

29. R.N. Thomas, H.M. Hobgood, G.W. Eldridge, D.L. Barrett, T.T. Braggins, B.L. Ta, and S.K. Wang in *Semiconductors and Semimetals*, (eds. R.K. Willardson and A.C. Beer) (New York, Academic, vol. 20;p. 1, 1984.

30. G.A. Antypas, *J. Cryst. Growth 33*;174, 1976.

31. G. Müller, J. Völkl, and E. Tomzig, *J. Cryst. Growth 64*;40, 1983.

32. S. Shinoyama, C. Uemura, A. Yamamoto and S. Tohno, *J. Electron. Mater. 10*;941, 1981.

33. D. Rumsby, R.M. Ware, and M. Whitacker, *J. Cryst. Growth 54*;32, 1981.

34. M. Morioka, K. Tada and S. Akai, *Ann. Rev. Mater. Sci., 17*;75, 1987.

35. L. Pekarek, *Czech. J. Phys. B20*;857, 1970.

36. C.A. Stolte in *Semiconductors and Semimetals*, (eds. R.K. Willardson and A.C. Beer), New York; Academic, vol. 20, p. 89, 1984.

37. T.R. AuCoin, R.L. Ross, M.J. Wade, and R.O. Savage, *Solid State Tech., 22*;59, 1979.

38. Y. Sasaki, J. Nakagawa, and K. Kurata, $46^{th}$ Autumn Meet. Jpn. Soc. Appl. Phys., Extended Abstr. 4a-E-4 (1985).

39. T. Inada, S. Ozawa, M. Eguchi, and T. Fukuda, $33^{rd}$ Spring Meet. Jpn. Soc. Appl. Phys. Related Soc., Extended Abstr. 2p-X-13 (1986).

40. H.P. Utech and M.C. Flemings, *J. Appl. Phys. 37*;2021, (1966).

41. K. Terashima, T. Katsumata, F. Orito, and T. Fukuda, *Jap. J. Appl. Phys. 23*;L302, 1984.

42. J. Kafalas, loc. cit Ref. 11.

43. T. Fukuda, S. Washizuka, Y. Kodubun, J. Ushizawa, and M. Watanabe, in *Gallium Arsenide and Related Compounds*, Inst. Phys. Conf. Ser. *63*;43, 1981.

44. A.S. Jordan, R. Caruso, A.R. Von Neida, and J.W. Nielsen, *J. Appl. Phys. 52*;3331, 1981.

45. G.M. Martin, A. Mitonneau and A. Micrea, *Electron. Lett. 13*;191, 1977.

46. H. Temkin and B.V. Dutt, *Mat. Res. Soc. Symp. Proc.* Vol. 14;253, 1983.

47. W.A. Bonner, *J. Cryst. Growth 54*;21, 1981.

48. W.C. Dash, *J. Appl. Phys. 28*;882, 1957.

49. A.S. Jordan, G.T. Brown, B. Cockayne, D. Brasen, and W.A. Bonner, *J. Appl. Phys. 58*;4383, 1985.

50. A.R. Von Neida and A.S. Jordan, *J. Metals 38*;35, 1986.

51. A.S. Jordan, R. Caruso, and A.R. Von Neida, *Bell Sys. Tech. J. 62*;477, 1983.

52. A.S. Jordan, *J. Cryst. Growth 49*;631, 1980.

53. K. Tomizawa, K. Sassa, Y. Shimanuki, and J. Nishizawa, in *Defects and Properties of Semiconductors: Defect Engineering* (eds. J. Chikawa, K. Sumino, and K. Wada), Tokyo, KTK Scientific Publishers, Tokyo, p. 25, 1987.

54. A.S. Jordan and J.M. Parsey, Jr., *J. Cryst. Growth 79*;280, 1986.

55. G. Antypas, Mountain View, CA; Crysta Comm. Inc., (personal communication).

56. A.R. Von Neida, R. Caruso, and A.S. Jordan, unpublished.

57. A.G. Elliot, C.L. Wei, R. Farraro, G. Woolhouse, M. Scott, and R. Hiskes, *J. Cryst. Growth 70*;169, 1984.

58. T. Fukuda, *Jap. J. Appl. Phys. 22*;413, 1983.

59. K. Katagiri, S. Yamazaki, A. Takagi, O. Oda, H. Araki, and I. Tsuhoya, in *GaAs and Related Compounds*, Inst. Phys. Conf. Ser. *79*;67, 1986.

60. S. Shinoyama, C. Uemura, A. Yamamoto, and S. Tohno, *Jap. J. Appl. Phys. 19*;L331, 1980.

61. S. Shinoyama, S. Tohno, and A. Katsui, in *GaAs and Related Compounds*, Inst. Phys. Conf. Ser. *79*;55, 1986.

62. G. Muller, J. Voelkl, and E. Tomzig, *J. Cryst. Growth, 64*;40, 1983.

63. H. Nakani, K. Yamada, H. Kohoda, and K. Hoshikawa, $16^{th}$ Int. Conf. on Solid State Devices and Materials, Extended Abstr. p. 63; 1984.

64. J. Osaka and K. Hoshikawa, in *Semi-insulating III-V Materials*, (eds. D. C. Look and J. S. Blakemore, Nantwich, England; Shiva Publishing, p. 126, 1984.

65. H. Kohda, K. Yamada, H. Nakanishi, T. Kobayashi, J. Osaka, and K. Hoshikawa, *J. Cryst. Growth, 71*;813, 1985.

66. S. Ozawa, H. Miyairi, M. Nakayima, and T. Fukuda, in *GaAs and Related Compounds*, Inst. Phys. Conf. Ser. *79*;19, 1986.

67. T. Kamejima, S. Matsui, Y. Seki, H. Watanabe, *J. Appl. Phys. 50*;3312, 1979.

68. G. Jacob, in *Semi-insulating III-V Materials*, (eds. S. Makram-Ebeid and B. Tuck,) Nantwich, England; Shiva Publishing, 1982, p. 2, 1982.

69. H.M. Hobgood, R.N. Thomas, D.L. Barrett, G.W. Eldridge, M.M. Sopira, and M.C. Driver, in *Semi-insulating III-V Materials*, (eds. D.C. Look and J.S. Blakemore,) Nantwich, England, Shiva Publishing, 1984) p. 149, 1940.

70. S. Tohno, E. Kubota, S. Shinoyama, A. Katsui, and C. Uemura, *Jap. J. Appl. Phys. 23*;L72 1984.

71. S.Z. Ye, X.L. Liu, J.H. Jiao, B.H. Yang and J.Q. Zhao, in *GaAs and Related Compounds,* Inst. Phys. Conf. Ser. *79*;61, 1986.

72. A. Katsui, S. Tohno, Y. Homma and T. Tanaka, *J. Cryst. Growth 74*;221, 1986.

73. T. Inoue, S. Nishine, M. Shibata, T. Matsutomo, S. Yoshitake, Y. Sato, T. Shimoda and K. Fujita, in *GaAs and Related Compounds,* Inst. Phys. Conf. Ser. *79*;7, 1986.

74. M.S.S. Young, A.S. Jordan, A.R. Von Neida and R. Caruso, unpublished.

75. J.P. Hirth and J. Lothe, *Theory of Dislocations,* New York; Wiley, 2$^{nd}$ Edition, p. 306, 1982

76. L.R. Weisberg, J. Blanc, and E.J. Stofko, *J. Electrochem. Soc. 109*;642, 1962.

77. A. Steinemann and V. Zimmerli, *Solid State Electron* 6;597, 1963.

78. S. Shinoyama, in *Defects and Properties of Semiconductors: Defect Engineering,* (eds. J. Chikawa, K. Sumino and K. Wada) Toyko; KTK Scientific Publishers, p. 87, 1987.

79. H.M. Hobgood, T.T. Braggins, D.L. Barrett, G.W. Eldridge and R.N. Thomas, 5$^{th}$ Int. Conf. Vapor Growth Epitaxy/5$^{th}$ American Conf. Cryst. Growth, San Diego, Calif.

80. S. Mahajan, W.A. Bonner, A.K. Chin and D.C. Miller, *Appl. Phys. Lett. 35*, 165 (1979).

81. K.J. Bachmann and E. Buehler,*J. Electron. Mat. 3*, 279 (1974).

82. J.D. Oberstar, B.G. Streetman, J.E. Baker, P. Williams, R.L. Henry and E.M. Swiggard, *J. Cryst. Growth 54*;443, 1981.

83. G.T. Brown, B. Cockayne, C.R. Elliott, J.G. Regnault, D.J. Stirland and P.D. Augustus, *J. Cryst. Growth 67*;495, 1984.

84. P.D. Augustus and D.J. Stirland, *J. Electrochem. Soc. 129*;614, 1982.

85. S.N.G. Chu and C.M. Jodlauk, unpublished.

86. W.G. Pfann, *Zone Melting*, New York;Wiley, 1958.

87. R.K. Willardson and W.P. Allred, in *GaAs and Related Compounds,* Inst. Phys. Conf. Ser. *3*;35, 1967.

88. J.B. Mullin, A. Royle and S. Benn, *J. Cryst. Growth*, 50;625, 1980.

89. Compilation of data in Landelt-Börnstein, Numerical Data and Functional Relationships in *Science and Technology,* vol. 17 Semiconductors, (eds. O. Madelung, M. Schulz, and H. Weiss,) Subvol. d New York;Springer-Verlag, 1984.

90. S.N.G. Chu, C.M. Jodlauk and A.A. Ballman, *J. Electrochem. Soc. 129*;352, 1982.

91. A.K. Chin, H. Temkin and S. Mahajan, *Bell Syst. Tech. J. 60*;2187, 1981.

92. M.S. Abrahams and C.J. Buiocchi, *J. Appl. Phys. 36*;2855, 1965.

93. J.G. Grabmaier and C.B. Watson, *Phys. Stat. Solidi 32*;K13, 1969.

94. H. Richter and M. Schulz, *Krist. Techn. 9*;1041, 1974.

95. A. Huber and N.T. Linh, *J. Cryst. Growth 29*;80, 1975.

96. R.C. Clarke, D.S. Robertson and A.W. Vere, *J. Mat. Sci. 8*;1349, 1973.

97. K. Akita, T. Kusunoki, S. Komiya and T. Kotani, *J. Cryst. Growth 46*;783, 1979.

98. F. Kuhn-Kuhnenfeld, *J. Electrochem. Soc. 117*;1063, 1972.

99. J. Nishizawa, Y. Oyama, H. Tadano, K. Inokuchi and Y. Okuno, *J. Cryst. Growth 47*;434, 1979.

100. V. Gottschalch, *Krist. Tech. 14*;939, 1979.

101. T. Saitoh, S. Matsubara and S. Minagawa, *J. Electrochem. Soc. 122*;423, 1975.

102. V. Gottschalch, R. Smanek and G. Wagner, *J. Mat. Sci. Lett. 1*;358, 1982.

103. L. Ekstrom and L.R. Weisberg, *J. Electrochem. Soc. 109*;321, 1962.

104. E.C. Lightowlers, *J. Electron. Mat. 1*;39, 1972.

105. E.S. Johnson, *J. Cryst. Growth 30*;249, 1975.

106. B.L. Ta, H.M. Hobgood and R.N. Thomas, *Appl. Phys. Lett. 41*;1091, 1982.

107. J.B. Clegg, in *Semi-insulating III-V Materials,* (eds. S. Makram-Ebeid and B. Tuck), Nantwich, England, Shiva Publishing Ltd., p. 80, 1982.

108. C.G. Kirkpatrick, R.T. Chen, D.E. Holmes, P.M. Asbeck, K.R. Elliott, R.O. Fairman and J.R. Oliver, in *Semiconductors and Semimetals,* (eds. R.K. Willardson and A.C. Beer) New York; Academic, vol. 20, p. 159, 1984.

109. H.C. Casey, Jr. and M.B. Panish, *Heterostructure lasers Part B: Materials and Opeating Characteristics* New York; Academic, p. 109, 1978.

110. H. Nelson, *RCA Rev. 24*;603, 1963.

111. J.M. Woodall, H. Rupprecht and W. Reuter, *J. Electrochem. Soc. 116*;899, 1969.

112. H.F. Lockwood and M. Ettenberg, *J. Cryst. Growth 15*;81, 1972.

113. H.C. Casey, Jr., M.P. Panish, W.O. Schlosser and T.L. Paoli, *J. Appl. Phys. 45*;322, 1974.

114. J. Steininger and T.B. Reed, *J. Cryst. Growth* 13/14;106, 1972.

115. G.H.B. Thompson and P.A. Kirkby, *J. Cryst. Growth 27*;70, 1974.

116. M.C. Tamargo and C.L. Reynolds, Jr., *J. Cryst. Growth 57*;349, 1982.

117. S.Y. Leung and N.E. Schumaker, *J. Cryst. Growth 67*;458, 1984.

118. S.Y. Leung and N.E. Schumaker, *J. Cryst. Growth 60*;421, 1982.

119. Zh.I. Alferov, B.Ya. Ber, K.Yu. Kizhaev, S.A. Nikitin and E.L. Portnoi, *Sov. Tech. Phys. Lett. 11*;397, 1985.

120. J.L. Zilko, S.Y. Leung and N.E. Schumaker, unpublished.

121. Y. Horikoshi, *Jap. J. Appl. Phys. 15*;887, 1976.

122. L.R. Dawson, *J. Appl. Phys. 48*;2485, 1977.

123. J. Heinen, *J. Cryst. Growth 58*;596, 1982.

124. B.V. Dutt, D.D. Roccasecca, H. Temkin and W.A. Bonner, *J. Cryst. Growth 66*;525, 1984.

125. D.J. Lawrence and L.F. Eastman, *J. Cryst. Growth 30*;267, 1975; *J. Electron. Mat. 6*;1, 1976.

126. L. Jastrzebski, H.C. Gatos and A.F. Witt, *J. Electrochem. Soc. 123*;1121, 1976.

127. J. Daniele, *Appl. Phys. Lett. 27*;373, 1975.

128. J. Daniele, D.A. Commack and P.M. Asbeck, *J. Appl. Phys. 48*;914, 1977.

129. A. Addul-Fadl and E.K. Stefanakes, *J. Cryst. Growth 39*;341, 1977.

130. L.J. Vieland, *Acta Metall. 11*;137, 1963.

131. C.D. Thurmond, *J. Phys. Chem. Solids 26*;785, 1965.

132. M.B. Panish and M. Ilegems, *Prog. in Solid State Chem.* vol. 7 (eds. H. Reiss and J. O. McCaldin), New York; Pergamon, p. 39, 1972.

133. G.B. Stringfellow and P.E. Greene, *J. Phys. Chem. Solids 30*;1779, 1969.

134. M. Ilegems and M.B. Panish, *J. Phys. Chem. Solids 35*;409, 1974.

135. A.S. Jordan and M. Ilegems, *J. Phys. Chem. Solids, 36*;329, 1975.

136. R.A. Swalin, *Thermodynamics of Solids,* 2nd ed., New York; Wiley, 1972.

137. Ref. 109, p. 74.

138. B.D. Lichter and P. Sommelet, *Trans. AIME 245*;99 and 1021, 1969.

139. G.A. Antypas, *J. Electrochem. Soc. 117*;700, 1970.

140. R.F. Brebrick, *Metall. Trans. 2*;1657, 3377, 1971.

141. M.V. Rao and W.A. Tiller, *J. Phys. Chem. Solids 31*;191, 1970.

142. E.A. Guggenheim, *Thermodynamics,* 5th ed. Amsterdam; North-Holland, p. 197, 1967.

143. M.B. Panish, *J. Cryst. Growth 27*;6, 1974.

144. Ref. 109, p. 81.

145. E.H. Perea and C.G. Fonstad, *J. Electrochem. Soc. 127*;313, 1980.

146. M. Ilegems and G.L. Pearson, Proceedings of 1968 Symposium on GaAs, The Institute of Physics and Phys. Soc. London, *7;3*, 1968.

147. A.S. Jordan, *J. Electrochem. Soc. 119*;123, 1972.

148. M.B. Panish and I. Hayashi, in *Applied Solid State Science* (ed. R. Wolfe) vol. 4, New York; Academic, p. 235, 1974.

149. Ref. 109, p. 87.

150. E. Kuphal, *J. Cryst. Growth 67*;441, 1984.

151. T.P. Pearsall, M. Quillac and M.A. Pollack, *Appl. Phys. Lett. 35*;342, 1979.

152. J.C. Phillips, *Phys. Rev. Lett. 20*;550, 1968.

153. J.A. Van Vechten, *Phys. Rev. 182*;891, 1969.

154. J.A. Van Vechten, *Phys. Rev. 187*;1007, 1969.

155. J.C. Phillips and J.A. Van Vechten, *Phys. Rev. B2*;2147, 1970.

156. G.B. Stringfellow, *J. Phys. Chem. Solids 33*;665, 1972.

157. G.B. Stringfellow, *J. Phys. Chem. Solids 34*;1749, 1973.

158. G.B. Stringfellow, *J. Cryst. Growth, 27*;21, 1974.

159. R.F. Brebrick and R.J. Panlener, *J. Electrochem. Soc. 121*;932, 1974.

160. K. Nakajima, in *Semiconductors and Semimetals,* (eds. R.K. Willardson and A.C. Beer), New York; Academic, 1985, vol. 22, Lightwave Communications Technology, vol. ed. W.T. Tsang, Part A, p. 1.

161. M. Ilegems and M.B. Panish, *J. Cryst. Growth 20*;77, 1973.

162. E.H. Perea and C.G. Fonstad, *J. Appl. Phys. 51*;331, 1980.

163. P.K. Bhattacharya and S. Srinivas, *J. Appl. Phys. 24*;5090, 1983.

164. J.J. Hsieh, *IEEE J. Quant. Electron. QE-17*;118, 1987.

165. G.A. Antypas, Y.M. Houng, S.B. Hyder, J.S. Escher and P.E. Gregory, *Appl. Phys. Lett. 33*;463, 1978.

166. K. Nakajima, T. Tanahashi, K. Akita and T. Yamaoka, *J. Appl. Phys. 50*;4975, 1979.

167. K. Nakajima and J. Okazaki, *J. Electrochem. Soc. 132*;1424, 1985.

168. M.B. Panish, *J. Appl. Phys. 44*;2676, 1973.

169. J.J. Hsieh in *Handbook of Semiconductors,* (Ser. ed. T.S. Moss), North Holland, Amsterdam, Vol. 3 (ed. S.P. Keller), p. 415, 1980.

170. S. Christensson, D. Woodard and L. Eastman, *IEEE Trans. Electron Dev. ED-17*;732, 1970.

171. E.G. Dierschke, L.E. Stone and R.W. Haisty, *Appl. Phys. Lett. 19*;98, 1971.

172. J.J. Hsieh, *J. Cryst. Growth 27*;49, 1974.

173. I. Crossley and M.B. Small, *J. Cryst. Growth 15*;268, 1972.

174. J.J. Hsieh, in *GaAs and Related compounds*, Inst. of Phys. Conf. Ser. (London) 33b;74, 1977.

175. D.L. Rode and R.G. Sobers, *J. Cryst. Growth, 29*;61, 1975.

176. K. Nakajima, S. Yamazaki and K. Akita, *J. Cryst. Growth 56*;547, 1982.

177. G.H.B. Thompson and P.A. Kirkby, *J. Cryst. Growth, 27*;70, 1974.

178. R.E. Nahory, M.A. Pollock and J.C. De Winter, *Appl. Phys. Lett. 25*;146, 1974.

179. H. Nagai and Y. Noguchi, *Appl. Phys. Lett. 26*;108, 1975.

180. D.B. Holt, *J. Phys. Chem. Solids 27*;1053, 1966.

181. J.W. Matthews, "Coherent interfaces and misfit dislocations," in *Epitaxial Growth* (ed. J.W. Matthews) Part B, New York, Academic, p. 562, 1975.

182. K. Nakajima, S. Komiya, K. Akita, T. Yamaoka and O. Ryuzan, *J. Electrochem. Soc. 127*;1568, 1980.

183. K. Nakajima, S. Yamazaki, S. Komiya and A. Akita, *J. Appl. Phys. 52*;4575, 1981.

184. M.S. Abrahams and C.J. Buiocchi, *J. Appl. Phys. 45*;3315, 1974.

185. D.L. Rode, *J. Cryst. Growth 27*;313, 1974.

186. M.S. Abrahams, L.R. Weisberg, C.J. Buiocchi and J. Blanc, *J. Mat. Sci. 4*;223, 1969.

187. M.S. Abrahams, C.J. Buiocchi, and G.H. Olsen, *J. Appl. Phys. 46*;4257, 1975.

188. G.B. Stringfellow, *J. Appl. Phys. 43*;3455, 1972.

189. Y. Takeda and A. Sasaki, *J. Cryst. Growth 45*;257, 1978.

190. M.C. Joncour, J.L. Benchimal, J. Burgeat and M. Quillec, *J. Phys. Colloq. 43*;C5-3, 1982.

191. J.J. Coleman, N. Holanyak, Jr., and M.J. Ludowise, *Appl. Phys. Lett. 28*;363, 1976.

192. J.J. Hsieh, M.C. Finn, and J.A. Rossi, in *GaAs and Related Compounds, Inst. of Phys. Conf. 33b*;37, 1977.

193. K. Oe and K. Sugiyama, *Appl. Phys. Lett. 33*;449, 1978.

194. M.F. Gratton, R.G. Goodchild, L.Y. Juravel and J.C. Woolley, *J. Electron. Mat. 8*;25, 1979.

195. R.E. Nahory, M.A. Pollack, E.D. Beebe and J.C. De Winter, *J. Electrochem. Soc. 125*;1053, 1978.

196. L.M. Dolginov, D.G. Eliseev, A.N. Lapshin and M.G. Milvidskii, *Kristall und Technik, 13*;631, 1978.

197. M. Quillec, C. Daguet, J.L. Benchimol and H. Launois, *Appl. Phys. Lett. 40*;3225, 1982.

198. F. Glas, M.M.J. Treacy, M. Quillec and H. Launois, *J. Phys. Colloq. 43*;C5-1, 1982.

199. B. De Cremoux, *J. Phys. Colloq. 43*;C5-19, 1982.

200. G.B. Stringfellow, *J. Cryst. Growth, 58*;194, 1982.

201. G.B. Stringfellow, *J. Electron. Mat. 11*;903, 1982.

202. G.B. Stringfellow, *J. Appl. Phys. 54*;404, 1983.

203. K. Onabe, *Jap. J. Appl. Phys. 22*;663, 1983.

204. H. Ijuin and S. Gonda, *J. Cryst. Growth, 33*;215, 1976.

205. I. Crossley and M.B. Small, *J. Cryst. Growth, 15*;275, 1972.

206. K. Hiramatsu, K. Tomita, N. Sawaki and I. Akasaki, *Jap. J. Appl. Phys. 23*;68, 1984.

207. H. Nagai and Y. Noguchi, *Appl. Phys. Lett. 32*;234, 1978.

208. S. Kondo, T. Amano and K. Akita, *J. Cryst. Growth 61*;8, 1983.

209. K. Nakajima, S. Yamazaki and A. Akita, *J. Cryst. Growth 61*;535, 1983.

210. M.B. Panish, S. Sumski and I. Hayashi, *Metall. Trans. 2*;795, 1971.

211. P.E. Brunemeier, K.C. Hsieh, D.G. Deppe, J.M. Brown and N. Holonyak, *J. Cryst. Growth 71*;705, 1985.

212. C.R. Elliott, M.M. Faktor, J. Haigh and M.R. Taylor, *Solid State Electron. 22*;446, 1979.

213. S. Mahajan, D. Brasen, M.A. DiGiuseppe, V.G. Keramidas, H. Temkin, C.L. Zipfel, W.A. Bonner, and G.P. Schwartz, *Appl. Phys. Lett. 41*;266, 1982.

214. V. Swaminathan, W.R. Wagner, N.E. Schumaker and R.L. Miller, *Thin Solid Films 93*;195, 1982.

215. B. Wakefield, *Appl. Phys. Lett. 33*;408, 1978.

216. P. Besomi, R.B. Wilson, and R.J. Nelson, *J. Electrochem. Soc. 132*;176, 1985; R.B. Wilson, P. Besomi, and R.J. Nelson, *J. Electrochem. Soc. 132*,;172, 1985.

217. B.W. Hakki, C.A. Gaw, W.R. Holbrook, and N.E. Schumaker, unpublished, 1981.

218. B.I. Miller, E. Pinkas, I. Hayashi, and R.J. Capik, *J. Appl. Phys. 43*;2817, 1972.

219. R.C. Peters, in *GaAs and Related Compounds*, Inst. Phys. Conf. Ser. (London), *17*;55, 1973.

220. S.G. Napholtz, unpublished.

221. Y. Nishitani, K. Akita, S. Komiya, K. Nakajima, A. Yamaguchi, O. Ueda and T. Kotani, *J. Cryst. Growth 35*;279, 1976.

222. B.V. Dutt, S. Mahajan, R.J. Roedel, G.P. Schwartz, D.C. Miller, and L. Darick, *J. Electrochem. Soc. 128*;1573, 1981.

223. P.D. Greene, in *The Chemistry of the Semiconductor Industry,* (eds. S.J. Moss and A. Ledwith) Glasgow, London; Blackie, p. 157, 1978.

224. P.D. Green and E.J. Thrush, *J. Cryst. Growth 72*;563, 1985.

225. S. Komiya, T. Tanahashi and I. Umebu, *Jap. J. Appl. Phys. 24*;1053, 1985.

226. G.A. Antypas, *Appl. Phys. Lett. 37*;64, 1980.

227. V. Wrick, G.J. Scilla, L.F. Eastman, R.L. Henry and E.M. Swiggard, *Electron. Lett. 12*;394, 1976.

228. T. Nishinaga, K. Pak and S. Uchiyama, *J. Cryst. Growth 42*;315, 1977.

229. K. Shima, N. Takagi, K. Segi, H. Imai, K. Hori and M. Takusagawa, *Appl. Phys. Lett. 36*;395, 1980.

230. E. Bauser, M. Frik, K.S. Lochner, L. Schmidt and R. Ulrich, *J. Cryst. Growth 27*;148, 1974.

231. D.L. Rode, *J. Cryst. Growth, 27*;313, 1974.

232. K. Hess, N. Stath and K.W. Benz, *J. Electrochem. Soc. 121*;1209, 1974.

233. E. Bauser and K.W. Benz, *Microelectronics J. 13*;10, 1982.

234. R. Messham and A. Majerfeld, loc. cit. Ref. 124.

235. R.C. Peters, in *GaAs and Related Compounds,* Inst. Phys. Conf. Ser. (London) *17*;55, 1973.

236. M.B. Small, A.E. Blakeslee, K.K. Shih and R.M. Potemski, *J. Cryst. Growth 30*;257, 1975.

237. F.R. Nash, W.R. Wagner, and R.L. Brown, *J. Appl. Phys. 47*;3992, 1976.

238. F.R. Nash, R.W. Dixon, P.A. Barnes, and N.E. Schumaker, *Appl. Phys. Lett. 27*;234, 1975.

239. R.A. Logan, N.E. Schumaker, C.H. Henry and F.R. Merritt, *J. Appl. Phys. 50*;5970, 1979.

240. T.P. Lee, C.A. Burrus, Jr., and A.G. Dentai, *IEEE J. Quant. Electron. QE-17*;232, 1981.

241. T. Martin, C.R. Stanley, A. Iliadis, C.R. Whitehouse, and D.E. Sykes, *Appl. Phys. Lett. 46*;994, 1985.

242. K. Nakajima, S. Yamazaki, T. Takanohashi and K. Akita, *J. Cryst. Growth 59*;572, 1982.

243. M.E. Weiner, *J. Electrochem. Soc. 119*;496, 1972.

244. P.D. Greene, *J. Phys. D6*;1550, 1973.

245. P.D. Greene and S.A. Wheeler, *Appl. Phys. Lett. 35*;78, 1979.

246. J.D. Oliver, Jr., and L.F. Eastman, *J. Electron. Mat. 9*;693, 1979.

247. E. Kuphal, *J. Cryst. Growth, 54*;117, 1981.

248. S.H. Groves and M.C. Plonko, in *GaAs and Related Compounds,* Inst. Phys. Conf. Ser. (London) *45*;71, 1979.

249. T. Amano, K. Takahei and H. Nagai, *Jap. J. Appl. Phys. 20*;2205, 1981.

250. W.F. Finch and E.W. Mehal, *J. Electrochem. Soc. 111*;814, 1964.

251. D. Effer, *J. Electrochem. Soc. 112*;1020, 1965.

252. R.C. Clarke, B.D. Joyce and H.W.E. Wilgoss, *Solid State Commun. 8*;1125, 1970.

253. J.A. Long, R.A. Logan, R.F. Karlicek, Jr., in *Optical Fiber Telecommunications II*, (ed. S.E. Miller and J.P. Kaminow), New York; Academic Press, Chap. 16, 1988.

254. S.K. Ghandi and I.B. Bhat, *MRS Bulletin 23*;37, 1988.

255. L. Hollan and J. Hallais, in *GaAs FET Principles and Technology* Dedham, MA; Artech, 1982, (ed. J.V. DiLorenzo and D.D. Khandelwal).

256. R.C. Clarke, *J. Cryst. Growth 54*;88, 1981.

257. D.W. Shaw, *J. Cryst. Growth 8*;117, 1971.

258. H.M. Cox, M.A. Koza, V.G. Keramidas, M.S. Young, *J. Cryst. Growth 73*;523, 1985.

259. P. Vohl, *J. Cryst. Growth 54*;101, 1981.

260. D.W. Shaw, *J. Cryst. Growth 35*;1, 1976.

261. H.M. Cox, A.S. Prior, V.G. Keramidas, in: Proc. 10$^{th}$ Int. Symp. on *GaAs and Related Compounds,* Albuquerque, NM, 1982. Inst. Phys. Conf. Ser. 65 (ed. G.E. Stillman) London-Bristol;Inst. Phys., p. 133, 1983.

262. J.V. DiLorenzo, *J. Cryst. Growth 17*;189, 1972.

263. D.W. Shaw, *J. Electrochem. Soc. 117*;683, 1970.

264. J. Komeno, M. Nogami, A. Shibatomi, and S. Ohkawa, *GaAs and Related Compounds 1980*, Inst. of Phys. Conf. Ser. No. 56, (ed. H.W. Thim), p. 9, 1981.

265. W.D. Johnston, Jr., M.A. DiGiuseppe, and D.A. Wilt, *AT&T Technical Journal, 60*;53, 1989.

266. M.A. DiGiuseppe, private communication.

267. J.J. Tietjen and J.A. Amick, *J. Electrochem. Soc. 113*;724, 1966.

268. G.H. Olsen, ''Vapour-phase Epitaxy of GaInAsP,''in *GaInAsP Alloy Semiconductors*, (ed. T.P. Pearsall), Chap. 1, New York; Wiley, 1982.

269. S.B. Hyder, *J. Cryst. Growth 54*;109, 1981.

270. R.F. Karlicek, Jr., and A. Bloemeke, *J. Cryst. Growth 73*;364, 1985.

271. D.N. Buckley, SOTAPOCS session of the Electrochemical Society Fall Meeting, Chicago, Illinois, held Oct. 1988, paper number 770 S0A.

272. L.M. Zinkiewicz, T.R. Lepkowski, T.J. Roth, and G.E. Stillman, *Gallium Arsenide and Related Compounds* (Inst. Phys. Conf. Ser. No. 56), p. 19, 1980.

273. H. Kanbe, Y. Yamauchi, and M. Susa, *Appl. Phys. Lett. 35*;603, 1979.

274. C.J. Nuese, D. Richman, and R. B. Clough, *Metall. Trans. 2*;789, 1971.

275. R.E. Enstrom, D. Richman, M.S. Abrahams, J.R. Appert, D.G. Fisher, A.H. Sommer, and B.F. Williams, *Gallium Arsenide and Related Compounds* (Inst. Phys. Conf. Ser. No. 9), p. 30, 1970.

276. M. Yoshida and H. Watanabe, *J. Electrochem. Soc. 132*;1733, 1985.

277. R.E. Enstrom, C.J. Nuese, V.S. Ban, and J.R. Appert, in *Gallium Arsenide and Related Compounds* 1972 (Inst. Phys. Conf. Ser. No. 17), p. 37.

278. W.D. Johnston, Jr., and K.E. Strege, 38th Annual IEEE Device Research Conf. Abstracts, Cornell University, vol. IVB-3, June 1980.

279. K. Sugiyama, H. Kojima, H. Enda, and M. Shibata, *Jpn. J. Appl. Phys. 16*;2197, 1977.

280. H. Enda, *Jpn. J. Appl. Phys. 18*;2167, 1979.

281. V.S. Ban, *J. Electrochem. Soc. 118*;1473, 1971.

282. D.W. Shaw, *J. Chem. Phys. Solids 36*;111, 1975.

283. H. Nagai, *J. Cryst. Growth 48*;359, 1980.

284. A. Koukitu and H. Seki, *J. Cryst. Growth 49*;325, 1980.

285. T.M. Mizutani and H. Watanabe, *J. Cryst. Growth 59*;507, 1982.

286. P.A. Longeway and R.T. Smith, *J. Cryst. Growth 89*;519, 1988.

287. R.F.C. Farrow, *J. Phys. D 7*;2436, 1974.

288. W.Y. Lum and A.R. Clawson, *J. Appl. Phys. 50*;5296, 1979.

289. C.R. Bayliss and D.L. Kink, *J. Phys. D9*;233, 1976.

290. H. Watanabe, M. Yoshida, and Y. Seki, Electrochem. Soc. Extended Abstracts, 151st Meeting, Philadelphia, May 1977, p. 255.

291. T. Mizutani, M. Yoshida, A. Usai, H. Watanabe, T. Yuasa and I. Hayashi, *Jpn. J. Appl. Phys. 19*;L113, 1980.

292. G.H. Olsen and T.J. Zamerowski, *Progress in Crystal Growth and Characterization,* vol. 2, (ed. B.R. Pamplin) London; Pergamon, p. 309, 1979

293. M.A. DiGiuseppe, H. Temkin, L. Peticolas, and W.A. Bonner, *Appl. Phys. Lett. 43*;906, 1983.

294. G. Beuchet, M. Bonnet, and J.P. Duchemin, Proc. 1980 NATO Conf. on *InP,* Rome Air

Development Center Tech. Memo. RADC-TM-80-07, Hanscom Air Force Base, MA, 1980, p. 303.

295. H.M. Cox, *J. Cryst. Growth 69*;641, 1984.

296. A.T. Macrander and K.E. Strege, *J. Appl. Phys. 59*;442, 1986.

297. A.T. Macrander, B.M. Glasgow, E.R. Minami, R.F. Karlicek, D.L. Mitcham, V.G. Riggs, D.W. Berreman, and W.D. Johnston, Jr., MRS Symp. Proc. *90*, 225 (1987).

298. R.F. Karlicek, B. Hammarlund, and J. Ginocchio, *J. Appl. Phys. 60*;794, 1986.

299. R.F. Karlicek, Jr., D. Mitcham, J.C. Ginocchio, and B. Hammarlund, *J. Electrochem. Soc. 134*;470, 1987.

300. H.M. Manasevit, *Appl. Phys. Lett. 11*;156, 1968.

301. R.D. Dupuis, *J. Cryst. Growth 55*;213, 1981.

302. R.M. Lum, J.K. Klingert, A.S. Wynn, and M.G. Lamont, *Appl. Phys. Lett. 52*;1475, 1988.

303. G.B. Stringfellow, *J. Cryst. Growth 68*;111, 1984.

304. W.D. Johnston, Jr., *J. Cryst. Growth 39*;117, 1977.

305. J.L. Zilko, in ''Handbook of thin-film deposition processes and techniques'' (ed. by K. K. Schuegraf), Park Ridge, NJ; Noyes, Chap. 7, p. 234, 1988.

306. S.K. Ghandi and R.J. Field, *J. Cryst. Growth 69*;619, 1984.

307. H. Itoh, H. Tanaka, T. Ohori, M. Takikawa, K. Kasai, and J. Komeno, *SOTAPOCS* session of the Electrochemical Society Fall Meeting, Honolulu, Hawaii, held Oct. 1987, paper number 1776 SOA.

308. A. Mircea, A. Ougazzaden, P. Dasté, Y. Gao, C. Kazmierski, J.C. Bouley and A. Carenco, *J. Cryst. Growth 93*;235, 1988.

309. J.P. Hirtz, M. Razeghi, M. Bonnet, and J.P. Duchemin, in ''GaInAsP Alloy Semiconductors'', (ed. T.P. Pearsall), New York; Wiley, Chap. 3, p. 64, 1982.

310. H.M. Manasevit and W.I. Simpson, *J. Electrochem. Soc. 116*;1968, 1969.

311. V. Gottschalch, W.H. Petzke, and E. Butter, *Kristall Tech. 9*;209, 1974.

312. M.R. Leys and H. Veenvliet, *J. Cryst. Growth 55*;145, 1981.

313. H. Krautle, H. Rochle, A. Escobosa, and H. Beneking, *J. Electron. Mater. 12*;215, 1983.

314. P.B. Chinoy, P.D. Agnello, and S.K. Ghandi, MRS Meeting, Boston 1988; MRS Symp. Proc. *144* (in press).

315. J.A. Long, V.G. Riggs, and W.D. Johnston, Jr., *J. Cryst. Growth 69*;10, 1984.

316. A.T. Macrander, J.A. Long, V.G. Riggs, A.F. Bloemeke, and W.D. Johnston, Jr., *Appl. Phys. Lett. 45*;1297, 1984.

317. J.T.S. Andrews and E.F. Westrum, Jr., *J. Organomet. Chem. 17*;349, 1969.

318. J.A. Long, V.G. Riggs, A.T. Macrander, and W.D. Johnston, Jr., *J. Cryst. Growth 77*;42, 1986.

319. N.K. Dutta, J.L. Zilko, T. Cella, D. A. Ackerman, T.M. Shen, and S.G. Napholtz, *Appl. Phys. Lett. 48*;1572, 1986.

320. D.P. Wilt, J.A. Long, W.C. Dautremont-Smith, M.W. Focht, T.M. Shen, and R.L. Hartman, *Electron. Lett. 22*;869, 1986.

321. J. Cheng, R. Stall, S.R. Forrest, J.A. Long, C.L. Cheng, G. Guth, R. Wunder, and V.G. Riggs, IEEE Electron. Dev. Lett. EDL-6, 384 (1985).

322. R.F. Karlicek, Jr., *J. Cryst. Growth 91*;33, 1988.

323. K. Tanaka, K. Nakai, and S. Yamakoshi, *SOTAPOCS* session of the Electrochemical Society Fall Meeting, Chicago, Illinois, held Oct. 1988, paper number 769 SOA.

324. F. Capasso, in *Physics and Applications of Quantum Wells and Superlattices,* (eds. E.E. Mendez and K. Von Klitzing), New York; Plenum, p. 377, 1987.

325. K.G. Günther, Z. Naturforsch. *Teil A 13*;1081, 1958.

326. J.E. Davey and T. Pankey, *J. Appl. Phys. 39*;1941, 1968.

327. J.R. Arthur, *J. Phys. Chem. Solids 28*;2257, 1967.

328. J.R. Arthur, *J. Appl. Phys. 39*;4032, 1968.

329. J.R. Arthur and J.J. Lepore, *J. Vac. Sci. Technol. 6*;545, 1969.

330. A.Y. Cho and J.R. Arthur, *Prog. Solid State Chem. 10*;157, 1975.

331. A.Y. Cho, *J. Vac. Sci. Technol. 16*;275, 1979.

332. W.T. Tsang, in *Semiconductors and Semimetals,* (eds. R.K. Willardson and A.E. Beer) New York; Academic vol. 22, *Lightwave Communications Technology,* vol. (ed. W.T. Tsang), Part A, p. 95, 1985.

333. L. L. Chang and R. Ludeke, in *Epitaxial Growth,* (ed. J.W. Matthews), New York; Academic, Part A, p. 37, 1975.

334. L.L. Chang in *Handbook on Semiconductors,* vol. 3 p. 565 North Holland, Amsterdam, 1980.

335. K. Ploog, *Ann. Rev. Mat. Sci. 11*;171, 1981.

336. C.T. Foxon and B.A. Joyce, in *Current Topics in Materials Science,* (ed. E. Kaldis) North Holland, Amsterdam, 1980.

337. C.T. Foxon, *J. Vac. Sci. Technol. B1*;293, 1983.

338. K.Y. Chang and A.Y. Cho, *J. Appl. Phys. 53*;441, 1982.

339. K. Ploog, in *Physics and Applications of Quantum Wells and Superlattices,* (eds. E.E. Mendez and K. Von Klitzing), New York; Plenum, p. 43, 1987.

340. J.R. Arthur, *Surf. Sci. 43*;449, 1974.

341. C.T. Foxon and B.A. Joyce, *Surf. Sci. 50*;434, 1975.

342. K. Ploog, A. Fischer, and H. Künzel, *Appl. Phys. 18*;353, 1979.

343. K.Y. Cheng, A.Y. Cho, W.R. Wagner, and W.A. Bonner, *J. Appl. Phys. 52*;1015, 1981.

344. A.Y. Cho, *J. Appl. Phys. 42*;2074, 1971.

345. C.T. Foxon and B.A. Joyce, *Surf. Sci. 50*;434, 1975.

346. C.T. Foxon and B.A. Joyce, *Surf. Sci. 64*;293, 1977.

347. B.A. Joyce and C.T. Foxon, *J. Cryst. Growth 31*;122, 1975.

348. R. Fischer, J. Klem, T.J. Drummond, R.E. Thorne, W. Kopp, H. Morkoc, and A.Y. Cho, *Appl. Phys. Lett. 54*;2508, 1983.

349. C.T. Foxon and B.A. Joyce, *J. Cryst. Growth 44*;75, 1978.

350. C.T. Foxon, *J. Vac. Sci. Technol. B1*;293, 1983.

351. T. Sakamoto, H. Funabashi, K. Ohta, T. Nakagawa, N.J. Kawai, T. Kogima, and K. Bando, *Superlattices and Microstructures 1*;347, 1985.

352. J.H. Neave, B.A. Joyce, P.J. Dobson, and N. Norton, *Appl. Phys. A31*;1, 1983.

353. J.M. Van Hove, C.S. Lent, P.R. Pukite, and P.I. Cohen, *J. Vacuum Sci. Technol. B1*;741, 1983.

354. P. Chen, T.C. Lee, N.M. Cho, and A. Madhukar, in *Growth of Compound Semiconductors*, (eds. R.L. Gunshor and H. Morkoc), Proc. SPIE 796, p. 139 1987.

355. A.Y. Cho, *J. Vac. Sci. Technol. 8*;531, 1971.

356. L.L. Chang, W.E. Esaki, W.E. Howard, R. Ludeke, and G. Schul, *J. Vac. Sci. Technol. 10*;655, 1973.

357. J.J. Harris, B.A. Joyce, and P.J. Dobson, *Surf. Sci. 103*;L90, 1981.

358. P.K. Larsen, P.J. Dobson, J.H. Neave, B.A. Joyce, B. Bolger, and J. Zhang, *Surf. Sci. 169*;176, 1986.

359. B.A. Joyce, P.J. Dobson, J.H. Neave, and J. Zhang, in *Two-Dimensional Systems: Physics and New Devices*, (eds. G. Bauer, F. Kuchar, and H. Heinrich) Berlin; Springer-Verlag, p. 42, 1986.

360. Y. Namba, R.W. Vook, and S.S. Chao, *Surf. Sci. 109*;320, 1981.

361. K.D. Grönwald and M. Henzler, *Surf. Sci. 117*;180, 1982.

362. B.A. Joyce, J. Zhang, J.H. Neave, and P.J. Dobson, Appl. Phys. A45, 255 (1988).

363. B.A. Joyce, P.J. Dobson, J.H. Neave, K. Woolbridge, J. Zhang, P.K. Larsen and B. Bölger, *Surf. Sci. 168*;423, 1986.

364. Y. Suzuki and H. Okamoto, *J. Appl. Phys. 58*;3456, 1985.

365. M. Tanaka, H. Sakaki, and J. Yoshino, *Jap. J. Appl. Phys. 25*;L155, 1986.

366. F. Voillot, A. Madhukar, J.Y. Kim, P. Chen, N.M. Cho, W.C. Tang and P.G. Newman, *Appl. Phys. Lett. 48*;1009, 1986.

367. M.B. Panish and H. Temkin, Ann. Rev. Mat. Science, in press (1989).

368. W.T. Tsang, "Chemical Beam Epitaxy", in *Beam Processing Technologies*", (ed. N.G. Einspruch, S.S. Cohen, and R.N. Singh) Academic Press, in press.

369. W.T. Tsang, *IEEE Circuits and Devices Magazine 4*;18, 1988.

370. M.B. Panish, "Gas Source Molecular Beam Epitaxy" in Proceedings of the Workshop on *Mechanisms of Reactions of Organometallic Compounds with Surfaces,* St. Andrews, Scotland, June 1988.

371. M.B. Panish, H. Temkin, and S. Sumski, *J. Vac. Sci. Technol. B3*;657, 1985.

372. M.B. Panish and S. Sumski, *J. Appl. Phys. 55*;3571, 1984.

373. W.T. Tsang, "Chemical Beam Epitaxy", in *Beam Processing Technologies*, (ed. N.G. Einspruch, S.S. Cohen, and R.N. Singh) New York; Academic Press, 1989.

374. A. Robertson, Jr., T.H. Chiu, W.T. Tsang, and J.E. Cunningham, *J. Appl. Phys.*, 64; 877, 1988.

375. E. Tokomitsu, Y. Kudou, M. Konagai, and K. Takahashi, *J. Appl. Phys. 55*;3163, 1984.

376. N. Pütz, E. Veuhoff, H. Heinecke, M. Heyen, H. Lüth, and P. Balk, *J. Vac. Sci. and Technol. B3*;671, 1985.

377. N. Kobayashi and T. Fukui, *Electron. Lett. 20*;887, 1984.

378. K. Kondo, H. Ishikawa, S. Sasa, Y. Suguyama, and Y. Hiyamizu, *Jap. J. Appl. Phys. 25*;L52, 1986.

# CHAPTER 3

# X-RAY STRUCTURAL CHARACTERIZATION

Structural information is conveniently obtained by x-ray diffraction. Samples can be easily mounted directly after they are grown without additional preparation and, except in the case of topography, can be quickly analyzed, often in just minutes. The ease and speed of analyses make x-ray techniques particularly well suited as a diagnostic tool for epitaxial crystal growth.

We present in this chapter techniques crystal growers are most likely to need and use. These techniques are double-crystal diffractometry which is used for measuring the lattice mismatch between an epitaxial layer and substrate and for assessing crystalline quality, the back reflection Laue method which is used to orient crystals, the Bond method for measuring lattice parameters, and x-ray topography.

These techniques can all be implemented using one of three x-ray sources. The most commonly used source and the one of primary concern here is an evacuated tube having a fixed anode and known as a Coolidge tube. These are typically rated at 1.5 kW and provide lines characteristic of the anode (such as Cu or W) superimposed on a white background as a result of Bremstrahlung radiation. Rotating anodes emit the same spectrum but can provide roughly an order of magnitude greater power. Synchrotron radiation which is continuous over a wide spectral range can also be used but is not considered here.

## 3.1 X-RAY DOUBLE CRYSTAL DIFFRACTOMETRY

### 3.1.1 Introduction

Since it was first used in 1917 by Compton [1] the double crystal diffractometer (DCD) has proved to be a powerful characterization tool for single crystals. This is especially true of semiconductor single crystals. GaAs and InP based heterostructure of such a high quality can now be produced that a double crystal arrangement is necessitated to obtain a reasonable assessment of the crystalline quality. A DCD is needed because the instrumental line broadening inherent in single crystal diffractometry (where the single crystal is the sample) obscures the intrinsic reflectivities of these crystals.

The DCD produces increased resolution because wavelength dispersion is eliminated as illustrated in Fig. 3.1. All x-ray sources produce radiation covering a finite wavelength range and with some finite divergence. The divergence is determined by the size of collimating slits and the distance between them. Unless this distance is made many meters long [2], which is impractical, this divergence will result in instrumental broadening which will be too large to permit a reasonable assessment of the reflection width of high quality crystals. The width of a rocking curve made using a DCD in the parallel configuration is, however, independent of the wavelength range and of the horizontal angular divergence (the divergence in the plane of Fig. 3.1). This is so because the same "rocking curve" results for each wavelength independently, and these are superimposed in the final measured rocking curve. The Bragg condition for each wavelength is satisfied simultaneously when the two crystals are exactly parallel. This is the situation shown in Fig. 3.1. The wavelength components are separated in space, however. If both $K\alpha$ components are present, the $K\alpha_2$ component can be blocked by a slit edge in between the two crystals. It is, furthermore, possible to show that rocking curves are independent of any vertical divergence for the parallel configuration [3].

A convenient illustration related to the double crystal arrangement is shown in Fig. 3.2 which is due to Dumond [4]. This plot is simply a graphical representation of Bragg's law. The spectral range of the x-ray source limits the range of interest on the wavelength axis, and the angular divergence of the x-ray beam in the plane of the diffractometer limits the range of interest on the angle axis. For conventional sources, the spectral range usually encompasses both $K\alpha_1$, and $K\alpha_2$ wavelengths. The inset in Fig. 3.2 is for such a case and contains Darwin bands for two crystals in the (+,−) or parallel configuration. (The (+,−) notation arises from a vector cross product convention between outgoing and ingoing beams, that is, if the cross product between outgoing and ingoing beams is up with respect to the plane of the diffractometer, then the reflection is denoted as +). The Darwin solution for the (400) reflection of InP is shown in Fig. 3.3a, and

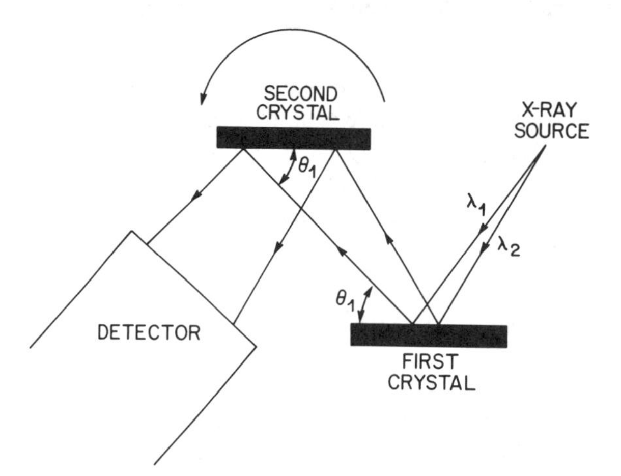


**Figure 3.1** Double crystal (+, −) geometry illustrating its wavelength dispersion free property.

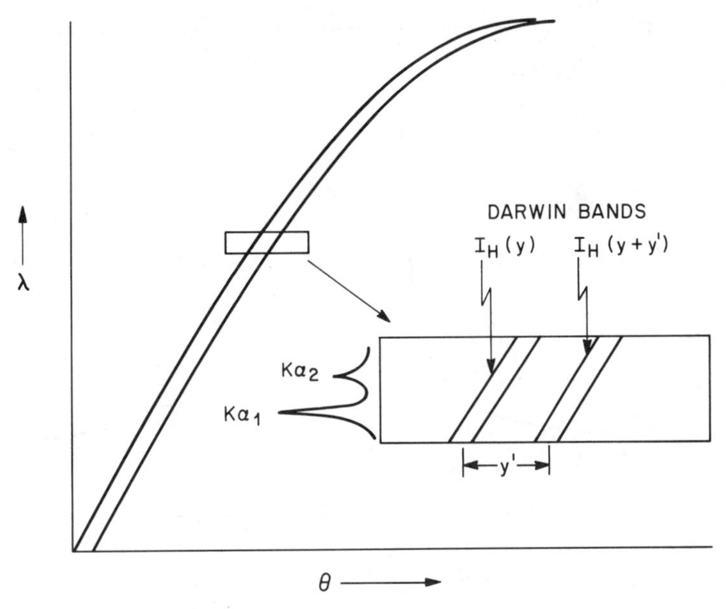


**Figure 3.2** Dumond diagram for the (+, –) double crystal configuration.

represents the intrinsic reflecting power of a thick crystal in the case of negligible absorption. It can be simply expressed [5, 6]:

$$\frac{P_H}{P_O} = 1, \; |y| \leq 1$$

$$\frac{P_H}{P_O} = \left( |y| - \sqrt{y^2-1} \right)^2, \; |y| \geq 1 \tag{3.1}$$

Here y is a reduced variable for the angular deviation from the Bragg angle [5]. If absorption is included, then the Darwin-Prins solution [6] is obtained. This is shown for the (400) reflection of InP in Fig. 3.3b. The two curves in Fig. 3.2 for each crystal are for $y = \pm 1$. The final result for the intensity in the case of a double crystal rocking curve is a convolution between the reflecting powers of the two crystals and contains contributions for both σ (electric field normal to the diffraction plane) and π (electric field in the diffraction plane) polarizations. However, the σ polarization usually predominates. Furthermore, in the (+,–) setting the required integral over wavelengths cancels in the normalization integral (which is the same as saying that wavelength dispersion is cancelled). Ignoring the rest of the normalization integral and considering only the σ polarization, we obtain the following expression for the rocking curve amplitude

$$P(y') = \int_{-\infty}^{\infty} P_H(y) P'_H(y'+y) dy \tag{3.2}$$

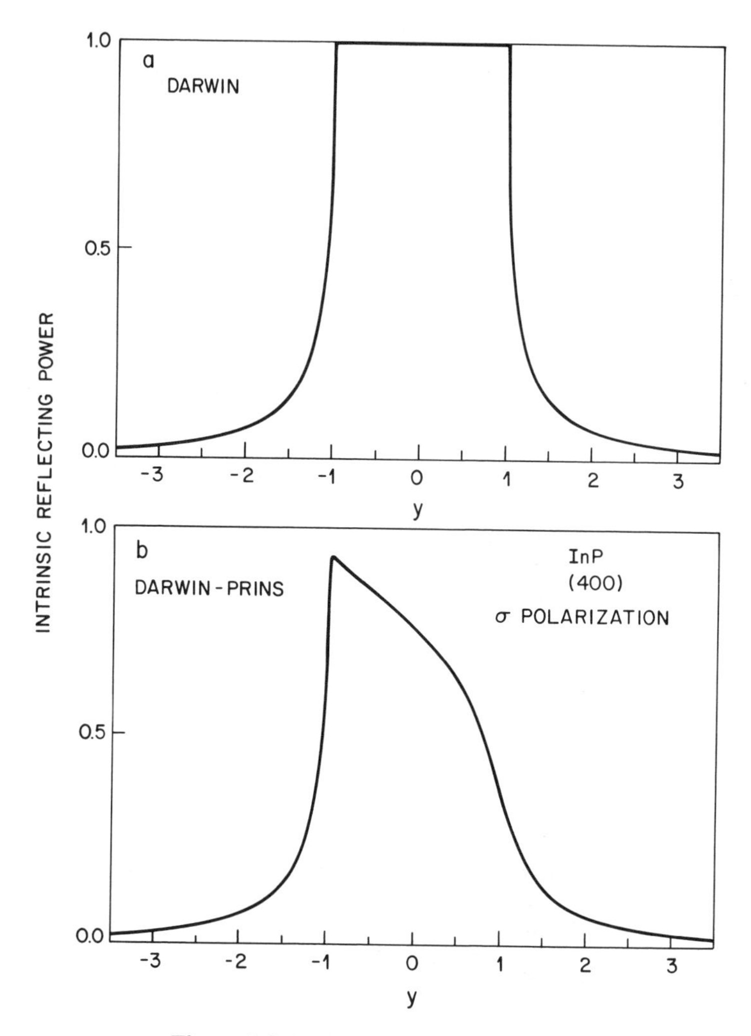


**Figure 3.3** Intrinsic reflecting power of the (400) reflection of a bulk (100) oriented InP crystal. a) Darwin solution (negligible absorption); b) Darwin-Prins solution (absorption included).

Here primed values refer to the second crystal and both reflections are symmetric. The complete expression can be found in Zachariasen's book [5]. The situation shown in the inset of Fig. 3.2 is such that there is no overlap between the two Darwin bands. For reduced values of $y'$, however, the bands overlap and $P(y')$ becomes nonzero. From this construction it is evident that the resulting $P(y')$ is independent of wavelength, that is, at each wavelength the two Darwin bands in Fig. 3.2 overlap identically so that the same convolution integral applies.

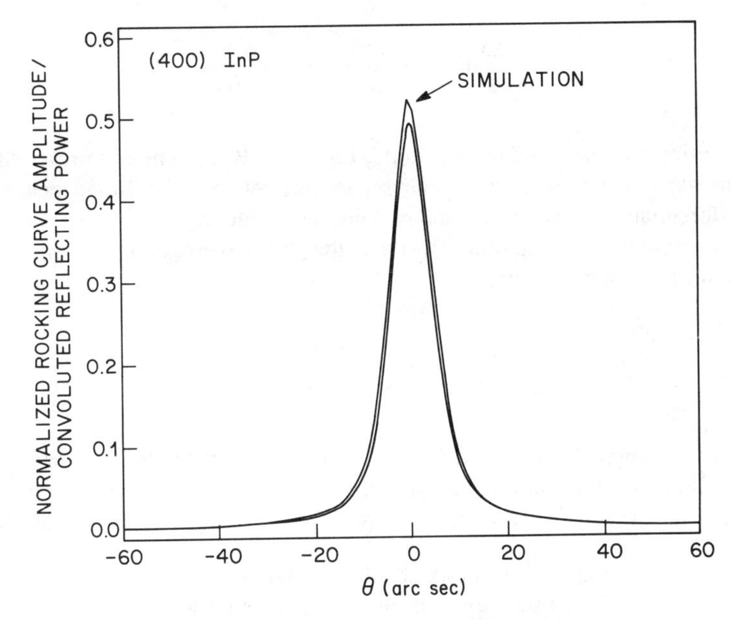


**Figure 3.4** (400) double crystal rocking curve data for a bulk (100)-oriented InP crystal and a dynamical diffraction theory simulation.

### 3.1.2 Indium Phosphide Substrates

The (400) reflection is convenient for (100) substrates because it is intense and the planes are parallel to the surface. A (400) rocking curve obtained using CuKα radiation and two large (100) InP crystals (~1 $in^2$) 4 mm thick with dislocation densities of $10^3$–$10^4$ $cm^{-2}$ is shown in Fig. 3.4. The observed value of 10 arcsec for the full width at half maximum (FWHM) is in excellent agreement with the value of 9.8 arcsec which is calculated from x-ray dynamical diffraction theory [7]. Noticeable influences of dislocations on rocking curves occur only at higher dislocation densities, that is, for etch pit densities $\gtrsim 2 \times 10^5$ $cm^{-2}$ [8]. Note that the tilt alignment of the sample (rotation of the reciprocal lattice vector into the diffraction plane) is crucial for obtaining linewidths of high quality samples for comparison to dynamical diffraction theory; the tilt between the two crystals must be less than 30 arcsec.

### 3.1.3 Single Epitaxial Layer Measurements and Comparisons to Dynamical Diffraction Theory

Straightforward analyses of rocking curve Bragg peak separations can be made in the case of the (400) reflection for epitaxial layers grown on (100) oriented substrates since in that case the Bragg spacing is given by $a_\perp/4$ where $a_\perp$ is the lattice parameter in the [100] direction. If the orientation is directly on [100] then tilt will normally not be present and the mismatch of the epitaxial layer is given by

$$\left(\frac{\Delta a}{a}\right)_{\perp} \equiv \frac{a_{\perp} - a_s}{a_s} = -\frac{\Delta\theta}{\tan\theta_B} \tag{3.3}$$

Here $a_s$ is the lattice parameter of the substrate, $\theta_B$ is the Bragg angle corresponding to $a_s$ and $\Delta\theta = \Delta\omega$ where $\Delta\omega$ is the measured angular separation between the Bragg peaks. Eq. (3.3) is obtained by differentiating Bragg's law and making the small angle approximation for $\Delta\theta$. If $\Delta\theta$ is not small compared to $2/(\tan\theta_B+2)$, then the full expression without a small angle approximation must be used, namely,

$$\left(\frac{\Delta a}{a}\right)_{\perp} = \frac{\sin\theta_B}{\sin(\theta_B + \Delta\theta)} - 1 \tag{3.4}$$

A schematic rocking curve is shown in Fig. 3.5. In this figure, either a $\theta$ or $a_{\perp}$ scale can be applied to the abscissa. The minus sign in Eq. (3.3) means the increasing $\theta$ corresponds to decreasing $a_{\perp}$. Thus the schematic situation shown in Fig. 3.5 is for positive mismatch.

**Figure 3.5** A (400) double crystal rocking curve for a 1.0μm thick mismatched coherent GaInAsP layer on a (100) oriented InP substrate. A computer simulation for a bandgap of 0.953 eV ($\lambda$ = 1.3μm) and for a mismatch of $(\frac{\Delta a}{a})_{\perp} = -0.070\%$ is shown.

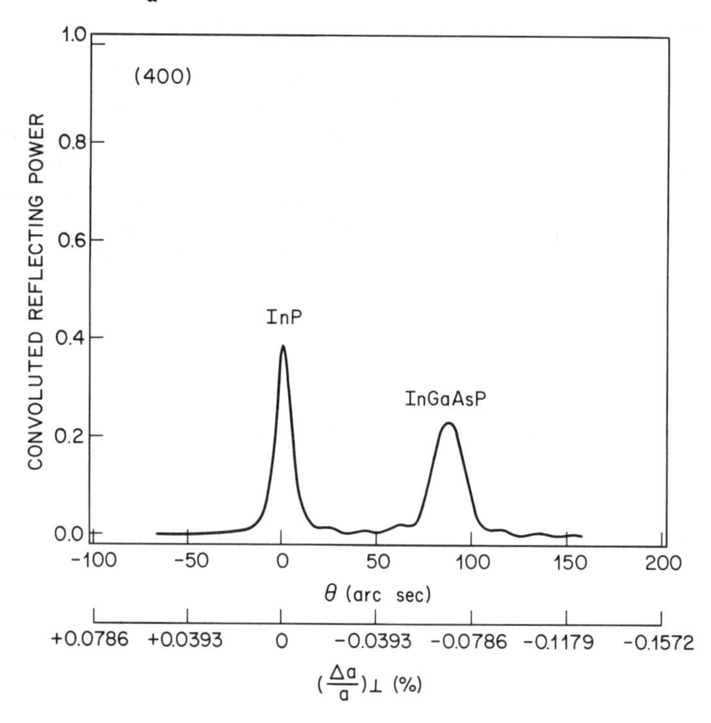

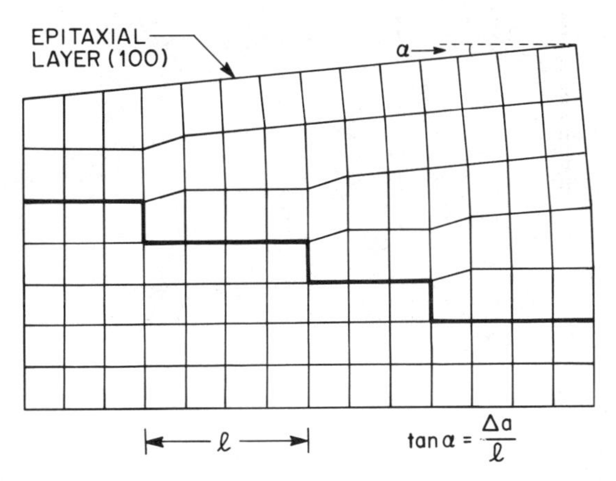


**Figure 3.6** Schematic illustration of the tilt which develops for a mismatched and tetragonally distorted coherent epitaxial layer grown on a stepped surface.

If the epitaxial layer is tilted with respect to the substrate, then two rocking curves are needed to obtain $(\Delta a/a)_\perp$. Such a situation is shown in Fig. 3.6 for an epitaxial layer grown on an off-[100] substrate, that is, the stepped substrate surface was vicinal to the (100) plane [9]. Note that Fig. 3.6 is a purely speculative and schematic illustration. If the angular tilt is denoted as $\Delta\phi$, and introduce another angle, namely, the azimuthal rotation of the sample (i.e., rotation axis normal to the sample), $\alpha$, then the separation of the Bragg peaks as a function of $\alpha$ will be as shown in Fig. 3.7. For any given azimuthal angle some fraction $\eta$ of $\Delta\phi$ will contribute to $\Delta\omega$. If,

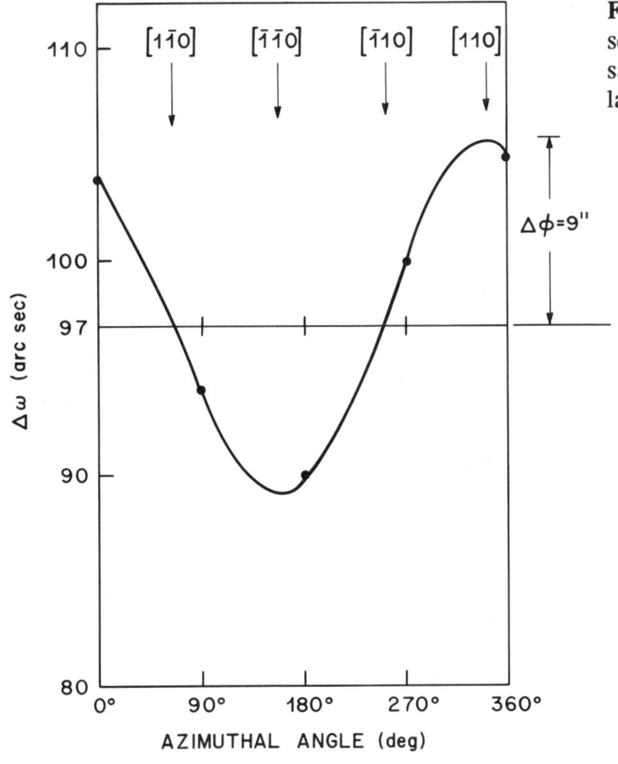


**Figure 3.7** Rocking curve peak separation as a function of azimuthal sample rotation for a tilted epitaxial layer.

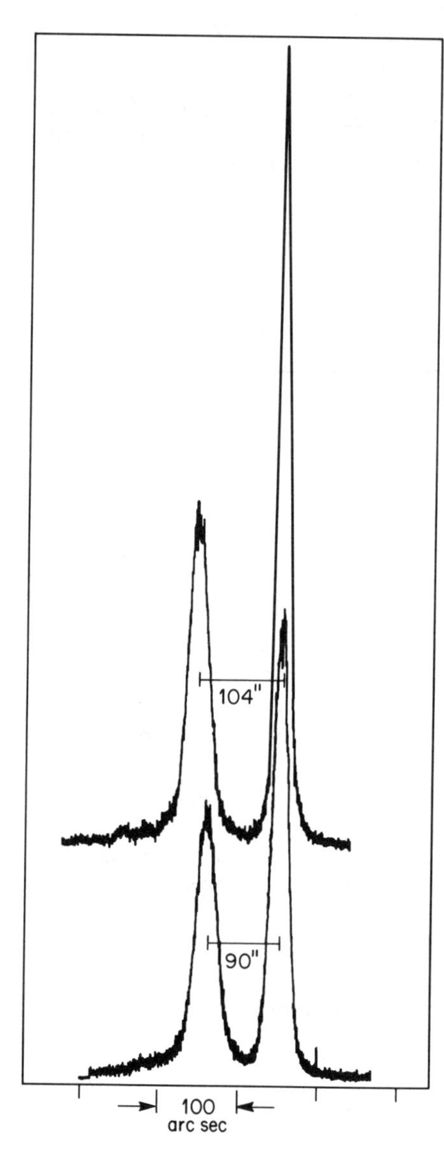


**Figure 3.8** Rocking curves at azimuthal angles of 0° and 180°.

however, a second rocking curve with the sample rotated aximuthally by 180 degree is made, then the two rocking curves will yield

$$\Delta\omega_1 = \Delta\theta + \eta\, \Delta\Phi \tag{3.5}$$

$$\Delta\omega_2 = \Delta\theta - \eta\, \Delta\Phi$$

and $\Delta\theta = (\Delta\omega_1 + \Delta\omega_2)/2$ can be obtained. This is the most practical way to obtain $\Delta\theta$ and to check for the presence of tilt. The reason that $\Delta\omega_1$ and $\Delta\omega_2$ are different can best be understood by considering the hypothetical case of tilt in the absence of mismatch ($\Delta\theta = 0$). In that case, because of tilt the substrate peak will appear before the epilayer peak in one case and vice versa for the other case so long as the sample is rocked in the same way. An example of this shown in Fig. 3.8 for an $Ga_xIn_{1-x}As$ sample grown on a substrate oriented by 3 degrees towards the nearest <110> from [100]. The tilt was found to be 9 arcsec towards this nearest <110> [10]. This value corresponds closely to that expected from the mismatch of $(\Delta a/a)_\perp$, $= 8.3 \times 10^{-4}$ based on the model of Nagai et. al., shown in Fig. 3.6.

Unless an epitaxial layer is perfectly lattice matched to the substrate its lattice will, in general, not have the same cubic unit cell as that of the substrate. In general, it will be distorted because of the mismatch stress. For layers grown on symmetry planes such as (100), (011), and (111) the lattice of the epitaxial layer will be tetragonally distorted with a different lattice parameter in the growth plane, $a_\parallel$, than in the growth direction, $a_\perp$ [11]. For other planes (such as the (311)) the lattice deformation is not symmetrical and has shear character [11]. The detailed nature of the distortion can be obtained using x-ray DCD in all cases and was found to agree with calculations for $Al_xGa_{1-x}As$ layers grown on (100), (011), $(\overline{111})$, and (311) oriented GaAs substrates [12].

For the case of tetragonal distortion, expressions will now be derived for the mismatch in the growth direction, $(\Delta a/a)_\perp$, and for the mismatch parallel to the interface, $(\Delta a/a)_\parallel$, in terms of angular separations between Bragg peaks observed on rocking curves.

Consider a general (hkl) reflection which is asymmetric as shown in Fig. 3.9. Note that coherency $a_\parallel = a_\perp$ is assumed in this figure, but not in the following derivation. Bragg's law for the substrate is given by

**Figure 3.9** The misorientation angle between epitaxial layer and substrate Bragg planes which are oblique to the surface (i.e., for an asymmetric x-ray reflection) for the case of a coherent and tetragonally distorted epitaxial layer.

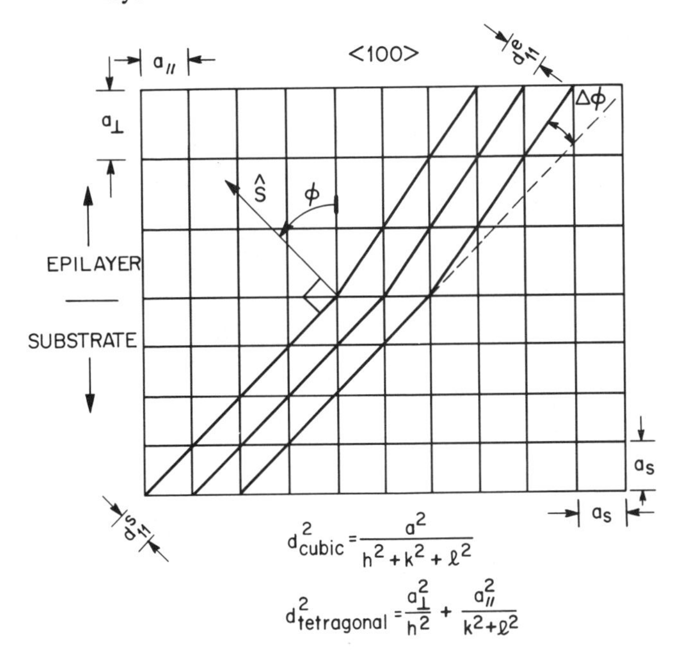

$$\lambda = \frac{2a_s \sin \theta_B}{\sqrt{h^2 + k^2 + \ell^2}} \tag{3.6}$$

For the epitaxial layer, the Bragg angle is written as $\theta_B + \Delta\theta$ and then obtain for Bragg's law,

$$\lambda = 2 \sin(\theta_B + \Delta\theta) \left[ \frac{h^2 + k^2}{a_\parallel^2} + \frac{\ell^2}{a_\perp^2} \right]^{-1/2} \tag{3.7}$$

where $\ell$ is taken to be the index along the growth direction which is taken to be along $\hat{z}$. Next the normals to (hkl) planes of the substrate, $\hat{s}$, and of the epitaxial layer, $\hat{e}$, are given by

$$\hat{s} = \frac{h\hat{x} + k\hat{y} + \ell\hat{z}}{\sqrt{h^2 + k^2 + \ell^2}} \tag{3.8}$$

$$\hat{e} = \frac{\frac{h}{a_\parallel}\hat{x} + \frac{k}{a_\parallel}\hat{y} + \frac{\ell}{a_\perp}\hat{z}}{\sqrt{\left(\frac{h}{a_\parallel}\right)^2 + \left(\frac{k}{a_\parallel}\right)^2 + \left(\frac{\ell}{a_\perp}\right)^2}}$$

We define $\phi$ as the angle between $\hat{s}$ and the surface normal, $\hat{z}$, and we define $\phi + \Delta\phi$ is defined as the angle between $\hat{e}$ and $\hat{z}$. Then we obtain directly

$$\cos\phi = \hat{s}\cdot\hat{z} = \frac{1}{\sqrt{h^2 + k^2 + \ell^2}} \tag{3.9}$$

and

$$\sin\phi = \hat{s}\cdot\hat{x} = \frac{h}{\sqrt{h^2 + k^2 + \ell^2}} \tag{3.10}$$

and

$$\cos(\phi+\Delta\phi) = \hat{e}\cdot\hat{z} = \frac{\frac{\ell}{a_\perp}}{\sqrt{\frac{h^2 + k^2}{a_\parallel^2} + \frac{\ell^2}{a_\perp^2}}} \tag{3.11}$$

and

$$\sin(\phi + \Delta\phi) = \hat{e}\cdot\hat{x} = \frac{\frac{h}{a_\parallel}}{\sqrt{\frac{h^2 + k^2}{a_\parallel^2} + \frac{\ell^2}{a_\perp^2}}} \tag{3.12}$$

Combining Eqs. (3.7) and (3.11) and substituting for $\lambda$ from Eq. (3.6) and for $\ell$ from Eq. (3.9) yields

$$a_{\perp} = a_s \frac{\sin \theta_B}{\sin (\theta_B + \Delta\theta)} \cdot \frac{\cos \phi}{\cos (\phi + \Delta\phi)} \tag{3.13}$$

and combining Eq. (3.7) and Eq. (3.12) and substituting for $\lambda$ from Eq. 3.6 and for h from Eq. 3.10 yields

$$a_{\|} = a_s \frac{\sin \theta_B}{\sin (\theta_B + \Delta\theta)} \cdot \frac{\sin \phi}{\sin (\phi + \Delta\phi)} \tag{3.14}$$

These are the desired expressions for the lattice parameters of the tetragonal epitaxial layer in terms of the measurable quantities $\Delta\theta$ and $\Delta\phi$. A convenient and better known [13, 14] form for the mismatches results if one makes the small angle approximation. The mismatches are then given by

$$\left(\frac{\Delta a}{a}\right)_{\perp} = -\Delta\theta \cot \theta_B + \Delta\phi \tan\phi \tag{3.15}$$

$$\left(\frac{\Delta a}{a}\right)_{\|} = -\Delta\theta \cot \theta_B - \Delta\phi \cot\phi$$

Because equivalent asymmetric planes of the substrate and of the epitaxial layer are not parallel in the case of tetragonal distortion, $\Delta\phi$ contributes to rocking curve Bragg peak separations just as if the epitaxial layer were tilted with respect to the substrate. Thus, just as in the case of a tilted epitaxial layer, one can obtain both $\Delta\theta$ and $\Delta\phi$ by making a second rocking curve with the sample rotated by 180 degrees around the surface normal. The two cases are depicted in Fig. 3.10. If the sample is rocked clockwise in this figure, then $\Delta\omega_1 = \Delta\theta - \Delta\phi$ for case 1 and $\Delta\omega_2 = \Delta\theta + \Delta\phi$ for case 2 is obtained. Therefore, both $\Delta\theta$ and $\Delta\phi$ can be obtained from the measured quantities $\Delta\omega_1$ and $\Delta\omega_2$.

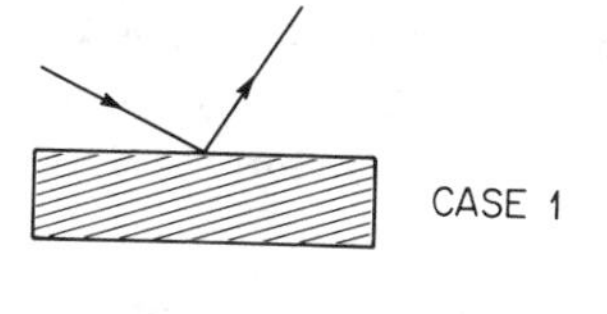


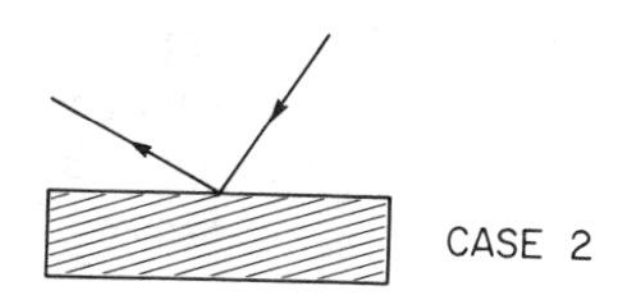


**Figure 3.10** Two possible cases for an asymmetric x-ray reflection.

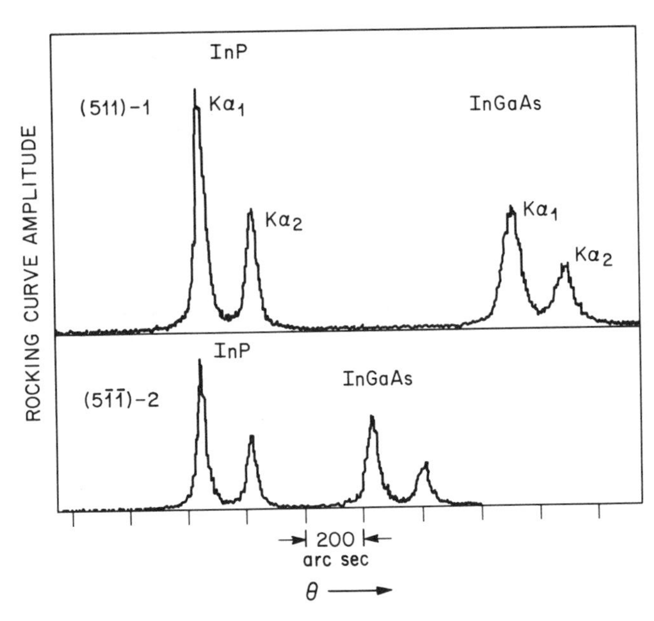


**Figure 3.11** Double crystal rocking curves for coherent GaInAs/InP for the (511) reflection (Case 1) and the $(5\bar{1}\bar{1})$ reflection (Case 2). Because the first crystal is a (100) oriented InP crystal set for the (400) reflection, wavelength dispersion is present and both $K\alpha_1$ and $K\alpha_2$ peaks occur.

An example for $Ga_xIn_{1-x}As/InP$ is shown in Fig. 3.11. The rocking curves shown are for the (511) and $(5\bar{1}\bar{1})$ reflections. (In practice it is convenient to use the $(5\bar{1}\bar{1})$ reflection instead of rotating the sample by 180 degrees). The measured angular separations were $\Delta\omega_1 = +1068$ arcsec and $\Delta\omega_2 = +580$ arcsec. These yielded $\Delta\theta = +824$ arcsec and $\Delta\phi = -244$ arcsec. Applying Eq. (3.15) with $\theta_B = 43.000$ degrees and $\phi = 15.793$ degrees we obtained the values $(\frac{\Delta a}{a})_\perp = -0.462$ percent and $(\frac{\Delta a}{a})_\parallel = -0.010$ percent which is less than the estimated uncertainty of $\pm 0.015$ percent for $(\Delta a/a)_\parallel$. Therefore, this epitaxial layer was coherent with the substrate to within the resolution of the measurement. A quick way to verify whether or not a layer is coherent is available from the ratio of $\Delta\omega_1$ to $\Delta\omega_2$. If one assumes $(\Delta a/a)_\parallel = 0$, then for the {511} reflection of InP, Eq. (3.15) implies $\Delta\omega_1/\Delta\omega_2 = 1.87$. The observed ratio for the case of Fig. 3.11 was 1.84.

The methods applied to measure the in-plane mismatch of coherent layers can also be applied to incoherent layers, that is, when there are misfit dislocations present. An example is shown in Fig. 3.12 for $Ga_xIn_{1-x}As/InP$. In this case the $K\alpha_2$ component was blocked by a slit edge between the two crystals because the $K\alpha_1$ and $K\alpha_2$ peaks were not resolved for the epitaxial layer. (If the $K\alpha_2$ component is not blocked, the epitaxial layer Bragg peak is asymmetric which

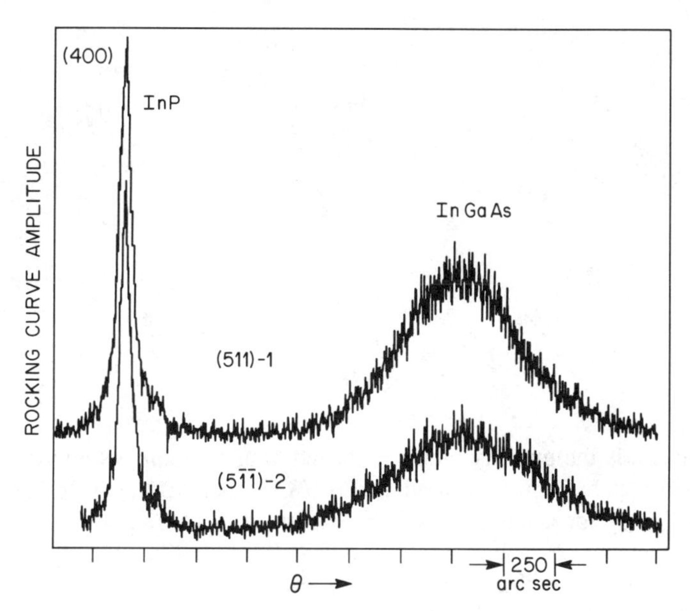


**Figure 3.12** Double crystal rocking curves for incoherent GaInAs/InP for the (511) reflection (Case 1) and for the $(5\bar{1}\bar{1})$ reflection (Case 2). The $K\alpha_2$ component was blocked by a slit in between the InP first crystal and the sample. A vestige of the $K\alpha_2$ substrate peak can still be discerned in both rocking curves.

hinders the measurement of $\Delta\omega$.) The result for the mismatches is $(\Delta a/a)_{\perp} = -0.851$ percent and $(\Delta a/a)_{\parallel} = -0.870$ percent. This layer was almost cubic and showed little tetragonal distortion. In such a case (see Fig. 3.9), $\Delta\phi$ is very small with the result that $\Delta\omega_1 \quad \Delta\omega_2$ in Fig. 3.12.

If misfit dislocations are the only lattice defects present which accommodate the misfit, then [15]

$$\left(\frac{\Delta a}{a}\right)_{\parallel} = \rho b \tag{3.16}$$

where $\rho$ is the linear density of misfit dislocations lying in the interface, and b is the misfit component of the Burgers vector. By modeling the interface in this way one can carry out a calculation of the stresses in the layer and in the substrate and derive expressions for the radius of curvature, R, and for the quantity $(\frac{\Delta a}{a})_r$ which is the relaxed mismatch. The relaxed mismatch is the mismatch in the absence of coherency strain. These expressions are, [15]

$$\frac{1}{R} = \frac{6t_1}{t_o^2} \cdot \frac{1}{1+6\frac{t_1}{t_o}} \cdot \frac{1-\nu}{1+\nu} \cdot \left[ \left(\frac{\Delta a}{a}\right)_\perp - \left(\frac{\Delta a}{a}\right)_{\parallel} \right] \tag{3.17}$$

and

$$\left(\frac{\Delta a}{a}\right)_r = \frac{1-\nu}{1+\nu} \cdot \left(\frac{\Delta a}{a}\right)_\perp + \frac{2\nu}{1+\nu} \cdot \left(\frac{\Delta a}{a}\right)_{\parallel} \tag{3.18}$$

Note that $(\Delta a/a)_r$ is the quantity which is related to alloy composition via Vegard's law and that measured mismatches must be used in Eq. (3.18) before Vegard's law is applied. For example, Vegard's law for $Ga_xIn_{1-x}As$ can be expressed as $(\Delta a/a)_r = 0.03227 - 0.06907x$ [16]. For coherent samples this becomes $(\Delta a/a)_\perp = 0.06857 - 0.14677x$ after application of Eq. (3.18) with $\nu = 0.36$. By applying this to the case of Fig. 3.8 for which $(\Delta a/a)_\perp = 0.076$ percent ($\Delta\theta = 97$ arcsec) one obtains $Ga_{0.462}In_{0.538}As$ for the alloy composition.

Eq. (3.17) has been tested by comparing calculated values of R to measured values for $Ga_xIn_{1-x}As/InP$ [15]. Results are given in Table 3.1. In all cases except one reasonable agreement between the two values was found. For the single exception stacking faults which were not incorporated in the model as a means of relaxing the misfit stress were also found to be present [15]. Therefore, by measuring both $(\Delta a/a)_\perp$ and $(\Delta a/a)_{\parallel}$ and, in addition, measuring the radius of curvature of a wafer, one can ascertain whether or not misfit dislocations are the only defects present which have relaxed the misfit stress.

Radius of curvature measurements are also conveniently made using x-ray techniques including DCD. However, for sufficiently well resolved peaks, single crystal diffractometry will suffice. The procedure in either case essentially consists of the measurement of the rotation of the sample required to maintain the Bragg condition as the sample is translated. A technique which makes this very convenient is known as automatic Bragg angle control [17, 18].

X-ray DCD also affords a means of monitoring the purity of epitaxial layers since the strain fields around impurities because of atomic misfit can result in measurable changes in the lattice parameter. DCD has been used to investigate changes in lattice parameter of epitaxial layers of $Al_{0.39}Ga_{0.61}As$ grown by LPE as a function of growth solution doping [19]. Because a change in alloy composition can also affect the lattice parameter, great care must be exercised to control the alloy composition in making such measurements. Wagner et al., [19] found that the presence of either $O_2$ in the growth ambient or Ge in the growth solution dilated the lattice. Dilatations caused by $O_2$ in the ambient were found to arise principally from a reduction in the background S contamination. Also, growth solution doping with Si was found to dilate the lattice for [Si] $<3 \times 10^{16} cm^{-3}$ but to contract the lattice at higher concentrations.

**Table 3.1**

Summaries of the measured and predicted results on six InGaAs/InP heterostructure samples.

| | Sample No. | $t_1$ (μm) | $t_0$ (μm) | $(\frac{\Delta a}{a})_\perp$ $(10^{-3})$ | $(\frac{\Delta a}{a})_\parallel$ $(10^{-3})$ | $(\frac{\Delta a}{a})_r$ $(10^{-3})$ | $R_{meas}$ (m) | $R_{cal}$ (m) | $\frac{(\Delta a/a)_r}{(\Delta a/a)_\perp}$ | $-\frac{\sigma^1_{xx}}{E_1}$ $(10^{-3})$ | $-\frac{\sigma^0_{xx}}{E_0}$ $(10^{-5})$ |
|---|---|---|---|---|---|---|---|---|---|---|---|
| VPE | 131 | 1.4 | 300 | 4.85 | 0.63 | 2.6 (2.6) | 5.6 | 5.6 (4.8) | 0.54 | 3.0 (3.5) | -5.6 (-5.8) |
| VPE | 389 | 2.2 | 231 | 3.60 | 1.80 | 2.7 (2.6) | 4.8 | 5.1 (2.5) | 0.75 | 1.4 (2.6) | -5.1 (-6.3) |
| VPE | 394 | 2.4 | 238 | -1.70[a] | ~0 | 0.8 (0.89) | -5.8 | -5.2 | -0.47 | -1.2 | 4.5 |
| LPE | 1-72 | 2.3 | 245 | -0.74[a] | ~0 | 0.35 (0.31) | -12 | -13 | -0.47 | 0.53 | -2.1 |
| VPE | 203 | 4.6 | 335 | 4.08 | 6.50 | 5.4 (5.5) | -3.5 | -3.9 (-2.3) | 1.32 | -1.8 (2.9) | 5.1 (-12) |
| VPE | 47 | 4.1 | 379 | 4.0 | 1.30 | 2.6 (3.7) | 21 | 4.9 (3.3) | 0.65 | 2.0 (2.7) | -3.6 (-45) |

(a) The minus sign indicates that the lattice constant of the epilayer is smaller than that of the substrate.

A powerful aspect of DCD is the ability to calculate theoretical rocking curves using dynamical diffraction theory. By making a comparison between such a calculated rocking curve and data, an assessment of the degree of crystalline perfection of a semiconductor wafer can be obtained. For thin layers, it is often convenient to apply the simpler kinematical theory, but if the substrate Bragg peak is also included theoretically, then dynamical theory must be applied.

The basic formulas used to obtain calculated rocking curves are presented here. The treatment of Zachariasen [5] is followed using his notation for the case of a thin layer and including anomalous dispersion. The dielectric constant is written as

$$\varepsilon = 1 + \psi \tag{3.19}$$

The Fourier components of $\psi$ obtained by expanding over the reciprocal lattice are related to the structure factor by

$$\psi_H = -\frac{r_e \lambda^2}{\pi a_{lp}^3} F_H \tag{3.20}$$

Here $a_{lp}$ is the lattice parameter, $\lambda$ is the wavelength, $r_e = 2.818 \times 10^{-5}$ Å is the classical electron radius. $F_H$ is written as

$$F_H = \Sigma_n \left[ f + \Delta f' + i\,\Delta f'' \right]_n \exp\left[ +2\pi i\, \vec{H} \cdot \hat{r}_n \right] . \tag{3.21}$$

Here $\hat{r}_n$ denotes an atomic site in the unit cell, $\vec{H}$ is a reciprocal lattice vector, $f_n$ is the atomic scattering factor, and $\Delta f_n'$ and $\Delta f_n''$ are the Hönl corrections for resonance and absorption. These corrections are also known as the anomalous dispersion corrections to the atomic scattering factor.

The full expression for the dynamical reflecting power of a thin layer is given by

$$\frac{P_H}{P_o} = \frac{1}{|b|} \frac{I_H^e}{I_o} = |b|\,|\psi_H|^2 \left[ \sinh^2 a w + \sin^2 a v \right] \tag{3.22}$$

$$\times \Bigg\{ |q+z|^2 + \left[ |z|^2 + |q+z|^2 \right] \sinh^2 a w$$

$$+ \left[ |z|^2 - |q+z^2| \right] \sin^2 a v$$

$$- \left[ v\,\mathrm{Re}(z) + w\,\mathrm{Im}(z) \right] \sinh 2 a w$$

$$- \left[ v\,\mathrm{Im}(z) - w\,\mathrm{Re}(z) \right] \sin 2 a v \Bigg\}^{-1}$$

Here

$$z = \frac{(1-b)\psi_o}{2} + \left[\frac{b}{2}\right]\alpha \text{ and } q = b\psi_H\psi_{\bar{H}}$$

where $\alpha$ is the variable which contains the deviation from the Bragg angle, namely, $\alpha = 2(\theta_B-\theta)\sin 2\theta_B$. Also, $a = \pi t/\lambda\gamma_o$, where t is the crystal thickness and $\gamma_o$ is dot product $\hat{n}\cdot\hat{s}_o$. Here $\hat{n}$ is the inward surface normal and $\hat{s}_o$ is a unit vector along the incident direction. Defining $\gamma_H = \hat{n}\cdot\hat{s}_H$ where $\hat{s}_H$ is a unit vector along the diffracted beam direction, then b is given by $b = \gamma_0/\gamma_H$. Last, v and w are the real and imaginary parts of $(q+z^2)^{1/2}$, respectively. Eq. (3.22) is a corrected version of Eq. (3.139) from Zachariasen's well-known book [5]. Zachariasen's equation is not correct in that the sign of radicals in the denominator must be taken as positive in some cases and minus in others. If used, the equation produces sharp features near $y = \pm 1$, where y is the reduced variable in which dynamical reflecting powers are naturally expressed [5] and is given by

$$y = \frac{\mathrm{Re}(z)}{K\,|\Psi'_H|\,\sqrt{|b|}} \tag{3.23}$$

where $\Psi'_H$ is caused by the $f_n$ and $\Delta f'_n$ terms in Eq. (3.21) and $K = 1$ for $\sigma$ polarized x-rays and $|\cos(2\theta_B)|$ for $\pi$ polarized x-rays. A comparison between a measured rocking curve and a theoretical one for an ideal $Ga_{0.462}In_{0.538}As/InP$ layer 0.78 μm thick grown by hydride-VPE is shown in Fig. 3.13. The theoretical rocking curve was obtained by convoluting intrinsic reflecting powers according to Zachariasen's Eq. (3.216) with reflecting powers for the first crystal (the (400) reflection of a (100) oriented InP piece in this case) for both $\sigma$ and $\pi$ polarizations of the x-rays. The calculated intrinsic reflecting powers of a 0.78 μm thick layer for the two polarizations are shown in Fig. 3.14. For the InP first crystal, the solutions for the reflecting powers of Cole and Stemple (Eq. (12) of ref. [20]) were applied. Note, however, that for the (400) reflection of InP the reflecting power expressions for centrosymmetric crystals first given by Hirsch and Ramachandran [21] can be applied (i.e., Eq. (12) of ref. [20] reduces to Eq. (16) of ref. [21]).

Because of the complexity of the dynamical expressions approximations are usually made to facilitate their application. First, since in many instances $\pi-2\theta_B$ is close to $\pi/2$, contributions to the total reflecting power for $\pi$ polarized x-rays (electric field vector in the diffraction plane) are small and are often ignored. Second, if absorption is ignored, if only symmetric reflections from thick crystals are considered, and if centrosymmetry of the lattice is assumed, then the Darwin curve given by Eq. (3.1) and shown in Fig. 3.3a results for the intrinsic reflecting power, and y reduces to

$$y = \left[\frac{-(\theta-\theta_B)\sin 2\,\theta_B + \psi_o}{\psi_H}\right] \tag{3.24}$$

The full width at half maximum (FWHM) is often quoted as a figure merit for a crystal. A

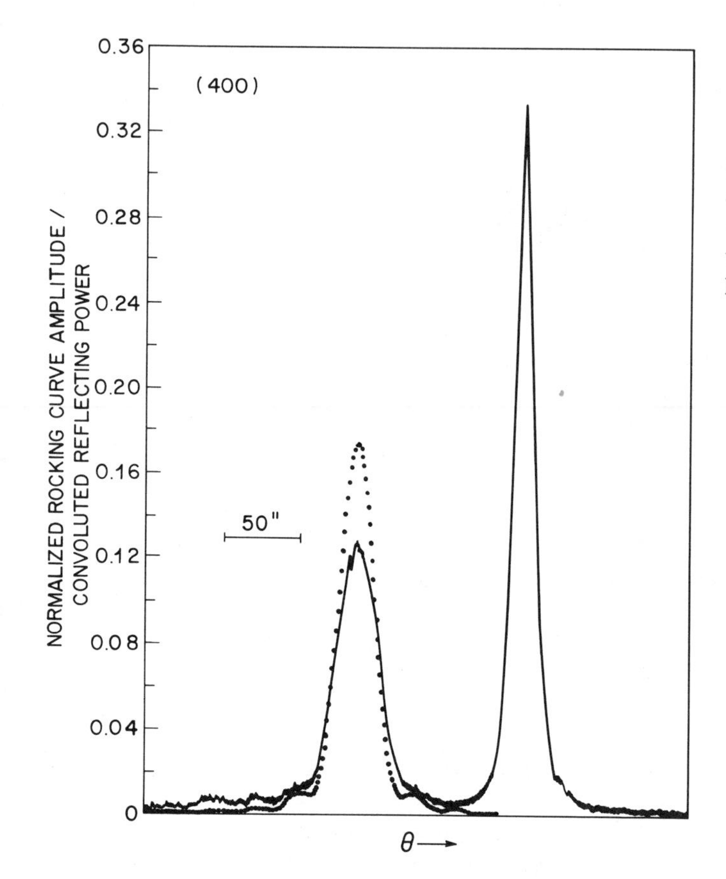


**Figure 3.13** Measured rocking curve and calculated convoluted reflecting power (dotted curve) for a highly perfect 0.78μm thick GaInAs/InP epitaxial layer grown by hydride-VPE.

**Figure 3.14** Calculated intrinsic reflecting power for the highly perfect 0.78μm thick GaInAs/InP layer of Fig. 3.13.

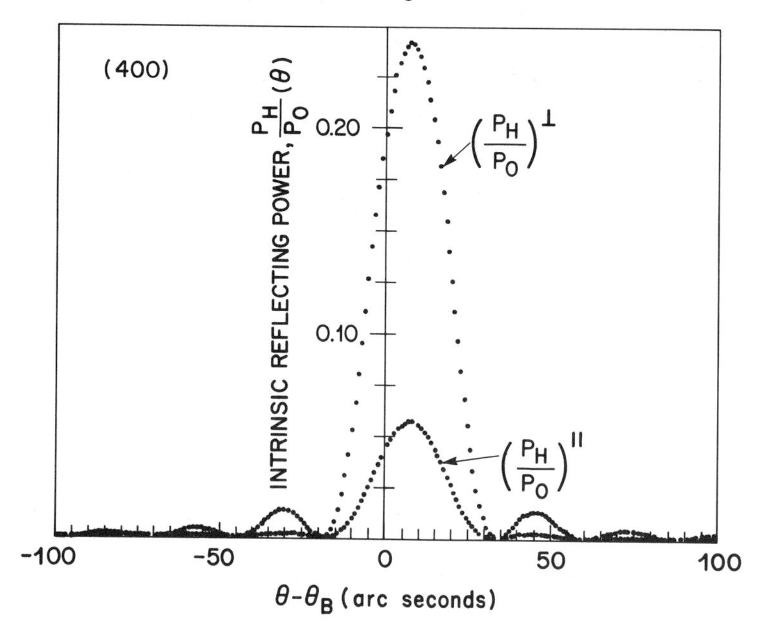

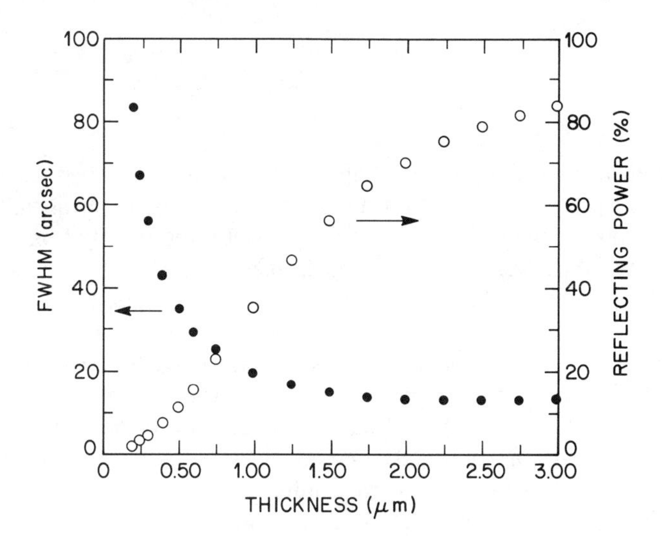


**Figure 3.15** Calculated results for the full width at half maximum and peak value of the intrinsic reflecting power of $Ga_{0.466}In_{0.534}As$ layers as a function of thickness.

simple expression for the FWHM results if the approximations made, which resulted in the Darwin curve, are made. The value of y at which $P_H/P_o \cong 1/2$ in Eq. (1) are $y = \pm 3\sqrt{2}/4 = \pm 1.06$, and the FWHM, $\beta$, becomes [14]

$$\beta = \frac{2.12 \mid \psi_H' \mid}{\sin 2\theta_B} \quad . \tag{3.25}$$

Here $\psi_H'$ is obtained from Eqs. (3.20) and (3.21) by using only the terms arising from f and $\Delta f'$.

However, if thin layers are considered, then one cannot apply Eq. (3.25). The dependence of the FWHM on layer thickness is shown for the case $Ga_{0.466}In_{0.534}As$ in Fig. 3.15. In this figure, the peak intrinsic reflecting power is also shown. These results were obtained by applying the full dynamical expression without any approximations Eq. (3.22). Note that these results for the (400) reflection are not very sensitive to the composition and that they apply approximately to all the $Ga_xIn_{1-x}As_{1-y}P_y$ alloys which have $1-x \cong 0.5$.

The critical thickness at which the FWHM becomes dependent on the thickness is known as the extinction depth and is given by [14],

$$t_{ext} = \frac{(\gamma_0 \gamma_H)^{1/2}}{\mid \psi_H' \mid} \tag{3.26}$$

For $Ga_{0.466}In_{0.534}As$ we have $t_{ext} = 1.6$ μm. If the layer thickness is less than $t_{ext}$, then the FWHM is given approximately by [14]

$$\beta = \frac{\lambda \gamma_H}{t \sin 2\theta_B} \tag{3.27}$$

Since the FWHM of a rocking curve Bragg peak may be broadened caused by crystalline imperfection such as a pronounced mosaic structure, measurement of this quantity does not afford a practical means of obtaining the thickness of an epitaxial layer directly from rocking curves. A way in which a layer thickness assessment made using the FWHM can be corroborated is, however, also available directly from a rocking curve if Pendellosung fringes are present. These fringes ought to occur for high quality thin layers, and the layer thickness can be obtained from the fringe spacing. An example of such fringes is shown in Fig. 3.16 which corresponds to the low angle part of the rocking curve shown in Fig. 3.13. The layer thickness deduced from the fringe spacing was 0.78 μm [10]. These fringes arise because of an interference between two allowed solutions for the wavefields, and they do not occur for a thick crystal because in that case the lack of a lower surface causes one of these solutions to be absent [22]. In addition to the case of GaInAs/InP discussed, they have been observed for AlGaAs/GaAs [12], and GaInAsP/InP [23].

For III-V ternary and quaternary semiconductor alloys, the lattice parameter is a function of composition which implies that if the stoichiometry wanders during epitaxial growth, mismatch grading can occur. Unless the stoichiometry wanders in the precise way required to preserve lattice matching (e.g., if both x and y in $Ga_xIn_{1-x}As_{1-y}P_y$ vary with $1 - y \cong 2.1x$) mismatch grading will occur. If we consider the ternary alloy $Ga_xIn_{1-x}As$, then this complication is avoided and the lattice parameter will be graded if x is not constant throughout the thickness of the epitaxial layer.

In many instances such grading exhibits itself in a rocking curve as intense fringes. An example is shown in Fig. 3.17 [24]. A calculated rocking curve based on the mismatch grading

**Figure 3.16** Measured and simulated rocking curves of the same sample as in Fig. 3.13 but at lower angles and showing Pendellosung fringes.

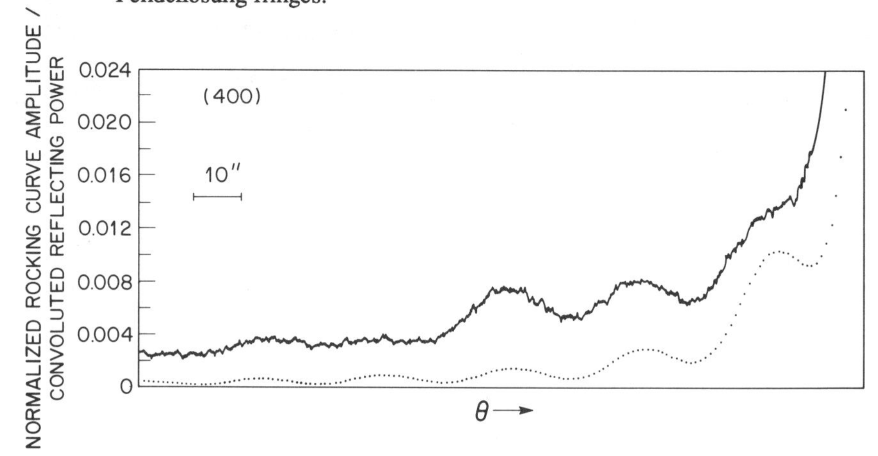

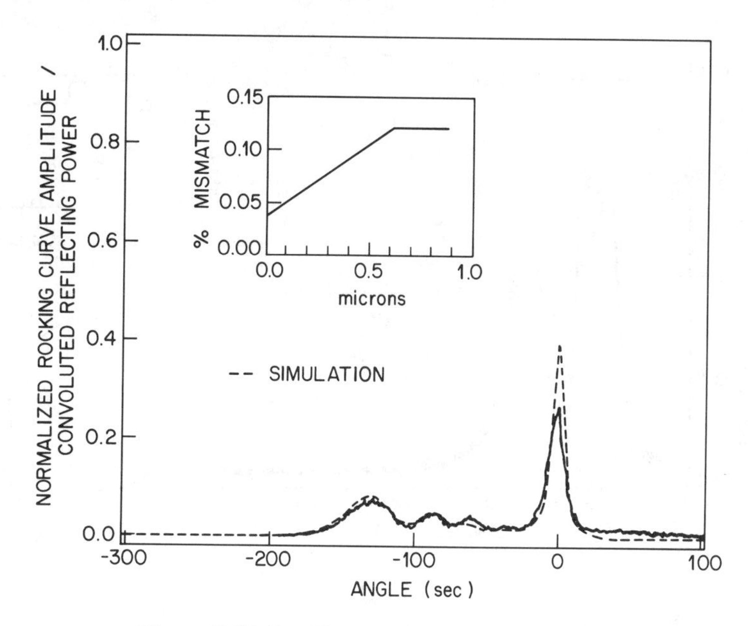


**Figure 3.17** Double crystal rocking curve data and computer simulation for a graded GaInAsP layer. The mismatch grading assumed for the simulation is shown in the inset.

shown in the inset is also shown. Because the lattice parameter is not constant, the dynamical expressions given cannot be applied. Instead computer simulations are used. The simulation shown in Fig. 3.17 [24] was generated using Abeles' method which is based on a $2 \times 2$ matrix formulation of Maxwell's equations [25]. Another technique which has been applied to this problem [26] involves making the electromagnetic field amplitudes position dependent and making a series of approximations until tractable expressions known as the Takagi-Taupin equations result [27, 28, 29].

### 3.1.4 Double Heterostructures

Rocking curves for double heterostructures (DH) can be considerably more complicated than those for single epitaxial layers because diffraction from the various layers can overlap. Two examples of rocking curves for a $\lambda = 1.3\ \mu m$ light-emitting-diode structure are shown in Fig. 3.18 [30]. In this figure, the insets show the mismatches and layer thicknesses which produced the dynamical diffraction theory fits (Abeles' method) overlaid on the data. The two rocking curves are for two locations on a 2-inch diameter round as shown in the figure. A broadening of the InP peak was observed which could be reproduced in the fits by assuming a slight Ga contamination of the cladding layer grown on the active layer. That is, the cladding layer was assumed to be $Ga_xIn_{1-x}P$ where x was $5.14 \times 10^{-4}$ and $5.20 \times 10^{-4}$ for Fig. 3.18a and 3.18b, respectively. In the first case, a pronounced second peak resulted for the cladding layer because in that case this layer was somewhat thinner ($1.55\ \mu m$) than in the other case ($1.70\ \mu m$). The active layer is clearly visible in Fig. 3.18b as a pronounced shoulder on the capping layer peak but not in Fig. 3.18a because diffraction features from the capping and active layer

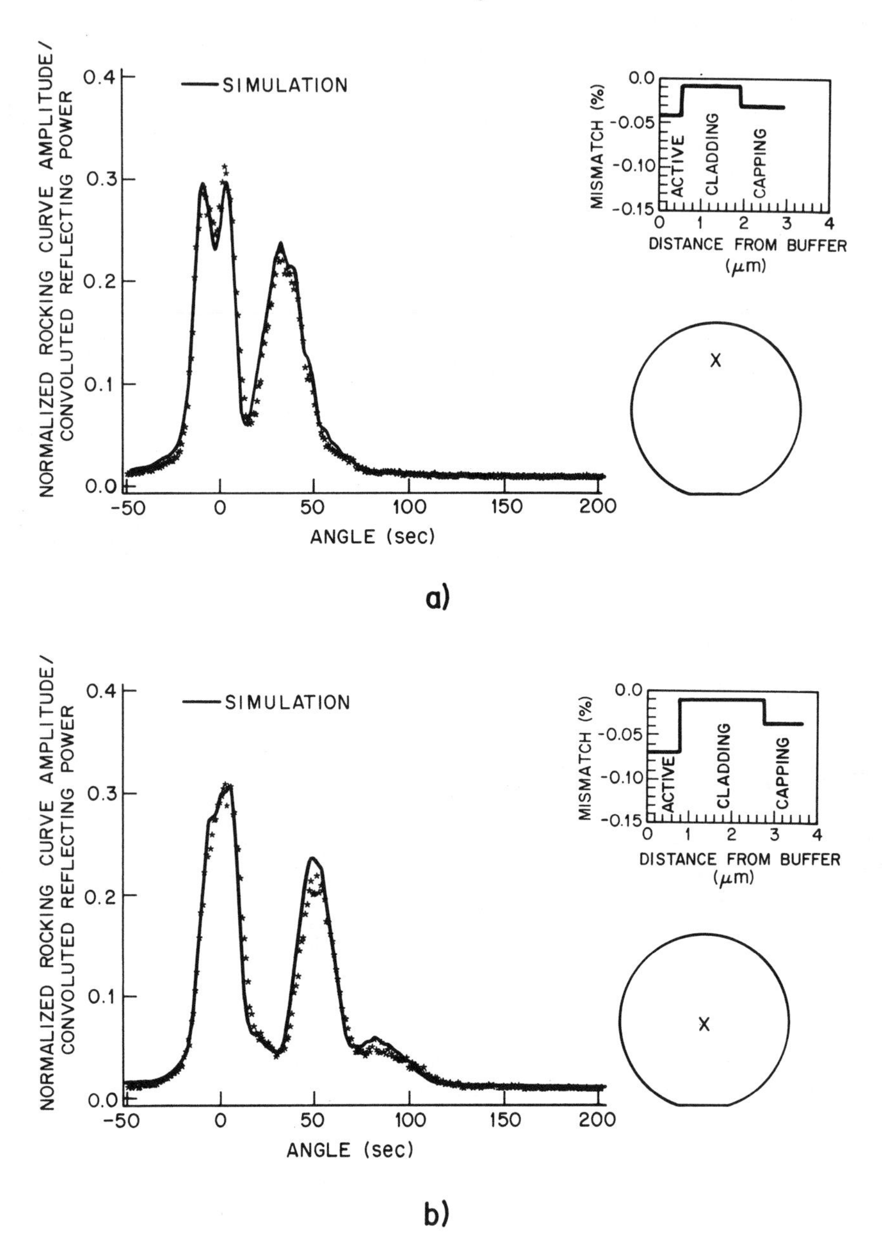


**Figure 3.18** Double crystal rocking curve data and computer simulations for a light-emitting-diode structure consisting of an InP/GaInAsP/InP double heterostructure with an GaInAsP capping layer.

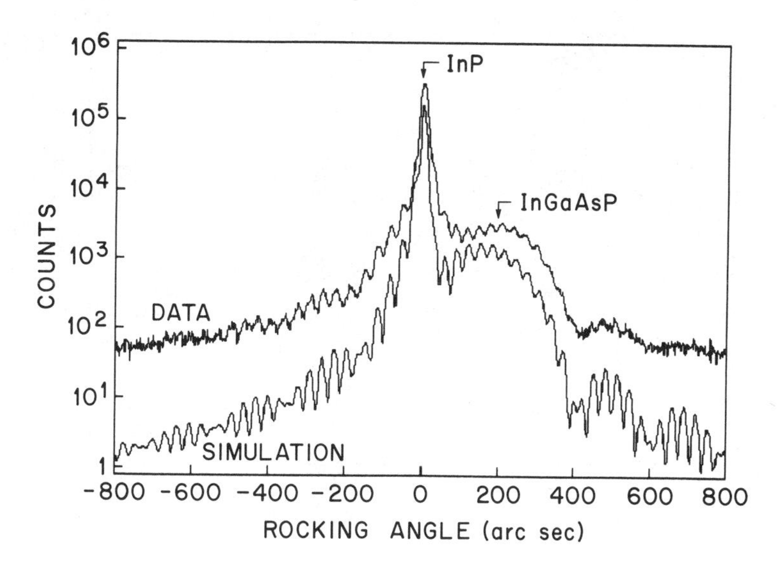


**Figure 3.19** X-ray double crystal rocking curve of the (400) reflection using Cuk$\alpha$ radiation of a double heterostructure and a best fit simulation. The intensity of the simulation is scaled arbitrarily.

completely overlap. Nuances in the shape of the capping layer (which could be fit very well as shown) are the only evidence for the presence of the active layer in this case.

For DHs having thinner layers, a semilog plot of rocking curve data is generally more useful than a linear one since weak intensity modulation caused by fringes is brought out. This is demonstrated in Fig. 3.19 which shows a rocking curve of an InP/GaInAsP/InP DH grown by hydride vapor phase epitaxy [31].

Next, the results displayed in Fig. 3.19 will be discussed at length since the analyses are illustrative of the method whereby layer thicknesses are measured from x-ray fringe spacings for multilayer structures. X-ray fringes of two types are visible in Fig. 3.19. The first have a period of ~31 arcsec, and the second have a period of ~187 arcsec. A Bragg peak for the negatively mismatched quaternary layer is also clearly seen. This rocking curve was simulated using Abeles' method [24], and the resultant fit is also shown in Fig. 3.19. The parameters producing the fit were 1000 Å and $-0.170$ percent for the thickness and mismatch of the quaternary layer, and 5500 Å for the thickness of the InP top cladding layer. The quaternary layer was assumed to be coherent and tetragonally distorted in proportion to the mismatch [15].

The period of the short-period fringes is related inversely to the InP top layer thickness by the well known relationship [12]

$$\Delta\theta = \frac{\lambda}{(2t\cos\theta_B)} \tag{3.28}$$

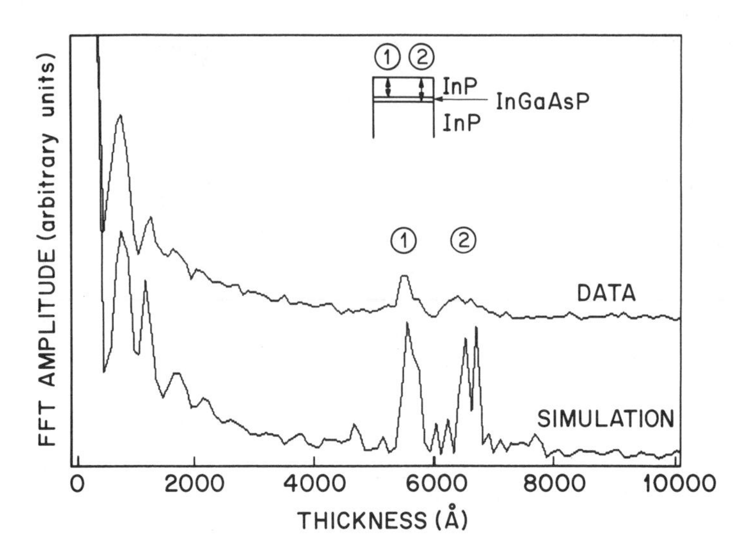


**Figure 3.20** Fourier transform spectra of the x-ray data in the previous figure and of the simulation.

where $\Delta\theta$ is the fringe period, $\lambda$ is the x-ray wavelength, t is the layer thickness, and $\theta_B$ is the Bragg angle. The short period fringes are similar to those reported previously by Bartels and Nijman [12] and to those reported by Chu and Tanner [32] for AlGaAs/GaAs/AlGaAs laser structures.

The origin of these fringes is clarified by taking the Fourier transform. The results obtained by applying a fast Fourier transform (FFT) algorithm to both the data and the simulation are shown in Fig. 3.20. The Fourier transform was applied to the logarithm of the intensities (2 arcsec between points over a range of 2000 arcsec), and the abscissa was converted to units of length using Eq. (3.28). The peaks labeled 1 and 2 in the transform (2 is actually a doublet in the simulation) both correspond to the short-period fringes with fringe spacings of 34.0 and 28.8 arcsec, respectively. The two peaks have a separation corresponding to the quaternary layer thickness, and they correspond to interference of x-rays scattered from the InP top layer/quaternary layer interface and from the quaternary layer/InP buffer layer interface, respectively. We note that the beat spacing for peaks 1 and 2 is 188 arcsec and that the widely spaced fringes can be viewed simply as resulting from a beating between the two sets of short-period fringes.

### 3.1.5 Superlattices

X-ray diffraction is a powerful tool with which to characterize semiconductor superlattices [33, 34, 35, 36]. Detailed analyses involve careful fitting of the rocking curves which can yield the strain and composition modulation. A quantity which is quite straightforwardly obtained, however, is the superlattice periodicity. Also, the intensity of the superlattice Bragg peaks is a good indicator of the quality of the superlattice, that is, of the degree of the crystallinity and of the uniformity of the layer thicknesses. If misfit dislocations are present or if the layer thicknesses do not repeat well, then the x-ray peaks will be broadened and have reduced intensity.

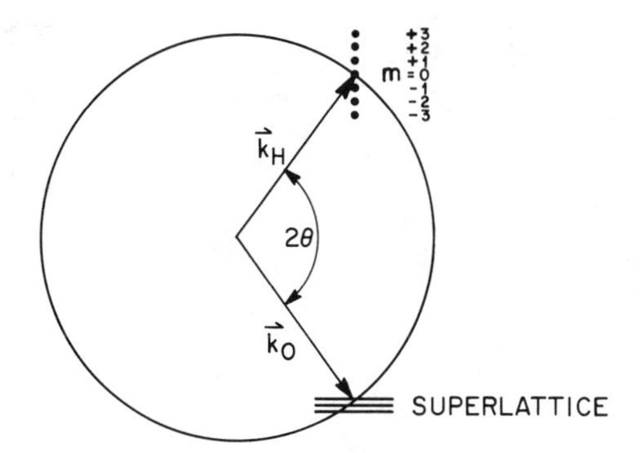


$$\vec{k}_H - \vec{k}_O = \vec{g} = \frac{1}{a} + \frac{m}{\Lambda}$$

**Figure 3.21** Ewald's sphere and reciprocal lattice for a superlattice.

In a superlattice, an additional periodicity is superimposed on the lattice periodicity. Ewald's construction in reciprocal space for an unstrained superlattice is shown in Fig. 3.21. Here $\vec{k}_o$ and $\vec{k}_H$ are the wavevectors of the incident and diffracted x-rays, respectively, and $|\vec{k}_o| = |\vec{k}_H| = 1/\lambda$. Additional reciprocal lattice points lie along the direction normal to the superlattice, and the Bragg condition becomes

$$\frac{2 \sin \theta_m}{\lambda} = |\vec{g}| = \frac{1}{d} + \frac{m}{\Lambda} \tag{3.29}$$

Here $\vec{g}$ is a reciprocal lattice vector, d is the interplanar spacing, $\Lambda$ is the superlattice period, and m is the superlattice order. Note that if $m = 0$, we obtain the familiar Bragg's law for the lattice. An expression for $\Lambda$ in terms of the separation between superlattice Bragg peaks can be obtained if we write $\theta_{m+1} = \theta_m + \delta\theta$ and then Taylor expand, $\sin\theta_{m+1} = \sin\theta_m + \cos\theta_m\, \delta\theta$. The result is given by

$$\Lambda = \frac{\lambda}{2(\cos\theta)\delta\theta} \tag{3.30}$$

An example of a rocking curve for an GaInAs/InP SLS grown on an InP substrate is shown in Fig. 3.22 [35]. The superlattice period was $\Lambda = 220$ Angstroms as obtained from the angular separation of the superlattice peaks.

The angular separation between the $m = 0$ peak and the substrate peak provides a measure of the average strain of the epitaxial layers [33]. The intensity of individual superlattice peaks is a function of both the strain and the composition modulation. Superlattice peaks result, in general, even without strain modulation if the structure factor is periodic. In such cases the Bragg peak intensities are related to the Fourier components of the structure factor modulation and can be used to measure interdiffusion within the superlattice [34].

Using kinematical theory an expression can be derived for the x-ray intensity diffracted from a periodic superlattice. The rocking curve of a superlattice can be demonstrated that it is symmetrical around the zero order peak in the case of zero strain (i.e., for pure structure factor modulation), but that for strained layer superlattices an asymmetric envelope function occurs.

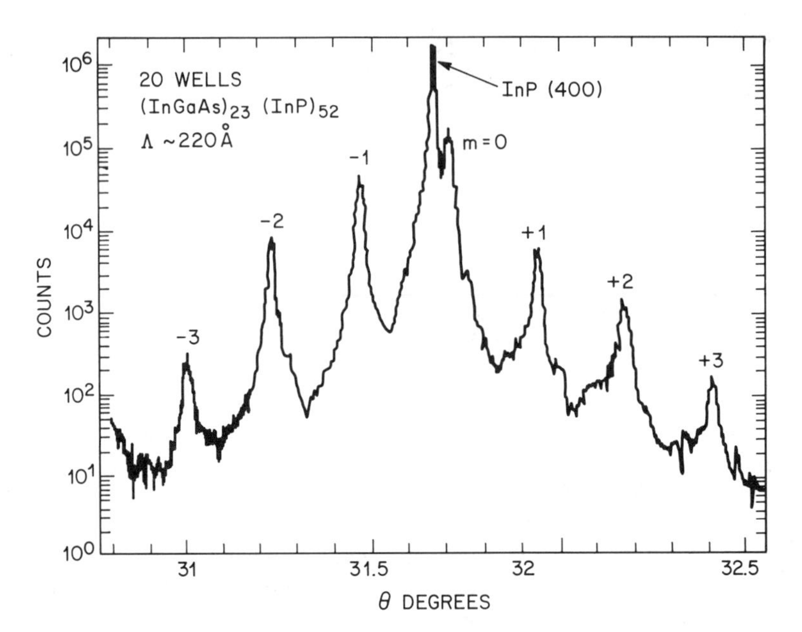


**Figure 3.22** Diffraction scan for the (400) reflection of an GaInAs/InP superlattice. The mismatch of the average lattice was determined to be $(\frac{\Delta a}{a})_{\perp} = -0.12\%$ from the angular difference between the zero order superlattice peak and the InP substrate peak.

Superlattices with layer thicknesses will be considered which are sufficiently large that a step model [37] for strain is appropriate. Thus the strain will be treated as changing abruptly at the interfaces between the layers which comprise the superlattice period. For superlattice layers which are only a few atomic layers thick, a frozen phonon model for the strain may be more appropriate [34, 37]. Diffraction from the substrate will be ignored for which kinematical theory would not be appropriate. For the sake of simplicity, we will ignore absorption and limit the treatment to symmetric (h00) reflections of the zincblende structure (i.e., (200), (400), and (600)). Furthermore, only a bilayer period will be considered with layer thicknesses of $t_a = N_a a_{\perp}^a$ and $t_b = N_b a_{\perp}^b$ (see Fig. 3.23) where $a_{\perp}^a$ and $a_{\perp}^b$ are the lattice parameters of the layers along the (100) growth direction. The structure is assumed to be coherent and free of dislocations so that the lateral lattice parameters of all the layers are identical and equal to the cubic lattice parameter of the substrate (i.e., $a_{\parallel}^a = a_{\parallel}^b = a_s$). Last, we also ignore changes in strain caused by bowing of the wafer.

The appropriate starting point for the kinematical diffraction treatment is an expression given by Warren [38] for the instantaneous field at an observation point P which is a distance R from the crystal (see Fig. 3.23) where R is sufficiently large that the scattered wave can be treated as a

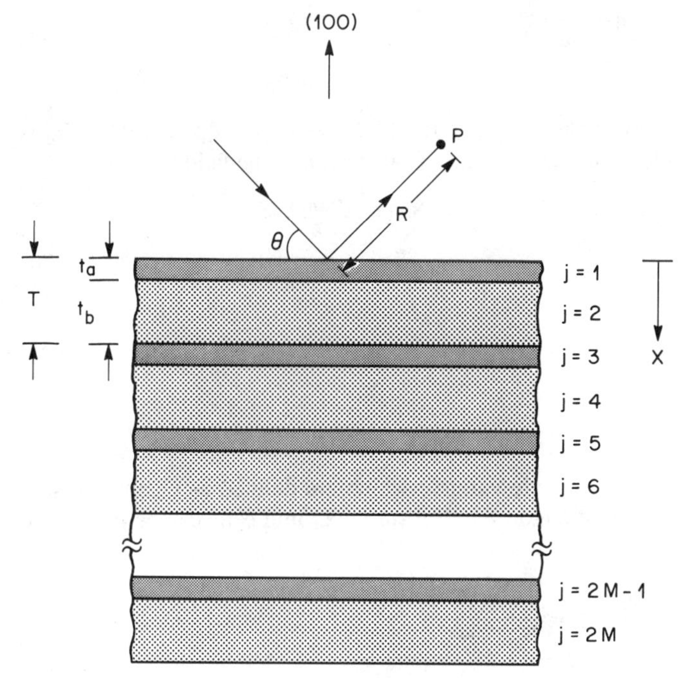


**Figure 3.23** Schematic superlattice.

plane wave. Here an incident plane wave with the electric field polarized perpendicular to the plane of the paper is assumed. This expression is [38]

$$E_p = \frac{E_o r_e}{R} e^{2\pi i[\nu t - R/\lambda]} \sum_n f_n e^{(2\pi i/\lambda)(\vec{s}-\vec{s}_o)\cdot\vec{r}_n} \tag{3.31}$$

Here $E_o$ is the electric field amplitude of the incident wave, $r_e$ is the classical electron radius, $\nu$ is the frequency, $\lambda$ is the wavelength, $\vec{s}$ and $\vec{s}_o$ are unit vectors along the scattered and incident directions, respectively, and $f_n$ is the atomic scattering factor and $\vec{r}_n$ is the position of the $n^{th}$ atom in the structure. For the present structure, phase terms caused by the atomic periodicity lateral to the superlattice direction can be factored out and summed over independently. They contribute factors of $N_y$ and $N_z$ to $E_p$ where $N_y a_s$ and $N_z a_s$ represent the lateral dimensions of the structure. The remaining x coordinate of an atom in the $j^{th}$ layer can be written as

$$X^j_{n,m} = \left\{ \sum_{\ell=1}^{j-1} t_\ell \right\}_{j\geq 2} + m a_j + x_n \tag{3.32}$$

where m is the number of lattice spacings ($a_j = a^a_\perp$ or $a^b_\perp$) down to the unit cell that the atom is in and $x_n$ is the coordinate of that atom within the unit cell. With these definitions the phase inside the sum in Eq. (3.31) reduces to

$$\frac{4\pi}{\lambda}\,(ma_j + x_n)\sin\theta \tag{3.33}$$

We can also leave out the traveling wave phase factor in Eq. (3.31) since it drops out when $E_p$ is multiplied by its complex conjugate to yield the intensity. Consequently, the field becomes

$$E_p = \frac{E_o r_e}{R}\,\frac{N_y N_z}{4}\sum_{j=1}^{2M} e^{\,i\frac{4\pi\sin\theta}{\lambda}\sum_{\ell=1}^{j-1} t_\ell}\sum_{m=0}^{N_j-1} e^{\,i\frac{4\pi\sin\theta}{\lambda} ma_j}$$

$$\times \sum_{\substack{n \\ \text{unit cell}}} f_n^j e^{\,i\frac{4\pi}{\lambda}\sin\theta\, x_n} \tag{3.34}$$

Next, the coefficients are lumped into a single variable which are defined as C, and the sum over the unit cell is the usual x-ray structure factor, $F_j$. The sum over unit cells can be performed easily to yield

$$\sum_{m=0}^{N_j-1} e^{\,i\frac{4\pi}{\lambda}\sin\theta\, ma_j} = \frac{1-e^{i2ga_jN_j}}{1-e^{i2ga_j}} \tag{3.35}$$

$$= \frac{\sin(ga_jN_j)}{\sin(ga_j)}\, e^{iga_j(N_j-1)}$$

where we have defined g as $2\pi\sin\theta/\lambda$. Thus we arrive at the following expression for the field amplitude,

$$E_p = C\sum_{j=1}^{2M} F_j\,\frac{\sin(ga_jN_j)}{\sin(ga_j)}\, e^{\,ig[a_j(N_j-1)+2\sum_{\ell=1}^{j-1}A_\ell N_\ell]} \tag{3.36}$$

It is convenient at this point to insert mismatches and angular deviations into the formulation. We define $\Delta\omega_j$ to be the angular deviation of the incident beam from the Bragg angle which corresponds to $a_j$. Thus

$$\Delta\omega_j = \theta - [\theta_B - (\frac{\Delta a}{a})^j_\perp \tan\theta_B] \tag{3.37}$$

where $\theta$ is the angle between the incident beam and the crystal planes, $\theta_B$ is the Bragg angle of the substrate, and $(\Delta a/a)^j_\perp$ is the mismatch with respect to the substrate. By Taylor expanding and by keeping only terms which are first order in the small quantities $\Delta\omega_j$ and $(\Delta a/a)^j_\perp$, we obtain in a straightforward manner the following relationship,

$$ga_j = \pi h + 2\pi h\cot\theta_B\,\Delta\omega_j \tag{3.38}$$

Here we have also used Bragg's law for the substrate. Since h is even, all terms in Eq. (5) caused by the $\pi$th term in Eq. (3.38) drop out, and we obtain for the field

$$E_p = C\sum_{j=1}^{2M} F_j \frac{\sin(u\,\Delta\omega_j N_j)}{\sin(u\,\Delta\omega_j)}\, e^{iu[\Delta\omega_j(N_j-1)+2\sum_{\ell=1}^{j-1}\Delta\omega_\ell N_\ell]} \tag{3.39}$$

Here we have defined u as $2\pi h \cot\theta_B$.

For the superlattice of Fig. 3.23, the sum over j separates into sums for layers of type a (j odd) and layers of type b (j even). For j odd the sum in the exponential factor of Eq. (3.39) becomes

$$\sum_{\ell=1}^{j-1} \Delta\omega_\ell N_\ell = (\Delta\omega_a N_a + \Delta\omega_b N_b)(\frac{j-1}{2}) \tag{3.40}$$

and for j even it becomes

$$\sum_{\ell=1}^{j-1} \Delta\omega_\ell N_\ell = \Delta\omega_a N_a \frac{j}{2} + \Delta\omega_b N_b(\frac{j}{2} - 1) \tag{3.41}$$

By performing the following sums,

$$\sum_{j\,\text{even}>0}^{2M} e^{i(\alpha_a N_a+\alpha_b N_b)j} = e^{i2(\alpha_a N_a+\alpha_b N_b)}\, \frac{(1-e^{i2(\alpha_a N_a+\alpha_b N_b)M})}{(1-e^{i2(\alpha_a N_a+\alpha_b N_b)})} \tag{3.42}$$

and

$$\sum_{j\,\text{odd}}^{2M-1} e^{i(\alpha_a N_a+\alpha_b N_b)j} = e^{i(\alpha_a N_a+\alpha_b N_b)}\, \frac{(1-e^{i2(\alpha_a N_a+\alpha_b N_b)M})}{(1-e^{2(\alpha_a N_a+\alpha_b N_b)})} \tag{3.43}$$

where $\alpha_a = u\Delta\omega_a$ and $\alpha_b = u\Delta\omega_b$, we obtain for the diffracted field amplitude,

$$E_p = Ce^{i[M(\alpha_a N_a+\alpha_b N_b)-\alpha_b N_b-\alpha_a]}\, \frac{\sin(M(\alpha_a N_a+\alpha_b N_b))}{\sin(\alpha_a N_a+\alpha_b N_b)}\, F_s \tag{3.44}$$

where $F_s$ is given by

$$F_s = F_a \frac{\sin(\alpha_a N_a)}{\sin\alpha_a} + F_b e^{i(\alpha_a N_a+\alpha_b N_b+\alpha_a-\alpha_b)} \frac{\sin(\alpha_b N_b)}{\sin\alpha_b} \tag{3.45}$$

Now the diffracted intensity can be expressed as

$$I_p = E_p^* E_p = |\,C\,|^2 \frac{\sin^2[M(\alpha_a N_a+\alpha_b N_b)]}{\sin^2(\alpha_a N_a+\alpha_b N_b)}\, |\,F_s\,|^2 \tag{3.46}$$

Superlattice peaks are produced by the $\sin^2[M(\alpha_a N_a + \alpha_b N_b)]/\sin^2(\alpha_a N_a + \alpha_b N_b)$ factor and occur at $\alpha_a N_a + \alpha_b N_b = m\pi$. By substituting for $\Delta\omega_a$ and $\Delta\omega_b$ from Eq. (8) in this condition, the following condition is obtained for $\theta_m$, the incident angle for the $m^{th}$ superlattice peak

$$\frac{2\sin\theta_m}{\lambda} = \frac{h}{<a>} + \frac{m}{\Lambda} \quad (3.47)$$

where $\Lambda = (N_a a_\perp^a + N_b a_\perp^b)$ and $<a> = \Lambda/(N_a + N_b)$. We recognize this as Eq. (3.29).

The quantity $F_s$ given in Eq. (3.35) is the superlattice structure factor. The intensity of the $m^{th}$ peak is given by

$$I_p^m = |C|^2 M^2 |F_s^m|^2 \quad (3.48)$$

where it is readily shown that

$$F_s^m = \sin(\alpha_a N_a)\left[\frac{F_a}{\sin\alpha_a} - \frac{F_b}{\sin\alpha_b} e^{i(\alpha_a - \alpha_b)}\right]. \quad (3.49)$$

The envelope function to be applied to the superlattice peak intensities can now be described. Two cases must be distinguished, namely, strained and unstrained superlattices. For the strain free case the lattice parameters are equal and, consequently, $\alpha_a$ equals $\alpha_b$. In this case Eq. (3.49) reduces to

$$F_s^m \cong \frac{N_a + N_b}{m\pi}(F_a - F_b)\sin\left(m\pi\frac{N_a}{N_a + N_b}\right) \quad (3.50)$$

Here $\sin\alpha \cong \alpha$ is also approximated. We note that Eq. (3.50) has a number of interesting features. First, it correctly indicates that superlattice peaks do not occur if $F_a$ equals $F_b$ or if either $N_a$ or $N_b$ vanishes. Second, the intensity of higher order peaks diminishes as $1/m^2$ and must be symmetric around $m = 0$ since $F_s^m$ equals $F_s^{-m}$. A simulated example of a strain free superlattice which exhibits these features is shown in Fig. 3.24. This simulation is for a lattice matched superlattice of GaInAs/InP and does not include a substrate.

Rocking curves for strained layer superlattices are not symmetric around $m = 0$. A simulated example for a superlattice of GaInAs/GaAs is shown in Fig. 3.25. The essential features of this rocking curve also follow from Eq. (3.49). In this case $\alpha_a$ and $\alpha_b$ are not equal. For an incident angle such that $\alpha_a \cong 0$ the term containing the factor $F_b$ can be ignored in Eq. (3.49), and the superlattice structure factor part of the intensity is given by

$$|F_s^m|^2 = N_a^2 |F_a|^2 \quad (3.51)$$

Similarly, for an incident angle such that $\Delta\omega_b \cong 0$, we have

$$|F_s^m|^2 = N_b^2 |F_b|^2 \quad (3.52)$$

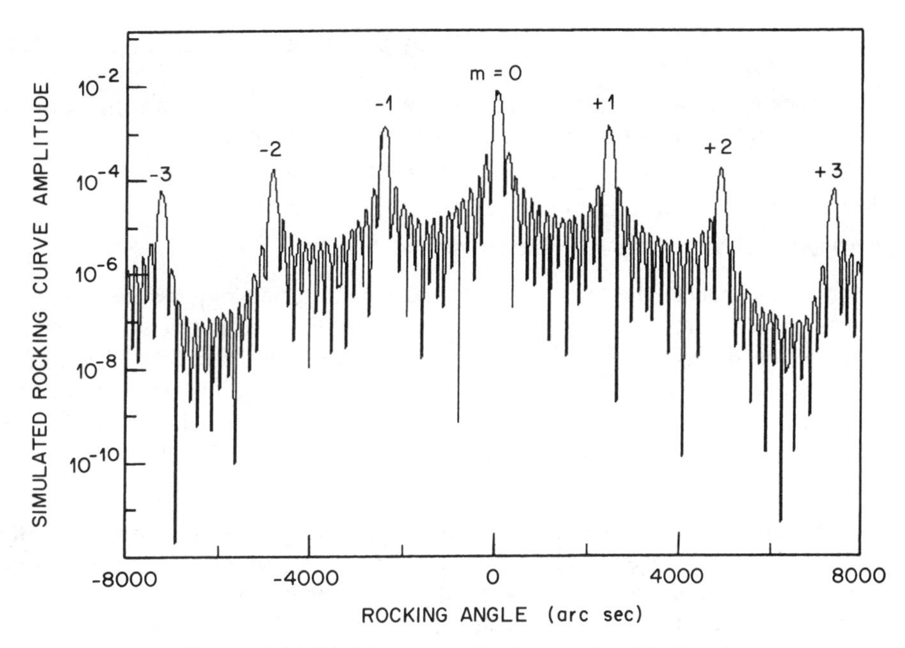


**Figure 3.24** Rocking curve simulation of an ideal strain free superlattice.

**Figure 3.25** Rocking curve simulation of an ideal highly strained coherent superlattice.

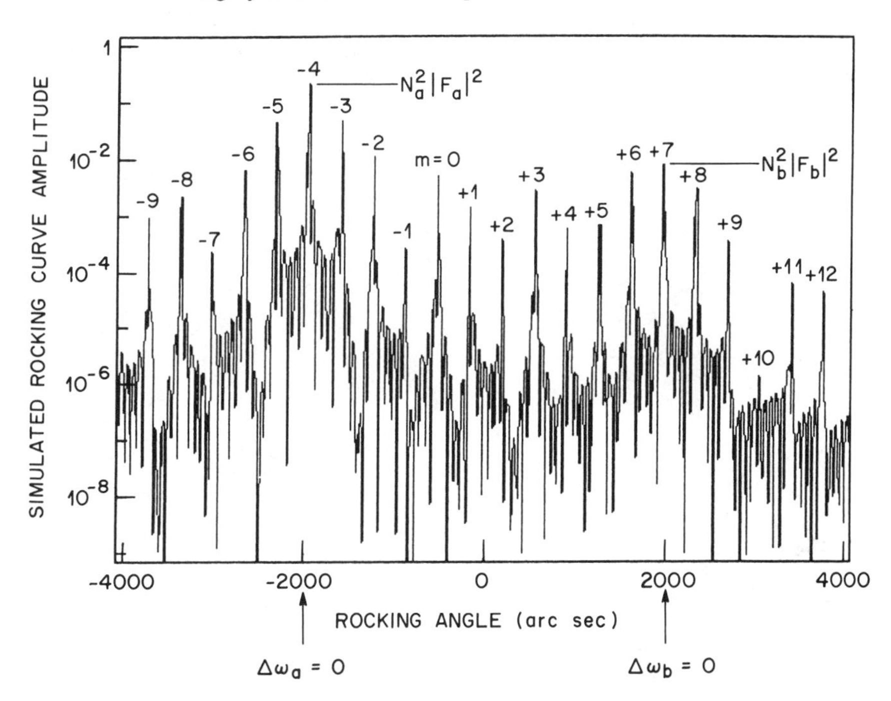

Therefore, the rocking curve has an envelope function with two peaks which correspond to each of the two layers. In the limit of a single superlattice period only a single a-type and a single b-type layer are present and the envelope function reduces to the two Bragg peaks which are diffracted from them.

### 3.1.6 Heteroepitaxial Structures with Large Mismatches

Traditional semiconductor heteroepitaxial crystal growth technology has been focussed on alloy systems which can be grown lattice matched to a convenient substrate (e.g., $Al_{1-x}Ga_xAs/GaAs$ and $Ga_xIn_{1-x}As_{1-y}P_y/InP$). As a result of the deleterious effects of crystal defects which plastically relax the misfit stress in devices, epitaxy with inherent large mismatches has not been considered viable. However, as a result of reports for field effect [39, 40] and bipolar transistors [41] and for lasers [42] made using GaAs grown on Si substrates, this may be changing.

X-ray diffraction provides a powerful tool with which to study these structures. In particular, x-ray DCD with a large area open detector provides an extremely sensitive means with which to monitor the mosaic structure since the entire range of reflection is obtained. This has been appreciated for many years [43] and is illustrated in Fig. 3.26. When a mosaic crystal is rocked, mosaic pieces with the same Bragg spacing but different orientations contribute to the observed rocking curve amplitude at different rocking angles, and the FWHM obtained using DCD is a very sensitive measure of the mosaic spread.

In spite of this, however, some workers use x-ray powder diffractometers to characterize epitaxial layers. Conventional powder diffractometers use Bragg-Brentano focussing and achieve high resolution via a slit aperture positioned in front of the detector [44]. However, diffractometers of this type are designed specifically for structured analyses of powdered samples for which the mosaic spread is not of interest. Powder diffractometers are designed to cancel the

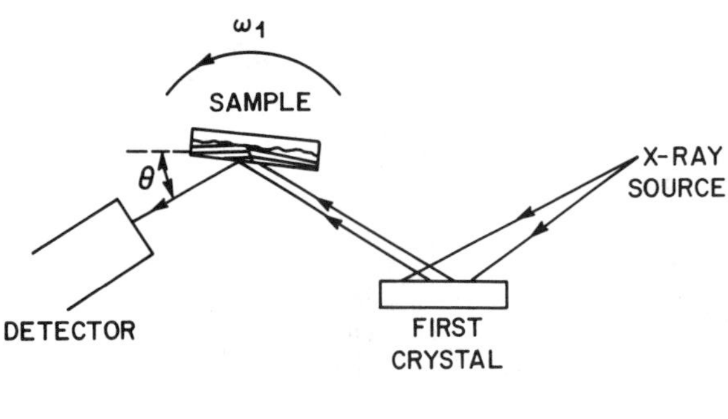


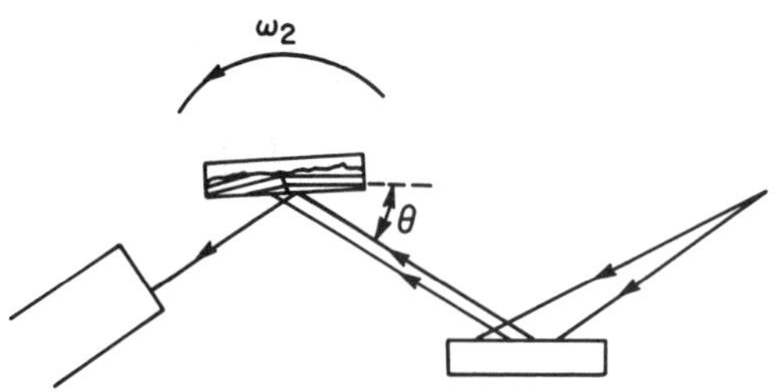


**Figure 3.26** Schematic illustration of sensitivity to mosaic structure in the double crystal arrangement.

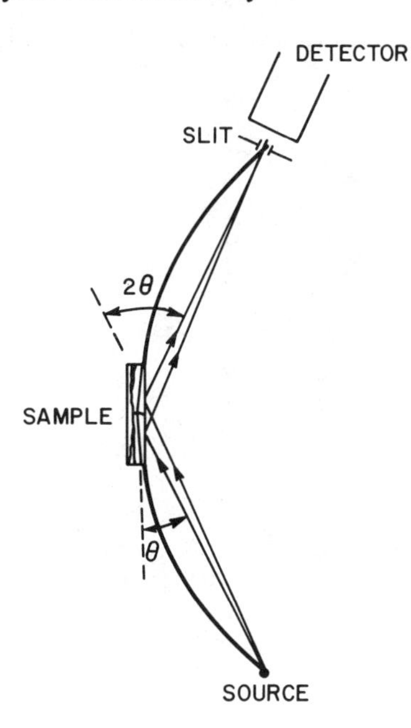


**Figure 3.27** Schematic illustration of the lack of sensitivity to mosaic structure in the Bragg-Brentano focussing arrangement.

mosaic misorientation effects as shown in Fig. 3.27. For diffractometers of this type there is a one-to-one correspondence between Bragg spacing and angle. At any given θ–2θ position, mosaic pieces with a range of orientations but all with the same Bragg spacing contribute simultaneously to the diffracted intensity observed at the detector. At a different θ–2θ position, other mosaic pieces with a different lattice parameter contribute. The FWHM observed for such diffractometers, therefore, is not a measure of mosaic spread, but instead is a measure of the size of the mosaic pieces and of the strain within them [44, 45].

Results for the (400) reflection of 3.0 μm thick InP layer grown by MOCVD on a (100) GaAs substrate [46] obtained using a powder diffractometer (Siemens model D500) and using a double crystal diffractometer ((100) InP first crystal) are shown in Figs. 3.28 and 3.29, respectively. The FWHM of the $K\alpha_1$, peak of the InP epitaxial layer in the powder diffractometer scan was 92 arcsec which is similar to that of the GaAs substrate peak, 72 arcsec. However, the crystallinity of the InP layer is not nearly as good as that of the substrate. This is evident from the double crystal rocking curve shown in Fig. 3.29. In spite of the fact that there is some wavelength dispersion for the GaAs substrate peak ($\beta = 35$ arcsec), the GaAs peak is still much narrower than that of the epitaxial InP (350 arcsec).

Approximate expressions for an average dislocation density in terms of the DCD FWHM were given by Hirsch [47] and can be applied in this case. The most appropriate expression is

$$D = \frac{\beta^2}{9b^2} \tag{3.53}$$

where D is the density of dislocations ($cm^{-2}$), β is the FWHM in radians, and b is the length of the Burgers vector of the dislocations. This expression is an estimate of the upper limit of the

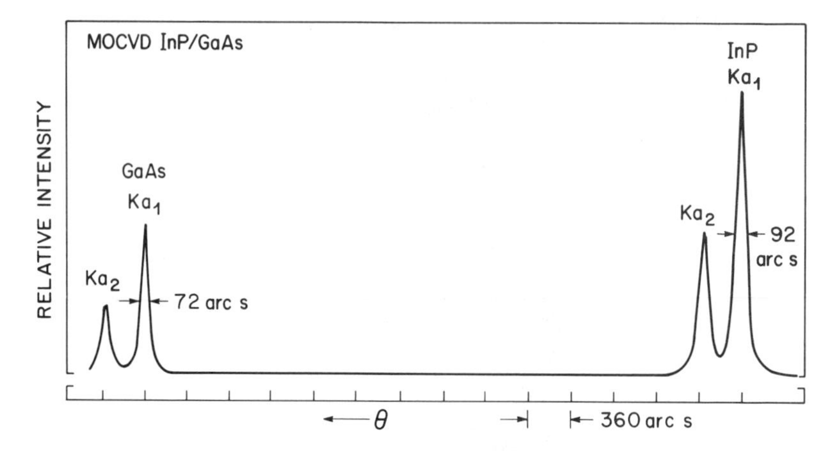


**Figure 3.28** A θ–2θ diffraction scan of the (400) reflection of epitaxial InP grown on a GaAs substrate using a Siemens D500 powder diffractometer which employes the Bragg-Brentano focussing geometry.

**Figure 3.29** A (400) double crystal rocking curve of the same sample as in Fig. 3.28. The first crystal was (100) InP set for the (400) reflection.

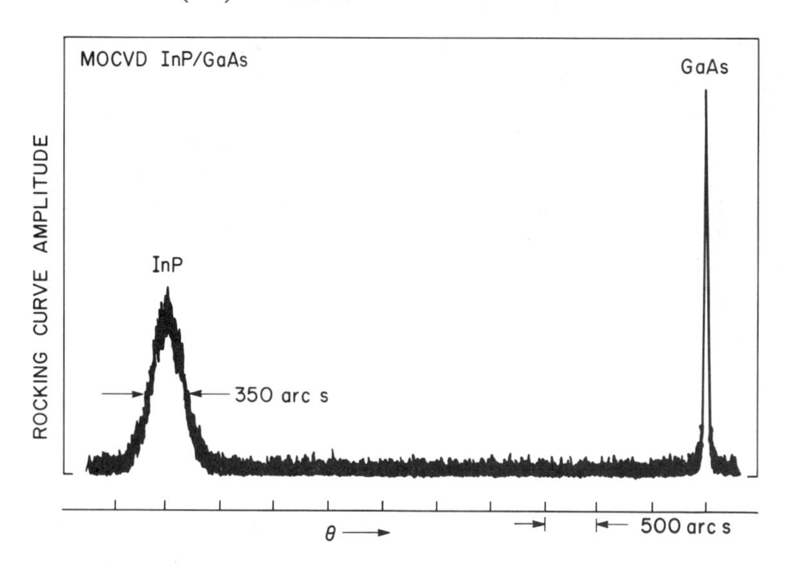

dislocation density [47]. Using the value $b=a_o/\sqrt{2}$ for 60 degree dislocations, where $a_o$ is the lattice parameter of InP, a value of $D = 5 \times 10^7 \text{ cm}^{-2}$ is obtained for the case of Fig. 3.29. We note that dislocation densities obtained from Eq. (3.53) for epitaxial $Ge_{1-x}Si_x$ alloys grown on Si substrates have been corroborated by plan-view transmission electron micrographs [48].

The utility of DCD rocking curves in assessing the degree of crystallinity of epitaxial Ge layers grown on (100) oriented Si substrates is demonstrated in Fig. 3.30. A substantial improvement in the quality of the layer (grown at 550°C by MBE) was revealed upon annealing it in a $H_2$ ambient at 750°C for 30 minutes. DCD rocking curves ((400) reflection) made before and after annealing shown in Fig. 3.30a and b, respectively, show a reduction in the FWHM from

**Figure 3.30** a) A double crystal rocking curve of a Ge layer grown at 550°C by MBE directly on a (100) Si substrate. The first crystal was a (100)-oriented InP crystal set to the (400) reflection, and the $K\alpha_2$ component was blocked by a slit edge. The wavelength dispersion contribution to the Ge peak widths is much smaller than that due to mosaic effects. This is evidenced by the narrow width of the Si peak for which there is more wavelength dispersion than for the Ge peak. b) After annealing at 750°C for 30 minutes.

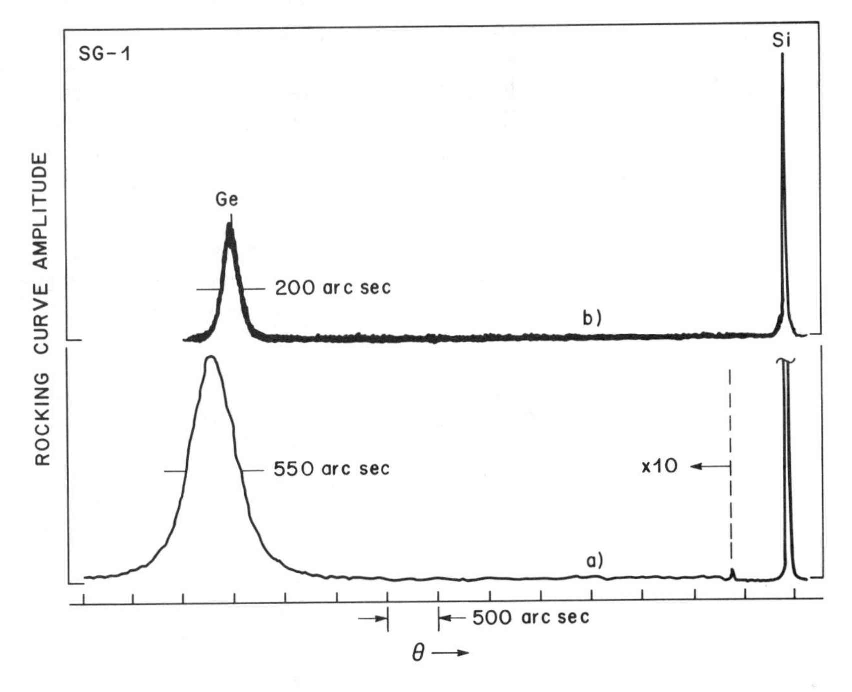

550 arcsec to 200 arcsec. This corresponds to a reduction in the dislocation density by a factor of 7.6 [46]. Similar results were found for $Ge_{1-x}Si_x$ alloys [49] and for GaAs [50] grown directly on Si substrates.

In general, mismatched epitaxial layers are tetragonally distorted caused by the misfit stress. Heterostructures with large mismatches exhibit tetragonality not according to the whole lattice mismatch but only to that fraction of the mismatch generated as a result of the thermal expansion mismatch [45]. The tetragonality can be measured using Eqs. (3.13) and (3.14) in which the small angle approximation must not be made [46].

## 3.2 OTHER X-RAY CHARACTERIZATION METHODS

### 3.2.1 The Back Reflection Laue Method

A crystal can be readily oriented using the Laue method [51], and this method is widely used to orient boules for wafer sawing. The back reflection method is illustrated in Fig. 3.31, and employs diffraction in the Bragg geometry (historically, the original arrangement was in transmission which is known as the Laue geometry). Bragg's law which relates the angle of a diffraction peak, θ, the x-ray wavelength, λ, and the interplanar spacing, d, is given by

$$n\lambda = 2\,d\,\sin\theta \tag{3.54}$$

Here n is the order of diffraction. This equation will suffice for an explanation of the method. A crystal positioned as in Fig. 3.31 will produce spots on a film for all those planes which produce a sufficiently intense diffraction peak. The angle of incidence with respect to all the planes is fixed since the crystal is not moved. Each plane uses that portion of the incident spectrum which is dictated by Bragg's law. For most planes, the wavelengths needed fall in the Brehmstrahlung part of the spectrum for all orders (which superimpose). However, occasionally, an atomic plane will be positioned properly for Bragg diffraction from a line characteristic of the anode and these are much more intense than the white radiation. In that case a very bright spot is produced. A Laue photograph of [100] InP obtained with W radiation is shown in Fig. 3.32. A (711) spot is

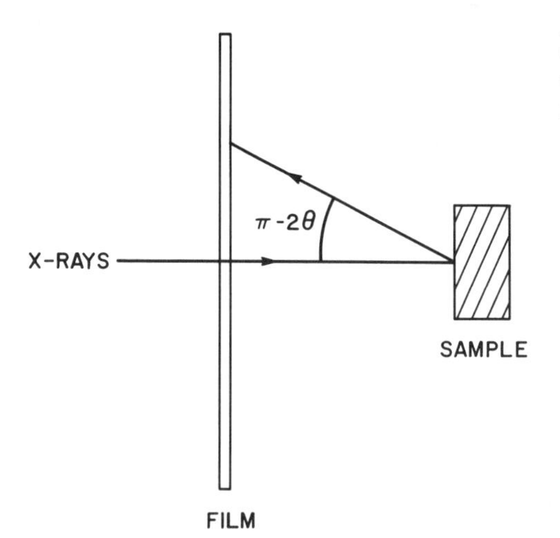


**Figure 3.31** X-ray beam, film, and sample arrangement for the back-reflection Laue method.

**Figure 3.32** Laue photograph for (100) oriented InP using a tungsten x-ray tube.

very bright because these planes were positioned properly for the $K\beta$ line of W. Note that the precision of the orientation depends on the precision of alignment between the incident x-ray beam and the crystal as it is mounted.

### 3.2.2 The Bond Method

Lattice parameters of single crystal semiconductors can conveniently be measured using the method devised by W. L. Bond [52]. The basic arrangement is shown in Fig. 3.33. The crystal is first positioned to diffract into one of the two detectors shown and is then rotated until it diffracts into the other detector. One measures the angle through which the crystal must be rotated to accomplish this, and since this angle is $180°-2\theta$, the Bragg angle, $\theta$, is thereby obtained.

In this procedure several of the errors which normally plague lattice parameter measurements are avoided. The Bragg angle can also be measured, of course, by simply measuring the angle that the diffracted beam makes with respect to the incident beam. This can be done by positioning the detector on a goniometer and noting its position at the Bragg condition. However, if the incident beam is not exactly at the goniometer zero an error is introduced in the measurement of $\theta$. This so called zero error is eliminated in the Bond method. In addition, the Bond method also avoids errors arising from eccentric placement of the crystal on the goniometer because the angle between two crystal positions is measured. If the angle between two detector positions were measured instead, the beam angles would be incorrectly measured. That eccentric crystal placement does not prevent the accurate determination of $\theta$ if crystal positions are measured, is illustrated in Fig. 3.34. Here we have changed over to using the surface normal to define the crystal's position. Large windows on the detector are needed, however, and, since the beam walks over the sample, good lateral uniformity is required for large eccentricities. Finally, a correction for absorption is also avoided since absorption displaces the beam but not the crystal angle at reflection.

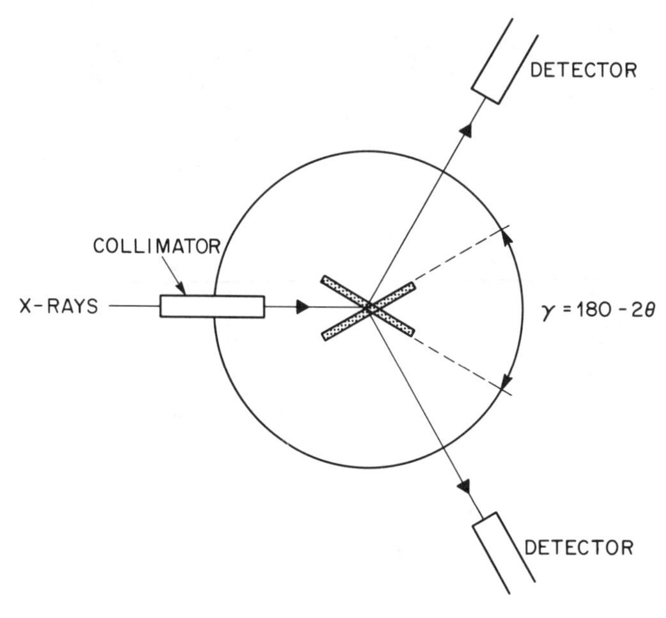


**Figure 3.33** X-ray beam, detector, and sample arrangement for the Bond method for precise lattice parameter measurements.

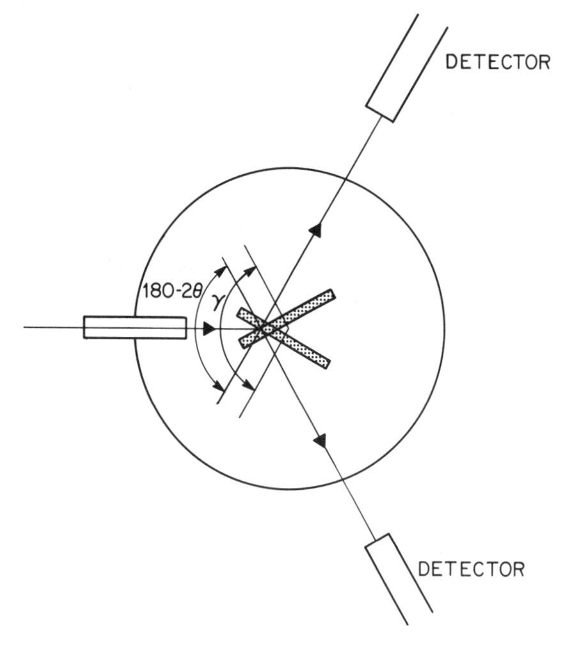


**Figure 3.34** Demonstration that the crystal rotation angle between Bragg reflection positions, $\gamma$, is equal to $\pi-2\theta$ even for eccentric placement of the crystal.

Other errors remain, unfortunately. For precise work, the atomic planes must be parallel to the axis of rotation and a deviation will result in a crystal tilt error. This is eliminated in the Bond method via a special crystal adjusting fixture designed to progressively reduce the tilt via a repeated procedure. A similar tilt error occurs if the collimator is not perpendicular to the goniometer axis. Because of this latter tilt, limits are placed on the divergence out of the plane of Figs. 3.33 and 3.34 and, as a result, on the collimator slit length. Finally, corrections for refraction, wavelength dispersion, and the Lorentz-polarization error are required, as usual.

### 3.2.3 Point Defect Influences on Lattice Parameters and X-Ray Measurements

There are strain fields associated with point defects in crystals and these strain fields can influence the lattice parameter of a semiconductor crystal [53]. In general, in addition to the lattice parameter, the size of the crystal will also be changed on the introduction of point defects. Such length and lattice parameter changes have been analyzed by Simmons and Balluffi [54] by conceptually introducing centers of dilatation using a sphere-in-hole model. Such dilatations were analyzed using the methods of Eshelby who had shown using elasticity theory that the dilatations of a random distribution of centers result in a strain which is uniform and isotropic [55]. If more than one kind of center is present, then the length and lattice parameter changes are simply additive unless the nonelastically strained cores of the defects overlap.

Only the dilatational strain causes changes in the lattice parameter, but an accounting of the number of atomic sites is also required to describe changes in length of a crystal. When a vacancy is formed, the number of sites is increased by one and when an interstitial is formed (without an accompanying vacancy) the number of sites is decreased by one. If $\Omega$ denotes an atomic volume and $x_v\Omega$ and $x_i\Omega$ are the dilatation volumes of a single vacancy and interstitial, then the length and lattice parameter changes are given by

$$3\frac{\Delta\ell}{\ell} = C_v - C_i + C_v x_v + C_i x_i \tag{3.55}$$

$$3\frac{\Delta a}{a} = C_v x_v + C_i x_i \quad , \tag{3.56}$$

where $C_v$ and $C_i$ are concentrations of vacancies and interstitials. These two expressions apply to elemental semiconductors. For compound semiconductors, antisite disorder is ignored if one applies them. Note that if there are no lattice distortions around vacancies or interstitials, then the dilation volumes are zero and there will be no change in the lattice parameter. We note, furthermore, that the difference between Eqs. (3.55) and (3.56) produces the relation

$$3\frac{\Delta\ell}{\ell} - 3\frac{\Delta a}{a} = C_v - C_i \tag{3.57}$$

This is a significant and useful result, and Eq. (3.57) is the basic relation on which the simultaneous length and lattice parameter method is based [53].

In the absence of simultaneous length and lattice parameter measurements, defect analyses in irradiated semiconductors and in semiconductors quenched from high temperature have been carried out using either Eq. (3.55) or Eq. (3.56) separately.

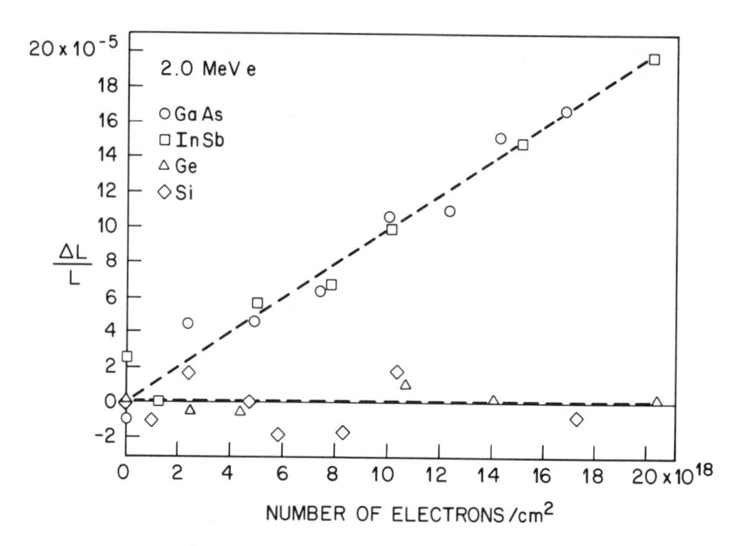


**Figure 3.35** Length changes for GaAs, InSb, Ge, and Si crystals irradiated with 2 MeV electrons [57].

Low temperature length change measurements have been made for germanium [56], silicon [56], indium antimonide [57], and gallium arsenide [57] bombarded with 2 MeV electrons using direct optical observations. The results are shown in Fig. 3.35, and show that the two compound semiconductors InSb and GaAs produced qualitatively different results than did the elemental semiconductors, Si and Ge. These results were analyzed by assuming that the concentrations of vacancies was equal to the concentration of interstitials in Eq. (3.55), that is, that $C_v = C_i = C$. This assumption is based on the general result that electron irradiations near 1 MeV produce mostly isolated defects [53]. With this assumption, Eqs. (3.55) and (3.56) become

$$3\frac{\Delta \ell}{\ell} = 3\frac{\Delta a}{a} = C\ (x_i + x_v) \tag{3.58}$$

An assessment of the defect concentrations based on radiation damage estimates involving a displacement threshold then lead to values of $(x_i + x_v)$ [53, 56, 57]. The results are $(x_i + x_v)_{Ge} < 0.02$, $(x_i + x_v)_{Si} < 0.01$, $(x_i + x_v)_{InSb} \cong 1.0$, and $(x_i + x_v)_{GaAs} \cong 0.94$.

Light ion bombardment will, in general, produce a very different kind of damage than will electron bombardment. This is evident from the data shown in Fig. 3.36 for Ge bombarded with 10.2 MeV deuterons [58]. Unlike for electron irradiations, a dilatation effect is readily seen in the $\Delta\ell/\ell$ measurement which implies a qualitatively different kind of damage. An effect on the lattice parameter was also observed for deuteron bombarded Ge [59]. These results were interpreted to imply that the observed dilatations for deuteron bombarded Ge are primarily caused by defects in large clusters [53].

Another way in which to obtain large concentrations of intrinsic point defects in

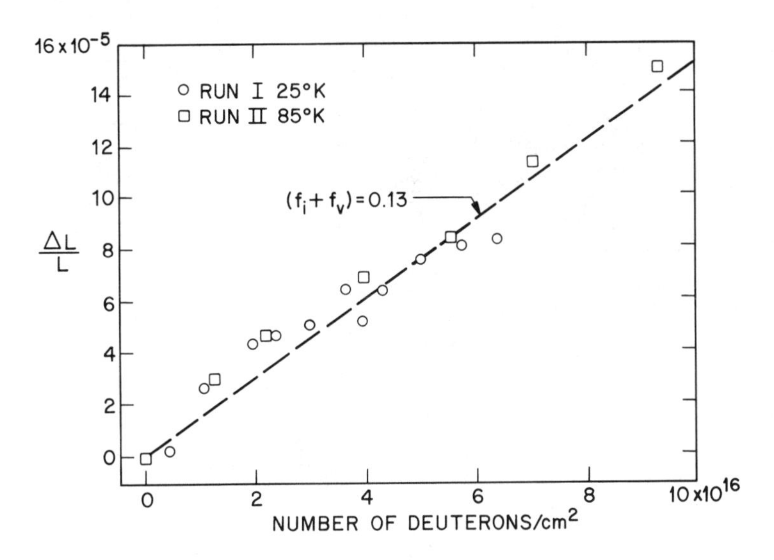


**Figure 3.36** Length changes for a Ge crystal irradiated with 10.2 MeV deuterons [58].

semiconductors is by quenching from a high temperature. In an elemental crystal, the equilibrium concentration of a thermal defect (e.g., a vacancy) at a temperature T is given by [60]

$$C = \frac{n}{N} = \exp\left(-\frac{g}{kT}\right) \tag{3.59}$$

Here k is Boltzman's constant, n is the number of defects, N is the number of possible defect sites, and g is the Gibbs free energy of formation of the defect. If we write

$$g = h - Ts \tag{3.60}$$

where h is the enthalpy of formation and s is the entropy of formation, then Eq. (3.59) becomes

$$C = \exp\left(\frac{s}{k}\right) \exp\left(-\frac{h}{kT}\right) \tag{3.61}$$

Quenched-in defects will affect the lattice parameter according to Eq. (3.56). Results of a quenching study of GaAs [61] are shown in Fig. 3.37. Data are shown for GaAs quenched in ampoules both without and with excess arsenic. Since excess arsenic substantially reduced the observed value of $\Delta a/a$ obtained after quenching, vacancies on As sites were concluded to be the primary defects formed.

The lattice parameter data were obtained using the Kossel line technique. This technique involves measurements of a diffraction pattern (Kossel lines) produced when a beam of electrons is focussed on a crystal to produce a point source of x-rays within the crystal.

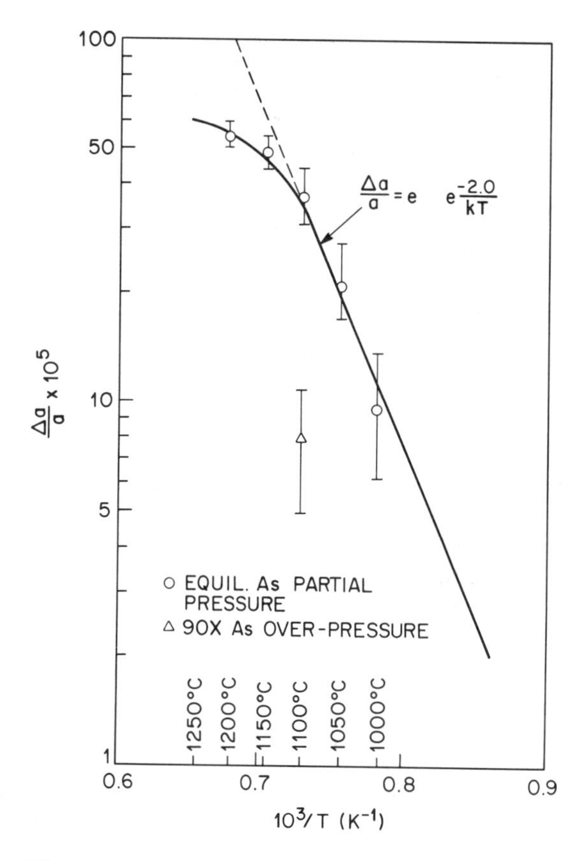


**Figure 3.37** Lattice parameter data for GaAs after quenching from high temperatures as a function of the quenching temperature. Data are shown for quenches made both with and without excess As in the ampoule [61].

The result for $\Delta a/a$ obtained by Potts and Pearson is $\Delta a/a = e^{8.8}e^{-2.0\text{eV}/kT}$. Applying Eq. (3.59) to this result and invoking Vook's values for the dilatation volumes, the values $s = 8.8\ k$ and $h = 2.0\ \text{eV}$ are obtained for As vacancies in GaAs. The quenched-in As vacancy concentration deduced in this way were as high as $5 \times 10^{-4}$ ($1 \times 10^{19}$ $\text{cm}^{-3}$).

A similar experiment was carried out by Driscoll and Willoughby [62] but using the Bond method to measure lattice parameters of quenched GaAs. These workers also found positive $\Delta a/a$ values after quenching but their values were an order of magnitude smaller than had been reported by Potts and Pearson [61]. This difference was ascribed to differences in quenching rates between the two experiments. Annealing data [62] are shown in Fig. 3.38 for samples quenched without additional As or Ga in the ampoule. Except for the quench made from the highest temperature, 1164°C, annealing was found to proceed in two stages. The first, stage I, is more rapid than the second and occurred up to about 26 hours annealing time at room temperature. The second stage, stage II, occurred from then on. This behavior confirmed the annealing results obtained by Potts and Pearson [61], and provided evidence for more than a single quenched-in defect.

Both As monovacancies and Frenkel disorder consisting of As-site monovacancy-As interstitial pairs were invoked to explain these annealing results. The interpretation of the data is that annealing occurs first by recombination of the Frenkel pairs until all the interstitials have

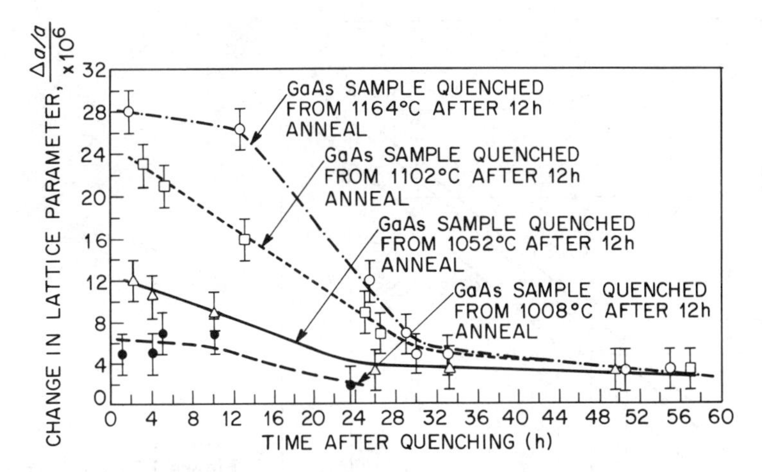


**Figure 3.38** Lattice parameter changes for GaAs as a function of time at room temperature after quenching subsequent to high temperature anneals [62].

been depleted. Stage II annealing occurs when the remaining vacancies diffuse to a sink such as the crystal surface, an internal boundary, or a dislocation.

A closely related set of measurements which also illustrate the usefulness of precise lattice parameter determinations were carried out by Straumanis and Kim [63]. These workers prepared samples at the limits of the phase extent of GaAs by taking highly stoichiometric starting material and putting it into quartz tubes containing in addition either elemental Ga or As and heating the tubes to 1000°C for 18 hours. The tubes were then furnace cooled. Lattice parameter determinations were subsequently made at various temperatures near room temperature from x-ray powder patterns using the (422) reflection and Cr K$\alpha_1$ radiation. The results are shown in Fig. 3.39. At 25°C the GaAs phase extends from 5.65325 Å at the Ga rich side to 5.65300 Å at the As rich side. Although the difference is only 0.00025 Å, it exceeds greatly the precision of the determinations (±0.00003 Å).

A general composition in this phase can be determined from precise lattice parameter values if the density is also known precisely. Precise density determinations were also made by Straumanis and Kim [63], and they report that the GaAs phase extends from 49.998 to 50.017 atomic percent As (using $6.0225 \times 10^{23}$ for Avogadro's number).

Characterization of the deviation from stoichiometry of GaAs can also be done by measurements of the x-ray intensity of a quasi-forbidden Bragg reflection [64]. For the zinc-blende lattice the structure factor for (200) and (600) reflections is given by

$$F = 4 \left[ f_{III} \exp(-M_{III}) - f_V \exp(-M_V) \right] \qquad (3.62)$$

where $f_{III}$ and $f_V$ are the atomic scattering factors (including the anomalous dispersion corrections, $\Delta f'$ and $\Delta f''$) for the atoms on the III and V sublattices and $M_{III}$ and $M_V$ are the Debye-Waller factors. Since the Ga and As factors are almost equal, these reflections are quasi-

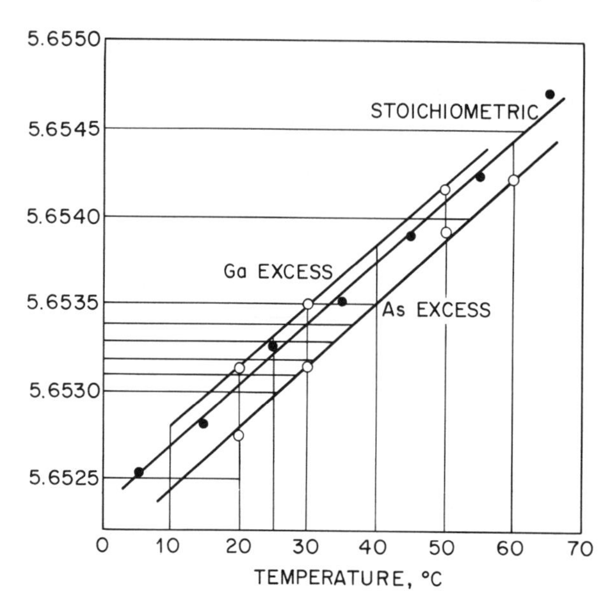


**Figure 3.39** Lattice parameters as a function of temperature for Ga rich, stoichiometric, and As rich GaAs samples [63].

forbidden in GaAs. Furthermore, Eq. (3.62) applies to a perfectly stoichiometric compound; for nonstoichiometric compounds this equation should be modified to

$$F = 4\left[C_{III} f_{III} \exp(-M_{III}) - C_V f_V \exp(-M_V)\right] \tag{3.63}$$

where $C_{III}$ and $C_V$ are effective atomic concentrations. Because of the near cancellation in the case of GaAs, small deviations of $C_{Ga}$ or $C_{As}$ from unity result in a large variation of F and consequently a large variation in diffracted x-ray intensity. Results obtained for an undoped semi-insulating liquid-encapsulated Czochralski (LEC)-grown crystal are shown in Fig. 3.40. Large values for $C_{Ga} - C_{As}$ were found relative to those obtained for Horizontal Bridgeman (HB), vapour phase epitaxy (VPE), and liquid phase epitaxy (LPE)-grown samples. If the HB, VPE, and LPE samples are assumed to have had a nearly stoichiometric composition, the observed deviation from stoichiometry in the LEC wafer corresponds to defect concentrations of $10^{18} \sim 10^{19}$ cm$^{-3}$. Since a primary use of LEC GaAs wafers is for integrated circuits, these results were plotted along with etch pit densities by Fujimoto. The well-known W shape is evident in both sets of data which suggests that the stoichiometric disturbances have a similar origin as do the dislocations [65].

A mapping of the deviation from stoichiometry for LEC GaAs wafers has also been done by making a precise lattice parameter map [66].

Strain fields associated with impurities can also effect the lattice parameter of a semiconductor crystal. This was demonstrated by Fewster and Willoughly [67] who reported lattice parameter measurements for GaAs heavily doped with Si made using the Bond technique. As shown in Fig. 3.41, they measured a large lattice contraction in p-type LPE material compared to n-type VPE and n-type gradient-freeze material. The lattice contraction for LPE material was found to $6 \times 10^{-4}$ Å at a hole concentration of $6 \times 10^{18}$ cm$^{-3}$. The measurements

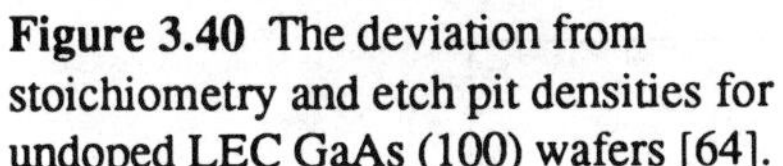

**Figure 3.40** The deviation from stoichiometry and etch pit densities for undoped LEC GaAs (100) wafers [64].

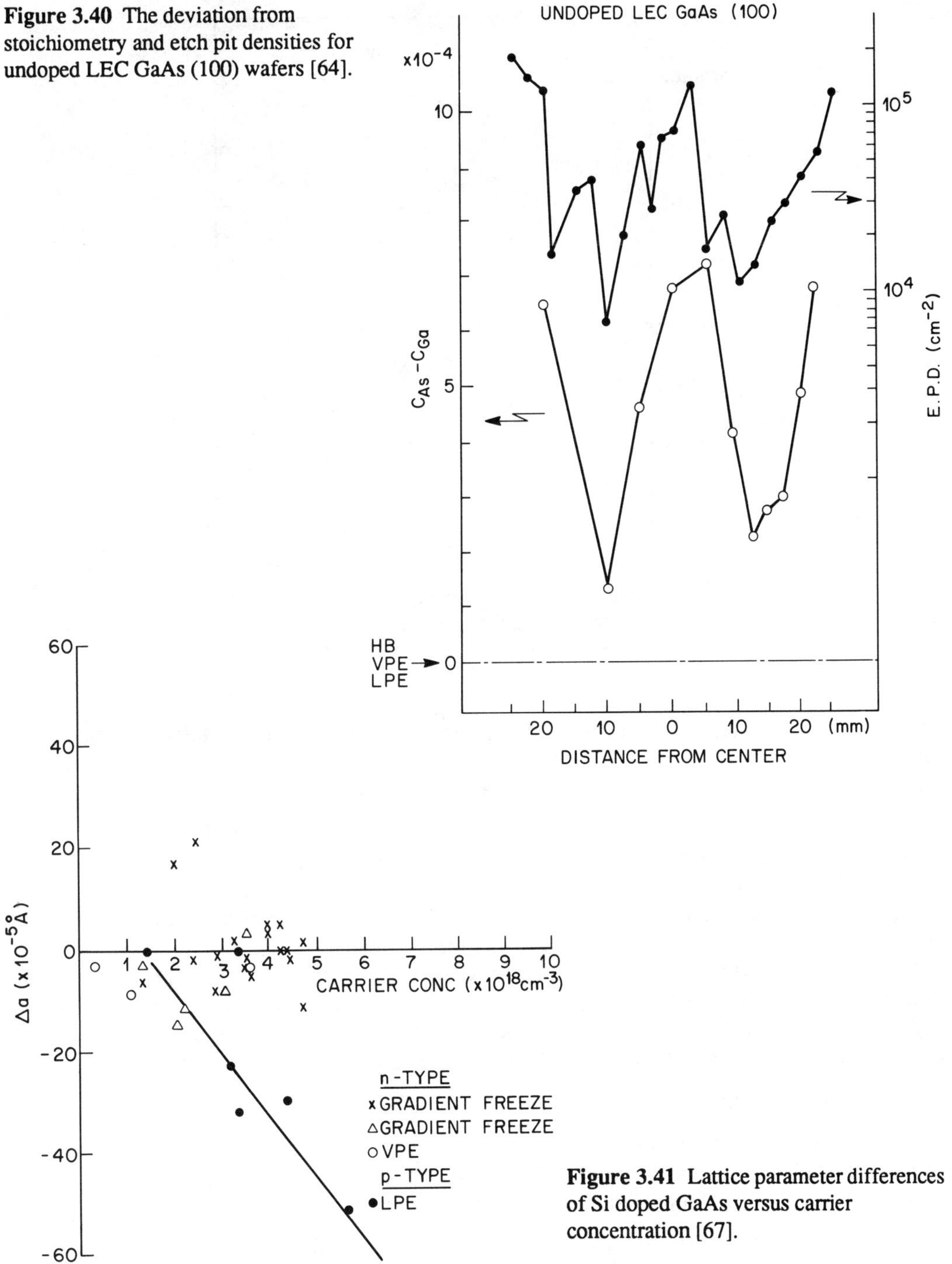


**Figure 3.41** Lattice parameter differences of Si doped GaAs versus carrier concentration [67].

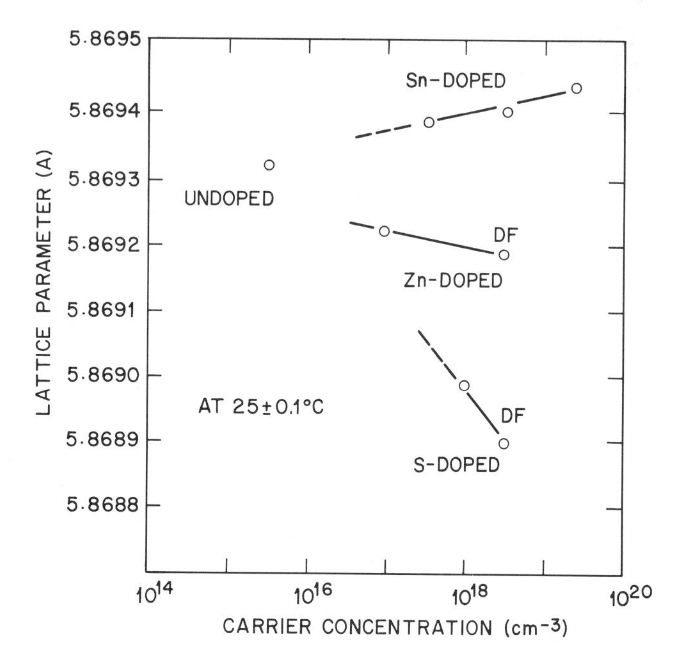


**Figure 3.42** Lattice parameter of LEC InP as a function of carrier concentration for undoped, Zn doped, S doped, and Sn doped crystals [68].

were made using the (800) reflection with Cuk$\beta$ radiation which yielded a lattice parameter with accuracy believed to be 1 ppm. The most likely defect responsible was considered to be Si on a Ga site.

The Bond method has also been applied to study LEC InP heavily doped with Sn, Zn, and S. Sugii et. al., [68] used the (600) reflection and Cu K$\alpha_1$, radiation and obtained the results shown in Fig. 3.42. The lattice parameter of a covalently bonded crystal is correlated to the bond strength [69]. The lattice contractions observed for Zn and S doping of InP imply an increased bond strength and the lattice expansion observed for Sn doping implies decreased bond strength. This result agrees with measurements of Knoop hardness number which were also found to depend on the type of dopant with the decreasing order of the hardness being S-doped, Zn-doped, undoped, and Sn-doped InP crystals, and these results have been used to explain the relative effectiveness of these dopants in reducing dislocation densities in LEC crystals [70].

### 3.2.4 X-Ray Topography

Techniques whereby images of crystals are obtained by recording diffracted x-ray beams are collectively called x-ray topography [71]. The contrast which results is informative concerning crystalline perfection, wafer curvature, and wafer thickness.

Two classes can be defined. These are distinguished according to whether topographs are obtained in reflection or in transmission. Diffraction in the two situations is often referred to as

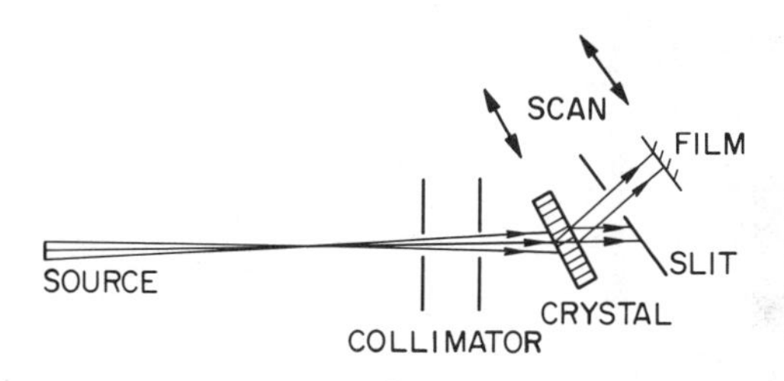


**Figure 3.43** Schematic arrangement for Lang Transmission topography. The sample and the film are scanned in tandem across a fixed beam.

Bragg case or Laue case for reflection and transmission, respectively. Which case is preferred depends on the type of information desired and on the structure of the sample.

Laue case topography is commonly carried out in the manner first described by Lang [72, 73]. A schematic illustration of Lang's technique is shown in Fig. 3.43. Characteristic radiation of an x-ray tube is employed, and the incident beam is sufficiently well collimated to allow the $K\alpha_1$ and $K\alpha_2$ components to be separated. In this way double images are avoided. Nuclear emulsion plates are commonly used to record the topographs, and the sample and plate are moved in tandem as the sample is traversed across the x-ray beam. A slit is incorporated between sample and plate to block all but one of the beams exiting from the back of the sample. This method can be applied even to samples which are much thicker than $1/\mu$, where $\mu$ is the x-ray absorption coefficient, if the samples are highly perfect. For such samples there is anomalous transmission which was discovered by Borrmann [74] and is known as the Borrmann effect. To understand the origin of this effect one must invoke dynamical diffraction theory (which has been excellently reviewed by Batterman and Cole [6]) and the topographic images observed in such cases are referred to as dynamical. The dynamical theory properly includes the interactions between the incident and diffracted beams and provides a complete description in all cases. However, for crystals thinner than $1/\mu$ or for highly imperfect crystals possessing a pronounced mosaic structure the simpler kinematic diffraction theory applies. The kinematical theory assumes that the beams do not interact and are affected only by normal absorption.

An example of a Lang topograph for which $\mu t=16.2$ for GaInAs/InP is shown in Fig. 3.44. Misfit dislocations exist as a cross grid in [110] and $[\bar{1}10]$ directions [15]. The contrast from misfit dislocations depends on the diffraction conditions, on the type of dislocation, and on the direction of the dislocation line [17]. An example of contrast analyses of misfit dislocations is provided in the work of Bartels and Nijman [75] for heteroepitaxial layers grown on (100) GaAs substrates. These workers found that there was an asymmetry in the character of the misfit dislocations which were running in the two orthogonal <110> directions in the (100) growth plane. In the [110] direction edge-type dislocations with their Burgers vector in the growth plane were observed. In the perpendicular $[\bar{1}10]$ direction 60 degree-type dislocations with their Burgers vector inclined at 45 degrees to the growth plane were visible.

Besides dislocations it is also possible to obtain contrast caused by other defects such as precipitates [76, 77].

A severe problem arises if one attempts to make a topograph of a large sample which is curved. Almost all samples will be sufficiently curved to cause the loss of the Bragg condition once the sample is moved along its traverse. The result will be that only a thin section of the sample has been imaged on the plate. Three different schemes are commonly applied to

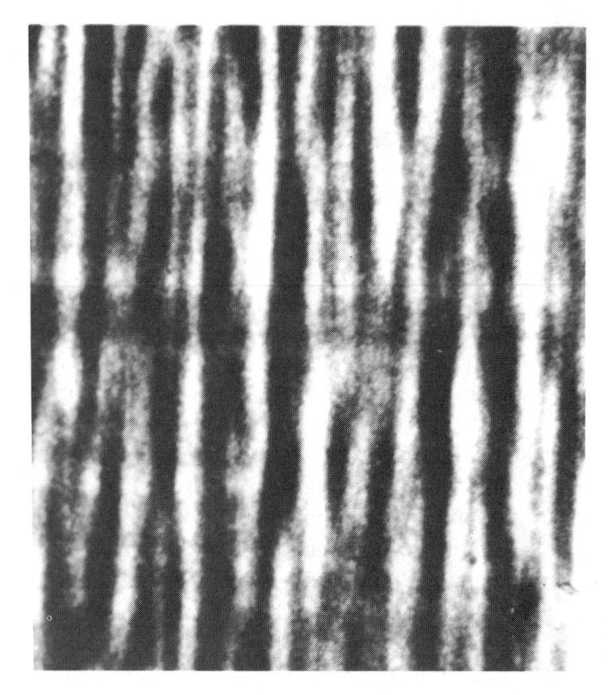

**Figure 3.44** Lang topograph for incoherent GaInAs/InP showing misfit dislocations.

overcome this problem. By oscillating the sample through the Bragg condition as it is traversed across the beam (as in Lang topography) Schwuttky found that large wafers (at least 2 in. in height) can be completely imaged in a single topograph [78]. A more efficient way in which to accomplish the same thing is to automatically stay on the Bragg peak during the traverse [79]. This has come to be known as automatic Bragg angle control [17] A third and different way in which to get complete exposures of large wafers is known as Hirst topography [80] and involves bending a wafer in a divergent x-ray beam in such a way that all points on the wafer diffract simultaneously.

Bragg case topography derives its usefulness from the fact that the region probed by the x-rays lies roughly within a depth of $1/\mu$ below the surface for imperfect samples or within one extinction depth for perfect samples. (Extinction is the phenomenon whereby x-rays are prevented from going deeper into a single crystal because they are back diffracted). Bragg case topography is therefore the topographical technique of choice for imaging diffraction from epitaxial layers only, that is, free from contrast as a result of defects in the bulk of the substrate.

The type of Bragg case topography which is simplest to implement is Berg-Barrett topography [81, 82]. However, it is hindered by the fact that double images caused by $K\alpha_1$ and $K\alpha_2$ components usually result and that resolution for rocking curve peaks is low because of wavelength dispersion. The problem of wavelength dispersion can be overcome by using a double crystal arrangement [83, 84], but then one still obtains double images because of the two $K\alpha$ components. Not only high resolution for rocking curve peaks but also single images can be obtained if one employs a multiple crystal arrangement [85, 86]. It has been possible to simulate the contrast observed for misfit dislocations in a GaAlAsP epilayer grown on a GaAs substrate using a multiple crystal monochromater [87].

**REFERENCES**

1. A.H. Compton, *Phys. Rev. 10*;95, 1917.

2. W. Ehrenberg and H. Mark, *Zeit. f. Physik 42*;807, 1927.

3. R.W. James, *The Optical Principles of the Diffraction of X-rays*, Ithaca, Cornell Univ. Press, Ch. VI, p. 3 309.

4. J.W.M. Dumond, *Phys. Rev. 52*;872, 1937.

5. W.H. Zachariasen, *Theory of X-ray Diffraction in Crystals* New York; Dover, 1967.

6. B. Batterman and H. Cole, *Rev. Mod. Phys. 36*;681, 1964.

7. J. Matsui, W. Watanabe, and Y. Seki, *J. Cryst. Growth 46*;563, 1979.

8. A.T. Macrander, W.A. Bonner, and E.M. Monberg, *Materials Letters 4*;181, 1986.

9. H. Nagai, *J. Appl. Phys. 45*;3789, 1974.

10. A.T. Macrander and K. Strege, *J. Appl. Phys. 59*;442, 1986.

11. J. Hornstra and W.J. Bartels, *J. Cryst. Growth 44*;513, 1978.

12. W.J. Bartels and W. Nijman, *J. Cryst. Growth 44*;518, 1978.

13. J. Matsui, K. Onabe, T. Kamijima and I. Hayashi, *J. Electro chem. Soc. 126*;664, 1979.

14. W.J. Bartels, *J. Vac. Sci. Technol. B*1;338, 1983.

15. S.N.G. Chu, A.T. Macrander, K.E. Strege, and W.D. Johnston, Jr., *J. Appl. Phys. 57*;249, 1985.

16. R.E. Nahory, M.A. Pollack, W.D. Johnston, Jr., and R.L. Barns, *J. Appl. Phys. 33*;659, 1978.

17. G.A. Rozgonyi and D.C. Miller, *Thin Solid Films 31*;185, 1976.

18. V. Swaminathan, J. Lopata, and J.W. Lee, Proc. Mat. Res. Soc., Boston 1986.

19. W.R. Wagner, N.E. Schumaker, J.L. Zilko, P.J. Anthony, and V. Swaminathan, *J. Electrochem. Soc. 130*;670, 1983.

20. H. Cole and N.R. Stemple, *J. Appl. Phys. 33*;2227, 1962.

21. P.B. Hirsch and G.N. Ramachandran, *Acta Cryst. 3*;187, 1950.

22. B.W. Batterman and G. Hilderbrandt, *Acta Cryst. A24*;150, 1968.

23. C. Bocchi, C. Ferrari, P. Franzosi, *J. Elec. Mat.*

24. A.T. Macrander, E.R. Minami, and D.W. Berreman, *J. Appl. Phys. 60*;1364, 1986.

25. D.W. Berreman, *Phys. Rev. B14*;4313, 1976.

26. M.A.G. Halliwell, M.H. Lyons, and M.J. Hill, *J. Cryst. Growth 68*;523, 1984.

27. S. Takagi, *J. Phys. Soc. Jpn. 26*;1239, 1969.

28. D. Taupin, *Bull. Soc. Frang. Miner. Crist. 87*;469, 1964.

29. J. Burgeat and D. Taupin, *Acta Cryst. A33*;137, 1977.

30. A.T. Macrander, B.M. Glasgow, E.R. Minami, R.F. Karlicek, D.L. Mitcham, V.G. Riggs, D.W. Berreman, and W.D. Johnston, Jr., Mat. Res. Soc. Proc. Boston 1986.

31. A.T. Macrander, S. Lau, K. Strege, and S.N.G. Chu, *Appl. Phys. Lett. 52*;1985, 1988.

32. X. Chu and B.K. Tanner, *Appl. Phys. Lett. 49*;1773, 1986.

33. V. Speriosu and T. Vreeland, Jr., *J. Appl. Phys. 56*;1591, 1984.

34. R. Fleming, D. McWhan, A.C. Gossard, W. Wiegmann, and R.A. Logan, *J. Appl. Phys. 51*;357, 1980.

35. J. Vandenberg, R.A. Hamm, A.T. Macrander, M.B. Panish, and H. Temkin, *Appl. Phys. Lett. 48*;1153, 1986.

36. A.T. Macrander, G.P. Schwartz, and G.J. Gualtieri, *J. Appl. Phys. 64*;6733, 1988.

37. (IBM early paper).

38. B.E. Warren, *X-ray Diffraction*, Reading, Mass; Addison-Wesley, Chap. 1, p. 8, 1969.

39. R. Fischer, T. Henderson, J. Klem, W.T. Masselink, W. Kopp, H. Morkoc and C.W. Litton, *Elect. Lett. 20*;945, 1984.

40. H. Morkoc, C.K. Peng, T. Henderson, W. Kopp, R. Fischer, L.P. Erickson, M.D. Longerbone, and R.C. Youngman, *IEEE Electron Device Letters* EDL-6;318, 1985.

41. R. Fischer, N. Chanel, W. Kopp, H. Morkoc, L.P. Erickson, and R.C. Youngman, *Appl. Phys. Lett. 47*;397, 1985.

42. T.H. Windhorn, G.M. Metze, B-Y Tsaur, and J.C. Fan, *Appl. Phys. Lett. 45*;309, 1984.

43. P.B. Hirsch, *"Mosaic Structure"*, Chapt. 6 in *Progress in Metal Physics*, B. Chalmers and R. King, New York;Pergamon, p. 278, 1956.

44. H.P. Klug and L.E. Alexander, *X-ray Diffraction Procedures For Polycrystalline and Amorphous Materials*. New York; John Wiley & Sons, 1974.

45. D.A. Neumann, X. Zhu, H. Zabel, T. Henderson, R. Fischer, W.T. Masselink, J. Klem, C.K. Peng, and H. Morkoc, *J. Vac. Sci and Tech. B4*;642, 1986; D.A. Neumann, H. Zabel, R. Fischer, and H. Morkoc, *J. Appl. Phys. 61*;1023, 1987.

46. A.T. Macrander, R.D. Dupuis, J.C. Bean, and J.M. Brown, Proceedings of the TMS/MRS 1986 Northeast Regional Meeting, May 1-2, 1986, Murray Hill, (ed. M.L. Green etal.), p. 75.

47. P.B. Hirsch, "Mosaic Structure", Chap. 6 in *Progress in Metal Physics*, (eds. B. Chalmers and R. King) New York; Pergamon, 1956.

48. D.C. Houghton, J.-M. Baribeau, P. Maigńe, T.E. Jackman, I.C. Bassignana, C.C. Tan, and R. Holt, Proc. Mat. Res. Soc. Boston 1986, in press.

49. R. Dupuis, J.C. Bean, J.M. Brown, A.T. Macrander, R.C. Miller, and L.C. Hopkins, *J. Electronic. Mat. 16*;69, 1987.

50. M. Chand, K.W. Wecht, R. People, F.A. Baiochi, F. Capasso, J. Allam, A.T. Macrander, and A.Y. Cho, *J. Vac. Sci. and Tech.*, in press.

51. E. Preuss, B.K. Urban, R. Butz, *Lane Atlas*, Halsted Press, New York; John Wiley & Sons, 1974.

52. W.L. Bond, *Acta Cryst. 13*;814, 1960.

53. F.L. Vook, 7$^{th}$ Int. Conf. of Phys. of Semiconductors vol. 3: Radiation Damage in Semiconductors New York; Academic Press, 1964, p. 51.

54. R.O. Simmons and R.W. Balluffi, *J. Appl. Phys. 30*;1249, 1959; R.W. Balluffi and R.O. Simmons, *J. Appl. Phys. 31*;2284, 1960.

55. J.D. Eshelby in *Solid State Physics*, (ed. F. Seitz and D. Turnbull) New York; Academic Press, vol. 3, p. 79, 1956.

56. F.L. Vook, *Phys. Rev. 125*; 855, 1962.

57. F.L. Vook, Proc. of the Int. Conf. on *Crystal Lattice Defects*, Kyoto, 1962, Conf. *J. Phys. Soc. Japan 18* Suppl. II, 190, 1963.

58. F.L. Vook and R.W. Balluffi, *Phys. Rev. 113*;62, 1959.

59. R.O. Simmons, *Phys. Rev. 113*;70, 1959.

60. R.A. Swalin, *Thermodynamics of Solids* New York; John Wiley and Sons, Chap. 12, 13 and 14, 1972.

61. H.R. Potts and G.L. Pearson, *J. Appl. Phys. 37*;2098, 1966.

62. C.M.H. Driscoll and A.F.W. Willoughby in *Defects in Semiconductors,* 1972: *Radiation Damage and Defects in Semiconductors,* Conf. Series No. 116, p. 377.

63. M.E. Straumanis and C.D. Kim, *Acta Cryst. 19*;256, 1965.

64. I. Fujimoto, *Jap. J. of Appl. Phys.* 23;L287, 1984.

65. A. Jordan, A. Von Neida, R. Carson, and C. Kim, *J. Electrochem. Soc. 121*;153, 1974.

66. Y. Takano, T. Ishiba, N. Matsunaga, N. Hashimoto, *Jap. J. Appl. Phys. 24*;L239, 1985.

67. P.F. Fewster and A.F.W. Willoughby, *J. Cryst. Growth 50*;648, 1980.

68. K. Sugii, H. Korzuma, and E. Kubota, *J. Elec. Mat. 12*;701, 1983.

69. L. Pauling, *The Chemical Bond* Ithaca, NY Cornell Univ. Press, 1967

70. S. Mahajan and A.K. Chin, *J. Cryst. Growth 54*;138, 1981.

71. B.K. Tanner, *X-ray Diffraction Topography* New York; Pergamon, 1976.

72. A.R. Lang, *J. Appl. Phys. 5*;358, 1958.

73. A.R. Lang, *Acta Cryst. 12*;249, 1959.

74. G. Borrmann, *Z. Phys. 127*;297, 1950.

75. W.J. Bartels and W. Nijman, *J. Cryst. Growth 37*;204, 1977.

76. J.R. Patel, *J. Appl. Phys. 44*;3903, 1973.

77. J.R. Patel, *J. Appl. Cryst. 8*;186, 1975.

78. G.H. Schwuttky, *J. Appl. Phys. 36*;2712, 1965.

79. L.J. van Mettaert and G.H. Schwuttke, *Phys. Stat. Sol.(a) 3*;687, 1970.

80. C.A. Wallace and R.C.C. Ward, *J. Appl. Cryst. 8*;28181, 1975.

81. W.F. Berg, *Naturwissenschaften 19*;391, 1931.

82. C.S. Barrett, *Trans. Metall. Soc. AIME 215*;483, 1959.

83. W.L. Bond and J. Andrus, *Am. Mineralogist 37*;622, 1952.

84. U. Bonse and E. Kappler, *Z. Naturforschung 13a*;348, 1958.

85. H. Hashizume, A. Iida, K. Kohra, *Jap. J. Appl. Phys. 14*;1433, 1975.

86. J.F. Petroff, M. Sauvage, P. Riglet, and H. Hashizume, *Phil. Mag. A42*;319, 1980.

87. P. Riglet, M. Sauvage, J.F. Petroff, and Y. Epelboin, *Phil. Mag A42*;339, 1980.

# CHAPTER 4

# ELECTRICAL CHARACTERIZATION

## 4.1 INTRINSIC SEMICONDUCTORS

Bonding in III-V compounds is a mixture of covalent and ionic bonding. Three electrons from a group III atom combine with five electrons from a group V atom to fill valence band states formed from hybridized $sp^3$ orbitals. In the absence of defects or impurities the valence band is filled at a temperature of zero degrees Kelvin (T = 0 K), and the conduction band is empty. For T above 0 K electrons are excited across the band gap into the conduction band. Since every such excitation leaves behind a hole in the valence band, the concentration of electrons in the conduction band, n in the valence band, equals the concentration of holes, p. This carrier concentration is known as the intrinsic concentration, $n_i$, and gives

$$n = p = n_i \quad . \tag{4.1}$$

Since electrons have spin 1/2 and are Fermions, they occupy states in the conduction band according to Fermi-Dirac statistics. The chemical potential which is introduced into the quantum statistics as the derivative of the Helmholtz free energy with respect to the total number of particles is called the Fermi energy, $E_F$ [1]. The Fermi energy is determined by the condition that the total number of physical particles, that is, electrons is fixed. For semiconductors one speaks in terms of electrons and holes, and the condition for the Fermi energy, is given by Eq. (4.1). The probability of occupancy of a conduction band state at energy E by electrons is given by the Fermi function,

$$f(E) = \frac{1}{1 + \exp\left(\frac{E - E_F}{kT}\right)} \tag{4.2}$$

where k is Boltzmann's constant. The electron concentration is given by

$$n = 2 \int_{E_c}^{\infty} \rho_c(E)\, f(E)\, dE \tag{4.3}$$

where $\rho_c(E)$ is the density of states in the conduction band and $E_c$ is the energy of the conduction band minimum. The factor of 2 arises from the spin degeneracy of the conduction band states. The density of states is derived from band structure calculations and is usually based on the assumption of parabolic conduction bands, that is,

$$E = E_c + \frac{\hbar^2 k^2}{2m_e^*} \tag{4.4}$$

where $\hbar k$ and $m_e^*$ are the momentum and the effective mass of the electron. Furthermore, if only nondegenerate semiconductors are considered, that is, if $(E_c - E_F) \gg kT$, then Eq. (4.3) leads to

$$n = N_c \exp - \left( \frac{E_c - E_F}{kT} \right) \tag{4.5}$$

Here $N_c$ is given by

$$N_c = \frac{1}{\sqrt{2}} \left( \frac{m_e^* kT}{\pi \hbar^2} \right)^{3/2} \tag{4.6}$$

and is known as the effective density of states for the conduction band. Similar relationships hold for holes in the valence band. The hole concentration is given by

$$p = N_v \exp - \left( \frac{E_F - E_v}{kT} \right), \tag{4.7}$$

where $E_v$ is the energy at the top of the valence band and $N_v$ is the effective density of states for the valence band which is obtained by substituting $m_h^*$ for $m_e^*$ in Eq. (4.6). For group IV and for all III-V semiconductors the valence band is degenerate for light and heavy holes and $(m_h^*)^{3/2}$ is given by

$$\left( m_h^* \right)^{3/2} = \left( m_{eh}^* \right)^{3/2} + \left( m_{hh}^* \right)^{3/2}$$

The intrinsic carrier concentration can be calculated from Eqs. (4.5) and (4.6) since

$$n_i^2 = np = N_c\, N_v \exp\left[-\frac{Eg}{kT}\right] \tag{4.8}$$

where $E_g$ is the band gap energy.

## 4.2 EXTRINSIC SEMICONDUCTORS

All real semiconductor crystals contain some electrically active impurities or defects which introduce energy levels in the band gap and must be included in the solution for the Fermi energy. These are classified as being either donors or acceptors. When ionized donors release electrons into the conduction band and become positively charged and acceptors accept electrons from the valence band (i.e., release holes) and become negatively charged. By considering the reaction equilibrium between the charge states of a donor impurity [2], one obtains to the following expressions for the number of electrons bound to donor sites,

$$n_D = \frac{g_D\, N_D}{g_D + \exp\left[\frac{E_c - \Delta E_D - E_F}{kT}\right]} \tag{4.9}$$

Here $\Delta E_D$ is the energy difference between the conduction band and the energy level caused by the defect and $g_D$ is the degeneracy of the donor level. This degeneracy arises from the number of internal degrees of freedom of the defect. In the simplest cases this is just the spin degeneracy, but it may also reflect other degrees of freedom such as, for example, the number of equivalent orientations of a defect in the unit cell [3]. Similarly, we have for the number of holes bound to acceptor sites

$$p_A = \frac{g_A\, N_A}{g_A + \exp\left[-\frac{E_v + \Delta E_A - E_F}{kT}\right]} \tag{4.10}$$

The Fermi level is determined by the condition that charge be conserved. When both donors and acceptors are present charge conservation demands that

$$n + N_A^- = p + N_D^+ \tag{4.11}$$

where $N_D^+ = N_D - n_D$ and $N_A^- = N_A - p_A$ are the numbers of ionized donors and ionized acceptors. These are given by

$$N_D^+ = \frac{N_D}{1 + g_D \exp\left[-\frac{E_c - \Delta E_D - E_F}{kT}\right]} \tag{4.12}$$

and

$$N_A^- = \frac{N_A}{1 + g_A \exp\left[\frac{E_v + \Delta E_A - E_F}{kT}\right]} \tag{4.13}$$

The Fermi energy is uniquely determined by Eq. (4.11) in which Eqs. (4.5), (4.7), (4.12), and (4.13) are substituted for n, p, $N_D^+$, and $N_A^-$. Intuition regarding the solution for the Fermi energy can be obtained from Shockley's graphical solution method [4]. An example of this method in which both donors and acceptors are included is shown in Fig. 4.1. The two sides of Eq. (4.11) are graphically constructed by adding the terms for the carriers to those for the ionized impurities, and the intersection of the two resulting curves gives the solution for $E_F$.

Figure 4.1 applies to InP at 300 K with $\Delta E_D = 4$ meV, $N_D = 1 \times 10^{16} cm^{-3}$, $\Delta E_A = 20$ meV, and $N_A = 5 \times 10^{15} cm^{-3}$. The solution for $E_F$ is at 120 meV below $E_c$ and the resulting value for n is $4 \times 10^{15}$ $cm^{-3}$. This example illustrates the compensation which occurs generally in semiconductors, that is, that n is given approximately by $N_D - N_A$.

Another example of the utility of Shockley's graphical method is found in the consideration of deep levels which can make semiconductors semi-insulating (S.I.). In many applications it is useful to have electrical isolation between layers of a device structure or to have isolation between devices on a wafer. To achieve this, one needs a semi-insulating semiconductor which is characterized by carrier concentrations approaching the intrinsic concentration given by Eq. (4.9).

**Figure 4.1** Graphical solution for the Fermi energy for InP. $N_c = 4.5 \times 10^{17}$ $cm^{-3}$, $N_v = 7.0 \times 10^{18}$ $cm^{-3}$, Eg = 1.351 eV, T = 300K. Donor parameters are $E_D = 1.347$ eV, $N_D = 1 \times 10^{16}$ $cm^{-3}$. Acceptor parameters are $E_A = 0.020$ eV, $N_A = 5 \times 10^{15}$ $cm^{-3}$.

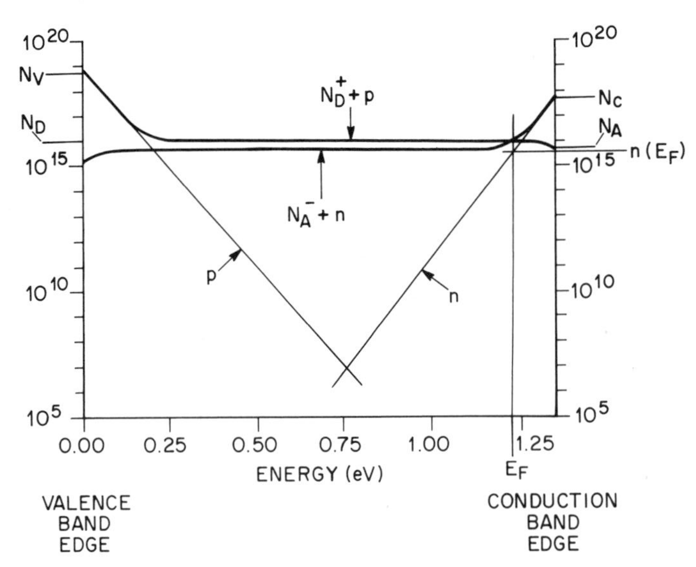

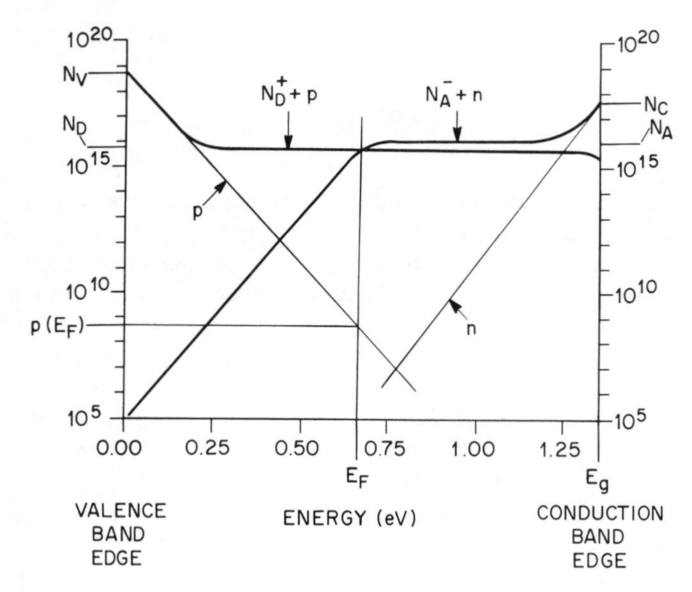


**Figure 4.2** Graphical solution for the Fermi energy for semi-insulating InP:Fe. InP parameters as in Fig. 1 and T=300K. Donor parameters are $E_D = 1.347$ eV, $N_D = 5 \times 10^{15}$ $cm^{-3}$. Acceptor parameters are $E_A = 0.671$ eV, $N_A = 1 \times 10^{16}$ $cm^{-3}$. The solution yields $E_F = 0.69$ eV below the conduction band edge, and $p = 5 \times 10^8$ $cm^{-3}$.

In practice this is accomplished via a deep level which pins the Fermi level near midgap. Here deep means that the level is not near a band edge. This is illustrated in Fig. 4.2 for Fe doped InP. The Fe in this hypothetical example for T = 300 K acts as a deep acceptor. The concentration of Fe, $1 \times 10^{16}$ $cm^{-3}$, chosen for the example exceeds that of the background donor concentration, which is taken to be $5 \times 10^{15}$ $cm^{-3}$. Note that the concentration of Fe need only be slightly higher than that of the background donor concentrations to pin the Fermi level near the Fe acceptor level.

## 4.3 MOBILITY AND THE HALL EFFECT

A very useful electrical figure of merit of an undoped semiconductor is its mobility. This quantity is the proportionality between the drift velocity of a carrier and the electric field that is,

$$v = \mu E \tag{4.14}$$

The mobility is strictly defined only for fields sufficiently low that the drift velocity and the field are linearly related. In this range of fields Ohm's law is satisfied since the current density is given by

$$J = qn\, v_n + qpv_p = \left[qn\mu_n + qp\mu_p\right] E \tag{4.15}$$

where q is the charge of the electron. Therefore, the conductivity, $\sigma$, and resistivity, $\rho$, is given by

$$\frac{1}{\rho} = \sigma = qn\mu_n + qp\mu_p \tag{4.16}$$

If one type of carrier predominates, then both the carrier concentration and the mobility are conveniently measured using the Hall effect [5]. Classically this is done by placing contacts on a bar or on a sample shaped in the well known bridge shape, but in practice it is much more convenient to apply the method of van der Pauw [6] which can be performed on samples of arbitrary shape. This method requires that the contacts be at the circumference of a sample, that the contacts be small, and that the sample be homogeneous in thickness and not perforated. Since GaAs and InP are conveniently cleaved along crystallographic <110> planes, the preferred sample geometry is a square with contacts in the corners as shown in Fig. 4.3. If we define $R_{AB,CD}$ to be the resistance obtained by dividing the voltage applied from contact C to contact D by the current which enters the sample at contact A and leaves it through contact B and if we define similarly the resistance $R_{BC,DA}$, then van der Pauw's theorem states that the resistivity is given by

$$\rho = \frac{\pi d}{\ell n 2} \frac{(R_{AB,CD} + R_{BC,DA})}{2} f \left( \frac{R_{AB,CD}}{R_{BC,DA}} \right) \tag{4.17}$$

where d is the thickness of the sample and f is a function which is unity if $R_{AB,CD} = R_{BC,DA}$. If a magnetic field is applied normal to the sample surface and if current enters and leaves the sample at diagonally opposite contacts in Fig. 4.3, then a Hall voltage, $\Delta V$, is built up between the other two contacts. The Hall voltage arises because the Lorentz force, given by $q\vec{v} \times \vec{B}$, separates electrons and holes which then give rise to an electric field. If j denotes the current entering at contact A and leaving through contact C, then the Hall voltage between contacts B and D is given by

$$\Delta V = \frac{\mu_H B j \rho}{d} \tag{4.18}$$

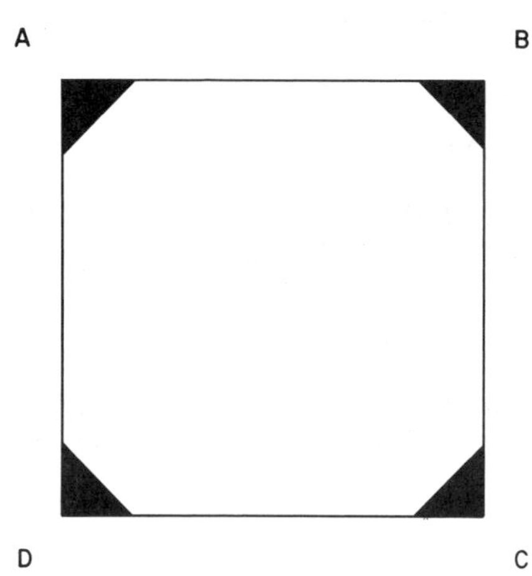


**Figure 4.3** An electrical contact configuration for a sample prepared for Hall effect measurements using van der Pauw's method.

Here $\mu_H$ is the Hall mobility which is closely related to the drift mobility [7]. The Hall mobility is obtained by measuring $\Delta R_{BD,AC}$ which is the change in $R_{BD,AC}$ caused by the magnetic field. Since the result of such a measurement is $\Delta V/j$, the Hall mobility can be computed from the expression

$$\mu_H = \frac{d}{B} \frac{\Delta R_{BD,AC}}{\rho} \tag{4.19}$$

where $\rho$ is measured using Eq. (4.17). The carrier concentration is subsequently obtained by computing the quantity $(q\rho\mu_H)^{-1}$. These results apply to ideal samples and do not include effects because of real contacts. Corrections caused by the size of the contacts have been calculated in the case square samples [8].

## 4.4 CARRIER EMISSION AND CAPTURE

A primary reason that semiconductors are so useful technologically is that their electrical properties can be varied over wide ranges. As we have seen in Sec. 4.2, the carrier concentration and conductivity of InP can be varied over many orders of magnitude with the additon of only ~1 ppm Fe. Not only extrinsic defects such as InP:Fe but also intrinsic defects such as vacancies and interstitials and antisite defects can introduce states within the band gap and can have a profound effect on electron-hole recombination and generation and on the carrier lifetimes. For an energy level situated deep within the band gap as shown in Fig. 4.4 interaction with both conduction and valence band states is possible. We denote by $e_n$ and $c_n$ the equilibrium probability per unit time of emission and capture of an electron, and the equivalent probabilities for holes are denoted by $e_p$ and $c_p$. These quantities are rates and have dimensions of $sec^{-1}$. The capture rates can be expressed as follows,

$$c_n = \sigma_n v_n n \tag{A}$$

$$c_p = \sigma_p v_p p \tag{B}$$

where $\sigma_n$ and $\sigma_p$ are capture cross sections and $v_n$ and $v_p$ are average thermal velocities. These velocities are related to the thermal energy by $1/2\, m^* v^2 = 3/2\, kT$. The effective mass for each type of carrier, $m^*$, is determined by the band structure. The energy level in Fig. 4.4 is schematically drawn at $E_T$ in an independent electron (one electron) picture. Such a picture is usually too simplistic and must be modified to include possible interactions between the charge

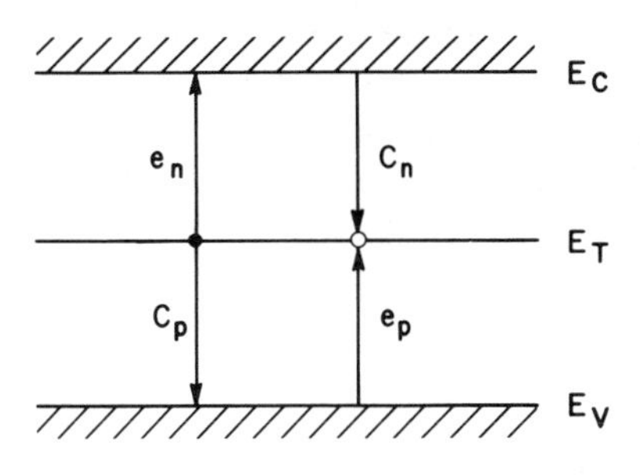


**Figure 4.4** Schematic representation of electron and hole emission and capture by a deep level situated in the middle of the band gap of a semiconductor. The arrow directions represent electron transistions.

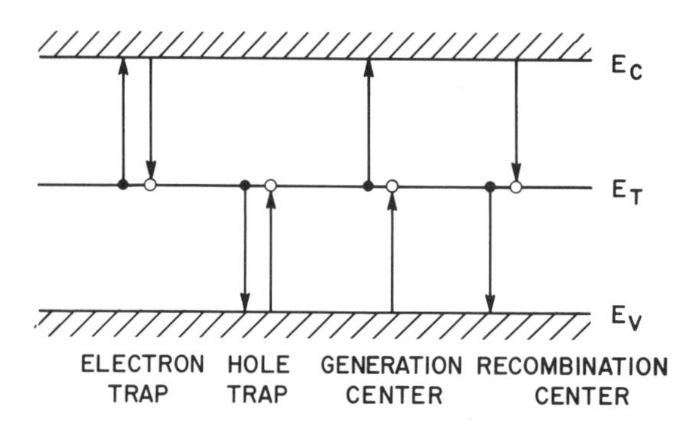


**Figure 4.5** Schematic representation of the predominant transistions applying to electron traps, hole traps, generation centers, and recombination centers.

state of the defect and lattice distortions such as Jahn-Teller distortions. However, we can retain the simple representation of the defect energy level shown in Fig. 4.4 if we add the additional proviso that the energy separations of $E_T$ from $E_c$ and from $E_v$ represent total defect electron and hole ionization energies. The directions of arrows in Fig. 4.4 correspond to electron transitions.

The term carrier trap is sometimes loosely used. A defect can also be referred to as a recombination center or as a generation center depending on the relative magnitudes of $e_n$, $c_n$, $e_p$, and $c_p$. We adopt here the definitions given by C.T. Sah [9] for these terms. The four types of centers are defined as follows (see Fig. 4.5):

a. electron traps, if $c_n \gg e_p$ and $e_n \gg c_p$,

b. hole traps, if $e_p \gg c_n$ and $c_p \gg e_n$,

c. generation centers, if $e_n \gg c_p$ and $e_p \gg c_n$,

d. recombination centers, if $c_n \gg e_p$ and $c_p \gg e_n$.

Thus a center which prefers to capture an electron from the conduction band rather than emit a hole and which then prefers to emit that electron back to the conduction band rather than to capture a hole is defined to be an electron trap. On the other hand, a center which prefers to capture an electron from the conduction band rather than emit a hole but which then prefers to capture a hole instead of emitting the captured electron back to the conduction band is a recombination center. The definitions for hole traps and generation centers are analogous.

Emission and capture rates are related by the principle of detailed balance which requires that at thermal equilibrium the net defect occupancy be constant in time. The rate of electron emission into the conduction band per unit volume is given by $e_n n_T$ where $n_T$ is defined to be the number of electrons per cm$^3$ trapped by the defect state and is given by Eq. (4.9) (where we have substituted T for D subscripts), and the rate of electron capture per unit volume by the defect state is given by $c_n N_T^+$ where $N_T^+$ represents the ionized defect concentration and is given by Eq. (4.12) (again with T instead of D subscripts). Equating these two rates and substituting from n from Eq. (4.5) yields

$$e_n = \frac{\sigma_n v_n N_c}{g_T} \exp\left[-\frac{\Delta E_T}{kT}\right] \tag{4.20}$$

Here $\Delta E_T$ is the free energy of ionization of the defect. An identical expression for holes results from the detailed balance of hole emission and capture rates given by $e_p p_T = c_p N_T^-$, where $p_T$ and $N_T^-$ must be taken from Eqs. (4.10) and (4.13) and p must be taken from Eq. (4.7).

Defects which introduce energy levels near the middle of the band gap can strongly affect carrier recombination rates if they are recombination centers as defined. In light emitting diodes and semiconductor lasers the recombination rate via deep centers competes with photon generation. Such a nonradiative recombination rate reduces the radiative efficiency, and, since the band gap energy is then put into the phonon system, enhanced degradation can result. In light sources, therefore, recombination centers are deleterious. On the other hand, caused by a reduction in carrier lifetimes, the speed of a photodetector can be enhanced because of the presence of defect centers. In such cases defect centers are desirable.

With the definitions given, we are now in a position to describe the recombination rate via defect centers quantitatively [10, 11, 12]. Consider the nonequilibrium situation in which there is carrier injection (which can be either electrical injection as, for example, in the case of a biased diode or optical injection achieved by illuminating the semiconductor). We denote the generation rate of carriers by injection as G. Then under steady state conditions we have

$$\frac{dn}{dt} = G - U = G - \left[c_n\, N_T(1-f) - e_n\, N_T f\right] = 0 \tag{4.21}$$

and

$$\frac{dp}{dt} = G - U = G - \left[c_p\, N_T f - e_p\, N_T(1-f)\right] = 0 \tag{4.22}$$

Here f is used to denote the occupation probability function. By equating Eqs. (4.21) and (4.22) and eliminating G, we can solve for f (which depends on the injection level). This yields

$$f = \frac{c_n + e_p}{c_n + e_n + c_p + e_p} \cong \frac{c_n}{c_n + c_p} \tag{4.23}$$

where we have applied the definition of a recombination center to simplify f. By substituting this result for f into U we obtain the net rate of recombination via the defect center

$$U = \frac{c_n c_p}{c_n + c_p}\, N_T \tag{4.24}$$

This is an important relationship from which one can, for example, obtain the minority carrier lifetime. If an n-type semiconductor under low-level injection is considered, then the lifetime of holes is defined by $U = p/\tau_p$, and since n>>p we have from Eq. (4.24) that

$$U = c_p N_T = \sigma_p v_p N_T\, p$$

Therefore the minority carrier lifetime is given by

$$\tau_p = \frac{1}{\sigma_p v_p N_T} \tag{4.25}$$

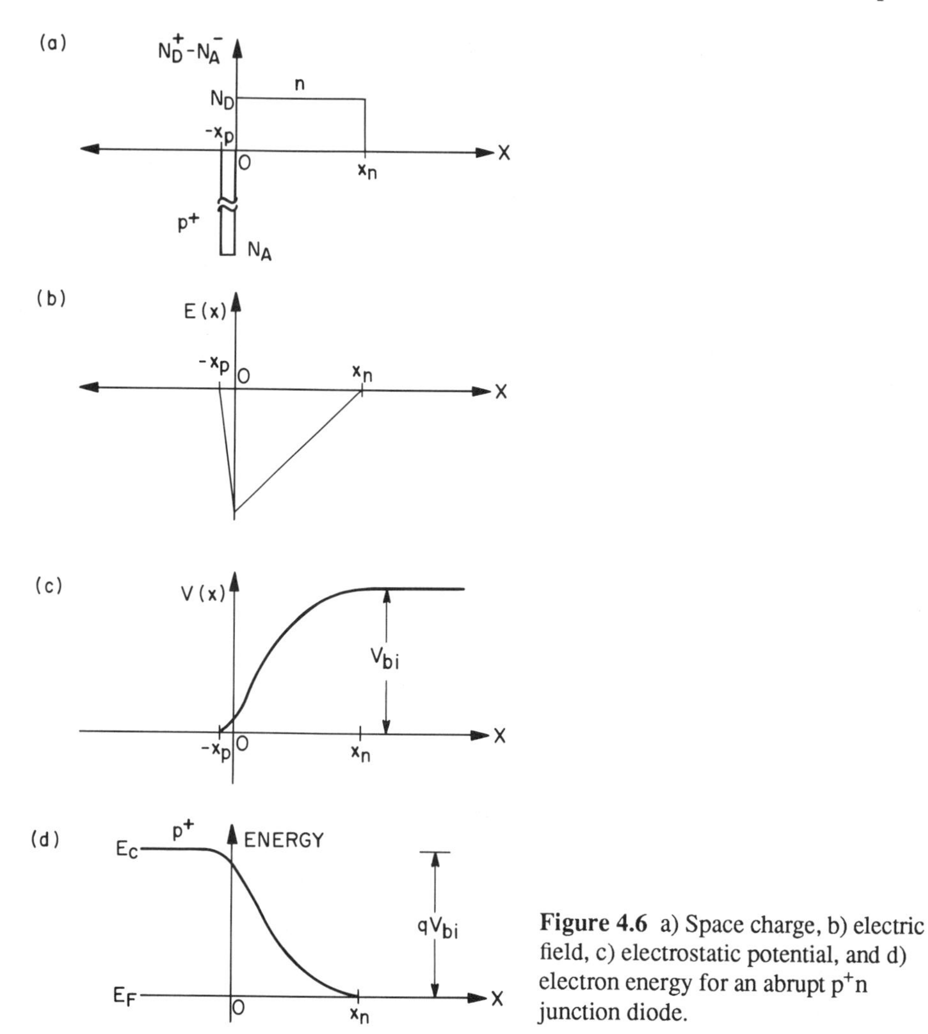


**Figure 4.6** a) Space charge, b) electric field, c) electrostatic potential, and d) electron energy for an abrupt $p^+n$ junction diode.

## 4.5 DEPLETION CAPACITANCE

Electrical properties of semiconductors relating to doping and to states which exist deep in the band gap are conveniently measured from the depletion capacitance of a diode. Both p-n junctions and Schottky diodes will serve this purpose. However, minority carrier effects can be studied only with p-n junctions since Schottky diodes are majority carrier devices. The present treatment to $p^+n$ junctions will be limited, and note that the result for the depletion capacitance for Schottky diodes is identical. By abrupt junctions we mean junctions which contain ideally abrupt doping profiles as shown in Fig. 4.6a, and by one sided we mean that the concentration of dopant is much larger on one side. Any variation lateral to the junction is ignored and the overall junction must be neutral which implies that $N_A x_p = N_D x_n$. The solutions for the electric field and electrostatic potential are shown in Fig. 4.6b and c. These are straightforwardly obtained by integrating Poisson's equation which for the n side is given by (mks units),

$$-\frac{\partial^2 V}{\partial x^2} = \frac{\partial E}{\partial x} = \frac{\rho(x)}{\varepsilon} = q\,\frac{N_D}{\varepsilon} \tag{4.26}$$

Here $\varepsilon$ is the dielectric constant and q is the charge of an electron. The result for the electric field is

$$E(x) = \frac{q\,N_D}{\varepsilon}\,(x-x_n) \quad \text{for } 0 < x \le x_n \tag{4.27}$$

and for the electrostatic potential is

$$V(x) = V_c - \frac{q\,N_D}{2\varepsilon}\,(x_n-x)^2 \tag{4.28}$$

where $V_c$ is the integration constant which corresponds to the total electrostatic potential across the diode. For a one-sided abrupt junction, $V(x = 0) \cong 0$ which implies that

$$V_c = V_{bi} + V = \frac{q\,N_D x_n^2}{2\varepsilon} \tag{4.29}$$

Here $V_{bi}$ is the "built-in" potential and V is any applied bias. For one-sided and abrupt junctions the depletion depth is given by

$$W = x_n + x_p \cong x_n \tag{4.30}$$

The solution for the electrostatic potential is often drawn superimposed on the bandstructure as shown in Fig. 4.6d in a "band-bending diagram." Here the Fermi level does not vary across the junction if no bias is applied since the Fermi level is the chemical potential for the carriers which must be constant in equilibrium. The crossing point between the Fermi level and the donor level determines the depletion depth.

If a slight bias, $\Delta V$, is applied, then according to Eq. (4.29) we have that $\Delta V = q\,N_D W \Delta W/\varepsilon$. Since the amount of donor charge which is uncovered at the edge of the depletion region is given by $\Delta Q = N_D A\, \Delta W$, where A is the area of the diode, the differential capacitance is given by

$$C = \frac{\Delta Q}{\Delta V} = \frac{\varepsilon\,A}{W} \tag{4.31}$$

This is a basic result which is easy to remember because it is identical to that of a parallel plate capacitor. If there is a quiescent applied bias, V, which increases the depletion depth (i.e., reverse bias), then the capacitance is decreased, and this affords a convenient way in which to ascertain the polarity of a diode. It follows from Eqs. (4.29) and (4.31) that

$$\frac{A^2}{C^2} = \frac{W^2}{\varepsilon^2} = \frac{2}{q\varepsilon N_D}\left[V_{bi} + V\right] \tag{4.32}$$

This key result implies that by plotting $A^2/C^2$ as a function of reverse bias the space charge density can be obtained from the slope. It can be shown that a similar procedure can be valid even if the space charge density is not uniform [13]. That is, even if $N_D$ varies with x, it is still true that

$$\frac{d}{dV}\left[\frac{A^2}{C^2}\right] = \frac{2}{q\varepsilon N_D(x)} \tag{4.33}$$

which is often stated as

$$N(x) = -\frac{C^3}{q\,\varepsilon A^2}\left[\frac{dC}{dV}\right]^{-1} \tag{4.34}$$

Here N(x) is the net space charge density which exists in the depletion region. These relations are the basis for a popular doping profiler known as the ''Miller Profiler'' which allows placement of a diode at the end of coaxial cables of essentially arbitrary length [14].

The charge distribution in this treatment was assumed to change abruptly from $N_D$ to zero at $x_n$. Actually the electrons spill over into the depletion region according to [15]

$$n(x) \approx n\, e^{-(x_n - x)^2/\ell_D^2} \tag{4.35}$$

Here $\ell_D$ is the Debye length given by

$$\ell_D = \sqrt{\frac{kT\varepsilon}{q^2 N_D}} \tag{4.36}$$

where k is Boltzmann's constant and T is the temperature. The Debye length limits the resolution of doping profiles obtained from C-V data. If in addition to space charge, the majority carriers are included in the treatment of the capacitance-voltage relationship, then for doping profiles with large gradients the derivative procedure of Eq. (4.34) yields a profile which corresponds more closely to the majority carrier distribution in the undepleted semiconductor [16]. Computer simulations have subsequently shown that the apparent doping profile near a step in the doping profile (as might occur, for example, at the interface between an undoped epitaxial layer and a heavily doped substrate) is considerably more complex than this and that a step in the doping profile cannot be resolved satisfactorily to less than several Debye lengths of the highly doped side of the profile [17].

The diodes most commonly used for capacitance-voltage (C-V) profiling are simple Schottky barriers formed by placing a metal on the semiconductor surface. This is usually accomplished by evaporating the metal through a shadow mask or by drawing liquid Hg into contact with the semiconductor surface using vacuum. In both cases the interface region usually has only a poorly defined chemical and structural composition and has surface states which can significantly perturb C-V profiles, especially at low biases [14]. In practice, therefore, one usually ignores profile features close to the zero bias depletion depth and considers as significant only those portions of a profile which occur at a high enough bias that surface related features are substantially stabilized [14].

A serious drawback associated with C-V profiling using metal-semiconductor diodes arises from the fact that for many semiconductors the profiling depth is limited by the current leakage which sets in at high biases. This is especially true for diodes with small Schottky barrier heights such $In_{1-x}Ga_xAs$ grown lattice matched to InP. In such cases C-V profiling is conveniently performed using an electrolyte instead of a metal to form an electrostatic potential barrier at the interface between the electrolyte and the semiconductor. The situation is completely analogous to that of a metal-semiconductor diode with the Fermi level of the metal replaced with the re-dox potential of the electrolyte [18]. The advantage of using the electrolyte arises from the fact that in-situ photo-chemical etching can be carried out in the same apparatus [19]. Thus doping profiles can be obtained to essentially unlimited depths by etching.

## 4.6 DEEP LEVEL INFLUENCES ON DIODE CAPACITANCE

The charge density in the space charge layer which exists at a pn junction or Schottky barrier affects the capacitance as discussed. If charged defects are present in addition to a dopant, then the capacitance will reflect this, and, consequently, by monitoring the capacitance one can monitor free carrier emission from and capture by defects. Defects can seriously affect the properties of devices built on GaAs or InP substrates, and, therefore, these defect properties can be instructive concerning the usefulness of the starting material before device processing. Calculation of the effect of the defect charge on the capacitance of a diode is complicated by the fact that the edge of the space charge region is not abrupt. For the sake of simplicity we assume in the following treatment that the edge of the space charge region is abrupt and that the defect energy level lies sufficiently high in the band gap that electron processes dominate and that the electron quasi-Fermi level (and not the hole quasi-Fermi level) is active on the n side of a $p^+n$ junction in the determination of defect occupation under reverse bias. (For a discussion of which quasi-Fermi level is applicable see ref. [20]). This situation is shown schematically in Fig. 4.7. We assume that the defect is a deep donor, that is, it is positively charged after emitting an electron, $n_T \rightarrow N_T^+ + \Delta n$ where the total defect concentration is $N_T = n_T + N_T^+$. The donor charge density of the uniformly distributed dopant is denoted by $N_D$ as it was in Sec. 4.5 and we assume $N_D \gg N_T$. We assume furthermore that the capacitance of the diode is measured using an AC signal at a frequency (1 MHz in most cases) which is sufficiently high that the defect does not change its charge state. In that case the parallel plate capacitance relation given by Eq. (4.31) holds in spite of the fact that there is defect charge present. The basic reason that defect parameters can be obtained follows from the fact that there is a change in depletion width, $\Delta W$, between the two reverse biased situations shown in Fig. 4.7. By invoking Eq. (4.31) for small changes we obtain the relation

$$\frac{\Delta C}{C} = -\frac{\Delta W}{W} \tag{4.37}$$

where $\Delta C$ is the transient capacitance change which takes place after the reverse bias is applied.

This transient results from the thermally activated emission of electrons from defects in the space charge region between x = W(x) – λ and W(0) – λ. Here λ is determined by the crossing point between the defect level and the electron quasi-Fermi level and is given by

$$\lambda = \sqrt{\frac{2\varepsilon(E_{Fe} - E_T)}{q^2 N_D}} \tag{4.38}$$

**Figure 4.7** Electron energy diagram for the n side of an abrupt $p^+n$ junction with a deep level donor a) in equilibrium (zero applied bias), b) directly after applying a step function in reverse bias with a step height V, and c) for a quiescent reverse bias condition.

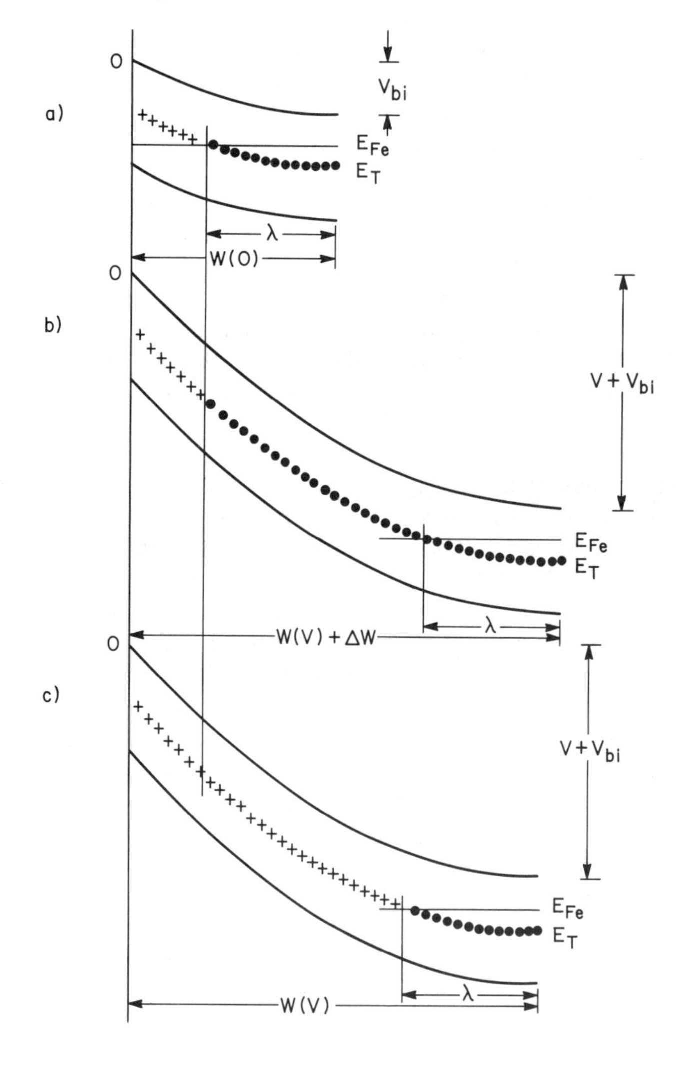

After such a transient has occurred, the whole process can be repeated by removing the bias for a sufficiently long time period that all the defects in this same region have captured an electron.

The potential drop across the diode during a capacitance transient is given by

$$V = \frac{1}{\varepsilon} \int_0^W x\,\rho(x)\,dx \tag{4.39}$$

where $\rho(x)$ is the total charge density. Since the bias is held fixed during the transient, we have

$$0 = \Delta V = \frac{1}{\varepsilon} \left[ W(V)\rho(W)\Delta W + \int_0^{W(V)} x\,\Delta\rho(x)\,dx \right] \tag{4.40}$$

which yields

$$\frac{\Delta W}{W} = \frac{1}{N_D[W(V)]^2} \int_{W(o)-\lambda}^{W(V)-\lambda} x n_T\,dx \tag{4.41}$$

Consequently, we have that the capacitance change in the case of a uniformly distributed defect is given by [21],

$$\frac{\Delta C}{C} = -\frac{n_T}{2N_D} \left[ 1-2\frac{\lambda}{W(V)} \left( 1-\frac{C(V)}{C(0)} \right) - \left( \frac{C(V)}{C(0)} \right)^2 \right] \tag{4.42}$$

For diodes which are biased sufficiently that $W(V) \gg \lambda$ and $W(V) \gg W(0)$ this reduces to the extremely simple and well known relationship [21, 22],

$$\frac{\Delta C}{C} = -\frac{n_T}{2N_D} = -\frac{N_T}{2N_D} \ . \tag{4.43}$$

The time dependence of the capacitance transient is considered next. The rate equation is very simple since we are ignoring hole processes and capture during the emission transient, and it is given by

$$\frac{dn_T}{dt} = -e_n n_T \tag{4.44}$$

which integrates to

$$n_T(t) = n_{T,o} \exp(-e_n t) \ . \tag{4.45}$$

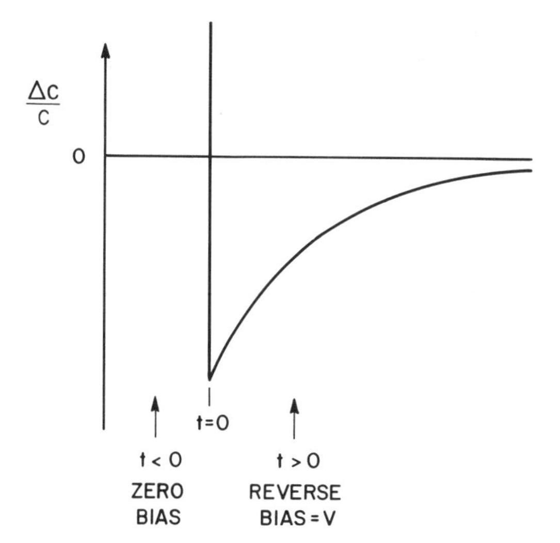


**Figure 4.8** Schematic capacitance transient of an abrupt $p^+n$ junction as electrons are thermally excited from a deep level to the conduction band.

The resulting capacitance transient is thus a simple exponential as shown in Fig. 4.8. Although it was derived for a deep donor, the same result applies to deep acceptors which interact with the conduction band, that is, $n_T \rightarrow N_T^o + \Delta n$, where the total defect density is $N_T = n_T + N_T^o$.

If we had considered a defect which interacted with only the valence band, that is, a hole trap, then a similar result would be obtained for hole emission from deep donors, $p_T \rightarrow N_T^o + \Delta p$, and for deep acceptors, $p_T \rightarrow N_T^- + \Delta p$. The transient is again exponential with $e_p$ replacing $e_n$ in Eq. (4.44), but it has the opposite sign, that is, $\Delta C/C > 0$.

If a defect interacts with both bands during an emission transient, that is, if it is a generation center, then both $e_n$ and $e_p$ appear in the expression for the capacitance transient, both in the time constant and in the equilibrium level occupation ratio [23].

It is clear from these discussions that defects which affect the capacitance of a diode also can affect the effective space charge density, $N_{eff}$, versus depth profile commonly obtained using Eq. (4.34) from C-V profiles. If the trapped charge density is a significant fraction of or exceeds the dopant concentration, then severe complications will result. The $N_{eff}$ profile will depend on whether or not the emission rate of the defect exceeds the frequency of the applied oscillatory bias used to measure the capacitance. If we consider point-by-point (static) C-V measurements made at a low enough temperature that the defect does not change its charge state, then only space charge at the edge of the depletion layer is uncovered, and the usual parallel plate capacitance relationship, $C = \varepsilon A/W$, applies. In the case of uniformly distributed deep donors in a space charge region having a uniform donor doping density, the result is [24]

$$N_{eff}(W-\lambda) = N_D + \left(1-\frac{\lambda}{W}\right)N_T \tag{4.46}$$

which for deep depletion approaches $N_T+N_D$. If now the temperature is raised sufficiently that the defect can change its charge state during the period of the bias oscillation used for the capacitance measurement, then space charge at both W (due to $N_D$) and W–λ (due to $N_T$) is uncovered and the apparent depth is given by

$$\frac{\varepsilon A}{C} = W - \frac{N_T}{N_T+N_D}\,\lambda \tag{4.47}$$

In this case $N_{eff}$ is just $N_D + N_T$. Different results apply to deep acceptors in n-type semiconductors [24]. Results for nonuniform distribution are considerably more complex. In summary, profiling results are complicated by deep levels present in significant concentrations, that is, $N_T \gtrsim N_D/10$. Because capacitance effects in these cases are large, such situations were discovered first, and many treatments of the various possible capacitance relationship have been given in the literature [20]. However, it is usually much simpler to deal with deep level densities which are small compared to the doping density.

## 4.7 DEEP LEVEL CHARACTERIZATION TECHNIQUES-TSCAP AND DLTS

If a diode is first cooled sufficiently to suppress carrier emission from deep levels, is then reverse biased, and subsequently warmed, the space charge capacitance will exhibit steps of a magnitude given by Eq. (4.42) when the temperature has risen sufficiently for carrier emission to take place. This type of deep-level data is known as thermally stimulated capacitance (TSCAP) [25]. An example is shown in Fig. 4.9 which applies to a gold on InP Schottky diode [26]. The InP was grown by vapor phase epitaxy. Since the capacitance of a diode can change with temperature in the absence of carrier emission from deep levels, two capacitance versus temperature (C-T) curves are needed to obtain TSCAP data. The C-T curve for which there is no change in deep level charge (empty traps) is then compared to one obtained as stated above (full traps) to reveal the steps caused by the carrier emission from a deep level. TSCAP is very useful as a first probe to see what the major deep levels are. However, carrier emission activation enthalpies from a deep level cannot be clearly extracted. An assumption must be made concerning the pre-

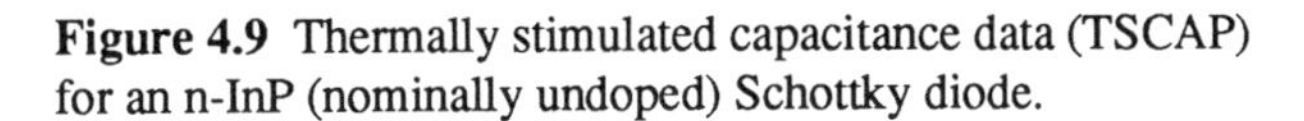
**Figure 4.9** Thermally stimulated capacitance data (TSCAP) for an n-InP (nominally undoped) Schottky diode.

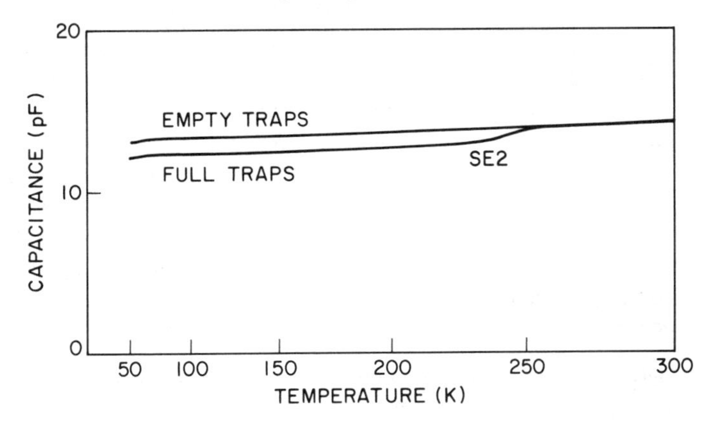

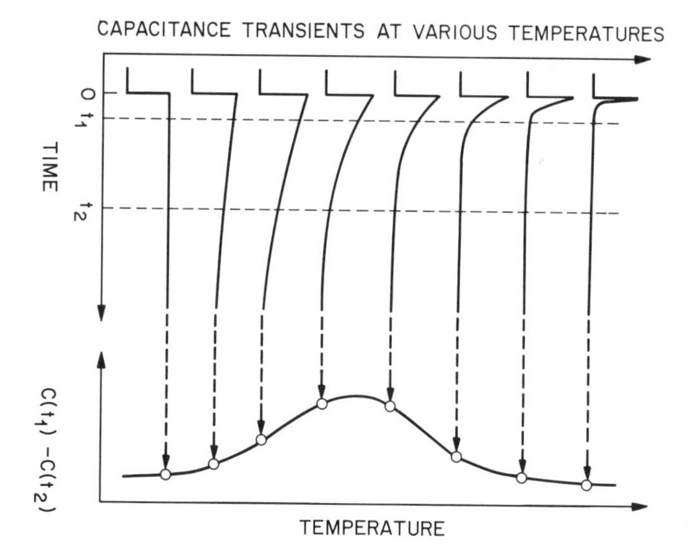


**Figure 4.10** Schematic illustration of a deep level transient spectroscopy (DLTS) peak generated by sampling capacitance transients at two sampling points.

exponential in the emission rate in order to extract the activation enthalpy, and carrier capture cross sections cannot be obtained.

A technique which overcomes these limitations is deep level transient spectroscopy (DLTS) [22]. The basic principle of this technique is illustrated in Fig. 4.10 and involves sampling a capacitance transient at two times and taking the DLTS signal to be the difference. This can be done conveniently using a boxcar averager with two integrators. The two sampling times define an emission rate window over which a DLTS signal is obtained. When the time constant of the capacitance transient is equal to $\tau_{max}$ where this quantity is given by

$$\tau_{max} = \frac{t_1 - t_2}{\ell n(t_1/t_2)} \tag{4.48}$$

a maximum in a DLTS peak occurs. Two other means of implementing a rate window are in common use. These involve using either a lock-in amplifier or an exponential correlator [23]. Of the three techniques, the correlator method provides the best signal to noise and has the additional advantage of eliminating drift in the baseline since a baseline restorer is implemented [27]. There is, of course, no better way to obtain data than to measure the entire transient at many temperatures. The advantage of DLTS is that it quickly generates the deep level parameters at the expense of not recording all of the possible data. This can have severe drawbacks especially when nonexponential transients are found. If one has an automated digital data acquisition capability and a computer, then obtaining the full capacitance transient must certainly be preferred. DLTS spectra can then still be generated from the acquired digitized transients.

Unlike TSCAP DLTS is a repetitive technique. By reducing the reverse bias during a "filling pulse" as shown in Fig. 4.11 the diode can be repeatedly prepared with majority carriers trapped

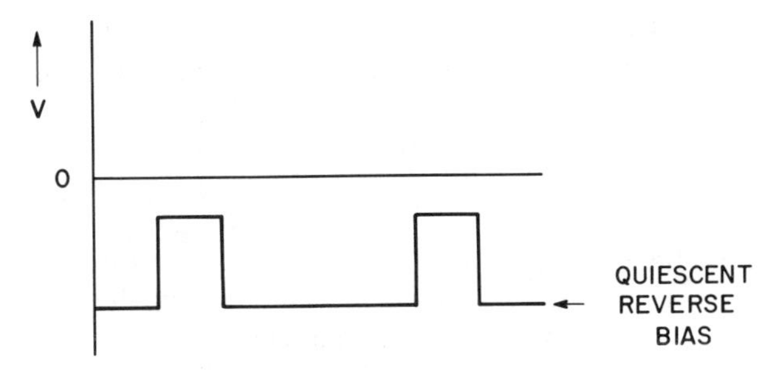


**Figure 4.11** Diode bias pulsing scheme used to implement DLTS.

at the defect. This permits signal averaging and has additional advantages. Majority carrier capture rates can be measured by varying the width of the filling pulse, and depth profiling can be accomplished by varying the height of the filling pulse.

The activation enthalpy can be obtained by making a number of DLTS scans at different rate windows as shown in Fig. 4.12. In this way one obtains data for the emission rate as a function of temperature. The slope of an Arrhenius plot of $\tau_{max}$ as a function of the inverse of the temperature at which a DLTS peak occurs yields the activation enthalpy. Here we have assumed that the capture cross section is not thermally activated and that ionization entropy is zero. In general these cannot be neglected [21].

We have also neglected the effect of the edge region of the space charge layer in our discussion of the filling pulse and majority carrier capture. The carrier concentration in this

**Figure 4.12** DLTS spectra of two hole traps in LPE n-GaAs. The transient sampling times and peak maxima are indicated. Arrhenius plots of the temperature at which these maxima occurred versus the rate window yielded the activation energies indicated.

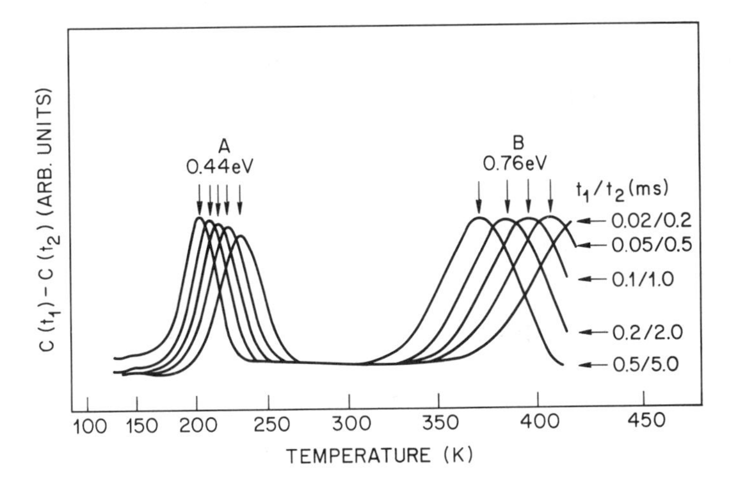

region diminishes over a distance of a Debye length which implies that the capture rate during the pulse also varies. The carrier concentration in this edge region in the case of Fig. 4.7a is given by

$$n(x) = N_c \exp - \frac{[E_c(x) - E_{Fe}]}{kT} \tag{4.49}$$

(This equation plus a quadratic dependence of $E_c$ on the distance in from W leads to Eq. (4.35) [24]). Since the electron capture rate is given by $c_n(x) = \sigma_n v_n n(x)$, the capture rate has the same spatial dependence as n(x). Capture and emission are related via detailed balance as previously discussed, and this relationship leads to Eq. (4.20) for $e_n$. The result is that during the filling pulse the number of capture events per unit volume per unit time exceeds the number of emission events for $x > W - \lambda$. For $x < W - \lambda$ emission events predominate and at $x = W - \lambda$ the two are equal. The fact that there is a spatially varying capture rate has significant consequences for majority carrier pulsing experiments [20, 28]. Correct values for the capture cross section, the enthalpy and entropy of ionization, and for the spatial profile of the defect all depend on the proper interpretation of the results. In practice it is usually convenient to reduce the effect of the edge region by using large depletion depths, W. However, this implies large values for the quiescent reverse bias which are not always achievable since the reverse breakdown voltage of the diode may be low.

For $p^+n$ and $n^+p$ junctions (but not Schottky diodes) it is also possible to inject minority carriers into the space charge region from which deep level capacitance effects arise. This is done by forward biasing the diode during the "filling" pulse. Minority carrier trapping and emission can be investigated with injection pulsing experiments as illustrated schematically in Fig. 4.13. To illustrate the difference between capacitance transients which result from majority carrier pulsing and injection pulsing, consider a $p^+n$ junction and a deep level for which $e_n \doteq e_p$.

The transients during and after a majority carrier pulse are schematically shown in Fig. 4.13a. During the pulse the rate of change of the trapped electron concentration is given by

$$\frac{dn_T}{dt} = c_n(N_T - n_T) \ , \tag{4.50}$$

and for a pulse of sufficient duration that $dn_T/dt \rightarrow 0$ the deep level defect state becomes completely filled with electrons ($n_T = N_T$). During the emission transient the rate equation is given by

$$\frac{dn_T}{dt} = -e_n n_T + e_p(N_T - n_T) \ . \tag{4.51}$$

Therefore, in the quiescent reverse bias state the occupation is given by $n_T = e_p N_T/(e_n + e_p)$.

If we now consider injection pulsing, then the rate equation during the pulse is given by

$$\frac{dn_T}{dt} = c_n(N_T - n_T) - c_p\, n_T \tag{4.52}$$

For an injection pulse of sufficient duration that $dn_T/dt \rightarrow 0$ the occupation is given by $c_n N_T/(c_n + c_p)$. The resulting transient for $c_n > c_p$ is shown schematically in Fig. 4.13b. Since we have assumed $e_n \doteqdot e_p$, the condition $c_n > c_p$ defines an electron trap according to the definitions given in Sec. 4.4. Similarly, the condition $c_p > c_n$ defines a hole trap, but now the transient has changed sign as shown in Fig. 4.13c. This shows that the nature of the trap can be observed from the sign of the capacitance transient during an injection pulsing experiment. An example of a single spectrum showing both electron and hole traps is shown in Fig. 4.14. We note that this discussion is completely symmetric with respect to an interchange of electron traps

**Figure 4.13** Schematic capacitance transients occurring during capture and emission for a) a majority carrier (reverse bias) pulse, b) an injection (forward bias) pulse in the case of an electron trap, c) an injection pulse in the case of a hole trap.

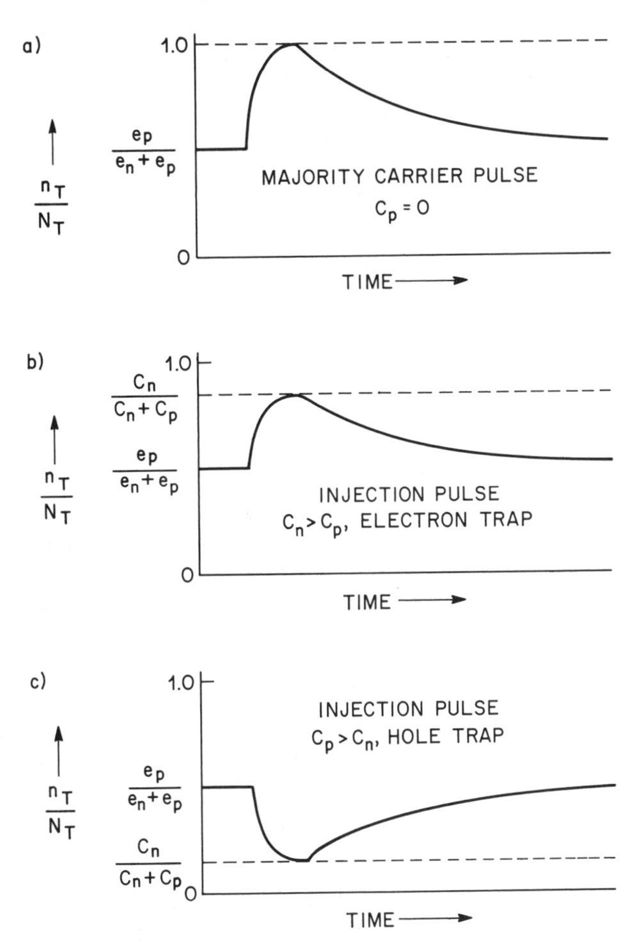

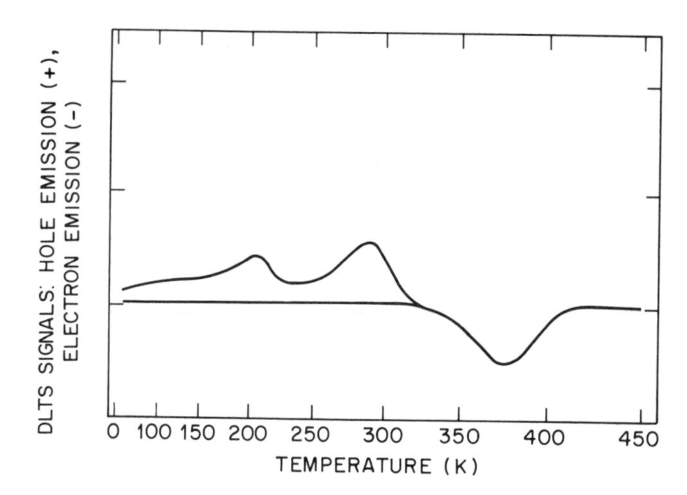


**Figure 4.14** A DLTS spectrum of a p$^+$n GaAs diode obtained in an injection pulsing experiment which reveals the presence of both electron (negative signal) and hole traps (positive signal).

and hole traps in n$^+$p junctions, and as a result the nomenclature of majority carrier trap and minority carrier trap is often used.

## 4.8 OTHER DEEP LEVEL CHARACTERIZATION TECHNIQUES-ADMITTANCE SPECTROSCOPY, CONSTANT CAPACITANCE BIAS TRANSIENTS, AND CURRENT TRANSIENTS

If the capacitance of a diode which contains a majority carrier trap (where $N_T \ll N_D$) is measured as a function of temperature, then there will be a step in the capacitance at a temperature such that the emission rate at the Fermi-level crossing point coincides with the frequency of the oscillatory bias voltage used to measure the capacitance. There is also a peak in the conductance at this temperature. The frequency in this case is akin to the DLTS rate window and thus such a temperature scan is also spectroscopic for deep-level ionization enthalpies. This technique is known as admittance spectroscopy [29], and an example for deuteron bombarded InP [26] is shown in Fig. 4.15. Only majority carrier traps can characterized using admittance spectroscopy and one is restricted to having concurrent carrier capture and emission. However, it is easier to study deep levels with small activation enthalpies ($\lesssim$0.2 eV) with admittance spectroscopy than it is with DLTS since the conductance peaks typically occur at more convenient (higher) temperatures. In addition, admittance spectroscopy allows one to measure emission rates decades faster than can be measured with DLTS. A typical admittance spectroscopy frequency is 100 KHz compared to a typical DLTS rate window of 50 Hz.

Another deep-level characterization technique which can be implemented using p$^+$n, n$^+$p, or Schottky diodes consists of measuring the bias voltage required to keep the capacitance constant. To observe the bias transient, one takes a zero biased diode at a fixed temperature and applies an initial bias to set the depletion depth. Then as majority carriers are emitted thermally, a steadily increasing bias is required to keep the capacitance constant. The bias transient which results will

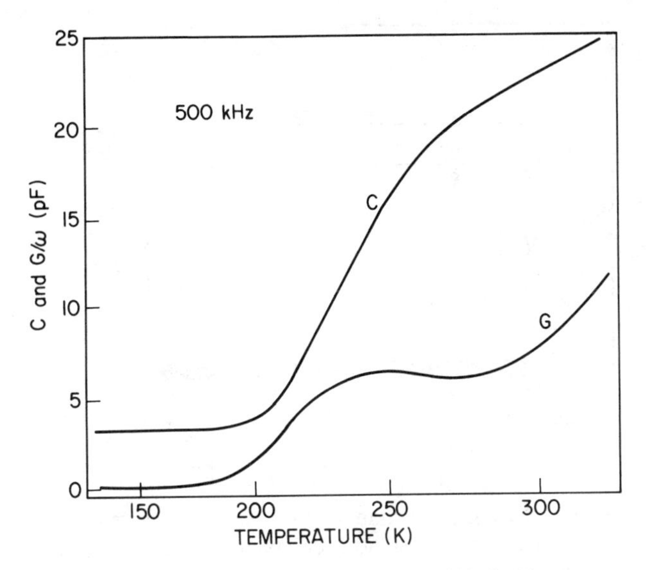


**Figure 4.15** Admittance spectroscopy data for deuteron irradiated InP.

be exponential even if $N_T \geq N_D$, $N_A$ [30], and this feature provides a significant advantage in such cases over DLTS for which nonexponential transients occur in the concentrated trap limit. A feedback loop is needed in the capacitance measurement apparatus to keep the capacitance constant, and this is conveniently accomplished using a Miller profiler set for constant distance profiling [14]. Typical constant capacitance bias transients have a time constant of 10 to 100 seconds. Thus constant capacitance bias transient measurements, DLTS, and admittance spectroscopy collectively provide the means to measure emission rates over ~7 decades from $\sim 10^{-2}$ Hz to $\sim 10^{5}$ Hz.

Finally, we briefly mention deep level detection techniques involving current transients [21, 31, 32, 33]. These can be used to carry out another form of deep level spectroscopy referred to as current transient spectroscopy [34]. One measures the diode current in place of the capacitance. Capacitance based DLTS has low sensitivity to traps at the diode junction, but for current transients and for majority carrier emission (but not for minority carrier emission) the sensitivity is a maximum at the junction [21]. However, majority and minority carrier emission produce current transients of the same sign, and this is a disadvantage compared to standard capacitance DLTS.

## 4.9 SEMI-INSULATING MATERIAL

As we showed in Sec. 4.2 for the case of InP:Fe, III-V crystals can be made semi-insulating (SI) by creating an electrically active midgap deep level. Such levels may be formed during the crystal growth without doping as in the case of EL2 in GaAs [35], may be introduced as an extrinsic dopant as in the case of GaAs:Cr [36] and InP:Fe [37], or may be formed as a result of ion bombardment [38]. In any case it is possible to evaluate the resistivity of SI material with simple current transport measurements.

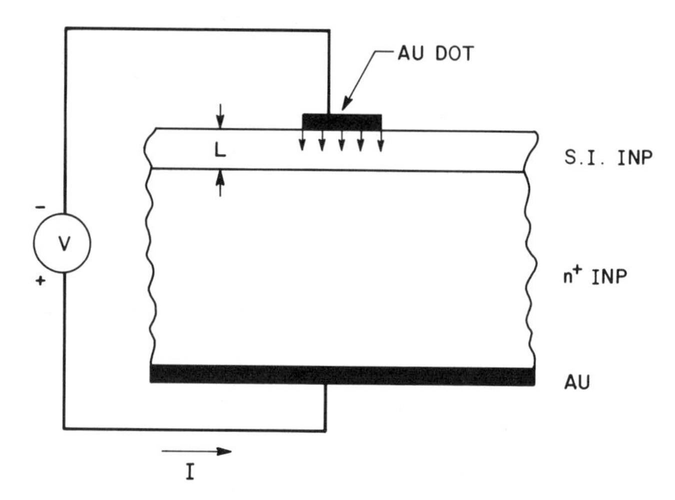


**Figure 4.16** Device configuration used to measure the resistivity of semi-insulating epitaxial layers grown on a conducting substrate.

All that is needed to make such measurements is a device with two ohmic contacts and a well defined volume with a known area over which current can be injected. These conditions are met by the device shown in Fig. 4.16 which applies to the case of an epitaxial SI layer grown on an $n^+$ substrate. An ohmic contact is made to the surface of the layer by a metal dot of known area which has been deposited by evaporation through a shadow mask. If the Schottky barrier height of this contact is low, then the contact resistance may be sufficiently low that an alloying step is not needed. The back side contact is made by the uniform evaporation of a metal layer to the bottom of the $n^+$ substrate. The $n^+$ doping in the substrate must be high enough to ensure good electron injection or collection at the SI-$n^+$ interface. Fig. 4.16 is drawn for anodic biasing of the dot for which electrons are injected into the SI-layer from the top contact. There will, in general, be some current spreading, but if the epitaxial layer thickness is much less than the thickness of the substrate then this can be ignored in the determination of the resistance.

To illustrate the basic I-V behavior, consider the case of InP:Fe which is SI caused by an Fe deep acceptor level. The deep level concentration must exceed the shallow background donor concentration as shown in Fig. 4.2. The negative charge density in the neutral SI material can be written as

$$\rho = q\,(n_o + N_A^- - N_D^+ - p_o) \tag{4.53}$$

where q is the charge of the electron, $n_o$ and $p_o$ are the equilibrium electron and hole concentrations, $N_A^-$ is the ionized deep level acceptor concentration and $N_D^+$ is the ionized background donor concentration. Let's assume in the following treatment that the background donors are completely ionized ($N_D^+$=$N_D$). Under applied bias, primarily electrons are injected in the case of Fig. 4.16 because of boundary conditions at the interface with the $n^+$ substrate [39, 40], and the quasi-Fermi level is moved up towards the conduction band. As a result the electron concentration becomes

$$n = n_o \exp \frac{\delta E_F}{kT} \tag{4.54}$$

where $\delta E_F$ is the difference between the quasi-Fermi and the Fermi level of the neutral material. This result follows from Eq. (4.5). Similarly, the hole concentration becomes

$$p = p_o \exp \frac{-\delta E_F}{kT} \tag{4.55}$$

The ionized deep level concentration becomes

$$N_A^- = \frac{N_A}{1 + g_A \exp\left[\frac{E_v + \Delta E_A - E_F}{kT}\right] \cdot \frac{n_o}{n}} \tag{4.56}$$

where Eq. (4.13) is used. If we denote the injected electron concentration by $\delta n = n - n_o$ and expand Eq. (4.56) to first order in $\delta n/n_o$ we obtain for the charge density when bias is applied,

$$\rho = q\left\{\left(n_0+\delta n\right) + \left[N_D + \frac{N_D^2}{N_A} g_A \exp\left(\frac{E_v+\Delta E_A-E_F}{kT}\right)\frac{\delta n}{n_o}\right] - N_D - \left(p_0 - \frac{p_0}{n_0}\delta n\right)\right\} \tag{4.57}$$

Here we have substituted $N_D$ for the equilibrium value of $N_A^-$ to simplify Eq. (4.57). If no bias is applied ($\delta n$=0), then the material must be neutral ($\rho$=0), and therefore Eq. (4.57) simplifies to

$$\rho = q\,\theta\,\delta n \tag{4.58}$$

where $\theta$ is given by

$$\theta = 1 + \frac{N_D}{n_o}\left(1 - \frac{n_o - p_o}{N_D}\right)\left(1 - \frac{N_D}{N_A} + \frac{n_o - p_o}{N_A}\right) + \frac{p_o}{n_o} \tag{4.59}$$

Therefore, Poisson's equation is given by

$$\frac{dE}{dx} = \frac{q}{\varepsilon}\,\theta\left(n(x) - n_0\right) . \tag{4.60}$$

where x is the distance into the SI layer from the top (dot) contact in Fig. 4.16.

The current density is taken to be due only to drift and thus

$$J = q\,n(x)\,\mu\,E(x) \tag{4.61}$$

where $\mu$ is the mobility of electrons. Corrections caused by the diffusion current affect mostly the electrostatic potential near the contacts without appreciably changing the I-V characteristics [41]. Furthermore, the cathode is taken to be an infinite reservoir of electrons which are available for injection. Thus the field must vanish at the cathode so that J remains finite. To solve Eqs. (4.60) and (4.61) it is convenient to employ dimensionless variables [42]. These are defined as follows:

$$u = \frac{n_o}{n(x)} = \frac{q\, n_o \mu E(x)}{J}$$

$$w = \frac{\theta q^2 n_o^2 \mu x}{\varepsilon J} \tag{4.62}$$

$$v = \frac{\theta q^3 n_o^3 \mu^2 V(x)}{\varepsilon J^3}$$

where V(x) is the electrostatic potential in the SI layer. In terms of these variables, Poisson's Eq. becomes

$$dw = \frac{udu}{1-u} \tag{4.63}$$

which integrates to

$$w = -u - \ell n(1-u) \tag{4.64}$$

Note that u=0 at v=0 which is the condition that the field vanish at the cathode. The voltage integral is given by

$$v = \int_0^W udw = \int_0^u u\frac{dw}{du}\, du = \int_0^u \frac{u^2}{1-u}\, du \tag{4.65}$$

Since $u \cong 1 - \frac{\delta n}{n_o}$, the $-\ell n(1-u)$ term in the above two equations dominates so that $v \cong w$. Substituting for v and w from Eq. (4.62) evaluated at the anode (x=L) we obtain the following relation between the applied voltage and the current density,

$$V = \frac{L}{qn_o\mu} J \tag{4.66}$$

This is just Ohm's law where the resistivity is that of the equilibrium SI material. Therefore, the resistivity can be obtained from current voltage measurements under low injection conditions.

Under larger injection currents than are used in the ohmic regime in the case of SI InP:Fe a regime will be reached where $N_A^- = N_A$ (see Fig. 4.2). In that case the hole concentration can be neglected and if the ionized donor concentration does not change (i.e. $N_D^+ = N_D$) Poisson's equation becomes.

$$\frac{dE}{dx} = \frac{q}{\varepsilon}\left[n + N_A - N_D\right] \tag{4.67}$$

The solution for the I-V behavior of this equation together with the drift equation Eq. (4.61) is conveniently done in the following reduced variables [42]:

$$u = \frac{N_A - N_D}{n(x)} = \frac{q(N_A - N_D)\mu E(x)}{J}$$

$$w = \frac{q^2(N_A - N_D)^2 \mu x}{\varepsilon J} \tag{4.68}$$

and

$$v = \frac{q^3(N_A - N_D)^3 \mu^2 V(x)}{\varepsilon J^2}$$

Poisson's Eq. now becomes

$$dw = \frac{u\,du}{1+u} \tag{4.69}$$

which integrates to

$$w = u - \ell n(1 + u) \tag{4.70}$$

Note that again u=0 at w=0 so that the field is zero at the cathode. The potential integral is given by

$$v = \int_0^w u\,dw = \frac{1}{2}u^2 - u + \ell n(1+u) \tag{4.71}$$

We examine this solution for sufficiently large injection currents that $u \ll 1$. By Taylor expanding the $-\ell n(1-u)$ term in these two equations around u=0, we obtain $v \cong \frac{1}{3}u^3$ and $w \cong \frac{1}{2}u^2$.

Therefore, $v \cong \frac{1}{3}(2w)^{3/2}$, and after substituting from Eq. (4.67), for anode values we obtain the following relation between the current density and the applied voltage

$$J \cong \frac{q}{8} \frac{\varepsilon\mu}{L^3} V^2 \tag{4.72}$$

A quadratic dependence of the current on the voltage is characteristic of a space charge limited regime which is well known as the Mott-Gurney law [43].

One can grow epitaxial InP:Fe by MOCVD [37]. SI material can be grown with resistivities as high as $2.2 \times 10^8 \Omega$–cm [44]. Current-voltage data for this material from which this value for

**Figure 4.17** Current-voltage data obtained for semi-insulating Fe doped InP grown by MOCVD.

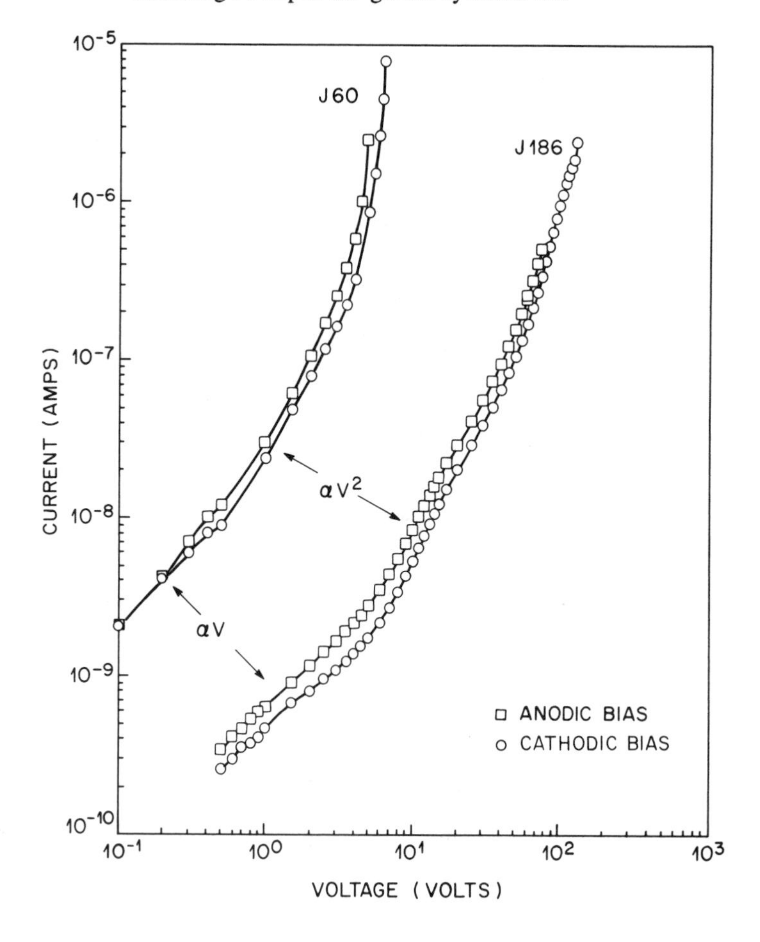

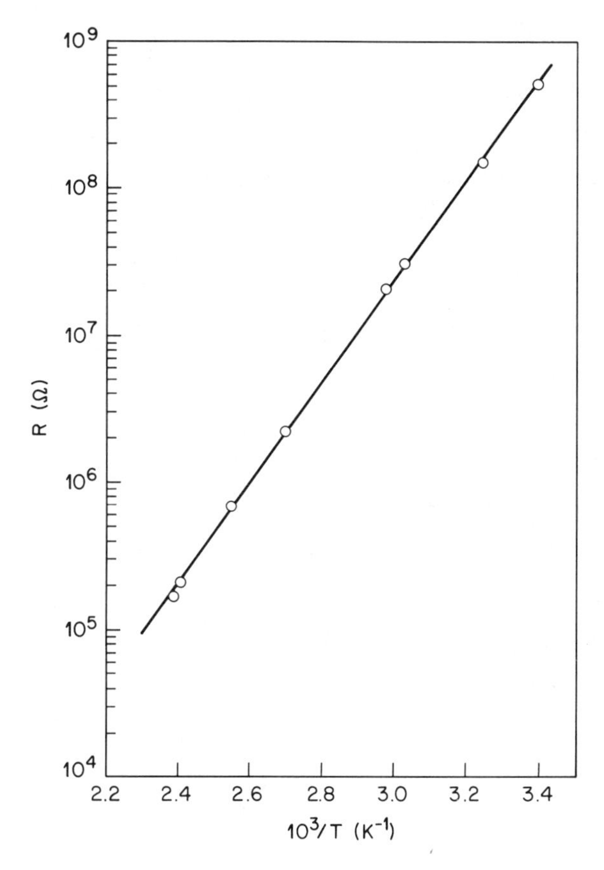


**Figure 4.18** Temperature dependence of the resistance of the device in Fig. 4.16 made with a MOCVD grown Fe-doped semi-insulating epitaxial layer.

resistivity was extracted are shown in Fig. 4.17. The ohmic regime as well as the space charge limited regime are clearly visible for two layers having thicknesses of L=0.8µm (J60) and L=9.0µm (J186).

Fe impurities in InP result in a deep level which causes the material to be SI. Fe acts as an acceptor to compensate the background donor impurities. The temperature dependence of the resistivity is Arrhenius-like (see Sec. 4.2), and the thermal ionisation energy of the Fe related deep level can be obtained from the slope of an Arrhenius plot. Such a plot of resistivity data for MOCVD grown InP:Fe obtained at various temperatures is shown in Fig. 4.18 [45]. An activation energy of 0.68 eV results from the slope of the Arrhenius plot. This places the Fe related deep level at the middle of the band gap of InP and demonstrates that Fe is extremely effective dopant for the production of SI InP. Deep levels which are not midgap would not be as effective.

**REFERENCES**


1. R. Reif, *Fundamentals of Statistical and Thermal Physics*, New York; McGraw-Hill, 1965.
2. J. Bourgoin and M. Lannoo, *Point Defects in Semiconductors II, Experimental Aspects*, New York; Springer-Verlag, Chapt. 5, 1983).
3. M. Lannoo and J. Bourgoin, *Point Defects in Semiconductors I, Theoretical Aspects*, New York; Springer-Verlag, Chapt. 6, 1981.
4. W. Shockley, *Electrons and Holes in Semiconductors* (New York; Krieger, 1976), Chapt. 16, 1981.
5. E. H. Hall, *Am. J. Math. 2*; 287 1879.
6. L. J. van der Pauw, *Philips Research Reports 13*; 1, 1958.
7. S. M. Sze, *Physics of Semiconductor Devices* New York; J. Wiley, Chapt. 2, 1969.
8. R. Chwang, B. J. Smith, and R. C. Crowell, *Solid State Electronics 17*; 1217, 1974.
9. C. T. Sah, *Proc. IEEE 55*; 654 1967.
10. C. T. Sah, R. N. Noyce, and W. Shockley, *Proc. IRE 45*; 1228, 1957.
11. R. N. Hall, *Phys. Rev. 87*; 387, 1952.
12. W. Shockley and W. T. Read, *Phys. Rev. 87*; 835, 1952.
13. J. Hilibrand and R. D. Gold, *RCA Review 21*; 245, 1960.
14. G. L. Miller, *IEEE Trans. on Electron Devices ED-19*; 1103, 1972.
15. S. Braun, H. G. Grimmeiss, J. W. Allen, *Phys. Stat. Solidi* (a) *14*; 527, 1973.
16. D. P. Kennedy, P. C. Murley, and W. Kleinfelder, *IBM J. Res. Develop. 12*; 399, 1968; D. P. Kennedy and R. R. O'Brein, *IBM J. Res. Develop. 13*; 212, 1969.
17. W. C. Johnson and P. T. Panousis, *IEEE Trans. on Elect. Dev. ED-18*; 965, 1971.
18. S. Roy Morrison, *Electrochemistry at Semiconductor and Oxidized Metal Electrodes*, New York; Plenum, 1980.
19. M. M. Faktor, T. Ambridge, C. R. Elliott, and J. C. Regnault, "Using Electrochemical Techniques," in *Current Topics in Materials Science*, (ed. E. Kaldis) New York; North Holland, Vol. 6, Chapt. 1, 1980.
20. J. Bourgoin and M. Lannou, *Point Defects in Semiconductors II, Experimental Aspects* (Springer-Verlag, New York, 1983). Chapt. 6.
21. D. V. Lang, "Space Charge Spectroscopy in Semiconductors", in *Thermally Stimulated Relaxation in Solids*, (ed. P. Braunlich), New York; Springer-Verlag, 1979.
22. D. V. Lang, *J. Appl. Phys. 45*; 3023, 1974.

23. G. L. Miller, D. V. Lang, and L. C. Kimerling, *Ann. Rev. Mater. Sci. 7*; 377 1977.

24. L. C. Kimerling, *J. Appl. Phys. 45*; 1839, 1974.

25. C. T. Sah, W. W. Chan, H. S. Fu, and J. W. Walker, *Appl. Phys. Lett. 20*; 193, 1972.

26. A. T. Macrander, B. Schwartz, and M. W. Focht, *J. Appl. Phys. 55*; 3595, 1984.

27. G. L. Miller, J. V. Ramirez, and D. A. H. Robinson, *J. Appl. Phys. 46*; 2638, 1975.

28. D. Pons, *Appl. Phys. Lett. 37*; 413, 1980.

29. D. L. Losee, *J. Appl. Phys. 46*; 2204, 1975.

30. G. Goto, S. Yanagisawa, O. Wada, H. Takanashi, *Jpn. J. Appl. Phys. 13*; 1127, 1974.

31. D. V. Lang, *J. Appl. Phys. 45*; 3014, 1974.

32. H. G. Grimmeiss, *Ann. Rev. Mater. Sci. 7*; 341, 1977.

33. B. W. Wessels, *J. Appl. Phys. 47*; 1131, 1976.

34. J. A. Borsuk and R. M. Swanson, *IEEE Trans. on Electron Devices ED-27*; 2217, 1980.

35. *Semi-insulating III-V Materials*, Kah-nee-ta 1984, (ed. D. C. Look and J. S. Blakemore) Shiva Publishing, Ltd.

36. *Semi-Insulating III-V Materials*, Nottingham 1980, (ed. by G. J. Rees), Shiva Publishing, Ltd.

37. J. A. Long, V. G. Riggs, and W. D. Johnston, Jr., *J. Cryst. Growth 69*; 10, 1984.

38. M. W. Focht, A. T. Macrander, B. Schwartz, and L. C. Feldman, *J. Appl. Phys. 55*; 3859, 1984.

39. D. L. Scharfetter, *Solid State Electron. 28*; 299, 1965.

40. S. Sze, *Physics of Semiconductor Devices*, $2^{nd}$ ed. New York; John Wiley and Sons, 1981, Sec. 5.4.5.

41. W. Shockley and R. C. Prim, *Phys. Rev. 90*: 753, 1953.

42. M. A. Lampert and P. Mark, *Current Injection in Solids*, New York; Academic Press; 1970. Sec. 4.2.

43. N. F. Mott and R. W. Gurney, *Electronic Processes in Ionic Crystals*, $2^{nd}$ ed. Oxford; Oxford Univ. Press; 1948.

44. A. T. Macrander, J. A. Long, V. G. Riggs, A. F. Bloemeke, and W. D. Johnston, Jr., *Appl. Phys. Lett. 45*; 1298, 1984.

45. J. A. Long, V. G. Riggs, A. T. Macrander, and W. D. Johnston, Jr., *J. Cryst. Growth 77*; 42, 1986.

# CHAPTER 5

# OPTICAL CHARACTERIZATION

Ever since the interest in semiconducting materials, optical processes in them have been an important field of study both from scientific and technological point of view. This is because the study of the interaction of light with semiconductor provides a plethora of information about the physical properties of the semiconductor. The numerous research articles as well as a few books which deal with this subject attest to its importance.

Absorption, emission, modulation, conversion, and scattering of light are the various phenomena which constitute optical processes in a semiconductor. Some of them can be studied as a function of intensity of the incident light, sample temperature, and under external perturbations such as magnetic field, electric field, uniaxial stress, and hydrostatic stress. All these techniques have been and are being used to characterize GaAs and InP. In fact, some of them are routinely employed in industrial laboratories to qualify wafers before processing them to devices and in some instances to monitor processing steps as well. In this section we emphasize those techniques which are commonly used for studying GaAs, InP, and their heterostructures and give only a brief treatment of others. However, appropriate references are given where more extensive coverage can be found.

## 5.1 ABSORPTION

Absorption is characterized by $\alpha$, the absorption coefficient, which is a measure of the attenuation of light as it traverses a distance x in the semiconductor. If the energy ($h\nu$) of the incident light is varied, so does $\alpha$ and $\alpha$ as a function of $h\nu$ carries information about the band structure of the semiconductor. Absorption is simply a process in which the incident light of a certain suitable energy excites an electron in the semiconductor from a lower to a higher energy state. The electrons that are excited can be (a) inner shell electrons, (b) valence band electrons , (c) free electrons (or free holes), and (d) bound electrons at impurities and defects. The absorption associated with the excitation of the valence band electrons, referred to as fundamental absorption, gives the most important information about the semiconductor.

### 5.1.1 Fundamental Absorption

Fundamental absorption involves excitation of an electron from the valence band to an empty state in the conduction band. This is simply measured by determining the transmission of light through the semiconductor as a function of photon energy, $h\nu$. As the photon energy approaches the band gap, $E_g$, the absorption increases rapidly by as much as 3 to 4 orders of magnitude and $h\nu \approx E_g$ represents the fundamental absorption edge. The position and shape of the absorption spectrum near this edge reveal the details about the band structure.

Consider a direct band gap semiconductor such as GaAs and InP with an E versus k dispersion relation shown schematically in Fig. 5.1(a). When $h\nu \sim E_g$, an electron from a state with a wave vector $\bar{k}_i$ in the valence band is excited to a state with wave vector $\bar{k}_f$ in the conduction band. If $\bar{q}$ is the photon wave vector, then conservation of momentum dictates $\bar{k}_f - \bar{k}_i = \bar{q}$, since $\bar{q}$, the momentum of the photon, is very small compared to the crystal momentum, $\bar{k}_f \approx \bar{k}_i$. In other words, for absorption in a direct band gap semiconductor only vertical transitions are allowed, that is, an electron in the valence band with a certain wave vector is transferred to a state in the conduction band with the same wave vector.

When the minimum in the energy gap does not occur at k=o but at a different point in k space as shown in Fig. 5.1(b) then vertical transitions cannot occur. This will be the case for the alloy $Al_xGa_{1-x}As$ when $x \gtrsim 0.40$. In such cases momentum is conserved by either phonon emission or absorption.

**Direct transitions**

For the direct transition shown in Fig. 5.1(a) connecting states $k_i$ and $k_f$, the photon energy required is

$$h\nu = E(k_f) - E(k_i) \tag{5.1}$$

For parabolic bands this gives

$$h\nu = E_g + \frac{\hbar^2 k_f^2}{2m_e^*} + \frac{\hbar^2 k_i^2}{2m_h^*} \tag{5.2}$$

where $m_e^*$ and $m_h^*$ are, respectively, the electron and hole effective masses. Since $k_f = k_i = k$, Eq. (5.2) becomes

$$h\nu = E_g + \frac{\hbar^2 k^2}{2}\left(\frac{1}{m_e^*} + \frac{1}{m_h^*}\right) \tag{5.3}$$

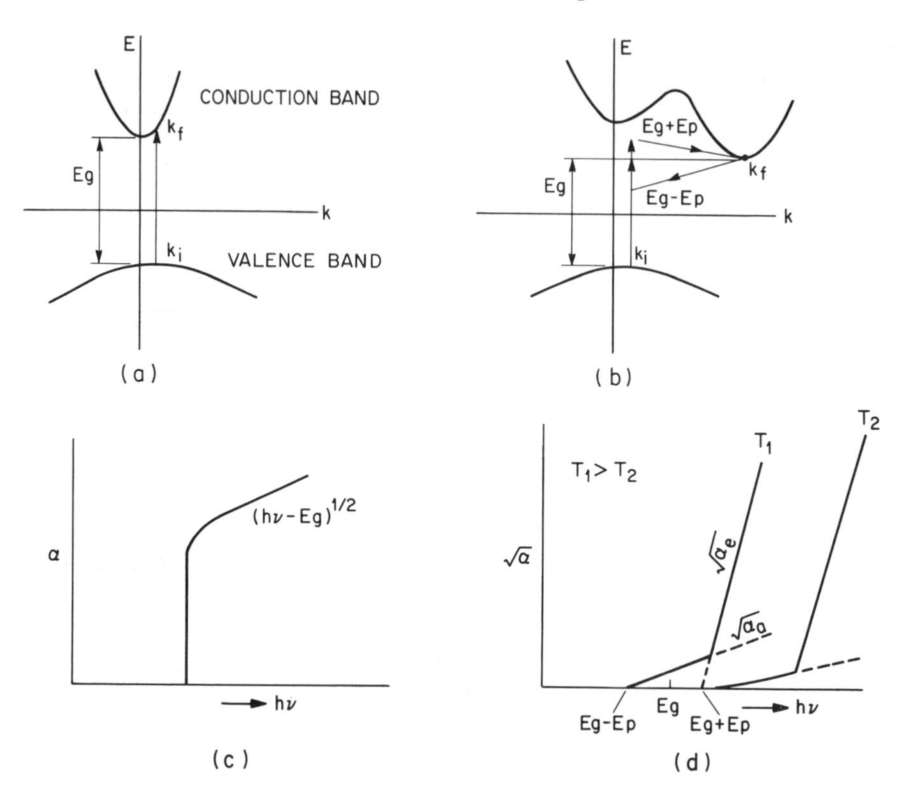


**Figure 5.1** E versus k relation for a direct band gap (a) and an indirect band gap (b) semiconductor showing direct optical transition and indirect optical transition, respectively. The corresponding absorption spectrum for a direct (c) and an indirect band gap (d) semiconductor are shown.

The absorption coefficient $\alpha(h\nu)$ is given by [1]

$$\alpha = A \mid M \mid^2 N(h\nu) \tag{5.4}$$

where A is a constant, M is the optical matrix element associated with the transition, and $N(h\nu)$ is the joint density of states, that is, the density of energy-level pairs per unit energy interval.

The number of states in the interval $h\nu$ and $h\nu + d(h\nu)$ can be written as [1, 2]

$$N(h\nu)d(h\nu) = \frac{8\pi k^2 dk}{(2\pi)^3} \tag{5.5}$$

From Eq. (5.3) it would follow that

$$N(h\nu)d(h\nu) = \frac{\sqrt{2}}{\pi^2 \hbar^3} \; (m_r^*)^{3/2} \; (h\nu - E_g)^{1/2} \; d(h\nu) \tag{5.6}$$

where $m_r^*$ is the reduced mass; $1/m_r^*=1/m_e^*+1/m_h^*$. Substituting Eq. (5.6) in Eq. (5.4) would yield

$$\alpha = A \mid M \mid^2 (h\nu - E_g)^{1/2} \tag{5.7}$$

Since $\mid M \mid^2$ is a slowly varying function of k and can be taken to be approximately constant near $k=o$ $\alpha$ is determined by the $(h\nu - E_g)^{1/2}$ term alone. Thus

$$\begin{aligned} \alpha &\sim (h\nu - E_g)^{1/2} \quad \text{for} \quad h\nu > E_g \\ &\sim 0 \quad \text{for} \quad h\nu \leq E_g \end{aligned} \tag{5.8}$$

as shown in Fig. 5.1(c).

Not all transitions at k=o are allowed according to parity selection rules. Transitions between any two of the three valence bands in a semiconductor would be examples for these forbidden transitions. However, such transitions become allowed at k#o and the transition probability is proportional to $k^2$ or $(h\nu - E_g)$. Hence $\alpha$ $(h\nu)$ varies as $(h\nu - E_g)^{3/2}$ for the direct forbidden transitions.

**Indirect transitions**

For the transition shown in Fig. 5.1(b) momentum is conserved with phonon cooperation. Usually the longitudinal acoustic and transverse acoustic phonons are involved. Transitions take place via virtual states with the emission or absorption of one or several phonons. Accordingly,

$$\begin{aligned} h\nu &= E_f - E_i + nE_p \quad \text{(emission)} \\ h\nu &= E_f - E_i - nE_p \quad \text{(absorption)} \end{aligned} \tag{5.9}$$

where $E_p$ is the phonon energy. Further, if we write

$$\alpha\,(h\nu) = \alpha_e + \alpha_\alpha \tag{5.10}$$

$\alpha_e$, $\alpha_a$ being the contributions as a result of emission and absorption of phonons respectively, then

$$\left.\begin{aligned} \alpha_e &= o \qquad h\nu < E_f - E_i + E_p \\ \alpha_a &= o \qquad h\nu < E_f - E_i - E_p \end{aligned}\right\} \tag{5.11}$$

For $h\nu > E_g - E_p$, $\alpha(h\nu)$ is given by

$$\alpha_a(h\nu) = \frac{A(h\nu - E_g + E_p)^2}{\left(\exp \dfrac{E_p}{kT} - 1\right)} \tag{5.12}$$

For $h\nu > E_g + E_p$, $\alpha\,(h\nu)$ is

$$\alpha_e(h\nu) = \frac{A\,(h\nu - E_g - E_p)^2}{1 - \exp\left[\dfrac{-E_p}{kT}\right]} \tag{5.13}$$

The temperature dependence of $\alpha_e$ and $\alpha_a$ are shown in Fig. 5.1(d) where the dotted lines intersects the energy axis at $E_g - E_p$ and $E_g + E_p$. As before, if transitions are forbidden by parity selection rules $\alpha$ in Eqs. (5.12) and (5.13) varies as $(h\nu - E_g \pm E_p)^3$.

**Absorption tail at hν , $E_g$**

In a direct band gap semiconductor $\alpha$ should vary with hν as depicted in Fig. 5.1c. That is, no absorption is to be expected below $E_g$. However, many times $\alpha$ decreases much more slowly below the fundamental edge than suggested by theory. This is illustrated in the room temperature absorption edge of GaAs [3] shown in Fig. 5.2. This absorption tail is sometimes referred to as Urbach's tail [4]. Urbach reported that the tail obeys a relationship $d(\ln\alpha)/d(h\nu) = 1/kT$. This led to the interpretation that the absorption tail involves phonon-assisted transitions [5]. Pankove [6] found in GaAs that the slope of the absorption tail varies with impurity concentration more so than with temperature. He related the slope to the presence of tail states in the conduction band for p-type material and in the valence band for n-type material. The slope was given by $E_o^{-1}$,

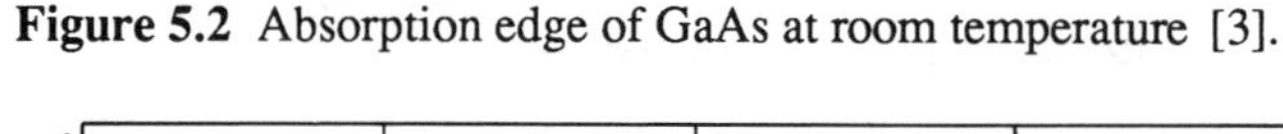

**Figure 5.2** Absorption edge of GaAs at room temperature [3].

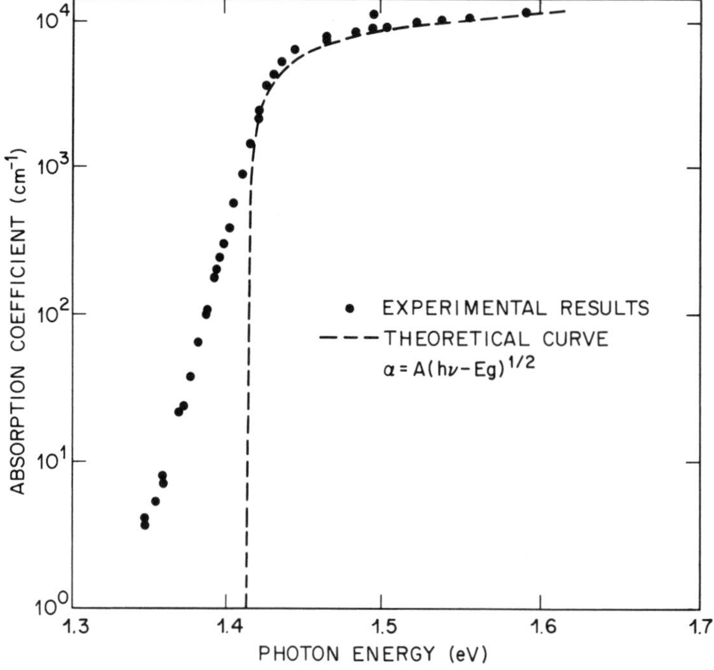

where $E_o$ is an empirical parameter having the units of energy and represents the distribution of the tail states. $E_o$ correlated well with impurity concentration.

Redfield [7] proposed that the internal electric fields associated with the charged impurities causes the broadening of the absorption edge in a manner analogous to the Franz-Keldysh effect [8]. According to this effect, in the presence of an electric field the probability of a valence electron reaching the conduction band via tunneling increases. The tunneling may or may not be assisted by photons. A consequence of electron tunneling, enhanced by the electric field, is an exponential tail in the absorption below $E_g$. The measurements by Redfield and Afromowitz [9] on compensated GaAs provided further evidence that the exponential broadening of the absorption edge is caused by charged impurities. When they looked at under-compensated p-type GaAs, $d(\ln\alpha)/d(h\nu)$ decreased with increasing temperature, consistent with increased ionization of the acceptors and hence impurity electric fields. However, in slightly over-compensated GaAs the slope was independent of temperature as expected since the density of ionized centers does not change with temperature.

While we are on the discussion of impurity effects on the absorption edge, it is worthwhile to mention yet another one. As the impurity concentration increases, the absorption edge shifts to higher energies. This is simply caused by the band filling effect otherwise known as the Burstein-Moss [10] shift. With increasing doping, the Fermi level moves inside the respective band and optical transitions connect states above the Fermi level. As a result, absorption edge occurs at an energy greater than $E_g$. The exact magnitude of the shift is, however, the Fermi level shift minus the band gap shrinkage induced by the high doping levels [11].

### 5.1.2 Measurement Techniques

Absorption near the band edge is easily determined by measuring the transmission through the sample in the standard light in and light out configuration. If the specimen has a thickness d and reflectivity R, then the transmission $T_r$ under normal incidence is given by [2]

$$T_r = \frac{(1-R)^2 \exp(-\alpha d)}{1-R^2 \exp(-2\alpha d)} \tag{5.14}$$

This expression takes into account multiple reflections but neglects interference. Equation (5.14) can be written in terms of optical density, OD, as

$$\begin{aligned} OD = -\log_{10} T_r &= \alpha d \log_{10} e - 2\log_{10}(1-R) \\ &+ \log_{10}\left[1 - R^2 \exp(-2\alpha d)\right] \end{aligned} \tag{5.15}$$

If R is constant over the spectral range the measurement is made and if $\alpha d \gg 1$ so that the third term in Eq. (5.15) is negligible, OD becomes

$$OD = \alpha d \log_{10} e + A \tag{5.16}$$

where A is a constant instrumental background including the second term and the value of the third term at $\alpha = o$ in Eq. (5.15).

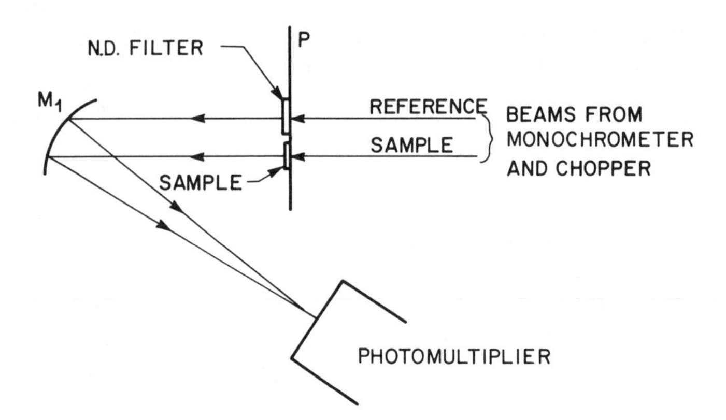


**Figure 5.3** Schematic representation of the optical configuration for a double-beam transmission measurement [12].

Absorption is generally measured in a double-beam configuration as shown in Fig. 5.3. This can be done using commercial absorption spectrometer or alternatively, the equipment can be set-up easily in a laboratory. The light source and detector must be selected depending on the spectral range investigated. For the wavelength range covered by III-V semiconductors and their alloys a tungsten lamp or a globar would be employed as a light source, the former for wavelengths in the visible to 2 μm and the latter for wavelengths greater than 2 μm. Similarly, one can use for detectors photomultipliers for the visible, photoconductive or photovoltaic detectors for the near infrared and thermocouples or bolometers for the far infrared spectral range. The spectral response of these various detectors and other relevant optical data are given by Pankove [2].

To measure absorption at low temperature the sample has to be mounted suitably in a cryostat. Sometimes it may be difficult to use such cryostats in the commercial absorption spectrometers without extensive modifications. In such cases absorption measurement can be made in a simple way using an appropriate lamp and focussing the transmitted light through the sample onto the entrance slit of a spectrometer which on its exit slit has the appropriate detector. Depending on the spectrometer used to disperse the transmitted light, very high spectral resolution can be achieved in the measurement. When the transmission through the sample is small, calibrated neutral density filters have to be used to minimize the amount of stray light entering the spectrometer. The spectrometer and the detector can be easily interfaced to a computer such that a plot of $\alpha$, calculated from Eq. (5.16), versus wavelength is obtained directly.

If the sample whose absorption is to be measured is in the form of an epitaxial film on a substrate, which will invariably be the case for the alloy semiconductors, then the substrate may have to be removed. For example, to measure the absorption of $Al_xGa_{1-x}As$ films grown on GaAs substrates, the substrate is removed by using selective etches whose etch rate for GaAs is many times greater than that for $Al_xGa_{1-x}As$. The substrate is removed over an area sufficient enough for the absorption measurement and at the same time to support the bare film. The latter aspect is important since otherwise strain in the film would affect the measurement. The availability of selective etches to remove GaAs with respect to $Al_xGa_{1-x}As$ has facilitated the

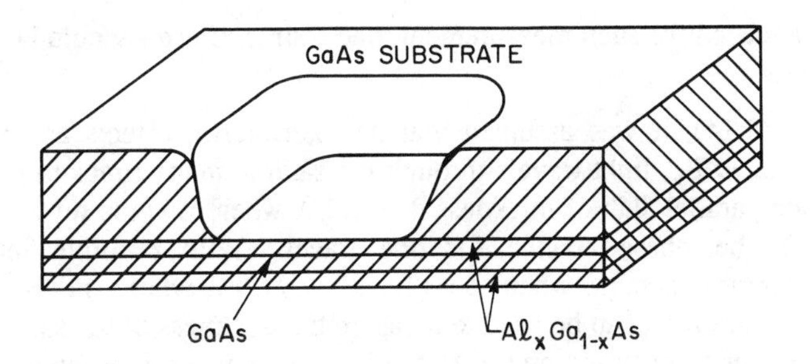


**Figure 5.4** Schematic cross-sectional view of the $Al_xGa_{1-x}As/GaAs$ heterostructure sample in which the GaAs substrate is removed to measure the absorption in the central GaAs layer [12].

measurement of a thin epitaxial layer of GaAs sandwiched between two nonabsorbing $Al_xGa_{1-x}As$ layers [12]. The specimen for the measurement is prepared as shown in Fig. 5.4.

When the epitaxial layer is grown on a nonabsorbing substrate, then the substrate need not be removed. For this case, the transmission coefficient given by Eq. (5.14) is modified to [13]

$$T_r = \frac{(1-R_{11})(1-R_{12})(1-R_{23})\exp(-\alpha d)}{1-R_{12}R_{23}-\exp(-2\alpha d)R_{01}\left\{R_{12}+R_{23}-2R_{12}R_{23}\right\}} \tag{5.17}$$

where α and d are, respectively, the absorption coefficient and thickness of the epitaxial layer, and $R_{01}$, $R_{12}$, and $R_{23}$ are the reflectivities at the air-substrate, substrate-epitaxial layer and epitaxial layer-air interfaces. For the usual single layer transmission, $R_{01} = R_{12} = R$, $R_{23} = o$ which gives Eq. (5.14). Equation (5.17) may be applicable for example, to measure α(hν) of a $Ga_{0.47}In_{0.53}As$ epitaxial layer on an InP substrate. However, if the sum of the transmittance and reflectance for the substrate is not unity in the spectral range of the epitaxial layer, which it should be to use Eq. (5.17), then computing α(hν) of the layer is not straightforward. The $\alpha d$ term in Eq. (5.17) then becomes $(\alpha_s d_s + \alpha_e d_e)$, the sum for substrate and epitaxial layer. Humphreys et al., [14] have found this to be the case for GaInAs/InP. They determined α(hν) of $Ga_{0.47}In_{0.53}As$ in the spectral range 1 to 1.7μm by assuming $R_{12} = o$ in Eq. (5.17).

We noted earlier that if $\alpha d \gg 1$, OD takes the simplified form given by Eq. (5.16). This can be used to calculate α if R is not known. By measuring the transmission of two samples having thicknesses $d_1$, and $d_2$, α is obtained from

$$\frac{T_2}{T_1} = \exp \alpha \left[d_1 - d_2\right] \tag{5.18}$$

Evtikhiev et al., [15] made use of Eq. (5.18) to determine α of $Ga_{0.47}In_{0.53}As$ and GaInAsP films grown by liquid phase epitaxy and of compositions corresponding to $E_g \sim 0.94$ and 0.78 eV.

When thick films are used for such measurement, one source of error would be the compositional gradients in the films.

In deriving Eq. (5.14), it was assumed that no interference effects are present. However, when $\alpha$ is small at $h\nu \ll E_g$, light travels through the sample many times to give an interference pattern. For a plane parallel slab of thickness d, $2\pi nd/\lambda$ where n is the refractive index and $\lambda$ is the wavelength, is the phase change for one travel. With multiple internal reflections, constructive interference occurs when $d = (2m+1)\lambda/(2n)$ where m is an integer. The interference pattern, therefore, can be used to measure the thickness of the sample very accurately [16]. The equations for transmitted and reflected intensities when interference occurs are given by Stratton [17].

### 5.1.3 Exciton absorption

The absorption curve shown in Fig. 5.1(c) and Eqs. (5.8), (5.12) and (5.13) are for a simple band-to-band absorption. When excitons are involved in the absorption process, the absorption spectrum can be characterized by sharp lines near the fundamental edge. According to Elliott [18], the energies of the lines are given by

$$h\nu = E_g - E_x \tag{5.19}$$

where $E_x$ is the exciton binding energy proportional to $1/n^2$ where n is an integer. The intensity of the lines varies as $n^{-3}$. A similar set of lines is expected for indirect transitions except that the first line is absent. Even for photon energies near and above $E_g$, the absorption process is affected by electron-hole Coulombic interaction. For direct allowed transitions [18],

$$\alpha(h\nu) = \alpha_o(h\nu)\; 2\pi A \frac{1}{1-\exp(-2\pi A)} \tag{5.20}$$

where $A = \sqrt{E_x/h\nu - E_g}$ and $\alpha_o(h\nu)$ is the absorption coefficient (Eq. (5.8)) in the absence of Coulomb interaction. It can be seen from Eq. (5.20) that the modification of the fundamental absorption edge by exciton absorption is greatest at $h\nu \approx E_g$. For higher values of $h\nu$ the exciton effects becomes small. Figures 5.5 and 5.6 show $\alpha$ versus $h\nu$ for GaAs [19] and InP [20],

**Figure 5.5** Exciton absorption in GaAs at 294, 186, 90 and 21K [19].

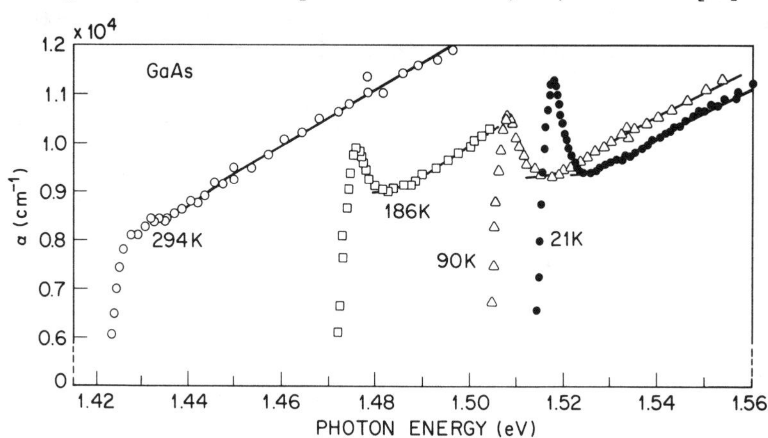

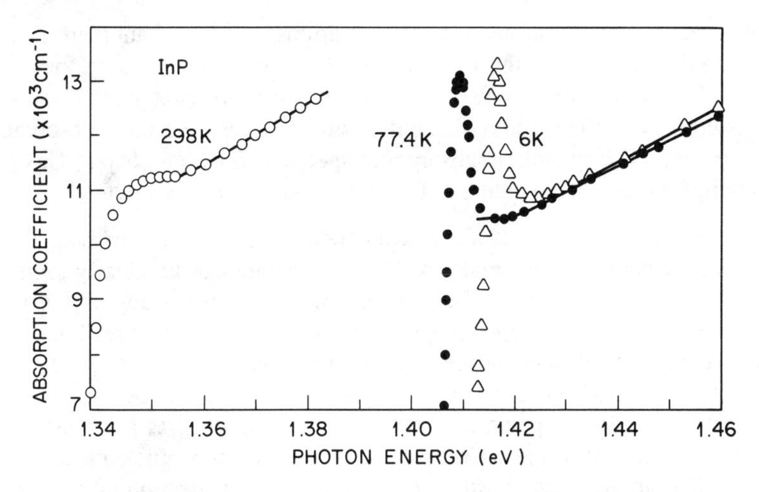


**Figure 5.6** Exciton absorption in InP at 298, 77 and 6K [20].

respectively, at different temperatures. Only the n = 1 (Eq. (5.19)) exciton is observed in Figs. 5.5 and 5.6 while the others merged with the absorption spectrum. Figure 5.7 shows the data of Monemar et al., [16] on the absorption spectrum of $Al_xGa_{1-x}As$ at 4K and 293K for $x \leq 0.55$. The shape of the absorption edge of the alloy is very similar to that of GaAs and shows the n = 1 exciton peak for x = 0.31 and x = 0.37 curves. As the band gap becomes indirect for x > 0.40, the exciton line is not observed and that the absorption curve below the edge broadens

**Figure 5.7** Optical absorption of GaAs and $Al_xGa_{1-x}As$ samples with x = 0.31, 0.37, 0.49 and 0.55 at 293 and 4K. The energy axis for GaAs is shifted. For the samples with x = 0.31 and 0.37, the ground state free exciton transition is indicated by the arrow [16].

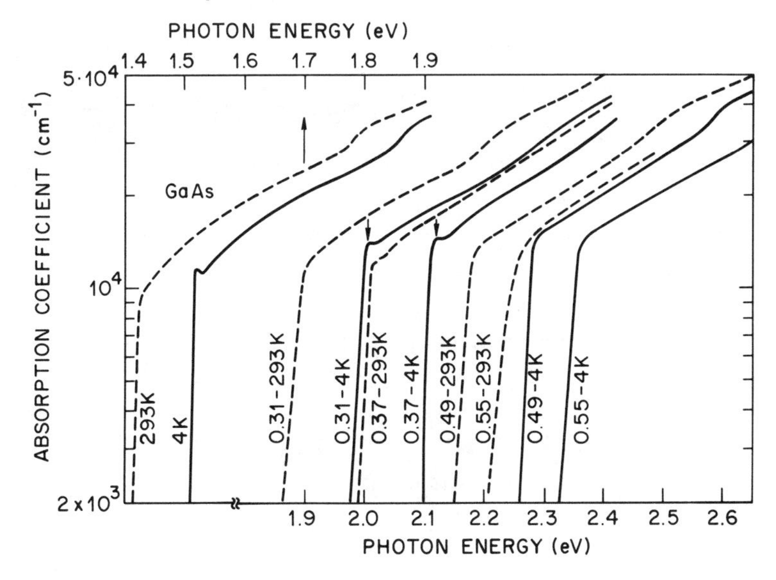

caused by the onset of phonon-assisted indirect transitions. An important point to note in Fig. 5.7 is that the absorption curves for the alloy for $x < 0.4$ are very similar to the one from GaAs illustrating that there is no additional broadening of the absorption edge because of potential fluctuations resulting from the random distribution of the atoms. A similar observation was made by Evtikhiev et al., [15] regarding absorption spectra from $Ga_{0.5}In_{0.5}P$, $Ga_{0.47}In_{0.53}As$ and GaInAsP of compositions corresponding to $E_g \approx 0.94$ and 0.78 eV, respectively.

In the case of epitaxial layers of alloy semiconductors, the effect of biaxial stress in them caused by the lattice parameter mismatch with the substrate can be clearly seen in the exciton absorption spectra. In the presence of a biaxial strain the degeneracy of the heavy-hole and light-hole valence bands is lifted depending on the magnitude and the direction of the strain [21]. The absorption spectra reveals the valence band splitting caused by stress [12, 22, 23]. Figure 5.8 shows the transmission spectrum of a GaAs sample sandwiched between two $Al_xGa_{1-x}As$ layers (Fig. 5.4). Although the lattice parameters of GaAs and $Al_xGa_{1-x}As$ are nearly the same at the growth temperature, because of the differences in the expansion coefficients of the two materials, the GaAs layer has an in-plane tensile stress. The exciton absorption edge splits into two components separated by 6.8meV. The higher energy component has nearly three times the optical transition strength of the lower energy component, as expected by crystal symmetry constraints [24, 25]. For GaAs the valence band splitting in eV is $6.0 \times 10^{-12}$ X where X is the stress in dyne $cm^{-2}$. Thus from the observed splitting in the absorption spectrum the in-plane epitaxial stress can be obtained. Zielinski et al., [23] discusses in detail the structure caused by

**Figure 5.8** Optical transmission at 2K of a 1.2μm thick GaAs sample sandwiched between the $Al_xGa_{1-x}As$ ($x \sim 0.60$) layers. The exciton absorption edge splits into two components separated by 6.8 meV due to splitting of the valence band caused by the in-plane tensile stress in GaAs [12].

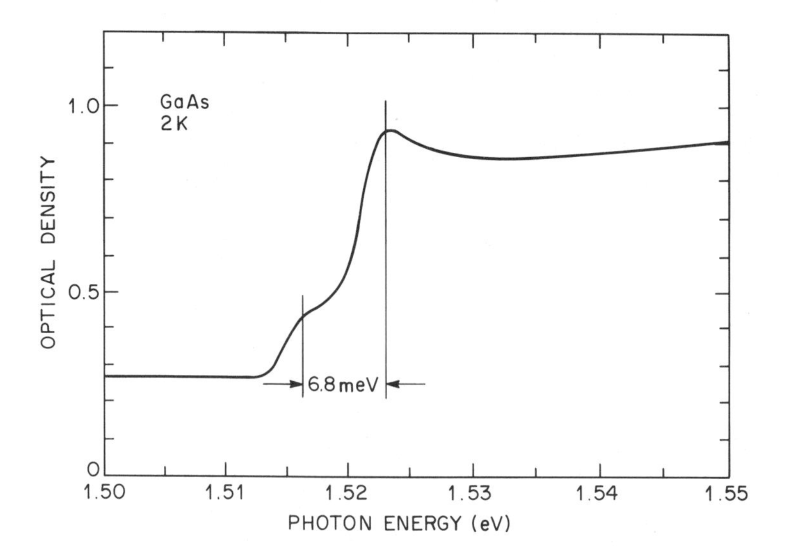

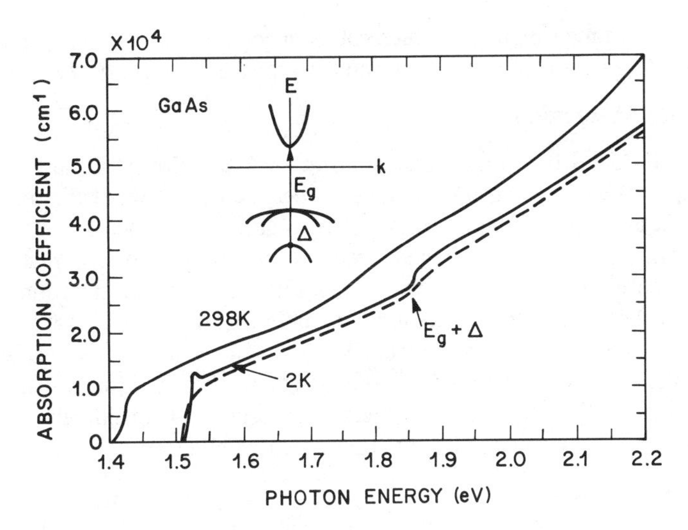


**Figure 5.9** Absorption of GaAs at 2 and 298K showing the spin-orbit split valence band. The dashed curve is the 298K data shifted 95 meV to higher energy. The inset shows the transition from the spin-orbit split valence band to the conduction band [12].

the lattice mismatch stress in the absorption spectrum from GaInAs layers grown on InP substrate.

**Absorption above the fundamental edge**

In the energy band diagram of III-V compounds shown in Fig. 1.3 direct transitions that are allowed can occur at all $k_s$ and at energies greater than $E_g$. One such transition that is observed as $h\nu$ increases beyond $E_g$ is the one from top of the spin-orbit split valence band to the bottom of the conduction band. The onset of this absorption gives the energy of the spin-orbit splitting as shown in Fig. 5.9 for GaAs [12]. The inset shows the associated transition. Note this transition in the absorption curves from $Al_xGa_{1-x}As$ in Fig. 5.7. Unlike the absorption edge near $h\nu = E_g$, no exciton peak is observed corresponding to the $E_g + \Delta$ transition. This is because the excitons associated with the $E_g + \Delta$ edge have short life time and as a result severe broadening of the absorption edge occurs [16].

A word should be mentioned about the experimental difficulty in measuring $\alpha$ at high energies. For $h\nu \gg E_g$, since $\alpha$ is rather large, very thin (0.1-0.2 μm) samples are needed for transmission measurement. Such thin samples are difficult to handle and are invariably strained if they have to be prepared from bulk material by mechanical polishing. This problem to some extent is simplified when thin epitaxial films sandwiched between nonabsorbing layers can be used. The GaAs data shown in Fig. 5.7 were obtained from such epitaxial layers where transmission measurements were made through a window etched in the substrate as shown in Fig. 5.4. Although sample preparation and handling are made easy with these samples, strain can still be present in the layer, either caused by heteroepitaxial mismatch or by differential thermal expansion. Reflectance measurements which can be made on bulk samples are more easily suited to study the high-energy transitions than transmission measurements [26, 27].

### 5.1.4 Intervalence Absorption

We noted in Sec. 5.1.1 that not all transitions at k=o are allowed because of parity selection rules. An example of such a transition is the one between any two of the three valence bands. The optical matrix element for this transition is zero at k=o but increases for non zero k. The intervalence absorption requires holes and hence is seen only in p-type material. Figure 5.10 shows the intervalence absorption peaks in p-GaAs with a carrier concentration of $2.7\times10^{17}cm^{-3}$ [28]. The three peaks in the absorption curve are assigned to $V_3\rightarrow V_1$ (0.42 eV), $V_3\rightarrow V_2$ (0.31 eV), and $V_2\rightarrow V_1$ (0.15 eV) transitions, where $V_1$, $V_2$, and $V_3$ denote the heavy hole, light hole, and spin-orbit valence bands. Figures 5.11(a) and 5.11(b) show the intervalence absorption data for p-InP [29, 30] and p-GaInAs [30]. Intervalence absorption is considered to be one of the loss mechanisms in the long wavelength (1.3µm–1.5µm) GaInAsP lasers [31, 32].

### 5.1.5 Impurity Absorption

At $h\nu < E_g$ impurities participate in the absorption process. There are four possible transitions involving shallow impurities as shown in Fig. 5.12. Absorption corresponding to transitions from the ground and excited states of the impurity to the respective band edge takes place at $h\nu \approx E_i$, the ionization energy of the impurity, and hence it occurs in the far infrared region of the absorption spectrum. More about this will be discussed in Sec. 6.3.5. The other

**Figure 5.10** Free carrier absorption due to holes in p-type GaAs at 84, 197, 295 and 370K. The peaks at 0.42 eV and 0.31 eV are due to inter valence band transitions from the spin-orbit valence band to heavy hole band and spin-orbit valence band to light hole band, respectively. The band at 0.15 eV is attributed to transition from the light hole to heavy hole band [28].

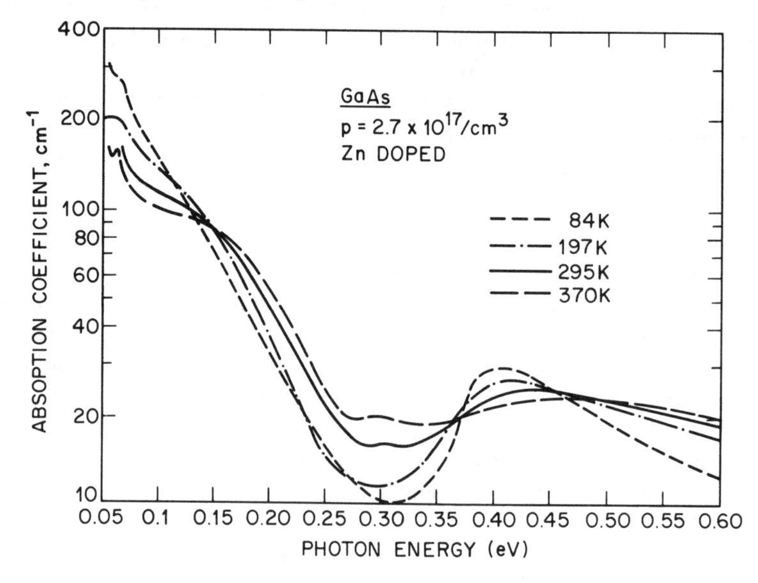

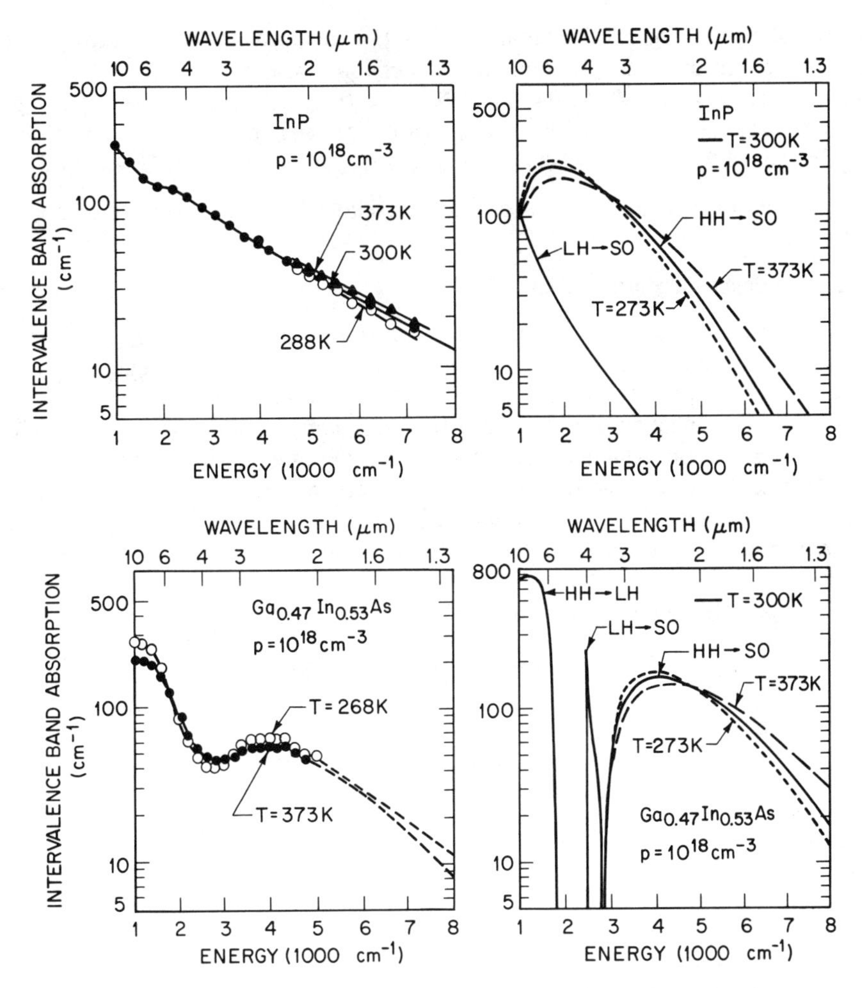


**Figure 5.11** Measured and theoretical inter valence absorption in p-type InP (a) and p-type $Ga_{0.47}In_{0.53}As$ (b) [30].

**Figure 5.12** Schematic band diagram showing optical transitions involving shallow donors and acceptors.

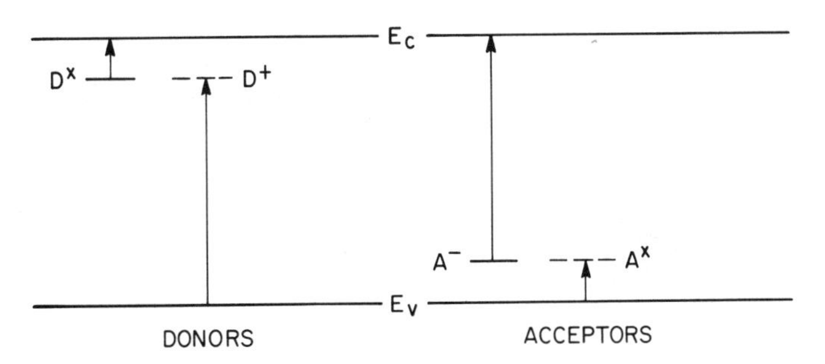

two transitions shown in Fig. 5.12 occur at $h\nu \sim E_g - E_i$ and appear as shoulders in the absorption spectrum near the fundamental edge. Impurity absorption in GaAs was observed by Sturge [19] who identified three levels at 0.70, 0.47, and 0.30 eV (Fig. 5.13). The 0.70 eV assigned to a donor level was originally suspected to be responsible for causing high resistivity in GaAs [33]. The absorption near 1.2 eV has now been associated with the EL2 center in GaAs [34].

Impurity absorption also includes that caused by excitons bound to isoelectronic traps and that by acceptor-to-donor transitions [2].

### 5.1.6 Free Carrier Absorption

Free carrier absorption occurs at hν much lower than the band edge absorption process. This corresponds to transition of the free carrier between states in a single band. Accordingly, it cannot occur in a completely filled band. In an n-type semiconductor, the absorption is proportional to the free electron density. Since free carrier absorption connects states of different k, momentum conservation requires phonon cooperation or scattering from ionized impurities. In the classical Drude theory on the behavior of free carriers in a periodic electric field free carrier absorption coefficient, $\alpha_f$, increases as the square of the wavelength ($\lambda^2$). The quantum mechanical treatment gives $\alpha_f$ proportional to $\lambda^{3/2}$, $\lambda^{5/2}$, and $\lambda^3$ or $\lambda^{7/2}$ depending upon whether

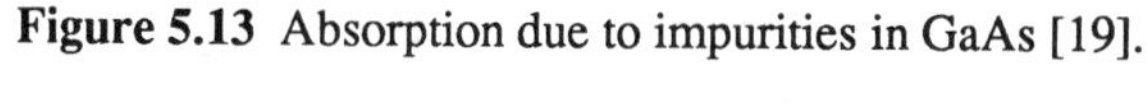

**Figure 5.13** Absorption due to impurities in GaAs [19].

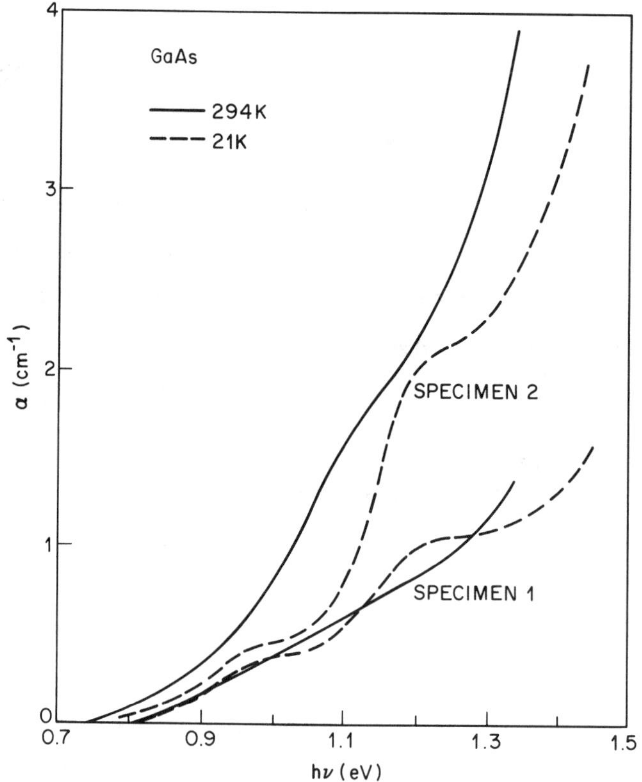

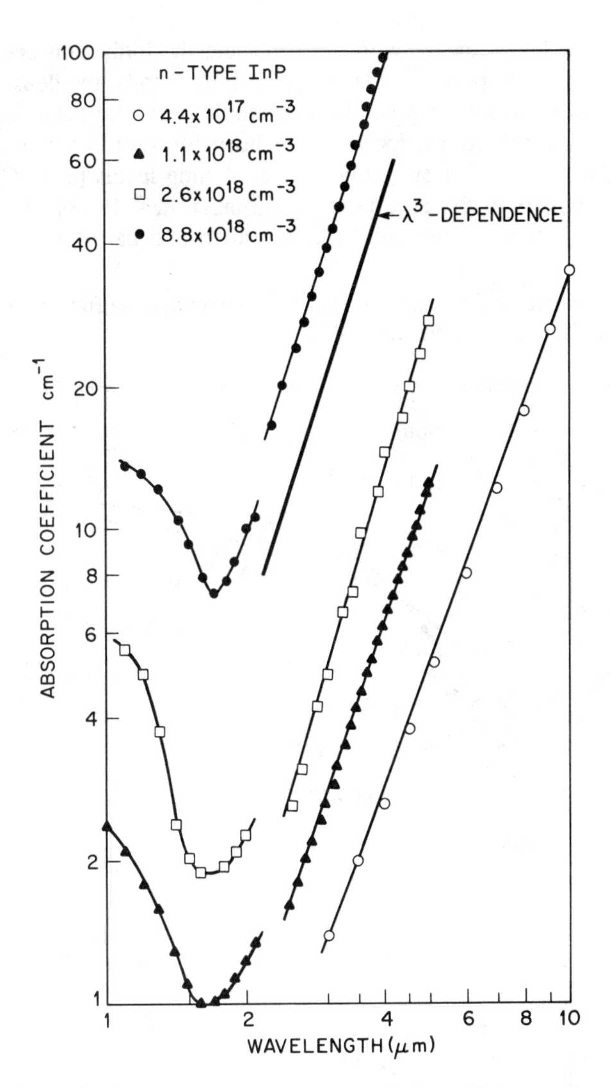


**Figure 5.14** Absorption spectrum for n-type S-doped InP at various carrier concentrations [38].

acoustic phonon scattering [35], optical phonon scattering [36], or ionized impurity scattering [37] is involved in the momentum conservation process. Kim et al., [38] measured the free carrier absorption in n-InP as a function of electron concentration. Their data is shown in Fig. 5.14.

### 5.1.7 Optical Absorption in Heavily Doped Material

As the doping level increases, interesting changes occur in the band structure of the semiconductor. With increasing concentration of the donors or acceptors the spacing between the impurities decreases leading to overlap of the wave functions of the carriers on adjacent sites.

Eventually the impurity levels merge with the band and the ionization energy of the impurity becomes zero. Also, as the impurity levels merge with the bands, the density of states function undergoes changes and is no longer given by the simple parabolic band function. Further, the k-selection rule for optical transitions, applicable in the undoped or low doped semiconductor, are broken down in the band tails that are found at high doping levels [39]. Casey and Stern [40] have worked out the details of the absorption spectrum in heavily doped n- and p-type GaAs. Figures 5.15(a) and (b) show respectively, the measured and calculated absorption spectra in

**Figure 5.15** Experimental (a) and calculated (b) absorption coefficient at 297K of p-type GaAs at various hole concentrations [40].

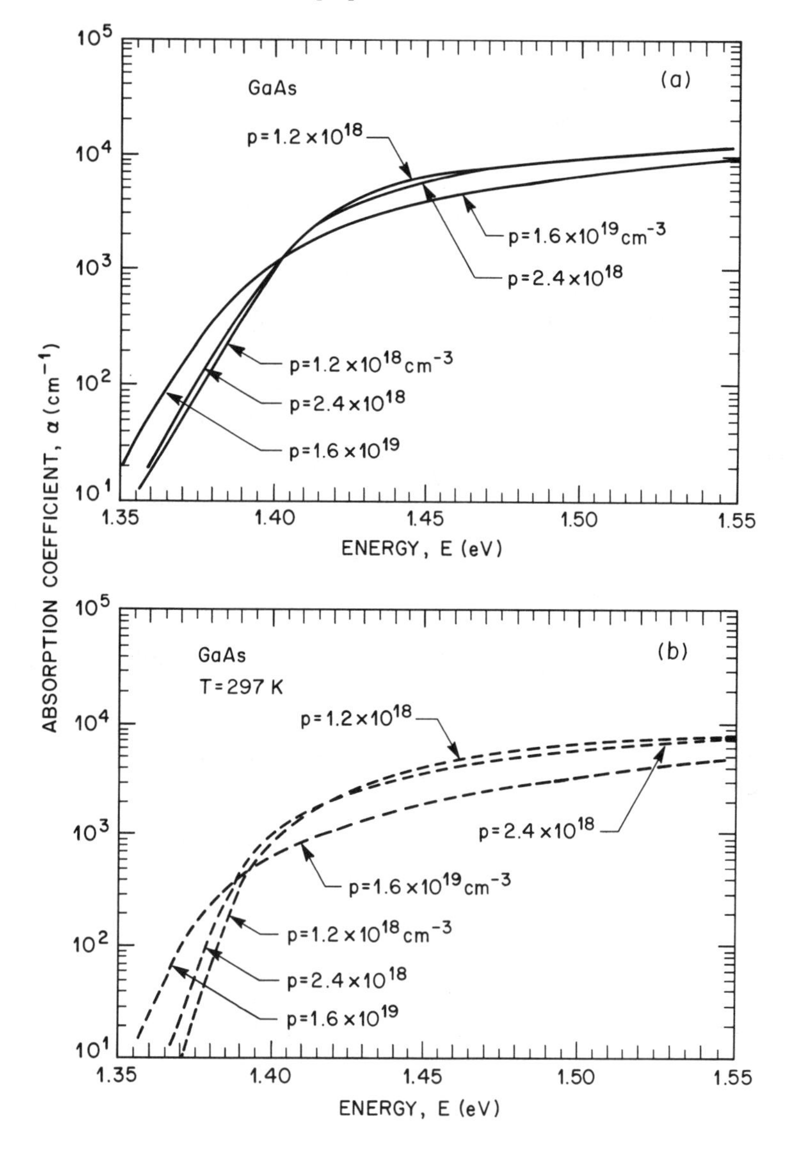

p-GaAs with carrier concentrations $1.2\times10^{18}\text{cm}^{-3}$, $2.4\times10^{18}\text{cm}^{-3}$, and $1.6\times10^{19}\text{cm}^{-3}$. In calculating the absorption spectra Casey and Stern [40] have included concentration dependent density of states and energy dependent matrix element.

## 5.2 PHOTOLUMINESCENCE

Of all optical techniques photoluminescence (PL) is perhaps the most widely used technique in characterizing III-V semiconductors and their alloys. This is because PL is a simple and an elegant technique that provides a wealth of information on the material for minimum investment in equipment and time. PL is employed both to understand the fundamental recombination process in the semiconductor as well as to characterize it for its "quality." In this latter role, PL has become a valuable nondestructive technique for routine characterization of GaAs, InP and the alloy semiconductors based on them. PL is used to study bulk or epitaxial films for their inter- and intra wafer variation and to compare material grown by different techniques. Information obtained by PL on the optical quality of the semiconductor becomes relevant when it is going to be used to make an opto-electronic device. In this section we discuss the fundamental process pertaining to PL, the experimental details of the measurement, and the type of information obtained with specific examples from GaAs, InP, and related material.

In Sec. 5.1 we dealt with the absorption processes in a semiconductor. The different transitions wherein an electron goes from a filled to an empty state by the absorption of photons of appropriate energy can, in principle, occur in the reverse direction also. The return of the electron to its ground state, or the recombination of the electron with a hole, can give rise to a photon. That is, the recombination process can occur radiatively.

Radiative recombination takes place when the semiconductor is under a nonequilibrium condition. That is, excess electron-hole pairs are created in the material by some external excitation to cause departure from equilibrium and the return to equilibrium occurs by the recombination of the excess carriers. When the excitation is achieved by optical means the resulting radiative recombination is photoluminescence. Excitation by current produces electroluminescence and excitation by an electron beam gives cathodoluminescence. Apart from the difference in the excitation process, the principles underlying each luminescence are the same.

The electron-hole recombination can also occur without the release of a photon, that is, nonradiatively. The nonradiative recombination competes with the radiative process and together they determine the recombination kinetics. In the application of the semiconductor as an optical device one is interested in reducing the nonradiative recombination relative to the radiative recombination.

### 5.2.1 Radiative Recombination in Semiconductors

After electron-hole pairs in excess of their equilibrium number are created in the crystal by means of one of the external sources discussed, they relax rapidly to states near the minimum of the conduction band in the case of electrons and near the maximum of the valence band in the case of holes. The carriers may be considered to be in quasi-thermal equilibrium and their distribution can be represented by invoking quasi-Fermi levels. The excess carrier density decays by recombination via one of several means as illustrated in Fig. 5.16 [41, 42]. There is band-to-band (B-B) recombination of an electron in the conduction band with a hole in the valence band (Fig. 5.16(a)). Recombination also occurs via shallow impurities. Ionized donors

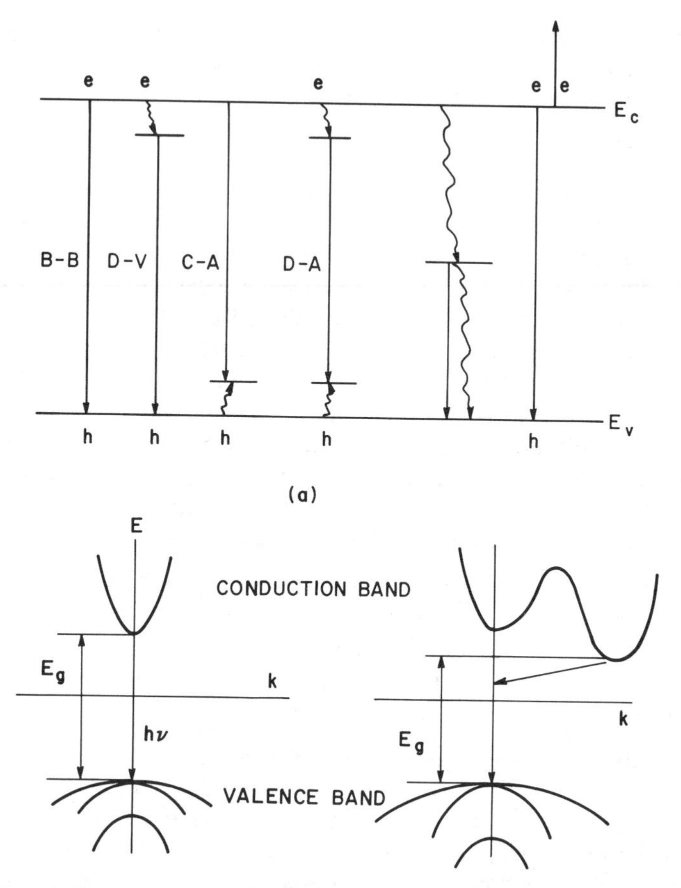


**Figure 5.16** Schematic band diagram showing the various electron-hole recombination paths (a). E versus k relation showing band to band electron-hole recombination in a direct band gap and an indirect band gap semiconductor (b).

and acceptors have a large capture cross-section for the photoexcited electrons and holes, respectively. If the captured electron recombines with a free hole in the valence band, a donor-to-valence (D-V) recombination ensues. Similarly recombination between the captured hole with a free electron in the conduction band gives conduction band-to-acceptor (C-A) transition. Both D-V and C-A transitions are referred to as free-to-bound (F-B) transitions. There is also donor-to-acceptor (D-A) pair transition where two bound particles recombine with each other. These transitions are radiative type. In contrast to these transitions recombination via a deep center is likely to be nonradiative. Also shown in Fig. 5.16 is another nonradiative process, the band-to-band Auger recombination process which involves three particles. In the example shown, the energy released by an electron-hole recombination kicks a third electron deep in the conduction band. These different decay paths will be discussed in latter sections.

Also shown in Fig. 5.16 are the recombination in direct and indirect semiconductors. As in the absorption process, momentum has to be conserved in the recombination process also. Since the photon has very small momentum, for direct transitions $k_e = k_h$. On the other hand, indirect transitions require phonon cooperation to conserve momentum in the transition; $k_e = k_h \pm q$, where q is the phonon wave vector. As a consequence indirect transitions have lower transition probability than direct transitions.

To calculate the radiative recombination rate in a direct band gap semiconductor under equilibrium, without recourse to detailed quantum mechanical calculations [43], the principle of detailed balance [44] can be used. The principle of detailed balance relates emission and the inverse process of absorption and from the experimentally measured absorption spectrum the shape of the emission spectrum can be calculated. From the principle of detailed balance it follows that the rate of radiative recombination at a frequency $\nu$ in an interval $d\nu$ at thermal equilibrium is equal to the rate at which electron-hole pairs are created. Thus R, the recombination rate, is given by [44]

$$R(\nu)\,d\nu = N(\nu)\,P(\nu)\,d\nu \tag{5.21}$$

where $N(\nu)\,d\nu$ is the density of photons of frequency $\nu$ in an interval $d\nu$ and $P(\nu)$ is the probability per unit time that a photon of frequency $\nu$ is absorbed. $N(\nu)\,d\nu$ is given by the well known Planck's formula

$$N(\nu)\,d\nu = \frac{8\pi\,\nu^2 [n^2 \dfrac{d(n\nu)}{d\nu}]\,d\nu}{c^3\,[\exp\dfrac{h\nu}{kT} - 1]} \tag{5.22}$$

where n is the refractive index and c is the speed of light. The probability of photon absorption is simply the inverse of its mean life time, $\tau$.

$$P(\nu) = \frac{1}{\tau(\nu)} \tag{5.23}$$

For an absorption coefficient $\alpha(\nu)$, the mean free path of a photon, before recombination is $\alpha^{-1}$. If $v_g$ is the group velocity of the photon, then $\tau(\nu) = (\alpha v_g)^{-1}$. Substituting for $v_g$ which is $c\,d\nu/d(n\nu)$ we have

$$P(\nu) = \alpha\,c\,\frac{d\nu}{d\,(n\nu)} \tag{5.24}$$

From Eqs. 5.21 to 5.24 it follows

$$R(\nu)\,d\nu = \frac{8\pi\,\nu^2\,n^2\,\alpha(\nu)}{c^2\left[\exp\dfrac{h\nu}{kT} - 1\right]}\,d\nu \tag{5.25}$$

The total radiative recombination rate R is obtained by integrating Eq. (5.25) over all values ν. Making a change of variable $u = h\nu/kT$, Eq. (5.25) gives

$$R = \frac{8\pi n^2 (kT)^3}{c^2 h^3} \int_0^\infty \frac{\alpha(\nu)\, u^2}{\exp(u) - 1}\, du \tag{5.26}$$

Equation (5.26) relates $\alpha(\nu)$ and the luminescence spectrum and thus from the experimentally obtained $\alpha(\nu)$ and knowing n, one can derive the emission spectrum. This is illustrated in Fig. 5.17, where the experimental absorption spectrum and the calculated emission spectrum using Eq. (5.25) for a p-type GaAs($p \sim 1.2 \times 10^{18}\,\mathrm{cm}^{-3}$) are shown [40]. In Eq. (5.26) it can be seen that the value of the integral is small both for high ν, since exp(u) in the denominator increases, and for low ν, since $\alpha(\nu)$ decreases. In other words, the emission spectrum corresponding to direct recombination will span a narrow frequency range corresponding to that near the absorption edge. Equation (5.26) is applicable to derive the recombination spectrum between any two states.

We noted that Eq. (5.26) is derived for thermal equilibrium conditions. We now calculate R under nonequilibrium conditions when excess carriers are generated by external means. The nonequilibrium recombination rate R′ is proportional to the concentration of the electrons (n) and holes (p). Thus

$$R' = B\, np \tag{5.27}$$

where B is the probability for radiative recombination. Equation (5.27) for equilibrium conditions becomes

$$R_o = B\, n_o p_o = B n_i^2 \tag{5.28}$$

where $n_o$ and $p_o$ are the concentration of electrons and holes at equilibrium and $n_i$ is the intrinsic concentration. From Eqs. (5.27) and (5.28)

$$R' = \frac{R_o}{n_i^2}\, np \tag{5.29}$$

If we write $n = n_o + \Delta n$ and $p = p_o + \Delta p$ where $\Delta n$ and $\Delta p$ are the excess values under non-equilibrium, then Eq. (5.29) becomes

$$R' = \frac{R_o}{n_i^2}\, (n_o + \Delta n)\, (p_o + \Delta p) \tag{5.30}$$

Neglecting the product $\Delta n\, \Delta p$, which is true at low levels of excitation, and after minor algebra Eq. (5.30) becomes

$$\frac{R' - R_o}{R_o} = \frac{\Delta R}{R_o} = \frac{\Delta n p_o + \Delta p n_o}{n_o p_o} \tag{5.31}$$

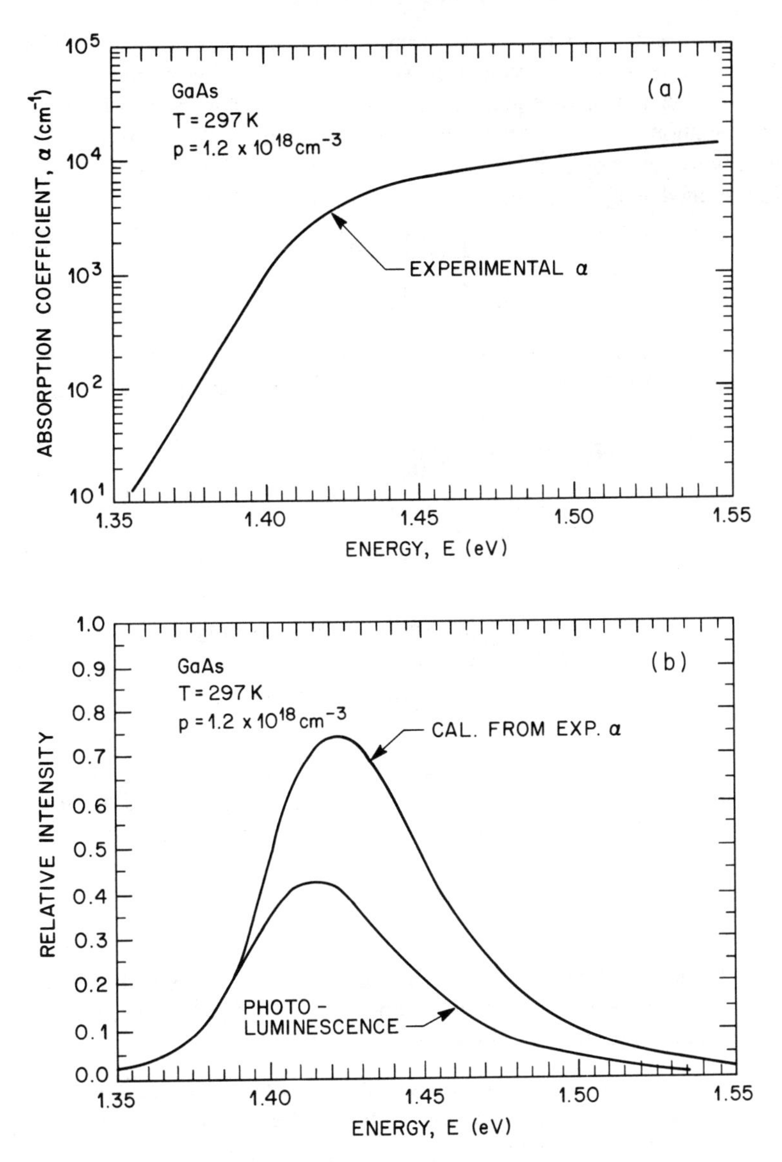


**Figure 5.17** Experimental absorption spectrum (a) and the calculated emission spectrum (b) for a p-type GaAs at 297K [40]. The experimental emission spectrum is also shown.

### 5.2.2 Minority Carrier Lifetime and Radiative Efficiency

When carriers in excess of their equilibrium value are generated in the semiconductor by some excitation source, after the excitation is removed, the system returns to equilibrium in certain time $\tau$. If the return to equilibrium is achieved predominantly by an increase in the radiative recombination rate, then $\tau \approx \tau_r$, the radiative lifetime. However, carrier decay can also occur nonradiatively as shown in Fig. 5.16 and each of the nonradiative path will have a characteristic $\tau$ defined as $\tau_{nr}$. The time $\tau$ is now given by

$$\frac{1}{\tau} = \frac{1}{\tau_r} + \sum_i \frac{1}{\tau_{nr}} \tag{5.32}$$

where the summation includes all nonradiative lifetimes. By definition $\tau_r = \Delta n/\Delta R$. From Eq. (5.31) and taking $\Delta n = \Delta p$, we obtain, respectively, $\tau_r$ for n- and p- type material

$$\begin{aligned} \tau_r \text{ (n–type)} &= \frac{n_o p_o}{R_o(n_o + p_o)} \approx \frac{p_o}{R_o} = \frac{1}{B n_o} \\ \tau_r \text{ (p–type)} &= \frac{n_o p_o}{R_o(n_o + p_o)} \approx \frac{n_o}{R_o} = \frac{1}{B p_o} \end{aligned} \tag{5.33}$$

Note in Eq. (5.31) that $\Delta R$ is inversely proportional to $p_o$ for n-type material and $n_o$ for p-type material; that is, the minority carrier density in each case. Similarly for other recombination paths also $\tau$ is decided by the minority carriers. For this reason $\tau$ is referred to as the minority carrier life time.

In deriving Eq. (5.33) we have assumed that $\Delta n, \Delta p \ll n_o, p_o$. This is satisfied for low levels of excitation conditions and is referred to as the small-signal approximation. However when $\Delta n, \Delta p \gg n_o, p_o$, a situation realized under high excitation conditions, $\tau_r$ varies as

$$\tau_r \approx \frac{1}{B\Delta n} \tag{5.34}$$

That is, $\tau_r$ is no longer constant as indicated by Eq. (5.33) but varies with $\Delta n$ as n decays to its equilibrium value. Thus $\tau_r$ assumes an instantaneous value and it corresponds to $\Delta R = B(\Delta n)^2$ which is referred as the bimolecular recombination region.

Given that excess carrier decay occurs via both radiative and nonradiative process, the efficiency of the radiative process, $\eta$, can be written as the ratio of the radiative recombination rate and total recombination rate. Thus

$$\eta = \frac{R_r}{R_T} = \frac{\frac{1}{\tau_r}}{\frac{1}{\tau}} = \frac{1}{1 + \frac{\tau_r}{\tau_{nr}}} \tag{5.35}$$

It is readily seen from Eq. (5.35) that a long $\tau_{nr}$ such that $\tau_{nr} \gg \tau_r$ results in high $\eta$.

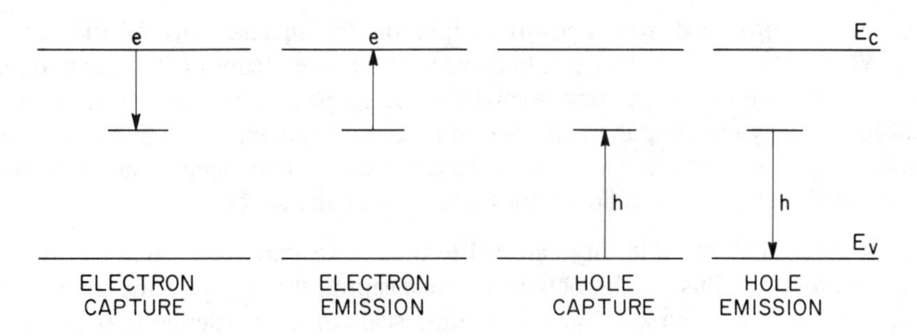


**Figure 5.18** Schematic band diagram illustrating capture and emission of carriers from a deep energy level.

We noted earlier that there are several processes which contribute to $\tau_{nr}$. Recombination via deep centers (Fig. 5.16) is generally a nonradiative process. Hall [45] and Shockley and Read [46] have worked out the kinetics of recombination through a deep level. The basic processes involved in the recombination are illustrated in Fig. 5.18. The recombination rate is given by

$$R = \frac{C_n C_p \,(np - n_i^2)}{C_n\,(n + n_1) + C_p\,(p + p_1)} \tag{5.36}$$

where $C_n$ is the rate at which an electron in the conduction band is captured into the level when it is empty, $C_p$ is the capture rate for a hole when the level is filled with the electron, and $n_1$ $(p_1)$ is the electron (hole) concentration in the conduction (valence) band when the Fermi level coincides with the deep level $E_t$. From Eq. (5.36) it can be shown that under low level of excitation for n-type material with $n_o \gg p_o$ and also $n_o \gg n_1$

$$\frac{1}{\tau_p} = C_p = N_t\,\sigma_p\,v_{th} \tag{5.37a}$$

and for p-type material with $p_o \gg n_o$ and $p_o \gg p_1$

$$\frac{1}{\tau_n} = C_n = N_t\,\sigma_n\,v_{th} \tag{5.37b}$$

where $N_t$ is the density of traps and $v_{th}$ is the thermal velocity of the holes or the electrons. $\sigma_{p(n)}$ are known as the capture cross-sections.

### 5.2.3 Experimental Techniques

When electron-hole pairs are created in the semiconductor by exciting it optically, the resulting radiative recombination is called photoluminescence. Therefore to study photoluminescence one needs a source of light of energy greater than the band gap of the semiconductor so as to excite electrons across the energy gap. One then needs equipment to collect, disperse and detect the luminescence from the sample. Additionally, the environment of the semiconductor such as its ambient temperature, or the presence of a magnetic field [47], electric field [48], hydrostatic pressure [49], uniaxial stress [50], and so on can alter the luminescence in a profound way and shed light on the mechanisms of the recombination processes. When photoluminescence has to be used as a process control technique to monitor the quality of a wafer, one has to perform spatially resolved photoluminescence [51]. This can be accomplished either by rastering the excitation source on the sample [51 - 53] or by rastering the sample with respect to the source [54]. Also one can combine both spectral and spatial resolution to obtain the distribution of a specific luminescence line on the wafer [55].

We noted earlier that the minority carrier life time is an important parameter describing the recombination process. This can be obtained by measuring the decay of the luminescence after terminating the excitation source [56]. As the time scale of the particular optical process under investigation decreases, the experimental complexity increases. Apart from the decay time, with repetitive excitation pulse the time-resolved spectrum can be measured after a certain time delay [57]. All the above experiments, from the most simple to the most complex, have been and are being performed to study GaAs, InP, and related materials. Many of the excellent reviews on photoluminescence characterization of semiconductors and the texts by Pankove [2], Berg and Dean [58], and Williams and Hall [59] would provide further details of the experimental techniques. The article by Hamilton et al., [60] gives a general introduction to luminescence instrumentation.

Figure 5.19 shows schematically the equipments and their layout for performing photoluminescence measurement in GaAs, InP, and their alloys. Table 5.1 lists the light sources, spectrometers, and detectors used to cover the different spectral regions. Calibrated neutral density filters (Fig. 5.19) are used to vary the excitation intensity on the sample. A mechanical chopper typically at a frequency of 300 to 500 Hz is used if lock-in detection of the luminescence signal is employed. Alternatively one can use photon counting techniques [61]. The exciting light, typically a laser, can be focussed on the sample to a diffraction limited spot size (~2 μm) which combined with x-y motion on the sample provides high spatial resolution. Instead of rocking the sample, the laser beam can be rastered on the sample with scanning mirrors. The sample can be kept in a dewar with suitable windows which have optimum transmission in the spectral range investigated to permit PL measurement as a function of temperature in the range 2 to 300K. The dewar can be kept between the poles of a magnet so that magneto-optical studies can be made. Also fixtures to provide hydrostatic or uniaxial stress on the sample can be incorporated into the dewar for piezo-optical measurements. The use of tunable light source permits photoluminescence excitation spectroscopy [62] which is an useful complementary technique to PL.

The luminescence from the sample has to be collected and focussed onto the slit of the spectrometer for dispersion and detection. It can also simultaneously be focussed onto a camera for quickly assessing the uniformity of the emission and thus facilitating PL topography. Appropriate combination of spectrometers, gratings, and detectors helps to achieve desired resolution as well as to cover the different spectral ranges.

An important application of PL is in the characterization of GaAs and InP based opto-electronic structures. In the analysis of laser structures, for demonstrating lasing action, optical pumping is a convenient technique. Even for routine characterization of laser structures

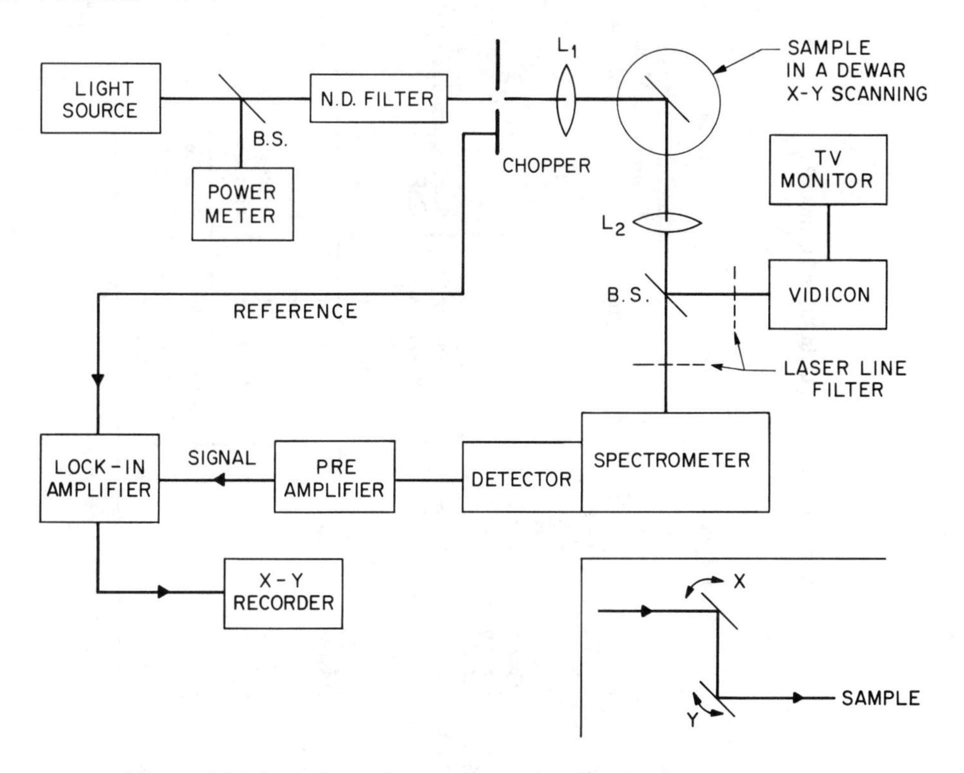


**Figure 5.19** Experimental arrangement for photoluminescence measurement. The inset shows the use of scanning mirrors to raster the exciting laser beam on the sample for spatially resolved photoluminescence.

for quality, in terms of lasing wavelength, lasing threshold, and so on, optical pumping provides much faster feedback than the conventional electrical pumping which involves several processing steps. In this regard optical pumping has become a standard experimental technique. Figure 5.20 shows the schematics of an optical pumping experiment [63]. Figures 5.21 and 5.22 show the correlation between optical pumping and electrical pumping of AlGaAs [63] and GaInAsP [64] laser structures, respectively.

### 5.2.4 Carrier Generation and Diffusion

When a semiconductor is excited by light of energy greater than the band gap, electron-hole pairs are created only near the surface of the sample depending upon the absorption coefficient of the semiconductor. That is, there will be an inhomogeneous carrier distribution near the surface of the semiconductor. With excess carrier concentration near the surface compared to the bulk, diffusion of carriers occurs to reduce the concentration gradients. The steady state distribution of excess carriers is determined by the minority carrier lifetime. The photon that is generated inside the sample by the recombination of the electron-hole pair can in turn get re-absorbed to create

**Table 5.1**
Light source, spectrometers and detectors suitable for photoluminescence measurements in GaAs, InP, GaInAs, and GaInAsP.

| Material | Spectral Range (nm) | Spectrometer | Light Source | Detector | Remarks |
|---|---|---|---|---|---|
| GaAs<br>Kr | 800-820<br>(Excitonic lines)<br>820-900<br>(Edge emission bands) | 1m or 3/4m spectrometer with 1200 groove/mm grating | 488, 514.5nm (Ar)<br>632.8nm (He-Ne) laser<br>647.1nm (Kr) lines<br>Tunable light source (e.g., dye laser) is desired for resonant excitation | GaAs<br>Photomultiplier | Quartz or borosilicate glass lenses and windows |
| | >900<br>(Deep center emission) | 1/2m spectrometer with 600 groove/mm or 300 groove/mm grating | Ar, He-Ne or Kr laser lines at 40-100mW | S1 photomultiplier<br>Ge photodiodes<br>InSb or PbS detectors | Sapphire windows<br>$CaF_2$ lenses |
| InP | 870-900<br>(Excitonic lines)<br>900-980<br>(Edge emission bands) | 1m or 3/4m spectrometer with 1200 groove/mm grating | Same as for GaAs | GaAs photomultiplier<br>S1 photomultiplier<br>Ge photodiode | Quartz or borosilicate glass lenses and windows |
| | >980<br>(Deep center emission) | 1/2m spectrometer with 600 groove/mm or 300 groove/mm grating | | Ge photodiode<br>PbS or InSb photodetectors | Sapphire windows<br>$CaF_2$ lenses |

**Table 5.1 (cont'd)**
Light source, spectrometers and detectors suitable for photoluminescence measurements in GaAs, InP, GaInAs, and GaInAsP.

| Material | Spectral Range (nm) | Spectrometer | Light Source | Detector | Remarks |
|---|---|---|---|---|---|
| GaInAs | 1600-1700 (Near band edge) | 1/2m spectrometer with 600 groove/mm grating | Ar, He-Ne Kr or Nd:Yag laser lines at 50-100 mW. | Ge photodiodes | Quartz or borosilicate glass lenses and windows |
| | >1700 | 1/2m spectrometer with 300 groove/mm grating | Tunable color center laser is desired for resonant excitation | PbS or InSb photodetectors | Sapphire windows $CaF_2$ lenses |
| GaInAsP 1.3μm or 1.55μm bandgap composition | Near band edge | 1/2m spectrometer with 600 groove/mm grating | Same as for GaInAs | Ge photodiode | Quartz or borosilicate glass lenses and windows |
| | Deep center emission bands | 1/2m spectrometer with 300 groove/mm grating | | PbS or InSb photodetector | Sapphire windows $CaF_2$ lenses |
| AlGaAs | Near band edge | 1m or 3/4m 1800 or 1200 groove/mm grating | Ar, He-Ne or Kr laser lines at 50-100mW. Tunable light source (e.g., dye laser) is desired for resonant excitation | GaAs photomultiplier | Quartz or borosilicate glass lenses and windows |
| | Deep center emission bands | 1200 groove/mm grating | | S1 photomultiplier | |

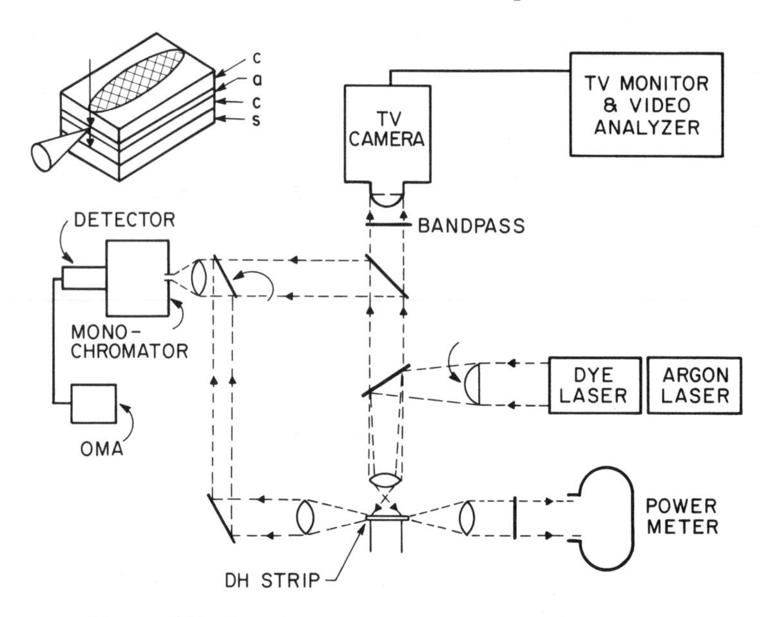


**Figure 5.20** Experimental arrangement for photo-pumping a laser structure shown in the inset where c, a, s correspond to the cladding layer, active layer and substrate, respectively [63].

**Figure 5.21** Device threshold current density versus optical current density (predicted from photo-pumping experiment). Devices which do not lase electrically cannot be photo-pumped to lase either. Also shown is the electrical lasing wavelength versus optical (predicted) lasing wavelength [63]. Both the data correspond to a $Al_xGa_{1-x}As$ laser structure.

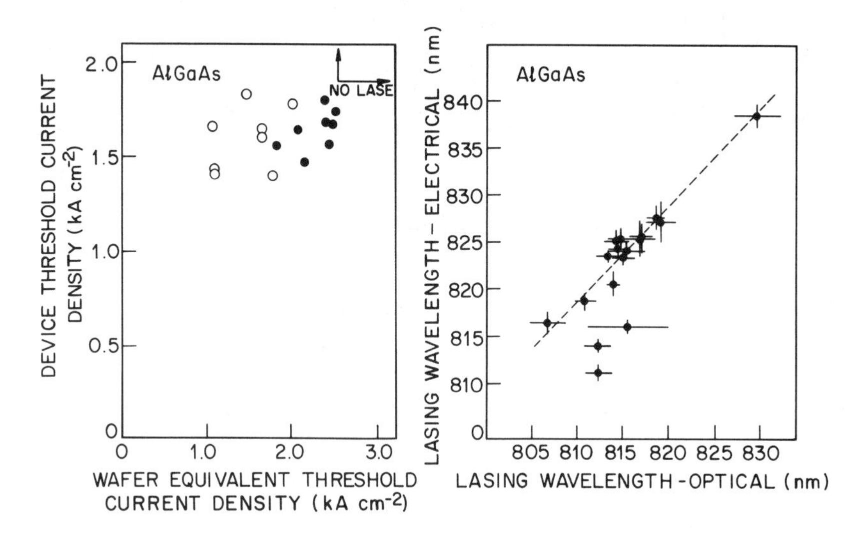

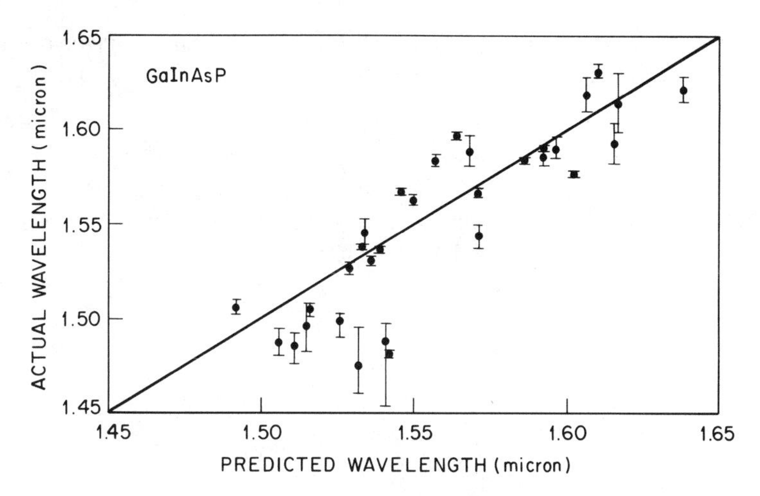


**Figure 5.22** Electrical lasing wavelength versus optical (predicted) lasing wavelength for a GaInAsP laser structure [64].

another pair or leave the sample. There is yet one another factor, namely, the reflection of the exciting or the recombination radiation. In order to relate the recombination rate calculated in Eq. (5.25) to the luminescence measured outside the sample, the carrier generation, carrier diffusion, recombination, re-absorption of the recombination radiation and reflection have to be analyzed in detail.

### Generation of carriers

Let a photon flux of $\phi$ photons per second of energy $h\nu > E_g$ be incident on an unit area of the sample. Each photon creates an electron-hole pair and the photon flux varies as

$$\phi = \phi(o) \exp(-\alpha x) \tag{5.38}$$

The generation rate G will be given by the photon removal rate

$$G = -\frac{dN}{dt} = \frac{dn}{dt} = \frac{dp}{dt} \tag{5.39}$$

where N is the photon density such that $\phi = v_g N$ and $x = v_g t$. The value of the flux incident on the sample at $x = 0$ is also reduced by the reflectivity, $R$, so that

$$G = -\frac{dN}{dt} = (1 - R)\, \alpha\, \phi(o) \exp(-\alpha x) \tag{5.40}$$

### Diffusion of carriers

Photoexcitation creates electron-hole pairs in excess of their equilibrium concentration near the surface of the semiconductors. Since a concentration gradient is set up, carriers will diffuse into the semiconductor to reduce the gradient. The diffusion of carriers gives rise to a diffusion current, J, which is related to the concentration gradient. For electrons, J is given by

$$J_e = D_e q \frac{dn}{dx} \tag{5.41}$$

where D is the diffusion constant and dn/dx is the gradient. The continuity equation imposes a condition that there should be no accumulation of charge. It is written as

$$\nabla J_e = -q \frac{\partial n(x)}{\partial t} \tag{5.42}$$

Including the generation (G) and recombination (R) rates Eq. (5.42) becomes

$$\frac{\partial n}{\partial t} = G_n - R_n + \frac{1}{q}\frac{dJ_e}{dx} \tag{5.43}$$

In order to solve Eq. (5.43) one assumes a semi-infinite solid and that R can be written in terms of the minority carrier lifetime as

$$R = \frac{\Delta n}{\tau} \tag{5.44}$$

substituting Eqs. (5.41) and (5.44) in (5.43) one gets

$$\frac{\partial n}{\partial t} = G_n - \frac{\Delta n}{\tau} + D_e \frac{d^2 n}{dx^2} \tag{5.45}$$

At steady state $\partial n/\partial t = \partial p/\partial t = 0$ which then gives the desired solution to Eq. (5.45). If we assume $G_n = 0$, then the solution of

$$\frac{d^2 n}{dx^2} = \frac{\Delta n}{D\tau} \tag{5.46}$$

gives $\Delta n = \Delta n(o) \exp(-x/L_e)$ where $L_e = (D_e \tau)^{1/2}$ is defined as the diffusion length of electrons; $\Delta n$ decays to 1/e of $\Delta n(o)$ when $x = L_e$. The Einstein relation $qD_e = \mu_e kT$ relates $D_e$ and $\mu_e$, the mobility of electrons. The general solution of Eq. (5.45) with $\frac{\partial n}{\partial t} = 0$ is

$$\Delta n = A \exp \frac{x}{L_e} + B \exp\left(-\frac{x}{L_e}\right) - \frac{G}{D_e(\alpha^2 - {}^1/L_e^2)} \tag{5.47}$$

The constants A and B are evaluated by the boundary conditions

$$D_e \nabla n = Sn \quad \text{at} \quad x = 0 \tag{5.48a}$$

$$n = 0 \quad \text{at} \quad x = \infty \tag{5.48b}$$

where S is the surface recombination velocity which characterizes the recombination rate at the surface. Williams and Chapman [65] have shown that for these boundary conditions Eq. (5.47) becomes

$$\Delta n = \frac{(1 - R)\,\alpha\phi(o)\,L_e^2}{D_e\,(1 - \alpha^2 L_e^2)} \left[ \exp(\alpha x) - \left\{ \frac{\alpha + \frac{S}{D}}{\frac{1}{L_e} + \frac{S}{D}} \right\} \exp \frac{-x}{L_e} \right] \tag{5.49}$$

The term $(1-R)\,\alpha\phi(o)$ is G at $x = 0$ (Eq. 5.40). Figure 5.23 shows the generation rate and the excess carrier concentration as a function of x for the cases $S = 0$ and $S = \infty$. Note that for $S = \infty$, $\Delta n$ reaches its maximum not at $x = 0$ as in the case for $S = 0$ but at a value of x near $L_e$.

The luminescence generated inside the sample can be reabsorbed in the sample and can suffer reflection at the surface before it can be measured outside. Williams and Chapman [65] have shown that the PL intensity $I_i$ caused by $i^{th}$ radiative recombination process is given by

$$I_i = (1 - R'(\theta))\,\varepsilon_i \int_0^\infty \frac{\Delta n}{\tau_i} \exp(-\beta_i x)\,dx \ \text{(photons/cm}^2\text{/sec)} \tag{5.50}$$

where $\tau_i$ is the life time for the recombination process, $\beta_i$ is the absorption coefficient for this luminescence, $R'$ is the reflectivity for the luminescence emitted from the sample at angle $\theta$ and $\varepsilon_i$ includes the solid angle factor and the detector efficiency. Substituting Eq. (5.49) in Eq. (5.50) and taking $\tau_i$ as constant (valid under small-signal approximation) $I_i$ is obtained as

$$I_i = \frac{(1 - R'(\theta))\,\varepsilon_i\,G_{x=0}}{\tau_i\,D_e} \; \frac{\frac{S}{D} + \frac{1}{L_e} + \alpha + \beta_i}{\left(\frac{S}{D} + \frac{1}{L_e}\right)(\alpha + \beta_i)\left(\alpha + \frac{1}{L_e}\right)\left(\beta_i + \frac{1}{L_e}\right)} \tag{5.51}$$

where $G_{x=0}$ is $(1-R)\,\alpha\,\phi(o)$. In this equation, $\tau_i$, $D_e$, and $L_e$ correspond to the minority carriers which we have taken to be electrons. If the material is n-type, then the minority carriers are holes and accordingly $D_h$, $L_h$, and $\tau_h$ have to be used. A generalized form of Eq. (5.51) corresponding to spectral photon flux (photons/cm$^2$/sec/eV) has been given by Bebb and Williams [43]. Table 5.2 lists the absorption coefficients at select excitation laser wavelengths and diffusion lengths of electron and hole at room temperature for GaAs, InP, AlGaAs/GaAs, GaInAs/InP, and GaInAsP/InP.

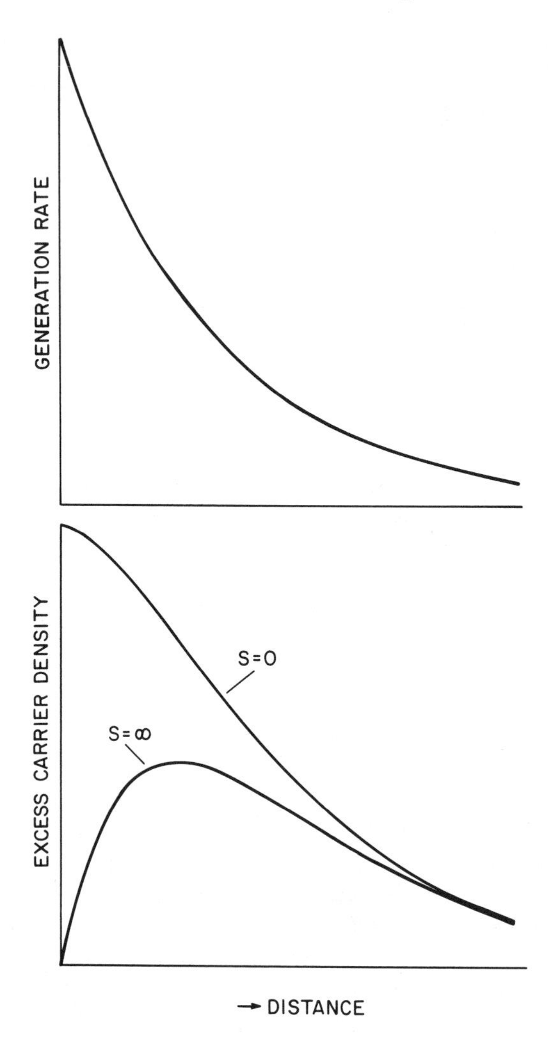


**Figure 5.23** Generation rate (a) and excess carrier concentration (b) as a function of distance for the cases $S=0$ and $S=\infty$ where S is the surface recombination velocity.

In using Eq. (5.51) several simplifications can be made depending on the recombination process. For example, if $I_i$ corresponds to band-to-impurity transition (D-V and C-A transitions in Fig. 5.16) and occurs below $E_g$, then $\beta_i$ will be small compared to $\alpha$ such that $\beta_i \ll \alpha$ in Eq. (5.51). On the other hand, for band-to-band recombination $\alpha \approx \beta_i$. Williams and Chapman [65] have made use of these assumptions in analyzing the temperature dependence of band-to-band and band-to- acceptor transitions in GaAs.

It should be emphasized again that Eq. (5.51) has been derived under the assumption that $\tau_i$ is constant, which is true only under low levels of excitation such that the injected carrier density is smaller than the equilibrium majority carrier density. As noted in Eq. (5.34) under high excitation conditions $\tau$ is no longer constant but becomes a function of the injected carrier density. Bebb and Williams [43] have given general expressions for $\tau$ for band-to-band and impurity-band recombinations which are appropriate for direct-gap III-V semiconductor. Simplified expressions applicable for small signal approximation have been given by Moss et al., [66]. These expressions have been used by Garbuzov et al., [67] to analyze the temperature dependence of the recombination life-times in GaAs.

### 5.2.5 Exciton Recombination

The band-to-band recombination process is generally observed in GaAs, InP and their alloys typically in the temperature range 75 to 300K. At low temperatures, for example, near liquid He temperature where PL measurements are generally made, and at low excitation intensities, the injected electrons and holes can combine to form free excitons or the electron-hole pair can be bound to an impurity to form bound excitons. Recombination then proceeds by the annihilation of these excitons rather than by band-to-band recombination.

#### Free excitons

In a direct band gap semiconductor when recombination occurs by the annihilation of a free exciton the transition energy is

$$h\nu = E_g - E_x \tag{5.52}$$

where $E_x$ is the binding energy of the exciton given by [5]

$$E_x = \frac{\mu q^4}{2\hbar \varepsilon^2} \frac{1}{n^2} \tag{5.53}$$

where $\mu$ is the reduced mass, $1/\mu = 1/m_e + 1/m_h$, $\varepsilon$ is the dielectric constant, q is the electron charge and n is an integer taking values 1, 2, and so on. From Eq. (5.53) it follows that the free exciton luminescence is characterized by a series of sharp lines. Since the intensity of the lines decreases as $n^{-3}$ [5, 18] one generally observes only a few transitions. Figure 5.24 shows the n = 1, 2, and 3 exciton lines in high purity GaAs epitaxial layer [68].

Free excitons can travel through the crystal and hence have kinetic energy. However, for recombination radiation caused by annihilation of free excitons, momentum conservation would impose that the free excitons have nearly zero momentum since the photon has very small momentum. Since there is no kinetic energy broadening, free excitonic transitions are sharp except for life time broadening [69, 70] as a result of the short lifetime of the zero momentum excitons.

Excitons with nonzero kinetic energy can take part in radiative transitions with the cooperation of phonons which satisfy momentum selection rules. Thus, one can observe phonon

**Table 5.2**

Absorption coefficient ($cm^{-1}$) at selected laser lines and electron ($L_n$) and hole ($L_p$) diffusion lengths at room temperature for GaAs, InP, AlGaAs/GaAs, GaInAs/InP, and GaInAsP/InP.

| Material | Laser line (nm) | Absorption Coefficient $(cm^{-1}) \times 10^4$ | Diffusion Length (μm) | |
|---|---|---|---|---|
| | | | $L_p$ | $L_n$ |
| GaAs | 514.5<br>647.1 | 10[a]<br>4[a] | 2 ($n \sim 10^{17}\ cm^{-3}$)[b]<br>0.3 ($n \sim 6 \times 10^{18}\ cm^{-3}$)[b] | 10 ($p \sim 10^{16}\ cm^{-3}$)[b]<br>1 ($p \sim 10^{19}\ cm^{-3}$)[b] |
| $Al_x\ Ga_{1-x}\ As$<br>x~ 0.3<br>x~0.37<br>x~0.55 | <br>514.5<br>514.5<br>514.5 | <br>5[a]<br>4[a]<br>2[a] | 2.0 ($n \sim 10^{17}\ cm^{-3}$, x~0.2)[a1] | 2.0 ($p \sim 5 \times 10^{19}\ cm^{-3}$, x ~ 0.2)[a1]<br>1.5 ($p \sim 9 \times 10^{16}\ cm^{-3}$, x ~ 0.28)[a2] |
| InP | 514.5<br>647.1 | 10[e]<br>5.5[e] | 1.5-2 ($n \sim 10^{17}\ cm^{-3}$)[f][g]<br>12 ($n \sim 10^{16}\ cm^{-3}$)[h] | 8 ($p \sim 10^{15}\ cm^{-3}$)[i]<br>1 ($p \sim 10^{18}\ cm^{-3}$)[i] |
| $Ga_{0.47}\ In_{0.53}\ As$ | 647.1<br>1060.0 | 8[e]<br>2.5[e] | 2 ($n \sim 5 \times 10^{15}\ cm^{-3}$)[j]<br>0.4 ($n \sim 2 \times 10^{16}\ cm^{-3}$)[k] | 2.5 ($p \sim 1.4 \times 10^{16}\ cm^{-3}$)[l]<br>0.83 ($p \sim 5 \times 10^{18}\ cm^{-3}$)[l] |
| GaInAsP<br>(Band gap 1.3μm) | 647.1<br>1060.0 | 7.5[e]<br>2[e] | 1.5 (undoped material)† (l) | 3.0 ($p \sim 10^{16}\ cm^{-3}$)† (l)<br>1.0 ($p \sim 10^{18}\ cm^{-3}$)† (l)<br>0.5 ($p \sim 5 \times 10^{18}\ cm^{-3}$)† †(m)<br>1.75 ($p \sim 2 \times 10^{18}\ cm^{-3}$)† † † (m) |
| GaInAsP<br>(Band gap 1.55μm) | 647.1<br>1060.0 | 8[e]<br>2.7[e] | | |

(a) B. Monemar, K-K Shih, and G.D. Pettit, *J. Appl. Phys. 47*;2604, 1976.
(a1) C. Amano, A. Shibukawa, and M. Yamaguchi, *J. Appl. Phys. 58*;2780, 1985.
(a2) M.J. Ludowise and W.T. Dietze, *J. Appl. Phys. 55*;4318, 1984.
(b) H.C. Casey, Jr., B.I. Miller, and E. Pinkas, *J. Appl. Phys. 44*;1281, 1973.
(e) H. Burkhard, H.W. Dinges, and E. Kuphal, *J. Appl. Phys. 53*;655, 1982.
(f) S. Li, *Appl. Phys. Lett., 29*;126, 1976.
(g) I. Umebu, A.N.M.N. Choudhury, and P.N. Robson, *Appl. Phys. Lett., 36*;302, 1980.
(h) V. Diadiuk, S.H. Groves, C.A. Armiento, and C.E. Hurwitz, *Appl. Phys. Lett., 42*;892, 1983.
This large value has been reported for a LPE grown InP.
(i) C.L. Chiang, S. Wagner, and A.A. Ballman, *Mater. Lett., 1*;145, 1983.
(j) K. Alavi, A.N.M.N. Choudhury, J. Vleek, N.J. Slater, C.G. Fonstad, and A. Y. Cho, *IEEE Electron. Dev. Lett., EDL-3*;379,
(k) Y. Takeda, M. Kuzuhara, and A. Sasaki, *Jap. J. Appl. Phys. 19*;899, 1980.
(l) M.M. Tashima, L.W. Cook, and G.E. Stillman, *J. Electron. Mat. 11*;831, 1982.
(m) G.H. Olsen, and T.J. Zamerowski, *IEEE J. Quant. Electron. QE-17*;128, 1981.
† These values are for a quarternary composition corresponding to band gap 1.55μm.
† † For a VPE grown quarternary of band gap 1.23μm.
† † † For a VPE grown quarternary of band gap 1.19μm.

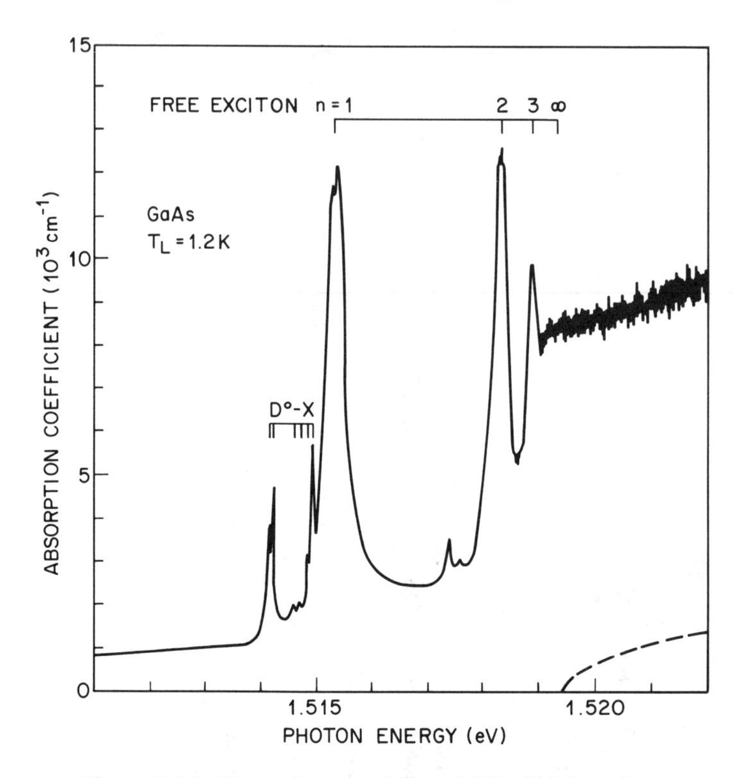


**Figure 5.24** Absorption spectrum at 1.2K of high purity GaAs near the band edge [68]. The n = 1, 2, 3 free exciton peaks are shown. Excitons bound to donors (D° − X) are also indicated. The rise at high energy is due to substrate absorption. The dashed line shows the $(E - E_g)^{1/2}$ behavior expected in the absence of electron-hole interacting (the absolute magnitude is chosen to fit the absorption far from the band edge).

replicas associated with the zero phonon line, [43]. The phonon replicas are broadened because of the kinetic energy of the excitons.

Because of the small binding energy of the free excitons, it is difficult to see them except at low temperatures. However, it turns out that excitonic or Coulomb effects can not be neglected even for optical transitions near room temperature. In fitting his absorption data from GaAs at 294K Sturge [19] found that Coulomb effects cannot be ignored. Nakashima et al., [71] measured the luminescence and photocurrent spectra of thin undoped GaAs epitaxial layer sandwiched between two $Al_xGa_{1-x}As$ layers at 77 and 300K. They found that at both temperatures exciton effects [18] together with lifetime broadening caused by acoustic phonon scattering [69, 72] have to be included to obtain good agreement between measured luminescence spectrum and the one calculated by principles of detailed balance [44].

The polarizing interaction between an exciton and a photon gives rise to a polariton [73]. This interaction is illustrated in Fig. 5.25 which is the polariton dispersion curve. The interaction between the particles is maximum near the point of intersection of the exciton and photon dispersion curves. Only the transverse exciton (polarization perpendicular to k) couples to the electromagnetic wave. The polariton above the knee behaves like an exciton and below it as a photon. When free excitons are formed, they can thermalize along the parabolic dispersion curve in Fig. 5.25 towards the knee at which point it can get scattered by an optical phonon and emit a photon. Polaritons have been seen in low temperature luminescence from GaAs [74].

**Bound excitons**

When impurities are present in the semiconductor, excitons can be bound to them analogous to the binding between two hydrogen atoms in the $H_2$ molecule. The complex, impurity and free exciton, is called a bound exciton (BE). Since the first experimental evidence of their presence in Si [75, 76] the study of bound excitons in both elemental and compound semiconductors has been a field of intense research [77, 78].

Radiative recombination through the annihilation of bound excitons is an efficient process. Since the exciton is generally bound only weakly to the impurity, its radius, $a_x$, is large. This means that the oscillator strength, proportional to $a_x^3$, is rather large [79]. Thus, the intensity of BE lines can be very large. Further, the bound exciton lines are narrow since no kinetic energy is

**Figure 5.25** E versus k dispersion curves for polaritons (dashed line) and free excitons and photon (solid lines).

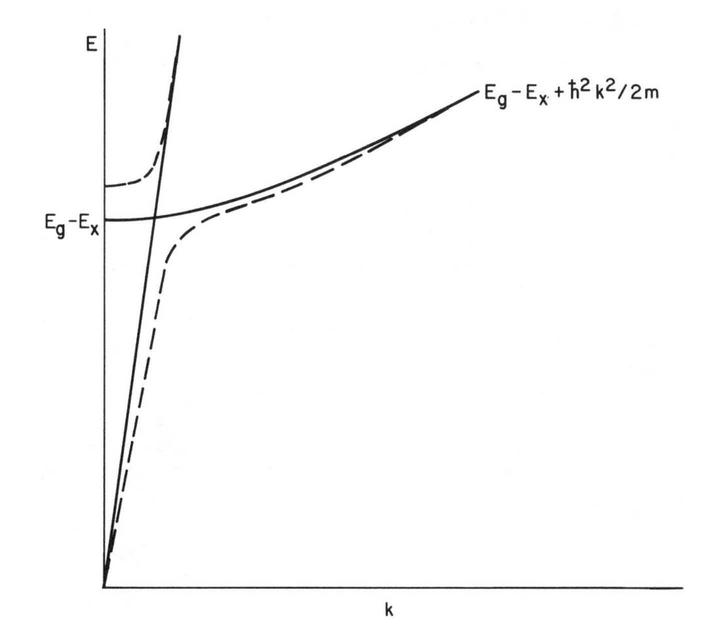

involved. The high intensity and narrow line width of the BE lines enable one to study their behavior under magnetic fields and stress which greatly elucidate the nature of the centers which bind the excitons.

The main reason why BE luminescence has been studied quite extensively in GaAs and InP is that it helps to identify the chemical nature of the impurities present in the material. It is generally found that the exciton localization energy to the impurity or the defect (obtained from the position of the BE line relative to the free exciton line in the spectrum) scales with the ionization energy of the center. The relationship first discovered in Si [76], is referred to as the Haynes' rule and is given as

$$E_{BX} = a + bE_i \tag{5.54}$$

where $a \neq 0$ in general and the value of b is different for donors and acceptors of ionization energy, $E_i$. However, in GaAs or InP the ionization energy between various donors and acceptors does not differ very much and the usefulness of Eq. (5.54) to identify the species is nil. In these cases one looks for the "two-electron" or "two-hole" transitions associated with the principal BE lines [77, 78]. These transitions occur when the BE recombination leaves the donor or the acceptor in one of its excited states as shown in Fig. 5.26 and they are shifted below the parent

**Figure 5.26** Relationship between the ground state and excited state binding energies for acceptors in GaAs including the theoretical datum T. The inset shows the experimentally observed transitions. The commonly observed two hole transition terminates in a 2S level [80].

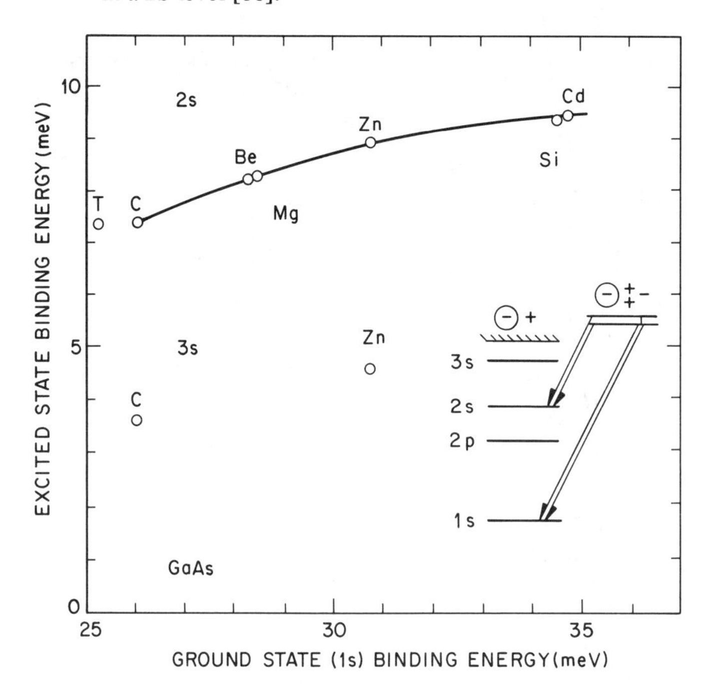

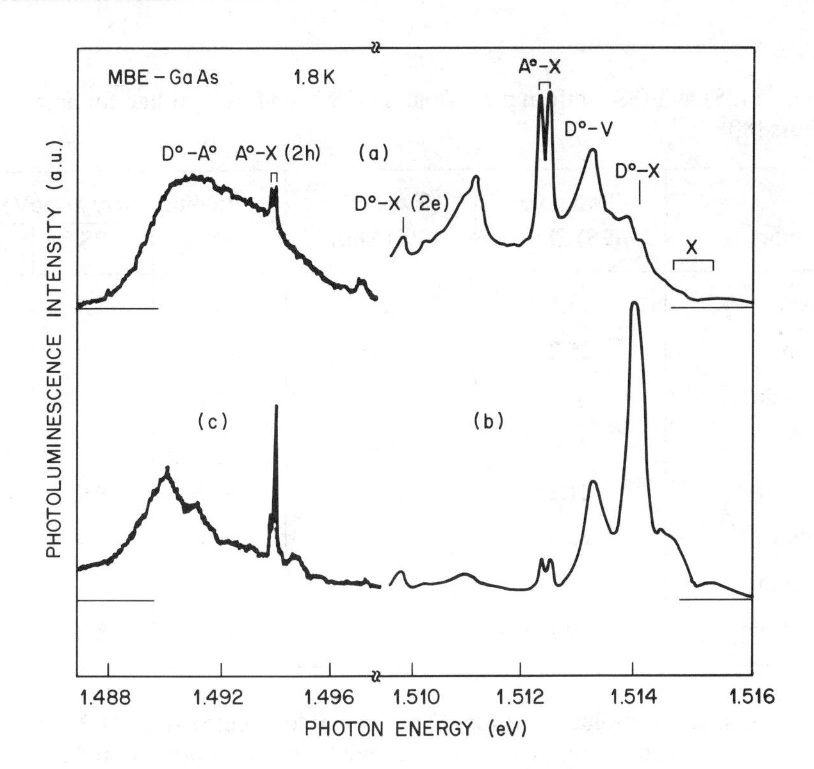


**Figure 5.27** Low temperatures photoluminescence spectra of MBE grown n-type GaAs under dye laser excitations. (a) under focussed non-resonant excitation at 1.63 eV. (b) under defocused non-resonant excitation at 1.63 eV. (c) under defocused resonant with the higher energy component of the $A^\circ-X$ doublet. $A^\circ-X$ and $D^\circ-X$ denote, respectively, acceptor bound and donor bound exciton transitions. The two hole $A^\circ-X$ (2h) and two electron $D^\circ-X$(2e) transitions associated with $A^\circ-X$ and $D^\circ-X$ are indicated. X, $D^\circ-V$, and $D^\circ-A^\circ$ denote free exciton, donor-to-valence band and donor-to-acceptor transitions, respectively [81].

BE lines by just the excitation energy of the level; the energy of the 2S level for the example shown in Fig. 5.26. The energy shift between the principal BE line and the first of the two-electron or two-hole replicas is a strong function of the chemical nature of the impurity since it carries most of the chemical shift or the central cell correction associated with the ground state [77, 78]. Thus, the two-hole transitions have served very well to identify the different acceptors in GaAs [80]. Figure 5.27 shows the principal BE line and the two-hole replicas of substitutional C acceptor in high purity GaAs layer grown by molecular beam epitaxy [81]. The doublet feature of the acceptor BE line and its two-hole replicas are due to j-j coupling between the holes in the excited state of the transition [80]. A catalogue of the two-hole energies for the different acceptors in GaAs is given in Table 5.3. The experimental requirements to observe the two-hole

**Table 5.3**

Two hole (2S) and (3S) shift in meV from the 1S bound exciton line for acceptors in GaAs [80].

| Acceptor | Two hole (2S) Shift | Two hole (3S) Shift | Binding Energy (meV) | | |
|---|---|---|---|---|---|
| | | | 1S | 2S | 3S |
| Carbon | 18.5 | 22.4 | 26.0 | 7.4 | 3.6 |
| Silicon | 25.2 | — | 34.5 | 9.3 | — |
| Germanium | — | — | 40.4 | — | — |
| Tin | — | — | 171 | — | — |
| Zinc | 21.8 | 26.1 | 30.7 | 8.9 | 4.6 |
| Cadmium | 25.3 | — | 34.7 | 9.4 | — |
| Beryllium | 19.7 | — | 28.0 | 8.2 | — |
| Magnesium | 20.1 | — | 28.4 | 8.3 | — |

replicas are high spectral resolution (0.1 meV or better), low temperature (4.2K or lower) and preferably resonant excitation conditions [81] and samples of high purity ($\lesssim 10^{15}$ cm$^{-3}$ for n-type and $\lesssim 10^{16}$ cm$^{-3}$ for p-type GaAs).

### 5.2.6 Band-to-Band Recombination

As mentioned earlier, exciton luminescence is observed only at low temperatures and in high purity crystals. As the temperature increases such that $kT > E_x$, the excitons break up into free carriers. Similarly, if doping increases, excitons tend to dissociate under local electric fields. Therefore, under these conditions electrons and holes recombine via the band-to-band process. In a direct band gap semiconductor with $\alpha(\nu) \sim (h\nu - E_g)^{1/2}$, the luminescence spectrum will exhibit a high energy tail characterized by $\exp(-h\nu / kT)$ and a sharp low energy cut off at $h\nu = E_g$. This is illustrated in the luminescence spectrum at 300K from a high purity ($n \sim 10^{15}$ cm$^{-3}$) GaAs epitaxial layer shown in Fig. 5.28.

As the doping level increases, the shape of the band-to-band recombination undergoes many changes. With increasing doping (a) the sharp low-energy cut off shown in Fig. 5.28 is no longer seen and the spectrum looks more symmetrical, (b) the peak shifts to higher energy in n-type material and to lower energy in p-type material and (c) the half-width increases for both n- and p-type material [82 - 90]. Effects (a) to (c) are illustrated in Figs. 5.29 to 5.31 using examples from GaAs [83, 84] or InP [11, 88]. Pearsall et al. [91] have given half-width versus carrier concentration for n- and p-type GaInAs and GaInAsP. Swaminathan et al., [92, 93] have given similar data for n- and p-AlGaAs. One can use Fig. 5.31 to obtain the carrier concentration from the half-width quite accurately without doing electrical measurements. In this sense, photoluminescence is a convenient nondestructive technique to obtain carrier concentration of individual epitaxial layers in multilayer structures or layers grown on n- or p-substrates where

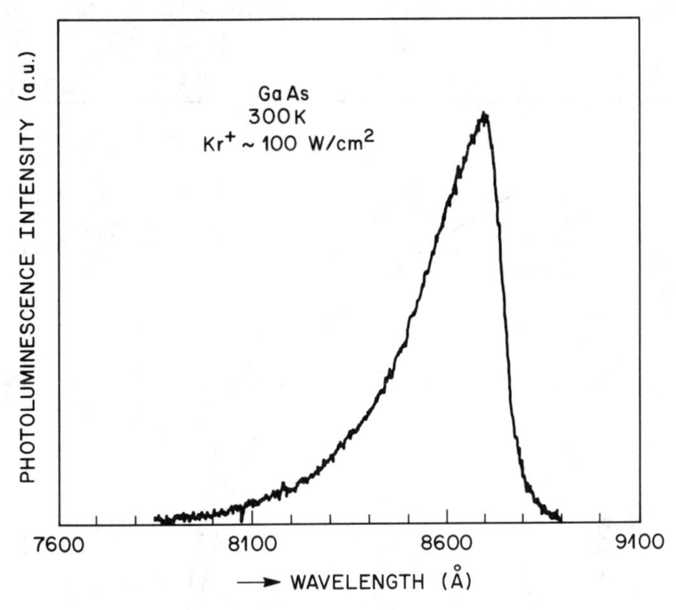


**Figure 5.28** Room temperature photoluminescence spectrum of a n-type high purity GaAs sample showing the high energy exponential tail and a sharp cut off at the low energy side.

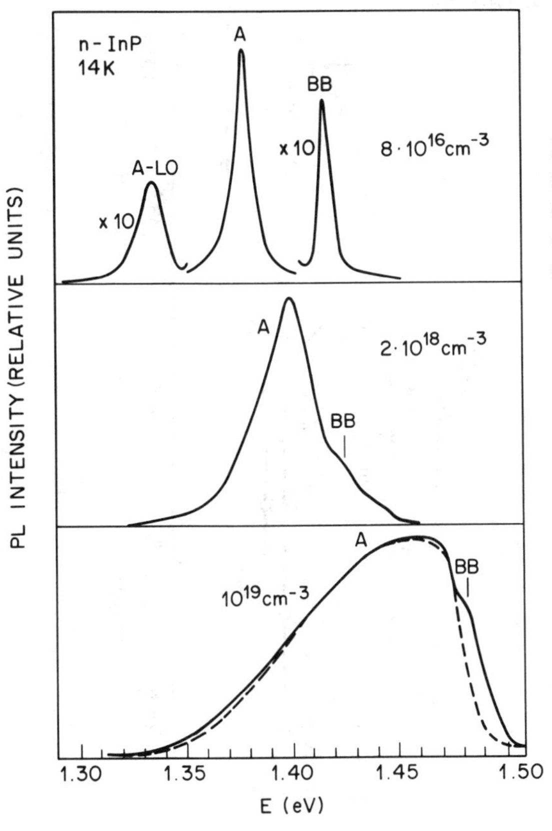


**Figure 5.29** Photoluminescence spectra of n-type InP as a function of free carrier concentration at 14K. Dashed line corresponds to spectrum taken at low excitation intensity [11].

**Figure 5.30** (a) Peak positions in cathodoluminescence at 77K and 300K from n- and p-type GaAs. The energy separations of 6 and 30 meV from the band gap as indicated correspond to donor and acceptor ionization energies. For the n-type samples the peaks shift to higher energy with increasing doping [84]. (b) Variation of band gap in n-type InP as a function of carrier concentration. The solid lines are calculated curves with the upper curve corresponding to k-conserving transitions and the lower curve to non k-conserving transitions [88].

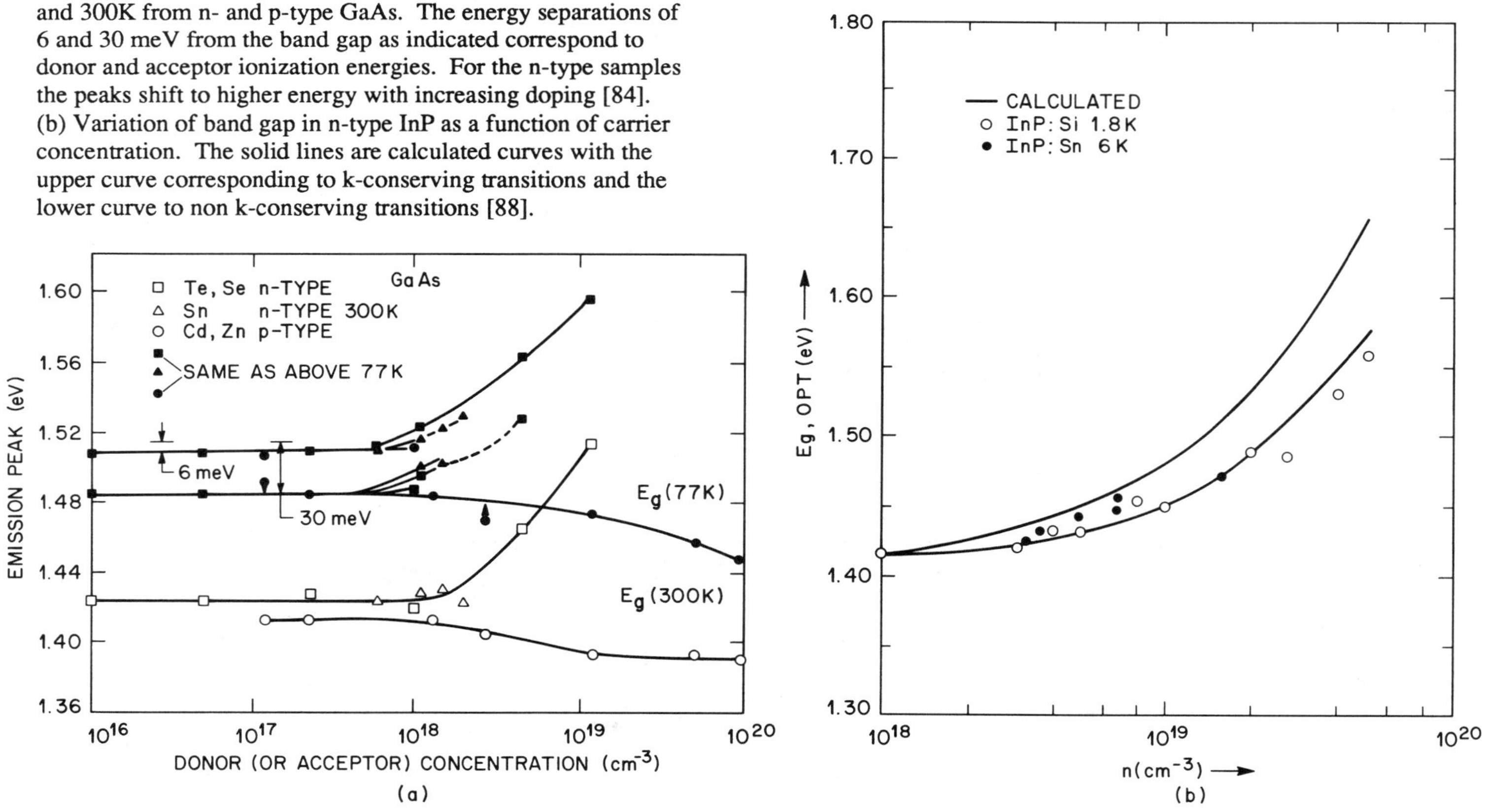

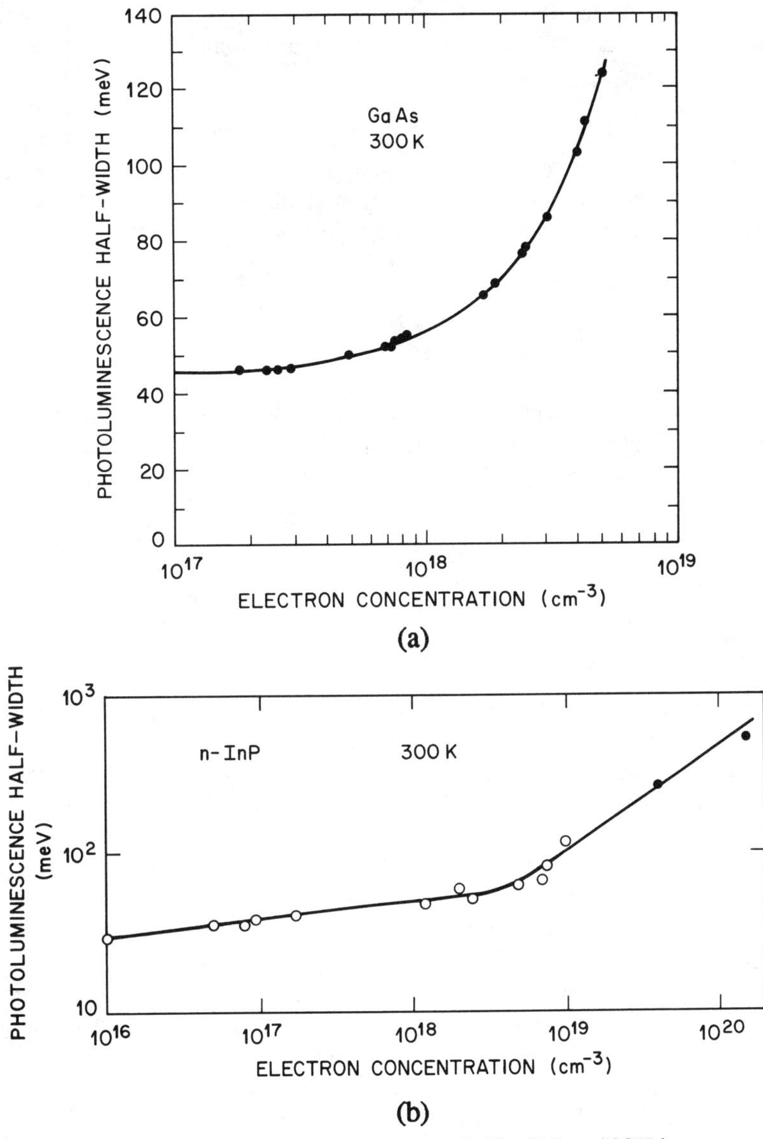


**Figure 5.31** (a) Photoluminescence half-width at 300K in n-type GaAs [83] and (b) n-InP [11] as a function of carrier concentration.

Hall measurements cannot be made readily. For an epilayer of an alloy semiconductor, lattice mismatch between the layer and the substrate and inherent random alloy composition fluctuations would contribute also to the line width of the band-to-band recombination. However, once a relationship between half-width and carrier concentration is established it can be used for other samples grown under identical conditions.

### 5.2.7 Free-to-Bound Transitions

Free-to-bound (F-B) transitions are the donor-to-valence band (D-V) and conduction band-to-acceptor (C-A) transitions in Fig. 5.16. They are typically seen at low temperatures when $kT < E_i$, the ionization energy of the donor or the acceptor, that is, when the centers remain neutral. The energy of the transition is given by [94]

$$h\nu = E_g - E_A\ (E_D) + \frac{1}{2}kT \tag{5.55}$$

The 1/2 kT term follows from Eagles calculation [94] of the photoabsorption cross section for scattering an electron initially bound to an isolated acceptor into a final conduction band state. The energy dependence of the F-B recombination spectrum is given by [94, 95]

$$I_{F\text{-}B}(h\nu) \propto (h\nu - E_g + E_{A,D})^{1/2} \exp - \frac{h\nu - E_g + E_{A,D}}{kT} \tag{5.56}$$

It follows from Eq. (5.56) that the low energy threshold of FB is $E_g - E_{A,D}$ and the high energy side is given by the Maxwell-Boltzman exponential factor in Eq. (5.56).

In direct band gap semiconductors such as GaAs and InP, the differences in $E_D$ between various donors is rather small which means that the D-V transitions are not distinguishable. On the otherhand, the acceptors are sufficiently separated in energy such that C-A transitions associated with each acceptor can be readily identified. In this sense C-A transitions complement the two-hole transitions of the neutral acceptor BE lines (Sec. 5.2.5) in identifying the unknown acceptor. When carrier concentrations are high ($\gtrsim 10^{16} cm^{-3}$) so that BE lines are no longer seen, one has to rely completely on the C-A transitions for acceptor identification. But, one has to watch for extraneous factors which affect the peak position and half-width of C-A lines besides chemically different acceptor species [80]. At high doping levels C-A lines could shift because of dielectric screening. There could be broadening effects caused by homogeneous or inhomogeneous strain. Also, the C-A luminescence should preferably be recorded at low excitation conditions and optimum temperatures which favor C-A emission over D-A pair recombination, to be discussed next. For these reasons, chemical identification of the acceptors based on C-A transitions alone should be done with care.

In fitting Eq. (5.56) to the experimentally observed C-A band one finds that generally the agreement on the high energy side is rather good. In fact, the carrier temperature, T, in the exponential in Eq. (5.56) can be used as a variable to obtain the best agreement with experiment. At high excitation conditions one can thus find that the carrier temperature can be higher than the lattice temperature as shown in Fig. 5.32 [96].

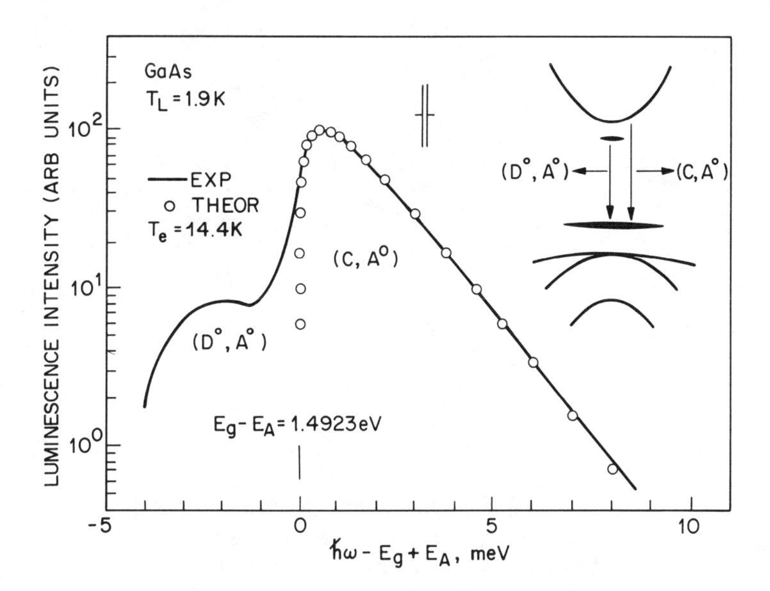


**Figure 5.32** Low temperature photoluminescence spectrum from lightly doped LPE GaAs recorded at an excitation intensity of 6W $cm^{-2}$. Note the logarithmic scale for intensity. The inset shows free-to-bound (C, A°) and donor-to-acceptor (D°, A°) transitions. The acceptor in these transitions is $C_{AS}$ with an ionization energy of ~26.9 meV. Theoretical fit to the (C, A°) transition using Eq. (5.56) indicates the electron temperature $T_e$ to be greater than the lattice temperature $T_L$ [96].

In contrast to the high energy side, agreement between Eagles equation and the low energy side of C-A band is rather poor as illustrated in Fig. 5.33 [97]. Dow et al., [98] attribute the origin of the discrepancy to small internal microfields of strength $10^3$ to $10^4$ V/cm created by one of several disorders such as piezoelectric phonons, ionized impurities, or excitons.

It can be seen from Eq. (5.55), that the peak energy of the C-A emission will follow the temperature variation of the band gap provided, of course, that the 1/2 kT term is taken into account. Williams and Chapman [65] have shown from Eq. (5.51) that for a C-A transition under the condition that $\beta \ll \alpha$, $\beta \ll L_e^{-1}$ and $S/D > \alpha > L_e^{-1}$

$$I_{C-A} \propto \frac{1}{\tau_{C-A}} (\tau/D)^{1/2} \frac{1}{\alpha^2} \tag{5.57}$$

where $\tau$ is lifetime of the minority electrons considering all recombination paths. Since $\alpha$ is nearly independent of temperature and since at low temperatures both $\tau$ and D usually increase

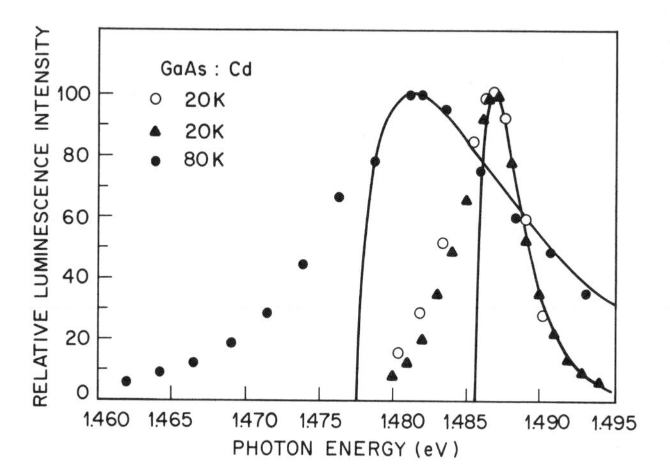


**Figure 5.33** Photoluminescence spectra from Cd-doped GaAs at 20 and 80K showing the free-to-bound transition involving the Cd acceptor. The solid lines are calculated using Eq. (5.56) showing the poor fit to the low energy side of the spectra [97].

with temperature which reduces the temperature dependence of $(\tau/D)^{1/2}$, $I_{C-A} \propto \tau_{C-A}^{-1}$. The lifetime $\tau_{C-A}$ has been shown to be [65]

$$\tau_{C-A} \propto \frac{1}{\Gamma(T)\, N^{\circ}_{A}} \tag{5.58}$$

where $\Gamma(T)$ is a factor which correct for the velocity distribution of the thermalized electrons and $N^{\circ}_{A}$ is the density of neutral acceptors. Williams and Chapman [65] found that for the C-A band caused by Cd acceptor in GaAs, Eqs. (5.57) and (5.58) fit the experimental data very well.

Since the C-A transition rate is proportional to $N^{\circ}_{A}$ (Eq. 5.58) its intensity decreases as $N^{\circ}_{A}$ decreases. For a given density of compensating impurities, $N^{\circ}_{A}$ would decrease as temperature increases because of the ionization of the hole. Thus the thermal quenching of the C-A emission would follow an Arrehenius type of process yielding an activation energy. Although it is not straightforward to relate the activation energy thus obtained with the electronic energy level of the impurity giving the C-A emission [99], the activation energy could correspond to the ionization energy of the acceptor. For example, the C-A transition involving substitutional $Cd_{In}$ acceptor in InP:Cd quenches above 100K as shown in Fig. 5.34 [100]. The activation energy obtained from the linear part of the curve in Fig. 5.34 is in good agreement with the ionization energy of $Cd_{In}$.

The intensity of the C-A band can also be related to $N^{\circ}_{A}$. However, unlike in optical absorption where the integrated absorption strength increases linearly with concentration at least up to concentrations before electronic states become broadened, in photoluminescence the intensity saturates once the concentration of the emission center equals the density of the excess carriers created by the absorbing light. In spite of this difficulty, PL intensity can be related to

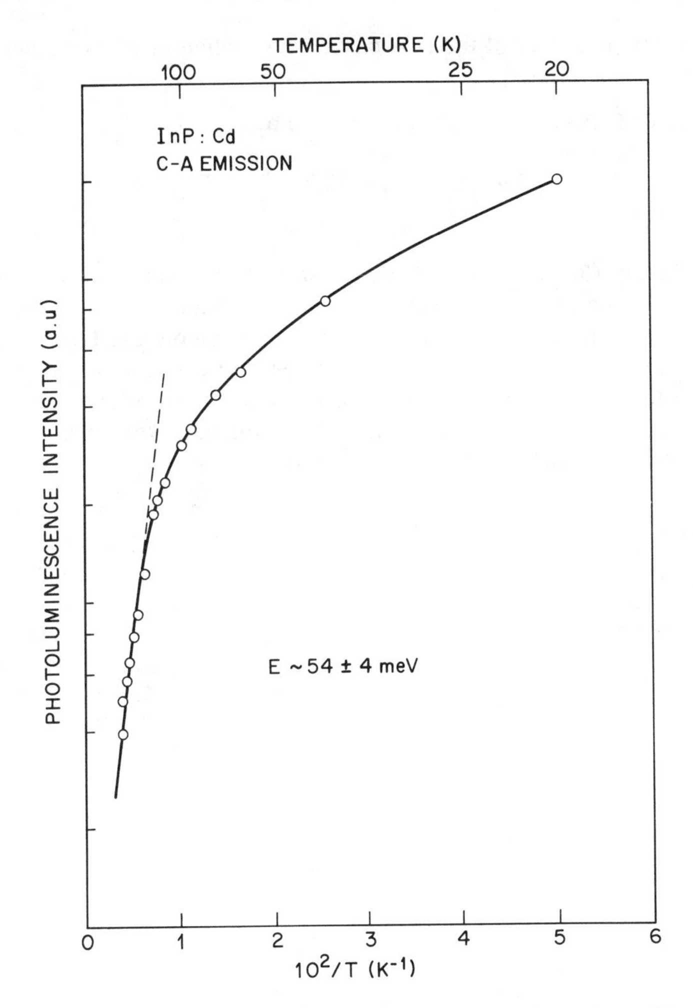


**Figure 5.34** Temperature dependence of the free-to-bound transition involving the Cd acceptor in InP. The activation energy of thermal quenching corresponds to the ionization energy of the acceptor [100].

impurity concentration by carefully maintaining the same experimental conditions as has been done for the D-V band in GaAs [101].

### 5.2.8 Donor-to-Acceptor Pair Recombination

When both donors and acceptors are present in the semiconductor, as is invariably the case, D-A pair recombination (Fig. 5.16) becomes a dominant recombination channel and it competes strongly with BE and C-A recombinations. At low temperatures such that $kT < E_D$ and $E_A$, when the thermal ionization probability of the carrier once it is captured at the center is low, D-A pair recombination dominates over C-A emission. As temperature increases, the shallow center which is generally the donor in the case of GaAs and InP, becomes ionized thereby promoting the C-A transition. The many properties of the D-A pair recombination in all its details can be found in the excellent review by Dean [99].

The energy of the D-A pair recombination is given by

$$h\nu = E_g - (E_A + E_D) + \frac{e^2}{\varepsilon r} - E_{vdw} \tag{5.59}$$

The $e^2/\varepsilon r$ term in Eq. (5.59), where $\varepsilon$ is the static dielectric constant and r is the separation between the recombining carriers, accounts for the electrostatic interaction between the ionized donor and acceptor in the final state of the D-A pair transition and $E_{vdw}$ is a Van der Waals interaction term in the initial state of the transition. Since r can take only discrete values in the lattice, according to Eq. (5.59) the D-A pair spectra is characterized by a series of sharp lines, each associated with a value of r allowed by the lattice structure, which merge into a broad band. Such a spectra is one of the hallmarks of a D-A pair recombination.

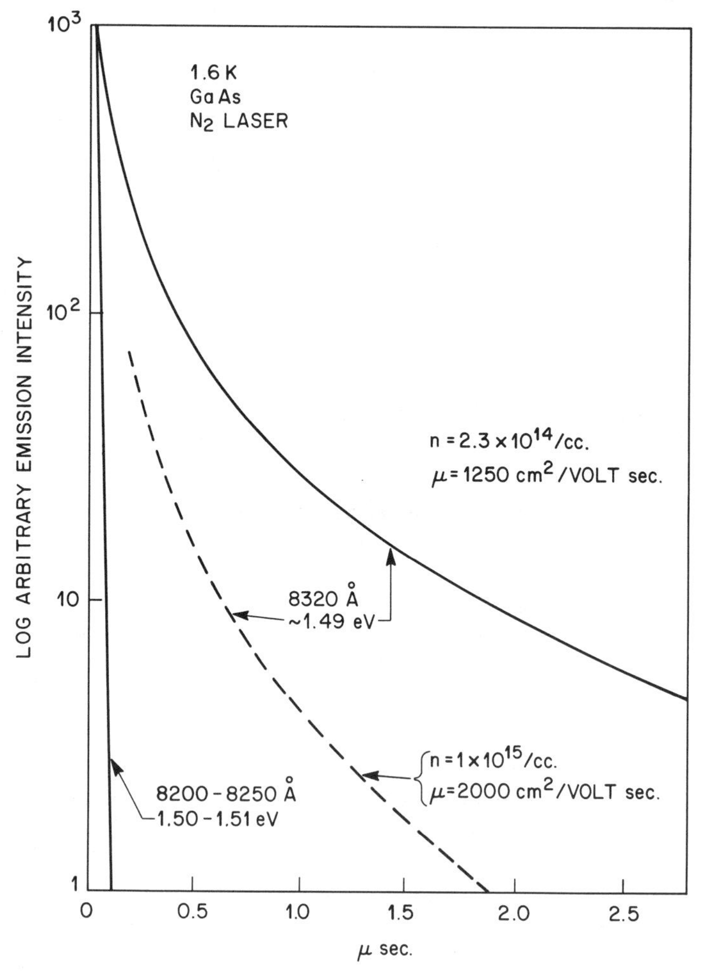


**Figure 5.35** Decay curves for the 1.49 eV ($D^\circ - A^\circ$) band and 1.50-1.51 eV (B-B) band in n-type GaAs. The (B-B) band decays very rapidly compared to the ($D^\circ - A^\circ$) band which is characterized by a longer non-exponential decay [57].

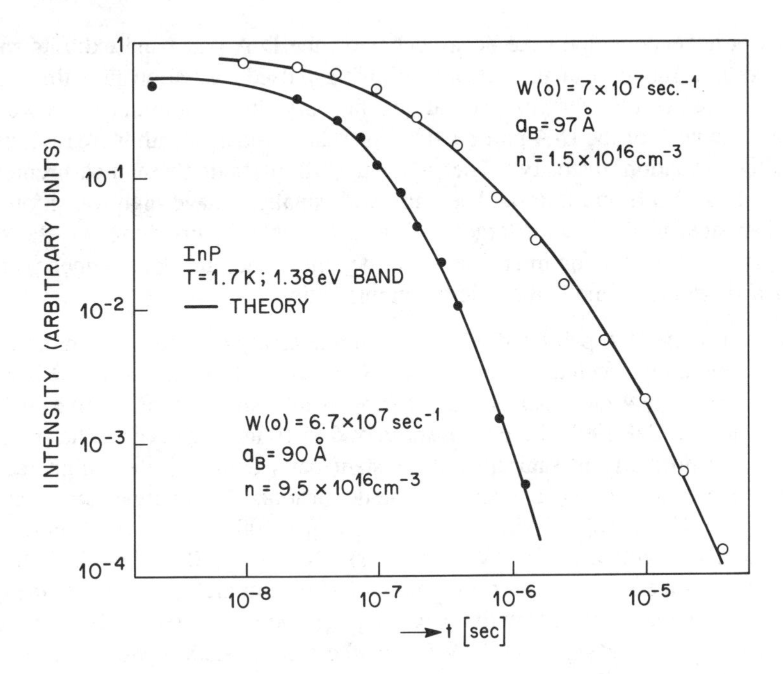


**Figure 5.36** Decay curves of the 1.38 eV ($D^\circ - A^\circ$) band in two InP samples with doping levels as indicated [102].

The D-A pair spectra in GaAs, InP or their alloys consist only of the broad band but not the discrete pair lines as discussed. The reason for this is believed to be the small ionization energies of the donor and the acceptor. For the most common donors and acceptors in GaAs and InP, $E_D \lesssim 10$ meV, and $E_A \sim 20$–$50$ meV. This means that only the distant pairs satisfy $E_A + E_D > e^2/\varepsilon r$ to give bound states. Further, the energy separation between the lines may become smaller than the width of the lines so that no fine structure can be resolved. Under these circumstances, the D-A pair spectrum has to be identified by its other characteristics, in particular the excitation and temperature dependence and the time decay.

While the energy of the D-A pair recombination is given by Eq. (5.59), the recombination rate between a given pair is proportional to the overlap integral between the electron and hole wave functions and is given by

$$W(r) = W(o) \exp \frac{-2r}{a_B} \tag{5.60}$$

where $a_B$ is the Bohr radius of the less tightly bound center in the pair. Equation (5.60) predicts that the distant pairs have longer recombination time or lower transition probability. The decay of D-A luminescence measured following pulsed excitation can be explained on the basis of Eq. (5.60) for a random distribution of noninteracting pairs. The values of W(o) and $a_B$ can be deduced from the characteristic shape of the decay curves. Figures 5.35 and 5.36 show the decay curves for a D-A pair at 1.49 eV in GaAs [57] and for one at 1.38 eV in InP [102], respectively.

Apart from a nonexponential time decay behavior, the D-A pair band exhibits spectral shift when measured as a function of time delay following pulsed excitation in a time-resolved PL measurement or as a function of excitation intensity in a continuous wave excitation measurement. The peak of the D-A pair band shifts to lower energy with increased time delay or with decreasing excitation intensity. The spectral shift in both these measurements can be understood via Eqs. (5.59) and (5.60). The pairs with smaller r have high transition probability and energy compared to those with larger r. Thus, the distant pairs dominate the spectrum at longer times and lower excitation intensities. On the other hand, the close-spaced pairs dominate the spectrum at shorter time and higher excitation intensities.

The peak shift of the D-A pair band with excitation intensity or time delay in direct-gap GaAs [57], InP [102], and alloys such as AlGaAs [103], GaInAs [104], GaInAsP [105] is rather small, of the order of only a few meV per decade change in intensity or time, unlike in indirect-gap semiconductor such as GaP [99] where peak shifts of 5 to 10 meV are seen. The small peak shift is attributed to the difficulty in saturating the distant pairs owing to their high recombination rates. [57, 106] For example, in GaAs and InP the donor ionization energies are small and hence $a_B$ is rather large in Eq. (5.60). Taking $r \sim 300\text{Å}$ and $a_B \sim 100\text{Å}$ in Eq. (5.60) corresponding to shallow donors in InP with $E_D \sim 7$ meV gives $W(r) \sim 10^7 \text{sec}^{-1}$ [102]. This is 3 to 4 orders of magnitude more than the recombination rate of distant pairs in GaP [99]. Since the exponential term in Eq. (5.60) decreases rapidly with decreasing $a_B$, one would expect that for a deep donor with $E_D > 7$ to 10 meV and $a_B < 100\text{Å}$, W(r) for the same $r \sim 300\text{Å}$ would be much smaller. Consequently, distant deep donor-acceptor pairs can be saturated giving rise to large peak shifts. In InP doped with Cd or Zn, an emission band on the low energy side of the shallow donor - $Cd_{In}$ or $Zn_{In}$ acceptor pair band showed large peak shifts with excitation intensity [100]. This was taken to imply that the band is caused by a deep donor-acceptor pair recombination.

While large peak shifts of emission bands near the band edge with excitation intensity could imply a D-A pair transition between deep centers, they are also seen for shallow D-A pair bands in compensated direct-gap semiconductors. Yu et al., [107] found in GaAs that the shift of the peak of the D-A pair band per decade change in excitation intensity, $\eta$, increases as the degree of compensation increases. The small value of $\eta$ observed in GaAs and InP for pair bands involving shallow donors and acceptors is applicable only for low to moderately doped and uncompensated crystals. For heavily doped, closely compensated crystals, the ideal picture of isolated noninteracting D-A pair as depicted by Eq. (5.59) is no longer valid. In these crystals, a recombination transition which in many respects resembles D-A pair recombination, takes place between carriers that are localized in spatially separated potential wells [108 - 110]. These wells are created if the number of free carriers is insufficient, as in the case of close compensation, to screen potential fluctuations generated by fluctuations in the concentration of charged impurities [111]. The energy of the photons ensuing from the recombination of electron-hole pairs trapped at these wells is given by [107, 108, 110]

$$h\nu = E_g - (E_A + E_D) - 2\Gamma \tag{5.61}$$

where $\Gamma$ is depth of the potential well. Since $\Gamma$ varies with the size of the fluctuation [111], deeper wells are separated by longer distances and hence have smaller overlap of the electron and hole wave functions and consequently smaller transition probability. In this respect, this

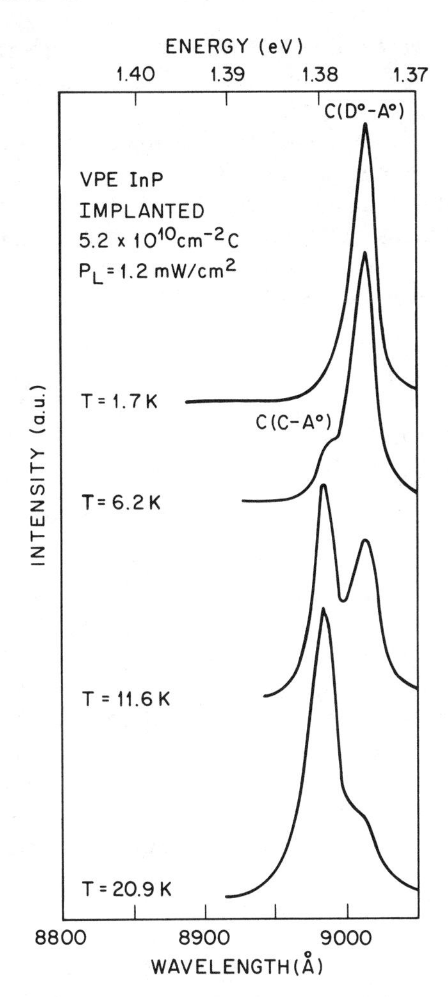


**Figure 5.37** Photoluminescence spectra in the temperature interval 1.7-20.9K from VPE grown InP implanted with $5.2 \times 10^{10}$ cm$^{-2}$ of carbon. The relative intensities of the carbon related $(D° - A°)$ and (C-A°) transitions change with temperature with the latter favored more at higher temperatures [116].

recombination is analogous to D-A pair recombination in that the peak shifts to higher energy (smaller $\Gamma$) with increasing excitation intensity. However, the magnitude of the shift can be very large and values of 10 to 20 meV for $\eta$ have been measured in compensated GaAs [107, 112, 113] and AlGaAs [114, 115].

The temperature dependence of the band-edge emission is also helpful in distinguishing F-B and D-A recombination. Generally in low to moderately doped samples at low temperatures, the D-A pair band dominates the spectra taken at low excitation intensities. As temperature increases the shallow center, which is the donor for most D-A pair bands in GaAs and InP, is ionized thereby quenching the D-A luminescence in favor of the F-B transition. This is illustrated in Fig. 5.37 which shows the gradual evolution of the F-B transition from the D-A transition [116]. Sometimes, with increasing temperature one can also observe a shift of the D-A band to higher energies. As the donor is ionized, $E_A$ of the acceptor is reduced by $e^2/\varepsilon r$ in Eq. (5.59). However, since $E_A > E_D$, $E_A - e^2/\varepsilon r > E_D$ and the acceptor is not ionized not withstanding the nearby ionized donor of the pair. The ionized electron will wander through the crystal and may get retrapped at a donor situated much closer to the acceptor than before and the resulting D-A emission will occur at higher energies. It should be noted that if $E_A$ and $E_D$ differ only slightly then as one of the centers is ionized, the reduced ionization energy of the other center may be less than the ionization energy of the shallower member. As a result, the deeper member is also ionized and both the carriers will move through the crystal and can get retrapped only at pairs with smaller $e^2/\varepsilon r$ or larger separations. Consequently, the D-A emission occurs at lower energies [117]. Evidence for excited state donor-acceptor pair recombination has also been cited in GaAs [118] and InP [118]. This manifests as a peak in between the D-A and F-B peaks as temperature increases.

To discriminate the F-B and D-A transitions, one can also rely on an externally applied electric field at low temperatures to cause impact ionization of the donor. This once again favors the F-B transition relative to the D-A band [119, 120].

The D-A pair band is also expected to shift to higher energies with increasing doping. This is readily seen by Eq. (5.59) since as the impurity concentration increases the average pair separation decreases giving higher photon energy [121].

Apart from serving to discriminate the different acceptor species, the D-A and F-B transitions have also been used to obtain the compensation ratio in GaAs [122] and InP [95]. The compensation ratio is usually determined from semiempirical formulas based on mobility measurements made at 77K for n-type samples with carrier concentration, $n \gtrsim 10^{15}\,cm^{-3}$. However, alternate methods are useful for $n < 10^{15}\,cm^{-3}$ and for semiconductors where an empirical mobility technique has not been developed. Kamiya and Wagner [122] proposed a technique which involves line shape analysis of the D-A and F-B bands to derive the compensation ratio. The analysis requires knowledge of n at 300K, easily obtainable from either Hall or C-V measurement, $E_D$, $E_A$, and $E_g$. The intensity of the broad D-A band is given by

$$I_{DA}(E)\,dE = P[r(E)]\,\frac{dr}{dE}\,W_{DA}[r(E)]\,P_A{}^{\circ}[r(E)]\,N_A\,dE \tag{5.62}$$

where E is the photon energy hν, $W_{DA}(r)$ is the D-A pair transition rate for separation r (Eq. 5.60), P(r) is the probability that a neutral donor will have a nearest-neighbor neutral acceptor at a distance r, $P_A{}^{\circ}(r)$ is the occupation probability under steady state conditions such that the total D-A pair intensity is proportional to the excitation intensity. The intensity of the F-B transition is given by

$$I'_{F-B}(E)\,dE = I_{F-B}(E)\,N_A{}^{\circ}\,dE \tag{5.63}$$

where $N_A{}^{\circ}$ is the stationary value of the neutral acceptor concentration and $I_{F-B}(E)$ is given by Eq. (5.56). Figure 5.38 shows the fit of D-A and F-B luminescence in InP [95] with the aid of Eqs. (5.60), (5.62) and (5.63).

Nam et al., [123] had developed an alternative technique for the analysis of the compensation. This technique also involves the analysis of the intensities of the D-A and F-B transitions obtained in the two opposite limits of low- and high-excitation intensities. They have applied this

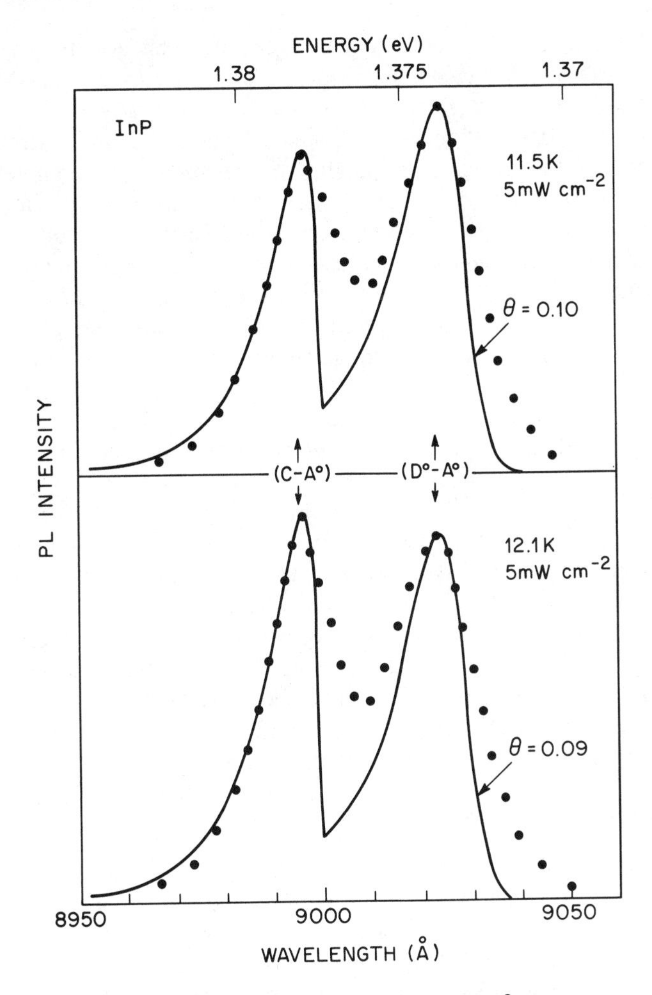


**Figure 5.38** Photoluminescence spectra from InP at two temperatures showing the (C-A°) and (D° – A°) transitions. The solid lines are theoretical fit to the experimental points using Eqs. (5.62) and (5.63) [95]. Θ indicates the compensation values obtained by the theoretical fit.

technique in p-type GaAs. Yu [124] derived compensation ratios for several n-type GaAs and found good agreement with the values determined from electrical measurements.

In connection with the discussion on D-A pair spectra it is important to mention two related optical techniques which reveal detailed structure even in the case of the broad D-A pair bands. The two techniques are photoluminescence excitation (PLE) spectroscopy and selectively excited photoluminescence (SPL) [62]. The former consists of selecting luminescence from only a narrow range of r by using a high resolution detecting monochromator and varying the energy of the excitation source. The latter technique involves resonantly exciting D-A pair band within only a narrow range of r. For both techniques the excitation is usually accomplished by means of a tuneable dye laser. The physics for both PLE and SPL processes are essentially the same. In the PLE process, photoluminescence corresponding to transitions from ground state of the donor to the ground state of the acceptor for a D-A pair within a certain r is excited by absorption processes which leave either the donor or the acceptor in a series of excited states. Excitation to both s and p states occurs and the return to ground state before recombination occurs very rapidly by emission of phonons. This absorption process manifests as a series of lines above the energy of the detected luminescence corresponding to the internal excitation energies of the donor or the acceptor. In SPL these lines occur below the excitation energy displaced by the same internal excitation energies. Such high resolution spectroscopic techniques provide rather precise value of the binding energies of acceptor excited states in GaAs [125 - 127], InP [128, 129], and AlGaAs [130] and more clear cut identification of the chemical nature of the acceptor.

### 5.2.9 Deep Level Transitions

Many impurities and defects give rise to deep energy levels in the forbidden energy gap of semiconductors. The term "deep" denotes that the ionization energies are of the order of $E_g$ itself. These deep states are efficient traps for excess carriers. However, not only the capture process but also the resulting recombination through such centers are generally nonradiative. But there are, however, several exceptions to this general rule in that many deep centers give rise to detectable emission bands. The knowledge gained about the deep levels by the detailed analysis of such luminescence is very useful since it is they which control the minority carrier lifetime, an important consideration for opto-electronic devices. We already showed (Eq. 5.37) that the trap limited lifetime is only a function of the deep-level concentration. Thus a reduction in that concentration increases carrier lifetime.

Among impurities, transition metals such as Fe, Mn, Cr, and so on, give rise to deep energy levels in GaAs and InP. Transitions associated with them occur at energies well separated from the band edge [62, 131]. Besides these transitions, several deep luminescence bands in GaAs [132 - 135], InP [136], and related alloy compounds [130, 137, 138] have been associated with native defects such as vacancy, vacancy clusters, and impurity-vacancy type complexes.

A common feature of all deep-level luminescence is the strong phonon participation in the optical transitions [139, 140]. This makes the deep-level PL quite broad and the no-phonon transition is quite weak compared to the total emission envelope. The concept of the configuration diagram is generally utilized to understand the deep-level luminescence. The configuration diagram as shown in Fig. 5.39 represents the energy of the ground state and the first excited state of an atom as a function of its position. The energy is the total energy of the atom including both the electronic and vibranic energies. Note in Fig. 5.39 that the minimum energy position for the excited state does not coincide with the minimum energy position for the ground state but is displaced by $\Delta x$ from it. Thus both the absorption from a to b and emission from c to d require atomic displacements of $\Delta x$ which is accomplished by phonon emission. The LO phonons are the ones which generally take part in this process since they give rise to largest change in energy per unit displacement because of their strong polarization. It can also be seen from Fig. 5.39 that the energy of the emission process $c \rightarrow d$ is lower than the energy of the

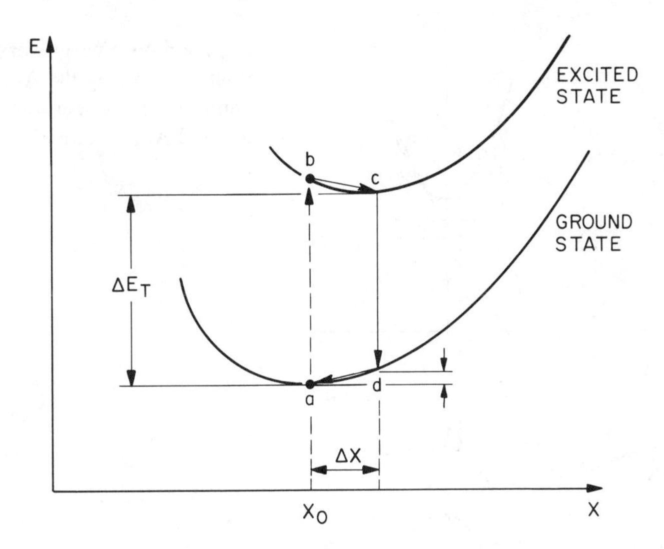


**Figure 5.39** Configuration co-ordination diagram showing the energy of the ground and excited states of an atom as a function of its position. The energy absorbed in the transition a to b is higher than the energy emitted in the transition c to d. The difference between the two energies is the Franck-Condon shift or Stokes shift. The thermal energy between the two states is $\Delta E_T$ [140].

absorption process a → b. This Stokes shift is referred to as the Franck-Condon shift [140]. A larger Franck-Condon shift implies a stronger phonon cooperation.

### 5.2.10 Nonradiative Recombination Processes

So far in this section we were concerned with radiative transitions. It is appropriate to end this section with a brief discussion of nonradiative transitions. Among the various types of nonradiative processes, the ones that control lifetimes in comparatively low to moderately doped crystals are the recombination at the surface and the Auger recombination of which there are several types. At high doping levels impurity clusters or impurity-defect complexes may limit radiative efficiency. In addition, point defects such as vacancies and line defects such as dislocations may also act as nonradiative centers.

The surface recombination is usually described by a surface recombination velocity, S. For excess electrons in p-type material the loss of carriers caused by recombination at the surface is given by the surface recombination current.

$$J_e = q\,\Delta n\,S \tag{5.64}$$

It is experimentally found that InP has several orders of magnitude lower S than GaAs. Queisser [141] suggested that a less strained interface between the native oxide and the semiconductor and

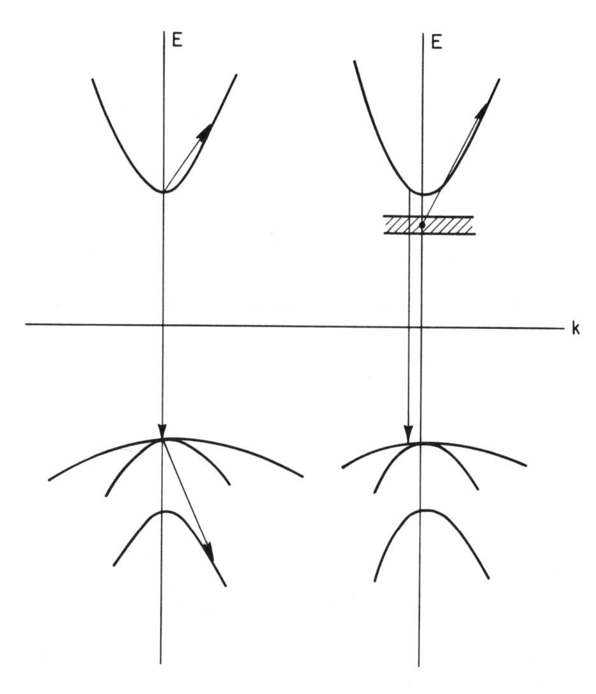


**Figure 5.40** Energy versus k dispersion relation showing the Auger transitions: (a) band to band Auger transition (b) impurity to band Auger transition.

the smaller piezoelectric coefficient of InP compared to GaAs could perhaps explain the lower S of InP.

Nelson et al., [142] reported a reduction in S of GaAs by surface protective coating with Ru. Also GaAs capped with AlGaAs has been used to determine lifetimes not limited by surface recombination [56].

Auger recombination involves three particles, the recombining electron and hole pair, and third electron or hole to which the recombination energy is transmitted [143]. In a band-to-band Auger process shown in Fig. 5.40 the recombination energy, equal to $\sim E_g$ is transmitted to an electron to transfer it to a state deep in the conduction band or alternatively to a hole to excite it to a state in the split-off valence band. The excited particle in turn relaxes nonradiatively to the extrema of the respective band. The rate of the band-to-band Auger process is proportional to $n^2p$ as opposed to np for the corresponding radiative process (Eq. 5.27). It is clear from momentum selection rules that the initial recombining e-h pair cannot lie near $k \approx o$ but must be separated by $\Delta k$ and the Auger probability decreases with increasing $\Delta k$. There exists an activation energy for the Auger process which is proportional to the width of the forbidden energy gap. The Auger process is expected to be important for narrow-gap semiconductors. The increased temperature dependence of current threshold of GaInAsP based semiconductor diode lasers operating at 1.3 to 1.5μm wavelength range compared to GaAs based lasers (wavelength 0.8–0.9 μm) has been suggested to be caused by the band-to-band Auger process [144].

Auger process can also occur associated with impurity related transitions. One such process is illustrated in Fig. 5.40. An impurity donor band is formed because of heavy doping. Electron in such a band is de-localized and the recombination of a bound exciton then kicks this electron to high energy conduction band state [145]. The impurity band Auger process is likely to be one of the processes to explain quenching of luminescence observed at high impurity concentrations in GaAs (concentration quenching) [146].

## 5.3 RAMAN SPECTROSCOPY

Raman Scattering (RS) is a useful characterization technique to study III-V semiconductors, both bulk and thin epitaxial films [147 - 149]. It has been employed to study ion-implantation damage, laser annealing effects, potential fluctuations in alloy semiconductors, composition dependence of electron effective mass, band offsets in quantum wells, interfaces (including semiconductor/vacuum, semiconductor/metal, heterointerfaces), and strain. RS has also been used to compare the crystalline quality of epitaxial films grown by different techniques [150]. Oxidation phenomena in GaAs and InP have been studied by RS [151]. Because of such wealth of information it provides, RS serves as a powerful complementary technique to other optical characterization techniques such as photoluminescence.

### 5.3.1 Principle of Raman Scattering

Raman scattering is an interaction of light with the optical modes of lattice vibration. It involves two photons, one incident and one scattered. The scattering event is inelastic, accompanied by the creation or annihilation of a phonon. The conservation of energy and momentum in the scattering process gives

$$h\nu_s = h\nu_i \pm h\nu_p \tag{5.65a}$$

$$k_s = k_i \pm q_p \tag{5.65b}$$

where subscripts s, i and p denote scattered radiation, incident radiation and phonon, k and q are the wavevectors of the photon and phonon, respectively. The minus sign denotes emission of a phonon and the plus sign denotes absorption of a phonon. The two processes are respectively called Stokes and anti-Stokes processes. The intensity of anti-Stokes modes is generally much weaker than that of the Stokes components since the probability of phonon absorption is less by $\exp(h\nu_p/kT)$ than phonon emission. As opposed to the first order RS depicted by Eq. (5.65), the second order RS involves two phonons.

Since the photon momentum is very small ($k = 2\pi/\lambda$) compared to phonon momentum (which can take values up to $k = 2\pi/a$, where a is the lattice spacing) from Eq. (5.65) it follows that q has to be small. This restriction does not apply for a second order Raman process for which Eq. (5.65b) takes the form

$$q_p + q'_p = k_s - k_i \tag{5.66}$$

where q and q′ are the wavevectors of the two phonons and they can have values throughout the Brillouin zone and yet satisfy Eq. (5.66).

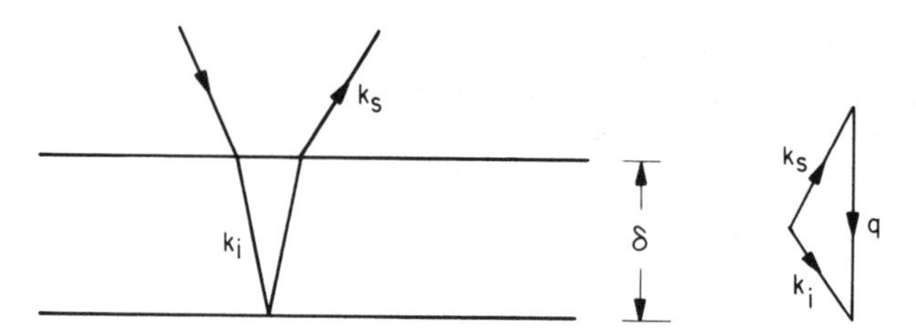


**Figure 5.41** Schematic view of the backscattering geometry for Raman Scattering showing momentum conservation for Stokes scattering [152].

### 5.3.2 Experimental Aspects

Raman Scattering experiments in GaAs and InP are done with Ar ion or Kr ion gas lasers. At these laser frequencies, GaAs and InP are opaque and hence the backscattering geometry is used for performing RS experiments. This is illustrated in Fig. 5.41 which also shows momentum conservation for Stokes scattering [152]. Note that the phonon wavevector is normal to the surface. In this surface-reflection RS, one can study the properties of phonons in the thin layer within the optical skin depth, $\delta$ near the surface of the crystal. The laser light is focussed on to the sample and the scattered light is collected normal to the sample surface and analyzed with a double monochromator. The dispersed light is usually detected by photon counting techniques. The scattered intensity can be measured for different polarization conditions of the incident and scattered light. Depending upon the sample orientation, selection rules would determine which lattice vibration can be observed. Table 5.4 summarizes the symmetry selection rules for backscattering geometry for crystals with diamond and zinc blende structures [147].

**Table 5.4**

Symmetry selection rules for backscattering from "ideal" semiconductors with diamond and zinc blende structures.

| Raman Active Mode | Surface Orientation | | |
|---|---|---|---|
| | <100> | <110> | <111> |
| LO | Allowed[a)] | Forbidden | Allowed[a)] |
| TO | Forbidden | Allowed[a)] | Allowed[a)] |

a) Allowed for at least one polarization configuration.

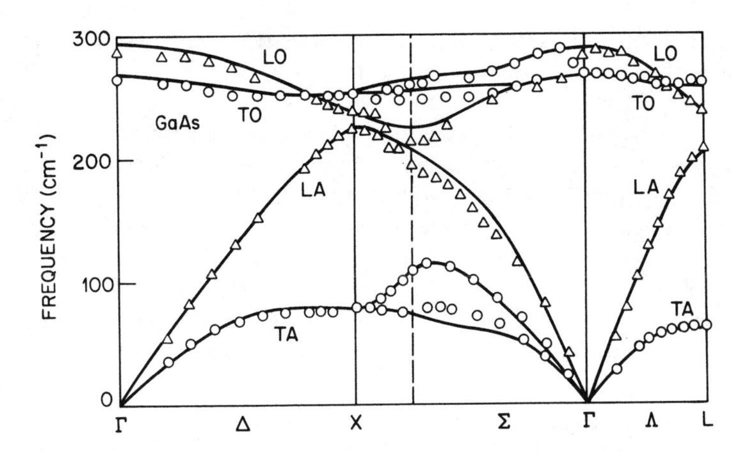


**Figure 5.42** Phonon dispersion relations for GaAs [153]. The experimental points are from neutron scattering studies.

### 5.3.3 First- and Second-Order Raman Scattering

The phonon dispersion curves along (100) and (111) for GaAs and InP are shown, respectively, in Figs. 5.42 and 5.43 [153, 154]. For the zinc blende type polar semiconductors the degeneracy at q = 0 of the optic phonons is lifted because of lack of inversion symmetry and one

**Figure 5.43** Phonon dispersion relations for InP [154]. The experimental points are from neutron scattering studies.

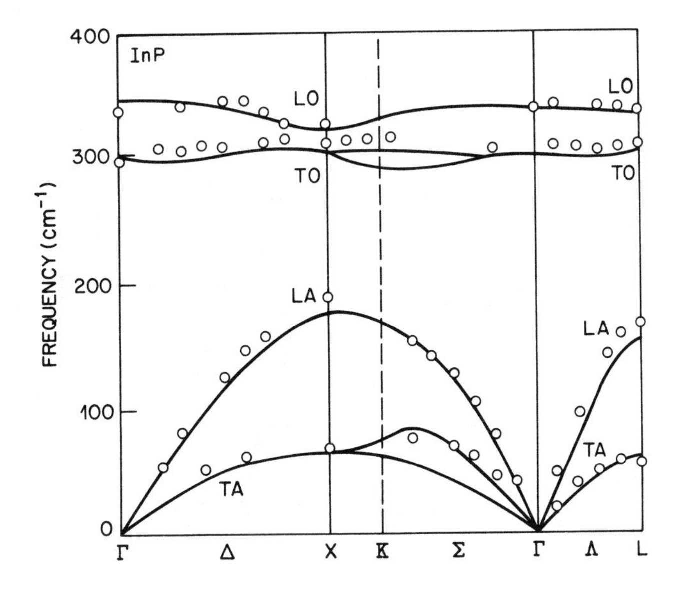

gets zone center phonons designated as longitudinal optic (LO) and transverse optic (TO) modes. Both LO and TO phonons are observed in the first order RS of GaAs and InP depending on crystal orientation. Figure 5.44 (top trace) shows the Raman spectrum obtained in the backscattering geometry from a (100) surface of a n-type ($n \sim 10^{15}$ $cm^{-3}$) GaAs layer grown by MOCVD [150]. The incident light was polarized in the plane of incidence and the scattered light was unpolarized. According to selection rules shown in Table 5.4, only the LO phonons are expected for the (100) orientation as observed. However, deviations from the selection rules do occur. For example, the Raman spectrum from a (100) MBE GaAs film shown also in Fig. 5.44 (bottom trace) contains both the LO and TO phonons though the latter is forbidden by selection rules. In a comparative study of GaAs films grown by LPE, MOCVD, and MBE, the incidence of TO lines was noted only in some of the MBE grown films. Abstreiter et al., [149] found that the

**Figure 5.44** Raman spectra from high purity and GaAs films grown by MBE (MB58) and MOCVD (OMA 326). For the (100) orientation only the LO phonon is allowed for the backscattering geometry [150].

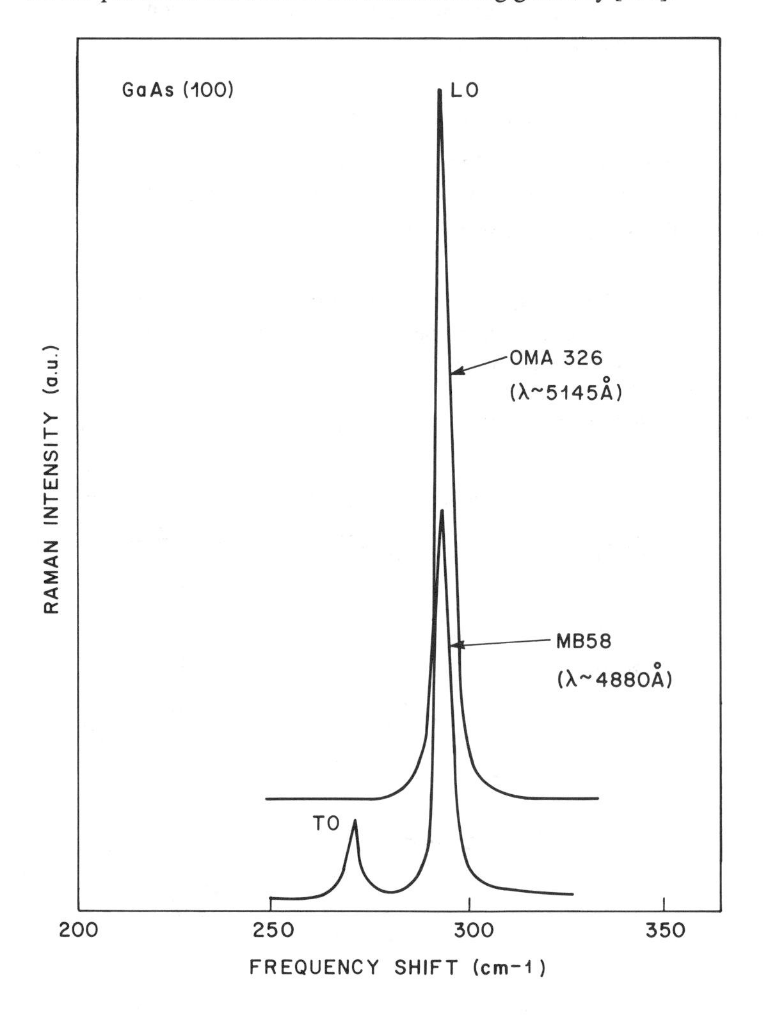

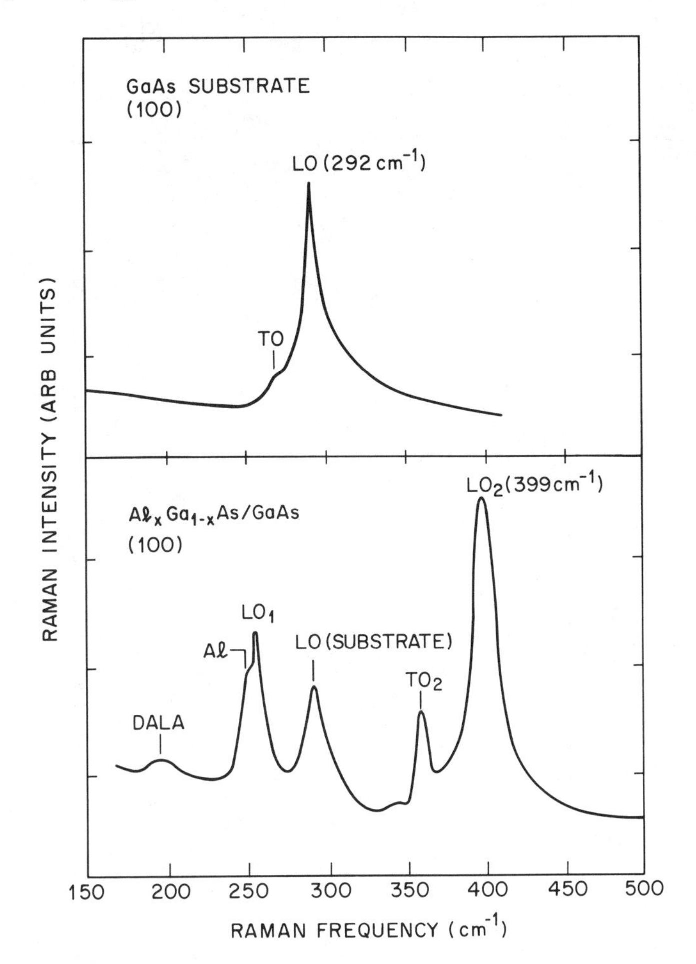


**Figure 5.45** (a) Raman spectrum of a (100) GaAs substrate and (b) Raman spectrum from a $Al_xGa_{1-x}As$ layer grown by LPE on the GaAs substrate in (a) [156]. DALA is disorder activated longitudinal acoustic mode. Al is an Al local mode. The GaAs like ($LO_1$) and AlAs like ($LO_2$) modes are indicated.

strength of the TO line depends strongly on the carbon contamination of the initial surface. They attributed the TO line to a slight misorientation of the growing film resulting from contamination with >0.3 monolayer of carbon. Biellmann et al., [155] have considered in detail sample misorientations and the relative intensities of the LO and TO lines.

It should be noted that the presence of the TO line in the MBE films does not necessarily suggest that they are of lower quality, since in terms of other measurements such as low-temperature exciton spectra, the MBE films are comparable to the ones grown by other techniques. However, in some instances, the appearance of the TO line may reflect the poor quality of the material. Figure 5.45 shows the Raman spectra from a GaAs substrate, used in the growth of an $Al_xGa_{1-x}As$ film by LPE, and from the grown film [156]. Note the symmetry

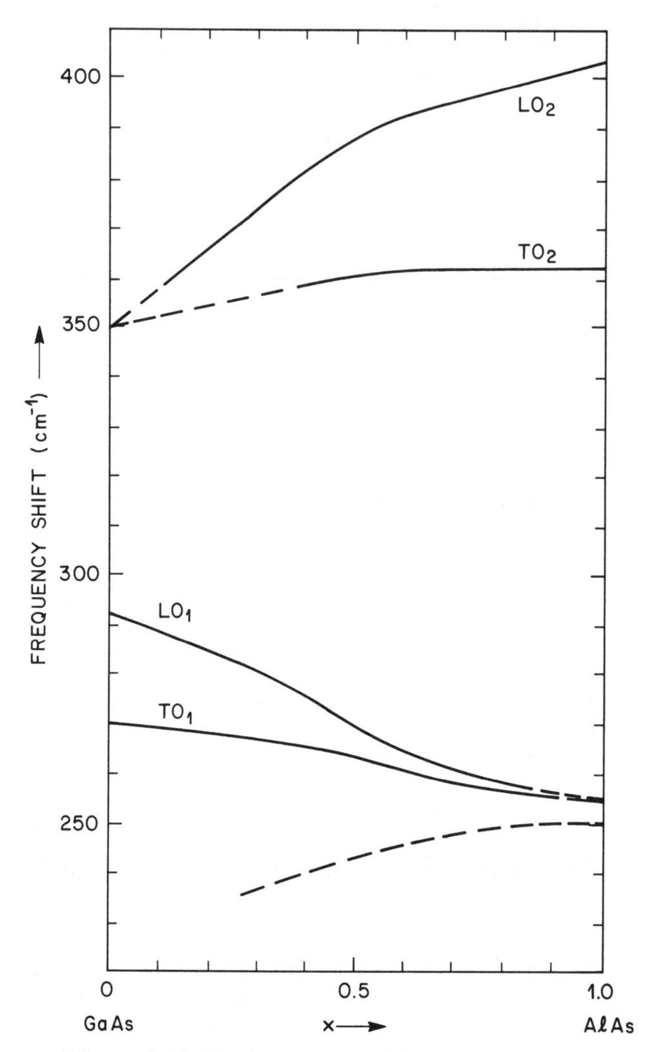


**Figure 5.46** The frequencies of the LO and TO phonons of $Al_xGa_{1-x}As$ as a function of x [149]. $LO_1$ and $TO_1$ are GaAs like and $LO_2$ and $TO_2$ are AlAs like.

forbidden TO line in the substrate as well as in the film. Also the linewidth of the LO lines were broader in comparison to other LPE films that were grown on substrates which did not show the TO line. The broad LO line and the appearance of the TO line may be indicative of poor quality of the epitaxial film.

The symmetry forbidden TO line has also been seen in the Raman spectrum taken near the interface of an $Al_xGa_{1-x}As$ and GaAs layer suggesting breakdown of selection rules, caused by perhaps the interfacial disorder [150]. TO line can also be seen in heavily p-doped GaAs where impurities are believed to be responsible for the breakdown of the selection rules [157].

Nakamura et al., [158] observed the TO line in heat-treated semi-insulating GaAs, though its presence did not have any effect on the thermal conversion of the sample.

In alloy semiconductors such as AlGaAs, two mode behavior is expected in RS. That is, there will be GaAs-like and AlAs-like TO and LO phonons, [149, 159] as illustrated in Fig. 5.46. The frequency variation of the LO and TO modes with Al mole fraction can be used to determine the composition of the alloy. Abstreiter et al., [149] used RS to determine the depth variation of Al composition of a 30 μm thick AlGaAs LPE sample. They obtained Raman spectrum as a function of position on a wedge shaped sample and from the frequencies of the AlAs like LO modes determined the Al mole fraction (Figs. 5.47). One can also use different frequencies of the exciting laser light, so as to vary the penetration depth to probe different layers, as for example, in a multilayer structure. The Al concentrations in a AlGaAs-GaAs-AlGaAs 3-layer structure have been obtained in this manner [149].

In alloy semiconductors, in addition to the LO and TO modes other forbidden modes are also seen. The Raman spectra from a (100) $Al_{0.75}Ga0_{25}As$ VPE film, shown in Fig. 5.48 contains the disorder activated longitudinal-acoustic (DALA) structure and disorder-activated transverse-acoustic (DATA) structure [159]. The spectra were taken at resonant excitation conditions with

**Figure 5.47** (a) Raman spectra from an angle lapped $Al_xGa_{1-x}As$ LPE sample at different depths in the layer as shown in the inset; (b) Al composition of the sample in (a) as a function of depth determined from the mode frequencies and using Fig. 5.46 [149].

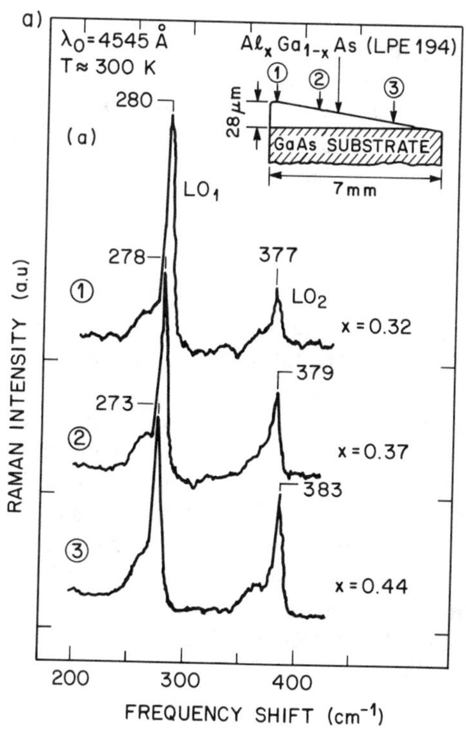


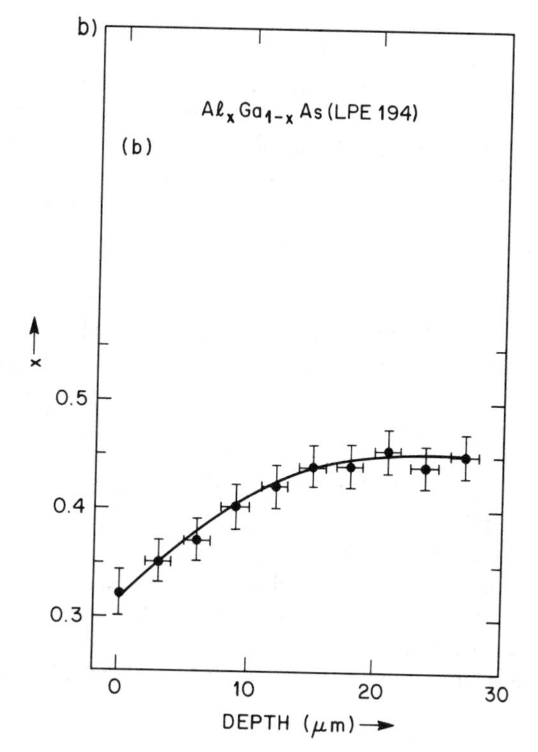

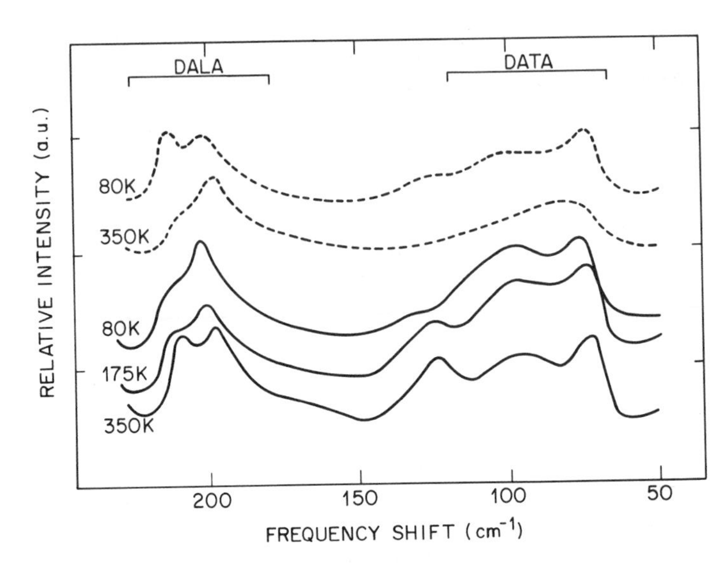


**Figure 5.48** Disorder-activated longitudinal acoustic structure (DALA) and disorder-activated transverse acoustic structure (DATA) in the Raman spectra of $Al_{0.25}Ga_{0.75}As$ at several temperatures. Excitation with the 5145Å (solid lines) and the 4880Å (dashed lines) laser lines [159].

the polarization of the scattered light parallel to that of the incident light. The lines around 200 cm$^{-1}$ are generally attributed to DALA and the ones between 70 and 150 cm$^{-1}$ to DATA [159]. The energy and temperature dependence of these lines cannot be explained within the frame work of RS theory for perfectly ordered materials. They arise due to one-phonon RS by short wavelength acoustic phonons which become Raman active because of the disorder in the lattice resulting from the substitution of Ga and Al in the Ga sub-lattice. Disorder modes have been seen in GaInAs [160], and GaInAsP [161].

**LO phonon line shape**

It has been observed in alloy semiconductors that the LO phonon lines are asymmetric and broader than the ones in the constituent binary semiconductors. The full width at half maxima of the LO mode is given by the sum of two parameters, $\Gamma = \Gamma_a + \Gamma_b$, with $\Gamma_a$ being greater than $\Gamma_b$. The broadening occurs on the low energy side as illustrated in Fig. 5.49 [162]. The broadening $\Gamma$ and asymmetry $\Gamma_a/\Gamma_b$ of the LO phonon mode are shown in Fig. 5.50 for a number of AlGaAs/GaAs and GaInAs/InP samples [163]. The broadening and asymmetry can be quantitatively explained by q-vector relaxation induced by the alloy potential fluctuations [163]. The second ordinate in Fig. 5.50 is a variable L in the LO line shape equation which can be taken as a microscopic spatial correlation length over which the material is compositionally ordered. Note in Fig. 5.50 that for Ga-rich AlGaAs the GaAs mode is narrower and more symmetric than the AlAs mode. The reverse is true for AlAs mode in the Al-rich AlGaAs. The reason for this is attributable to the composition dependence of L.

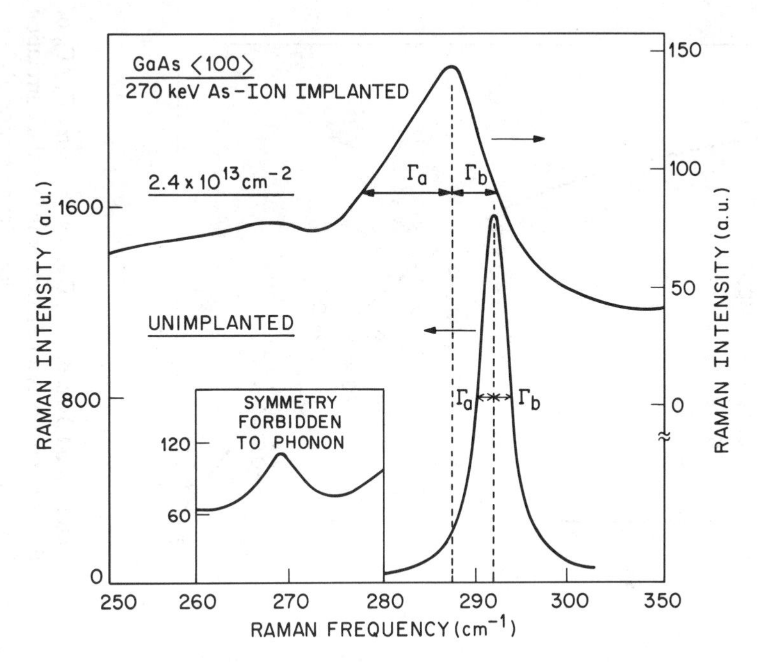


**Figure 5.49** Raman spectra from an unimplanted and C implanted (100) GaAs showing the broadening and asymmetry of the LO phonon line in the implanted sample [162].

**Second-order Raman scattering**

From Eq. (5.66) it follows that for the two phonon Raman process the phonons involved are those for which the sum of their wavevectors is very nearly equal to zero. Phonons throughout the Brillouin zone can take part in the second order RS. Near-zone edge phonons for example, the X and L phonons are favored because of density of states considerations. The properties of the two-phonon RS can be utilized to characterize alloy semiconductor.

### 5.3.4 Superlattices and Quantum Well Structures

Raman scattering has been used quite extensively to study superlattices and quantum well structures. One of the areas of great interest is light scattering by two-dimensional electron or hole gas in selectively doped superlattice structures. Menendez et al., [164] have determined conduction band offset in GaAs-AlGaAs quantum wells using a light scattering method. Also in a superlattice, the periodicity in the direction perpendicular to the layers gives rise to a folding of the Brillouin zone [165]. The manifestation of zone-folding effect is evidenced by the presence of new Raman lines as shown in Fig. 5.51 [166] which shows the Raman spectrum from a GaAs-AlAs superlattice consisting of 1720 layers each of GaAs wells of 13.6Å and AlAs barriers of 11.4Å. The sharp lines, at about 65cm$^{-1}$ labelled $A_1^{(1)}$ and $B_2^{(1)}$ are from LA phonons folded to the zone center from $q = 2\pi/d$ in the extended zone, where d is the superlattice period. The energy position of the zone-folded lines depends on d and hence can be used to determine the

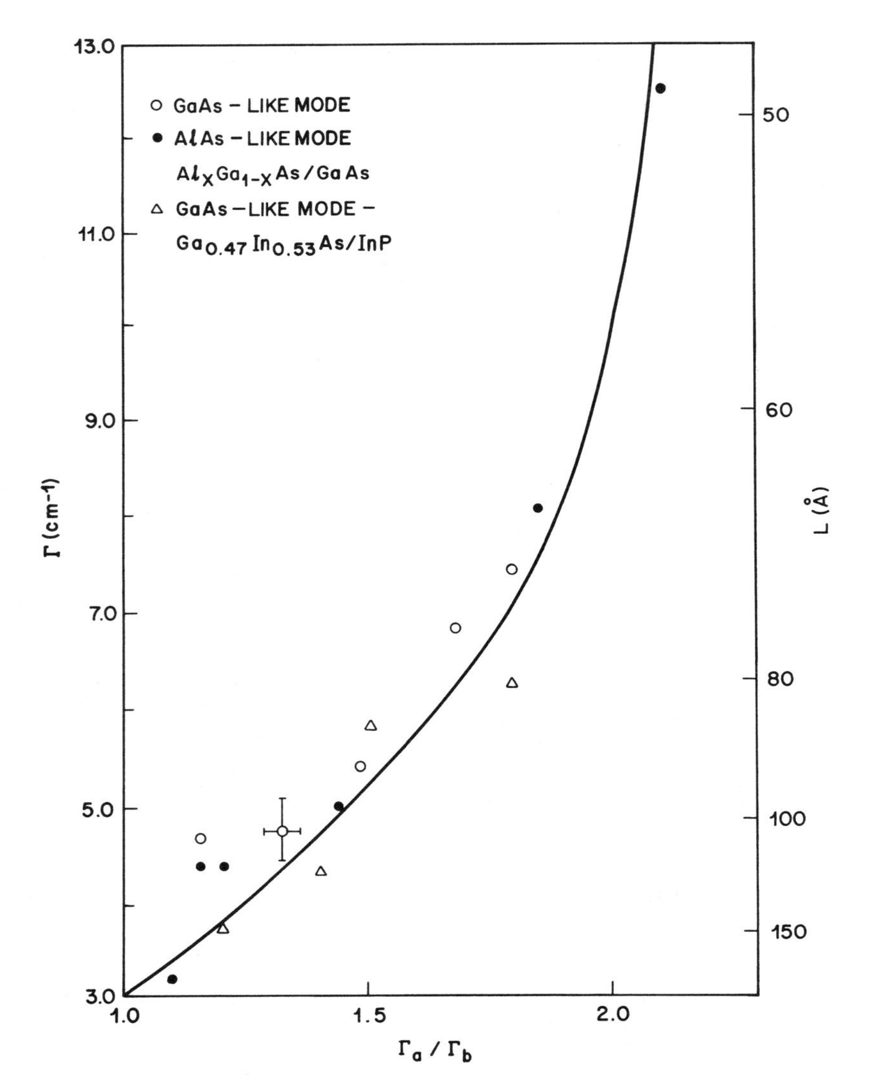


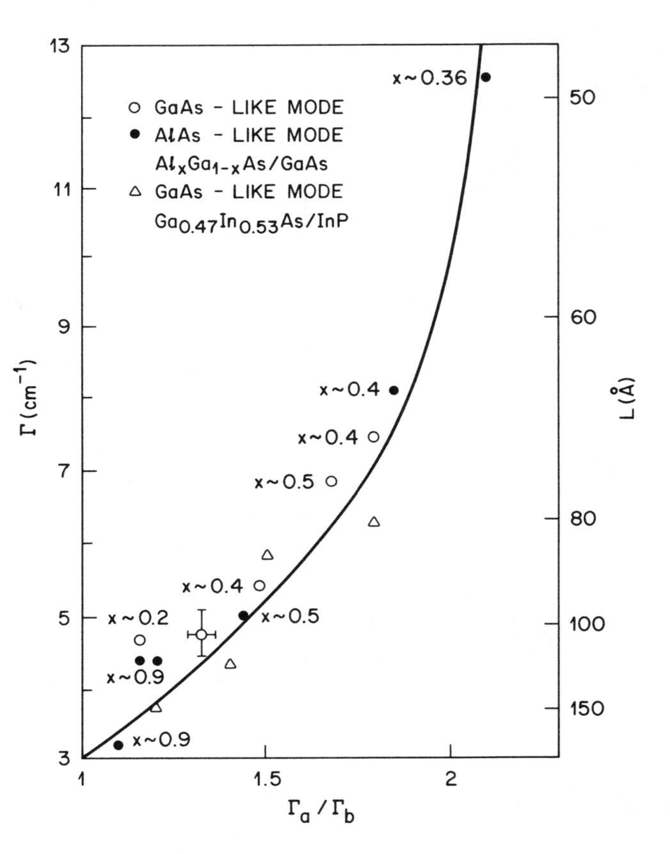


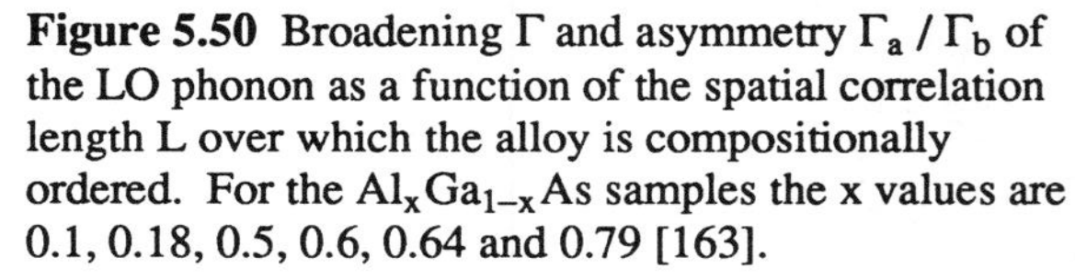

**Figure 5.50** Broadening $\Gamma$ and asymmetry $\Gamma_a / \Gamma_b$ of the LO phonon as a function of the spatial correlation length L over which the alloy is compositionally ordered. For the $Al_xGa_{1-x}As$ samples the x values are 0.1, 0.18, 0.5, 0.6, 0.64 and 0.79 [163].

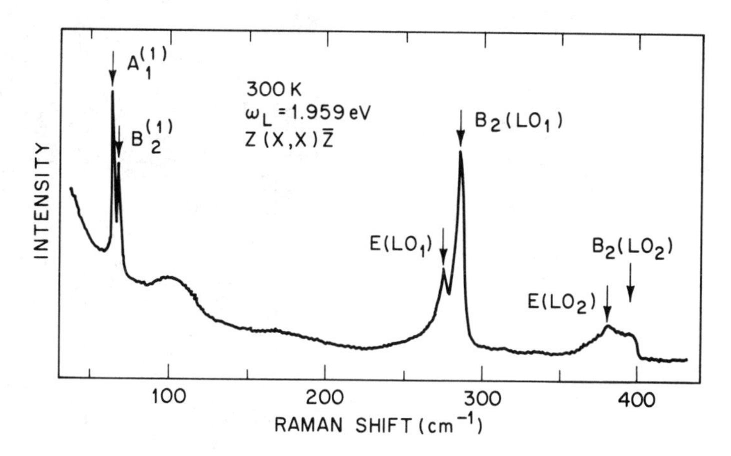


**Figure 5.51** Raman spectrum of a superlattice consisting of 1720 periods of 13.6Å GaAs and 11.4Å AlAs grown on (100) GaAs [166]. The sharp features denoted $A_1^{(1)}$ and $B_2^{(1)}$ at ~65 $cm^{-1}$ are from LA phonons folded to the zone center from $q = 2\pi / d$ in the extended zone where d is the superlattice period. $B_2$ ($LO_1$) is the GaAs-LO phonon. E($LO_1$) and E($LO_2$) are due to scattering from zone center LO phonons propagating parallel to the layers, i.e., with E symmetry.

superlattice period. Pollak and Tsu [147] suggested that the width of the zone-folded lines could be a good indicator of superlattice quality.

In quantum well structures resonant RS can give information about the band-offsets [164] as well as on the transport properties of the confined electrons [167]. In the GaAs-AlGaAs system the band-offsets have been determined by electronic light scattering studies on photoexcited samples [164]. For a ten period GaAs (334Å) and $Al_xGa_{1-x}As$ ($x \sim 0.06$, 459Å) layer the excitation energy was chosen to be near $E_o + \Delta_o$ gap of GaAs, for which resonant enhancements of the electronic light scattering occurs. The spectra show (Fig. 5.52) peaks which are assigned to different intersubband transitions (shown in the inset) between the first four levels in the conduction band of the GaAs quantum wells. To ensure that the measured transition energies coincide with the actual intersubband energies, the spectra were taken in the depolarized configuration, that is, with the polarization of the incident and scattered light perpendicular to each other. By calculating the energy level structure in the conduction band using the conduction band discontinuity, $\Delta E_c$, as an adjustable parameter to match the measured values, gave $\Delta E_c$ for the best fit. To determine $\Delta E_c / \Delta E_g$ where $\Delta E_g$ is the total band gap discontinuity, resonant RS at delocalized excitons was used. Figure 5.53 shows the resonance behavior of the GaAs like LO phonon of the AlGaAs layer in the superlattice as a function of laser energy. The Raman spectra were taken in the polarized configuration (parallel incident and scattered polarizations). Two resonances can be seen in Fig. 5.53 one in which the incident laser energy is equal to the excitonic transition and another in which the scattered photon energy is equal to the excitonic transition. Given LO ~36.25 meV for the sample in question, the exciton energy is obtained from

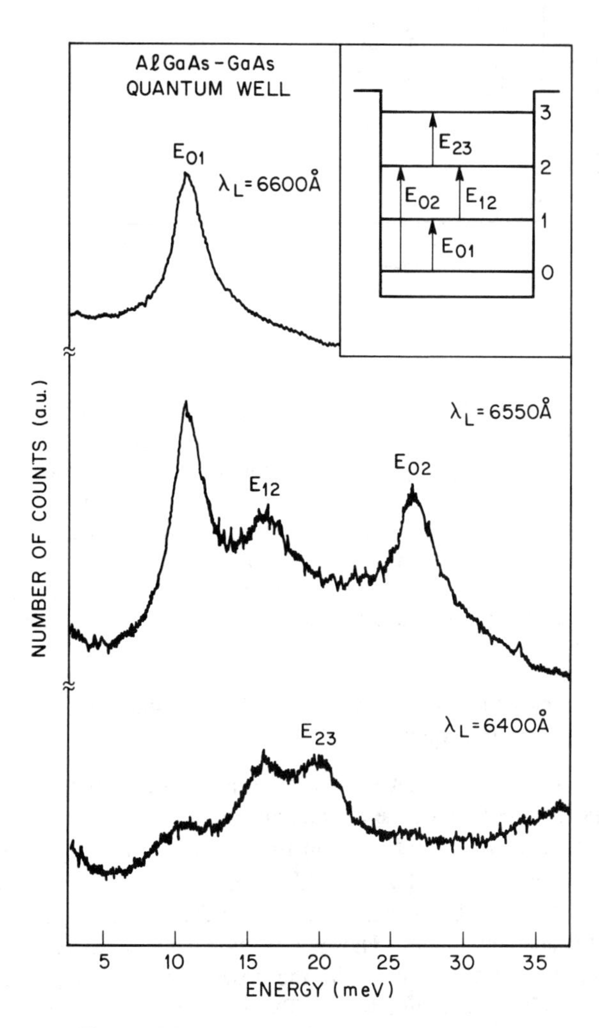


**Figure 5.52** Depolarized (polarization of the incident and scattered light perpendicular to each other) Raman spectra for three different excitation wavelengths from a multiquantum well sample consisting of 10 periods of 334 ± 3Å GaAs and 459 ± 3Å $Al_xGa_{1-x}As$ layers (x ~ 0.06). The inset shows schematically the labelled transitions [164].

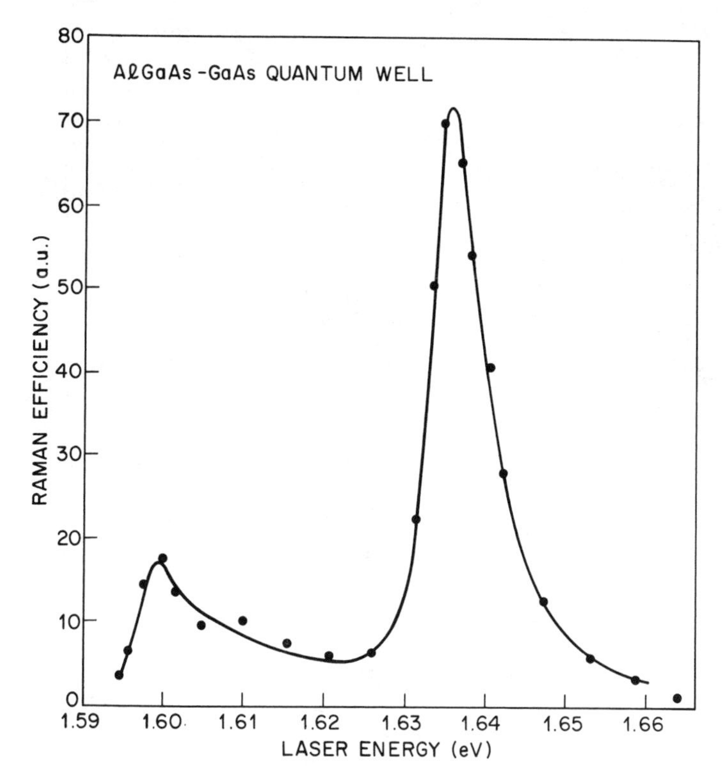


**Figure 5.53** Resonance of the Raman efficiency of the $LO_1$ phonon of $Al_xGa_{1-x}As$ in sample of Fig. 5.52 as a function of laser energy. The Raman spectra were taken in the polarized configuration [164].

Fig. 5.53 to be $1.599 \pm 0.001$ eV. To obtain the band gap from this value, two corrections have to be made. One is the difference between the delocalized levels and the barrier heights which has to be subtracted and the other is the exciton binding energy which has to be added. These two values are roughly of equal magnitude and tend to cancel each other. Thus, a band gap of $1.599 \pm 3$ meV is estimated for the AlGaAs in the structure. From the $\Delta E_c$ and $\Delta E_g$ thus measured $\Delta E_c/\Delta E_g$ is found to be $0.69 \pm 0.03$ which is in agreement with many other experimental data for different x values. Thus, RS offers a unique nondestructive way for determining the band gap discontinuity and conduction band offsets.

The FWHM of intersubband transitions in resonant RS spectra from modulation-doped GaAs-AlAs quantum well structures has been correlated with Hall mobilities [167]. Figure 5.54 shows the FWHM of the $E_{01}$ transition (see the inset in Fig. 5.52) of modulation doped GaAs-$Al_{0.12}Ga_{0.88}As$ samples with Hall mobilities 28,000 $cm^2/Vsec$ (Sample 1), 62,000 $cm^2/$ Vsec (Sample 2), and 93,000 $cm^2/Vsec$ (Sample 3). Note the increase in FWHM with decreasing Hal mobility. Apart from this dependence between mobility and FWHM, in the low mobility sample there is a photon energy dependence of FWHM. This effect has been attributed to wavevector

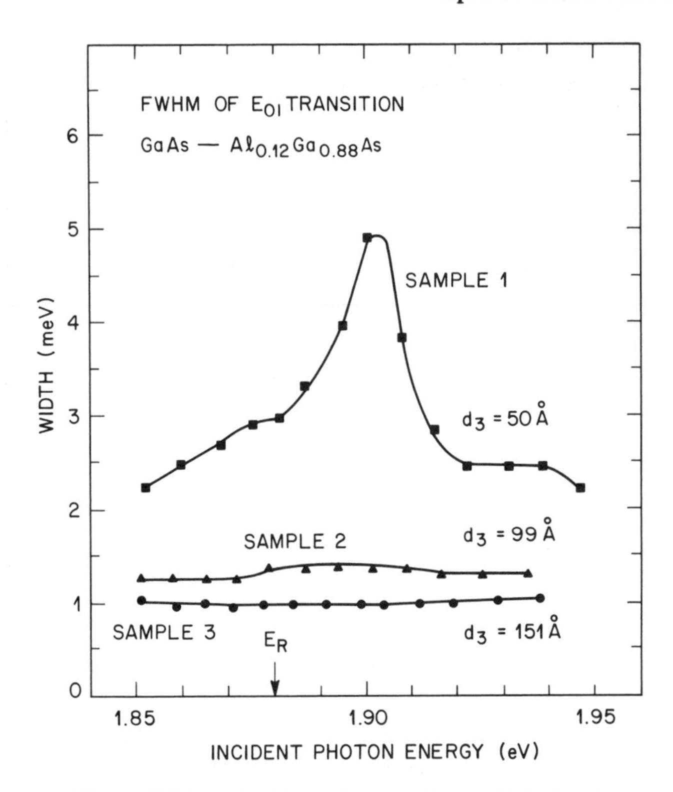


**Figure 5.54** Full width at half maxima (FWHM) of the $E_{01}$ intersubband transition (see inset Fig. 5.52) as a function of the incident phonon energy for three $Al_{0.12}Ga_{0.88}As$/GaAs single quantum well samples. Sample 1 has a mobility of 28,000 $cm^2$ V $sec^{-1}$, Sample 2, 62,000 $cm^2$ V $sec^{-1}$ and Sample 3, 93,000 $cm^2$ V $sec^{-1}$ [162]. $E_R$ is the photon energy of the maximum in resonance enhancement.

relaxation caused by scattering of electrons by the Coulomb potential of the ionized donors, which also limits electron mobilities.

### 5.3.5 Applications of RS

#### Ion-implanted samples

When doping of the semiconductor is achieved by ion-implantation the implanted region is rendered disordered and to activate the dopants and to restore crystallinity of the implanted region the semiconductor is suitably annealed. Since RS depends on the lattice structure, it has been used by several authors to study ion-implantation damage and its recovery with annealing in GaAs and InP. Figure 5.55 shows the unpolarized Raman spectrum of a <100> GaAs sample in the unimplanted condition and the Raman spectra obtained for various polarization configurations

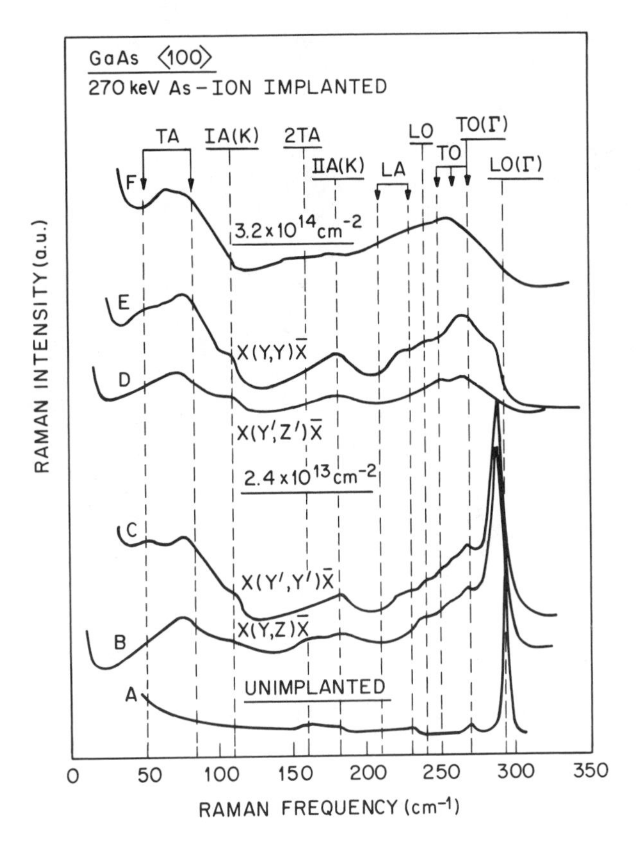


**Figure 5.55** Raman spectra os (100) GaAs before and after implantation with As to various fluences up to $3.2\times10^{14}$ $cm^{-2}$ for different polarization configurations B, C, D and E [162].

after 270 keV As-ion implantation to a fluence of $2.4\times10^{13}$ $cm^{-2}$ (curves B, C, D and E) or $3.2\times10^{14}$ $cm^{-2}$ (curve F) [162]. The unimplanted sample shows a strong LO phonon line at 290 $cm^{-1}$ with a half-width of 3.0 $cm^{-1}$ (after correcting for instrument resolution). A weak symmetry forbidden TO line and other features at 150 to 180 $cm^{-1}$ caused by second order modes are also seen. The polarization conditions are denoted by the standard notation $x(y,z)\bar{x}$ where x, y, z correspond to [100], [010] and [001] directions; y′ and z′ denote [011] and [0$\bar{1}$1]. For a fluence of $2.4\times10^{13}$ $cm^{-2}$, the polarization selection rules are still obeyed, as evidenced by the absence of LO phonon which is forbidden for the polarization conditions $x(y', z')\bar{x}$ (curve D) and x (y, y) x (curve E). When the fluence is $3.2\times10^{14}$$cm^{-2}$, the spectrum approaches that of an amorphous material. The additional features seen below 250 $cm^{-1}$ can be related to zone edge phonons. They are activated because of the disorder which destroys the q-vector conservation.

The major effect of ion-implantation on the Raman spectrum is the broadening and softening of the LO phonon line. This is illustrated in Fig. 5.56 which shows the broadening of the LO line

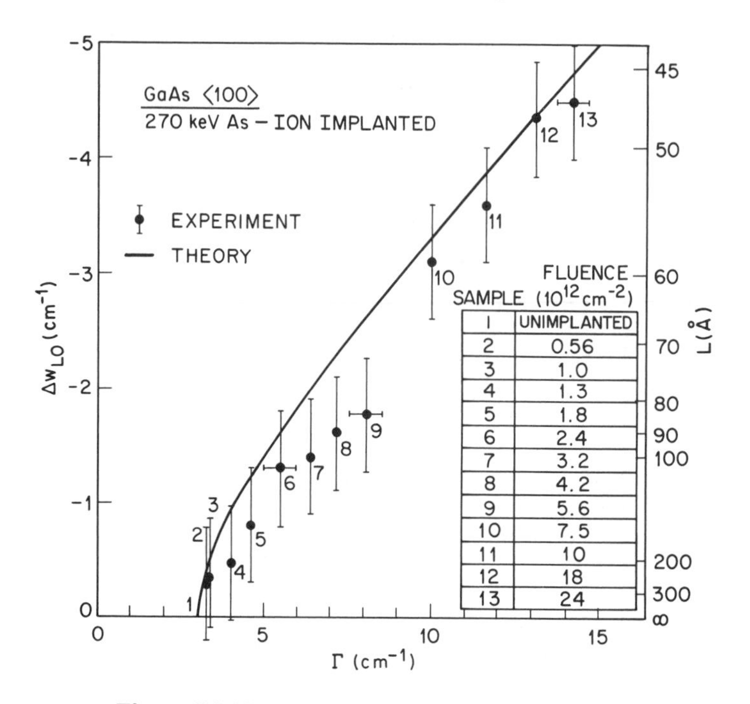


**Figure 5.56** Shift of LO phonon and broadening Γ as a function of correlation length L for different fluences of As implanted GaAs. Γ is corrected for spectrometer resolution [162].

(Γ) and the red shift ($\Delta\omega_{LO}$). Besides the increase in Γ, the asymmetry $\Gamma_a / \Gamma_b$ also increases with fluence. The TO line shows only a small symmetric red shift. Tiong et al., [162] have quantitatively explained the behavior of the LO and TO modes by a "spatial correlation" (SC) model [163] similar to the one proposed for RS in microcrystalline Si. The physical basis of the SC model is that as the crystal is damaged by ion-implantation, the mode correlation functions of the phonons become finite which relaxes the q-selection rules, whereas in an "ideal" crystal the correlation function is infinite in extent which leads to the usual plane-wave phonon eigenstates and the q=0 selection rules for first order RS. For the damaged crystal, a Gaussian attenuation factor, $\exp(-2r^2/L^2)$, where L is the diameter of the correlation region, is assumed for the phonon wavefunction. This function upon Fourier transformation leads to an average over q with a similar weighting factor, $\exp(-q^2L^2/4)$.

Based on the SC model, Tiong et al., [162] give the intensity, I, of the Raman line at frequency, ω

$$I(\omega) \propto \int_0^1 \exp\frac{-q^2L^2}{4} \frac{1}{[\omega-\omega(q)]^2 + \frac{\Gamma_o}{2}^2} d^3q \tag{5.67}$$

where q is in units of $2\pi/a$, a being the lattice constant, $\Gamma_o$ is the width of the Raman line in the unimplanted crystal and $\omega(q)$ represents the phonon dispersion relation. For $\omega$ (q) Tiong et al., [162] use

$$\omega(q) = A + B \quad \cos(\pi q) \tag{5.68}$$

where A and B are constants. For GaAs they are 269.5 $cm^{-1}$ and 22.5 $cm^{-1}$, respectively. Using Eqs. (5.67) and (5.68), the shift $\Delta\omega_{LO}$ and width $\Gamma$ of the LO phonon line as a function of L are calculated for various fluences for GaAs and compared with experimental values (Fig. 5.56). The asymmetry $\Gamma_a/\Gamma_b$ of the LO line can also be evaluated for various L. Thus, the red shift and asymmetric broadening of the LO phonon line can be related to the size of the undamaged region. The SC model also explains the small red shift of the TO line and the appearance of the zone-edge phonons with increasing fluence.

**Effect of strain**

Strain gives rise to broadening and shift of the optic phonons and as such can be measured from Raman spectrum. Several authors have used RS to measure stress in Si/ $Si0_2$ or Si/ $Al_20_3$ interfaces [147]. In <100> GaAs and InP, Raman spectra obtained for various surface polishing conditions show that the LO phonon line broadens and shifts to higher frequency when the surface damage increases [152, 168]. The increased linewidth is caused by the increased anharmonic decay and scattering channels caused by reduced local symmetries and/or shifting of the peak frequency caused by inhomogeneous strain. Shen et al., [168] investigated LO phonon line shapes in (100) GaAs and InP which have been polished with differing size polishing grit for various laser excitation frequencies. They explained the line shapes by a model based on the convolution of the penetration depth of the light and the skin depth of the mechanical damage induced strain. In their theory, the frequency shift of the LO phonon caused by strain and the skin depth of the strain were used as adjustable parameters. Figure 5.57 shows the Raman spectra from surface damaged (100) GaAs and InP. The calculated line shape according to this model agrees well with the experiment. The frequency shift is related to the strain by

$$\frac{\Delta\omega}{\omega_o} = \left[\frac{q}{\omega_o^2} + \frac{p}{\omega_o^2}\left(\frac{S_{12}}{S_{11} + S_{12}}\right)\right]\varepsilon \tag{5.69}$$

where $\varepsilon$ is the strain, p and q are coefficients which describe the strain-induced shift of the optic phonons, $S_{11}$ and $S_{12}$ are elastic compliance coefficients and $\omega_o$ is the unperturbed Raman frequency. The parameters p and q are not very different for different III-V materials. The term in the bracket in Eq. (5.69) for GaAs is $-1.5$ and for InP is $-1.3$.

In the growth of alloy semiconductor films, lattice mismatch between the epitaxial film and the substrate can give rise to strain and hence broadening of the LO phonons. Kakimoto and Katoda [160] measured the Raman spectra from an angle lapped 1 μm thick $Ga_{0.9}In_{0.1}As$ layer grown on a GaAs substrate as a function of distance from the epi-substrate interface and found that the LO phonon line near the interface is broader compared to those far from the interface.

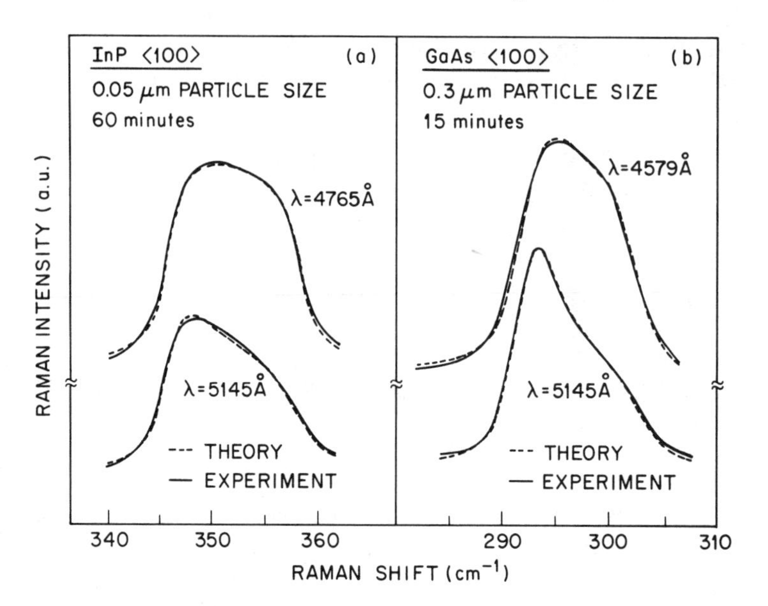


**Figure 5.57** Comparison of experiment and theory for (100) InP damaged with 0.05μm particle size for 60 min (a) and for (100) GaAs damaged with 0.3μm particle size for 15 min (b). The excitation wavelengths are indicated [168].

### Determination of carrier concentration and mobility in n-type material

In polar semiconductors it is possible to have a coupled LO phonon-plasmon mode because of the lack of inversion symmetry and the resulting net electric dipole moment. The frequencies of the coupled modes (generally labelled $L^+$ and $L^-$) are given by [169]

$$\omega = \left\{ \frac{\omega_{L_o}^2 + \omega_p^2(q)}{2} \pm \left[ \frac{[\omega_{L_o}^2 + \omega_p^2(q)]^2}{4} - \omega_p^2(q)\omega_{TO}^2 \right]^{\frac{1}{2}} \right\}^{\frac{1}{2}} \tag{5.70}$$

where $\omega_p(q)$ is the wavevector dependent plasma frequency

$$\omega_p^2(q) = \omega_p^2 + \frac{3}{5}(qV_F)^2 \tag{5.71}$$

with

$$\omega_p^2 = \frac{4\pi ne^2}{\varepsilon_\infty m^*} \tag{5.72}$$

n is the carrier concentration, $V_F$ the Fermi Velocity and $m^*$ is the effective mass of the carriers. The + or − sign in Eq. (5.70) give $L^+$ or $L^-$ mode frequencies, respectively. The value of $\omega_{L^+}$ depends strongly on n for $n \geq 5\times10^{17}\,cm^{-3}$ whereas $\omega_{L^-}$ depends strongly on n for $1\times10^{16} \leq n \leq 5\times10^{17}\,cm^{-3}$. Generally, $\omega_{L^+}$ is more frequently used than $\omega_{L^-}$ to measure n. Figure 5.58 gives the dependence of $\omega_{L^+}$ mode on n in GaAs for various laser lines [149].

The carrier concentration determined from $\omega_{L^+}$ agrees well with the values determined from electrical measurements. The attractive feature of the Raman technique is that it requires no contacts, that it can be used for thin films irrespective of the substrate unlike Hall measurements which require semi-insulating substrates and that it can be performed with high spatial resolution. Olego and Serreze [170] have compared the n values obtained from $L_+$ modes in RS and from the linewidth of the emission band in PL in ion-implanted and annealed InP.

**Figure 5.58** The frequency of the $L_+$ mode versus the carrier concentration for different excitation wavelengths. The open and closed circles are experimental points [149].

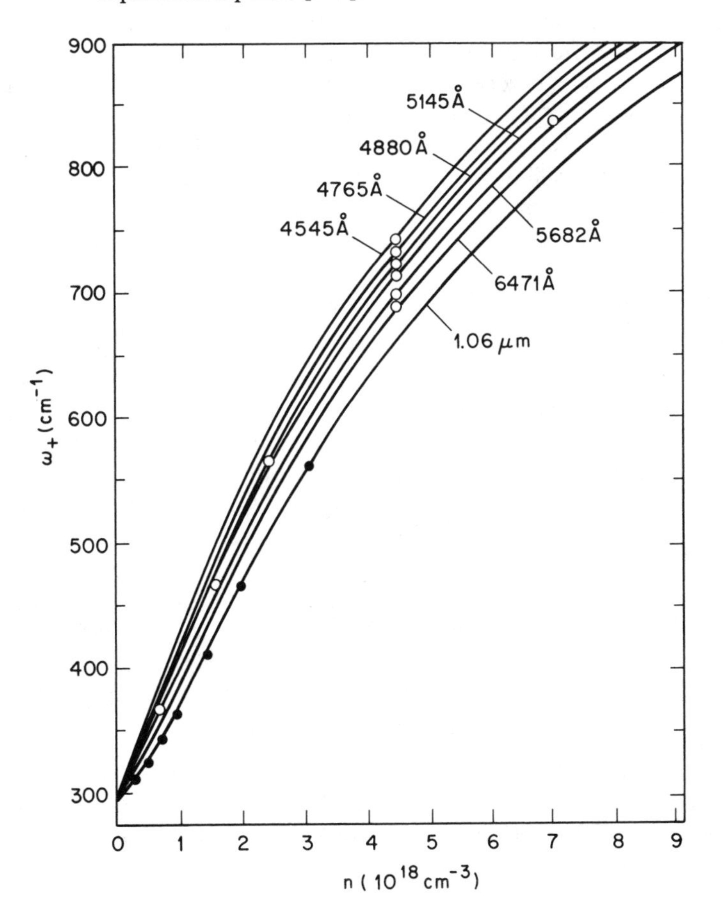

One can also use the coupled mode to determine electron effective mass if n in Eq. (5.72) is known from other means. Olego et al., [171] thus determined the compositional dependence of the electron effective mass in AlGaInAs lattice matched to InP.

Besides n and $m^*$ obtained from the position of the $L_+$ mode, its linewidth ($\Gamma_{L_+}$) can give information about electron scattering time ($\tau$) since $\Gamma_{L_+} \sim 1/\tau$ [149]. Since $1/\tau = e/m^*\mu$, the mobility $\mu$ can be obtained from $\Gamma_{L_+}$. If $\Gamma_{L_+}$ is larger than what is expected from mobility, other scattering mechanisms can be suspected. In GaAs for $n \sim 7\times10^{17}\text{cm}^{-3}$, $1/\tau$ determined from $\Gamma_{L_+}$ is about a factor of two larger than calculated from $\mu$. Surface scattering is thought to be the additional scattering mechanism. For $n > 10^{18}\text{cm}^{-3}$ better agreement between $\Gamma_{L_+}$ and $\tau$, calculated from $\mu$, is seen [149].

In p-type material the LO phonon-plasmon mode is not observed since because of increased hole effective mass $\omega_p$ decreases (Eq. (5.72)). As a result the frequency of the coupled mode is very close to the unscreened LO frequency.

**Surface band bending and barrier heights**

We saw in this section that at high carrier concentrations one observes coupled LO-plasmon modes. Under certain circumstances both the coupled LO phonon-plasmon mode and the unscreened LO phonon can be observed. For example, at a real surface of a semiconductor there normally exists a depletion region. The width of this layer varies with carrier concentration, decreasing with increasing carrier concentration. Therefore, in order to observe the coupled modes in the bulk, the penetration depth of the laser should be greater than the depletion width. However, by varying the laser line one can observe both the coupled and the unscreened mode, the former coming from the bulk and the latter from the surface depletion region. Thus, by measuring the relative intensities of the two lines surface band bending and barrier heights can be determined. This has been done for free and covered GaAs surfaces [172].

Surface or interface electric fields caused by surface or interface space charge layer can give rise to breakdown of symmetry selection rules. Since the intensity of the symmetry forbidden lines is proportional to the square of the electric field, barrier heights can be determined. The usefulness of this approach to determine Fermi level pinning on clean and covered (110) GaAs surfaces has been demonstrated [172]. For this orientation the back scattering geometry allows only the TO phonons (Table 5.4) in first order RS. The LO phonon is forbidden by symmetry. In GaAs, to calibrate the intensity of the LO phonon with electric field, its dependence was studied as a function of oxygen exposure and compared with the results obtained from ultraviolet photoelectron spectroscopy and contact potential difference measurements. Since the intensity of the allowed TO line, $I_{TO}$, is independent of the surface electric field, the LO phonon intensity, $I_{LO}$, is normalized to $I_{TO}$. The results are shown in Fig. 5.59 for (110) GaAs with $n \sim 7\times10^{17}\text{cm}^{-3}$. The saturation value of the barrier height for an oxygen exposure of about $10^5$ has been used to relate $I_{LO}/I_{TO}$ directly to the barrier height.

The same approach has also been used to study the changes of the barrier height for Ge coverage on GaAs. The Ge layer was grown on UHV cleaved (110) GaAs at 675K. For small Ge coverages, only the TO phonon of the GaAs substrate was observed as expected from selection rules. The optical phonon of the Ge film was seen even for a Ge layer thickness of only six

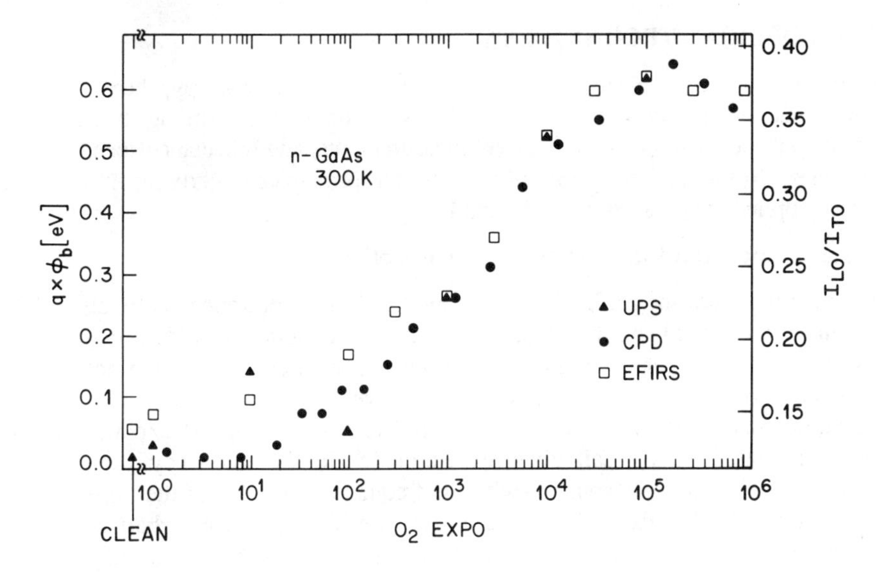


**Figure 5.59** Barrier height ($q\phi_b$) in n-GaAs as a function of oxygen exposure. The relation between electric field induced (EFIRS) LO phonon intensity $I_{LO}$ normalized to $I_{TO}$, $I_{LO} / I_{TO}$, and $q\phi_b$ is also indicated [172]. The barrier heights are obtained from ultraviolet photoelectron spectroscopy (UPS) and contact potential difference (CPD) measurements.

monolayers, its intensity increasing strongly with film thickness. When the layer thickness was greater than 40Å, the bulk Ge phonon energy of 37.4 meV was obtained. For layers thinner than 40Å, the energy was slightly less than this value. The barrier height in GaAs, determined from $I_{LO}/I_{TO}$, increases rapidly with Ge coverage at submonolayer level, almost independent of the growth temperature. For crystalline films the barrier height reaches a maximum and decreases after about 10Å while for amorphous films it also reaches saturation after ~10Å. The saturation has been explained as a result of Fermi-level pinning at interface states.

Zahn et al., [173] studied the electric field induced LO phonons in (110) InP on to which Sb was deposited at 80K in a UHV chamber. The RS experiments were also done at 80K *in situ* after each evaporation stage. The ratio $I_{LO}/I_{TO}$ decreased with Sb coverage indicating that the initial band bending at the cleaved InP surface is also reduced. By calibrating $I_{LO}/I_{TO}$ rates against the surface potential of an InP/air interface determined by contact potential difference, it was determined that for a cleaved clean surface of InP at 80K the Fermi level is pinned between 330 and 440 meV below the conduction band.

In addition to the study of band bending at clean surfaces, RS can also be used to investigate real surfaces. Pinczuk et al., [174] used RS to study the Schottky barrier of Ag on InP. The sample was a <111> B InP surface and a semitransparent Ag was evaporated on it. From RS it was deduced that the band bending is 0.55 eV compared to the value of 0.2 eV before evaporation.

## 5.4 OTHER OPTICAL TECHNIQUES

Besides optical absorption, photoluminescence and Raman spectroscopy, there are a few other optical techniques which are used to study III-V semiconductors. Among them, infrared local mode vibrational spectroscopy, photo current measurements, and reflectance spectroscopy are the important ones. In this section we briefly review the principles underlying these methods and how they are applied to characterize GaAs and InP.

### 5.4.1 Infrared Localized Vibrational Mode Absorption

The normal modes of the perfect lattice are modified in the presence of defects and the extent of the modification depends on the nature of the defect. In some cases defects give rise to new modes with frequencies greater than the maximum phonon frequency of the defect-free lattice. Unlike the modes in the perfect lattice, which are travelling plane waves, the defect modes are localized vibrational modes with the amplitude of vibration falling off exponentially from the impurity. Sometimes defects can give rise to modes in the frequency gap in the phonon dispersion curve or modes in resonance with the frequency spectrum of the unperturbed crystal. In semiconductors the localized vibrational modes are used for defect characterization [175 - 177]. These modes are infrared active and are observed as bands in infrared absorption measurements. They can also be seen in Raman scattering experiments.

If a substitutional impurity of mass $M^1$ replaces a host atom of mass M but with no change in the force constants, that is, in the isotopic assumption, the frequency $\omega_L$ of the local mode is $\omega_L \approx (k/M^1)^{1/2}$ where k is the force constant and $\omega_L$ is greater than $\omega_m$ the maximum phonon frequency of the lattice for $M^1 \ll M$. The impurity has tetrahedral symmetry and has the potential of a spherical harmonic oscillator for small displacements of the impurity from equilibrium. Thus vibrations in x, y, and z directions are equivalent and the local mode is triply degenerate.

If the substitutional impurity which gives rise to the local mode is also a dopant, then there is free carrier absorption which can interfere with the local mode measurement. Since the absorption cross section of the localized vibrational mode is much smaller than that of the free carriers it is necessary to reduce the free carriers before the measurement of the localized modes. The free carrier concentration has to be reduced by electrical compensation that is, without reduction of the impurity concentration. In local mode measurement compensation has been achieved either by electron irradiation or by diffusion of a compensating impurity. In either case, however, there is the risk of interaction of the defect under study either with the radiation induced defects or with the diffusing impurity, depending on the method used for compensation. As for techniques measuring the local mode frequencies infrared absorption is the mostly used technique. With the availability of Fourier transform infrared spectrometer, measurements can be made at high spectral resolution. Local modes can also be observed in Raman measurements. They have the advantage over infrared measurements in that electrical compensation is not required since measurements are made at photon energies near the energy gap such that free carrier absorption is low.

In the case of GaAs, defects associated with Si impurity have received a considerable amount of attention in local mode experiments [175, 176]. Table 5.5 lists the local mode frequencies for Si, B, C, P, and Al in GaAs and B in InP measured at 77K. A defect pair such as the $Si_{Ga} - Si_{As}$ complex is expected to have four modes as sketched in Fig. 5.60. The approximate values of these modes are 419, 390, 369, and 327 $cm^{-1}$. The experimentally observed lines at 367 and 393

**TABLE 5.5**

Local mode frequencies (cm -1) for substitutional impurities in GaAs and InP (77K) [176]

| | Fundamental (Theory) | Fundamental (Experiment) | Second Harmonic |
|---|---|---|---|
| $^{31}$P(As) | 351 | 355.4 | 709.7 |
| $^{27}$Al(Ga) | 369 | 362 | 722 |
| $^{28}$Si(Ga) – $^{28}$Si (As) | 369 | 367.2 | |
| $^{28}$Si(Ga) – $V_{Ga}$ | | 368.5 | |
| $^{30}$Si(Ga) | 352 | 373.4 | |
| $^{29}$Si(Ga) | 357.5 | 378.5 | |
| $^{28}$Si(Ga) | 363 | 383.7 | 766.5 |
| $^{28}$Si(Ga) – $^{28}$Si(As) | 390 | 393 | |
| $^{28}$Si(Ga) | 365 | 398.2 | |
| $^{28}$Si(Ga) – $^{28}$Si(As) | 419 | 464.0 | |
| $^{11}$B(Ga) | 542 | 517.0 | |
| $^{10}$B(Ga) | 570 | 540.2 | |
| $^{13}$C(As) | 500 | 561.2 | |
| $^{12}$C(As) | 519 | 582.4 | |
| $^{10}$B(In) | 558 | 543.5 | |
| $^{11}$B(In) | 536 | 522.8 | |

**Figure 5.60** Schematic view of the four localized vibrational modes of the ($Si_{Ga} - Si_{As}$) pair in GaAs.

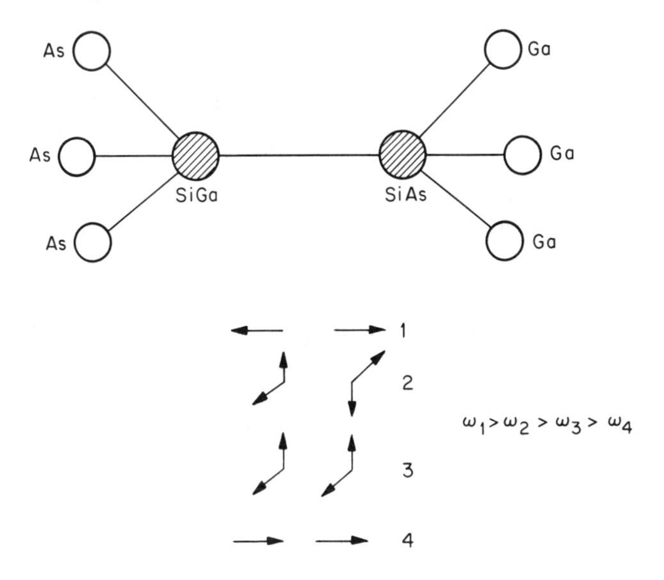

$cm^{-1}$ are close to the predicted 369 and 390 $cm^{-1}$. The mode at 464 $cm^{-1}$ is larger than the calculated value of 419 $cm^{-1}$, but this is the out-of-phase axial mode whose frequency is extremely sensitive to the $Si_{Ga} - Si_{As}$ force constant which is not included in the calculation. The line expected at 327 $cm^{-1}$ is not observed owing to large lattice absorption in this frequency range.

The assignment of the C modes in GaAs to $C_{As}$ has been made by Theis et al., [178] based on the splitting of the $^{12}C$ mode produced by the differences in the isotopic masses of nearest-neighbor host atoms. The $^{12}C$ local mode was measured under high resolution using a Fourier transform infrared spectrometer. Figure 5.61 shows the localized vibrational mode absorption

**Figure 5.61** Localized vibrational mode absorption band for $^{12}C$ in GaAs as measured at 80K with three different instrumental resolutions. The values in parentheses are estimated band strengths [178].

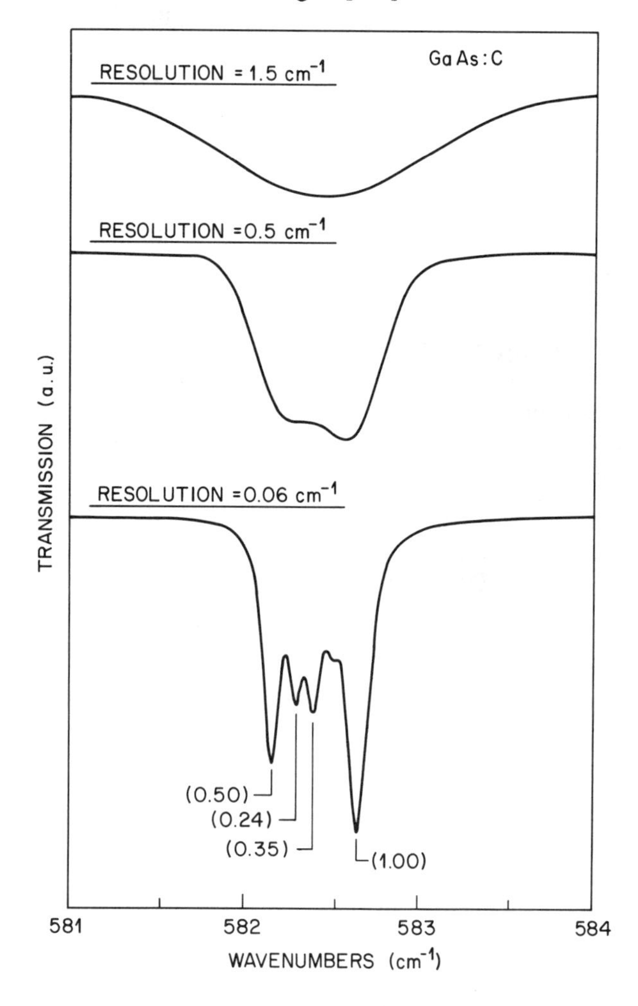

band for $^{12}C$ at 80K measured at three different instrumental resolutions. Note the splitting of the 582 $cm^{-1}$ line into at least four and perhaps five overlapping bands at high resolution. Theis et al.. [178] explained the multiple lines as due to the different nearest-neighbor isotopic arrangements. Though the C-local mode is highly localized, some of the kinetic energy of the mode is in the vibration of the nearest neighbors and hence different isotopes would be expected to give rise to small frequency shifts. The $C_{As}$ has four Ga in its nearest neighbor positions and Ga has two isotopes with masses 69 and 71 with 60.4 percent and 39.6 percent relative abundances, respectively. Depending on the arrangement of the two Ga isotopes in the nearest neighbor positions, different mode frequencies occur. No fine structure is expected if the mode were $C_{Ga}$ since there is only one As isotope.

One of the advantages of the local mode absorption measurement is that impurity concentrations can be obtained from the strength of the local mode. In the case of GaAs:C the detection limit for [C] is $5\times10^{14}cm^{-3}$ [179]. Kitagawara et al. [179] have observed a shift of the $582cm^{-1}$ $^{12}C_{As}$ mode towards low energy and a broadening of the band when the GaAs crystals are doped with In to reduce the dislocation density. They found a shift of $1cm^{-1}$ for $2.7\times10^{20}cm^{-3}$ In. They also related the absorption strength to carbon concentration. Theis [177] discusses the various factors that would affect the calibration of the local mode absorption to actual concentration.

### 5.4.2 Photocurrent Measurements

Junction photocurrent measurement can be a useful characterization technique in structures which contain a p-n junction. In many instances it serves as a complementary technique to photoluminescence. Acket et al., [180] combined both photoluminescence and photocurrent measurements to characterize GaAs-AlGaAs double heterojunction structures used in the fabrication of injection lasers. Both measurements were used to estimate the radiative efficiency of the lasing active region. A schematic of their experiment is shown in Fig. 5.62. The photoluminescence from the lasing active layer excited by the He-Ne laser irradiation was

**Figure 5.62** Schematic representation of the photocurrent and photoluminescence measurements on a AlGaAs-GaAs-AlGaAs double heterostructure wafer [180].

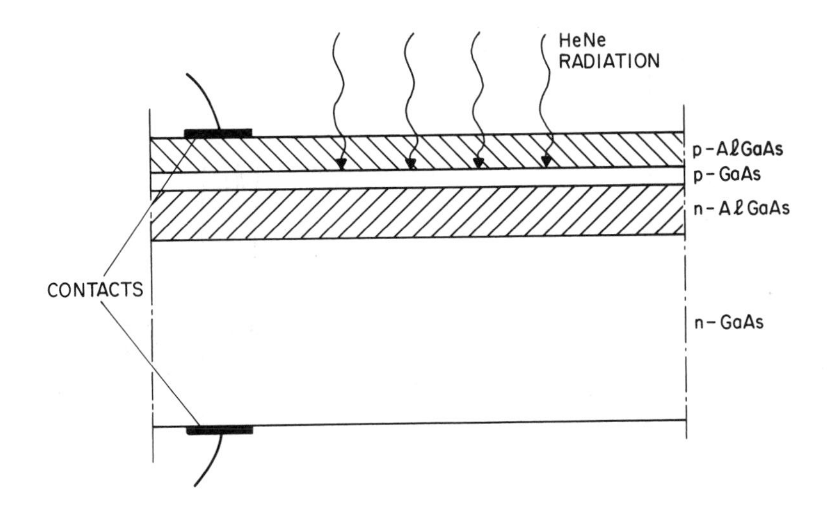

measured when the sample was in the open circuit condition and when it was reverse biased. In the latter condition the photocurrent flowing across the function was also measured. Under a reverse bias the net generation of carriers in the active layer is given by [180]

$$G = \frac{P_{abs}}{h\nu} - \frac{I_{ph}}{q} \tag{5.73}$$

where $P_{abs}$ is the amount of He-Ne laser of energy $h\nu$ absorbed, $I_{ph}$ is the photocurrent and q is the electron charge. The external luminescence efficiency is therefore given by

$$\eta_{ex} = \frac{P_{lum}}{G} \tag{5.74}$$

where $P_{lum}$ is the luminescence intensity. If the photo excited carriers are able to cross the depletion layer of the n-AlGaAs -p-GaAs heterojunction to give a photocurrent before they recombine, one would expect the photoluminescence efficiency to be low. This was indeed observed when both $I_{ph}$ and $P_{lum}$ were measured as a function of position on the wafer. In regions where $I_{ph}$ was high $P_{lum}$ was low and vice versa as shown in Fig. 5.63. In other words, low $P_{lum}$ need not necessarily suggest low radiative efficiency of the active layer but could be caused by the minority carriers crossing the depletion layer. The depletion layer effect was also confirmed by the improved uniformity of photoluminescence in isotype (all p-type layers in Fig. 5.62) heterostructures. Acket et al., [180] interpreted the variation in $I_{ph}$ in Fig. 5.63 in terms of a variation of the height of the conduction band barrier the electrons have to overcome in order to reach the depletion layer and obtained information regarding the grading at the n-AlGaAs - p-

**Figure 5.63** Variation of the photoluminescence intensity (curve A) and photocurrent (curve B) as a function of position parallel to a cleaved edge of a AlGaAs-GaAs-AlGaAs double heterostructure [180].

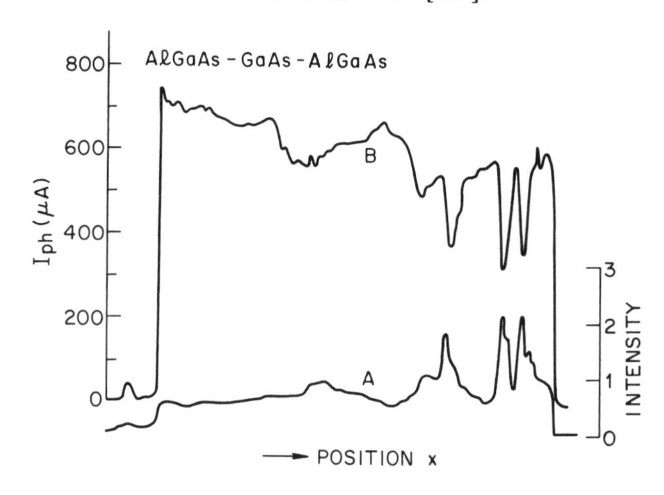

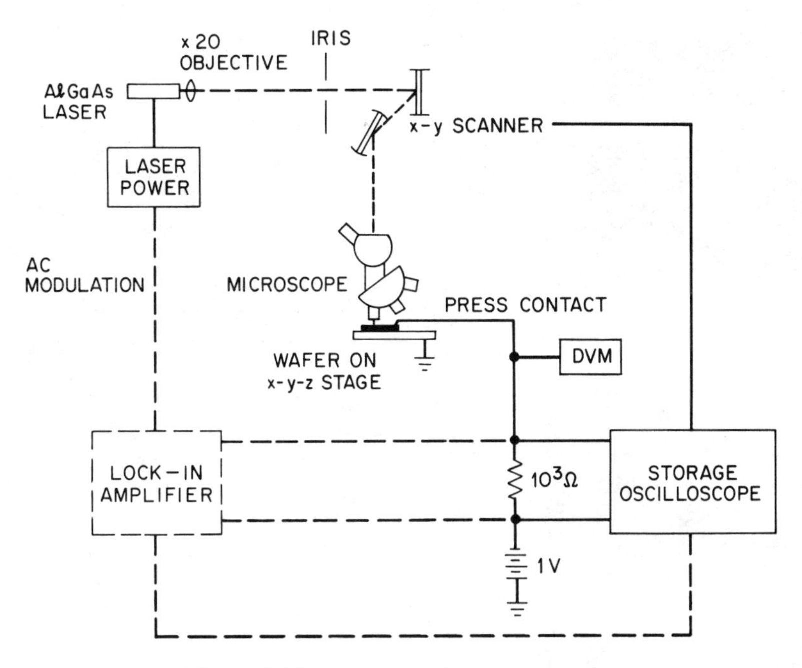


**Figure 5.64** Experimental arrangement for scanning photocurrent measurement [181].

GaAs interface. Further, they found that $\eta_{ex}$ obtained from Eq. (5.74) is a better way to qualify the laser wafers than a straightforward photoluminescence measurement.

In the prequalification of wafers used for certain device application, it is often necessary to obtain information about the spatial uniformity of the material. Simply by scanning the laser beam in Fig. 5.62 over the wafer, it is possible to obtain information about the spatial distribution of the photocurrent which in turn reflects the material quality. This scanning photocurrent technique has been used to characterize AlGaAs heterostructure wafers [181, 182]. A schematic of the apparatus is shown in Fig. 5.64. Thickness variations on a microscopic scale in the GaAs active layers of the AlGaAs-GaAs-AlGaAs laser structures grown by LPE have been recorded using the scanning photocurrent technique [181].

Figure 5.65 shows the scanning photocurrent image and a luminar photograph of a heterostructure wafer consisting of a n-type GaAs substrate, n-type ($2 \times 10^{17}$ cm$^{-3}$, 2μm thick) $Al_{0.36}Ga_{0.64}As$ layer, p-type ($2 \times 10^{17}$ cm$^{-3}$, 0.1 – 0.2μm thick) $Al_{0.08}Ga_{0.92}As$ lasing active layer, p-type ($2 \times 10^{17}$ cm$^{-3}$, 1.2μm thick) $Al_{0.36}Ga_{0.64}As$ layer and a p-type ($1 \times 10^{18}$ cm$^{-3}$, 0.5μm thick) GaAs layer. All the layers were grown by LPE. The photocurrrent signal (Fig. 5.65(a)) varies by a factor of three across the wafer. This variation in photocurent has been associated with variation in an "effective thickness" of the active layer [181].

Thickness and/or composition variations in the active layer which manifest as dark lines in photoluminescence topograph also show up as modulations in the photocurrent signal. This is illustrated in Fig. 5.66 where the modulation in the photocurrent signal corresponds to the dark

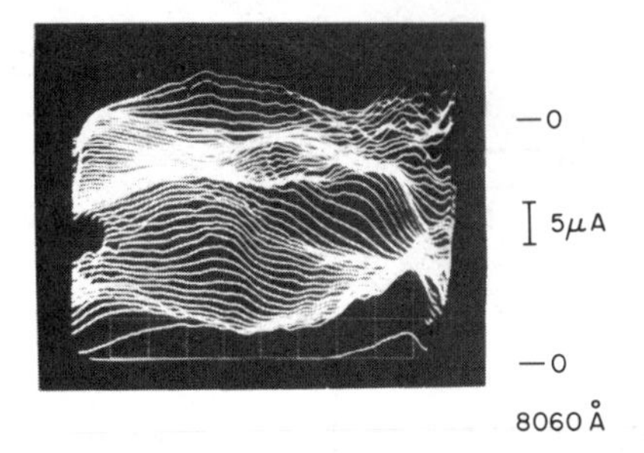


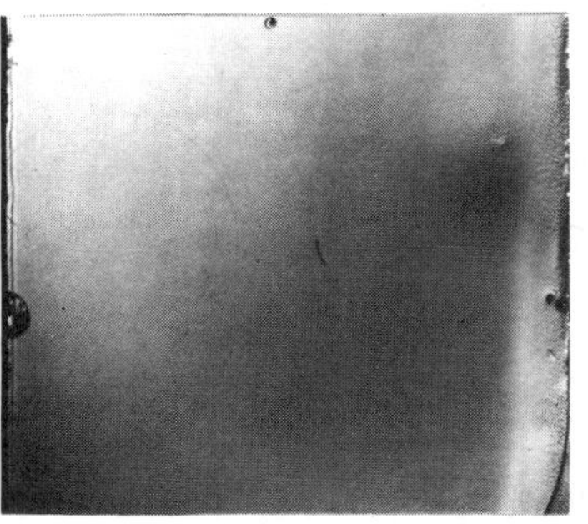


**Figure 5.65** Photocurrent scan of a wafer (a) and its top surface luminar photograph (b). The zero marks are for the first and last scan lines. The variations in the photocurrent represent variations in the "effective thickness" of the active layer [181].

**Figure 5.66** Thickness and/or composition variations (rake lines) in the thin GaAs active layer of a AlGaAs-GaAs-AlGaAs double heterostructure seen by (a) photocurrent (b) transmission cathodoluminescence and (c) photoluminescence measurements. The backscattered electron image from the same area as the cathodoluminescence image is also shown.

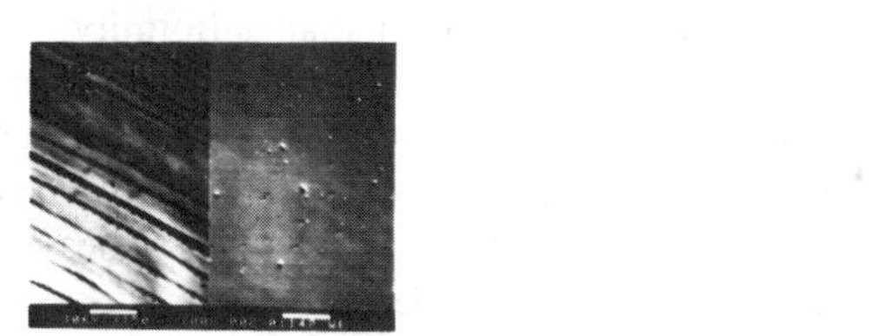


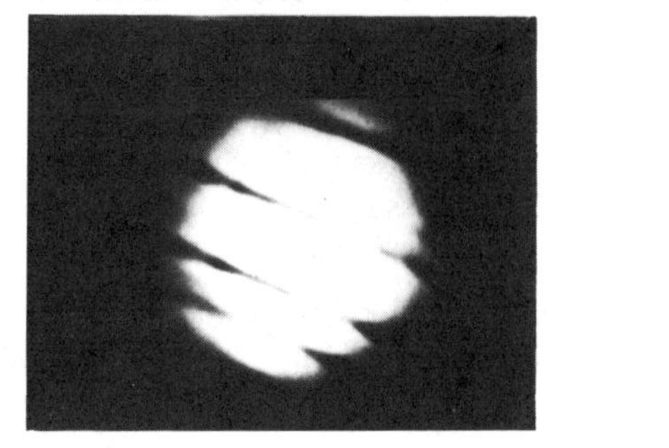

lines in the photoluminescence topograph. To obtain the photoluminescence topograph shown in Fig. 5.66 the top GaAs layer of the four-layer heterostructure has to be removed so as to facilitate absorption of the excitation source in the active layer. In this sense, the photoluminescence technique is destructive. However, the photocurrent measurement can be made with the top GaAs layer and hence has the advantage over PL for nondestructive analysis of the wafer at comparable spatial resolution.

Lang and Henry [183] have developed the technique of scanning photocurrent microscopy to study inhomogeneously distributed recombination centers. The technique is analogous to the EBIC microscopy performed in the SEM except that the junction current is generated by a laser beam as opposed to an electron beam. The essential features of experimental set-up are similar to that shown in Fig. 5.64. The amplified photocurrent signal is used to modulate the intensity of the CRT display of an oscilloscope in synchronism with the laser scanning mirrors. Also, sub-band gap excitation is used so that the generation of the photocurrent depends on the two-step excitation process via deep energy levels in the gap. This technique has been applied to study localized nonradiative defects in $Al_xGa_{1-x}As/GaAs$ double heterostructures [183]. Figure 5.67 shows a photocurrent line scan across three dark lines (labeled D) which have been introduced by catastrophic degradation of a $GaAs/Al_xGa_{1-x}As$ double heterostructure. The photocurrent signal for above band gap excitation decreases at the dark lines in contrast to below band gap excitation where it increases. This demonstrates that nonradiative recombination at the dark lines is caused by a large concentration of point defects. Note also in Fig. 5.67 that the irregular background in

**Figure 5.67** Line scan across three dark lines (labelled D) which have been induced by catastrophic degradation of a $GaAs/Al_xGa_{1-x}As$ double heterostructure. The upper trace shows the decrease in the photocurrent signal for above band gap excitation with 6328A laser line and the bottom trace shows the photocurrent signal for below band gap excitation with 1.06μm laser line [183].

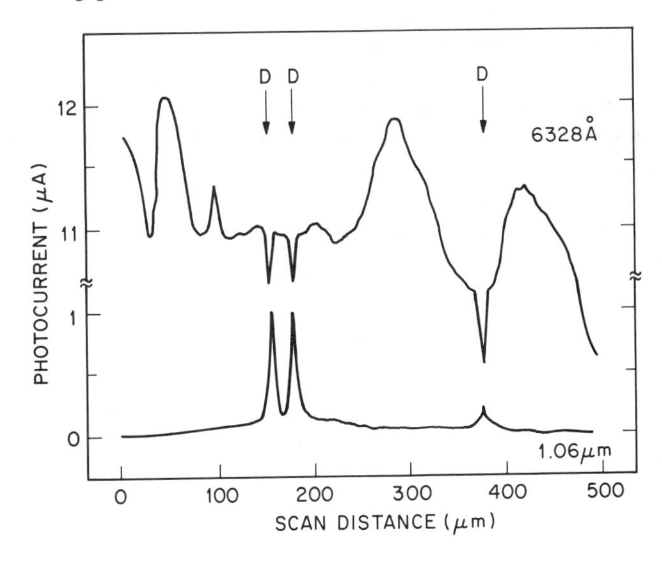

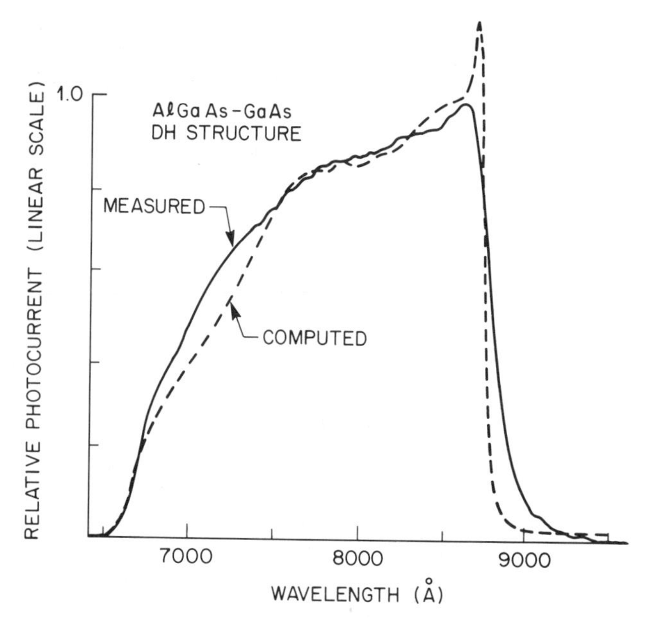


**Figure 5.68** Photocurrent spectrum for a AlGaAs-GaAs-AlGaAs double heterostructure. The dashed line is calculated using the parameters of Table 5.6 [184].

the signal obtained for above band gap excitation vanishes for below band gap excitation showing that this signal does not arise because of recombination centers.

In measuring the photocurrent, if the wavelength of the exciting light is varied from below the band gap to just above the band gap, then a photocurrent spectrum is obtained. The shape of the photocurrent spectrum resembles the absorption spectrum from which one can determine the band edge of the region where the light is absorbed. The photocurrent spectroscopic technique has been used to obtain the active layer composition and hence to predict the emission wavelength of the AlGaAs-GaAs laser heterostructure similar to the one giving the scanning photocurrent image in Fig. 5.65. Figure 5.68 shows the photocurrent spectrum obtained from a double heterostructure wafer with undoped GaAs active layer [184]. The cutoff at 6500Å occurs because of a long pass filter in the experimental set-up which blocks light with wavelengths shorter than 6600Å. The calculated photocurrent spectrum is also shown in Fig. 5.68. To calculate the photocurrent of the double heterostructure, Nygren [184] extended the model of Hovel [185] for a single heterostructure and this will be discussed.

The photocurrent from an n AlGaAs -p GaAs -p AlGaAs double heterostructure wafer consists of three components; (1) photogenerated carriers from the depletion region, (2) photogenerated minority carrier electrons from the p-type material, and (3) photogenerated minority carrier holes from the n-type material. That is, there will be photocurrent contributions from the active layer and the two $Al_{0.37}Ga_{0.63}As$ cladding layers. For all the three cases a quantum efficiency of unity is assumed.

In the depletion region the electric field is assumed to be high enough that all photogenerated carriers are collected before they recombine. Then, the photocurrent per unit bandwidth is given by

$$J_{dr}(\lambda) = q\,(F_1 - F_2) \tag{5.75}$$

where $J_{dr}$ is the current density in depletion region, q the electronic charge and $F_1$ and $F_2$ are respectively the light fluxes at boundary of layer nearest the surface and farthest from the surface. There will be a $J_{dr}$ for whatever portion of each region that is contained within the depletion layer. They will be the active layer, the n-AlGaAs cladding layer and the p-AlGaAs cladding layer if the depletion layer extends into it.

The second component to the photocurrent is the minority carrier hole current from the n-AlGaAs layer that is outside the depletion region. This is given by

$$J_p(\lambda) = \frac{q\,F_N\,\alpha_N\,L_N}{(\alpha_N^2\,L_N^2 - 1)} \times$$

$$\left[\alpha_N L_N - \frac{\dfrac{S_N L_N}{D_N}\left[\cosh\dfrac{t_N}{L_N} - \exp\left(-\alpha_N t\right)\right] + \sinh\dfrac{t_N}{L_N} + \alpha_N L_N \exp\left(-\alpha_N t_N\right)}{\dfrac{S_N L_N}{D_N}\sinh\dfrac{t_N}{L_N} + \cosh\dfrac{t_N}{L_N}}\right] \tag{5.76}$$

where $F_N$, $\alpha_N$, $L_N$, $D_N$, and $t_N$ are the light flux at boundary of depletion layer, the absorption coefficient, the minority carrier diffusion length, minority carrier diffusion coefficient, and the thickness of the undepleted region in the n-AlGaAs layer, respectively, and $S_N$ is the surface recombination velocity at n-AlGaAs-substrate interface.

The third component to the photocurrent comes from the minority carrier electron current in the p-type region outside the depletion region. If this region is entirely in the p-AlGaAs, then the minority carrier current is

$$J_n(\lambda) = \frac{q\,F_p \alpha_p L_p}{(\alpha_p^2 L_p^2 - 1)} \times$$

$$\left[\frac{\left[\dfrac{S_p L_p}{D_p} + \alpha_p L_p\right] - \exp\left(-\alpha_p t_p\right)\left[\dfrac{S_p L_p}{D_p}\cosh\dfrac{t_p}{L_p} + \sinh\dfrac{t_p}{L_p}\right]}{\dfrac{S_p L_p}{D_p}\sinh\dfrac{t_p}{L_p} + \cosh\dfrac{t_p}{L_p}} - \alpha_p L_p \exp\left(-\alpha_p t_p\right)\right] \tag{5.77}$$

where $F_p$ is the light flux at p-AlGaAs/p-GaAs interface, $\alpha_p$, $L_p$, and $D_p$ are, respectively the absorption coefficient, minority carrier diffusion length, and minority carrier diffusion coefficient in p-AlGaAs, $t_p$ is the thickness of the undepleted portion of p-AlGaAs and $S_p$ is the surface recombination velocity at the p-AlGaAs/p-GaAs interface.

If the depletion layer is within the p-GaAs active region then Eq. (5.77) has to be used with appropriate factors to calculate the current from the active layer. The p-AlGaAs current in such a case is obtained by attenuating the value calculated from Eq. (5.77) by $\exp(-t_a / L_a)$ where $t_a$ and $L_a$ are the undepleted thickness and minority carrier diffusion length in the active layer.

Nygren [184] gives the formulae for calculating the depletion layer width for two cases. First, when the depletion region is entirely within the active layer and the n-AlGaAs, the depletion layer widths are given by

$$W_d = \left[ \frac{2\varepsilon_d \psi_b}{q N_d} \left( \frac{\varepsilon_a N_a}{\varepsilon_d N_d + \varepsilon_a N_a} \right) \right]^{1/2} \tag{5.78}$$

$$W_a = \frac{N_d W_d}{N_a} \tag{5.79}$$

where $W_{d,a}$, is the depletion layer thickness, $\varepsilon_{d,a}$, is the dielectric constant, $N_{d,a}$ is the impurity concentration in n-AlGaAs or GaAs layers and $\psi_b$ is the built-in voltage of the junction.

Second, when the depletion layer extends into the p-AlGaAs, the Poisson's equation is solved assuming no excess charge at the active layer/p-AlGaAs interface.

$$W_p = \frac{1}{2a} \left[ -b + (b^2 - 4ac)^{1/2} \right] \tag{5.80}$$

with

$$a = \frac{q N_p}{2 \varepsilon_p} + \frac{q N_p^2}{2 \varepsilon_d N_d}$$

$$b = \left(\frac{q}{\varepsilon_a}\right) W_a N_p + \frac{q N_p N_a W_a}{\varepsilon_d N_d}$$

$$c = \frac{q N_a^2 W_a^2}{2 \varepsilon_d N_d} + \frac{q N_a W_a^2}{\varepsilon_a} - \left( \frac{q}{2 \varepsilon_a} \right) W_a^2 N_a - \psi_b$$

and

$$W_d = \frac{N_p W_p + N_a W_a}{N_d} \tag{5.81}$$

In Eqs. (5.80) and (5.81), $W_{p,a}$ is the depletion layer thickness in p-AlGaAs or active layers, $\varepsilon_p$ is the dielectric constant in p-AlGaAs layer and $N_p$ is the acceptor concentration.

**TABLE 5.6**

Material parameters used for photocurrent model [184]

| | p GaAs cap | p AlGaAs clad | p GaAs active | n AlGaAs clad |
|---|---|---|---|---|
| % Aluminum | 0 | 37 | 0 | 37 |
| Thickness (μm) | 1.2 | 1.8 | 0.13 | 2.2 |
| Carrier concentration ($x10^{18}cm^{-3}$) | 2.0 | 0.5 | 0.05 | 0.5 |
| Minority carrier diffusion length (μm) | | 0.5 | 0.5 | 0.5 |
| Minority carrier diffusion coefficient ($cm^2/sec$) | | 10 | 51 | 2 |
| Energy band gap (eV) | 1.424 | 1.867 | 1.424 | 1.867 |
| Dielectric constant | 13.10 | 11.99 | 13.10 | 11.99 |

Surface recombination velocity = 1 cm/sec at all interfaces.

To calculate the photocurrent shown in Fig. 5.68 Nygren [184] used the material parameters shown in Table 5.6. The refractive indices for GaAs are obtained from Casey [186]. For the absorption coefficients of GaAs, the data of Casey [186] and Sturge [19] are used. The same data is also used for AlGaAs after shifting the wavelength by the band gap change between AlGaAs and GaAs.

The wavelength where the slope of the photocurrent edge is the highest in the measured photocurrent spectrum is taken to be the active layer band edge. The wavelengths determined this way from AlGaAs-GaAs heterostructures have been correlated with the emission wavelengths of lasers fabricated from them. This relationship is shown as a probability plot in Fig. 5.69. The standard deviation of the distribution shown in Fig. 5.69 is 37Å and 90 percent of the lasers have their emission wavelengths within ±62Å of their value predicted by photocurrent spectroscopy. The accuracy of this prediction is comparable to what has been achieved on similar laser wafers by photoluminescence technique. But the major advantage of the photocurrent technique is that it is nondestructive and hence a whole wafer can be measured, unlike photoluminescence, which can be used only on a section of the wafer since it requires removal of the top GaAs layer of the heterostructure.

### 5.4.3 Reflectance Modulation

Reflectance modulation is an important experimental technique to probe the band structure of semiconductors [187 - 191]. Sharp structures related to the third derivative of the optical constants are obtained in this technique. Further it has high sensitivity even at room temperature and is a function of surface and interfacial electric fields. For these reasons reflectance modulation is increasingly used to study superlattices, quantum wells and heterostructures.

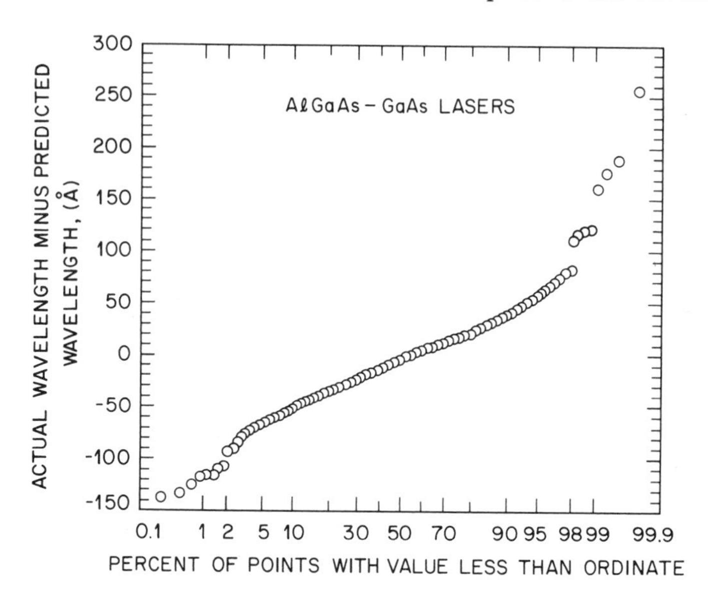


**Figure 5.69** Cumulative probability distribution showing the difference between actual emission wavelength and wavelength predicted by photocurrent spectral measurement of AlGaAs-GaAs-AlGaAs laser [184].

In reflectance modulation, the change in reflectance is measured as the optical properties of the semiconductor are modified by applying a periodic perturbation. There are two reflectance modulation techniques which have been proven to be very useful. One is electroreflectance and the other is photoreflectance. In the former a periodic electric field is applied to the sample and in the latter modulation is achieved by a means of a strongly absorbing radiation. A schematic drawing of the experimental arrangement used for the electrolyte electroreflectance technique is shown in Fig. 5.70 [191]. Light from a suitable lamp passes through a monochromator and is focussed on the sample. The light from the sample is collected by another lens and is focussed onto a photomultiplier or other appropriate detectors. The sample is immersed in an electrolyte which is contained in a cell with a window. An ac voltage at frequency $\omega_m$ is applied between the sample and another electrode, usually platinum. This modulates the space charge region of the semiconductor-electrolyte interface. The choice of the electrolyte is critical in that it should not dissolve the semiconductor or produce adverse chemical reactions at the surface. The disadvantage of this technique is that it can be used only at or near room temperature.

Electric field modulation can also be achieved with samples in MOS (metal-oxide-semiconductor) or Schottky barrier configuration consisting of semitransparent metal electrodes. Both methods can be used at low temperatures. For high resistivity samples ($>10^8\ \Omega$ cm) a transverse electric field is applied between two parallel metal electrodes evaporated onto the sample, usually about 1 mm thick. The maximum electric field that can be applied is limited by electric breakdown accross the gap and is of the order $10^5$ V cm$^{-1}$.

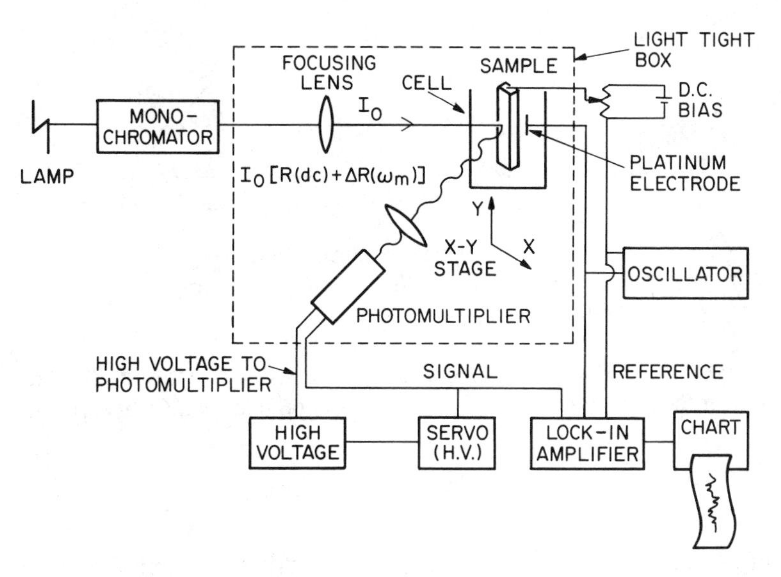


**Figure 5.70** Experimental arrangement for electrolyte electroreflectance [191].

The signal from the photomultiplier consists of a dc value, $I_oR$, and an ac value, $I_o \Delta R (\omega_m)$. The ac component is obtained from the lock-in amplifier. By eliminating $I_o$, the desired quantity $\Delta R/R$ can be obtained. In the set-up shown in Fig. 5.70 the dc output is applied to a servo power supply which adjusts the high voltage to the photomultiplier to keep $I_oR$ constant. Then the lock-in output is proportional to $\Delta R/R$.

The dispersion relation for the index of refraction n in terms of the absorption coefficient $\alpha(E)$ is given by the Kramers-Kronig relation

$$n(E) = 1 + \frac{hc}{2\pi^2} P \int_o^\infty \frac{\alpha(E')}{E'^2 - E^2} dE' \tag{5.82}$$

where P is the Cauchy principal value of the integral,

$$P \int_o^\infty \equiv \lim_{a \to o} \left[ \int_o^{E-a} + \int_{E+a}^\infty \right]$$

h is the Planck's constant and c is the velocity of light. Integrating Eq. (5.82) by parts one gets

$$n(E) = 1 + \frac{hc}{4\pi^2} P \int_o^\infty \ell n \frac{1}{E'^2 - E^2} \frac{d\alpha(E')}{dE'} dE' \tag{5.83}$$

The derivative term in Eq. (5.83) determines the value of the integral when the logarithmic factor

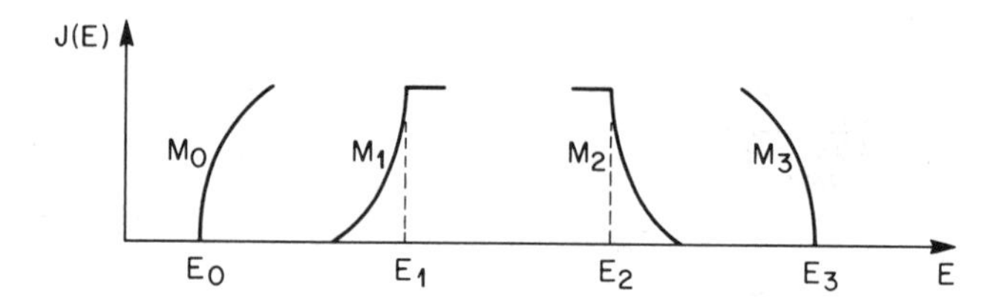


**Figure 5.71** Optical joint density of states distributions for the four main types of critical points [2].

becomes large. Since $\alpha(E')$ is proportional to the joint density of states, $d\,\alpha(E')/dE'$ will be large for transitions near the critical points in the E(k) dispersion relation. Critical points are defined as regions where

$$\nabla_k\,[E_c\,(k) - E_v\,(k)] = o \tag{5.84}$$

There are four general types of critical points whose corresponding optical joint density of states are shown in Fig. 5.71. The $d\,\alpha\,(E')/dE'$ term is large and positive above $E_o$ and below $E_1$ and is large and negative above $E_2$ and below $E_3$. As a result, n(E) will go through maximum at $E_o$ and $E_1$ and a minimum at $E_2$ and $E_3$. In other words, as $d\,\alpha(E)/dE$ goes through a large maximum or minimum a structure is obtained in n(E). The structure in n(E) is reproduced in the reflectance spectrum.

The structure in n(E) and hence in the reflectance spectrum can be enhanced by periodic modulation of a parameter that affects the joint density of states. Let us consider a modulating electric field. The change in n(E) caused by the applied field $\boldsymbol{E}$ is

$$\Delta n\,(E, \boldsymbol{E}) = \frac{hc}{\pi} \int_o^\infty \frac{\Delta\alpha(E', \boldsymbol{E})}{E'^2 - E^2}\, dE' \tag{5.85}$$

The variations in $\Delta n$ and $\Delta\alpha$ of Eq. (5.85) are similar to the changes in the real and imaginary parts of the dielectric constant, $\Delta\varepsilon_1$ and $\Delta\varepsilon_2$.

The contribution of the critical points to reflectivity can be described phenomenologically as

$$\frac{\Delta R}{R} = \alpha\,(\varepsilon_1, \varepsilon_2)\,\Delta\varepsilon_1 + \beta\,(\varepsilon_1, \varepsilon_2)\,\Delta\varepsilon_2 \tag{5.86}$$

where $\alpha$ and $\beta$ are the Seraphin coefficients. Through $\varepsilon_1$ and $\varepsilon_2$, $\alpha$ and $\beta$ are functions of energy. A knowledge of the sign and relative magnitude of $\alpha$ and $\beta$ helps to analyze the modulated reflectance spectrum in different spectral regions. For example, in the electroreflectance spectrum from GaAs shown in Fig. 5.72 below $E_1$, $\alpha$ is large and positive and $\Delta R/R$ is proportional to $\Delta\varepsilon_1$. At $E_2$, $\alpha$ is negative and $\beta$ is small and the spectrum is proportional to $-\Delta\varepsilon_1$.

Since electroreflectance spectrum readily gives the critical point energies of the semiconductor, as illustrated by the spectrum shown in Fig. 5.72, it can be used to determine the

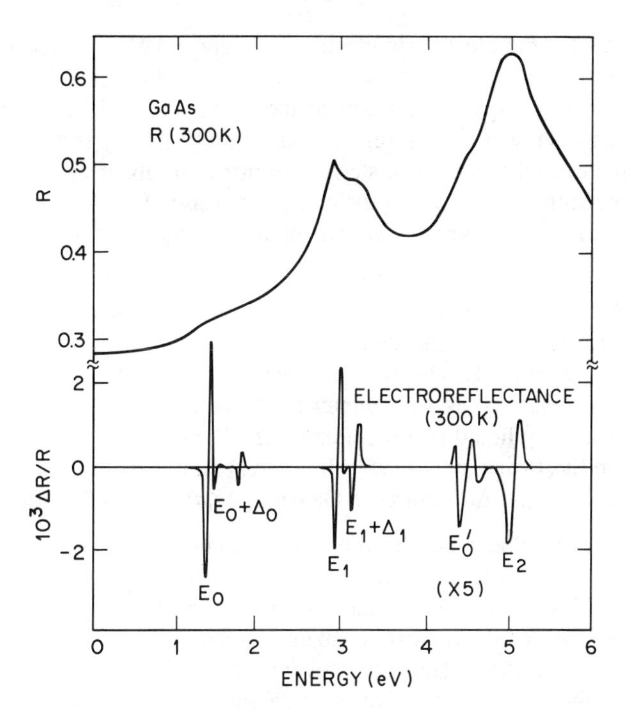


**Figure 5.72** Comparison of room temperature reflectivity and electroreflectance of GaAs [191].

band gap and hence the composition of alloy semiconductors. By using a small light spot and scanning it across the sample spatial compositional uniformity can also be assessed. The electrolyte used in the measurement can also be used to anodize and to remove systematically the material. Electroreflectance spectrum taken after each etching step will thus give composition information in the depth of the material [191].

Electroreflectance can also be used to obtain carrier concentration in the material [191]. Under conditions of small electric fields and when phase-sensitive detection is used, the change in reflectivity is given by

$$\frac{\Delta R}{R} = -\frac{2q}{\varepsilon_o} V N L(h\omega) \tag{5.87}$$

where $\varepsilon_o$ is the static dielectric constant, N the carrier density, V is the amplitude of the modulating ac voltage, and L ($\hbar\omega$) is a line shape factor characteristic of the material and transition being studied. To determine N using Eq. (5.87) measurements have to be made in fully depleted space charge layer; that is, $\Delta R/R$ should vary linearly with V. Also the intensity of light should be low enough so that the density of photo excited carriers is small. That is, $\Delta R/R$ should be independent of the intensity of light. Spatial variations in the carrier concentration of GaAs have thus been studied using electrolyte electro reflectance [192]. A type of depth profiling of doping inhomogeneity can also be achieved by looking at different features in the electrolyte reflectance spectra. For example, in GaAs near the $E_o$ region (8370Å) the penetration depth of light is much greater than near the $E_1$ region. As a result, the $E_o$ feature would reveal doping inhomogeneties much further into the material. Information about type of carriers can also be obtained by electroreflectance because of the change in phase of $\Delta R/R$ in going from n- to p-type. This has been used to identify type converted regions in implanted GaAs both spatially and in depth [193].

Small strains (~0.1%) at the interface between epitaxial layers have been detected in the AlGaAs-GaAs heterostructure by the electrolyte electroreflectance technique [194]. This was done by observing shifts of the $E_o$ and $E_o + \Delta_o$ spectra of the GaAs substrate. The broadening of the ER spectrum has been used as a parameter to characterize $Ga_xIn_{1-x}As_yP_{1-y}$ epitaxial layers grown on InP substrates by liquid phase epitaxy. The broadening was found to be larger on the As rich side than on the P-rich side. [195]. Interfacial strains have also been deduced from the line shapes of the $E_o$ and $E_o + \Delta_o$ features in $Ga_XIn_{1-x}As_yP_{1-y}$ on InP substrates [196].

Photoreflectance is another powerful modulation technique as electrolyte electroreflectance. This technique does not require any electrodes and can be used at low and high temperatures, different ambients and with other perturbations on the sample such as hydrostatic pressure. In this technique modulation of the optical constants is achieved by injection of electron-hole pairs by photo excitation [197, 198]. There are several possible ways by which the photo excited carriers can perturb the reflectivity. They include screening of excitons [199], Burstein-Moss effect [200], the creation of a Dember potential [201] and the reduction of the built-in surface field through recombination of minority carriers with surface charge states [201, 202]. Of these different mechanisms, the reduction in the surface band bending appears to be the dominant mechanism for reflectance modulation. This mechanism is schematically illustrated in Fig. 5.73

**Figure 5.73** Depletion region at a semiconductor surface before (solid line) and after (dashed line) excitation with intense above band gap light [198].

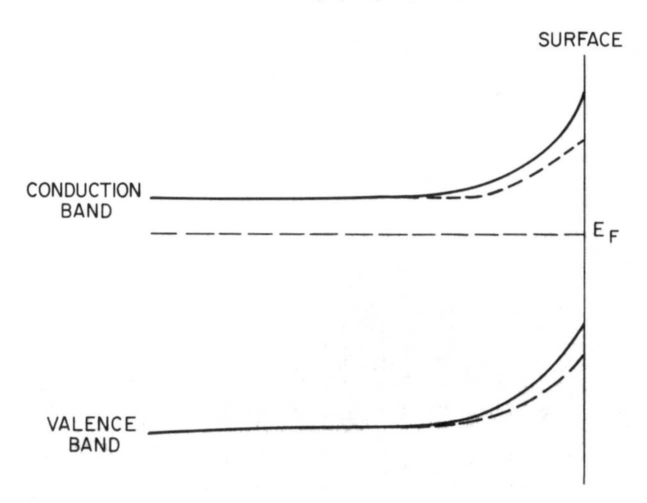

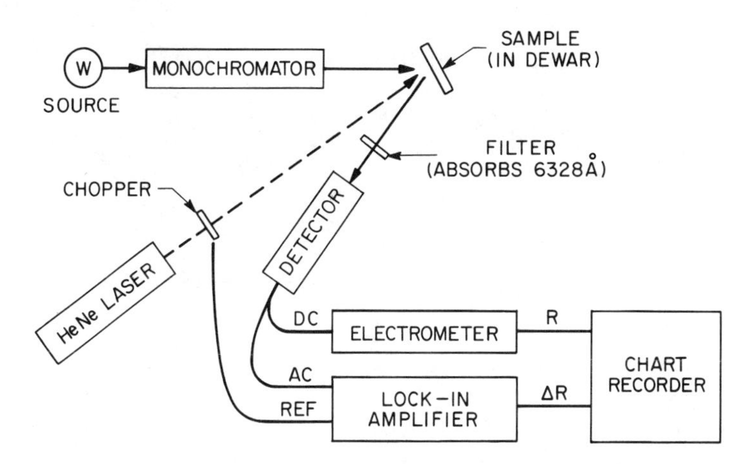


**Figure 5.74** Experimental arrangement for photoreflectance measurement [198].

for an n-type semiconductor. The situation depicted in Fig. 5.73 is what one encounters in a surface photovoltage experiment. Therefore Eq. (5.87) can be used to describe photo reflectance by replacing V with the surface photovoltage [203].

The experimental set-up used to measure photoreflectance is shown in Fig. 5.74 [198]. The modulation in the reflectivity is produced by mechanically chopping the laser beam used to create the electron-hole pairs. To obtain a uniform distribution of carriers the laser light is defocussed to a large spot (~1 cm in diameter) and is larger than that of the probe beam. Also the intensity of the probe beam is much less than that of the laser intensity.

The photoreflectance spectra are very similar to electroreflectance spectra discussed earlier and provide very much the same information about the semiconductor under study. Lately photoreflectance has become a popular technique to characterize quantum well structures. Figure 5.75 shows the photo reflectance spectra of a $GaAs/Al_xGa_{1-x}As$ (x ~0.24) quantum well sample grown by MBE with nominal thicknesses of 220 A and 650Å respectively for the wells and barriers [204]. Even at 300K it is possible to observe sharp structure from n = 5 confined quantum wells. The peak at about 1.42 eV corresponds to the direct gap of GaAs ($E_{01}$) and originates from the GaAs substrate. The features between 1.43 and 1.68 eV are the allowed ($\Delta n =$ o) heavy-hole-to-conduction band ($h_1, \ldots h_5$) and light-hole-to-conduction band ($\ell_1, \ldots \ell_4$) transitions where the subscript indicates the quantum number. The spectral AlGaAs peak A is related to the direct gap of the AlGaAs barrier layer. The solid lines in Fig. 5.75 are a least-square fit of the data to the Aspnes third-derivative line shape function for electromodulation spectra given by [190]

$$\frac{\Delta R}{R} = \mathrm{Re}\left[\sum_{j=1}^{p} C_j\, e^{i\theta_j} (E - E_{g,j} + i\Gamma_j)^{-m_j}\right] \tag{5.88}$$

where p is the number of spectral features to be fit, $C_j$, $\theta_j$, $E_{g,j}$, and $\Gamma_j$ are the amplitude, phase, energy and broadening parameter, respectively, of the jth structure while $m_j$ denotes the type of

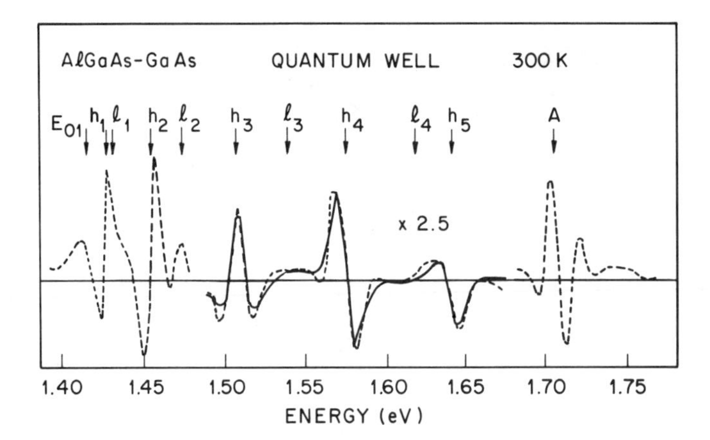


**Figure 5.75** Photoreflectance spectrum of a $GaAs/Al_xGa_{1-x}As$ (x ~ 0.24) multiquantum well at 300K using 6328Å pump beam. The solid line is calculated [204].

critical point. Because of Franz-Keldysh oscillations, the feature A could not be fit by Eq. (5.88). However, accurate energy value for A can be obtained from a three-point fit to the spectrum. Good agreement between the experimentally determined energies and theoretical calculations for the various confined transitions has been obtained.

Using the photoreflectance technique at room temperature, Parayanthal et al., [204] have investigated the topographical variations in barrier height and quantum well width (and hence confinement energy) with a spatial resolution of about 100μm. By employing Eq. (5.88), these authors have shown that variations in well width to one monolayer can be determined. A wavelength dependence of the photoreflectance signal has also been observed in GaAs/AlAs quantum wells suggesting that the electric field modulation depends on whether the excitation is below or above the wells [205].

In GaAs/AlGaAs quantum wells Shen et al. [206] have observed in room temperature photoreflectance spectra evidence for several symmetry forbidden transitions. Denoting the transitions in the notation nmH or nmL which represents transitions between the nth conduction subband and the mth valence subband of heavy hole (H) or light hole (L) character, allowed transitions have n = m while for symmetry forbidden transitions n ≠ m. Shen et al., [206] observed features corresponding to 12H, 13H, and 21L. These forbidden transitions have also been reported in the low temperature photoluminescence excitation spectra [207].

The doping dependence of photoreflectance which showed directly that it is caused by modulation of the surface electric field by photo injected carriers, has been used to obtain a correlation between Franz-Keldysh oscillation, and doping concentrations in Se implanted GaAs [208]. According to the Franz-Keldysh theory, the period of the oscillation in the reflectance structure ΔE is related to the electric field via $\Delta E \alpha \boldsymbol{E}^{\frac{2}{3}}$ [209]. The surface electric field $\boldsymbol{E}$ itself varies as $(N_d)^{1/2}$ for a constant barrier height where $N_d$ is the dopant density. Thus,

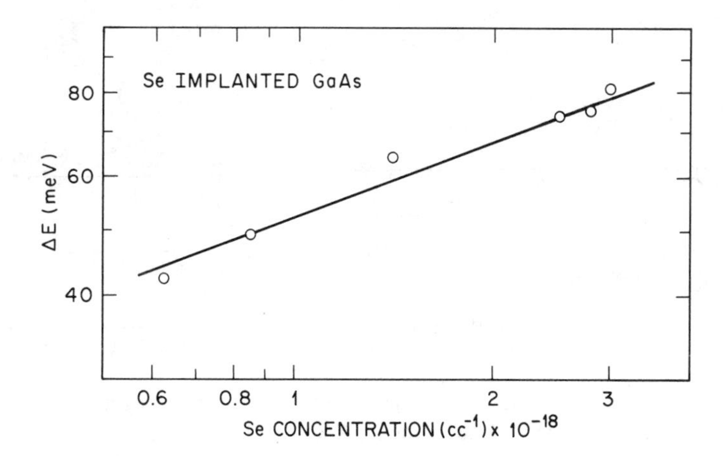


**Figure 5.76** Photoreflectance lineshape oscillation period ΔE versus surface Se concentration in Se implanted GaAs [208].

$\Delta E \; \alpha \; (N_d)^{1/3}$. This relationship is illustrated in Fig. 5.76 for Se implanted GaAs over the concentration range $0.6 - 3\times10^{18}$ $cm^{-3}$. Thus photoreflectance can be used to measure the dopant concentration at the surface of ion-implanted samples in a nondestructive manner. The applicability of this technique to lower concentration limits is dictated by the requirement that the extent of the space charge region must be smaller than the depth of implantation.

Glembocki et al. [203] have shown that the photoreflectance technique can be used to detect the presence of two-dimensional electron gas (2DEG) in modulation doped heterostructures. In the modulation doped GaAs/AlGaAs system, they observed an extra feature above the GaAs $E_o$ transition which was interpreted as caused by transitions between the valence band and subband states in the triangular potential well formed at the GaAs/AlGaAs interface. While the nature of this extra feature has not been completely understood, it can serve as a fingerprint of 2DEG in a simple room temperature measurement. InP doping superlattices have been studied by the photoreflectance technique by Gal et al., [210]. In these structures, photoreflectance signals arise because of modulation of the subbands of the conduction band.

### 5.4.4 Optical Detection of Magnetic Resonance (ODMR)

Another optical technique which has been used to study native defects in GaAs and InP is optically detected magnetic resonance (ODMR) [211]. Before discussing ODMR a brief discussion of the closely related electron paramagnetic resonance (EPR) [212] technique is appropriate. In EPR the sample is placed in a microwave cavity and a magnetic field is applied to the sample such that $H_{r.f}$ is normal to $H_{ext}$, as illustrated in Fig. 5.77. Microwaves are generated usually with a frequency stabilized klystron and the isolator prevents backward reflections. Since the ground state will be in thermal equilibrium, resonant absorption of microwave power occurs when the microwave energy is equal to the separation between the Zeeman electronic levels of the defect. The resonant absorption is measured by the detector. The resonance can be measured under illumination (photo EPR) which changes the charge state of the defect and/or under

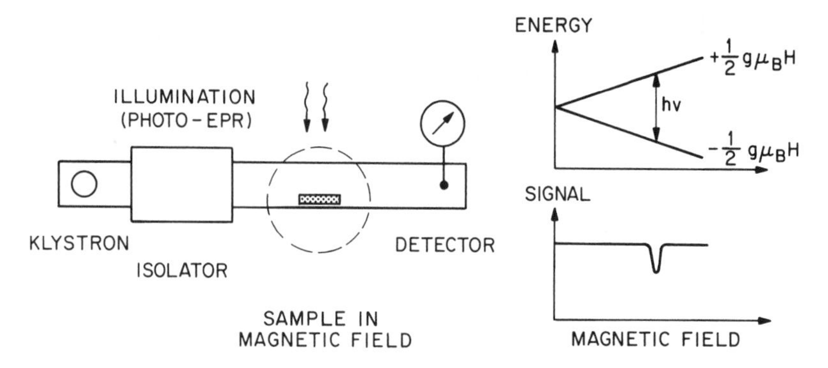


**Figure 5.77** Schematic view of the electron paramagnetic resonance (EPR) experiment [212].

uniaxial stress to obtain detailed information on the nature and symmetry of the defect. In the optically detected magnetic resonance technique Fig. 5.78 [213] the sample is excited by a laser to produce luminescence and the microwave induced changes in the luminescence intensity or polarization are measured at resonance. Alternatively, microwave induced changes in the absorption can be measured for the detection of the ground state resonance.

Semiconductors are diamagnetic and hence do not produce any EPR signal. However, impurities and defects depending on their charge state can have an unpaired electron and produce an EPR signal. Defects which do not have an unpaired electron are not accessible by EPR. In the presence of a magnetic field the Zeeman levels of an unpaired electron are split. Resonant transitions are set-up between these levels by the microwave field whose frequency satisfies the resonance condition

$$h\nu = g\,\mu_B\,H \tag{5.89}$$

where $\mu_B$ is the Bohr magneton and g is a constant determined by the Dirac theory of electron. The above equation is applicable to an istropic spin 1/2 center.

In the case of an electron localized at a defect, the EPR signal is affected by the hyperfine interaction of the electron spin with the nuclear magnetic moment of the surrounding nuclei and by the crystal field (fine structure). The quantities describing these interactions are no longer scalars but are tensors reflecting the symmetry of the electron wavefunction and thus of defect symmetry. In these cases, the EPR spectrum is generally described by the Spin Hamiltonian

$$\boldsymbol{H} = \mu_B\,S_i\,g_{ij}\,H_j + S_i\,A_{ij}\,I_j + S_i\,D_{ij}\,S_j \tag{5.90}$$

where $g_{ij}$, $A_{ij}$ and $D_{ij}$ are, respectively, the g-tensor, hyperfine interaction tensor and the fine structure tensor. Additional hyperfine interaction terms are added to Eq. (5.90) if interaction with more than one nucleus occurs. The analysis of an isotropic EPR spectra is frequently performed with the help of Eq. (5.90).

Since the EPR signal is proportional to the occupation of the Zeeman split levels, which is

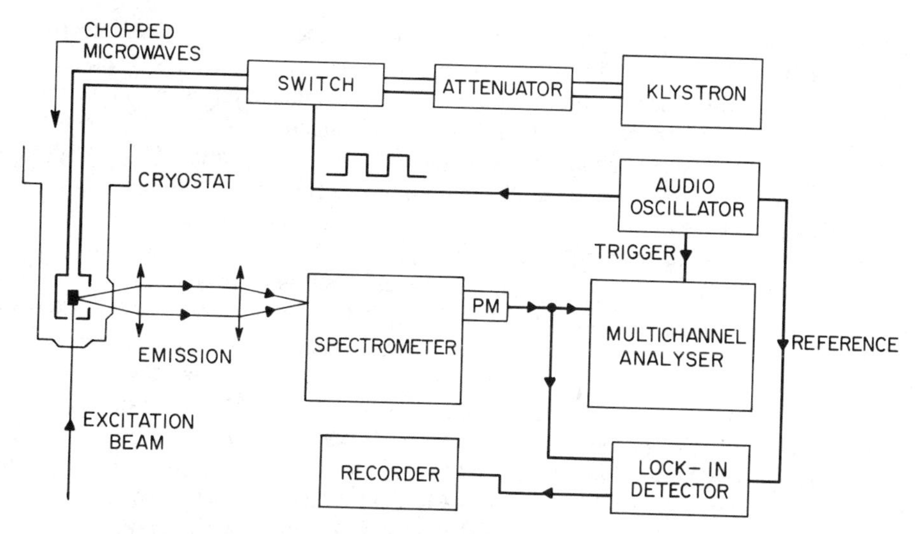


**Figure 5.78** Experimental arrangement for optically detected magnetic resonance (ODMR) experiment [213]. The sample in the cryostat at the left is held in a magnetic field which is slowly swept to obtain the ODMR spectrum through measurement of the change in total intensity or polarization of the luminescence in a given spectral range using lock-in detection. Alternatively, the spectral response of a given ODMR signal can also be measured by observing the luminescence through a spectrometer at a constant magnetic field.

given by the Boltzman distribution, it is advantangeous to do EPR measurement at low temperatures. On the other hand, at low temperatures there is not sufficient spin lattice relaxation which restores the population difference of the levels that is upset at resonance. Therefore, there is an optimum temperature for measuring the EPR spectrum.

A good example for the application of EPR to GaAs and InP is the study of anion antisite defects that is, anions occupying caption sites, $As_{Ga}$ or $P_{In}$. These defects are expected to behave as double donors since a pentavalent atom substitutes a trivalent atom. The antisite defect in the neutral charge state with both its electrons is diamagnetic. If the defect is singly ionized, then an EPR signal is observed. Figure 5.79 shows the EPR spectrum of $As_{Ga}$ in as-grown, plastically deformed and neutron-irradiated, liquid encapsulated Czochralski grown GaAs [212]. The quadruplet structure seen in the spectra is the fingerprint of $As_{Ga}$ [214]. The four lines are caused by an hyperfine interaction of the resonant electron with an $I = 3/2$ nucleus. Plastic deformation and electron or neutron irradiation enhance the EPR signal, indicating creation of $As_{Ga}$ centers under these conditions. An EPR signal caused by $P_{In}$ has been identified in electron irradiated InP [215, 216]. It has not, however, been yet observed in as-grown material. The EPR signal is a two line spectrum characteristic of the central $^{31}P$ atom with nuclear spin $I = 1/2$.

The advantage of ODMR over EPR is that increased magnetic resonance sensitivity is generally obtained. The reason for the increased sensitivity in ODMR, is that weakly allowed magnetic dipole transitions at resonance induce changes in the highly allowed electric dipole transitions of the optical process. The improved sensitivity and power of ODMR, particularly the detection of magnetic resonance via absorption, have been demonstrated by several experiments connected with the study of antisite defects in III-V semiconductors. These experiments involve observing the microwave induced changes in the magnetic circular dichroism (MCD) of the broad IR absorption associated with the antisite defect. Unlike the Zeeman splittings of a sharp line, magnetic field has no effect on a broad absorption band. However, the changes in the circularly polarized absorption and the refractive index components can result in large dichroism and Faraday rotation signals, respectively. The MCD signal which is a measure of $I_{\sigma^+} - I_{\sigma^-}$ usually expressed as a function of absorption wavelength, is dependent upon the population of the ground state. Hence magnetic resonance in the ground state is manifested as a microwave induced change in the MCD signal. The MCD-ODMR technique was first used by Meyer et al., [217] to observe the $As_{Ga}$ resonance in GaAs. Figure 5.80 shows the MCD of the absorption of as-grown semi insulating GaAs in the spectral range 0.8 to 1.4 eV measured at T = 4.2K and a magnetic field of 2T. The MCD-ODMR signal measured by monitoring the absorption at 1.35μm at T = 1.4K and 24.31GHZ for B || [100] is shown in Fig. 5.81. The resulting EPR spectrum is very much similar to the $As_{Ga}$ resonance shown in Fig. 5.79 in its four-line structure, the g-value and the $^{75}As$ hyperfine-splitting parameters. Deiri et al., [215] have observed the $P_{In}$ antisite resonance in electron irradiated InP by monitoring the MCD at 0.95μm. Kana-ah et al., [218] studied MCD-ODMR in e-irradiated n- or p-type InP doped with Sn or Zn and found different signatures for the $P_{In}$ resonance depending upon the conductivity type.

In the conventional EPR spectrum that has been assigned to the $As_{Ga}$ defect (Fig. 5.79) the quadruplet structure is caused by hyperfine interaction with the central As nucleus (I = 3/2). This

**Figure 5.79** EPR of $As_{Ga}^{4+}$ in as-grown, plastically deformed and n-irradiated LEC GaAs [212].

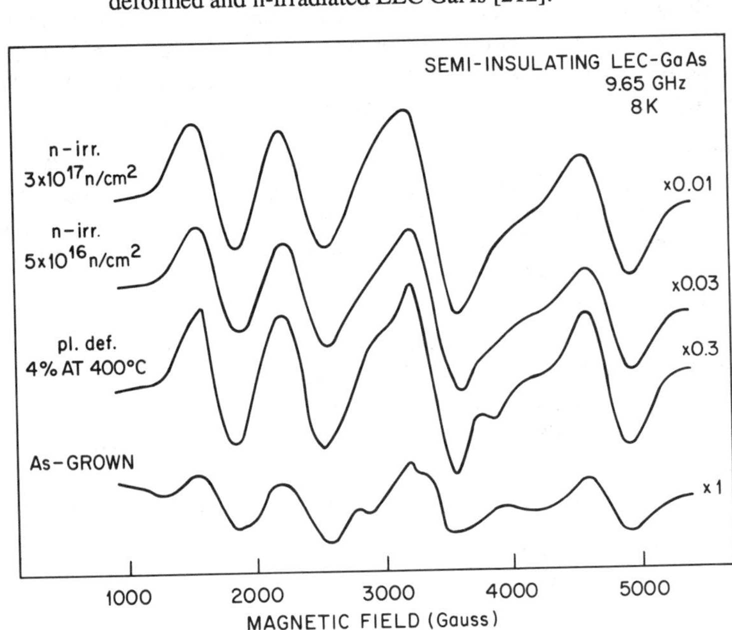

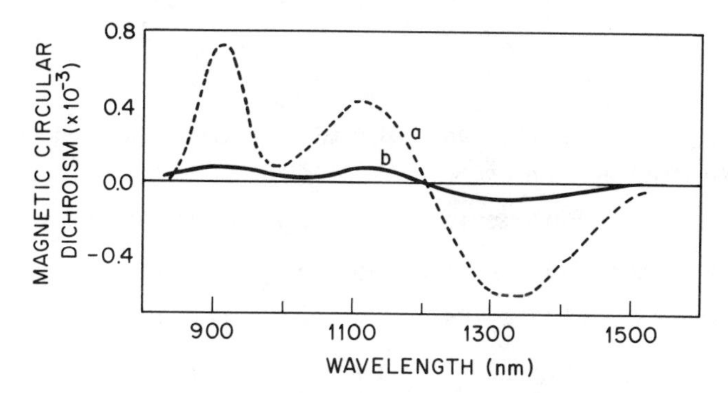


**Figure 5.80** Magnetic circular dichroism of the absorption of as-grown semi-insulating GaAs at 4.2K and a magnetic field of 2T (curve a). Excitation spectrum of the optically detected electron spin resonance line of the $As_{Ga}$ antisite defect (curbe b) [217].

spectrum does not resolve the ligand hyperfine interaction. Therefore, from EPR alone it is not possible to decide whether the central As atom in the antisite defect is surrounded by four As atoms or whether an impurity or vacancy exists in the nearest or next nearest neighbor positions. Electron nuclear double resonance (ENDOR) which is capable of resolving ligand hyperfine interactions has not been very successful in materials such as GaAs because of the large linewidth of the EPR lines. Hoffmann et al., [219] used optically detected ENDOR similar to ODMR, and because of the increased sensitivity of optical detection of magnetic resonance, were able to resolve and analyze ligand hyperfind interactions. These results showed that the $As_{Ga}$ antisite structure in deformed GaAs is different from the regular $As_{Ga}$ defect in as-grown material, even though they both give the same EPR spectrum thus suggesting the possibility that there are several $AsAs_4$ antisite complexes.

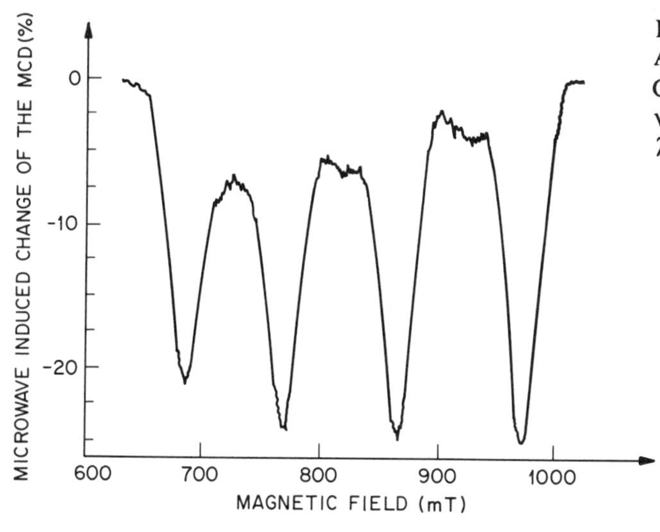


**Figure 5.81** ODMR spectrum of the $As_{Ga}$ antisite defect in semi-insulating GaAs · B || [100], T = 1.4K, $\nu_{ESR}$ = 24.31 GHz, measured at $\lambda$ = 1350nm [217].

## REFERENCES

1. T.S. Moss, G.J. Burrell, and B. Ellis, ''Semiconductor Opto-Electronics,'' London;Butterworths, p. 58, 1973.
2. J.I. Pankove, ''Optical Processes in Semiconductors,'' New York;Dover, p. 93, 1975.
3. T.S. Moss and T.D.F. Hawkins, *Infrared Phys. 1*;111, 1961.
4. F. Urbach, *Phys. Rev. 92*;1324, 1953.
5. R.S. Knox, ''Theory of Excitons,'' Suppl. 5, *Solid State Physics, Advances in Research and Applications* F. Seitz, D. Turnbull, and H. Ehrenreich, New York;Academic, 1963.
6. J.I. Pankove, *Phys. Rev.* 140;A2059, 1965.
7. D. Redfield, *Phys. Rev.* 130;916, 1963.
8. W. Franz, *Z Naturfarsch, 13a*;484, 1958; L.V. Keldysh, *Sov. Phys. JETP 7*;788, 1958.
9. D. Redfield, and M.A. Afromowitz, *Appl. Phys. Lett. 11*;138, 1967.
10. E. Burstein, *Phys. Rev. 93*;632, 1954; T.S. Moss, *Proc. Phys. Soc.* (London) B76, 775, 1954.
11. M. Bugajski, and W. Lewandowski, *J. Appl. Phys. 57*;521, 1985.
12. D.D. Sell, and H.C. Casey, Jr., *J. Appl. Phys. 45*;800, 1974.
13. S.H. Wemple, and J.A. Seman, *Appl. Optics* 12;2947, 1973.
14. D.A. Humphreys, R.J. King, D. Jenkins, and A.J. Moseley, *Electron. Lett. 21*;1187, 1985.
15. V.P. Evtikhiev, D.Z. Garbuzov, and A.T. Gorelenok, *Sov. Phys. Semicond. 17*;891, 1983.
16. B. Monemar, K.K. Shih, and G.D. Pettit, *J. Appl. Phys. 47*;2604, 1976.
17. J.A. Stratton, ''*Electromagnetic Theory*,'' New York;McGraw-Hill, 1941.
18. R.J. Elliott, *Phys. Rev. 108*;1384, 1957.
19. M.D. Sturge, *Phys. Rev.127*;768, 1962.
20. W.J. Turner, W.E. Reese, and G.D. Pettit, *Phys. Rev. 136*;A1467, 1964.
21. H. Asai and K. Oe, *J. Appl. Phys. 54*;2052, 1983.
22. R. Dingle and W. Wiegmann, *J. Appl. Phys. 46*;4312, 1975.
23. E. Zielinski, H. Schweizer and K. Streubel, *J. Appl. Phys. 59*;2195, 1986.
24. F.H. Pollak and M. Cardona, *Phys. Rev. 172*;816, 1968.
25. D.O. Sell, S.E. Stokowski, R. Dingle and J.V. DiLorenzo, *Phys. Rev. B7*;4568, 1973.

26. H.R. Phillip, and H. Ehrenreich, in ''*Semiconductors and Semimetals*'' eds., R.K. Willardson, and A.C. Beer, vol. 3, New York;Academic, p. 93, 1967.

27. M. Cardona, ''Semiconductors and Semimetals'' eds., R.K. Willardson, and A.C. Beer, vol. 3, New York;Academic, p. 125, 1967.

28. R. Braunstein, *J. Phys. Chem. Solids, 8*;280, 1959.

29. H.C. Casey, Jr., and P.L. Carter, *Appl. Phys. Lett. 44*;82, 1984.

30. C.H. Henry, R.A. Logan, F.R. Merritt and J.P. Luongo, *IEEE J. Quant. Elect. 19*;947, 1983.

31. A.R. Adams, M. Asada, Y. Suematsu and S. Arai, *Jap J. Appl. Phys. 19*;L621, 1980.

32. K. Nonomura, I. Suemune, M. Yamanishi and N. Mikoshiba, *Jap J. Appl. Phys. 22*;L556, 1983.

33. J. Blanc and L.R. Weisberg, *Nature 192*;155, 1961.

34. G.M. Martin, *Appl. Phys. Lett. 39*;747, 1981.

35. H.Y. Fan, *Rep. Prog. Phys. 19*;107, 1956.

36. S. Visvanathan, *Phys. Rev. 120*;376, 1960.

37. H.Y. Fan, W.G. Spitzer, and R.J. Collins, *Phys. Rev. 101*;566, 1956.

38. O.K. Kim and W.A. Bonner, *J. Electron. Mat. 12*;827, 1983.

39. J. Serre, A. Ghazali, and P. Hugon, *Phys. Rev. B 23*;1971, 1981.

40. H.C. Casey, Jr., and F. Stern, *J. Appl. Phys. 47*;631, 1976.

41. M. Voos, R.F. Leheny and J. Shah, in ''Handbook on Semiconductors'' (eds., M. Balkanski,) vol. 2, North-Holland, Amsterdam, p. 329, 1980.

42. M. Gershenzon in, ''Semiconductors and semimetals'' (eds., R.K. Willardson and A. C. Beer), vol. 2, New York;Academic, p. 289, 1966.

43. H.B. Bebb and E.W. Williams, in ''Semiconductors and semimetals'' (eds., R. K. Willardson and A. C. Beer), vol. 8, New York;Academic, p. 181, 1972.

44. W. Van Roosbroeck and W. Shockley, *Phys. Rev. 94*;1558, 1954.

45. R.N. Hall, *Phys. Rev. 87*;387, 1952.

46. W. Shockley and W.T. Read, *Phys. Rev. 87*;835, 1952.

47. D.C. Reynolds and C.W. Litton, in ''Optical Characterization Techniques for Semiconductor Technology'' (eds., D.E. Aspnes, S. So and R.F. Potter) *Proc. SPIE, 276*;11, 1981.

48. Y. Horikoshi, A. Fischer and K. Ploog, *Appl. Phys. A39*;21, 1986.

49. D.J. Wolford, J.A. Bradley, K. Fry, J. Thompson and H. E. King, in "GaAs and Related Compounds," Inst. of Conf. Ser. *65*;477, 1983.

50. R.N. Bhargava and M.I. Nathan, *Phys. Rev. 161*;695, 1967.

51. W.D. Johnston, Jr., G.Y. Epps, R.E. Nahory and M.A. Pollack, *Appl. Phys. Lett. 33*;992, 1978.

52. E. Fabre, *Solid State Comm. 9*;635, 1971.

53. M.J. Luciano and D.L. Kingston, *Rev. Sci. Instrum. 49*;718, 1978.

54. R.E. Frahm, Unpublished.

55. P.M. Petroff and R.A. Logan, *J. Vac. Sci. Tech. 17*;1113, 1980.

56. R.J. Nelson and R.G. Sobers, *J. Appl. Phys. 49*;6103, 1978.

57. R. Dingle, *Phys Rev. 184*;788, 1969.

58. A.A. Berg and P.J. Dean, "Light-Emitting Diodes," Oxford University Press, 1976.

59. E.W. Williams and R. Hall, "Luminescence and the light emitting diode," Pergamon, Oxford, 1978.

60. T.D.S. Hamilton, I.H. Munro and G. Walker, in "Luminescense Spectroscopy" (ed. M. D. Lumb), New York;Academic, p. 150, 1978.

61. R.Z. Bachrach, *Rev. Sci Instrum. 43*;734, 1972.

62. S.G. Bishop, in "Optical Characterization Techniques for Semiconductor Technology," (eds., D.E. Aspnes, S. So and R.F. Potter), *Proc. SPIE, 276*;1, 1981.

63. R.G. Waters, S.R. Chinn and B.D. Schwartz, in "Spectroscopic Characterization Techniques for Semiconductor Technology" (eds., F.H. Pollak and R.S. Bauer), *Proc. SPIE, 452*;17, 1984.

64. V. Swaminathan, C.A. Green, D.T.C. Huo, and M. Brelvi, *J. Mater. Res. 3*;309, 1988.

65. E.W. Williams and R.A. Chapman, *J. Appl. Phys. 38*;2547, 1967.

66. T.S. Moss, G.J. Burrell, and B. Ellis "Semiconductor Opto-Electronics," London;Butterworths, p. 204, 1973.

67. D.Z. Garbuzov, V.B. Khalfin, M.K. Trukan, V.G. Agafonov and A. Abdullaev, *Sov. Phys. Semicond. 12*;809, 1978.

68. R.G. Ulbrich and C. Weisbuch, 1976, unpublished.

69. Y. Toyozawa, *Prog. Theor. Phys. 20*;53, 1958.

70. Y. Toyozawa, *rog. Theor. Phys. 27*;89, 1962.

71. H. Nakashima, N. Chinone and R. Ito, *J. Appl. Phys. 46*;3092, 1975.

72. Y. Toyozawa, *J. Phys. Chem. Solids 25*;59, 1964.

73. J.J. Hopfield, in "II-VI Semiconducting Compounds," New York;Benjamin, p. 800, 1967.

74. R.G. Ulbrich, C. Weisbuch, *Phys. Rev. Lett. 38*;865, 1977.

75. J.R. Haynes, *Phys. Rev. Lett. 17*;860, 1966.

76. J.R. Haynes, *Phys. Rev. Lett. 4*;361, 1960.

77. P.J. Dean and D.C. Herbert, in "Excitons" (ed., K. Cho), Berlin;Springer-Verlag, p. 55, 1979.

78. E.I. Rashba and M.D. Sturge, "Excitons," Amsterdam;North-Holland, 1982.

79. E.I. Rashba and G.E. Gurgeneshvili, *Sov. Phys. Solid State 4*;759, 1962.

80. D.J. Ashen, P.J. Dean, D.T.J. Hurle, J.B. Mullin, A.N. White and P.D. Greene, *J. Phys. Chem. Solids 36*;1041, 1975.

81. R. Dingle, C. Weisbuch, H.L. Stormer, H. Morkoc, and A.Y. Cho, *Appl. Phys. Lett. 40*;507, 1982.

82. Ref. 58, p. 135.

83. H.C. Casey, Jr., and R.H. Kaiser, *J. Electrochem. Soc. 114*;149, 1967.

84. D.A. Casano, *Solid State Commn.* 2;353, 1964.

85. D.E. Hill, *Phys. Rev. A133*;866, 1964.

86. F.Z. Hawrylo, *Appl. Phys. Lett. 37*;1038, 1980.

87. P.M. Raccah, H. Rahemi, J. Zehnder, F.Z. Hawrylo, H. Kressel, and J.S. Helman, *Appl. Phys. Lett. 39*;496, 1981.

88. S. Bendapudi and D.N. Bose, *Appl. Phys. Lett. 42*;287, 1983.

89. J. De-Sheng, Y. Makita, K. Ploog, and H.J. Queisser, *J. Appl. Phys. 53*;999, 1982.

90. D. Olego and M. Cardona *Phys. Rev.* 22B;886, 1980.

91. T.P. Pearsall, L. Eaves, and J.C. Portal, *J. Appl. Phys. 54*;1037, 1983.

92. V. Swaminathan, N.E. Schumaker, J.L. Zilko, W.R. Wagner, and C.A. Parsons, *J. Appl. Phys. 52*;412, 1981.

93. V. Swaminathan, J.L. Zilko and S.F. Nygren, *Mat. Lett.* 2;308, 1984.

94. D.M. Eagles, *J. Phys. Chem. Solids 16*;76, 1960.

95. C. Pickering, P.R. Tapsten, P.J. Dean, and D.J. Ashen, in "GaAs and Related Compounds," *Inst. Phys. Conf. Ser. 65*;p. 469, 1983.

96. R. Ulbrich, *Phys. Rev. B8*;5719, 1973.

97. E.W. Williams and H.B. Bebb, *J. Phys. Chem. Solids 30*;1289, 1969.

98. J.D. Dow, D.L. Smith, and F.L. Lederman, *Phys. Rev. B8*;4612, 1973.

99. P.J. Dean, in "Proress in Solid State Chemistry" (eds., J. O. McCaldin and G. Somorjai), vol. 8, New York;Pergamon, p. 64, 1973.

100. V. Swaminathan, V.M. Donnelly, and J. Long, *J. Appl. Phys. 58*;4565, 1985.

101. V. Swaminathan, A.R. VonNeida, R. Caruso, and M.S. Young, *J. Appl. Phys. 53*;6471, 1982.

102. O. Roder, U. Heim, and M.H. Pilkuhn, *J. Phys. Chem. Solids 31*;2625, 1970.

103. V. Swaminathan, N.E. Schumaker, J.L. Zilko, W.R. Wagner, and C.A. Parsons, *J. Appl. Phys. 52*;412, 1981.

104. V. Swaminathan, R.A. Stall, A.T. Macrander, and R.J. Wunder, *J. Vac. Sci. Tech. 3B*;1631, 1985.

105. V. Swaminathan, unpublished.

106. J. Golka, *Solid State Comm. 28*;401, 1978; J. Golka, and H. Stall, *32*;479, 1979.

107. P.W. Yu, *J. Appl. Phys. 49*;5043, 1977.

108. V.P. Dobrego and I.S. Shlimak, *Phys. Stat. Solid, 33*;805, 1969.

109. D. Redfield, J.P. Wittke, and J. Pankove, *Phys. Rev. B2;1830, 1970.*

*110. Zh.I. Alferov, V.M. Andreev, D.Z. Garbuzov, and M.K. Trukan, Sov. Phys. Semicond. 6*;1718, 1973.

111. B.I. Shkolovskii and A.L. Efros, *Sov. Phys. JETP, 33*;468, 1971; *34*;435, 1972.

112. M.I. Nathan, T.N. Morgan, in "Proceeding of the International Conference on Quantum Electronics" (eds., P.L. Kelly, B. Lax, and P.E. Tannewald), New York;McGraw-Hill, p. 478, 1966.

113. V. Swaminathan, N.E. Schumaker, and J.L. Zilko, *J. Luminescence 22*;153, 1981.

114. V. Swaminathan, P.J. Anthony, J.L. Zilko, M.D. Sturge, and N.E. Schumaker, *J. Appl. Phys. 52*;5603, 1981.

115. J. Mazzaschi, J. Barrau, J.C. Brabant, M. Brousseau, H. Maaref, F. Voillot, and M.C. Boissy, *Rev. Phys. Appl. 15*;861, 1980.

116. B.J. Skromme, G.E. Stillman, J.C. Oberstar, and S.S. Chan, *J. Electron. Mat. 13*;463, 1984.

117. A.S. Kaminskii and Ya.E. Pokrovskii, Sov. Phys. Semicond. 3;1496, 1970; See also Ref. 94, p. 59.

118. B.J. Skromme and G.E. Stillman, *Phys. Rev. 29B*;1982, 1984.

119. W. Schairer and N. Stath, *J. Appl. Phys. 43*;447, 1972.

120. W. Bludau, E. Wagner, and H.J. Queisser, *Solid State Comm. 18*;861, 1976.

121. D. Bois and D. Beaudet, *J. Appl. Phys. 46*;3882, 1975.

122. T. Kamiya and E. Wagner, *J. Appl. Phys. 48*;1928, 1977.

123. S.B. Nam, D.W. Langer, D.L. Kingston, and M.J. Luciano, *Appl. Phys. Lett. 31*;652, 1977.

124. P.W. Yu, *Solid State Comm. 27*;1421, 1978.

125. A.T. Hunger and T.C. McGill, *Appl. Phys. Lett. 40*;169, 1982.

126. D.W. Kisker, H. Tews, and W. Rehm, *J. Appl. Phys. 54*;1332, 1983.

127. S.R. Hetzler, T.C. McGill, and A.T. Hunter, *Appl. Phys. Lett. 44*;793, 1984.

128. P.J. Dean, D.J. Robbins, and S.G. Bishop, *J. Phys. C. Solid State 12*;5567, 1979.

129. P.J. Dean, D.J. Robbins, and S.G. Bishop, *Solid State Comm. 32*;379, 1979.

130. J.C.M. Henning, J.J.P. Noijen, and A.G.M. deNijs, *Phys. Rev. 27B*;7451, 1983.

131. T. Nishino, Y. Fujiwara, A. Kojima, and Y. Hamakawa, in ''Spectroscopic Characterization Techniques for Semiconductor Technology,'' (eds., F.H. Pollak and R.S. Bauer), *Proc. SPIE,, 452*;2, 1984.

132. V. Swaminathan, *Bull. Mat. Sci. 4*;403, 1982; see also Ref. 60, p. 158.

133. B.V. Shanabrook, P.B. Klein, E.M. Swiggard, and S.G. Bishop, *J. Appl. Phys. 54*;336, 1983.

134. M. Tajima, in *Defects and Properties of Semiconductors: Defect Engineering,* (eds. J. Chikawa, K. Sumino, and K. Wada), Tokyo;KTK Scientific Publishers, p. 37, 1987.

135. P.W. Yu, *Phys. Rev. 29B*;2283, 1984.

136. H. Temkin and B.V. Dutta, in ''Defects in Semiconductors II'' (eds., S. Mahajan and J.W. Corbett), Materials Research Society Proceedings, *vol. 14*, North-Holland, Amsterdam, p. 253, 1983.

137. K.H. Goetz, D. Bimbrg, K.A. Brauctile, H. Jurgensen, J. Selders, M. Razeghi, and E. Kuphal, *Appl. Phys. Lett. 46*;277, 1985.

138. E.V.K. Rao, R.E. Nahory, B. Etienne, G. Chaminant, M.A. Pollack, and J.C. DeWinter, in *GaAs and Related Compounds, Inst. Phys. Conf. Ser. 56*;605, 1981.

139. B. Monemar and H.G. Grimmeiss, *Prog. Crystal Growth Charact. 5*;47, 1982.

140. Ref. 2, p. 113.

141. H.J. Queiisser, *Solid State Elect. 21*;1495, 1978.

142. R.J. Nelson, J.S. Williams, H.J. Leamy, B. Miller, H.C. Casey, Jr., B.A. Parkinson, and A. Heller, *Appl. Phys. Lett. 36*;76, 1980.

143. P.T. Landsberg, *Phys. Status Solids 41*;457, 1970.

144. N.K. Dutta and R.J. Nelson, *J. Appl. Phys. 53*;74, 1982.

145. J.C. Tsang, P.J. Dean, and P.T. Landsberg, *Phys. Rev. 173*;814, 1968.

146. Ref. 59, p. 157.

147. F.H. Pollak and R. Tsu, in "Spectroscopic Characterization for Semiconductor Technology" (eds., F. Pollak and R. Bauer), *Proc. SPIE 452*;p. 26, 1984.

148. R. Tsu, in "Optical Characterization Techniques for Semiconductor Technology," (eds., D.E. Aspnes, S. So, and R.F. Potter), *Proc. SPIE 276*;p. 78, 1981.

149. G. Abstreiter, E. Bauser, A. Fischer, and K. Ploog, *Appl. Phys. 16*;345, 1978.

150. V. Swaminathan, A. Jayaraman, J.L. Zilko, and R.A. Stall, *Mat Lett. 3*;325, 1985.

151. G.P. Schwartz, in "Optical Characterization Techniques for Semiconductor Technology," (eds., D. E. Aspnes, S. So and R. F. Potter), *276*;p. 72, 1981.

152. D.J. Evans and S. Ushida, *Phys. Rev. B1*;1638, 1974.

153. J L. Waugh and G. Dolling, *Phys. Rev. 132*;2410, 1963.

154. P.H. Borcherds, G.F. Alfrey, D.H. Saunderson, and A.D.B. Woods, *J. Phys. C8*;2022, 1975.

155. J. Biellmann, B. Prevot, and C. Schwab, *J. Phys. C16*;1135, 1983.

156. P. Parayanthal, F.H. Pollak, and J.M. Woodall, *Appl. Phys. Lett. 41*;961, 1982.

157. D. Olego and M. Cardona, *Phys. Rev. B24*;7217, 1981.

158. T. Nakamura, A. Ushirokawa, and T. Katoda, *Appl. Phys. Lett. 38*;13, 1981.

159. B. Jusserand and J. Sapriel, *Phys. Rev. B24*;7194, 1981.

160. K. Kakimoto and T. Katoda, *Appl. Phys. Lett. 40*;826, 1982.

161. R.K. Soni, S.C. Abbi, K.P. Jain, M. Balkanski, S. Slempkes, and J.L. Benchimol, *J. Appl. Phys. 59*;2184, 1986.

162. K.K. Tiong, P.M. Amirtharaj, F.H. Pollak, and D.E. Aspnes, *Appl. Phys. Lett. 44*;122, 1984.

163. P. Parayanthal and F.H. Pollak, *Phys. Rev. Lett. 52*;1822, 1984.

164. J. Menendez, A. Pinczuk, A.C. Gossard, J.H. English, D.J. Werder, and M.G. Lamont, *J. Vac. Sci. Tech. B4*;1041, 1986.

165. R. Tsu and S.S. Jha, *Appl. Phys. Lett. 20*;16, 1972.

166. C. Colvard, R. Merlin, M.V. Klein, and A.C. Gossard, *Phys. Rev. Lett. 45*;298, 1980.

167. A. Pinczuk, J.M. Worlock, H.L. Stormer, A.C. Gossard, and W. Weigmann, *J. Vac. Sci. Technol. 19*;561, 1981.

168. H. Shen and F.H. Pollak, *Appl. Phys. Lett. 45*;692, 1984.

169. A. Pinczuk, G. Abstreiter, R. Trommer, and M. Cardona, *Solid State Commun. 21*;959, 1977.

170. D.J. Olego and H.B. Serreze, *J. Appl. Phys. 58*;1979, 1985.

171. D. Olego, T.Y. Chang, E. Silberg, E.A. Caridi, and A. Pinczuk, *Appl. Phys. Lett. 41*;476, 1982.

172. G. Abstreiter, *J. Vac. Sci. Technol. B3*;683, 1985.

173. D. Zahn, N. Esser, W. Pletschen, J. Geurts, and W. Richter, *Surf. Sci. 168*;823, 1986.

174. A. Pinczuk, A.A. Ballman, R.E. Nahory, M.A. Pollack, and J.M. Worlock, *J. Vac. Sci. Technol. 16*;1168, 1979.

175. W. Spitzer, in ''Festkorper problem - Advances in Solid State Physics,'' (ed. O. Madelung), *vol. XI*, New York;Pergamon, p. 1, 1971.

176. R.C. Newman, ''Infrared studies of crystal defects.'' London;Taylor and Francis, 1974.

177. W.M. Theis, in ''Spectroscopic characterization techniques for semiconductor technology,'' *Proc. SPIE 524*;p. 45, 1985.

178. W.M. Theis, K.K. Bajaj, C.W. Litton, and W.G. Spitzer, *Appl. Phys. Lett. 41*;70, 1982.

179. Y. Kitagawara, I. Itoh, N. Noto, and T. Takenaka, *Appl. Phys. Lett. 48*;788, 1986.

180. G.A. Acket, W. Nijman, R.P. Tijburg, and P.J. deWaard, *Inst. Phys. Conf. Ser. 24*;181, 1975.

181. P.J. Anthony, N.E. Schumaker, and J.W. Lee, unpublished.

182. R.A. Logan, N.E. Schumaker, C.H. Henry, and F.R. Merritt, *J. Appl. Phys. 50*;5970, 1979.

183. D.V. Lang and C.H. Henry, *Solid State Electronics 21*;1519, 1978.

184. S.F. Nygren, *IEEE J. Q. Electron. QE-19*;898, 1983.

185. H.J. Hovel, in ''Semiconductors and Semimetals,'' (eds., R.K. Willardson and A.C. Beer), *vol. 11*. New York;Academic, 1975.

186. H.C. Casey, Jr., and M.B. Panish,'' Heterostructure lasers,'' part A, p. 44-47. New York;Academic, 1975.

187. M. Cardona, ''Modulation Spectroscopy,'' Supplement 11, of ''Solid State Physics-Advances in Research and Applications'' (eds., F. Seitz, D. Turnbull, and H. Ehrenreich), New York;Academic, 1969.

188. ''Semiconductors and Semimetals'' (eds., R.K. Willardson and A.C. Beer), *vol. 9*. ''Modulation Techniques,'' New York;Academic, 1972.

189. Y. Hamakawa and T. Nishino, in ''Optical properties of Solids: New Developments'' (ed., B. O. Seraphin), North-Holland, Amsterdam, p. 259, 1976.

190. D.E. Aspnes, in ''Handbook on semiconductors'' (T.S. Moss, ed.) *vol. 2*, North-Holland, Amsterdam, p. 198, 1980.

191. F.H. Pollak, in ''Optical Characterization Techniques for Semiconductor Technology'' (eds., D.E. Aspnes, S. So and R.F. Potter), *Proc. SPIE 276*;p. 142, 1981.

192. F.H. Pollak, C.E. Okeke, P.E. Vanier, and P.M. Raccah, *J. Appl. Phys. 50*;5375, 1979.

193. R.L. Brown, L. Schoonveld, L.L. Abels, S. Sundaram, and P.M. Raccah, *J. Appl. Phys. 52*;2950, 1981.

194. F.H. Pollak and J.M. Woodall, *J. Vac. Sci. Technol. 17*;1108, 1980.

195. Y. Yamazoe, Ph.D. Dissertation, Osada University, Japan, 1981.

196. K. Alavi, R.L. Aggrawal, S.H. Groves, and Z.L. Liau, Annual Report Francis Bitter National Magnet Lab., MIT, 1980.

197. R.E. Nahory and J.L. Shay, *Phys. Rev. Lett. 21*;1569, 1968.

198. J.L. Shay, *Phys. Rev. 2B*;803, 1970.

199. W. Albers, Jr., *Phys. Rev. Lett. 23*;410, 1969.

200. J.G. Goy and L.T. Klauder, *Phys. Rev. 172*;811, 1968.

201. D.E. Aspnes, *Solid State Commun., 8*;267, 1976.

202. F. Cerdeira and M. Cardona, *Solid State Comun. 7*;879, 1969.

203. O.J. Glembocki, B.V. Shanabrook, N. Bottka, W.T. Beard, and J. Comas, *Proc. SPIE 524*;86, 1985.

204. P. Parayanthal, H. Shen, F.H. Pollak, O.J. Glembocki, B.V. Shanabrook, and W.T. Beard, *Appl. Phys. Lett. 48*;1261, 1986.

205. H. Shen, P. Parayanthal, F.H. Pollak, M. Tomkiewicz, T.J. Drummond, and J.N. Schulman, *Appl. Phys. Lett. 48*;653, 1986.

206. H. Shen, P. Parayanthal, F.H. Pollak, A.L. Smirl, J.N. Schulman, R.A. McFarlane, and I. D'Haenens, *Solid State Commun. 59*;557, 1986.

207. R.C. Miller, A.C. Gossard, G.D. Sanders, Y.C. Chang, and J.N. Schulman, *Phys. Rev. 32*;8452, 1985.

208. W.M. Duncan and A.F. Schreiner, *Solid State Commun. 31*;457, 1979.

209. D.E. Aspnes, *Phys. Rev. 153*;972, 1967.

210. M. Gal, J.S. Yuan, J.M. Viner, P.C. Taylor, and G.B. Stringfellow, *Phys. Rev. 33B*;4410, 1986.

211. B.C. Cavenett, *Adv. Physics 30*;475, 1981.

212. E.R. Weber, *Proc. SPIE 524*;160, 1985.

213. D.J. Dunstan and J.J. Davies, *J. Phys. C 12*;2927, 1979.

214. R.J. Wagner, J.J. Krebs, C.H. Strauss, and A.M. White, *Solid State Commun. 36*;15, 1980.

215. M. Deiri, A. Kana-ah, B.C. Cavenett, T.A. Kennedy, and N.D. Wilsey, *J. Phys. C, Solid State Phys. 17*;L793, 1984.

216. T.A. Kennedy and N.D. Wilsey, *Appl. Phys. Lett. 44*;1089, 1984.

217. B.K. Meyer, J.M. Spaeth, and M. Scheffler, *Phys. Rev. Lett. 52*;851, 1984.

218. A. Kana-ah, M. Deiri, B.C. Cavenett, N.D. Wilsey, and T.A. Kennedy, *J. Phys. C, Solid State Phys. 18*;L619, 1985.

219. D.M. Hoffman, B.K. Meyer, F. Lohse, and J.M. Spaeth, *Phys. Rev. Lett. 53*;1187, 1984.

# CHAPTER 6

# IMPURITIES AND NATIVE DEFECTS

## 6.1 INTRODUCTION

The material quality of semiconductors, be they in bulk form or in the form of thin films, is often the overriding factor in determining device performance. Poor devices result from poor starting material and the quality of the material is generally determined by crystalline imperfections that are invariably present. Because of this the study of defects in III-V semiconductors has been an active field of research over the last two decades.

Crystalline imperfections include point defects, dislocations, stacking faults, grain boundaries, interfaces, and so on. Foreign impurities are generally intentionally added to the material during growth to obtain desired electrical properties. Even when the material is grown intentionally "undoped," satellite impurities such as C, Si, and S, are present because of contamination from graphite crucibles (C, S) and quartz containers (Si) used for crystal growth. In VPE and MOCVD, reaction gases are often impurity sources. Besides foreign impurities, point defects include native defects such as vacancies.

Dislocations can be present in the as-grown material caused by a variety of factors such as thermal stress, excess point defects precipitating to form prismatic dislocation loops, and so on. Dislocations are also introduced during the multivarious device processing steps, for example, diffusion and ion implantation. The presence of dislocations is, in general, harmful to the devices and often the crystal grower aims to grow "low" dislocation density material. Stacking faults are two dimensional defects which can be present in the as-grown material due to excess impurity atoms precipitating as ''extrinsic faults'' or introduced during processing steps such as diffusion and oxidation. The same as dislocations, stacking faults also can have deleterious effects on devices.

Interfaces between semiconductors as in multilayer structures or that between semiconductors and dielectric films that are deposited on them as part of device processing can also be source and sink for defects. In either case defect free interfaces are desired.

In this chapter we discuss these various defects in GaAs, InP, and some of their alloys. After outlining the notation we employ to denote the different imperfections in Sec. 6.2, in Sec. 6.3 we discuss those foreign impurities which give rise to energy levels in the band gap that are close to either band edge (shallow levels). Impurities such as Si, S, Te, C, Be, and Mg belong to this category. Some impurities have energy levels near the middle of the band gap (deep levels). Transition metal impurities belong to this category. These impurities are discussed in Sec. 6.4. Section 6.5 covers native point defects, that is, vacancies, interstitials, and antisite defects, with respect to their formation, structure and energy levels. The influence of the ambient such as the group V partial pressure on the formation of point defects is discussed within the framework of defect chemistry, the branch of chemistry which constitutes formulation of chemical reactions for defect incorporation and interaction. The incorporation of impurities during crystal growth is discussed in Sec. 6.6. Section 6.7 discusses dislocations, their types and structures, electrical properties, and their relation to the mechanical properties of the semiconductor. Also discussed in this section is the introduction of dislocations during crystal growth and the ways to reduce their density.

## 6.2 CLASSIFICATION AND NOTATION

The atomic imperfections are of three types:

1. Vacancies: sites which in the ideal structure should be occupied by atoms but are not,
2. Interstitials: atoms in sites where they should not be,
3. Misplaced atoms: atoms in sites which in the ideal crystal are assigned to atoms of different type, thus in compound AB, A atoms occupy sites belonging to B atoms and vice versa. These misplaced atoms are also called antisite defects.

Each imperfection is denoted by a symbol with a subscript indicating the site. $A_i$ means atom A at an interstitial site, $V_A$ a vacancy at an A site and $A_B$ atom A occupying a B site. This scheme can also be used to indicate the normal situation. That is, A atom at an A site is $A_A$ and $V_i$ is an unoccupied interstitial site. Foreign atoms occupying substitutional or interstitial sites can also be indicated by the same method; for example, $F_A$, $F_B$, and $F_i$ for a foreign atom F at site A, site B, and interstitial site, respectively.

Defects can exist in neutral or charged form. The difference in the charge of the crystal site with and without an imperfection gives the charge state of the defect. This represents the effective charge as opposed to the actual charge which is present inside the bounds of the defect [1]. However, it is the effective charge of the defect which is important in electrostatic interaction between defects. For an interstitial, the actual and effective charges are the same. To denote the effective charge on the defect, we use +, –, and x as superscripts for positive, negative and neutral charge, respectively. Thus, $A_i^+$ is a positively charged interstitial, $V_A^{--}$ is an A vacancy with a double negative effective charge, and $A_A^x$ is an A atom at A site with zero effective charge. Free electrons and holes will be indicated by e and h, respectively.

From a practical point of view, one would like to know what should be expected for the charge state of a particular imperfection, that is, whether the imperfection will act as a donor or an acceptor or will have no electrical activity. For substitutional impurities which give rise to shallow levels, one can determine the donor or acceptor behavior from the difference of the valency of the foreign atom and the lattice atom it replaces. For deep impurities and native defects, detailed calculations of the electronic structure are required before the electrical activity can be determined and in the case of native defects even such calculations are not always unambiguous. Not withstanding these difficulties, it is perhaps helpful to follow some general empirical rules, which are borne out by experiments in many semiconductors, regarding the charge state of the point defects [2].

(a) Misplaced atoms: For a III-V semiconductor, a group V atom on a group III site would be expected to behave the same as a pentavalent impurity on a group III site. Similarly, a group III atom on a group V site would behave the same as a trivalent impurity on a group V site. That is, $As_{Ga}$ and $P_{In}$ would behave as a doubly ionizable donor and $Ga_{As}$ and $In_P$ would behave as a doubly ionizable acceptor.

(b) Interstitial atoms act as donors when their outer electron shell is less than half-full and as acceptors in the opposite case. For example, one may expect $Ga_i$ to behave as donor.

(c) Vacancies have donor or acceptor action depending on whether the number of unpaired electrons near the vacancy is less than or more than half the number of valence electrons present near the same site in the perfect crystal. Thus one may expect cation vacancies ($V_{Ga}$, $V_{In}$) to be the same as acceptors and anion vacancies ($V_{As}$, $V_P$) to be the same as donors.

In a zinc blende structure semiconductor there are three nonbonding interstitial sites, the tetrahedral ($T_R$), the hexagonal (H), and the $T_x$ sites, as shown in Fig. 6.1 [3]. Unlike an elemental semiconductor, in a compound there are two different tetrahedral sites, one in which the interstitial is surrounded by group III atoms and another in which it is surrounded by group V atoms. Accordingly the electronic energy of the interstitial at the two sites can be different [4]. The interstitial can also exist in a bonded configuration as shown in Fig. 6.1(c) where the interstitial is nestled between two substitutional atoms. Such a configuration is possible for a small foreign impurity such as $Li_i$. Another configuration for the interstitial is the split-interstitial configuration, in which two atoms straddle a substitutional site. The axis of the defect is $\langle 100 \rangle$ and the associated bonds are $sp^2$ hybridized (Fig. 6.1(d)).

### 6.2.1 Native point defects

Native defects can be formed in a III-V compound AB by disorder, nonstoichiometry, or doping with foreign elements. As enumerated earlier, there are six possible defects; two types of vacancies ($V_A$, $V_B$), interstitials ($A_i$, $B_i$), and misplaced atoms ($A_B$, $B_A$). Since the A/B site ratio required by the crystal structure has to be maintained, in a stoichiometric crystal the imperfections occur in sets containing at least two types. These sets constitute the basic types of native atomic disorder [5].

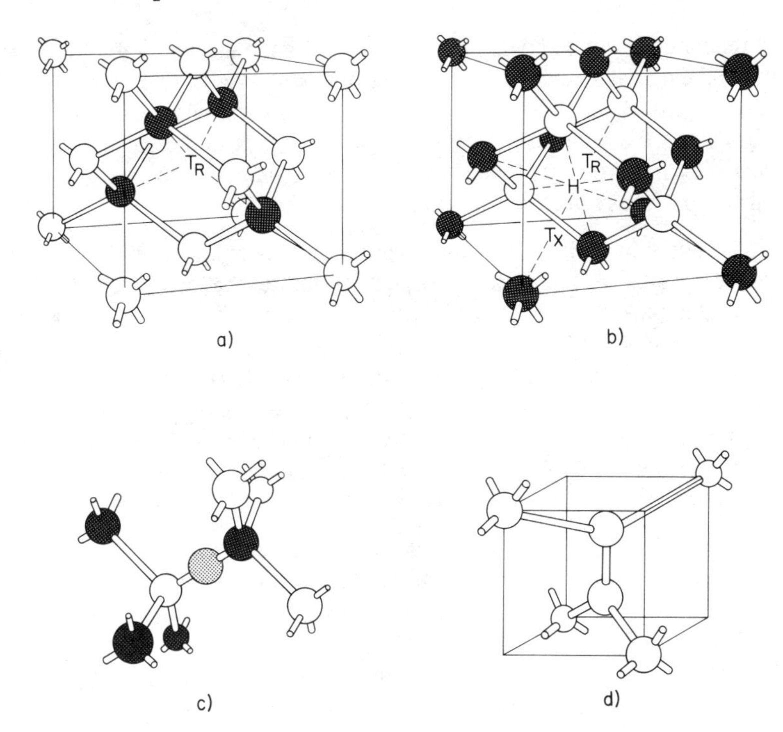


**Figure 6.1** The different interstitial configurations in the zinc blende lattice:

(a) tetrahedral interstitial site $T_R$ (b) tetrahedral interstitial site $T_x$; hexagonal interstitial site H shown joining $T_R$ and $T_x$ (c) the bond centered interstitial site (d) the [100] split or dumb-bell interstitial [3].

(a) Schottky disorder involving $V_A$ and $V_B$,

(b) interstitial disorder involving $A_i$ and $B_i$,

(c) antistructure disorder involving $A_B$ and $B_A$.

There is also the possibility of combinations of these three disorder types. A disorder involving vacancies and interstitials, for example, $V_A+A_i$ or $V_B+B_i$ is called Frenkel disorder. Another type involves $V_A+A_B$ or $V_B+B_A$. A third type of disorder involves $A_i+B_A$ or $B_i+A_B$. Thus there are nine possible types of disorder in a III-V compound.

## 6.3 SHALLOW LEVEL IMPURITIES

It is no exaggeration to say that the electrical properties of a semiconductor are controlled to a large extent by foreign impurities which are either intentionally added or present as contaminants. The effect of impurities arises from the fact that they introduce energy levels in the forbidden energy gap and the nature of these energy levels determines the electrical and in many instances, optical properties of the host semiconductor. Impurities give rise to two kinds of energy levels which are often denoted as shallow and deep energy levels. Shallow impurities have energy levels that are placed on the order of 0.1 eV or less with respect to the band edge. On the other hand, deep impurities have energy levels towards the middle of the energy gap. The deep impurities are also sometimes referred to as traps.

In the case of shallow impurities, a further classification is made depending on whether they are donors and acceptors. For impurities which substitute the cation or anion site in the compound semiconductor, donor or acceptor action is realized based on the chemical valence difference between the host and the impurity. If the valence difference is greater than zero, donors are created and if it is less than zero acceptors are created. If there is no valence difference then the impurity is called isovalent or isoelectronic, the latter being the common nomenclature even though it is a misnomer. The isovalent impurity can behave as a donor or an acceptor. An empirical guide line to predict doping type is that atoms much larger than the host they replace are donors whereas smaller atoms are acceptors [6].

For the GaAs and InP semiconductors, group VI impurities such as S, Se, and Te substitute for As or P and behave as shallow donors. Likewise, group II impurities such as Be, Mg, Zn, and Cd give rise to shallow acceptors from which it can be inferred that they substitute for the Ga or In site. Group IV impurities, on the other hand, show amphoteric behavior. That is, a group IV impurity on the anion site behaves as an acceptor while one on the cation site behaves as a donor. Among the group IV impurities, with the exception of carbon which is predominantly an acceptor, Si, Ge, and Sn behave amphoterically. As a result, when these dopants are used they tend to be self-compensated.

### 6.3.1 Effective Mass Theory

The energy level associated with a donor (acceptor) represents the binding energy of an electron (hole) to the impurity in the semiconductor host. For shallow impurities the binding of the electronic particle can be described by the so-called effective-mass approximation [6, 7]. Consider a shallow donor such as S in GaAs. Sulphur with six valence electrons substituting for the As site supplies five electrons to satisfy the co-valent bonds with Ga and the extra electron goes in the conduction band. Consequently $S_{As}$ assumes a positive charge and produces an electrostatic potential which is screened by the dielectric function of GaAs

$$U(r) = \frac{e}{\varepsilon r} \tag{6.1}$$

where $\varepsilon$ is the dielectric constant of the semiconductor. The electric field is screened both by electronic polarization and lattice polarization. Accordingly, $\varepsilon$ lies between the low frequency limit of the electronic contribution to the dielectric constant $\varepsilon_0$ and the static dielectric constant $\varepsilon_s$ [7]. If the motion of the carrier around the impurity is slow such that the ions will follow the particle, $\varepsilon = \varepsilon_s$ otherwise $\varepsilon = \varepsilon_0$. The extra electron in the conduction band is bound to the positively charged $S_{As}^+$ by the potential given by Eq. (6.1). If the orbit of the trapped electron around $S_{As}^+$ extends over many unit cell distances, then the energy state of the electron can be described by the Schroedinger equation

$$\left[-\frac{\hbar^2}{2m^*}\nabla^2 - \frac{e^2}{\varepsilon r}\right]F(r) = EF(r) \tag{6.2}$$

Equation (6.2) is called the effective mass equation where F(r) is hydrogen-like envelope function. The solution of Eq. (6.2) is similar to that of the hydrogen atom, except that $e^2$ is replaced by $e^2/\varepsilon$ and the free electron mass is replaced by the effective mass $m^*$. Hence the energies of the bound states are given by

$$E_n = \frac{m^* e^4}{n^2\, 2\hbar^2 \varepsilon^2} = \frac{1}{n^2}\left(\frac{m^*}{m}\right)\left(\frac{1}{\varepsilon^2}\right) Ry \tag{6.3}$$

where Ry is 13.6 eV, the ionization energy of the free hydrogen atom. For electrons (donors) $m^*$ is the electron effective mass and the energies are measured from the bottom of the conduction band. For holes (acceptors) $m^*$ is the hole effective mass and the energies are measured from the top of the valence band. The Bohr radius corresponding to the bound states is also modified from that of the hydrogen atom and is given by

$$a = \frac{\varepsilon \hbar^2}{m^* e^2} = a_H \frac{\varepsilon}{(m^*/m_0)} \tag{6.4}$$

where $a_H$ is the Bohr radius of the hydrogen atom.

For a direct gap semiconductor such as GaAs and InP with nondegenerate conduction band and isotropic electron effective mass, Eq. (6.2) is a good description for the binding of the electrons to donors. The binding energies and Bohr radius of donors in GaAs and InP obtained from Eqs. (6.3) and (6.4) are, respectively, $E_D = 5$ meV, $a_D = 102$ Å, and $E_D = 7$ meV, $a_D = 84$ Å with $\varepsilon_s = 12.9$ and $m_e = 0.067\, m_0$ for GaAs and $\varepsilon_s = 12.61$, $m_e = 0.08\, m_0$ for InP. For acceptors, the situation is somewhat complicated because of the complex valence band structure. But qualitatively it can be expected that the larger effective mass of the holes compared to that of the electrons gives rise to larger binding energies and smaller Bohr radius for the acceptor compared to that of the donor. The large Bohr radius means that the electron wave function is not localized at the donor but extends over many unit cells. Under this condition the validity of the effective mass approximation (Eq. 6.2) is increased and as a result the agreement between calculated and experimental ground state binding energies for donors is rather good. Excited states have even larger orbits and the effective mass approximation works better for them. On the other hand, acceptors have shallow orbits and the agreement between theory and experiment is generally poor.

### 6.3.2 Effective Mass Approximation for Acceptor States

The treatment of the acceptor states via the effective mass theory though similar in concept to the donor states, is much more complex owing to the complexity of the valence band [7]. The top of the p-like valence band in cubic semiconductors at $k = 0$ is six-fold degenerate if spin-orbit interaction is not included. In the presence of spin-orbit interaction, however, the six-fold degeneracy is lifted and the valence band consists of an upper set of bands which is four-fold degenerate and a lower set which is two-fold degenerate. When the effective mass approximation is considered for acceptors, the upper set of bands must be considered on an equal basis and the Hamiltonian is, therefore, a $4 \times 4$ matrix. If the spin-orbit splitting is comparable to the acceptor binding energy, then it may be necessary to include the two-fold degenerate split-off bands as well.

Baldereschi and Lipari [8, 9] have calculated the energy levels of shallow acceptor states in cubic semiconductors by reformulating the effective mass Hamiltonian (Eq. 6.2). In doing so, they made use of the similarity between acceptor centers in semiconductors and atomic systems with the spin-orbit interaction included and applied the methods of the angular momentum theory, which have been successfully used in the study of atomic systems. The acceptor Hamiltonian was separated into two terms with cubic and spherical symmetry. Since the contribution of the cubic term to binding is much less than that of the other, it is neglected in a first approximation. Hence the model is called a spherical model of shallow acceptor states. In the limit of strong spin-orbit interaction, that is, the acceptor binding energy is small relative to the spin-orbit splitting, the acceptor binding energy is given by [8],

$$E_a = Rf(\mu) \tag{6.5}$$

where R and $\mu$ are defined as

$$R = \frac{13.6\ \text{eV}}{\varepsilon^2 \gamma_1} \quad \text{(effective Rydberg)} \tag{6.6}$$

$$\mu = \left[ \frac{6\gamma_3 + 4\gamma_2}{5\gamma_1} \right] \tag{6.7}$$

and $\gamma_1$, $\gamma_2$, and $\gamma_3$ are the valence band parameters proposed by Luttinger [10]. The function $f(\mu)$ is the acceptor energy spectrum calculated as a function of $\mu$ using variational trial wave functions [7].

Table 6.1 lists the valence band parameters, $\mu$ and R for the III-V semiconductors [8, 11]. Figure 6.2 shows the calculated acceptor energy spectrum as a function of $\mu$. The ground state binding energy of the acceptor ($1S_{3/2}$ state where the subscript represents the $\bar{L} + \bar{J}$ value; for the S state $L = 0$, $J = \frac{3}{2}$) in GaAs and InP obtained from Fig. 6.2 are, respectively, 25.6 meV and 35.2 meV. Table 6.2 lists the ground state and the excited state energies of acceptors in III-V semiconductors.

### 6.3.3 Chemical Shifts and Central-Cell Corrections

The effective mass theory predicts the same energy levels for donors or acceptors irrespective of their chemical nature. But in reality the binding energies vary depending on the impurity and the variation is significant, especially for the acceptors in GaAs and InP. The difference in the effective mass binding energy and the experimental value is generally referred to as a "chemical

**TABLE 6.1**

Values of the valence band parameters, $\gamma_1, \gamma_2, \gamma_3$, and $\mu$ and the energy unit R for III-V semiconductors.

| | $\gamma_1^{(a)}$ | $\gamma_2^{(a)}$ | $\gamma_3^{(a)}$ | $\mu^{(b)}$ | $R^{(b)}$(meV) |
|---|---|---|---|---|---|
| AlAs | 4.04 | 0.78 | 1.57 | 0.621 | 45.1 |
| AlSb | 4.15 | 1.01 | 1.75 | 0.701 | 22.8 |
| GaP | 4.20 | 0.98 | 1.66 | 0.661 | 28.0 |
| GaAs | 7.65 | 2.41 | 3.28 | 0.767 | 11.3 |
| GaSb | 11.80 | 4.03 | 5.26 | 0.808 | 4.7 |
| InP | 6.28 | 2.08 | 2.76 | 0.792 | 14.1 |
| InAs | 19.67 | 8.37 | 9.29 | 0.907 | 3.2 |
| InSb | 35.08 | 15.64 | 16.91 | 0.935 | 1.2 |

a) [11]  b) [8]

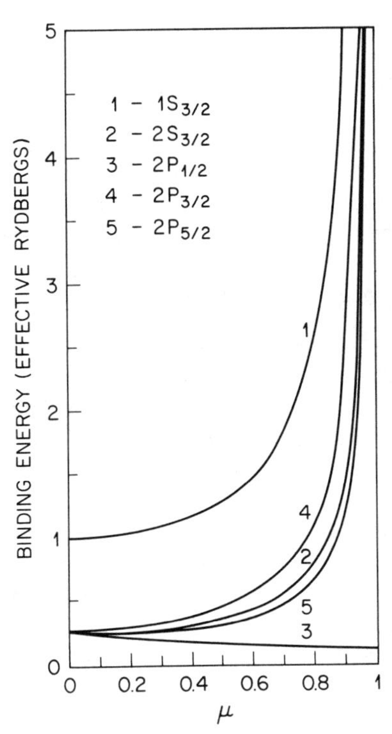


**Figure 6.2** Calculated acceptor ground state and excited state energies (in units of effective Rydberg R) as a function of $\mu$ in the strong spin-orbit coupling limit [8].

**TABLE 6.2**

Theoretical ground state and excited state energies in meV of acceptors in III-V semiconductors [8].

| | $1S_{3/2}$ | $2S_{3/2}$ | $2P_{1/2}$ | $2P_{3/2}$ | $2P_{5/2}$ |
|---|---|---|---|---|---|
| AlSb | 42.4 | 12.4 | 3.3 | 17.5 | 10.5 |
| GaP | 47.5 | 13.7 | 4.2 | 19.1 | 11.7 |
| GaAs | 25.6 | 7.6 | 1.6 | 11.1 | 6.5 |
| GaSb | 12.5 | 3.8 | 0.65 | 5.6 | 3.2 |
| InP | 35.2 | 10.5 | 2.0 | 15.5 | 8.9 |
| InAs | 16.6 | 5.1 | 0.4 | 7.9 | 4.4 |
| InSb | 8.6 | 2.7 | 0.2 | 4.2 | 2.3 |

shift'' [12]. In Eq. (6.2) the one-electron impurity potential, $e^2 / \varepsilon r$ should, therefore, be modified to include these chemical shifts and can be written as [6]

$$U(r) = \frac{\Delta Z\, e^2}{\varepsilon r} + U_{cc}(r) \tag{6.8}$$

where $\Delta Z$ is the difference in the valence between the impurity and the host and $U_{cc}(r)$ is the correction term giving the chemical shift. It is called the "central-cell" potential signifying the fact that the effects responsible for the chemical shift do not extend beyond a unit cell distance from the impurity but are confined to a "central-cell" region. Note that for isovalent impurities $\Delta Z = 0$ and $U_{cc}(r)$ is the only potential responsible for binding [13].

Phillips [14] proposed a dielectric model to explain the chemical shifts in terms of the differences in electronegativity between the substitutional impurity and the host atom that it has replaced. When the electronegativity difference, $\Delta X$, is large, the deviation of the binding energy from the effective mass value, $\Delta E$, is also large. Specifically, $\Delta E \propto (\Delta X)^2$ and the constant of proportionality is related to the probability that a valence electron will find itself in the atomic cell of the cation ($P_C$) or the anion ($P_A$) [15]. The probability ratio $P_C / P_A$ can be estimated from the ionic character of the covalent band and it is about 0.5 and 0.4, respectively, for GaAs and InP. It would, therefore, appear that for large values of $\Delta E$, the chemical shift for impurities on anion sites is larger than that for impurities on cation sites. This is qualitatively true for acceptors in GaAs like Zn and Cd which substitute on the Ga site and the same as Si, Ge, and Sn which substitute on the As site. A quantitative description of the chemical shifts would require the use of an appropriate impurity pseudopotential, which would already include the chemical effects in the effective mass theory [12].

### 6.3.4 Donor Levels Associated with Subsidiary Minima

The conduction band structure of GaAs and InP consists, besides the minimum at $k=0$, of other minima of higher energies occurring at L, X, and at other high symmetry points. The presence of these secondary minima can influence the bound states produced by an attractive potential by shifting their energies relative to the values calculated by the effective mass approximation or by producing new bound and resonance states [16]. The effects caused by the secondary minima naturally increase as the energy separation between the minima is reduced, as for example, by applying a hydrostatic pressure. Resonance states associated with the secondary minima at X in GaAs [17] and InP [18] containing donor impurities have been observed.

### 6.3.5 Experimental Methods to Determine Impurity Energy Levels

A quantity of practical importance regarding impurities in semiconductors is the position of their energy levels in the forbidden energy gap. This is usually obtained by observing the transitions of electrons from the impurity state to one of the band continua or to another impurity state or conversely from one of the band continua to the impurity state. The ground state and/or the excited states of the impurity may also be involved in such transitions. Optical measurements such as absorption and emission and transport measurements give information about the positions of the excited and ground states. Optical measurements can also give information regarding the symmetry of the energy states. Paramagnetic measurements give detailed information on the impurity wave functions which aid in the fundamental understanding of the energy states.

#### Transport Measurements

In transport measurements the carrier concentration, n, is obtained as a function of temperature by measuring the conductivity $\sigma$ or the Hall coefficient $R_H$. When only one type of carrier is present, the carrier concentration is given by

$$n = \pm \frac{r}{R_H e} \tag{6.9}$$

where r is the ratio of Hall mobility to drift mobility and its value is generally close to unity. If ionization of only one impurity contributes to n, it varies as $\exp(-E/kT)$ where E is related to the ionization energy. In the absence of compensation the ionization energy is $E/2$ otherwise it is E. The analysis becomes more complicated when more than one impurity contributes carriers or the semiconductor is degenerate. In such cases the equation describing the charge neutrality condition has to be solved numerically taking into amount density of states in the band, concentrations of impurities and the degeneracy of the impurity levels. The ionization energy obtained from such an analysis can be used to identify the impurity present, especially for acceptors in GaAs and InP. However, for donors which have small ionization energies, there is significant interaction between donor atoms even at low concentrations. This causes the ionization energy to decrease and the magnitude of the decrease is generally much greater than the maximum variation in the ionization energy between different donor species. Usually one relies on optical measurements to identify the chemical nature of the donor atom.

**Optical Measurements**

Absorption, photoconductivity, photocapacitance, luminescence, and electronic Raman scattering are the commonly used optical measurements to obtain energy levels of impurities. Another complementary optical technique which gives information on the site occupation and symmetry of the impurity is local mode vibrational spectroscopy (see Sec. 5.4.1).

In absorption the transitions from the ground state to excited states of hydrogenic impurities manifest as sharp lines in the far infrared and give an accurate estimate of the energy level positions. This method has been successfully used to obtain energy levels of donors and acceptors in elemental semiconductors [19]. In III-V semiconductors, fundamental lattice absorption falls in the same spectral range as the impurity transitions and renders interpretations difficult. For GaAs and InP, spectral photoconductivity measurements have proven to be useful in obtaining information about excited states of impurities, ionization energies and emission and capture rates. Even in the presence of strong background absorption it is possible to observe the 1s–2p transition which would otherwise be difficult in direct absorption measurement. Another variation of the far infrared photoconductivity technique is photothermal ionization spectroscopy which measures transitions between discrete states of an impurity [20]. This technique involves excitation of an electron from the ground state to one of the excited states of the impurity by absorption of a far infrared photon of appropriate energy and the subsequent ionization of the excited state by absorption of phonons. The resulting electron in the band gives photoconductivity. Since the ionization from the excited state occurs by phonons, the sample temperature should be such that there are sufficient phonons to excite an electron from the excited state to the band continuum but not enough to excite the electron directly from the ground state. For the study of donors in GaAs and InP where the photothermal ionization spectroscopy has been used most extensively, the temperature range for the experiment is about 1.5 to 6K [21]. Figure 6.3 shows a photoconductivity spectrum from high purity GaAs as a function of temperature [22]. The prominent peaks at 4.2K labelled A, B, and C correspond to the 1s–2p transitions of three different donors. The structure at higher energies corresponds to transitions from the 1s to other higher p-like excited states. These transition energies agree rather well with those predicted from the hydrogenic model which, as we noted earlier, is a good description of the shallow donor states. Note also in Fig. 6.3 that the excited state photoconductivity is reduced as temperature decreases indicating that the ionization probability of the excited state is decreasing with temperature.

## 6.4 DEEP IMPURITIES AND NATIVE DEFECTS

### 6.4.1 Overview of Theory

Unlike in the case of substitutional Zn, Cd, C, Si, Ge, Sn, S, Se, and Te which belong to columns II, IV, and VI of the periodic table, impurities from other columns which are incorporated substitutionally in a III-V crystal may have energy level(s) deep within the band gap. The electronic wave functions centered on these impurities are, in general, more localized than is the case for simple donors and acceptors [23, 24]. We saw in Sec. 6.3 that the effective mass

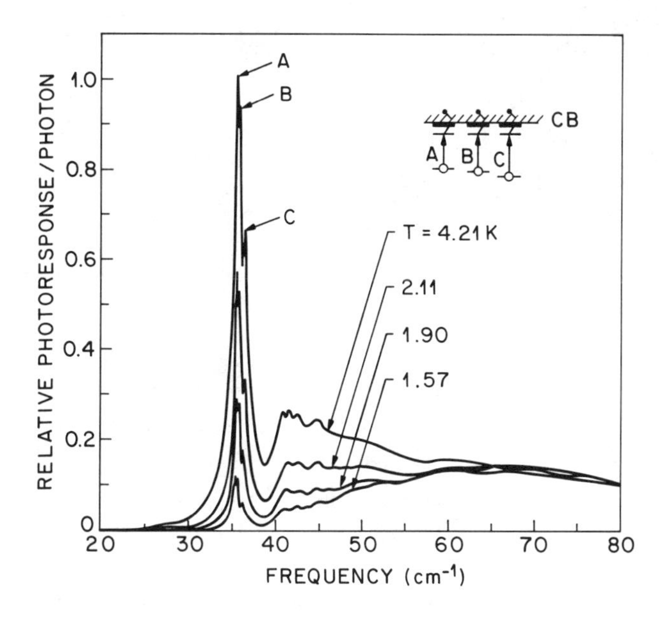


**Figure 6.3** Photoconductivity spectra from high purity GaAs at different temperatures [22]. The peaks labeled A, B and C are the 1s-2p transition from three different donors. The structure at higher energy is due to transitions from the 1s to other higher p-like excited states.

theory is a good approximation for shallow levels for which the extent of the electron wave function is over several unit cells. For deep level impurities, on the other hand, the wave functions fall off exponentially away from the impurity over a distance much less than a unit cell, and effective mass theory is not applicable [23, 24].

The technologically important substitutional deep level impurities are broadly divisible into two categories. The first of these encompasses those species for which only the s and p atomic orbitals participate in the bonding. To the second category belong those species for which d orbitals also participate in the bonding. We include in particular the transition metal atoms Ti, Cr, Mn, Fe, Co, Ni, and Cu in this second category.

For a physical description of the bonding for the first category a basis change from s and p orbitals to $sp^3$ hybridized orbitals is the first step since matrix elements of the Hamiltonian between these states are most important [25]. These orbitals derive an angular dependence from the p orbitals and are directed to the four corners of a tetrahedron as shown in Fig. 6.4. For diamond, Si, and Ge the atomic configuration is $s^2p^2$ and an s electron is promoted to a p orbital in purely covalent $sp^3$ bonds. For the eight orbitals participating in four covalent bonds in Fig. 6.4 there are eight electrons. For GaAs and InP the four corner atoms are different than the central atom (i.e., they are from different sublattices). However, since the atomic configurations are $s^2p^1$ and $s^2p^3$, we again have eight electrons to distribute over eight orbitals. Now, however, more than five electrons are concentrated near the group V atom to form an anion, and less than

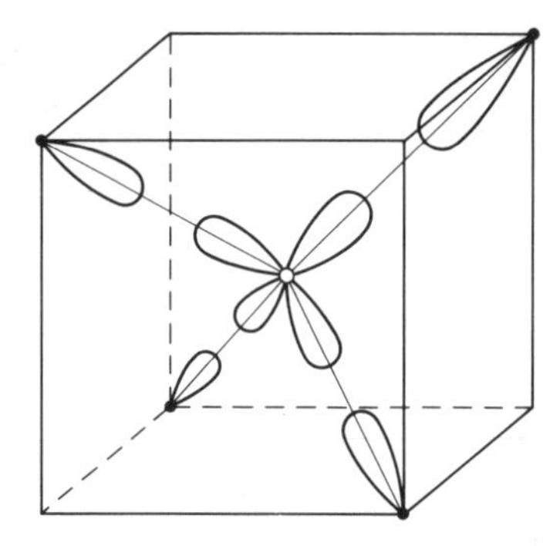

**Figure 6.4** The 8 orbitals participating in 4 covalent bonds in spref3 bonding.

three electrons are near the group III atom to form a cation. The bonding is partially covalent and partially ionic. For $sp^3$ impurities this same picture is applied but either fewer than or more than eight electrons are then distributed over the eight orbitals.

Native defects are usually also treated with the $sp^3$ basis states [24]. Isolated antisite defects are simply cases of substitutional $sp^3$ bonded impurities for which the central atom is the same as the four corner atoms in Fig. 6.4. Vacancies are another example. Energy levels for vacancies are obtainable from the solution of the general substitutional impurity problem (we shall elaborate on this later). Self interstitials, however, fall into a different category and require an expanded set of basis states.

If the basis is limited to the orbitals shown in Fig. 6.4, we have what is known as the defect molecule approximation [25]. In this case there is perfect tetrahedral symmetry and it follows that all states must belong either to the $A_1$ or to the $T_2$ irreducible representations of the tetrahedral point group, $T_d$. If other orbitals are included such as, for example, "back bonds" to the other orbitals of the four corner atoms, then this is no longer true.

In classical models of transition metal impurities in $sp^3$ bonded crystals, not all of the impurity valence electrons are used up in forming bonds, which leaves chemically unsaturated electrons responsible for the magnetic, electrical, and optical properties of the doped crystals. In $T_d$ symmetry there is a correspondence between the allowed crystalline levels and the free ion multiplets [26]. This correspondence is

$$S \rightarrow A_1$$

$$P \rightarrow T_1$$

$$D \rightarrow T_2 + E$$

$$F \rightarrow T_1 + T_2 + A_2$$

So, for example, $Fe^{2+}$ ($3d^6$) has a $^5D$ ground state multiplet of 25 states (s=2, l=2), and when incorporated in GaAs or InP this multiplet is split into a higher lying orbital triplet $^5T_2$ and a lower lying orbital doublet $^5E$ by the tetrahedral crystal field.

No single theoretical framework for the deep levels has been formulated which has been as successful as effective mass theory has been for shallow levels. The simplest approaches are tight binding theories in which the orbitals are taken to be linear combinations of atomic orbitals (LCAO) and overlap integrals between orbital wave functions centered on different atoms are completely ignored [24, 27]. If only nearest neighbor interactions are considered, we have the simple Huckel theory [24]. In the extended Huckel theory overlap integrals are not neglected, but to make the problem tractable, empirical approximations are made to matrix elements of the Hamiltonian. The extended Huckel approach was the first to be applied to defects in semiconductors, and was used to predict bond angles [28].

More sophisticated tight binding theories also include the charge state of a defect. Since the charge of the defect affects the electron density of the perfect host crystal and vice versa, the potential, screening, and defect charge must be determined self consistently [23, 24]. A model similar to this was applied to transition metal elements in covalent crystals by Haldane and Anderson [29]. They concluded that often several of the charge states of a particular impurity species should produce deep levels in the band gap which is found experimentally to be true in many instances. For example, $Fe^{+1}$ and $Fe^{+2}$ both produce levels within the band gap of InP.

With the incorporation of many electron effects we come to the class of the most sophisticated deep level theories being applied to defects in semiconductors [24]. These are based on finding the ground state wave function through variational methods and include Hartree, Hartree-Fock, and density-functional theory approaches. Variational wave functions are constructed from one electron wave functions and the potential is determined self consistently [23, 24].

Green's function techniques can be applied to both the simple tight binding approaches and to the more sophisticated self consistent approaches and are widely used [23, 24]. When they are applied to the problem of a substitutional impurity in a one band tight binding model we have the Koster-Slater problem [30].

Finally, we note that lattice distortions of the defect configuration may, in general, lower the energy [31, 32]. In addition, if the symmetry of the defect is such that the ground state is degenerate, then this state will split upon a lowering of the symmetry to produce a ground state of lower overall energy (i.e., including the nuclear configuration). Distortions driven by this effect are known as Jahn-Teller distortions and have been found to be very important for the vacancy in Si [31]. Lattice distortions have also been included in the theoretical treatments of defects in GaAs [33-35].

### 6.4.2 $sp^3$ Bonded Impurities

In spite of the lack of a comprehensive theory for deep level defects in III-V semiconductors, chemical trends and a number of physical insights are available from the simple tight binding theory applied to the defect molecule [36, 37]. We, therefore, present at length the basic ideas, relationships, and results of this theory in the case of a substitutional $sp^3$ bonded impurity in an otherwise ideal III-V crystalline host. This problem has been treated by Dow, Hjalmarson, and co-workers, and we present here mostly their approach and their results together with insights obtained from the volumes by Lannoo and Bourgoin [24] and Bourgoin and Lannoo [32].

An illustration of the defect molecule is shown in Fig. 6.4. For binary III-V semiconductors the central atom is either a group III atom, in which case the four corner atoms are group V atoms

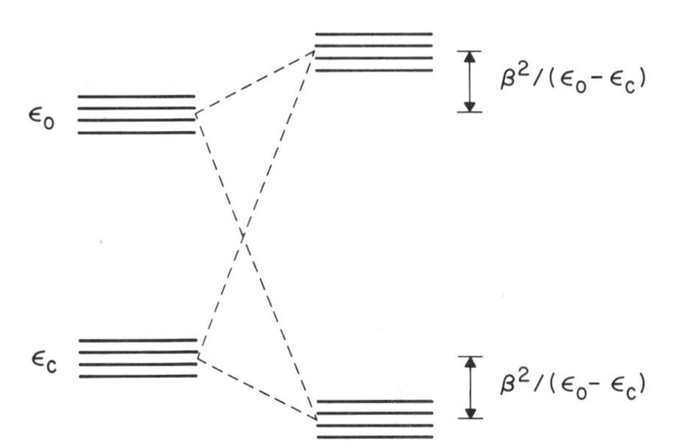


**Figure 6.5** Bonding and antibonding orbital energies.

or vice versa. If matrix elements of the Hamiltonian which account for the interaction between orbitals on different atoms are neglected, then the 4 $sp^3$ hybridized orbitals for the central atom are degenerate at energy $\varepsilon_c$ and the 4 orbitals for the other atoms are degenerate at $\varepsilon_o$. If now we turn on the interaction, $\beta$, between the bonding orbitals, that is, along [111], $[1\bar{1}\bar{1}]$, $[\bar{1}1\bar{1}]$, and $[\bar{1}\bar{1}1]$, then the eight orbitals will be separated into four bonding and four antibonding orbitals as shown in Fig. 6.5. Using a two state basis where the two states are any pair of bonding orbitals we can write the Hamiltonian as

$$\begin{bmatrix} \varepsilon_o & \beta \\ \beta & \varepsilon_c \end{bmatrix} \tag{6.10}$$

which has solutions

$$E_{A,B} = \frac{1}{2}(\varepsilon_o + \varepsilon_c) \pm \frac{1}{2}((\varepsilon_o - \varepsilon_c)^2 + 4\beta^2)^{1/2} \tag{6.11}$$

In the limit where $4\beta^2 \ll (\varepsilon_o - \varepsilon_c)^2$ we have $E_A = \varepsilon_o + [\beta^2/(\varepsilon_o - \varepsilon_c)]$ and $E_B = \varepsilon_c - [\beta^2/(\varepsilon_o - \varepsilon_c)]$. The bonding states correspond to the valence band of the whole crystal, and the antibonding states correspond to conduction band states.

So far we have not included an impurity. This is done by perturbing the central atom orbital energy by $V_o$ and by solving the Hamiltonian given by

$$\begin{bmatrix} \varepsilon_o & \beta \\ \beta & \varepsilon_c + V_o \end{bmatrix}. \tag{6.12}$$

The solutions are obtained by substituting $\varepsilon_c + V_o$ for $\varepsilon_c$ in Eq. (6.11) and are shown in Fig. 6.6 in units of $\beta$ with $\varepsilon_o = 2.0$ and $\varepsilon_c = 0.0$. We see that the effect of perturbing the central atom orbital energy is the formation of a pair of deep levels. If the perturbing potential, $V_o$, is very weak, then the levels are not deep, and we see that the expected energy level heirarchy of donors and acceptors is obtained in this very simple model. Thus for group II atoms, for which $V_o > 0$, a defect acceptor level lies near the valence band in III-V semiconductors, and for group VI impurities, for which $V_o < 0$, a defect donor level lies near the conduction band.

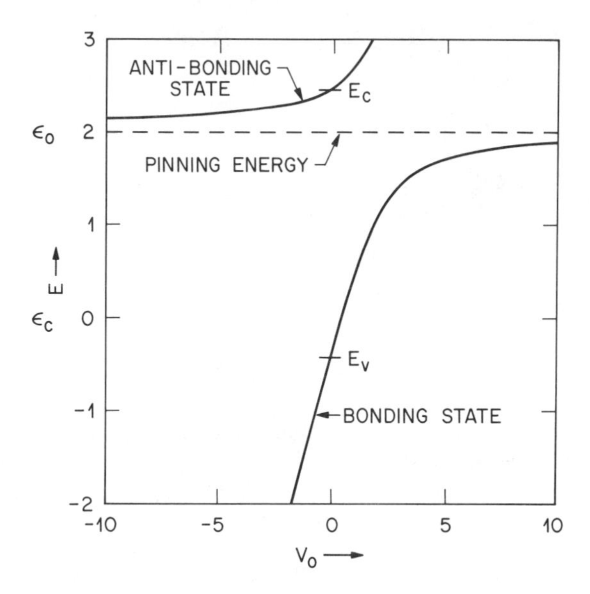


**Figure 6.6** Defect molecule energy solutions as a function of the potential energy perturbation due to an impurity at the central atom position.

Since there are four equivalent bonds in the defect molecule, each of the deep levels in Fig. 6.6 is four-fold degenerate. In the limit of $V_o \rightarrow \pm\infty$ we obtain the results for a vacancy. This is so because matrix elements of the Hamiltonian between orbitals which are far removed from one another energetically are very small. This means that the orbitals of the central atom can be voided in the limit of either an extremely positive or an extremely negative orbital energy for the $sp^3$ hybrids of the central atom. In both of these limits these four deep $sp^3$ hybrid levels are degenerate and are pinned at $\varepsilon_o$ (see Fig. 6.6).

A splitting of these four degenerate levels occurs if one includes next nearest neighbor interactions. For the vacancy only the orbitals shown in Fig. 6.7 are included in the defect molecule, and if we denote the interaction between any pair of these by $\gamma$, we obtain (in accordance with $T_d$ symmetry) a splitting into three degenerate higher lying $T_2$ states and a lower lying $A_1$ state as shown in Fig. 6.8.

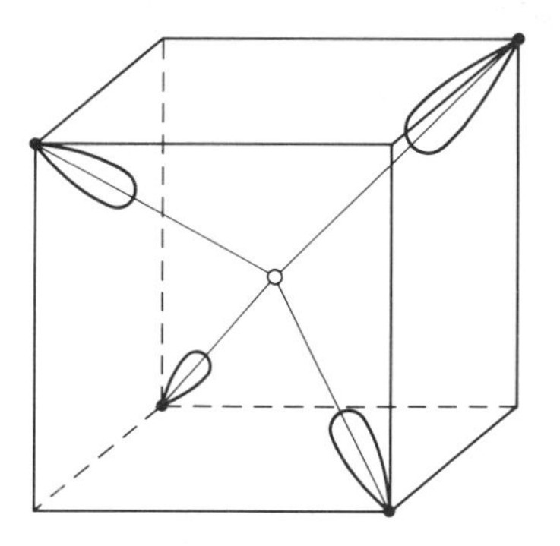

**Figure 6.7** Orbitals included in the defect molecule for a vacancy.

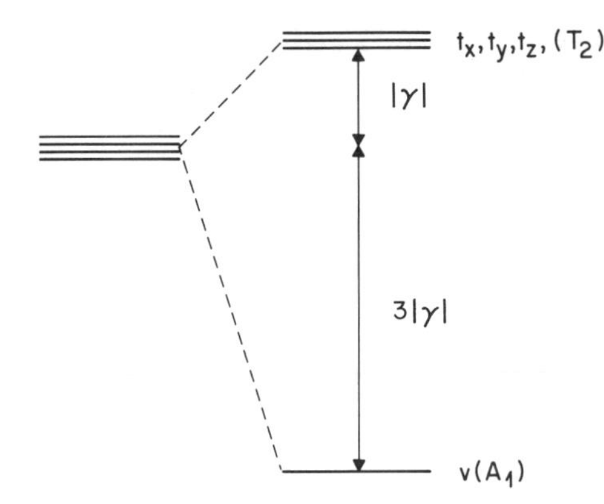


**Figure 6.8** Level splitting resulting from the interaction between orbitals in the defect molecule for a vacancy.

This splitting also occurs in the case of substitutional impurities and affords a means of categorizing the levels. The $A_1$ level has a large s-like (symmetric) character, and, thus, it is convenient to use s orbital energies to determine $V_o$ for the $A_1$ level. Similarly, since the $T_2$ levels have a p-like character, it is convenient to use p orbital energies to determine $V_o$.

We give now the results obtained by Hjalmarson using Green's function techniques applied to substitutional $sp^3$ impurities in GaAs and InP [36, 37]. Results for $A_1$ levels split off from the conduction band for anion site impurities ($V_o < 0$) are shown in Fig. 6.9 for GaAs. The corresponding $T_2$ levels lie at a higher energy above the conduction band edge as do both the $A_1$ and $T_2$ levels for InP. Results for $T_2$ states split off from the valence band for anion site impurities ($V_o > 0$) are shown in Fig. 6.10. Here the corresponding $A_1$ states lie below the valence band edge. Results for cation site impurities for states split off from the conduction band are shown for $T_2$ states in Fig. 6.11 and for $A_1$ states in Fig. 6.12. In Figs. 6.9 through 6.12 various impurities are noted along the abscissca near their orbital energy. A common feature is that there is an orbital energy threshold for the formation of a deep level, that is, $|V_o|$ must be larger than a threshold value to split off a deep level. (For cation site impurities this threshold is too large to split off valence band states in either InP or GaAs).

We note that the tight binding calculations of Hjalmarson et al [36, 37] were not done using solely a basis of four hybridized atomic orbitals per atom. An additional s state was also included for each atom because the calculations did not produce the correct band structure for the

**Figure 6.9** Results of Hjalmarson for $A_1$ levels resulting from anion site impurities [37].

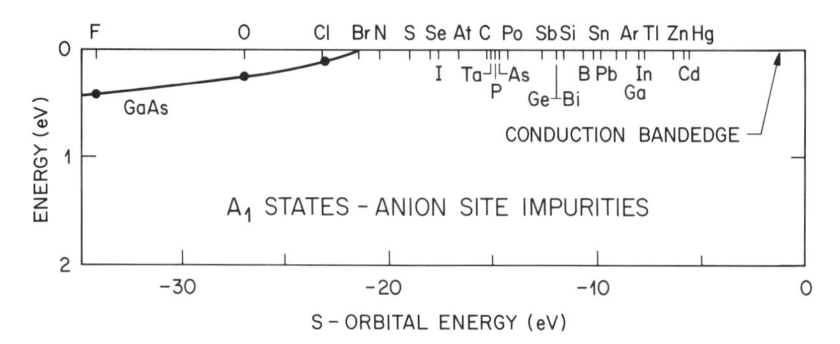

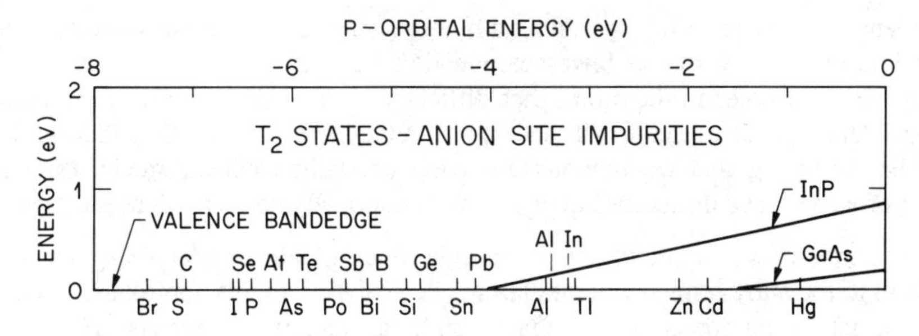


**Figure 6.10** Results of Hjalmarson for $T_2$ levels resulting from anion site impurities [37].

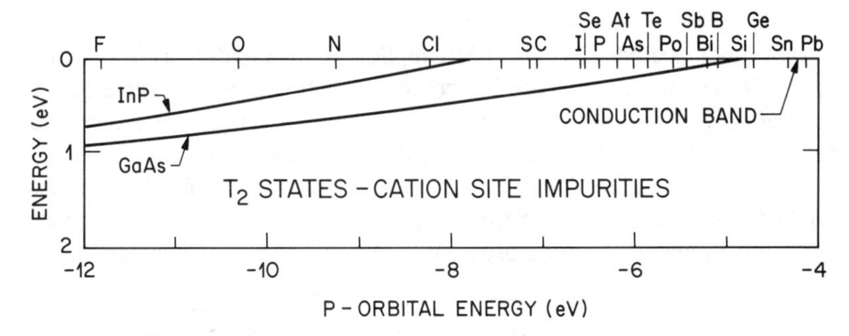


**Figure 6.11** Results of Hjalmarson for $T_2$ levels resulting from cation site impurities [37].

**Figure 6.12** Results of Hjalmarson for $A_1$ levels resulting from cation site impurities [37].

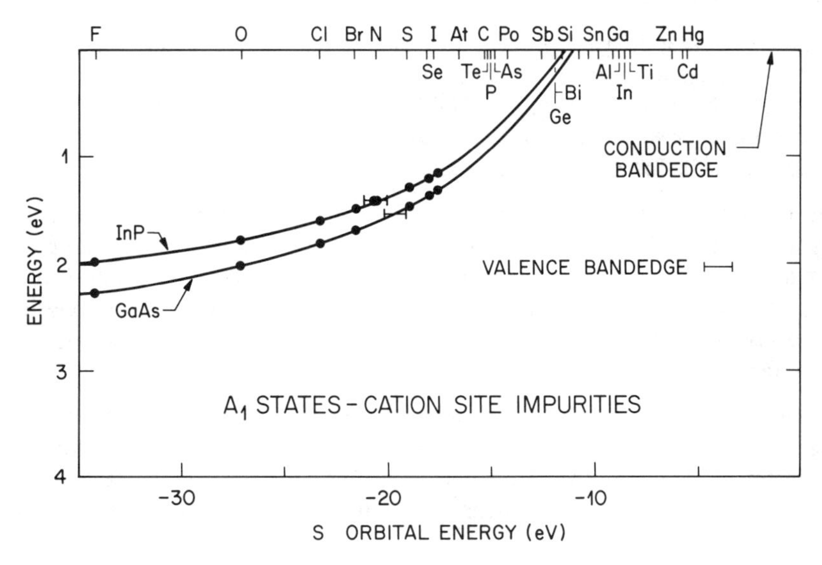

conduction bands otherwise [36]. By adding a higher energy s state, these workers were able to place the minimum energy of the lowest conduction band at the correct symmetry point in reciprocal space. We note, furthermore, that difficulty in predicting numerically correct band structures for ideal crystals occurs in all modern theoretical approaches. Considering this fact, it should not be surprising that calculations involving crystalline defects made using the same theoretical approaches have difficulties in predicting numerically correct defect energies.

Using the virtual crystal approximation, the tight binding defect molecule approach has been applied as well to the alloy semiconductor $GaAs_{1-x}P_x$ [36, 37]. In this approximation the alloy is treated as a virtual crystal composed of a Ga cation lattice and an average $As_{1-x}P_x$ anion lattice. $A_1$ defect levels for O and N impurities on the anion site were found not to follow the conduction band minimum (which changes from $\Gamma$ to X at $x \cong 0.45$), but it follows the anion vacancy level instead since $V_o$ is large and negative for N and O. This is in agreement with experimental data [36, 37].

The same tight binding model has also been applied to antisite defects in $Ga_xIn_{1-x}As_{1-y}P_y$ in a model where both the cation and anion lattices were occupied by average atoms in the virtual crystal approximation [38]. $A_1$ anion on cation site defects for the technologically important stoichiometries, that is, those lattice matched to InP, are predicted to lie 0.55-0.57 eV above the valence band. $T_2$ anion on cation site defects are predicted to lie within the conduction band for the lattice matched stoichiometries. For the cation on anion site defects the $A_1$ states lie within the valence band and $T_2$ states lie very close to or below the valence band edge [37].

Calculations have also been made for anion vacancies in $Ga_xIn_{1-x}As_{1-y}P_y$ [38]. Neither the $A_1$ nor $T_2$ levels are predicted to lie within the band gap for any of the alloys lattice matched to InP. For stoichiometries very close to $Ga_{0.47}In_{0.53}As$ the predicted levels lie very close to the conduction band edge.

Aside from the general observation that there are very few reports of deep levels in $Ga_xIn_{1-x}As_{1-y}P_y$, experimental corroborations of these theoretical predictions for antisite defects or vacancies are very scarce.

Using a theoretical approach similar to that of Hjalmarson et al., [36, 37] Sankey and Dow [39] have shown that pairing of $sp^3$ bonded defects dramatically affects the predicted deep level energies in InP. If pairs are thermodynamically favored, that is, if they are bound, then attempting to make an association between calculated deep level energies of isolated point defects and experimental results for as-grown InP would appear to be pointless.

Besides the simple tight binding treatment for which we have so far given some of the reported results, a large body of other theoretical work exists not only for $sp^3$ impurities but also for intrinsic defects and, more recently, for pairs of intrinsic defects. Many physical insights can be obtained from these calculations, and to provide some indication of the physics which is available, we give now the results of more sophisticated theoretical treatments for vacancies in GaAs due to Baraff, Schluter, and Bachelet.

A limitation of the simple tight binding treatment which we have discussed is the inability to treat charge state effects. These are undoubtedly important and can be explored using self-consistent treatments to obtain the defect potential. Even for neutral defects self-consistency is expected to affect results in polar semiconductors because the partially ionic character of the bonding in the perfect crystal is altered [40]. Bachelet, Baraff, and Schluter [40] have presented self-consistent results for the neutral $V_{As}$ and $V_{Ga}$ defects in GaAs obtained using Green's function techniques in a local pseudopotential approximation. A charge contour plot for $V_{Ga}$ along the (110) plane is shown in Fig. 6.13. The units can be deduced from Fig. 6.14 which shows the corresponding charge density profile along the [111] direction from a vacant Ga site to a near neighbor As atom. Similar results for the $V_{As}$ are shown in Figs. 6.15 and 6.16. Note that in both cases the short range nature of the vacancies is demonstrated since the range of the total charge disturbance is roughly one bond length.

The charge states of $V_{Ga}$ and $V_{As}$ as a function of the Fermi energy have also been treated theoretically by Baraff and Schluter using self-consistent Green's function techniques and local-density-functional theory [41]. Results for the total formation energies of the defects with charge N and of the electron reservoir with chemical potential $\mu$ (and ignoring the entropy term in the Gibbs free energy) are shown in Fig. 6.17. We see that $V_{Ga}$ is a triple acceptor and $V_{As}$ is a double donor. Note that these results are in agreement with the rule (see Sec. 6.2) that if more than half of the eight electrons of the four bonds in Fig. 6.4 are present after a simple neutral atom removal process, then the vacancy is an acceptor, and if fewer than half are present, then the vacancy is a donor.

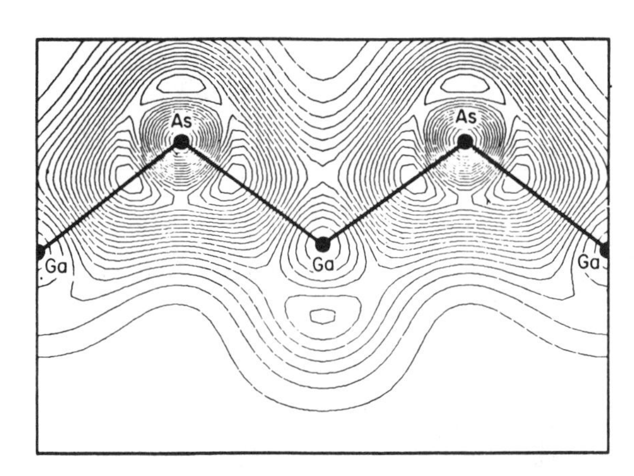


**Figure 6.13** The central panel shows the results of Bachelet, Baraff, and Schluter for charge density contours around a Ga vacancy in GaAs. The top panel shows their results for defect-free GaAs, and the lower panel shows the total charge disturbance introduced by the Ga vacancy (from ref. 40).

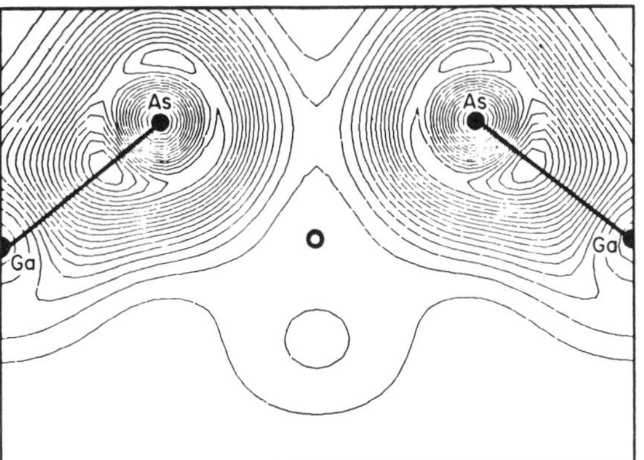


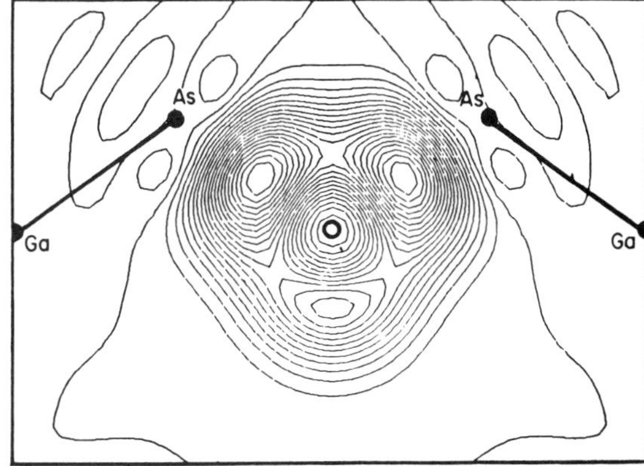

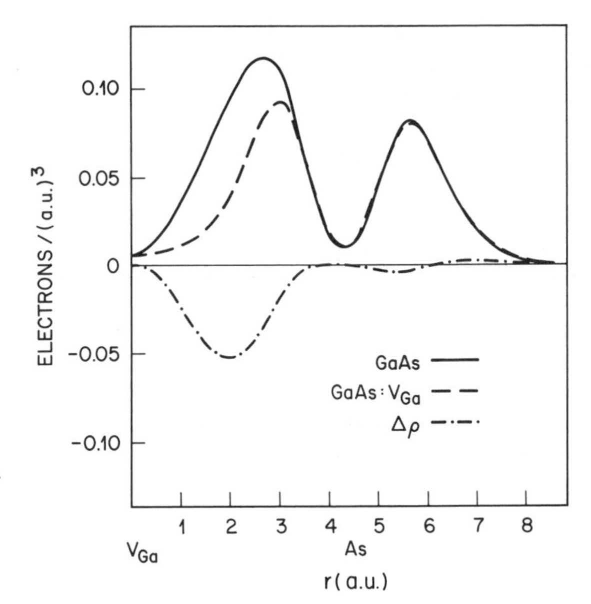


**Figure 6.14** Charge density profile along the [111] direction from a vacant Ga site to a near neighbor As corresponding to Fig. 6.13 (from ref. 40).

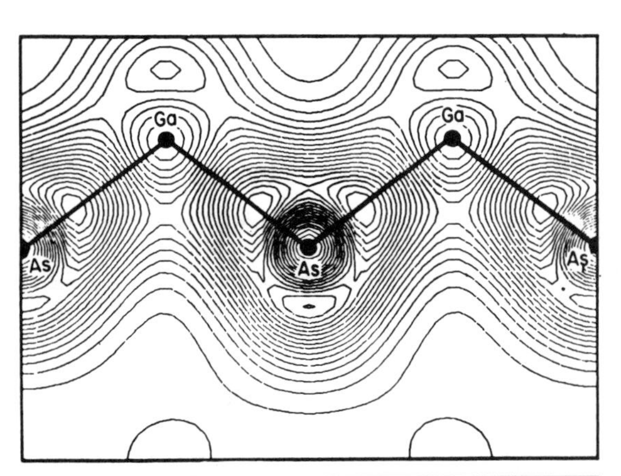


**Figure 6.15** The central panel shows the results of Bachelet, Baraff, and Schluter for charge density contours around an As vacancy in GaAs. The top panel shows their results for defect-free GaAs, and the lower panel shows the total charge disturbance introduced by the As vacancy (from ref. 40).

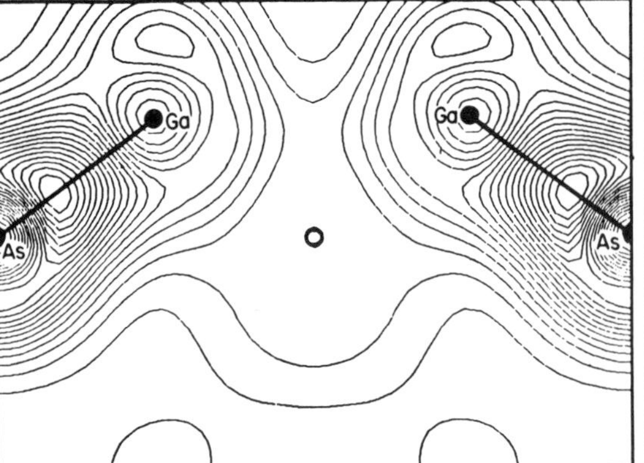


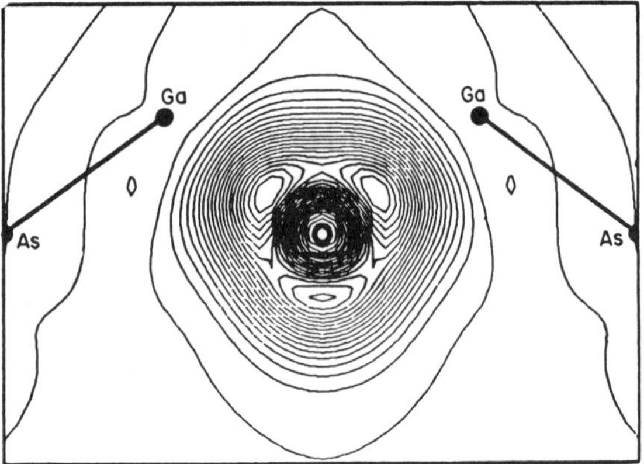

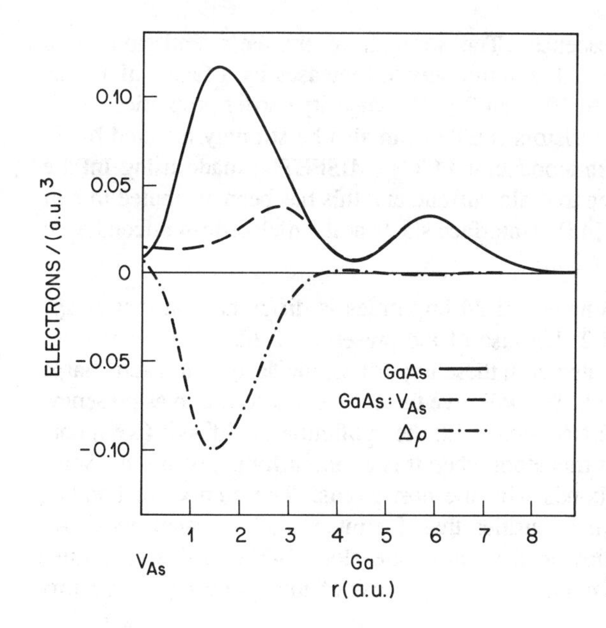


**Figure 6.16** Charge density profile along the [111] direction from a vacant As site to a near neighbor Ga site corresponding to Fig. 6.15 (from ref. 40).

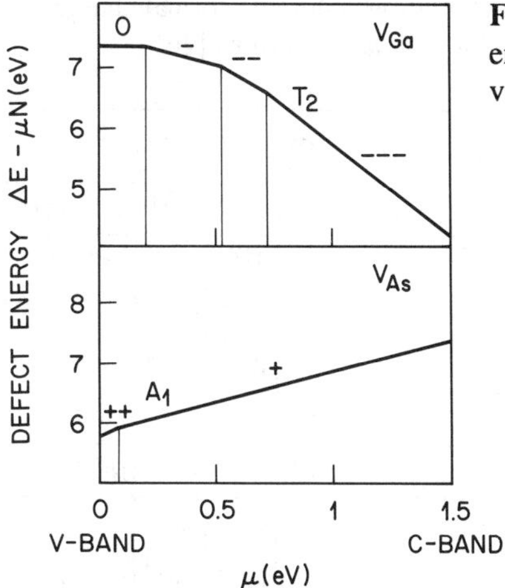


**Figure 6.17** Total formation energies of charged Ga and As vacancies [41].

### 6.4.3 3d Transition Metal Impurities

The 3d transition metal elements form a very important class of impurities in III-V semiconductors [42]. They are important because they can be used to make these semiconductors semi-insulating (SI) with carrier concentrations and resistivities akin to that of the intrinsic semiconductor. Also, transition metal impurities are usually present in crystal growth systems, and, as a result, they are often present as background impurities in the crystals. Because these impurities can act as efficient recombination centers and determine the minority carrier lifetime, they can limit the efficiency of near band edge emission and limit the performance of electroluminescent devices [43]. Due to internal d-to-d transitions, the impurities can emit radiation strongly in the near infrared (at wavelengths in the range 2-4 μm). Fe impurities in InP can dramatically affect band edge luminescence. The strength of the $Fe^{+2}$ emission (at a wavelength of ~3.5 μm) relative to the band edge luminescence increases by a factor of $10^5$ as the Fe concentration is increased from $10^{14}$ to $10^{17}$ $cm^{-3}$ [44]. Majority carrier properties which determine the performance of field-effect-transistors (FET's) can also be strongly affected by 3d impurities. For example, metal-insulator semiconductor FET's (MISFET's) made using InP:Fe substrates generally exhibit drift in the source to drain current, and this has been attributed in part to carrier recombination via an Fe impurity [45]. (Interface states at the dielectric-semiconductor interface may also play a role [46], see Sec. 7.2.2).

The electronic, optical, and magnetic behavior of 3d impurities is different than that of $sp^3$ impurities (which are discussed in Sec. 6.4.2) because of the presence of 3d valence electrons. Before we can discuss the deep levels associated with these impurities, the designation and charge state of 3d impurities first need to be clarified. We will give the classical description as presented by Zunger [26]. We consider first a neutral free atom with the configuration $3d^n4s^m$ (we ignore the core electrons). The nominal valence of this atom when it is an impurity is just n+m. Some of the valence electrons are used to form bonds with the host atoms. The number of bonding electrons will be different for substitutional impurities than for interstitially located ones. All reports place the substitutional 3d impurities on the cation site alone [42]. In this case three electrons are required by the bonds. We imagine that the process of incorporating an impurity

**Figure 6.18** Oxidation states and formal charge states of transition metal impurities [26].

| NON BONDING VALENCE ELECTRONS | OXIDATION STATE | | | | | |
|---|---|---|---|---|---|---|
| 1 | $K^0$ | $Ca^+$ | $Sc^{2+}$ | $Ti^{3+}$ | $V^{4+}$ | $Cr^{5+}$ |
| 2 | $Ca^0$ | $Sc^+$ | $Ti^{2+}$ | $V^{3+}$ | $Cr^{4+}$ | $Mn^{5+}$ |
| 3 | $Sc^0$ | $Ti^+$ | $V^{3+}$ | $Cr^{3+}$ | $Mn^{4+}$ | $Fe^{5+}$ |
| 4 | $Ti^0$ | $V^+$ | $Cr^{2+}$ | $Mn^{3+}$ | $Fe^{4+}$ | $Co^{5+}$ |
| 5 | $V^0$ | $Cr^+$ | $Mn^{2+}$ | $Fe^{3+}$ | $Co^{4+}$ | $Ni^{5+}$ |
| 6 | $Cr^0$ | $Mn^+$ | $Fe^{2+}$ | $Co^{3+}$ | $Ni^{4+}$ | $Cu^{5+}$ |
| 7 | $Mn^0$ | $Fe^+$ | $Co^{2+}$ | $Ni^{3+}$ | $Cu^{4+}$ | |
| 8 | $Fe^0$ | $Co^+$ | $Ni^{2+}$ | $Cu^{3+}$ | | |
| 9 | $Co^0$ | $Ni^+$ | $Cu^{2+}$ | | | |
| 10 | $Ni^0$ | $Cu^+$ | | | | |
| 11 | $Cu^0$ | | | | | |
| | | | | | | |
| SITE | FORMAL CHARGE STATE, q | | | | | |
| INTERSTITIAL | 0 | +1 | +2 | +3 | +4 | +5 |
| III - V CATION | -3 | -2 | -1 | 0 | +1 | +2 |

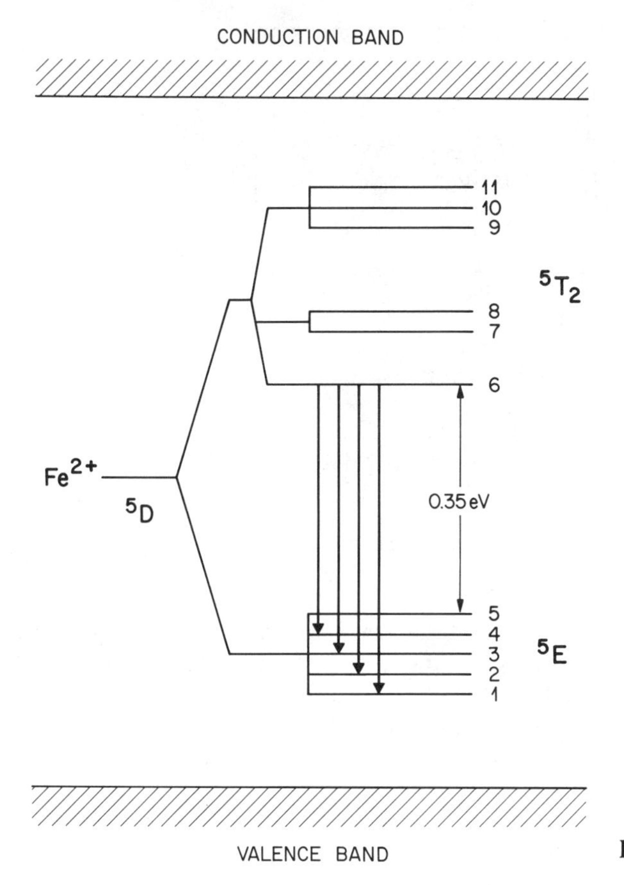


**Figure 6.19** Energy levels of $Fe^{2+}$ [42].

into the lattice starts with the removal of a cation followed by the insertion of the impurity. If the inserted impurity is not neutral, but is charged, we speak of the formal charge state, q, of the impurity. Including the three electrons lost to the bond, the net oxidation state becomes $3 + q$. For example, neutral atomic Fe has an electron configuration given by $3d^6 4s^2$. Three of the eight valence electrons are used in forming the bonds to yield an oxidation state of $Fe^{3+}$. This center has no formal charge (q=0), and if an additional electron is added to the impurity center, then the oxidation state becomes $Fe^{2+}$ which has formal charge of $q=-1$. This state is referred to [26, 47] as an occupied (with an electron) acceptor state since it acts as a Lewis base. Similarly, the $q = +1$ formal charge state is referred to as an empty donor state since it acts as a Lewis acid [26]. A chart due to Zunger [26] of the various oxidation states and the corresponding formal charge states is given in Fig. 6.18. An interstitial impurity is usually considered not to have any electrons given up to bonds.

In InP:Fe, the $Fe^{2+}$ internal d-to-d luminescence arises from the transitions between $^5T_2$ and $^5E$ multiplets shown schematically in Fig. 6.19 [42]. The photoluminescence spectrum corresponding to these transitions is shown in Fig. 6.20 [42]. The $^5E$ levels can be considered to lie very near the middle of the InP band gap (1.34 eV at 300K) because the thermal activation energy (at 300K) for the acceptor transition $Fe^{3+} \rightarrow Fe^{2+} + h^+$ is 0.68 eV. Taking the valence

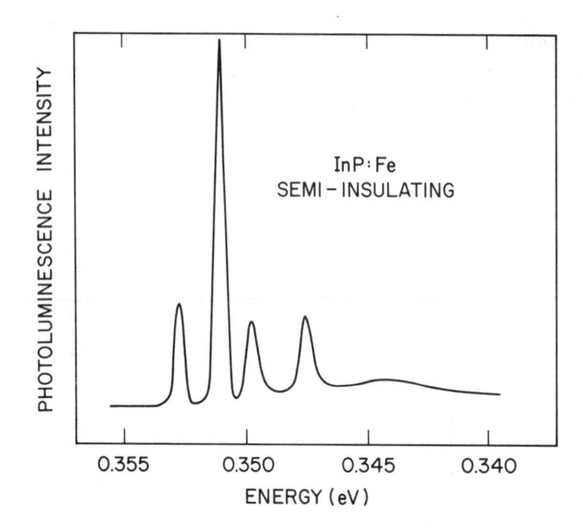


**Figure 6.20** Photoluminescence arising from $Fe^{2+}$ intra-center transitions [42].

band maximum as the zero of energy the energy level diagram shown in Fig. 6.21 is obtained [42]. Although no luminescence attributable to $Fe^{3+}$ intra-center transitions has been observed [48], electron spin resonance (ESR) signals from $Fe^{3+}$ in InP:Fe have been observed and are shown in Fig. 6.22 [49]. (We note, however, that these data did not unambiguously demonstrate that substitutional $Fe^{3+}$ was present since tetrahedral symmetry also applies to possible interstitial sites [42].

**Figure 6.21** Energy level diagram for transitions between the oxidation states of Fe in GaAs and InP [42].

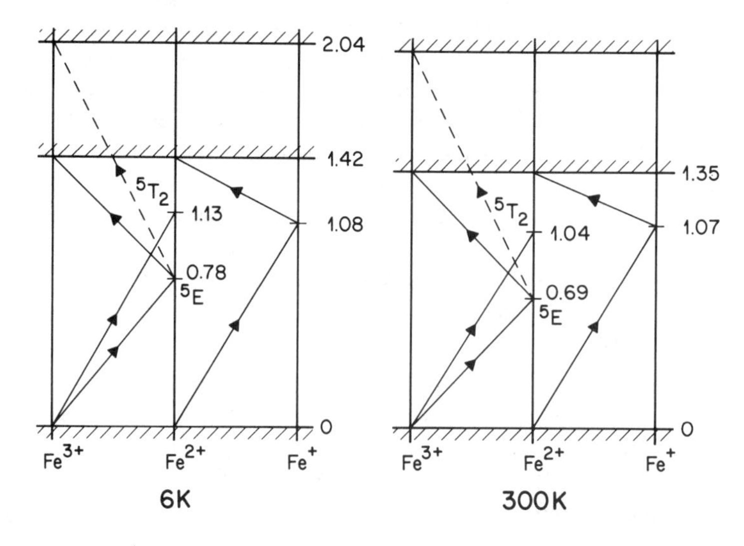

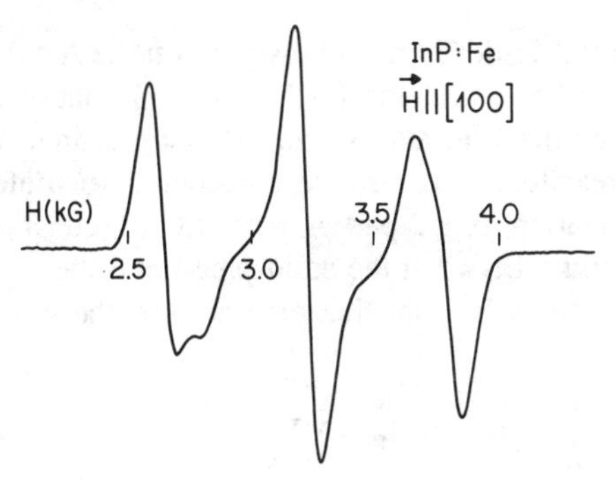


**Figure 6.22** Electron spin resonance (ESR) signal from $Fe^{3+}$ in InP:Fe [49].

Energy level diagrams of other 3d impurities are constructed in the same manner. Levels deduced from photoluminescence data for internal d-to-d transitions are placed in the band gap according to the transition energies between two oxidation states. Another example of this is the energy level diagram for GaAs:Cr shown in Fig. 6.23 [23]. The energy required to accomplish the transition from $Cr^{3+}$ to $Cr^{2+}$ by adding a single electron from the valence band to $Cr^{3+}$ is used to place the level labeled with the number 1 within the bandgap. Similarly, a second level for the $Cr^{2+}$ to $Cr^{1+}$ transition (representing a transfer of two valence band electrons to a $Cr^{3+}$ impurity) labeled with the number 2 is placed higher in the gap. Unlike the case of InP:Fe, however, still another transition, the $Cr^{4+}$ to $Cr^{3+}$ transition, can also be represented with a level in the gap as shown in Fig. 6.23 [23].

## 6.5 CHEMISTRY OF IMPERFECTIONS

The various disorder processes discussed in Sec. 6.2.1 can be described by simple chemical reactions. For an AB compound the following reactions can be written.

$$0 \rightleftarrows V_A^x + V_B^x \tag{6.13}$$

$$A_A + V_i \rightleftarrows A_i + V_A^x \tag{6.14}$$

$$A_A + B_B \rightleftarrows A_B + B_A \tag{6.15}$$

$$A_A + V_B^x \rightleftarrows A_B + V_A^x \tag{6.16}$$

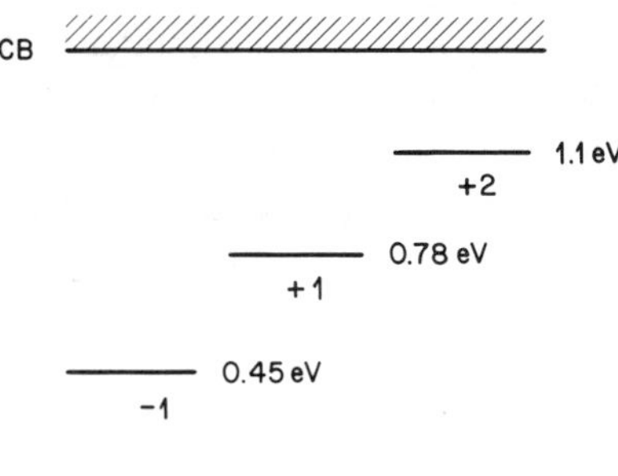


**Figure 6.23** Energy level diagram for transitions between the oxidation states of Cr in GaAs [42].

Equation (6.13) describes the Schottky disorder. Frenkel disorder on the A sublattice is described by Eq. (6.14). A similar reaction can be formulated for B atoms. Equations (6.15) and (6.16) describe, respectively, the antistructure disorder and the mixed disorder involving vacancies and antisite defects. Further chemical reactions can also be considered for defect formation as a function of the conditions of preparation (e.g., partial pressure of constituents). These are reactions which represent the equilibrium between the solid phase and the external liquid or gas phase. For example, the transfer of atoms B from the gas phase to the compound AB can be written as

$$\frac{1}{2} B_2(g) \rightleftarrows B_B + V_A^x \tag{6.17}$$

In Eqs. (6.13) to (6.17) all defects are treated as neutral. If the defects are charged, then one can write chemical reactions for the ionization of the defect.

$$V_B^x \rightleftarrows V_B^+ + e \tag{6.18}$$

$$V_A^x \rightleftarrows V_A^- + h \tag{6.19}$$

$$V_A^- \rightleftarrows V_A^{--} + h \tag{6.20}$$

Here $V_B$ is taken to be a single donor and $V_A$ as a doubly ionizable acceptor. In addition, there is the thermal excitation of electron-hole pair which can be described by the reaction [50,51].

$$0 \rightleftarrows e^- + h^+ \tag{6.21}$$

### 6.5.1 Mass Action Relations

For ideal behavior, that is, at sufficiently low concentrations of defects, the law of mass action for reactions Eqs., (6.13 to 6.17) relates the ratio of the product of the concentrations of products to the product of the concentrations of reactants to an equilibrium constant K having the form

$$K = \exp\frac{\Delta S}{k} \exp -\frac{\Delta H}{kT} \tag{6.22}$$

where $\Delta S$ and $\Delta H$ are the changes in the entropy and enthalpy for a given reaction, k is the Boltzmann constant and T is the absolute temperature. If concentration of defects are not low enough to apply the laws of dilute solutions, then activities instead of concentrations of defects have to be used in the mass action relations, as described later.

For reactions involving electrons and holes, at low concentrations one can use Boltzmann statistics. Then the application of law of mass action to the chemical Eq. (6.21) gives

$$K_i = np = N_c N_v \exp\frac{-E_g}{kT} \tag{6.23}$$

where $N_c$ and $N_v$ are the density of states in the conduction band and valence band, respectively,

$E_g$ is the band gap, and n and p are electron and hole concentrations, respectively in $cm^{-3}$. If concentrations are given as atom (or molar) fractions, for example, [e], [h] for electrons and holes where square brackets denote fractions, then $K_i' = [e][h] = K_i / N_0^2$, where $N_0$ is the number of atoms (or molecules) per $cm^3$. When the concentration of free carriers becomes equal to the density of states, Boltzmann statistics has to be replaced by Fermi-Dirac statistics. In other words, the laws of dilute solutions is no longer applicable for those reactions that involve charge carriers. In such cases, activities, **a**, instead of concentrations have to be used in the mass action laws. This is conveniently done by introducing activity coefficients, $\gamma$, which are related to **a** as $\mathbf{a} = \gamma \times$ (concentration). Thus, Eq. (6.23) becomes

$$K_i = \gamma_p \, p \, n \quad \text{(p-type material)} \tag{6.24a}$$

or

$$K_i = p \, \gamma_n \, n \quad \text{(n-type material)} \tag{6.24b}$$

where $\gamma_p$ and $\gamma_n$ are the activity coefficients for holes and electrons, respectively. Since in p-type material n is very low, $\gamma_n = 1$ and, similarly, in n-type material $\gamma_p = 1$. An ionization reaction such as Eq. (6.18) has the mass action relation

$$\frac{[V_B^+] \, n \gamma_n}{[V_B^x]} = K_{V_B} \, . \tag{6.25}$$

Several authors have derived $\gamma_s$ as a function of carrier concentration [52 - 57]. There are two effects to consider at high carrier concentrations. One is the Fermi degeneracy which gives $\gamma > 1$ when the concentration of free carriers exceeds one-tenth the density of states [53]. The other is the Debye-Huckel screening of the ionized donors or acceptors which give rise to the carriers [52]. The screening tends to shift the band edges opposite to that caused by Fermi degeneracy. The net result is that $\gamma$ is greater than unity only at very high-carrier concentrations and ideal behavior may be assumed for many practical situations [55].

### 6.5.2 Equilibrium of Imperfections: Brouwer's Approximation

With the help of the chemical reactions, and the associated mass action relations as discussed one can obtain the relation between defect concentrations and the conditions of the ambient. In doing so, the restriction of electrical neutrality has to be considered. Further, if equilibrium is not maintained as would be the case when crystals prepared at high temperatures are rapidly cooled, then an atom balance equation has to be considered for each defect whose high temperature concentration is frozen in the solid. As an example, let us consider in compound AB only vacancies $V_A$ and $V_B$ present as $V_A^x$, $V_A^-$, $V_B^x$, and $V_B^+$. The electroneutrality condition is given by

$$[e] + [V_A^-] = [h] + [V_B^+] \tag{6.26}$$

The atom balance relations for the vacancies are

$$[V_A]_{Total} = [V_A^x] + [V_A^-] \tag{6.27a}$$

$$[V_B]_{Total} = [V_B^x] + [V_B^+] \tag{6.27b}$$

The six unknown quantities, [e], [h], $[V_A^x]$, $[V_B^x]$, $[V_A^-]$, and $[V_B^+]$, can be obtained by solving the six equations (Eqs. (6.13) (6.17 to 6.19), (6.21), and (6.26)) which gives

$$[e] = \left\{ \frac{K_i \left[ K_i + \dfrac{K_S K_{V_B^+}}{K_{V_A} p_{B_2}^{1/2}} \right]}{\left[ K_i + K_{V_A} p_{B_2}^{1/2} K_{V_A^-} \right]} \right\}^{1/2} \tag{6.28}$$

where $K_S$, $K_{V_A}$, $K_{V_B^+}$, and $K_{V_A^-}$ are the reaction constants for the reaction Eqs. (6.13), (6.17), (6.18), and (6.19), respectively, and $p_{B_2}$ is the partial pressure of component B at the temperature of equilibrium. From [e], the concentrations of [h], $[V_B^+]$ and $[V_A^-]$ can be obtained.

Although the exact solution of Eq. (6.28) can be obtained, the problem becomes intractable when several defects are involved. One then resorts to approximate solutions by considering only the predominant defects in the electroneutrality and the atom balance equations. This type of solution is called Brouwer's method of approximation [58]. For example, if Eq. (6.26) is approximated by $[e] = [V_B^+]$, then $[e] = \left[ K_S K_{V_B^+} / K_{V_A} p_{B_2}^{1/2} \right]^{1/2}$.

When dopants are introduced into the semiconductor chemical reactions describing their incorporation and ionization can also be formulated in a similar manner and defect concentrations as a function of the activity of the dopant can be obtained. In general, the concentration of defect j is of the form $[j] = \prod K^m p_{B_2}^n a_F^o$ where K denotes the equilibrium constants, $a_F$ is the activity of the dopant, and m, n and o are small integers or simple fractions. At a given temperature, a log-log plot of [j] against $p_{B_2}$ or $a_F$ contains straight lines with slope n or o and at constant $p_{B_2}$ and $a_F$, a semilog plot of [j] versus $T^{-1}$ has slopes $\sum m \Delta H/k$.

Assuming different defect models several of these plots can be generated, which then serve to understand and/or to predict the physical properties of the crystals as a function of their preparation or treatment. Ruda et al., [59] proposed defect schemes involving native defects and the predominant impurity, viz., carbon, to explain the semi-insulating behavior of GaAs crystals as function of the arsenic activity in the melt [60, 61]. The native defects considered were $Ga_{As}$ for growth from Ga rich melts and $V_{Ga}$ and $As_{Ga}$ for growth from As rich melts. In one defect model, $As_{Ga}$ was taken to be the deep donor $EL_2$ which compensates the shallow $C_{As}$ acceptors to produce semi-insulating behavior. Also $V_{Ga}$, $Ga_{As}$ and $As_{Ga}$ were taken to be a single acceptor, a double acceptor and a double donor, respectively. After formulating defect reactions

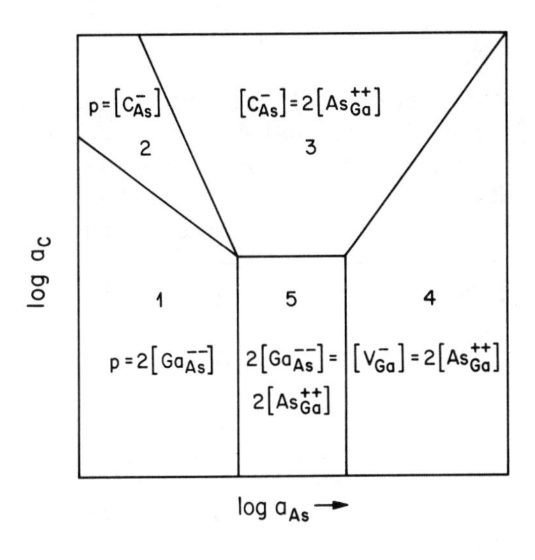


**Figure 6.24** Log-log plot of activity of carbon and activity of arsenic in the melt showing regions of different partial electroneutrality conditions taking $As_{Ga}$ as the $EL_2$ defect [59].

different partial neutrality conditions were assumed in order to construct a range diagram between activity of carbon in the melt, $a_C$, and activity of arsenic in the melt, $a_{As}^{-2}$. This is shown in Fig. 6.24. The slopes of the lines separating the regions as indicated in the figure can easily be calculated from the defect reactions. For example, in region 1, $[Ga^{--}_{As}]p^2 \approx a_{As}^{-1}$ which yields from the neutrality condition $p \approx a_{As}^{-2/3}$. Since $[C^-_{As}]p \approx a_C a_{As}^{-1}$, $[C^-_{As}] \approx a_C a_{As}^{-1/3}$. In region 2, $p = [C^-_{As}]$, which gives $[C^-_{As}] \approx a_C^{1/2} a_{As}^{-1/2}$. Equating $[C^-_{As}]$ at the boundary between regions 1 and 2 gives $a_C \approx a_{As}^{-1/3}$. Similarly, the other boundaries can be derived. Figure 6.24 shows that the crystal is p-type at low $a_{As}$ (regions 1 and 2) but becomes semi-insulating as $a_{As}$ increases (regions 3 - 5). Note also that semi-insulating behavior can be obtained even at low $a_{As}$ when the concentration of $C_{As}$ is above a critical level such that the neutrality corresponds to region 3.

### 6.5.3 Reaction Constants

In order to obtain a quantitative agreement between defect diagrams such as shown in Fig. 6.24 and experiments, one should know the values of the equilibrium constants for the various defect reactions. That is, the entropy and enthalpy changes (Eq. 6.22) associated with each reaction should be known. There have been only a few reactions for which the enthalpies have been calculated from first principles. In most cases the task involves fitting the various experimental data to arrive at a plausible value for the reaction constant.

Phillips and Van Vechten [62, 63] have shown that vacancies in semiconductors with the diamond structure may be treated as macroscopic, octahedral cavities. Accordingly, the enthalpy of formation of a neutral vacancy is the surface area of the cavity times its surface energy. It was proposed that the surface energy consists of two contributions, one from long-ranged metallic forces and another from short-ranged covalent forces. The former is estimated from the surface tension of the metallic liquid phase of the semiconductor after correcting for the difference in density between the solid and liquid phases. The latter is simply the energy per co-valent bond

TABLE 6.3

Formation enthalpies in eV of single neutral vacancies $H(V_A^x)$ and $H(V_B^x)$ and the Schottky disorder $H_S^x = H(V_A^x) + H(V_B^x)$ for III-V compounds [64]

| Compound | $H(V_A^x)$ | $H(V_B^x)$ | $H_S^x$ | Tetrahedral covalent radius |
|---|---|---|---|---|
| AlAs | 2.76 | 2.75 | 5.51 | Al = 1.230 Å |
| AlSb | 2.30 | 2.81 | 5.11 | |
| GaP | 2.98 | 2.64 | 5.62 | Ga = 1.225 Å |
| GaSb | 2.03 | 2.56 | 4.59 | |
| GaAs | 2.59 | 2.59 | 5.18 | In = 1.405 Å |
| InP | 3.04 | 2.17 | 5.21 | As = 1.225 Å |
| InAs | 2.61 | 2.07 | 4.68 | P = 1.128 Å |
| | | | | Sb = 1.405 Å |
| InSb | 2.12 | 2.12 | 4.24 | |

times half the number of broken bonds. Van Vechten [64] has extended the macroscopic cavity model to vacancies in zinc blende or wurtzite type crystals. Table 6.3 gives the enthalpies of formation of $V_A^x$ and $V_B^x$ and their sum which is the enthalpy of the Schottky disorder reaction Eq. (6.13). Since the size of the vacancy cavity is assumed to be proportional to the tetrahedral co-valent radius, the element which has the large co-valent radius has a higher formation enthalpy. The enthalpies shown in Table 6.3 are expected to be 5 to 20 percent overestimates [64].

The enthalpies $H(V_A^x)$ and $H(V_B^x)$ shown in Table 6.3 are those associated with the virtual reactions

$$0 \rightleftarrows V_A^x \; ; \quad H(V_A^x) \tag{6.29a}$$

$$0 \rightleftarrows V_B^x \; ; \quad H(V_B^x) \tag{6.29b}$$

In GaAs, combining Eq. (6.29b) with another virtual reaction

$$Ga(\ell) \rightleftarrows Ga_{Ga} \; ; \quad H_{Ga} \tag{6.30}$$

Hurle [65] estimated the enthalpy associated with the actual reaction

$$Ga(\ell) \rightleftarrows Ga_{Ga} + V_{As} \; ; \quad H = H(V_{As}^x) + H_{Ga} \tag{6.31}$$

Equation (6.31) together with the reaction for the formation of GaAs from the vapor

$$Ga(\ell) + \frac{1}{2} As_2(g) \rightleftarrows GaAs(s) \; ; \quad H_f \tag{6.32}$$

gives

$$\left[V_{As}^{x}\right] = K_{V_{As}} \frac{p_{As_2}^{-1/2}}{K_f} \tag{6.33}$$

where $K_{V_{As}}$ and $K_f$ are the constants for Eqs. (6.31) and (6.32), respectively and $p_{As_2}$ is the partial pressure of $As_2$.

Hurle [65] also estimated the values for $K_{V_{As}}$ and $K_f$ which require the entropy values for these reactions. For a reaction such as Eq. (6.29a), the entropy has two contributions, a change in the vibrational entropy upon the formation of the vacancy and a configurational entropy resulting from the Jahn-Teller distortion of the vacancy. In Si and Ge the latter term was estimated to be 1.1k [66, 67]. In analyzing the data of the quenching experiments in Si and Ge, Van Vechten and Thurmond [67] assumed that the vibrational entropy term is negligible and explained the observed high entropy values in terms of the entropy of ionized vacancies. On the other hand Lannoo and Allan [68] computed a value of 2.7k from their Green's function calculation of the formation entropy. They also showed that small changes in the force constants of the nearest neighbors of the vacancy can lead to large entropy values.

There are no quenching experiments in GaAs or InP to enable an analysis similar to the one made by Van Vechten and Thurmond in Si and to conclude whether the entropy of the vacancy is only the entropy of its Jahn-Teller distortion. Hurle [65] assumed the latter and the value of 1.1k for the formation entropy of $V_{As}$ and $V_{Ga}$ in GaAs. The values of the reaction constant $K_{V_{As}}$ and $k_f$ required to calculate $[V_{As}^{x}]$ in GaAs as derived by Hurle are given in Table 6.4. Also the values of the constants derived in a similar manner for InP, InAs, and GaP are listed in Table 6.4 [69]. For calculating $[V_{B}^{x}]$ under LPE growth conditions, the value of $p_{B_2}$ can be assumed to be the equilibrium pressure along the metal rich liquidus which is given in Table 6.4. The values of $[V_{B}^{x}]$ thus obtained at 600°C are shown in Table 6.5. If the formation entropy for the vacancy were higher because of the vibrational entropy, say by a factor of 4, then the vacancy concentrations would be 20 times higher than the values given in Table 6.5.

For the ionization reactions Eqs. (6.18 to 6.21), the calculation of the entropy terms is some what simpler. Consider the reaction Eq. (6.21) for which the forbidden energy gap $E_g$ is the Gibbs energy [50, 71]. It follows from familiar thermodynamic relationships that the entropy and enthalpy for reaction Eq. (6.21) are

$$E_g = \Delta G = \Delta H - T\Delta S$$

$$\Delta S = -\frac{dE_g}{dT} \tag{6.34}$$

$$\Delta H = \frac{dE_g}{d\frac{1}{T}}$$

**TABLE 6.4**

Values of enthalpy (in eV) and entropy (in units of k) associated with the reactions listed and the pressure of the group V species in equilibrium with the A-B liquidus which are required to calculate the concentration of neutral group V vacancies in InP, GaAs, InAs, and GaP

| Reaction | AB Compound | | | | | | | |
|---|---|---|---|---|---|---|---|---|
| | InP[b] | | GaAs[a] | | InAs[b] | | GaP[b] | |
| | $\Delta H$ | $\Delta S$ | $\Delta H$ | $\Delta S$ | $\Delta H$ | $\Delta S$ | $\Delta H$ | $\Delta S$ |
| $0 \rightarrow V_B^x$ | 1.9 | 1.1 | 2.3 | 1.1 | 1.8 | 1.1 | 2.3 | 1.1 |
| $A(\ell) \rightarrow A_A^x$ [c] | −0.49 | −1.9 | −0.53 | −2.82 | −0.33 | −3.9 | −0.55 | −1.5 |
| $A(\ell) + \frac{1}{2} B_2(g) \rightarrow AB(s)$ | −1.57 | −12.7 | −1.98 | −13.8 | −1.69 | −15.2 | −1.99 | −12.1 |
| $p_{B_2}^{1/2}$ [d] | 1.55 | 13 | 1.98 | 13.8 | 1.8 | 14.7 | 1.85 | 12.2 |

(a) Ref. 65
(b) Ref. 69
(c) In obtaining $\Delta H$ and $\Delta S$ values for this reaction we have apportioned equally the enthalpy and entropy values of the reaction $A(\ell) + B(\ell) \rightarrow AB(s)$.
(d) Ref. 70

TABLE 6.5

Neutral group V vacancy concentration calculated for LPE growth conditions at 600°C using the values listed in Table 6.4.

| | InP | GaAs | InAs | GaP |
|---|---|---|---|---|
| $[V_B^x]$ | $2.5\times10^{-9}$ | $1.9\times10^{-11}$ | $1.6\times10^{-9}$ | $7.5\times10^{-12}$ |

The temperature dependence of $E_g$ is usually described by the Varshni (72) equation

$$E_g(T) = E_g(T=0°K) - \frac{\alpha T^2}{T+\beta} \tag{6.35}$$

where $\alpha$ and $\beta$ are constants, the latter being approximately equal to the Debye temperature. Equations (6.34) and (6.35) give

$$\Delta S = \frac{\alpha T(T+2\beta)}{(T+\beta)^2} \tag{6.36}$$

At temperatures above the Debye temperature $\Delta S$ tends to a constant value of $\alpha$. For GaAs $\alpha \sim 5.405\times10^{-4}$ eV $K^{-1}$ which gives $\Delta S \sim 6k$. Thurmond [71] estimated the uncertainty in the value of $\Delta S$ obtained from Eq. (6.36) to be a few percent at 300°K and 5 to 8 percent at the melting point. Van Vechten [73] calculated $\Delta S$ using a simple point charge model for the effect of electron-hole pair creation on the phonon modes of the crystal and found it to be comparable to the value obtained from Varshni equation.

For charged point defects, the entropy of the ionization reactions can also be shown to be $-dE_i/dT$ just as in Eq. (6.34) where $E_i$ is the free energy of ionization. Van Vechten and Thurmond [74] have shown that for the ionization of vacancies the entropy is essentially the same as that given by Eq. (6.36). They have also shown that for hydrogenic impurities $\Delta S$ for the ionization is zero.

While the entropies of the ionization of vacancies can be taken to be the entropy of the formation of electrons and holes, the enthalpies are not known. Analyzing a large number of experimental data in GaAs, Hurle [65] deduced that the ionization enthalpy of $V_{As}^x$ to form $V_{As}^+$ is 0.27 eV. According to many theoretical calculations [75 - 77] the neutral undistorted group V vacancy has two energy levels, an $A_1$ level filled with two electrons and a $T_2$ level filled with one electron (Sec. 6.4). If the levels were not in resonance with the valence or conduction bands, their positions in the energy gap depending upon the model range from 0 to 0.6 eV above the valence band for the $A_1$ level and 0.05 to 0.5 eV below the conduction band for the $T_2$ level for the $V_{As}$ in GaAs. In InP, the $T_2$ level of $V_p$ is above the conduction band [77]. But the picture of an unrelaxed vacancy is an ideal situation. In reality there are energy lowering structural distortions around the vacancy which have profound effects on its energy levels [78] (Sec. 6.4).

No realistic numbers are yet available for the vacancy ionization energies such that one can calculate the reaction constants for the ionization reactions Eqs. (6.18) to (6.20).

### 6.5.4 Interstitial and Antistructure Disorder

We noted in Sec. 6.2 that in a compound besides Schottky disorder, interstitial and antistructure disorder could also exist. Using the dielectric two band model [79, 80] Van Vechten has calculated the energies of formation of individual antisite defects and antisite defect pairs [81]. Table 6.6 lists these values for some of the III-V compounds. The enthalpy for antistructure disorder with neutral defects and charged defects are, respectively,

$$H^x_{As} = H(A^x_B) + H(B^x_A) \tag{6.37a}$$

$$H^z_{As} = H^x_{As} - (E_g - E_a - E_d) - (E_g - E_{2a} - E_{2d}) \tag{6.37b}$$

where $E_{1,2a}$ and $E_{1,2d}$ are the ionization energies for the doubly ionizable $A_B$ and $B_A$ defects. The ionization enthalpies $H^z_{As}$ are calculated from Eqs. (6.37b) neglecting $E_a$ and $E_d$ in a first approximation. From Tables 6.3 and 6.6 it can be seen that the enthalpy of formation of the neutral antisite pair is considerably less than that of the Schottky disorder suggesting that the antisite pair will be the most common native defect in III-V compounds. Self-consistent Green's function calculations of defect pairs in GaAs give a value of ~ 1.7 eV for the $(A_B B_A)^x$ pair [82] which is higher than the value of 0.7 eV in Table 6.6, but nonetheless is still considered to be a dominant defect compared to Schottky disorder. Further, in GaAs the deep level $EL_2$ which is

**TABLE 6.6**

Enthalpy of formation in eV of antisite defect pairs $H(A_B B_A)^x$, single antisite defects $H(A^x_B)$, $H(B^x_A)$ and antistructure disorder $H^x_{AS}$ for III-V compounds [1,81].

| Compound | $H(A^x_B)$ | $H(B^x_A)$ | $H(A_B B_A)^x$ | $H^x_{AS}$ |
|---|---|---|---|---|
| AlAs | 4.90 | 5.05 | 0.75 | 9.95 |
| AlSb | 3.12 | 3.46 | 0.38 | 6.58 |
| AlP | 5.60 | 5.50 | 1.11 | 11.10 |
| GaP | 5.38 | 5.08 | 1.06 | 10.46 |
| GaAs | 3.21 | 3.21 | 0.70 | 6.42 |
| GaSb | 1.92 | 1.68 | 0.40 | 3.60 |
| InP | 3.59 | 3.12 | 1.30 | 6.71 |
| InAs | 1.27 | 1.03 | 0.90 | 2.30 |
| InSb | 0.61 | 0.61 | 0.54 | 1.22 |

present in semi-insulating material has been linked to $As_{Ga}$ (Sec. 7.6), thus giving experimental support to include antisite defects in formulating the equilibrium defect structure of III-V compounds.

Hurle [65] in his point defect model of GaAs considered Frenkel disorder in the As sublattice ($As_i$ , $V_{As}$) as the dominant disorder. This view was taken based on lattice parameter [83, 84] and density [85] measurements of GaAs crystals prepared under conditions of varying stoichiometry. The lattice parameter measurements indicated that the lattice parameter increased with increasing As concentration in the melt. The density measurements showed that on the As rich side there is an excess mass per unit cell. Both these measurements suggested the existence of $As_i$ or $As_{Ga}$ on the As rich side of GaAs. A concentration of $\sim 5 \times 10^{18}$ cm$^{-3}$ of $As_i$ or $\sim 7 \times 10^{19}$ cm$^{-3}$ of $As_{Ga}$ could account for the measured density changes [65]. Hurle [65] supposed that such a high concentration of $As_{Ga}$ would produce larger changes in lattice parameter than measured and thus chose to consider only $As_i$. Hurle [86], similarly, proposed Frenkel defects in the Ga sublattice producing $V_{Ga}$ [Eq. 6.14] which served to explain the compensation, the formation of interstitial loops and the dilation of the lattice in Te-doped GaAs.

Low-temperature electron irradiation experiments in GaAs indicates a threshold of 17 eV for Frenkel pair formation on the Ga sublattice [87] suggesting a rather large formation enthalpy for self-interstitials. Self-consistent Green's function calculations of the total energy of elemental defects in GaAs also suggest that single interstitials of $T_d$ symmetry are not thermodynamically favorable in either As or Ga rich crystals [88]. In view of these results the inclusion of Frenkel disorder in equilibrium defect structure of III-V compounds is questionable.

### 6.5.5 Partial Equilibrium; the Situation After Cooling

The analysis of the defect structure of a crystal at the temperature of preparation, say near the growth temperature, would require measurements at that temperature which are not always practicable. Measurements are generally made at room temperature and based on the results the high temperature equilibrium is reconstructed. In doing so it is usually assumed that the crystals are cooled suddenly from the high temperature such that atomic imperfections are frozen in. This is a reasonable assumption since a change in the concentration of the imperfection would require migration, a process generally supposed to have a large activation energy. The movement of electronic particles is, however, not affected by cooling and hence electronic equilibrium will be maintained at all temperatures. The concentrations of the defects can then be obtained in the same way as outlined in Sec. 6.5.1 with the additional restraint that the total concentration of atomic imperfections is a constant and given by the equilibrium situation prevailing at the high temperature [89].

### 6.5.6 Alloy Semiconductors

Since the defect structure of the individual III-V compounds themselves has not been established to any certainty, that of their alloys is very much unknown. Blom [90] extended the analysis of Van Vechten [64, 81] for binary compounds to ternary systems and calculated defect concentrations in AlGaAs assuming Schottky and antistructure disorder. The analysis involved calculating the concentrations of vacancies and antisites in each binary compound separately and adding them with appropriate weight factors to obtain the concentration in AlGaAs. The defect concentrations thus obtained are shown as a function of Al composition in Fig. 6.25 for LPE growth at a temperature of 800°C. The effect of increasing the Al composition is that the concentration of vacancy increases while that of anti-sites decreases.

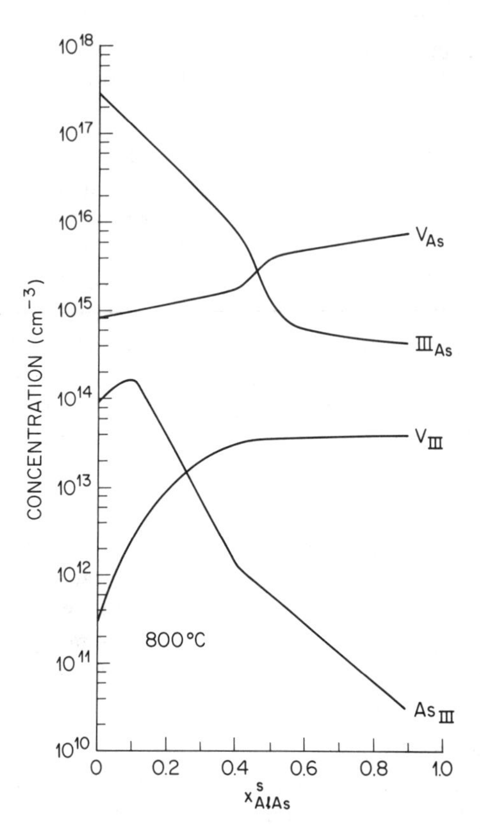


**Figure 6.25** Concentration of vacancies and antisite defects in $Al_xGa_{1-x}As$ at 800°C as a function of x [90].

Buisson et al., [38] calculated the anion vacancy levels in $Ga_x In_{1-x} As_{1-y} P_y$ quaternary alloys using the Hjalmarson et al., [36] theory of deep levels which is based on the empirical tight-binding model of semiconductor band structures [91] (Sec. 6.4). Their results are shown in Fig. 6.26 as contours of constant energy for anion vacancy energy levels of $A_1$ (Fig. 6.26(a)) and $T_2$ (Fig. 6.26(b)) symmetry in the gap. It should be noted that the accuracy of the level positions is not better than a few tenths of an eV but the results are useful in describing chemical trends. The zero of energy in Fig. 6.26 is the valence band maximum. The line marked $E_g$ corresponds to the predicted vacancy level coincident with the conduction band edge. In other words, alloys whose composition lie below this line do not have anion vacancy levels in the gap. In GaAs, the $T_2$ level of $V_{As}$ lies at ~ 1.45 eV and the $A_1$ level lies at ~ 0.6 eV. In $Ga_{0.47} In_{0.53} As$ there are no anion vacancy levels in the gap, so is the case for InP and InAs. These are in agreement with other similar calculations [77], except for the $A_1$ level in InP which is predicted to be about 1 eV above the valence band edge [92].

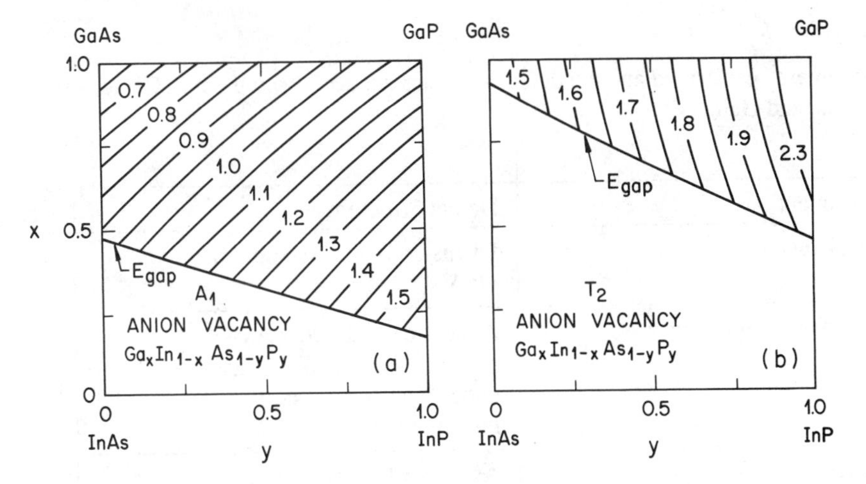


**Figure 6.26** Calculated contours of constant anion vacancy energy for states of $A_1$ and $T_2$ symmetry in $Ga_xIn_{1-x}As_{1-y}P_y$. The zero of energy is the valence band maximum. The contours are labeled by their energies in eV. The line marked $E_{gap}$ corresponds to the vacancy level being coincident with the conduction band edge [38].

### 6.5.7 Experimental Observation of Point Defects

Since native or foreign point defects, neutral or charged, affect the various physical properties of the semiconductor, the determination of the defect structure involves studying these properties as a function of temperature, component activities, and dopant concentration. The physical properties that are generally studied include electrical properties such as resistivity, Hall effect, capacitance (e.g., DLTS), optical properties, (e.g., photoluminescence), magnetic properties (EPR), lattice constant, density, diffusion constants, mechanical properties (e.g., internal friction), structural properties (e.g., transmission electron microscope) and so on. Generally, the chances of arriving at the correct defect structure for the crystal as a condition of preparation or the correct atomic model for a specific defect are greatly improved if several properties are studied preferably on the same samples. In fact, such a synergestic approach has become the norm of current studies of defects in semiconductors. Table 6.7 lists the material, the technique, and the defect structure that is deduced from the measurement. Only some key references for each technique are given but they would be helpful to the interested reader to delve further into the subject.

It would emerge from Table 6.7 that regarding defect structure, GaAs has been studied to a greater extent than InP and that the alloy semiconductors, especially the InP based ones have not received their share of attention. But even in the case of GaAs, only in a few instances have specific defects been identified from the physical measurements, as for example, the association of the four line spectrum in the electron paramagnetic resonance experiment with $As_{Ga}$. In the majority of cases, microscopic identification has not been possible but instead only suggestions for probable defect structures have been made based on the response of the crystal to a change in

**TABLE 6.7**

Experiments performed and the defect structure deduced from them in GaAs, AlGaAs, InP, GaInAs, and GaInAsP.

a) GaAs

| Experiment | Defect Structure | Reference |
|---|---|---|
| Self-diffusion | Diffusion in the As-sublattice via $V_{As}$ | 93, 94 |
| Hall effect, conductivity | Annealing under low As pressure leading to $V_{As}^{+}$ and under high As pressure leading to $V_{Ga}^{-}$ | 95 |
| density | $As_i$($\sim 5\times 10^{18}$ $cm^{-3}$ in as-grown crystals from the As-rich side) | 85 |
| Lattice Parameter | $As_i$ on the As-rich side; $V_{As}$ on the Ga-rich side. | 83, 84 |
| Internal friction | ($V_{Ga} - V_{Ga}$) pair<br>($Te_{As} - V_{Ga}$) pair | 96<br>97 |
| Deep level transient spectroscopy | Defects produced by e-irradiation due to displacements in the As-sub-lattice | 98-100 |
| Photoluminescence | Impurity-vacancy complexes | 101, 102 |
| Electron paramagnetic response | $As_{Ga}$, $V_{Ga}$ | 103, 104 |
| Positron life time | Impurity-vacancy complexes; the concentration of $V_{As}$ and $V_{Ga}$ in melt-grown crystals are $\sim (1–4) \times 10^{17}$ $cm^{-3}$ | 105 - 108 |
| Transmission Electron microscopy | Vacancies (interstitials) in crystals annealed at high temperature as evidenced by the presence of prismatic vacancy type (interstitial type) dislocation loops | 102 |
| Extended x-ray absorption fine structure (EXAFS) | Local structure of an impurity $- S_{As}$ | 109 |

**Table 6.7** (cont'd)

b) AlGaAs

| Experiment | Defect Structure | Reference |
|---|---|---|
| Deep level transient Spectroscopy | DX center<br>Defects produced by e-irradiation | 110, 111 |
| Photoluminescence | Impurity-vacancy complexes | 112, 113 |
| Transmission Electron Microscopy | Interstitials in epitaxial layers annealed at high temperature as evidenced by the presence of prismatic interstitial type dislocation loops | 114 |
| Ballistic phonon attenuation | Symmetry of DX centers | 115 |

c) InP

| Experiment | Defect Structure | Reference |
|---|---|---|
| Self-diffusion | Defect structure unknown | 93 |
| Positron life time | Vacancy defects in e-irradiated InP | 116 |
| Photoluminescence | $V_P$, $V_{In}$ related emission bands | 117 |
| Electron Paramagnetic Resonance | $P_{In}$, $V_{In}$ | 104 |
| Deep level transient spectroscopy | Configurationally bi-stable defect | 118, 119 |

d) GaInAs, GaInAsP

| Experiment | Defect Structure | Reference |
|---|---|---|
| Deep level transient spectroscopy | Several electron and hole traps- with their microscopic identities unknown | 120 |
| Transmission electron microscopy | Prismatic vacancy type loops | 121 |

the ambient such as the group V partial pressure. Even then a unified defect model for the III-V compounds has not emerged. One reason for this is that often impurity effects would dominate over the effects of the native defects leading to erroneous interpretation [1].

## 6.6 INCORPORATION OF IMPURITIES

### 6.6.1 Introduction

Impurities are sometimes added to the pure crystal to obtain a certain desired physical property. For example, one would add donor or acceptor impurities to obtain predominantly n-type or predominantly p-type conductivity. On the other side of this issue, often the inadvertent presence of impurities may also limit the lowest background carrier concentration that would be desired for certain application. When impurities are added intentionally one considers the ease of incorporation and the maximum solubility of the impurity in the semiconductor. When impurities are present as contaminants and their identity is known, one would like to choose the conditions of preparation of the crystal such that their incorporation is minimized. In either situation it is, therefore, very important to understand the factors governing impurity incorporation.

The mechanism of impurity incorporation is determined by the type of ions it is going to form, that is, whether it is a single or multiple donor or acceptor, by other charged imperfections, native or foreign, and by the conditions of preparation, that is, temperature and the partial pressures of the constituents. In the case of substitutional impurity, the incorporation will be variable depending on the site availability which is determined by the partial pressures of the constituents. The incorporation of impurities and interactions among them and those between impurities and native imperfections may be treated within the framework of defect chemistry as outlined in Sec. 6.5.

For an impurity F in a compound AB, the stability limits of the compound as defined by the A-B-F phase diagram would control the solubility limits of F. In the ternary phase diagram, solid solubility isotherms at temperature T along which the ternary liquid is in equilibrium with solid AB saturated with F, provide detailed information about the incorporation of F into AB [122]. Panish has constructed such isotherms for a few of the dopants in GaAs and AlGaAs; GaAs:Zn [123, 124], GaAs:Sn [125, 126], GaAs:Ge [125], GaAs:Si [127, 128], GaAs:Sn, Ge [129], and AlGaAs:Sn [130].

Let us now consider the incorporation of a substitutional impurity F from the external phase, say the liquid, into a compound AB. Let us assume that F substitutes for A and acts as a single acceptor. Zn, Cd, Mg, and Be would be examples for F in GaAs or InP. The chemical reactions which would describe the incorporation and ionization of F in AB are

$$F(\ell) + V_A^x \rightleftarrows F_A^x \tag{6.38a}$$

$$F_A^x \rightleftarrows F_A^- + h \tag{6.38b}$$

From Eq. (6.38) the following mass action relation is obtained

$$[F_A^-]\,[h] = K_{F_A^-}\, K_{F_A^x}\, [V_A^x]\, a_F \tag{6.39}$$

where $K_{F_A^x}$ and $K_{F_A^-}$ are the reaction constants, respectively for (6.38a) and (6.38b) and $a_F$ is the activity of F in the liquid. Because of the incorporation of $F_A^-$ into the semiconductor, let us assume that the electroneutrality is governed by

$$[F_A^-] = [h] \tag{6.40}$$

Note that if [h] exceeds the valence band density of states one may have to use the activity coefficient for holes in Eq. (6.39) (Sec. 6.5.1). Substituting Eq. (6.40) into Eq. (6.39) gives

$$[F_A^-] = [h] = \left\{ K_{F_A^x} K_{F_A^-} \ [V_A^x] \ a_F \right\}^{1/2} \tag{6.41}$$

Two important points can be made from this equation. First, if $F_A^-$ is the major species of F in the solid such that $[F]_{Total} \approx [F_A^-]$, then it follows from Eq. (6.41) that both $[F]_{total}$ and [h] would vary as the square root of the concentration of F in the liquid. Second, the concentration of F is proportional to $[V_A^x]$ which in turn is proportional to the activity of B in the liquid Eq. (6.17). In other words $[F_A]$ increases with increasing activity of B in the external phase from which growth occurs.

The square-root dependence of ionized dopant concentration or the carrier concentration in the solid on the impurity concentration in the external phase, under the simple case of the electroneutrality condition governed by only these two species, is a direct indication of the existence of equilibrium between the bulk of the semiconductor and the external phase. The incorporation of Zn into GaAs follows the square-root law. Figure 6.27 shows the Zn solid solubility in GaAs at 1000°C [124]. Note the square-root relationship between $[Zn_{Ga}^-]$ (= [h]) in the solid and the concentration of Zn in the liquid. The solid line is calculated using Eq. (6.41) and by including the activity coefficient for holes as well. Note the increase in $[Zn_{Ga}^-]$ in crystals grown from As rich solutions as compared to Ga rich solutions, consistent with the increase in $[V_{Ga}]$ for the former growth condition Eq. (6.41).

The incorporation of Zn in GaAs perhaps represents the simplest situation as depicted by Eq. (6.41). For most other common dopants, n- or p-type, in GaAs, InP, AlGaAs, GaInAs, and GaInAsP the square-root law is not observed. Generally the room temperature carrier concentration would show a linear dependence on the concentration of the dopant in the liquid. This is illustrated in Fig. 6.28 for Te in GaAs where it occupies the As site and acts as a donor [131]. To differentiate the linear dependence from the square-root dependence of the carrier (dopant) concentration on the activity of the dopant in the liquid it was proposed that the equilibrium between the external phase and the bulk of the semiconductor does not exist but instead equilibrium is established with only its surface. The essential feature of this proposal is that the liquid-semiconductor interface at the growth temperature acts as a Schottky barrier which pins the Fermi level at the interface. As a consequence the electron concentration is only a function of temperature and is given by

$$n_{gt} = N_c \exp \frac{-\phi}{kT} \tag{6.42}$$

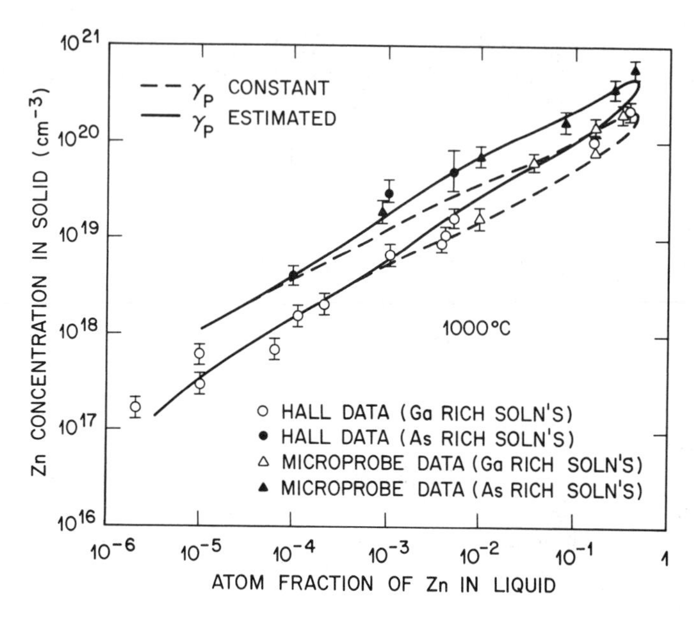


**Figure 6.27** Zinc concentration in the solid versus zinc concentration in liquid for LPE growth at 1000°C. $\gamma_p$ is the activity coefficient of holes [124].

**Figure 6.28** Tellurium or electron concentration in the solid versus Te fraction in liquid for growth at 1000°C [131].

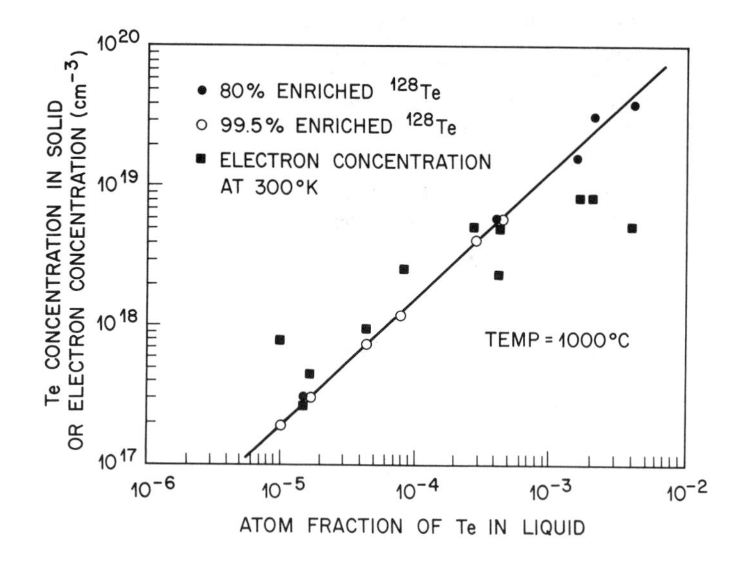

where $\phi$ is the barrier height and $n_{gt}$ is the electron concentration at the growth temperature. Applying Eq. (6.39) to the specific case of Te in GaAs gives

$$[Te_{As}^{+}]\, n_{gt} \propto a_{Te} \tag{6.43}$$

At room temperature the electroneutrality is given by

$$n_{rt} = [Te_{As}^{+}] \tag{6.44}$$

where $n_{rt}$ is the electron concentration at room temperature. Substituting Eqs. (6.42) and (6.43) into (6.44) would give

$$n_{rt} = [Te_{As}^{+}] \propto a_{Te} \tag{6.45}$$

Whether the dopant would follow the square root or linear dependence on the amount in the liquid, that is, whether equilibrium exists with the bulk or the surface, would be determined by the growth rate, $\nu$, and the diffusion coefficient of the impurity in the solid, D [132, 133]. Because of the band bending at the liquid-semiconductor interface, there exists a space charge layer of width **w** which is of the order of the Debye screening length, $(\varepsilon kT / 4\pi n_{gt} e^2)^{1/2}$, where $\varepsilon$ is the dielectric constant and e is the electronic charge. If $D/w \gg \nu$, then equilibrium with the bulk is established, otherwise, surface equilibrium prevails. Typically for LPE growth conditions $\nu \sim 10^{-4} - 10^{-7}$ cm sec$^{-1}$ and for a fast diffusing impurity such as Zn in GaAs, $D/w \gg \nu$ is easily satisfied thus obeying the square root law according to bulk equilibrium situation. On the other hand, for a slow diffusing impurity such as Te [131], Sn [126], and so on, $D/w \ll \nu$ thus leading to linear law according to surface equilibrium situation.

Barring the case of Zn in GaAs, most other dopants in both GaAs and InP grown from the melt or solution do not exhibit the square-root law. Generally for the group IV impurities which can substitute either the metal or the nonmetal site and act as donor or acceptors, and for some of the group II acceptor impurities, the incorporation may be consistent with the surface equilibrium picture. They have low values for $D_s$ such that $D/w \ll \nu$. Examples of these dopants are Te [131], Sn [126], Ge [134], Mg [135, 136], in GaAs; Sn [137, 138], Te [137, 138], Se [138], Ge [138] in InP; Sn [138], Te [138] in GaInAs. On the other hand, a few of the group II acceptor impurities which have comparatively high $D_s$ may satisfy the condition $D/w > \nu$. For example, D for Zn in InP is estimated to be about $10^{11}$ cm$^2$ sec$^{-1}$ at a LPE growth temperature of 630°C. For a carrier concentration of $\sim 10^{18}$ cm$^{-3}$ at the growth temperature, $w \sim 7 \times 10^{-7}$ cm which then gives $D_{Zn}/w \sim 1.4 \times 10^{-5}$ cm sec$^{-1}$. This is greater than $\nu$ used typically for growing p-type InP layers by LPE. Nevertheless, Zn incorporation does not show the expected square-root behavior. This is also true for other fast diffusing impurities such as Be, Mg, and Cd in InP [137 - 142] as well as in GaInAsP [141, 142] and GaInAs [138, 142 - 146].

Hurle [147 - 149] proposed an alternative model to the Schottky barrier model to explain the linear doping curves obtained for Te [147], Sn [148], and Ge [149] in GaAs in which he assumed that the electroneutrality at high temperatures is governed by $V_{As}^{+}$ at all but the highest doping levels. He also proposed complexes such as $(Te_{As}V_{Ga})$, $(Sn_{Ga}V_{Ga})$ to explain the saturation in the room temperature carrier concentration at high doping levels in Te or Sn doped GaAs. The support of his model comes from the apparent absence of orientation dependence of doping [150, 151] which otherwise be expected from the Schottky barrier model.

### 6.6.2 Amphoteric Dopants

Group IV elements in III-V compounds can occupy either the group III site or the group V site when they are incorporated substitutionally. Accordingly, they exhibit donor or acceptor behavior and in this regard, they are called amphoteric impurities. The distribution of the amphoteric impurity over the two sublattices can be influenced by the concentrations of the metal and nonmetal vacancies and by the interaction of charges (Fermi level effects). The two effects may act in unison to enhance the solubility on one site compared to the other. Consider, for example, Si in GaAs. If the crystal is heated in an As vapor leading to p-type behavior, the solubility of $Si_{Ga}^{+}$ is increased since $[Si_{Ga}^{+}] \propto 1/n$ where n is the electron concentration. Further, excess As condition gives $V_{Ga}$ which also favors $Si_{Ga}^{+}$. Although the distribution of the group IV atoms between two sites is determined by the availability of these sites, the situation obtained in experiments is very complex. Let us consider Si in GaAs. The reactions which describe the site distribution are the following:

$$Si(g, \ell) + V_{Ga} \rightleftarrows Si_{Ga}^{x} \tag{6.46a}$$

$$Si(g, \ell) + V_{As} \rightleftarrows Si_{As}^{x} \tag{6.46b}$$

where the doping occurs either from the gas phase or from the liquid. Applying the law of mass actions to these reactions one gets

$$\frac{[Si_{Ga}]}{[Si_{As}]} = \frac{K_{Ga}\,[V_{Ga}]}{K_{As}\,[V_{As}]} \tag{6.47}$$

where $K_{Ga}$ and $K_{As}$ are the reaction constants for 6.46a and 6.46b, respectively. When crystals are grown from the melt under stoichiometric conditions $[Si_{Ga}] \gg [Si_{As}]$ and n-type crystals are obtained. However, for growth from Ga-solutions since $[V_{As}] > [V_{Ga}]$, $[Si_{As}] > [Si_{Ga}]$, and p-type GaAs is obtained at least at low-growth temperatures. If the growth temperature is greater than a certain critical temperature (750 - 850°C for Si in GaAs) then n-type behavior is obtained. In fact, high-efficiency light-emitting diodes have been fabricated by solution grown Si-doped GaAs p-n junctions. Teramoto [152] considered this n-to-p-type transition during LPE growth and derived an expression for the transition temperature by calculating the equilibrium constants of Eq. (6.47). The calculated transition temperature was in reasonable agreement with the experiments. The transition in the site occupation during LPE growth has also been observed for Si in AlGaAs and the transition temperature in that case decreased logarithmically with increasing Al to As atom fraction ratio in the liquid [153].

### 6.6.3 Incorporation of Impurities During VPE Growth.

Unlike in melt and LPE growth, dopant incorporation during VPE and MBE growth is generally thought to be governed by kinetic factors. The reasoning stems from many observations that are contrary to thermodynamic expectation such as the doping dependence on crystal orientations and growth rate, the inconsistency in the solubility of the dopant with changes in the group III/group V ratio, and the increase in the concentrations of the dopant (e.g., Te in GaAs) in the solid with increasing growth temperature opposite to the situation in LPE or melt growth. Albeit these observations a thermodynamic description of dopant incorporations may still provide a framework to differentiate the equilibrium and the nonequilibrium effects.

As a rule, dopant incorporation during vapor phase growth which would include both the hydride / chloride process as well as the metal organic chemical vapor deposition process is characterized by the linear relationship between the dopant or electron / hole concentration in the solid and the partial pressure of the dopant species in the gas phase. The commonly used dopants in GaAs and InP grown by VPE and MOCVD are Si, Sn, S, Se, and Te as n-type dopants and Zn, Cd, Mg, Be as p-type dopants. Table 6.8 lists the dopants used in VPE and MOCVD growth of GaAs and InP.

**TABLE 6.8**

Dopants used in GaAs and InP prepared by vapor phase epitaxy (VPE) or metal organic chemical vapor deposition (MOCVD).

| Growth | Matrix | Dopant (Source in the gas phase) | Ref. |
|---|---|---|---|
| VPE | GaAs | Ge ($GeH_4$) | 154 |
| | | Ge (added to Ga Source) | 155 |
| | | Se ($H_2Se$) | 154 |
| | | S ($H_2S$) | 154, 156 |
| | | Te ($(C_2H_5)_2Te$) | 157 |
| | | Si ($SiCl_4$) | 158 |
| | | Sn (added to Ga source) | 155, 159 |
| | | Zn (added to Ga source) | 155 |
| | | Cd (added to Ga source) | 155 |
| MOCVD | GaAs<br>AlGaAs | Se ($H_2Se$) | 160-163 |
| | GaAs | S ($H_2S$) | 164 |
| | | Ge ($GeH_4$) | 162, 165 |
| | | Sn ($(CH_3)_4$ Sn) | 162 |
| | | Si ($SiH_4$) | 165 |
| | | Mg ($(C_5H_5)_2Mg$) | 166, 167 |
| | | Be ($(C_2H_5)_2Be$) | 168, 169 |
| | | Zn ($(CH_3)_3Zn$) | 163, 164 |
| | | Zn ($(C_2H_5)_2Zn$) | 162 |
| VPE | InP | Cd (metal source) | 170 |
| | | Zn (metal source) | 171 |
| MOCVD | | Si ($SiH_4$) | 172 |
| | | S ($H_2S$) | 173 |
| | | Se ($H_2Se$) | 174 |
| | | Zn ($(CH_3)_3Zn$) or ($(C_2H_5)_2Zn$) | 175 |
| | | Cd ($(CH_3)_3Cd$) | 176 |
| | | Mg ($CP_2$ Mg) | 177 |

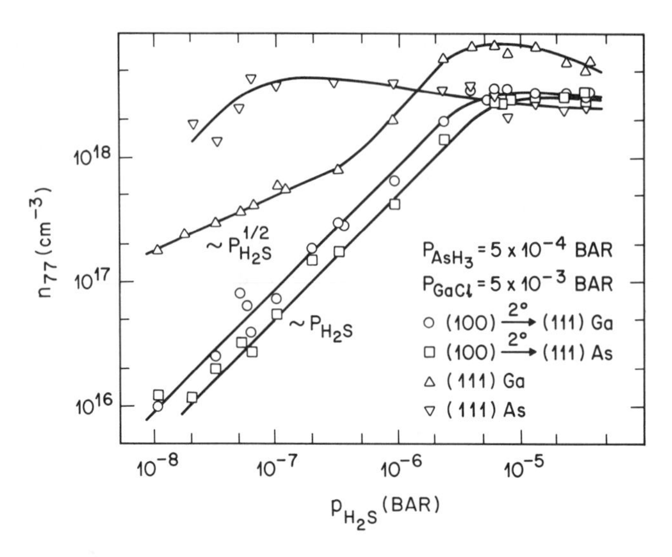


**Figure 6.29** Electron concentration at 77K versus partial pressure of $H_2S$ for differently oriented GaAs substrates during VPE growth [178].

Figure 6.29 shows the doping curve for the incorporation of S in GaAs grown by HCl-Ga-$AsH_3$-$H_2$ system as electron concentration at 77K, n, versus the partial pressure of $H_2S$, $p_{H_2S}$, for different orientations of the substrate [178]. The linear relationship between n and $p_{H_2S}$ for the (100) surfaces would be consistent with S incorporation occurring under equilibrium conditions wherein the electron concentration at the growth temperature is independent of the S concentration. As discussed in Sec. 6.6.1, such a situation can be brought about either by the electroneutrality at the growth temperature being governed by $V_{As}^{+}$ or by the pinning of the Fermi level at the growth interface. In the latter situation equilibrium is reached only at the surface. The data shown in Fig. 6.29 suggests that there are problems in choosing either model. The $V_{As}^{+}$ model assumes bulk equilibrium, and therefore, would not predict the observed orientation dependence. On the other hand, if the Fermi level is pinned at the surface then an orientation dependence can come about owing to the possibility of different surface states pinning the Fermi level for different orientations. But the quick saturation of n for the (111) As orientation and the surprising $p_{H_2S}^{1/2}$ dependence of n for (111) Ga orientation at low $H_2S$ pressure, that is consistent with an equilibrium situation as described by Eqs. (6.40) and (6.41), are not totally in accordance with the surface equilibrium model.

The thermodynamic equilibrium model can also only partially explain the doping dependence on the concentrations of the group III and group V gas phase components. For dopants which substitute on the group III site, the carrier concentration would be expected to increase with increasing group V component activity, since the group III vacancy concentration increases. A similar situation is true for the dopant on the group V site and the group III component activity. For Zn and Sn in GaAs where they occupy the Ga site as acceptor and donor, respectively, the carrier concentration increases as the $AsH_3$ pressure increases [162, 164]. Similarly, for S and Se which occupy the As site and act as donors, the carrier concentration decreases as $AsH_3$ increases since $V_{As}$ decreases [162,164]. But Si and Ge which also substitute on the Ga site do not behave the same as Zn and Sn. In their case, the electron concentration decreases with increasing $AsH_3$ concentration.

The effect of varying the group III component activity in the gas phase does not always produce results consistent with thermodynamic expectations of site availability. In the HCl-Ga-$AsH_3$–$H_2$ growth system when $H_2S$ is used for S-doping, the electron concentration showed an inverse quadratic dependence on GaCl pressure. GaCl is formed by reaction of metallic Ga with HCl according to

$$Ga + HCl \rightleftarrows GaCl + \frac{1}{2} H_2 \tag{6.48}$$

and the activity of Ga, $a_{Ga}$, is proportional to $p_{GaCl} / p_{HCl}$. Since $p_{GaCl}$ increases because $p_{HCl}$ increases, $a_{Ga}$ is constant. That is, $[V_{As}]$ is constant when $p_{GaCl}$ is changed and accordingly n should have been independent of GaCl pressure contrary to what is observed.

For MOCVD growth, n increases with increasing concentration of trimethyl Ga (TMG) for S, Se [162, 164] but decreases for Si, Ge, and Sn [162, 165]. In the case of Zn doping when dimethyl Zn is used as the source, the hole concentration increases as TMG increases [164]. However, the hole concentration is independent of TMG when diethyl Zn is used [162]. Increasing the group III concentration would increase the concentration of group V vacancy and decrease that of group III vacancy. Thus the behavior exhibited by S, Se, Si, Ge, and Sn are consistent with the site availability while that by Zn is not. Even for the former dopants site availability with the change in group III activity is only a partial picture. Dopant incorporation can also depend on growth rate. The growth rate increases as the group III activity increases and further it markedly depends on the substrate orientation [179].

As for the temperature dependence of doping, once again there is no uniform behavior among dopants. When $SiH_4$, $GeH_4$, and TMSn are used as the dopant sources, since dopant incorporation depends on their dissociation the carrier concentration increases with increasing temperature according to an Arrhenius type relationship [162, 165]. On the other hand, for elements like S, Se, and Zn which have high vapor pressures near the growth temperature, escape from the surface increases with increasing temperature giving rise to lower carrier concentration [162, 164].

From all respects it appears that a thermodynamic model explains only partially the doping phenomena in vapor phase growth and that kinetic factors play an important role. It is these factors which give rise to the marked orientation dependence on growth rate which in turn affects doping. For example, the decrease in the doping efficiency for S in GaAs with increasing misorientation of the substrate from the (100) orientation can be attributed to the increased growth rate caused by an increase in the density of lateral steps in the misoriented surface and thus less residence time for the dopant to allow surface migration and site occupation [178]. But a complete quantitative description of kinetically controlled dopant incorporation is not possible because of the lack of knowledge of gas chemistry [180]. For example, the partial pressure of the dopant adjacent to the growth interface may be higher or lower than the input value. One has to reckon with such factors as vapor transport which can be convective, laminar or turbulent flow, mixing of the dopant gas and the group III and group V carriers, the transport in the boundary layer in the immediate vicinity of the growth surface, reactions occurring on the crystal surface such as adsorption, desorption, chemical reactions, lateral diffusion, and finally the site availability and defect interactions in the crystal [180]. Jacobs [181] had developed a model which has the ingredients of both thermodynamic and kinetic factors and was able to explain most of the experimental observations in Te-doped GaP during Ga-HCl-$PH_3$-$H_2$ VPE growth. For the effects occurring on the surface he invoked ''sticking coefficients'' for both the host material and the dopant.

### 6.6.4 Incorporation of Impurities During MBE Growth

In the MBE growth of III-V compounds the commonly used n-type dopants are Sn, Si, and Ge [GaAs 182]; [AlGaAs 183]; [InGaAs 184]; [InGaAsP 185]. Although Be is the preferred p-type dopant [AlGaAs, GaAs 186]; [GaInAs, 187]; [GaInAsP 185], Mg [188], Mn [189], and Ge [190] have also been used. Heckingbottom et al., [191, 192] have considered in detail dopant incorporation, specifically in GaAs, and concluded that in most cases there are no major kinetic hindrances and that a thermodynamic description of doping is adequate. Their analysis consisted of estimating the dopant pressures required to achieve a low ($10^{16}$ $cm^{-3}$) and a high ($10^{19}$ $cm^{-3}$) doping level in the solid under the typical MBE operating conditions. For the common dopants such as Si, Ge, Sn, and Be, the dopant pressures required turn out to be very low, much less than the operating pressure, and under the assumption of unity sticking coefficients, which is the case for these dopants, the ratio of dopant and Ga beam fluxes gives the dopant incorporation. On the other hand, Zn and Cd have a much higher vapor pressure and thus have very low-sticking coefficients. For some dopants, the postulation of kinetic effect seemed inevitable. For example, the doping of S in GaAs [191] or InP [193] can be hindered by a competing reaction in which S reacts with Ga(In) to form volatile $Ga_2S$($In_2S$ or InS) on the surface. Similarly, incorporation of Mg is possibly hindered by the surface reaction between Mg and As to form $Mg_3As_2$ [192]. The accumulation of Sn at the surface during doping in GaAs is another situation which also requires a modification to the thermodynamic description of incorporation [194].

For MBE grown films, the doping curve is usually presented as carrier concentration at room temperature versus the dopant effusion cell temperature. Figure 6.30 shows such a curve for Be in GaAs [186]. The hole concentration increases exponentially with increasing Be cell temperature. If we consider an ideal Knudsen cell the arrival rate of the dopant at the surface is related to the vapor pressure according to

$$\Gamma = \frac{p(T)AN}{\pi \ell^2 (2\pi mRT)^{1/2}} \tag{6.49}$$

where p is the dopant vapor pressure, A is the aperture area of the cell, $\ell$ is the distance of the substrate from the aperture, m is the molecular weight of the dopant, N is the Avogadro's number, R is the gas constant, T is the cell temperature, and $\Gamma$ is the arrival rate in atoms $cm^{-2}$ $sec^{-1}$. Since p(T) varies exponentially with T, its temperature dependence dominates $\Gamma$. The solid line through the data points in Fig. 6.30 represents the Be vapor pressure. The good agreement between the slope of the vapor pressure curve and that of the carrier concentration suggests that

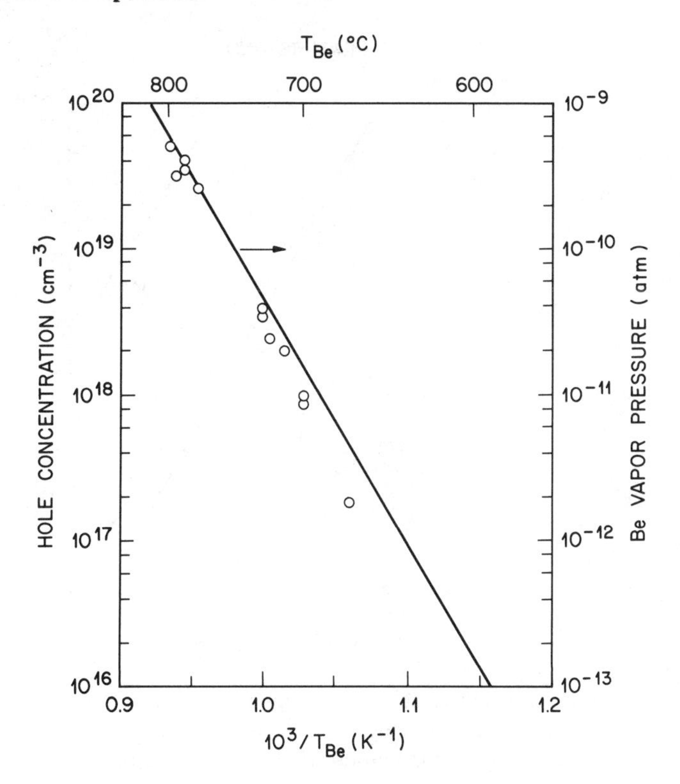


**Figure 6.30** Hole concentration versus Be oven temperature for MBE grown GaAs. The solid line is the Be equilibrium vapor pressure curve. The oven to substrate distance was ~5cm and the average growth rate was 1.4μm/h [186].

the Be doping level is proportional to the arrival rate of Be in accordance with Eq. (6.49). Figure 6.31 is an universal doping curve for III-V compounds as a function of dopant effusion cell temperature and for the group III element arrival rate of $3 \times 10^{15}$ $cm^{-2}$ $sec^{-1}$ [195].

The dependence of dopant incorporation on substrate temperature and the group V/group III flux ratio can be understood on the basis of the availability of the substitutional site that the dopant occupies. For example, a dopant occupying the group III site will show reduced incorporation under group III element-rich growth conditions. The importance of the group V/group III flux ratio is rather well demonstrated by the doping behavior of Ge in GaAs. Ge exhibits a strong amphoteric character in GaAs. Both n- and p-type material can be obtained depending on the substrate temperature and As/Ga flux ratio [196, 197]. In fact, p-n junctions can be prepared in a single growth step simply by changing the flux ratio during growth [190]. Both the flux dependence and the substrate temperature dependence can be explained by site availability. Ge on a Ga site acts as a donor and n-type material is obtained therefore under As-

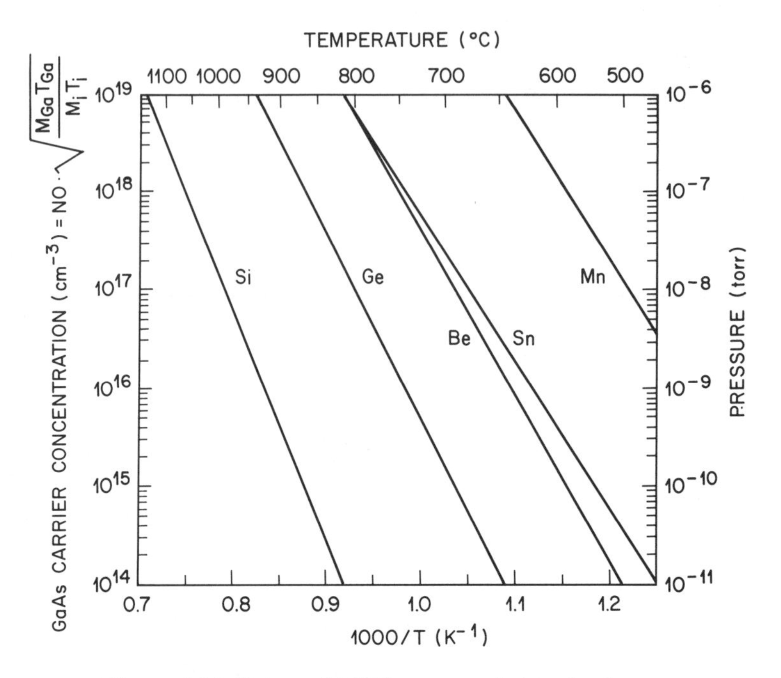


**Figure 6.31** Universal III-V compound chart for dopant concentration as a function of dopant effusion cell temperatures when the group III element arrival rate is $3 \times 10^{15}$ cm$^{-2}$ s$^{-1}$ [195].

stabilized growth conditions. Similarly, p-type material is obtained under Ga-stabilized growth conditions owing to Ge occupying As sites and acting as acceptors. At a constant flux ratio the concentration of Ge on As sites increases as substrate temperature increases simply due to the increased evaporation of As from the surface and thereby increasing $[V_{As}]$ [197]. Si in GaAs, though not a strong amphoteric dopant as Ge, also exhibits a similar tendency for site occupancy when temperature and As/Ga flux are varied [198]. The variation of the compensation ratio with substrate temperature can thus be related to the change in the ratio of $[V_{Ga}]$ and $[V_{As}]$ which can be obtained from defect chemistry principles outlined earlier. By adopting such an equilibrium approach Munoz-Yague and Baceiredo [199] have been able to show qualitative agreement with the experimental results. The different electrical behavior obtained for doping Si in GaAs of orientations other than (100) has also been explained based on the availability of the Ga or As site at the growing surface [200, 201].

### 6.6.5 Unintentional Impurities

The growth of a ''pure'' material, that is, one with as low a background carrier concentration as possible, is often a challenge common to all growth methods. The hindrance is caused by the incorporation of unintentional impurities which have a number of origins. Impurities can come from the various source materials including the substrate, the ambient gases, and the material used in the growth apparatus such as quartz and graphite. The solvents used in cleaning the substrate can be another source of impurities. The identification of their nature and especially their origin have indeed been a formidable task in the growth of III-V semiconductors.

The background doping in MBE grown GaAs is generally p-type which has been attributed to $C_{As}$ with the gases ($CH_4$, CO, $CO_2$) in the ambient perhaps being the source of carbon [202 - 205]. Mn has also been identified as another acceptor impurity coming from the hot stainless steel parts [202]. The high temperature heater used for the sources is found to be another major source of impurities in MBE films [206]. Employing a ''load-lock transfer system'' wherein samples are introduced without breaking the vacuum in the main chamber has greatly helped to improve the quality of the films.

A major residual impurity in halide VPE grown materials is Si which is introduced by the reaction of HCl with quartz to form chlorosilanes

$$n\,HCl(g) + SiO_2(s) + (4-n)H_2(g) \rightarrow SiCl_nH_{4-n}(g) + 2H_2O(g) \tag{6.50}$$

$$(n-2)H_2(g) + SiCl_nH_{4-n}(g) \rightarrow Si(s) + n\,HCl(g) \tag{6.51}$$

with $n=1$ to 4. It follows from these equations that the activity of Si is inversely proportional to the partial pressure of HCl.

Carbon is also a dominant residual impurity in MOCVD grown material [207 - 209]. The source of carbon is most likely to be the metal alkyls used in the growth process. These alkyls may also contain other impurities such as Si. Another source of Si is perhaps the SiC coated graphite susceptor. Zn has also been identified in many MOCVD grown GaAs films and is suspected to be introduced from the $AsH_3$ gas since $ZnAs_2$ is used as the starting material in the synthesis of high purity $AsH_3$ [207]. A judicious choice of growth temperature, group III/group V flux ratio, and reactor pressure is very important to achieve the highest purity in the grown films. A careful consideration of the details of the reactor design is also important besides the purity of the starting materials. The growth apparatus may also include previsions for repurification of the metal alkyls [210].

In melt growth and LPE growth also, the purity of the starting materials is an important factor limiting the purity of the product. Quartz and graphite parts used in the growth apparatus can act as sources for Si and C. Boron has been found as a contaminant in GaAs crystals grown by the Czochralski technique which uses $B_2O_3$ as an encapsulant. Sulphur has been identified as a residual donor, perhaps coming from the graphite boats, in LPE grown GaAs [211] film and AlGaAs [212, 213]. Prolonged baking of the melts helps to reduce the background doping levels [213] (see Sec. 2.2.6).

## 6.7 DISLOCATIONS

### 6.7.1 Types and Structures

#### Perfect Dislocations

As we have seen in Sec. 1.2, the close packed {111} planes of the fcc lattice become pairs of {111} planes in the zinc blende structure as shown in Fig. 6.32(a) or in a two dimensional projection on the {110} plane in Fig. 6.32b. Because of the presence of two type of atoms, in GaAs and InP, there is a polarity in stacking of the {111} planes. If ABCABC is the stacking order of {111} planes in the first sublattice and $\alpha\beta\gamma\alpha\beta\gamma$ in the second, then $\alpha B, \beta C, \gamma A$ are closely spaced pairs of planes connected by three times wider spaced $A\alpha, B\beta, C\gamma$ pairs of planes.

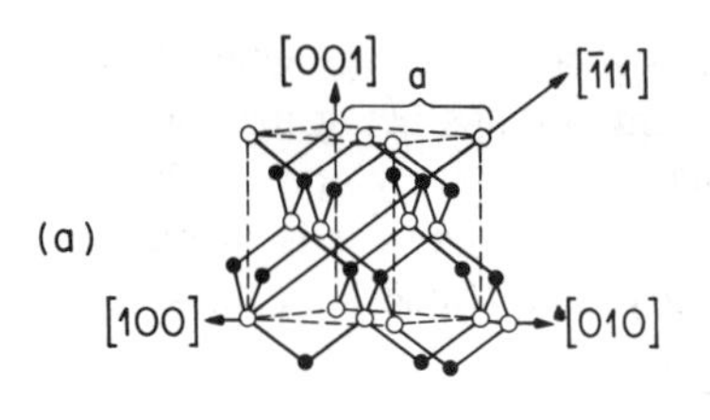


o-Ga
•-As

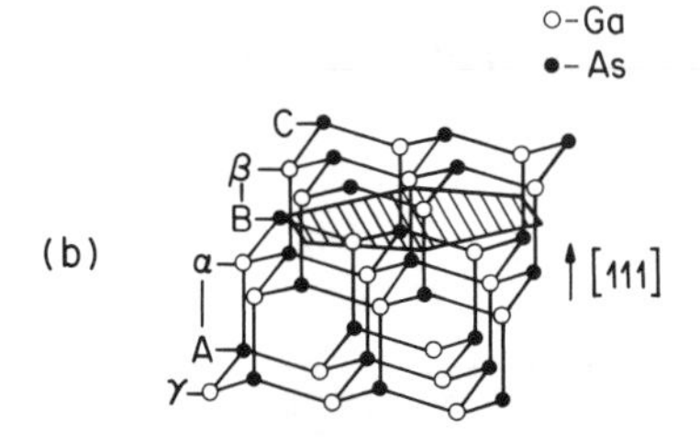


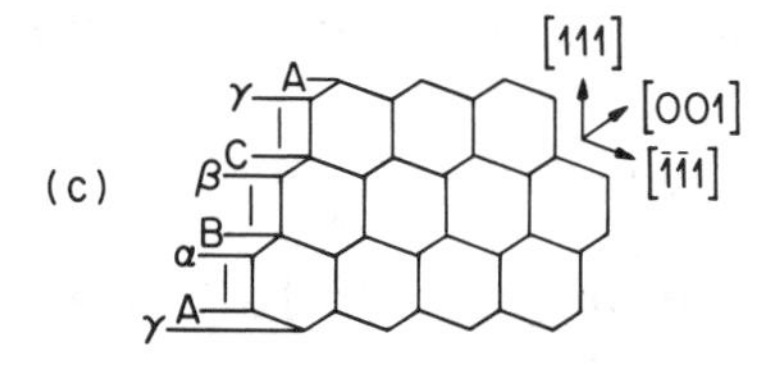


**Figure 6.32** The zinc blende unit cell (b) stacking of {111} planes (c) projection of (b) on the (110) plane [261].

Hornstra [214] first examined the possible structure of dislocations in crystals with diamond cubic structure and later Holt [215] extended the analysis to the zinc blende structure. The important types of dislocations are the 60-degree dislocation and the screw dislocation in the {111} plane. The creation of a 60-degree dislocation is illustrated in Fig. 6.33. For this dislocation the angle between the dislocation line and the Burgers vector is 60-degrees. The characteristics of perfect dislocations having Burgers vector a/2 [110] where a is the lattice constant are shown in Table 6.9.

Because of the lack of a center of symmetry in the zinc blende structure, Haasen first pointed out that there are two types of edge dislocations depending on whether the extra half-plane ends

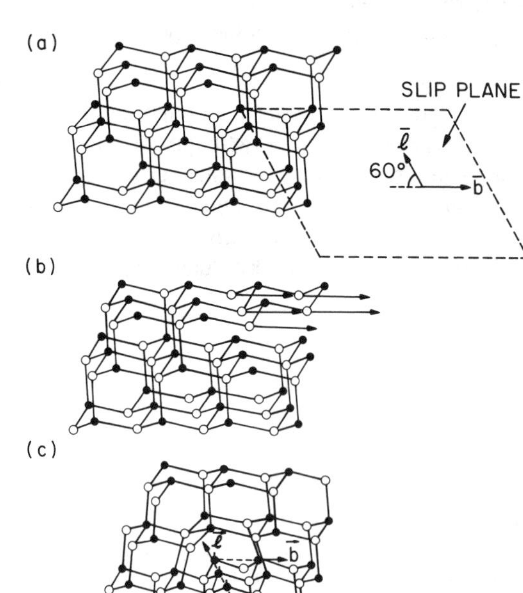


**Figure 6.33** (a) The GaAs structure with a [111] direction vertical. The dislocation line direction and the Burgers vector of a 60° dislocation are shown in a slip plane.
(b) The result of moving a 60° dislocation in the crystal from right to left is to produce the relative moments of the atoms indicated by the arrows.
(c) The 60° dislocation of the shuffle set [215].

**TABLE 6.9**

Types of dislocations having Burgers vector a/2 [110] in the zinc blende lattice [214]

| No. | Direction of dislocation axis | Angle between axis and Burgers vector | Glide Plane | Number of broken bonds per a cm† |
|---|---|---|---|---|
| 1 | <110> | 0° (screw) | — | 0 |
| 2 | <110> | 60° | {111} | 1.41 |
| 3 | <110> | 90° | {100} | 2.83 or 0 |
| 4 | <211> | 30° | {111} | 0.82 |
| 5 | <211> | 90° | {111} | 1.63 |
| 6 | <211> | 73°13′ | {311} | 2.45 or 0.82 |
| 7 | <211> | 54°44′ | {110} | 1.63 or 0 |
| 8 | <100> | 90° | {110} | 2.0 or 0 |
| 9 | <100> | 45° | {100} | 2.0 or 0 |

† For dislocations with {111} glide plane the number of broken bonds is proportional to the sin of the angle between the dislocation axis and Burgers vector [216]

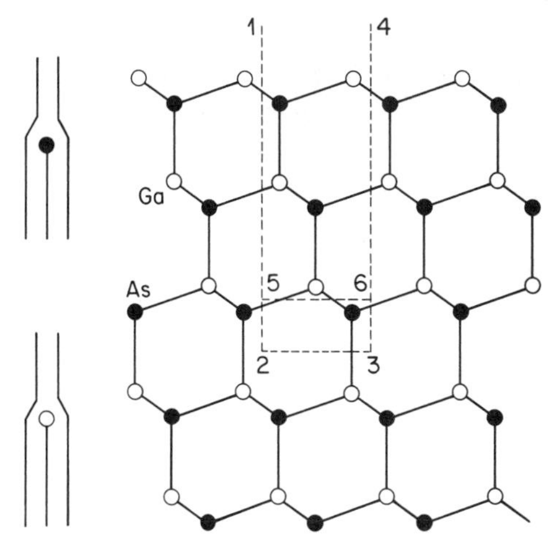


**Figure 6.34** The glide and shuffle set dislocation for the α-configuration in GaAs. The glide dislocation is formed by removing the extra half plane (1564) and the shuffle dislocation is formed by removing the extra half plane (1234) [218].

with a row of group III atoms or group V atoms [217]. However, the extra half-plane can terminate either between the Aα planes or between the αB planes, (Fig. 6.34). In the former case, if the dislocation ends on a row of group III atoms, in the latter it would end with a row of group V atoms. These two types of dislocations are referred to as the shuffle set (extra half-plane between Aα planes) and glide set (extra half-plane between αB planes) [218]. In a compound AB, the glide set is denoted as A(g) and B(g) and the shuffle set as α(or A(s)) and β(or B(s)) [219].

**Partial dislocations and stacking faults**

A stacking fault occurs in the zinc blende structure in the same way as it does in the fcc lattice. Since there are two {111} planes in the zinc blende lattice the formation of the stacking fault involves removal (intrinsic fault) or insertion (extrinsic fault) of a pair of planes to preserve the tetrahedral coordination. If the AαBβCγ stacking sequence of {111} planes is written simply as ABCABC, then extrinsic fault gives rise to ABC|B|ABC stacking and intrinsic fault to A|C ABC stacking. A low energy fault is formed only between the closely spaced αB planes (glide set) [218].

A stacking fault is a region bounded by two partial dislocations formed by the dissociation of a perfect dislocation. By the use of weak beam contrast analysis in the transmission electron microscope [220], it has now been established that the perfect a/2 [110] dislocations in III-V compounds are dissociated into Shockley partials [221 - 224] according to

$$\frac{a}{2}\,[0\bar{1}\bar{1}] \rightarrow \frac{a}{6}\,[121] + \frac{a}{6}\,[\bar{1}12] \tag{6.52}$$

Since the long-range strain energy associated with a dislocation is proportional to $|b|^2$ where b is the Burgers vector, it can be seen from Eq. (6.52) that dissociation is energetically favored. The repulsive elastic interaction between the two partials pushes them apart to form a stacking fault between them. Figure 6.35 shows a <110> projection of a 60-degree dislocation of the glide set dissociated into a 90-degree (edge) and a 30-degree [225,226] partial. As opposed to the

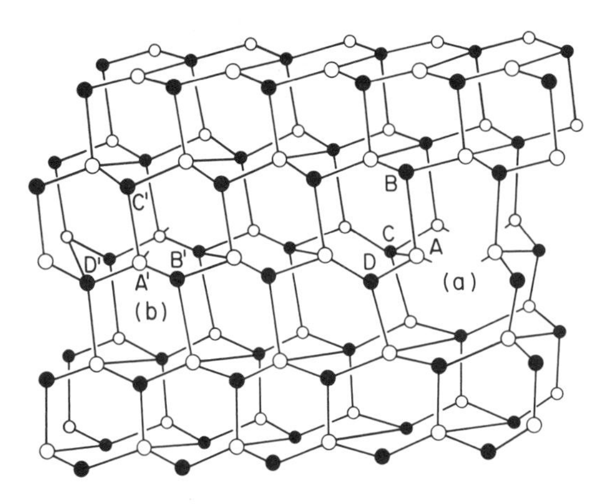


**Figure 6.35** Dissociation of a complete 60° Ga(g) (As(s)) dislocation into (a) one 90° and (b) one 30° glide partial separated by an intrinsic fault [241].

60-degree dislocation, the screw dislocation dissociates into two 30-degree partials and the edge dislocation into two 60-degree partials.

In GaAs [221] both A(g) and B(g) dislocations [227] introduced by bending the samples at 650°C, were found to be dissociated and to have the same separation suggesting that the different core structures of the two dislocations do not have any effect on the separation. Even in alloy semiconductors dissociation was observed [224]. From the separation of the partials, the intrinsic stacking fault energy in undoped n-type GaAs ($n \sim 2 \times 10^{16}$ $cm^{-3}$) was estimated to be 48±6 erg $cm^{-2}$ [221].

The observation that dislocations in semiconductors are dissociated raises doubt whether dislocations glide in the Aα planes (shuffle set) since dissociation in these planes produces a high energy fault. However, the motion of the αB dislocation needs breaking of three times more covalent bonds than does the motion of the Aα dislocation. The conflict between mobility and dissociation may be reconciled by a model proposed by Hornstra [214] for the dissociation of dislocation in the shuffle set. The idea is that the mobile Aα dislocation might associate itself on coming to rest with a pair of partials of opposite sign on the neighboring glide set planes. The corresponding reaction is

$$\frac{a}{2}[011] + \frac{a}{6}[1\bar{1}\bar{2}] + \frac{a}{6}[\bar{1}12]$$

$$= \frac{a}{6}[121] + \frac{a}{6}[\bar{1}12] \tag{6.53}$$

Note that the sum of the Burgers vectors of the perfect dislocation (a/2 [011]) and of the partial (a/6 $[1\bar{1}\bar{2}]$) bounding the stacking fault on the side of it is equal to the Burgers vector of the partial dislocation located on the same side in the case of splitting of a αB dislocation (Eq. (6.52)). Hirth and Lothe [218] proposed that a dislocation in the shuffle set with a stacking fault (Eq. 6.53) can be considered as a normal dissociated glide set dislocation with one of its partials having a row of either interstials or vacancies in its core. This row of point defects may be represented by a vertical dipole of two 60-degree dislocations with Burgers vectors a/2 $[0\bar{1}\bar{1}]$ and a/2 ~ [0 1 1] in αB and Aα planes, respectively. Then the following reaction can be considered:

B

α $\quad \frac{a}{2}[0\bar{1}\bar{1}] + \frac{a}{6}[121] + \frac{a}{6}[\bar{1}12]$

stacking fault

A $\quad \frac{a}{2}[011]$

---

B $\quad \frac{a}{6}[1\bar{1}\bar{2}] + \frac{a}{6}[\bar{1}12]$

≡ α

A $\quad \frac{a}{2}[011]$ (6.54)

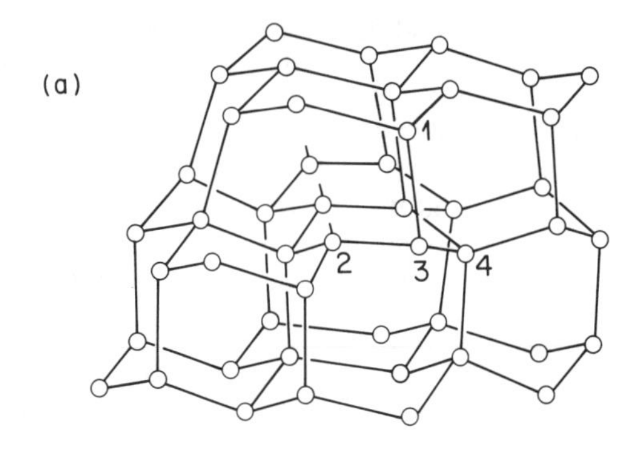


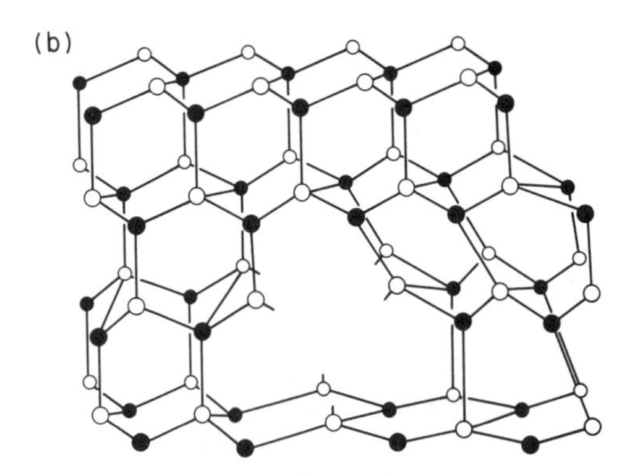


**Figure 6.36** Two possible geometries of the 30° Ga(s) partial in the shuffle set: the shuffle set partial is converted to a glide set partial with a row of interstitials (a) or with a row of vacancies (b) [218]. The motion of the partial in (a) involves the sheer of bonds between 1 and 3 to form bonds between 1 and 2, but in addition, it requires the rearrangement of the rows 3 and 4 by a shuffling process.

Note that the result is the same as in Eq. (6.53). One of the partials, that is, $a/6[\bar{1}12]$ is of the Shockley type while the other is of the complicated type a/6 [121] + $\{a/2[0\bar{1}\bar{1}]+a/2[011]\}$. The bracketed dislocation pair is a dislocation dipole equivalent to a row of point defects. The transformation of the a/6 [121] glide set partial into one in the shuffle set with a row of point defects is illustrated in Fig. 6.36 [218]. The Shockly partial glides rather easily while the other requires shuffling of the bonds (hence the name shuffle set) and therefore is less mobile. This model is attractive since it explains the source of point defects during plastic deformation [225].

The question of whether dislocations lie totally on the glide planes, or on the shuffle planes, taking up one of the configurations as proposed by Hornstra [214] or by Hirth and Lothe [218], has not yet been settled by any microscopic technique. However, great progress has been made with the use of high-resolution transmission electron microscope to determine the core structure of the dislocation and thus whether it belongs to the glide or shuffle set. These studies [228 - 231] involve the comparison of the experimental images with those calculated assuming a particular model structure and taking into account the aberrations of the electromagnetic lenses of the microscope. Figure 6.37 shows the high-resolution transmission electron microscope image of the core of a 30-degree partial dislocation in Si as well as the calculated images for 30-degree partial with broken bonds either between widely separated planes or between closely separated planes [229]. A comparison of Figs. 6.37(a) and 6.37(c) would suggest that the dislocation core

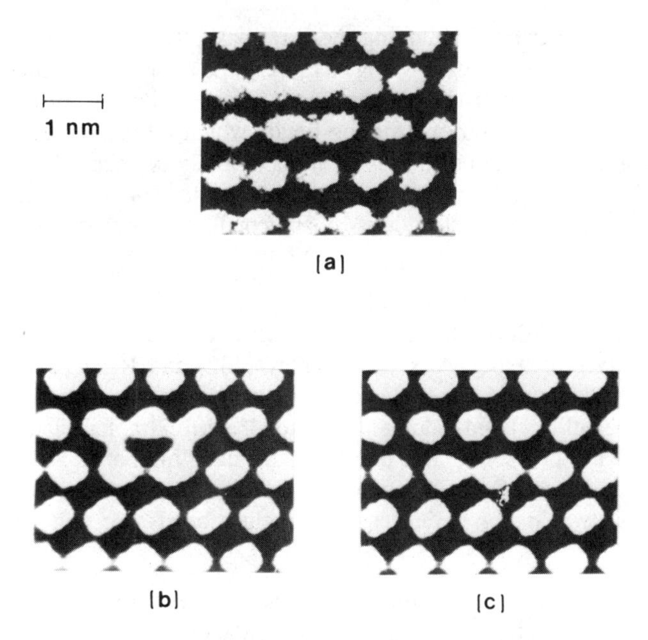


**Figure 6.37** (a) High resolution TEM image of the core of 30° partial dislocation in Si [229]. The dislocation is viewed with the electron beam parallel to the dislocation line direction. Each white dot corresponds to a pair of closely spaced atoms. (b) Computer simulated image of the 30° partial with broken bonds between widely separated (111) planes. (c) Computer simulated image of the 30° partial with broken bonds between closely separated (111) planes.

consists of dangling bonds between closely separated planes. However, it appears that agreement with the experimental image can still be obtained even if up to half of the dangling bonds lie between the widely separated planes [231]. Therefore, other arguments have to invoked [232] to decide whether dislocations are of the glide set or of the shuffle set.

Several studies [223, 225] seem to suggest that dissociated dislocations do not collapse during motion but remain dissociated. In other words, the need for association (Eq. (6.53)) as proposed by Hornstra [214] does not arise. The alternative model proposed by Hirth and Lothe (Eq. (6.54)) for the transformation of the shuffle set dislocation to glide set dislocation could explain the observation of EPR signal associated with point defects at the core of a 30-degree partial, the partial with the complex core structure in Eq. (6.54). But the small magnitude of the signal suggests that for most part the dislocation has simple glide character [233]. There is one more reason yet to favor the glide set dislocation. This is to do with bond reconstruction along the core of the partials. As illustrated in Fig. 6.34, a 60 degree shuffle dislocation is favored by removing the extra half plane 1234 and then joining the top of the crystal. This leaves dangling bonds in the Aα planes. On the other hand removing the extra half plane 1564 leaves dangling bonds in the αB planes. Since these are at shallow angles to the glide plane, reconstruction of the dangling bonds across and along the core can occur easily. At least in the case of the 30 degree partial, theory suggests that reconstruction of dangling bonds which minimizes the total energy (strain energy and dangling bond energy) of the dislocation, is likely to occur [234-236]. However, no direct experimental evidence is available to support this hypothesis. At present it appears that the most plausible interpretation for dislocations in semiconductors is that they are of the glide set configuration.

### 6.7.2 Electrical Properties of Dislocations

The electronic states of dislocations have been attributed to the dangling bonds associated with them. The early pioneering work of Shockley [237] and Read [216] on dislocations in diamond cubic structure showed that a 60-degree dislocation has a row of atoms, each with an unpaired electron (dangling bond) along the edge of the extra halfplane (Fig. 6.33(c). For dislocations in the {111} planes, the line density of the dangling bonds is equal to $2\ \mathrm{Sin}\theta / \sqrt{3}\,b$, where $\theta$ is the angle between the dislocation line and the Burgers vector b. Accordingly, a 60-degree dislocation shown in Fig. 6.33, has 1/b dangling bonds per unit length while a screw dislocation has none.

The unstaturated dangling bond along the dislocation core can either accept another electron to satisfy the bond or donate its electron. In this sense the dislocation can act as an acceptor or a donor. Because of the translational symmetry along the dislocation, the dangling electrons are expected to form an one-dimensional band with a bandwidth roughy of the same order of magnitude as that of the conduction and valence band. The band is half filled with one electron per dangling bond (because the spin of each electron can have one of two values) and the dislocation is neutral [238].

The one dimensional band model for the dislocation would lead to the expectation that an enhanced conductivity along the dislocation core could occur at low temperature. Such an expectation has not been verified. On the other hand, a Peierls transition at low temperatures (as in other one-dimensional systems [239]) when the dislocation is half filled could explain the absence of metallic conductivity along the dislocation core. In any event, the original treatment of dislocations in Ge as a row of acceptor centers closely spaced but having a single level in the energy gap [216] has long been abondoned in favor of the one-dimensional band model.

The total energy of a dislocation may be written as the sum of various terms [240]

$$E_{tot} = E_{ei} + E_{ii} + E_{ee} + E_{ex} \tag{6.55}$$

where, $E_{ei}$, $E_{ii}$ and $E_{ee}$ are the electron-ion, ion-ion and electron-electron interaction energies, respectively, and $E_{ex}$ represents quantum-mechanical effects such as exchange and correlation. The calculation of energy levels of the dislocation would involve minimizing $E_{tot}$. Since the atomic configuration at the core of the dislocation is not experimentally known, the calculations start with atomic coordinates first deduced from isotropic elasticity theory and then modified by letting the atoms around the core to relax in response to a valence potential. Also arbitrary displacements of the atoms having the broken bond (A and A′ in Fig. 6.35) were allowed so as to

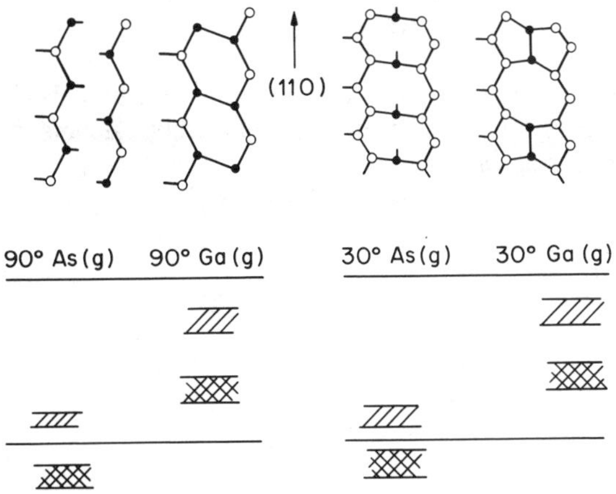

**Figure 6.38** Schematic diagram showing the energy levels of 90° and 30° partials of the As and Ga glide set dislocations [241]. The upper trace shows the unreconstructed and reconstructed 90° and 30° partials [240]. Empty level is shown hatched and the full level is shown double-hatched.

decrease ("in") or increase ("out") the height of these atoms above the planes BCD and B′C′D′, respectively [241]. The calculations neglect $E_{ex}$ and $E_{ee}$ terms and in this sense thay are semiempirical single-electron calculations.

Jones et al., [241] have calculated the energy levels of the 30- and 90-degrees glide partials of a 60-degree dislocation in GaAs. However, their results can also be applied to other dislocations since all partial dislocations can be considered to be made up of alternate segments of 90- and 30-degree partials [226]. According to Jones et al., [241] arsenic dislocations introduce bands into the lower half of the band gap and gallium dislocations into the upper half. Also introducing alternate ''in'' and ''out'' type relaxation along each dislocation resulted in splitting of these bands into a filled and an empty band. The latter result is expected from the Peierls instability of partially filled one-dimensional band against lattice distortions [239]. The energy bands of 90- and 30-degree partials thus obtained are schematically illustrated in Fig. 6.38 [241]. Table 6.10 gives the energy values obtained for two sets of parameters used in the solution of the Hamiltonian and in turn they represent extreme points of view.

On experimental side, information about the electronic states of dislocations in GaAs and InP are very scarce. The electrical properties of dislocations are generally deduced from measurements made on deformed samples. From Hall measurements on Cr-doped high resistivity GaAs bent at 580°C to produce an excess of As(g) dislocations [227], Lin and Bube [242] observed an increase in the electron concentration after deformation. On the other hand, in n-type GaAs ($n \sim 10^{16} cm^{-3}$) both carrier concentration and mobility decrease after deforming the crystals either by bending [243 - 245] or by uniaxial compression [246]. A similar effect was observed in p-type GaAs though not as pronounced as in n-type GaAs [244]. When deformation

**TABLE 6.10**

Energy bands of the alternately "in" and "out" reconstructed partials. Energies in eV are measured from the top of the valence band [241]. $E_o$ and $E_1$ denote the positions of the filled and empty bands, respectively.

| Type | Parameter I | Parameter II |
|---|---|---|
| 90° As(g) | $-0.35 < E_0 < -0.05$ | $-0.65 < E_0 < 0.05$ |
| | $-0.05 < E_1 < 0.05$ | $0.3 < E_1 < 0.7$ |
| 90° Ga(g) | $-0.20 < E_0 < 0.60$ | $0.95 < E_0 < 1.25$ |
| | $1.05 < E_1 < 1.35$ | $1.55 < E_1 < 2.05$ |
| 30° As(g) | $-0.20 < E_0 < 0.0$ | $0.05 < E_0 < 0.35$ |
| | $0.15 < E_1 < 0.25$ | $0.85 < E_1 < 1.15$ |
| 30° Ga(g) | $0.75 < E_0 < 0.9$ | $> E_c$ |
| | $1.4 < E_1 < 1.6$ | |

is produced by bending, for either direction of bending so as to produce an excess of either Ga or As dislocations [247], qualitatively similar results were obtained [244]. However, a larger decrease in carrier concentration (both n and p) was observed when the GaAs crystals were bent to produce an excess of Ga dislocations than when they were bent to produce an excess of As dislocations.

The results of the bending experiments suggested that in n-type GaAs dislocations show acceptor action while in p-type GaAs they show donor behavior. In light of the electronic states of the As(g) and Ga(g) dislocations derived by Jones et al., [241], the experimental results can be explained for p-type GaAs but not for n-type GaAs. As can be seen from Fig. 6.38, the Ga(g) dislocation has its filled band near the middle of the energy gap and hence can compensate acceptors in p-type material. On the other hand, As(g) dislocation with its filled band in resonance with the valence band cannot compensate acceptors and hence will be less effective in reducing hole concentration, consistent with the experimental observations. For n-type GaAs,

As(g) dislocation with its empty band near the valence band can compensate donors more effectively than Ga dislocations which has empty bands near the conduction band. This is, however, contrary to the experimental results.

The position of the energy levels of the dislocation has not been clearly established experimentally. From analyzing spectral response and decay process of photoconductivity in bent undoped n-type GaAs ($n \sim 10^{16} cm^{-3}$), Nakata and Ninomiya [248] deduced two levels, at 0.7 and 0.8 eV below the conduction band edge, and attributed them to the partial dislocations of As(s). Since dislocations taking part in deformation are of the glide set, one should really assign these levels to Ga(g). Then these levels are consistent with that derived by Jones et al., [241] for the Ga(g) partials (Table 6.10). Ishida et al., [249] studied deformation produced deep levels in deformed GaAs by using deep level transient spectroscopy. They found that the concentrations of two grown-in electron trap levels ($E_c - 0.65$ eV and $E_c - 0.74$ eV) and one grown-in hole trap level ($E_v + \sim 0.4$ eV) increased with plastic deformation, while that of a grown-in electron trap level ($E_c - \sim 0.3$ eV) first decreased with strain and after 4 percent compressive strain increased with strain. The electron trap at $E_c - 0.7$ eV and the hole trap at $E_v + 0.4$ eV have been associated with EL2 and level A. Ishida et al., [249] tentatively identified a broad hole trap spectrum consisting of several peaks corresponding to energy levels near $E_v + 0.46$ eV with dislocations. However, Kadota and Chino [245] found no trap levels that could be assigned to dislocations in bent epitaxial GaAs films. They also suggested that the hole trap seen by Ishida et al. [249] could arise because of Cu and/or Fe contamination.

Although the DLTS measurements failed to identify dislocation levels, they revealed an important aspect of plastic deformation namely that the concentration of point defects could increase on deformation. This is a consequence of dragging jogs on screw dislocations formed by dislocation intersection or cross slip. What this means is that some of the electrical effects in deformed GaAs could very well be caused by point defects. Also at the temperature of deformation, impurities in GaAs could diffuse and precipitate near dislocations and thus changes in resistivity would merely be a result of this gettering near dislocations. In order to isolate the effects solely caused by dislocations from that by point defects, one has to measure the electrical properties after careful post-deformation annealing experiments. The point defects usually anneal at moderate temperatures while relatively high temperatures are required to alter considerably the dislocation structure. As a result, the electrical effect which persists after the point defects anneal out may then be attributed somewhat unambiguously to dislocations. Such careful experiments are wanting in GaAs and InP.

### 6.7.3 Mechanical Properties and Impurity Hardening

#### Glide system and critical resolved shear stress

The mechanical properties of GaAs and InP have been studied by several authors by using either compression testing or micro hardness indentation technique. The former is used exclusively in the case of bulk crystals while the latter is used for epitaxial films as well as bulk crystals. Like other fcc structures, GaAs [250, 251] and InP [252 - 255] have {111} <110> slip system as their primary slip system. But these semiconductors are brittle at room temperature and become increasingly ductile at temperatures above $\geq 0.5 T_M$ where $T_M$ is the melting point in degrees Kelvin. In this temperature range, the plasticity of semiconductors is similar to that of

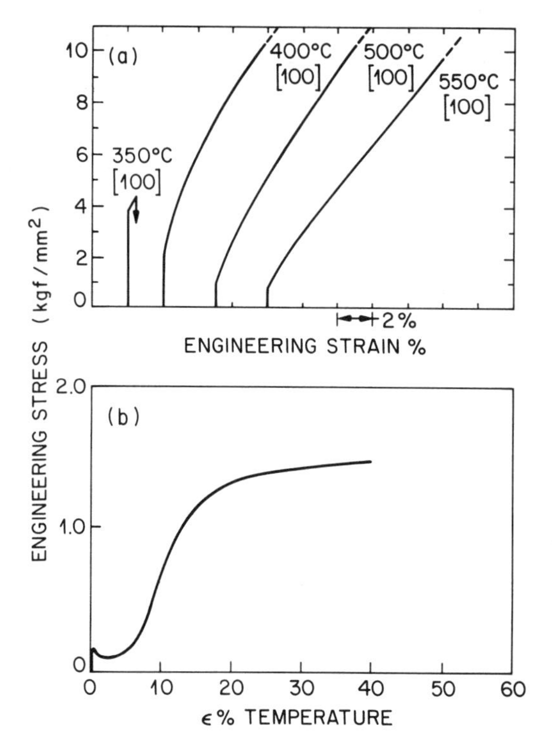


**Figure 6.39** (a) Stress-strain curves for Si-doped GaAs (n ~$1.8 \times 10^{18}$ $cm^{-3}$) single crystals compressed with [100] stress axis at temperatures indicated. The specimens were loaded at 0.14 MPa secref–1 [251]. The arrow indicates fracture. (b) Stress-strain curve for an undoped (n ~$3 \times 10^{16}$ $cm^{-3}$) GaAs deformed in tension with [100] stress axis at 850°C. The specimen was loaded at a constant strain rate $3.8 \times 10^{-3}$ $sec^{-1}$ [250]. 1kgf $mm^{-2}$ = 9.8 MPa.

fcc metals [256-258]. The lowest temperature at which plastic deformation under uniaxial loading conditions was observed is 250°C (0.2 $T_M$) in the case of GaAs and is 480°C (0.56 $T_M$) in the case of InP. Below these temperatures specimens failed by fracture which propagated along the {110} planes [251, 252]. In InP, plastic deformation occurred by a combination of slip and twinning in the temperature range 0.56 $T_M$ to 0.75 $T_M$ [252] and only by slip at higher temperatures. No twinning was observed in the case of GaAs, however, [250, 251] for high temperature deformation.

Figure 6.39 shows the stress strain curves for Si doped n-type (n~$1.8 \times 10^{18} cm^{-3}$) GaAs single crystals compressed with [100] stress axis at constant force rate (Fig. 6.39(a)) in the temperature range 350 to 550°C [251]. Also shown in Fig. 6.39(b) is the stress strain curve for an undoped GaAs crystal tested in tension at a constant displacement rate at 850°C [250]. For the first few percent of strain, a constant force rate or a constant displacement rate can be considered as a constant stress rate or a constant strain rate. Stress-strain curves for S-doped n-type (n~$1.4 \times 10^{18} cm^{-3}$) InP single crystals compressed at constant strain rate in the temperature range 475 to 675°C are shown in Fig. 6.40 [253]. The InP crystals were oriented such that the compression axis was parallel to [$32\overline{1}$] and the side faces were parallel to [111] and [$\overline{1}45$] (Fig. 6.41). The [$32\overline{1}$] stress axis facilitates the operation of only one of the possible {111} <110> slip systems in the initial stages of deformation. Also the specimens had an aspect ratio of 2:1 to ensure a purely uniaxial state of stress in a significant volume of the material [251].

The stress-strain curve measured under a constant strain rate exhibits typically three stages of deformation: stage I is characterized by a yield drop and followed by an easy glide region; stage II

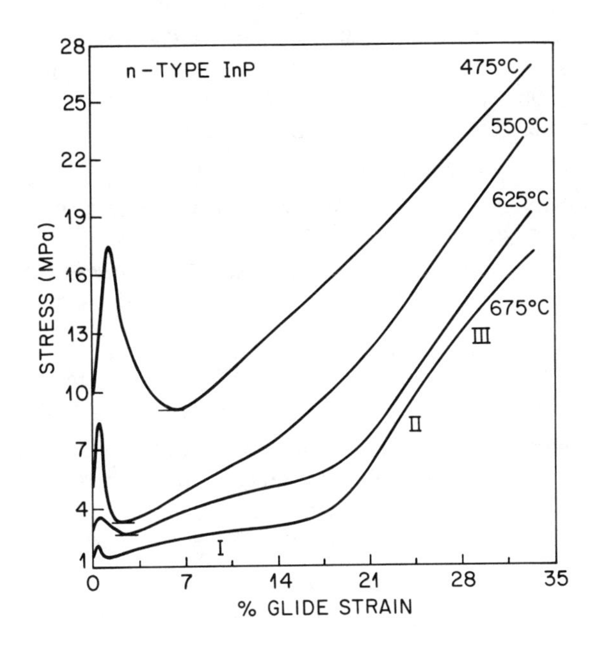


**Figure 6.40** Resolved shear stress-glide strain curves of S-doped InP ($n \sim 1.4 \times 10^{18}\ cm^{-3}$) at temperatures indicated. The compression axis was [123] and the loading was at a constant strain rate $6.7 \times 10^{-5}\ sec^{-1}$. The stress-strain curves exhibit pronounced yield behavior. The three stages of deformation I, II and III denote easy glide, work hardening and recovery, respectively [253].

**Figure 6.41** Crystallographic orientation of the specimens used for compression testing. With [321] stress axis only one of the possible {111} <110> slip systems operates. Therefore, deformation of the crystals occurs by single slip initially [255].

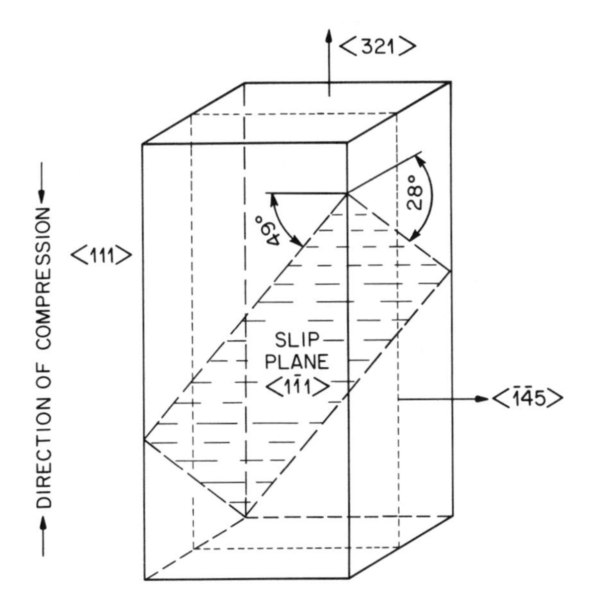

represents a hardening stage where the slope of the stress-strain curve is constant; stage III characterizes a recovery stage. The three stages are marked in the strain-strain curves from InP crystals Fig. 6.40. At high degree of deformation and at low strain rates, a second hardening stage (stage IV) and a second recovery stage have been seen in Si, Ge, and InSb [256, 257] and can be expected in InP and GaAs as well.

A mechanical parameter of great interest that is obtained from the stress-strain measurement is the critical resolved shear stress (CRSS). The CRSS corresponds to the minimum shear stress at which a glide system begins to operate [259]. In other words, it is the stress that is needed to initiate dislocation motion in the crystals. For a given stress axis, the resolved shear stress on a glide plane is

$$\sigma_{CRSS} = \frac{P}{A_0} \cos\phi \cos\lambda \tag{6.56}$$

where P is the applied load, $A_0$ is the cross-sectional area of the specimen, $\phi$ is the angle between the normal to the glide plane and the stress axis and $\lambda$ is the angle between the glide direction and the stress axis. The $\cos\phi \cos\lambda$ term is called the Schmid factor. When there are several crystallographically equivalent glide systems, the one with the maximum CRSS will operate first.

For Si-doped GaAs crystals, Swaminathan and Copley [251] obtained the CRSS values by multiplying the yield stress (taken as the stress corresponding to 0.2% strain in the stress-strain curves shown in Fig. 6.39a) by the appropriate Schmid factors (Eq. 6.56). When the stress-strain curves are measured under constant strain rate, Muller et al., [255] have shown that the onset of dislocation formation occurs at a stress level 10 to 20 percent lower than the lower yield point, $\sigma_{ly}$. Further, the upper yield point, $\sigma_{uy}$, is influenced much more by the initial dislocation density, being smaller for higher dislocation density, than $\sigma_{ly}$. For these reasons $\sigma_{ly}$ appears to be the proper value to determine CRSS. Figure 6.42 shows CRSS for GaAs as a function of

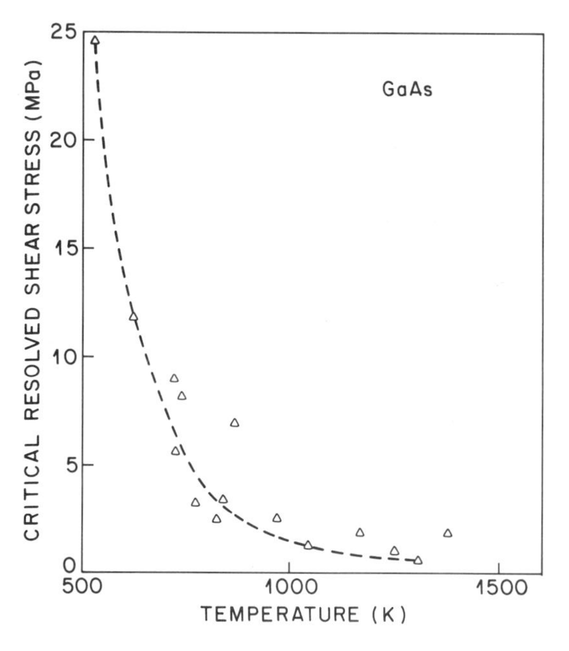


**Figure 6.42** Critical resolved shear stress versus temperature for GaAs single crystals [260].

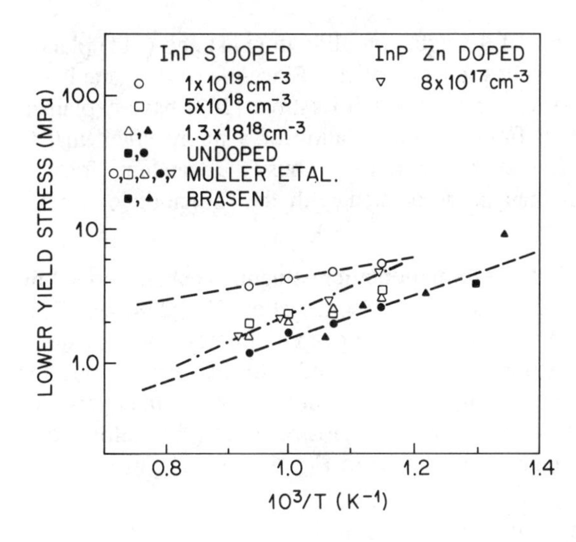


**Figure 6.43** Temperature dependence of lower yield stress (equated to CRSS) for S-doped, Zn-doped and undoped InP single crystals [255].

absolute temperature [260]. Figure 6.43 shows the CRSS values normalized to a strain rate of $10^{-4}$ $sec^{-1}$ for undoped InP [253] and S-doped InP [253, 255] for different doping levels. For data which are taken at different strain rates the normalization to one strain rate is done using the following relation between $\sigma_{ly}$ and $\dot{\varepsilon}$ [261].

$$\sigma_{ly} = C \left( \dot{\varepsilon} \exp \frac{E}{kT} \right)^{\frac{1}{m+2}} \tag{6.57}$$

where E is the activation energy for dislocation motion, C is a constant and the exponent m characterizes the stress dependence of dislocation velocity according to

$$v \propto \sigma^m \exp \frac{-E}{kT} \tag{6.58}$$

The value of m is close to unity in most cases [261, 262].

### Deformation micro structure

The distribution and nature of dislocations in deformed GaAs and InP have been studied using transmission electron microscopy after various stages of deformation. The dislocations in the deformed sample have Burgers vectors of the type a/2 <110> and most of them are of the 60-degree type with their axis lying along <110> direction. Pure edge dislocations with <112> axis have also been seen. In some cases Lomer dislocations with {100} slip plane have also been observed, especially in samples tested up to or beyond stage II [253, 258]. The Lomer dislocations are formed most likely as a result of the reaction of two glissile dislocations on different {111} planes according to

$$\frac{a}{2}[101] + \frac{a}{2}[01\bar{1}] \rightarrow \frac{a}{2}[110] \tag{6.59}$$

where the two dislocations on the left hand side of the equation glide in $(11\bar{1})$ and (111) planes, respectively, and the resultant Lomer dislocation lies in (001) plane. Since the (001) plane is not a glide plane, the Lomer dislocations are sessile. Abrahams and Ekstrom [263] have explained cleavage of III-V semiconductors as resulting from the propagation of a micro crack that is formed because of the coalescence of Lomer dislocations. Thus the random fracture characteristics of the crystals at high strains can be associated with the formation of Lomer dislocations.

For small amounts of strain, during stage I, the deformation microstructure is characterized by the pressure of long straight dislocations aligned along <110> direction [250,258] which is presumably the direction of low Peierls energy. With increasing degree of deformation (stage II and beyond) a highly disordered dislocation structure, which is similar to that in deformed fcc metals, is developed. Dislocation walls which lie along <110> directions form and the interior of the cell structure becomes relatively dislocation free. Such a microstructure in InP deformed at 475°C to approximately 34 percent strain (stage II) is shown in Fig. 6.44 [253]. In stage III,

**Figure 6.44** TEM micrograph showing the deformation microstructure in S-doped InP deformed at 475°C to ~34% strain (stage II). The planes of the micrographs are approximately (111). The operating reflections are (a) $(20\bar{2})$, (b) $(0\bar{2}2)$ and (c) $(2\bar{2}0)$. The scale bar represents 1.0μm [253].

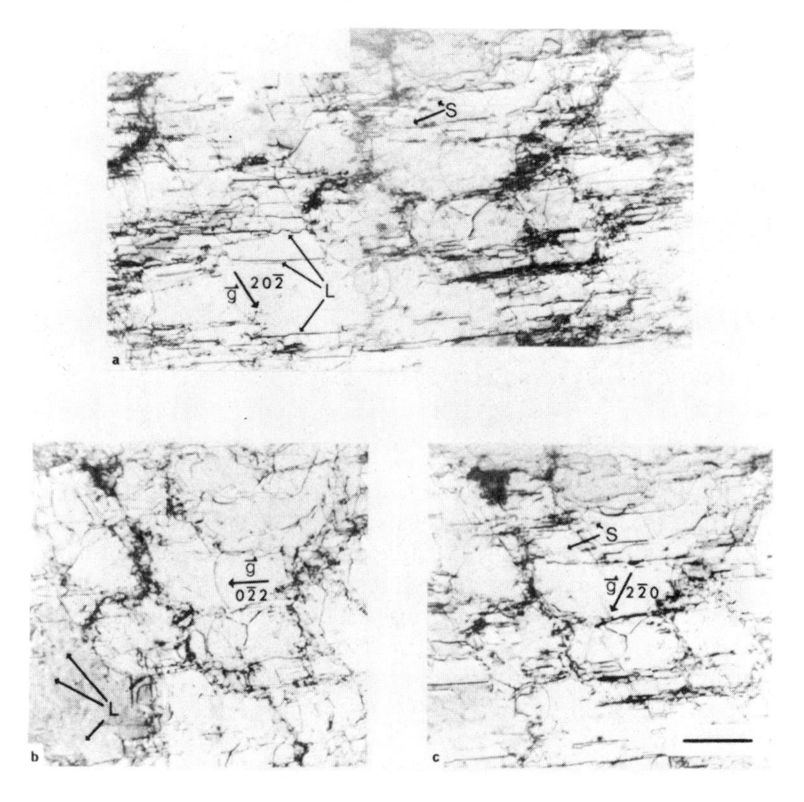

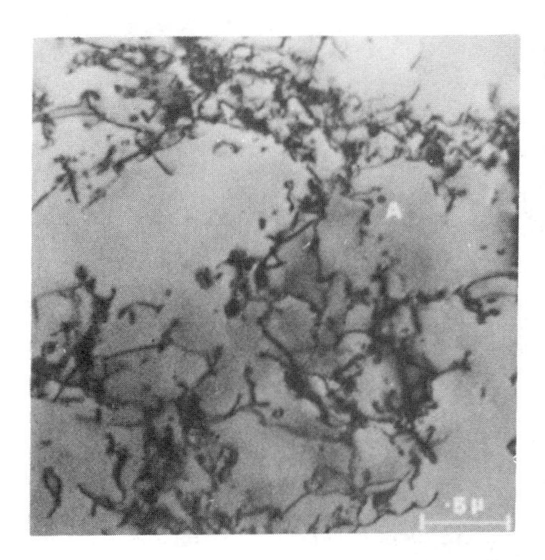


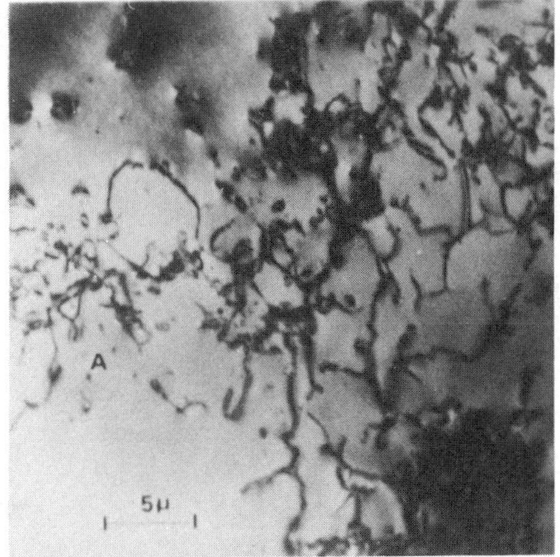


**Figure 6.45** TEM micrograph showing the dislocation substructures of Si-doped GaAs deformed at 550°C to 20% strain [258]. Dislocation subboundary formation is quite pronounced.

subboundary formation becomes quite pronounced as illustrated in the microstructure in Fig. 6.45 of Si-doped GaAs deformed at 550°C to 20 percent strain.

**Doping effect and impurity hardening**

The addition of dopants has a profound effect on the dislocation velocity and thus on $\sigma_{CRSS}$ in semiconductors. In this respect, the compound semiconductors differ from elemental semiconductors such as Si and Ge. In Si and Ge, addition of donors softens the lattice while that of acceptors hardens the lattice [264]. On the other hand, the effect of dopants on the mechanical strength of compounds is quite the opposite. Several investigations have shown that in GaAs the addition of donors such as Si or Te to give a carrier concentration $\gtrsim 10^{18} cm^{-3}$, increases the yield stress while the addition of acceptors such as Zn to give a hole concentration $10^{18} cm^{-3}$, decreases it [250,251,265]. The hardening caused by donors is generally more pronounced than the softening effect caused by acceptors as shown in Fig. 6.46 [251]. In fact, at high acceptor concentrations ($> 10^{18} cm^{-3}$) a certain hardening of the material was observed, though still less than that produced by donors [266, 267]. The mechanical hardening and softening produced by donors and acceptors, respectively, have also been verified in epitaxial GaAs and AlGaAs films via microhardness measurements [268].

In the case of InP, addition of S (donor) to give electron concentrations of $10^{18} - 10^{19}$ $cm^{-3}$, increases the yield stress compared to the undoped crystals (Fig. 6.43). Brown et al., [254] found for Ge which acts as a donor in InP that at doping levels corresponding to carrier concentrations of $10^{17} cm^{-3}$ the crystals are weaker compared to undoped ($n \sim 4 \times 10^{15}$ $cm^{-3}$) crystals. However, at high Ge doping levels ($n \sim 10^{19}$ $cm^{-3}$) $\sigma_{uy}$ was temperature independent over the range 823 to

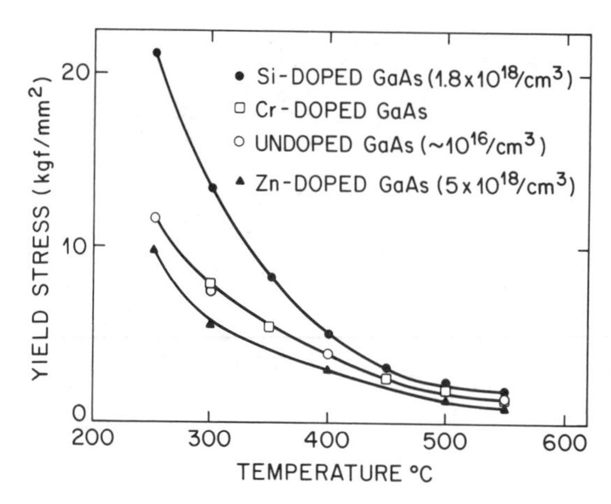


**Figure 6.46** Temperature dependence of yield stress obtained from repeated yielding experiments on single crystals of undoped and doped GaAs [251]. 1kgf mm$^{-2}$ = 9.8 MPa.

923K unlike undoped and low doped crystals where $\sigma_{uy}$ decreased with increasing temperature. At temperatures above 600°C highly doped crystals had high $\sigma_{uy}$. The improved mechanical strength in turn produced low dislocation density in Ge-doped crystals [269] (Sec. 6.7.4).

Zinc which acts as an acceptor in InP produces a hardening effect even at low doping levels ($<10^{18}$ cm$^{-3}$) unlike in GaAs as illustrated in Fig. 6.43. Further, Zn-doped crystals show a stronger dependence of $\sigma_{CRSS}$ on temperature than the S-doped crystals.

In the case of GaAs, the doping effect on yield stress is also reflected in dislocation velocity measurements [267, 270]. The velocity is measured by double etch technique in specimens which were stressed by three point bending in the temperature range 150 to 500°C [270]. Based on geometrical consideration on expanding dislocation loops, the dislocation type was determined and the velocities of As(g), Ga(g) or screw dislocations were determined at resolved stresses 0.2 to 10 kg/mm$^2$. The velocities and activation energies of As(g) and Ga(g) dislocations are shown as a function of carrier concentration in Fig. 6.47 [267]. Screw and Ga(g) dislocations were found to have roughly equal velocities in n-type material. The relation between velocity and stress as given in Eq. (6.58) was found to hold in a narrow range of stress (1–5 kg/mm$^2$) at a given temperature and the value of m was found to be ~1.5 [270]. The activation energy was found to be nearly independent of stress at stresses greater than 2 kg/mm$^2$. Note in Fig. 6.47 that the velocity of Ga(g) dislocations depend strongly on donor concentration, decreasing with increasing donor concentration.

The doping effect on yield stress can also be understood from the velocity data in Fig. 6.47 with the assumption that the slowest dislocations determine the yield stress. First, in p-type GaAs dislocations move faster than the slowest Ga(g) dislocations in n-type undoped and doped GaAs. Therefore, the yield stress of p-type material would be smaller than that of undoped or n-type material, as observed (Fig. 6.46). Second, at high doping levels ($>3 \times 10^{18}$ cm$^{-3}$) in p-type GaAs the velocity decrease rather suddenly which could explain the slight hardening observed in heavily Zn-doped GaAs [266].

Detailed dislocation velocity measurements in InP as a function of doping have not been done to confirm whether in both Zn- and S-doped crystals dislocation move slower compared to undoped crystals so as to be consistent with the yield stress data shown in Fig. 6.43. A

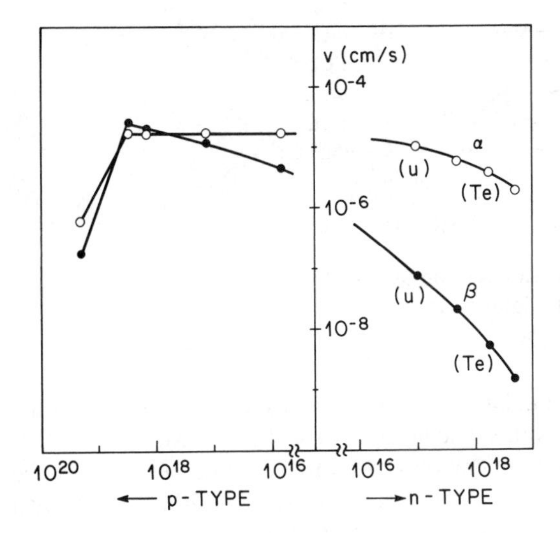


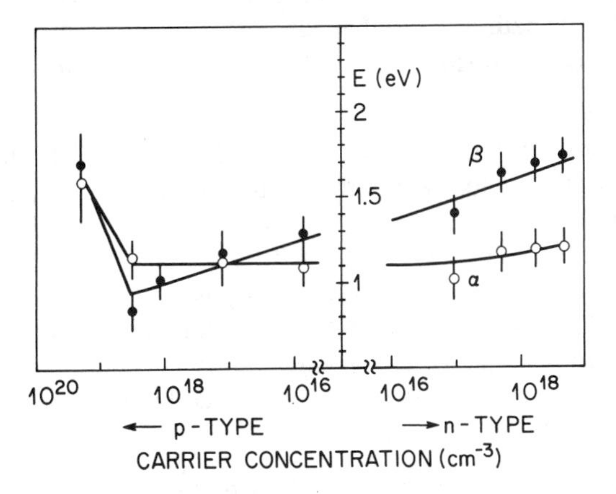


**Figure 6.47** Velocities (a) and activation energies (b) of α and β dislocations in GaAs as a function of carrier concentration at 300°C and $\tau = 10$ MPa. U-undoped, Te-doped. Other n-type samples are Si-doped and p-type samples are Zn-doped [267].

preliminary study by Nagai [271] indicated that P(g) dislocations in undoped InP have a similar velocity to As(g) dislocations in undoped GaAs. But surprisingly, in S-doped crystals they had a higher velocity compared to the undoped crystals. More experiments are needed to confirm this result. Further, the velocity of In(g) dislocations needs to be determined. If In(g) dislocations behave the same as Ga(g) dislocations in GaAs, they would be the slowest and the yield stress would be affected by them.

The motion of dislocations in semiconductors involves the formation and migration of double kinks on the dislocation [272, 273]. At 0K the equilibrium configuration of the dislocation is that it is straight and lies at the bottom of the Peierls valley. However, at finite temperatures, a number of randomly positioned double kinks are formed such that a part of the dislocation extends from one Peierls valley into the next as in Fig. 6.48. Under the action of an applied stress the positive and negative kinks move in opposite direction and are annihilated with the result that the whole dislocation moves to the next Peierls valley. The earlier theories of dislocation mobility considered perfect dislocations and calculated kink velocities assuming that the moving

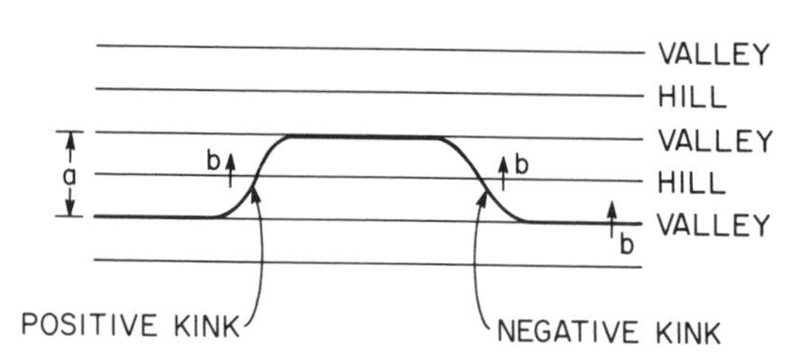


**Figure 6.48** Dislocation motion by double kink formation and migration from one Peierls valley to another.

kinks have to overcome barriers presented by randomly spaced obstacles, the exact nature of the obstacles being unspecified. Since dislocations in semiconductors are dissociated, Muller [274] analyzed dislocation motion by considering the generation and motion of a double kink on the two partials of the dissociated dislocation. It appears that at high stresses ($\sigma > 20\ \text{N/mm}^2$) when the nucleation of the double kinks on the two partials is uncorrelated, the velocity may be described by Eq. (6.58). At low stresses the deviation from the equation was qualitatively explained by the existence of "weak" obstacles on the dislocation line which must be overcome by thermal activation by the moving kinks.

Given that the motion of dislocations occurs via double kinks, several theories have been proposed to explain the doping effect on dislocation velocity in terms of the dependence of the formation and/or migration of double kinks on the Fermi level. A review of the various theories has been given by Hirsch [232, 275] and Jones [276]. Frisch and Patel [277] proposed a phenomenological theory in which they assumed that the dislocation velocity is proportional to the line charge on the dislocation. Haasen [278] proposed that the kink formation energy decreases with increasing charge on the dislocation because of the higher electrostatic energy of a straight than of a kinked dislocation. The charge on the dislocation is determined by the relative positions of the Fermi level and the dislocation levels.

Hirsch [232, 275] proposed that kinks with dangling bonds, formed on partial dislocations, have deep donor and acceptor levels. Thus kinks can be neutral or charged depending on the relative postion of the Fermi level and the kink energy levels. The change in the Fermi level brought about by doping can in turn affect the concentration and/or the migration energy of charged kinks and thus the velocity of the dislocation. The detailed theory proceeds along the lines developed by Hirth and Lothe [279] for uncharged kinks. Jones [280] assumed that the kinks are reconstructed and hence have no dangling bonds. This led to the suggestion that the migration energy is significantly lowered with doping.

The different theories of doping effect on dislocations have been applied to elemental semiconductors with limited success. The presence of two dislocations with different core structures introduces further complexities and uncertainties into the theories when applied to compound semiconductors. Besides the purely electronic effects of doping on the double kink formation and/or migration energy for the hardening observed in n-type GaAs, Swaminathan and Copley [281] proposed that interactions of moving dislocations with static defect complexes are an important contribution as well. To explain the magnitude of hardening, they proposed that the defect complexes were solute-vacancy pairs similar to those that are known to strengthen the alkali halide crystals [282]. Specifically, in Si-doped GaAs the solute-vacancy pair was supposed to be the ($Si_{Ga} - V_{Ga}$) pair. The elastic distortion of such a pair is expected to have deviatoric as well as hydrostatic parts and thus both edge and screw components of the glide dislocations will interact elastically with the pair. Also, in GaAs, $Si_{Ga}$ is a single donor and $V_{Ga}$ is expected to have an acceptor like behavior (Sec. 6.2). Thus, there may be a short-range electrostatic interaction between pairs and dislocations associated with the displacement of the oppositely charged defects relative to each other by the shearing action of the gliding dislocations. An interaction of this type has been proposed by Gilman to explain impurity hardening in ionic crystals [283]. The relationship between resolved shear stress and impurity-vacancy pairs was further established by the work of Chen and Spitzer [284] in Si-doped GaAs. Based on changes induced by various thermal treatments on carrier density, defect-induced infrarred localized vibrational mode absorption, microstructural characteristics as determined by transmission electron microscopy and the resolved shear stress, they showed that the resolved shear stress varies as the $1/4^{th}$ power of the concentration of the ($Si_{Ga}$–$V_{Ga}$) pairs.

Although the solute-vacancy pair model is shown to be applicable for the hardening, specifically in Si-doped GaAs, it can be extended to other donors and perhaps to other III-V compounds as well. For all donors in GaAs, a deep photoluminescence band at 1.2 eV that has been identified with ($D_{Ga}$ $V_{Ga}$) or ($D_{As}$ $V_{Ga}$) pairs where D is a donor atoms occupying either a Ga site or an As site, respectively, is invariably seen in melt grown crystals. Thus the hardening observed in GaAs doped with Te, S, Ge, and Sn could be explained by the presence of donor-vacancy complexes. It is conceivable that S-doped InP may also contain ($S_P$–$V_{In}$) pairs to cause hardening.

The solute-vacancy pair model can not, however, explain the softening observed with Zn-doping in GaAs. On the one hand, doping with Zn substituting for Ga sites should reduce the concentratin of $V_{Ga}$ so that the concentration of ($Si_{Ga}$ $V_{Ga}$) will be lower than that of the undoped sample, $Si_{Ga}$ being present as a contaminant in both cases. This would explain the decrease in the yield stress in the Zn-doped sample compared to the undoped sample. On the other hand, a complex such as ($Zn_{Ga}$ $V_{As}$) can exist in Zn-doped sample which would have the same effect as ($Si_{Ga}$ $V_{Ga}$) pairs. Another difficulty with the model is that it would predict less of an increase or no increase at all in the yield stress at high temperatures when the thermal stability of the complex decreases. But hardening with Te-doping in GaAs is observed even at 800°C [250] and further the decrease in dislocation density in Czochralski grown GaAs crystals containing high concentrations of Si, Te, or other donors has been attributed to the hardening of the lattice at very near the growth temperature. A comprehensive theory of the doping effect on the yield stress of GaAs and InP is needed which takes into account electronic effects such as kink formation and/or migration energy and elastic effects such as solute-vacancy pairs and precipitation hardening. The possibility of precipitation hardening is particularly appealing in the case of high Zn-doping both in GaAs [267] and in InP [285].

**Microhardness and plasticity at room temperature**

We noted earlier that the microhardness measurements are easily suited to determine the plasticity of epitaxial films. In addition, even in bulk crystals low-temperature deformation characteristics are studied by microhardness measurements. Although GaAs and InP are brittle at room temperature, plastic deformation may be induced by using indenters. This is because a large hydrostatic stress component around the indenter with a super-imposed shear stress inhibits

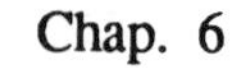

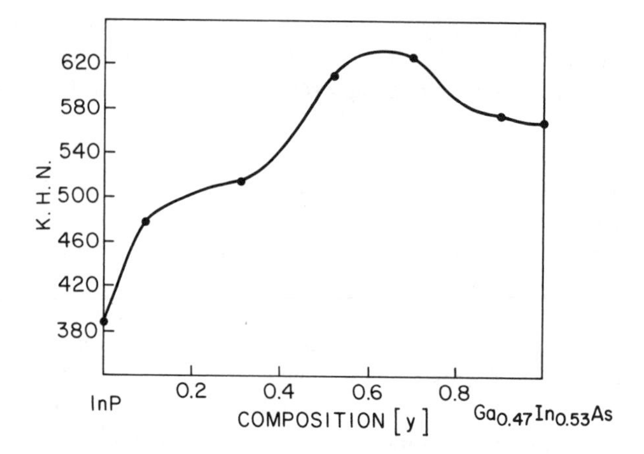


**Figure 6.49** The composition dependence of the knoop microhardness (KHN) of $Ga_xIn_{1-x}As_yP_{1-y}$ in the [100] direction [286].

the onset of fracture and plastic deformation may therefore occur at temperatures where uniaxial loading alone would lead to fracture. For this reason hardness testing has been applied quite extensively in III-V semiconductors.

Watts and Willoughby [286] have studied the composition dependence of the microhardness of $Ga_xIn_{1-x}As_yP_{1-y}$ lattice matched to InP. The quarternary layer was grown by liquid phase epitaxy and the indendations were made using a Knoop indentor under 25g load making sure that the plastic deformation caused by the indent is predominantly confined to the epitaxial layer. Their results are shown in Fig. 6.49. It can be seen that addition of Ga and As to InP increases the hardness rather significantly up to y ~ 0.09. After this composition, the hardness continues to increase but at a slower rate, and reaches a maximum at y ~ 0.7. Beyond y ~ 0.7, the hardness decreases somewhat but $Ga_{0.47}In_{0.53}As$ is still harder than InP. Watts and Willoughby interpret these results as caused by a combined hardening and softening effect caused by P and Ga additions, respectively.

In making hardness indentations, one should keep in mind the hardness anisotropy exhibited by semiconductors [287, 288] which is dependent upon both the orientation of the indentor and the plane of indentation. Brasen [287] found in InP that the hardness was highest on the (100) plane followed by (110) and (111) planes and in the (100) and (110) planes the hardness varied with the direction of indentation. The hardness anisotropy on the (100) plane, is shown in Fig. 6.50. Note the maximum along the [100] direction and two subsidiary minima closer to the [110] directions. Brasen [289] explains the variation in hardness with direction and plane of indentation in terms of the slip systems that are activated. Also shown in Fig. 6.50 is the hardness curve for a S-doped sample whose overall hardness is higher than that of the undoped sample consistent with the hardening of the lattice observed in the yield stress measurements. As for the two [110] directions on the (100) plane, they are found to be equivalent. In this regard, the

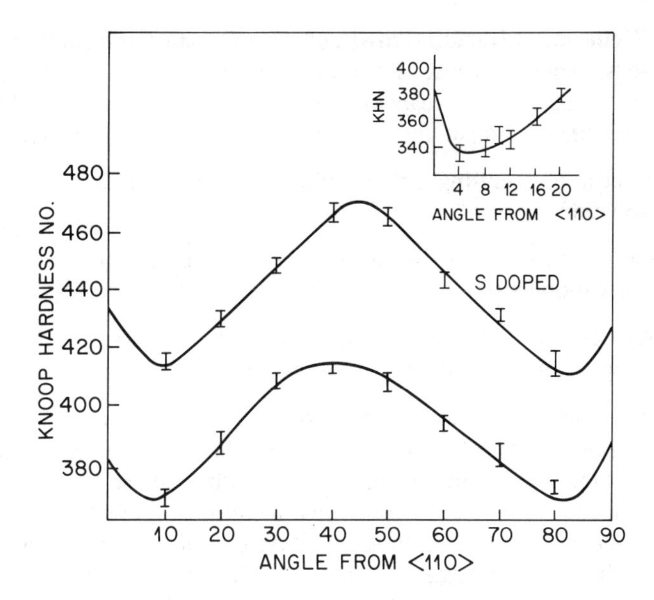


**Figure 6.50** Knoop hardness anisotropy of undoped (lower curve) and S-doped InP (upper curve) [288].

asymmetry observed between the <110> and <$\bar{1}$10> directions with respect to dislocations [290] and fracture [291] is not observed with respect to hardness.

The Knoop hardness anisotropy curves of GaInAs and GaInAsP exhibit two maxima and one minimum instead of one maximum and two minima as in Fig. 6.50 [288]. The different shape of the hardness anisotropy curves suggest that the low-temperature deformation behavior of the ternary and quaternary alloys differ significantly from that of the binary.

In addition to the anisotropy mentioned, there is also an effect of polarity in hardness on (111) faces of III-V compounds. In general, the hardness of (111) and $(\overline{111})$ faces is different. For n-type GaAs the surface terminating in Ga atoms is harder than that terminating in As atoms, while for p-type material, the As face is harder than the Ga face. This polarity of hardness has been explained in terms of the differences in the velocities of the As(g) and Ga(g) dislocations for n- and p-type material [292].

The hydrostatic stress around the indenter which prevents brittle failure at room temperature can also be used in compression experiments [293, 294]. Rabier et al., [293] performed uniaxial compression experiments at room temperature on n- and p-type GaAs under a confining hydrostatic pressure. Unlike the high temperature deformation characteristics, the n-type crystals were found to be softer than intrinsic or p-type crystals. This suggests that the influence of doping can be different at low temperature and high stress from that at high temperature and low stress. Further, microtwinning has been observed in deformed crystals indicating it as a deformation mode [294, 295].

### 6.7.4 Dislocation Generation During Growth of Bulk Crystals and Its Reduction

Since dislocations generally cause deleterious effects in devices (Chap. 7), considerable efforts have been made in the last 20 years to grow dislocation free GaAs and InP crystals. There are three main mechanisms leading to the generation of dislocations during growth [296]:

(a) nonuniform heat flow during solidification and the ensuing thermal stresses causing plastic deformation,

(b) condensation of the excess point defects present near the growth temperature to form prismatic dislocation loops,

(c) defective seed crystal or accidental introduction of macroscopic foreign particles during growth causing generation and multiplication of dislocations.

Of these three mechanisms, the first two are the most important to be considered. Brice and King [297] and Brice [298] in the late 60s noted that the partial pressure of As during growth of GaAs is a parameter to control the dislocation density in horizontal Bridgman crystals and pulled crystals. More systematic study later demonstrated the critical role of melt stoichiometry in the generation of dislocations [299 - 302]. In small GaAs crystals (< 1.5 cm) grown under reduced thermal stress by the horizontal Bridgman method, in which the stoichiometry was controlled by controlling the arsenic source temperature, $T_{As}$, it was found [301] that the dislocation density was a minimum at a $T_{As} \sim 617°C$ as shown in Fig. 6.51. Deviations from the minimum arsenic temperature gave high dislocation densities similar to those obtained in crystals grown under

**Figure 6.51** Dislocation density versus arsenic source temperature for lightly doped n- and p-type GaAs [302].

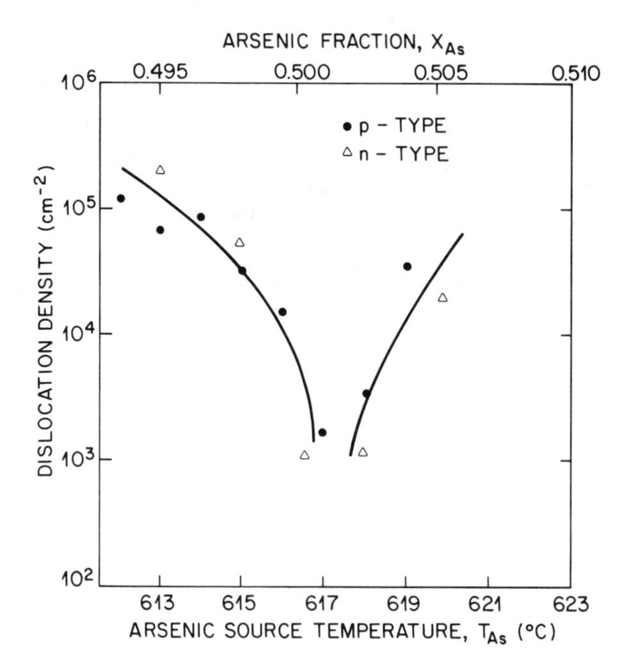

large thermal stresses. The dislocations generated as a result of deviations from stoichiometry are essentially the dislocation loops formed by the condensation of point defects.

The formation of dislocations by point defect condensation also showed unique dependence on conductivity type in GaAs. For n-type doping at levels of $10^{17}$ $cm^{-3}$, less than that required for impurity hardening, the dislocation density decreased. In contrast, with p-type doping the dislocation density increased. These results are interpreted as a Fermi level effect on point defect condensation [302]. Specifically, the effect is supposed to be a change in gallium vacancy concentration. The formation of prismatic vacancy loops requires the presence of vacancies from both sublattice. However, the presence of $V_{Ga}$ alone would be sufficient since $V_{As}$ can be created upon migration of $V_{Ga}$ [302]. Since $V_{Ga}$ behaves as an acceptor with an energy level close to the middle of the energy gap, the shift of the Fermi energy towards the valence band with p-type doping increases the concentration of neutral $V_{Ga}$ which promotes migration and condensation.

The As pressure effect seen in small crystals (up to ~2 cm in diameter) grown by the horizontal Bridgman technique has not been reproduced in crystals of large sizes of commercial interest. It is believed that for the commercial size material there is incomplete solid-vapor equilibrium [303] leading to an apparent insensitivity to As pressure. Nevertheless, it is believed that for dislocation density below 3000 $cm^{-2}$, stoichiometry control and point defect condensation are important considerations in the growth of commercial size crystals [304].

The primary cause for the large density of dislocations in most III-V semiconductor crystals grown from the melt by the Czochralski technique is the plastic deformation caused by thermal stresses [305]. Jordan et al., [304, 306] have reviewed some of the early work on thermal stress generated dislocations. Qualitatively, the thermal stress effect can be described [307]. The main cause of the thermal stress is the heat dissipation from the growing crystal. The heat enters the crystal at the solid-liquid interface and leaves through the external surfaces by radiation and convection. This leaves each cross-section of the crystal with a cooler periphery than core and, consequently, because of thermal contraction and expansion, the periphery is left under tension and the core under compression. If the corresponding resolved shear stress components of $\{111\}$ $\langle 1\bar{1}0\rangle$ system are exceeded, then the resulting plastic deformation would exhibit the same symmetry as the observed dislocation pattern.

Jordan et al., [308] developed a quasi-steady state heat transfer/thermal stress model of the Czochralski process which provided the fundamental description of dislocation generation in III-V compounds. The analysis consists of solving for the temperature distribution of the growing crystal (taken to be a cylinder) assuming planar solid-liquid interface at a temperature equal to the melting point, $T_f$, that the ambient temperature around the boule is a constant value $T_a < T_f$ and that the heat loss by natural convection and radiation from the lateral and top surfaces is proportional to the temperature difference between the surface and ambient fluid. The classical thermoelastic principles are used to obtain the radial, tangential and axial thermal stresses caused by the nonuniform temperature profile. The resolved shear stress for the $\{111\}$ $\langle 1\bar{1}0\rangle$ system is then determined. The stress in excess of the critical resolved shear stress gives rise to plastic flow and thus the overall dislocation density is proportional to the total glide strain which in turn is proportional to the total excess shear stress, $\sigma_{ex}$. In regions $\sigma_{ex} = 0$ the dislocation density will be zero.

Figure 6.52 shows the calculated $\sigma_{ex}$ or dislocation pattern in half a $\{100\}$ GaAs grown at an

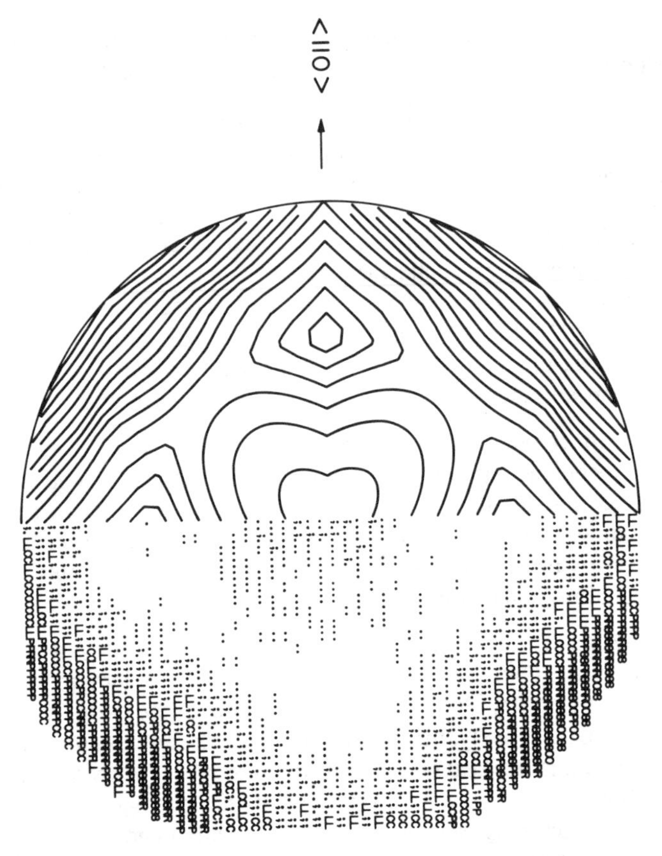

**Figure 6.52** Comparison of predicted dislocation density contour map (top) with KOH etched pattern of pits for (100) GaAs (bottom). A computerized scan generated the etch pit distribution [304].

average ambient temperature 200°K below the melting point ($T_f - T_a = 200°K$) [304]. The adjacent macrophotograph shows the dislocation structure, revealed by etching in fused KOH of a {100} wafer cut from a Te-doped boule. The idea that excessive thermal stresses is the cause of dislocation generation is supported by the good agreement between the calculated and observed patterns in Fig. 6.52.

The (100) dislocation distribution exhibits four-fold symmetry with a minimum in the <110> direction at ~ 60% of the radius value. The center and edge of the wafer are heavily dislocated with maximum disloction density at the <100> edge. The pattern across the diameter of the wafers thus exhibits the W shaped profile which has been reported by several authors in correlating physical properties and dislocation density. The predicted dislocation patterns for {111} wafers display a six-fold symmetry and are also in satisfactory agreement with the etch pit distribution obtained on both LEC grown GaAs [306] and InP [309] crystals (see Sec. 2.1.3). The agreement improves when axial gradients and displacements are also considered in the thermoelastic model besides only radial extension [310].

Jordan and co-workers [304] have investigated the effect of various parameters in the quasi steady state heat transfer/thermal stress model on dislocation generation. They find that decreasing the pull rate would have only a minor effect on reducing the dislocation density. On the other hand, the magnitude of the heat transfer coefficient has a profound influence on dislocation density. The higher the heat transfer coefficient the higher is the dislocation density, especially near the periphery. The incorporation of a radiation shield and/or an after-heater in the growth system would help lowering the heat transfer coefficient. Alternatively replacing the $B_2O_3$ liquid encapsulant with a gaseous ambient such as $As_2$ can also achieve the same goal. For the standard LEC growth conditions for small diameters, the dislocation density rises superlinearly with crystal diameter for $d \leq 2$ cm. With increasing size the density saturates at the periphery and declines in the interior [303]. To reduce the dislocation density in large diameter crystals, the ambient temperature must be high leading to low-temperature gradients. The beneficial effect of reduced temperature gradients on defect density is shown by the work of Shinoyama et al., [311] on the LEC growth of InP.

The low temperature gradients necessary for defect control are achievable in Bridgman growth but are difficult to realize in LEC growth because of the concomittant solid-liquid interface instability and diameter control problems. Although the horizontal Bridgman technique produces crystals of low dislocation density it has certain disadvantages compared to the LEC technique. First, the crystals are D-shaped instead of round which make them less attractive for IC applications. Second, the inevitable contamination with Si from the quartz boat hinders the making of semi-insulating crystals. In this regard, the horizontal gradient freeze technique employing pyrolytic boron nitride boats produced low dislocation density (1000-4000 $cm^{-2}$) and semi-insulating two-inch D-shaped GaAs crystals and appears promising for further scale up [312]. The Kyropoulous method which uses $B_2O_3$ encapsulation also permits low gradients and low dislocation density [313 - 315]. In the LEC method, the shape instability associated with small confining radial temperature gradients can be alleviated to some extent by employing automatic diameter control and/or a vertical magnetic field [316]. Alternatively, conventional "necking-in" procedure can be used but it does not prevent defect generation above a critical diameter for a given temperature gradient as shown in the case of InP [311].

The best solution to achieve stable growth of large diameter and low dislocation crystals by the LEC method appears to be a combination of moderate temperature gradient and impurity hardening of the lattice that would rise the CRSS for dislocation generation. We already discussed in Sec. 6.7.3 the increase in CRSS in the medium temperature range caused by the addition of donors in GaAs and of Zn and Ge besides S in InP. Since addition of these impurities results in reduced dislocation density it should be supposed that the hardening prevails even near the growth temperature. It has been found that addition of S [317, 318] Zn [285, 317], Te [317], and Ge [269] in InP and Si [319], Te [317], S [317], and Se [320] in GaAs at levels exceeding $10^{18} cm^{-3}$ reduces the dislocation density. However, such heavily doped substrates are not suitable for certain device applictions and further heavy doping generates other defects such as

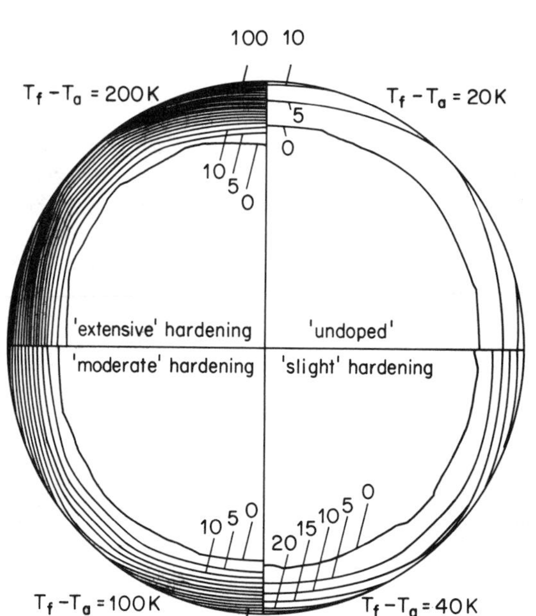


**Figure 6.53** Dependence of $\sigma_{ex}$ or dislocation density distribution on the degree of impurity hardening at the top end of a LEC pulled 75 mm diameter (100) GaAs boule. In each of the quadrants the temperature gradients are matched with the CRSS to yield approximately equal defect-free areas in the central region. The quadrants represent undoped, slight, moderate and extensive impurity hardening conditions. The pull rate is 0.0004 cm secref–1. Multiplication of the numerical labels by 1 MPa yield $\sigma_{ex}$ [303].

precipitates [285]. Interestingly enough several isovalent dopants have also been found to reduce dislocation density again presumably caused by lattice hardening effects. In this regard In, Sb, and N in GaAs and Sb, As and Ga in InP have been used successfully [314] to reduce dislocation density. Besides reducing dislocation density, the isovalent elements have no effect on the electrical properties of the crystals. This is particularly important in growing semi-insulating crystals.

Jordan et al., [304] considered the effects of varying degrees of hardening combined with different temperature gradients on the dislocation distribution maps. Their results for a 75 mm diameter (100) GaAs grown at five different temperature gradients from $T_f$–$T_a$ = 200°K to 10°K and at each gradient with light, moderate, and extensive hardening are shown in Fig. 6.53. The complementarity between temperature gradient and hardening is illustrated in Fig. 6.53 by the approximately equal dislocation-free areas achieved for high gradient combined with extensive hardening or low gradient combined with moderate or light hardening. A particularly note worthy point in Fig. 6.53 is that the periphery is not significantly affected by impurity hardening but the interior is virtually dislocation free even with high gradients. This phenomenon has sometimes been commericially used to grow a large diameter (say, ~6 cm) crystals by the high gradient LEC technique with large In additions and then "core out" a 5 cm diameter dislocation free crystal.

Among the isovalent dopants used to reduce dislocation density in GaAs, In has received considerable attention. The effectiveness of In doping is illustrated in Fig. 6.54 [321] which compares the radial etch pit density of an undoped and an In-doped (mid $10^{19}$ $cm^{-3}$ of In) LEC grown 50 mm diameter GaAs crystals. Note the nearly three orders of magnitude decrease in dislocation density except near the periphery of the crystal.

The reduction in dislocation density in In doping GaAs crystals is because of the hardening of the lattice as indicated by both hardness [322] and yield stress measurements [260, 323 - 325]. For In additions of $0.5 - 2 \times 10^{20}$ $cm^{-3}$, the CRSS values in the temperature range 700 to 1100°C are a factor of 2 to 4 higher compared to crystals containing no In [260,323]. This increase in CRSS is roughly in agreement with the hardening predicted based on a solid solution hardening model where an In atom with four nearest As neighbors (a solute molecule) is the hardening agent [326]. Also consistent with the hardening effect, Yonenaga et al. [327] observed that the velocity of the As(g) dislocations is greatly reduced under low stresses in the temperature range 350 to 750°C in In-doped ($2 \times 10^{20}$ $cm^{-3}$) GaAs and is lower than that of the usually slow moving Ga(g) dislocations. Extrapolating the CRSS versus temperature data to the melting point of GaAs shows that the factor of 2 to 4 higher CRSS in In doped crystals compared to the undoped crystals is still valid. According to the thermal stress model this amount of hardening should produce dislocation free 75mm diameter crystals [304].

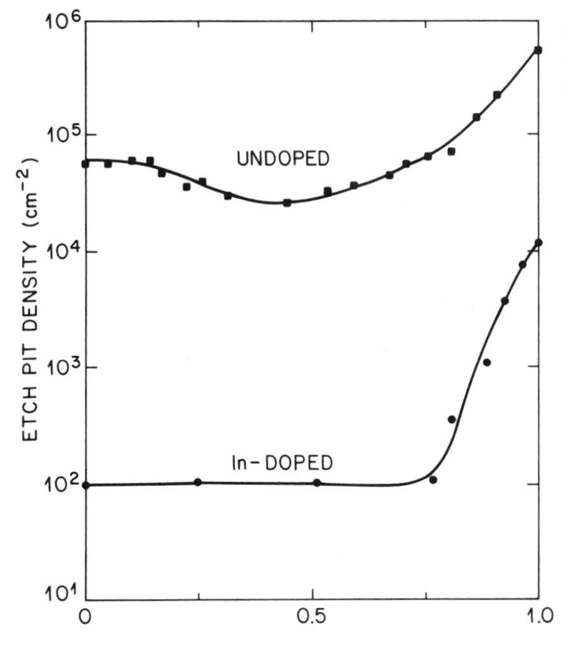


**Figure 6.54** Radial etch pit distribution for (100) GaAs crystals: undoped and In-doped ($5 \times 10^{19}$ $cm^{-3}$) [321].

## REFERENCES

1. F.A. Kroger, *Ann. Rev. Mater. Sci. 7*;449, 1977.

2. F.A. Kroger, "The Chemistry of Imperfect Crystals," Vol. 2, 2nd ed., Amsterdam;North Holland, p. 9, 1974.

3. J.W. Corbett, and J.C. Bourgoin, in "Point defects in Solids," (eds. J.H. Crawford, Jr., and L.M. Slifkin), vol. 2, New York;Plenum, p. 1, 1975.

4. P.J. Lin-Chung and Y. Li, in "Thirteenth International Conference on Defects in Semiconductors," eds. L. C. Kimerling and J. M. Parsey, Jr. (The Metallurgical Society of AIME, Pennsylvania, 1985), p. 1071.

5. F.A. Kroger, "The Chemistry of Imperfect Crystals," Vol. 2, 2nd ed., Amsterdam;North-Holland, p. 237, 1974.

6. M. Altarelli and F. Bassani and "Handbook on Semiconductors" (ed. T. S. Moss), vol. 1, (ed. W. Paul), Amsterdam;North-Holland, p. 169, 1982.

7. W. Kohn in "Solid State Physics-Advances in Research and Applications" (eds. F. Seitz and D. Twinbull), vol. 5, New York;Academic, p. 257, 1957.

8. A. Baldereschi and N. O. Lipari, *Phys. Rev. B8*;2697, 1973.

9. A. Baldereschi and N. O. Lipari, *Phys. Rev. B9*;1525, 1974.

10. J.M. Luttinger, *Phys. Rev. 102*;1030, 1956.

11. P. Lawaetz, *Phys. Rev. B4*;3460, 1971.

12. S.T. Pantelides in Festkorperprobleme XV (ed. H. J. Queisser), p. 149, 1975.

13. D.G. Thomas and J.J. Hopfield, *Phys. Rev. 150*;680, 1966.

14. J.C. Phillips, *Phys. Rev. B1*;1540, 1970.

15. J.C. Phillips, Bonds and Bands in Semiconductors, New York;Academic, p. 235-240, 1973.

16. F. Bassani, G. Iadonisi, and B. Preziosi, *Phys. Rev. 186*;735, 1969.

17. P.M. Adler, *J. Appl. Phys. 40*;3554, 1969.

18. A. Onton, Y. Yacobi, and R.J. Chicotka, *Phys. Rev. Lett. 28*;966, 1972.

19. R.L. Aggarwal and A.K. Ramdas, *Phys. Rev. 140*;A1246, 1965; J.K. Reuszer and P. Fisher, *Phys. Rev. 135*;A1125, 1964.

20. Sh.M. Kogan and B.I. Sedunov, *Sov. Phys. Solid State 8*;1898, 1967.

21. G.E. Stillman, L.W. Cook, T.J. Roth, T.S. Low, and B.J. Skromme in "GaInAsP Alloy Semiconductors" (ed., T.P. Pearsall), Chap. 6, New York;Wiley, p. 121, 1982.

22. G.E. Stillman, C.M. Wolfe, and J.D. Dimmock, in Semiconductors and Semimetals, (eds., R.K. Willardson and A.C. Beer New York;Academic, vol. 12, p. 169, 1977.

23. M. Jaros, ''Deep Levels in Semiconductors,'' Published by Adam Hilger Lts., Techno House, Red Cliffe Way, Bristol, printed by The Universities Press Ltd., Belfast.

24. M. Lannoo and J. Bourgoin, ''Point Defects in Semiconductors I: Theoretical Aspects,'' New York;Springer-Verlag, 1981.

25. C.A. Coulson and M.J. Kearsley, *Proc. R. Soc. A241*;433, 1957.

26. A. Zunger, ''Electronic Structure of 3d Transition-Atom Impurities in Semiconductors,'' in *Solid State Physics*, (eds., H. Ehrenreich and D. Turnbull), vol. 39, pp. 275-264, 1986.

27. M. Lannoo, *J. Phys. C: Solid State Phys., 17*;3137, 1984.

28. R.P. Messmer and G.D. Watkins in ''Radiation Damage in Semiconductors,'' (eds., J.W. Corbett and G.D. Watkins) New York;Gordon and Breach, p. 23, 1971.

29. F.D.M. Haldane and P.W. Anderson, *Phys. Rev. B13*;2553, 1976.

30. G.J. Koster and J.C. Slater, *Phys. Rev. 95*;1167, 1954.

31. G.D. Watkins, *J. Phys. Soc. Jpn. (Suppl. 2) 18*;22, 1963; in ''Radiation Effects in Semiconductors,'' (ed. P. Baruch), Dunod, Paris, p. 97, 1964.

32. J. Bourgoin and M. Lannoo, ''Point Defects in Semiconductors II: Experimental Aspects,'' New York;Springer-Verlag, 1983.

33. G.A. Baraff, M. Lannoo, and M. Schluter, *Phys. Rev. 38B*;6003, 1988; see also M. Lannoo, *Rev. Phys. AP 23*;817, 1988.

34. J. Dabrowski and M. Scheffler, *Phys. Rev. Lett. 60*;2183, 1988.

35. D.J. Chadi and K.J. Cheng, *Phys. Rev. Lett. 60*;2187, 1988; *61*; 873, 1988.

36. H.P. Hjalmarson, P. Vogl, D.J. Wolford, and J.D. Dow, *Phys. Rev. Lett. 24*;810, 1980.

37. H.P. Hjalmarson, ''Studies in the Theory of Solids,'' Ph.D. thesis, Univ. of Illinois, Urbana, 1979.

38. J.P. Buisson, R.E. Allen, and J.D. Dow, *Sol. St. Commun. 43*; 833, 1982; see also *J. Physique 43*;181, 1982.

39. O.F. Sankey and J.D. Dow, *J. Appl. Phys. 52*;5139, 1981.

40. G.B. Bachelet, G.A. Baraff, and M. Schuter, *Phys. Rev. B24*;915, 1981.

41. G.A. Baraff and M. Schluter, *Phys. Rev. Lett., 55*;1327, 1985.

42. S.G. Bishop, ''Experimental Studies of Iron Impurity Centers in III-V Semiconductors,'' Chap. 8, *Deep Centers in Semiconductors*, (ed., S.T. Pantelides), New York;Gordon and Breach, 1986.

43. P.J. Dean, A.M. White, B. Hamilton, A.R. Peaker, and R.M. Gibb, *J. Phys. D10*;2545, 1977.

44. O. Mizuno and H. Watanabe, *Electro. Lett. 11*;118, 1975.

45. L.G. Meiners, in "Dielectric Films on Compound Semiconductors," (eds., V.J. Kapoor, D.J. Connolly, and Y.H. Wong), *Electrochem. Soc. Proc.*, vol., 86-3, p. 3, 1986.

46. S.J. Prasad and S.J. Towen, in "Dielectric Films on Compound Semiconductors" (eds., V.J. Kapoor, D.J. Connolly, and Y.H. Wong), Electrochem Soc. Proc., vol. 86-3, p. 23, 1986.

47. S.M. Sze, "Physics of Semiconductor Devices," New York;Wiley, Interscience, 1969.

48. S.G. Bishop, P.B. Klein, R.L. Henry, and B.D. McCombe, "Semi-Insulating III-V Materials," (ed., G.J. Rees), Shiva, Onpington, p. 161, 1980.

49. G.H. Stauss, J.J. Krebs, and R.L. Henry, *Phys. Rev. B16*;974, 1977.

50. Ref. 5, p. 77.

51. C. D. Thurmond, *J. Electrochem. Soc., 132*;673, 1985.

52. W.W. Harvey, *J. Phys. Chem. Solids, 23*;1545, 1962.

53. A.J. Rosenberg, *J. Chem. Phys. 33*;665, 1960.

54. M.B. Panish, and H.C. Casey, Jr., *J. Phys. Chem. Solids* 28;1673, 1967; *29*;1719, 1968; H.C. Casey, Jr., M.B. Panish, and L.L. Chang, *Phys. Rev. 162*;660, 1967.

55. C.J. Hwang, and J.R. Brews, *J. Phys. Chem. Solids, 32*;837, 1971.

56. P.T. Landsberg and A.G. Guy, *Phys. Rev. B28*;1187, 1983.

57. M.E. Orazem and J. Newman, *J. Electrochem. Soc. 131*;2715, 1984.

58. Ref. 5, p. 152.

59. H.E. Ruda, J. Lagowski, H.C. Gatos, and W. Walukiewicz, in "Semi-insulating III-V materials," (eds., D.C. Look, and J.S. Blakemore), Shiva, Nantwich England, p. 263, 1984.

60. D.E. Holmes, R.T. Chen, K.R. Elliott, and C.G. Kirkpatrick, *Appl. Phys. Lett. 40*;46, 1982; D.E. Holmes, R.T. Chen, K.R. Elliott, C.G. Kirkpatrick, and P.W. Yu, *IEEE Trans. Electron Devices, ED-29*;1045, 1982.

61. L.B. Ta, H.M. Hobgood, A. Rohatgi, and R.N. Thomas, *J. Appl. Phys. 53*;5771, 1982.

62. J.C. Phillips and J.A. Van Vechten, *Phys. Rev. Lett. 30*;220, 1973.

63. J.A. Van Vechten, *Phys. Rev. B. 10*;1482, 1974; *Phys. Rev. B, 11*;3910, 1975.

64. J.A. Van Vechten, *J. Electrochem. Soc. 122*;419, 1975.

65. D.T.J. Hurle, *J. Phys. Chem. Sol. 40*;613, 1979.

66. R.A. Swalin, *J. Phys. Chem. Sol. 18*;290, 1961.

67. J.A. Van Vechten and C.D. Thurmond, *Phys. Rev. B 14*;3551, 1976.

68. M. Lannoo and G. Allan, *Phys. Rev. B. 25*;4089, 1982.

69. V. Swaminathan, unpublished; the necessary thermochemical data were taken from Landolt-Bernstein Numerical data and functional relationships in science and technology, vol. 17, subvol. A, (eds. O. Mdelung, M. Schulz, and H. Weiss), Berlin;Springer-Verlag, 1982.

70. M.B. Panish, *J. Cryst. Growth 27*;6, 1974.

71. C.D. Thurmond, *J. Electrochem. Soc. 122*;1133, 1975.

72. Y.P. Varshni, *Physica, 39*;149, 1967.

73. J.A. Van Vechten, in *Lattice Defects in Semiconductors* 1974, p. 212, London;Institute of Phys., 1975.

74. J.A. Van Vechten, and C.D. Thurmond, *Phys. Rev. B 14;3539, 1976.*

75. *G.A. Baraff, in "Semi-insulating III-V materials," (eds. D. C. Look and J. S. Blakemore), Shiva, Nantwich England p. 416, 1984.*

76. *J.S. Blakemore and S. Rahimi, in "Semiconductors and Semimetals," vol. 20, (eds. R.K. Willardson and A.C. Beer), New York;Academic, p. 233, 1984.*

77. *W. Potz and D.K. Ferry, Phys. Rev. 31B*;968, 1985.

78. G.A. Baraff, E.O. Kane, and M. Schluter, *Phys. Rev. B21*;3563, 5662, 1980.

79. J.C. Phillips, "Bonds and Bands" in *Semiconductors* New York;Academic, 1973.

80. J.A. Van Vechten, *Phys. Rev. 182*;891, 1969; *187*, 1007, 1969.

81. J.A. Van Vechten, *J. Electrochem. Soc. 122*;423, 1975.

82. G.A. Baraff and M. Schluter, *Phys. Rev. 33B*;7346, 1986.

83. A.F.W. Willoughby, C.M.H. Driscoll, and B.A. Bellamy, *J. Mater. Sci. 6*;1389, 1971.

84. C.M.H. Driscoll, and A.F.W. Willoughby, in "Defects in Semiconductors" p. 377. Inst. of Phys. Conf. Series, no. 16, 1973.

85. V.T. Bublik, V.V. Karataev, R.S. Kulagin, M. Milvidskii, G. Osvenskii, V.B. Stolyarov, and L.P. Kholodnyi, *Sov. Phys. Crystallogr., 18*;218, 1973.

86. D.T.J. Hurle, *J. Phys. Chem. Solids 40*;627, 1979.

87. J.A. Grimshaw and P.C. Banbury, Proc. Phys. Soc., (London) *84*;151, 1964.

88. G.A. Baraff and M. Schluter, *Phys. Rev. Lett., 55*;1327, 1985.

89. Ref. 5, p. 165.

90. G.M. Blom, *J. Cryst. Growth 36*;125, 1976.

91. P. Vogl, H.P. Hjalmarson, and J.D. Dow, *J. Phys. Chem. Solids, 44;365, 1983.*

92. *P.J. Lin-Chung and T.L. Reinecke, Phys. Rev. B27;1101, 1983.*

93. *B. Goldstein, Phys. Rev. 121*;1305, 1961.

94. H.D. Palfrey, M. Brown, and A.F.W. Willoughby, *J. Electrochem. Soc., 128*;2224, 1981; *J. Electron. Mat., 12*;863, 1983.

95. S.Y. Chiang, and G.L. Pearson, *J. Appl. Phys. 46*;2986, 1975.

96. B.K. Chakraverty and R.W. Dreyfus, *J. Appl. Phys. 37;631, 1966.*

97. *V.B. Osvenskii, L.P. Kholodnyi, and M.G. Milvidskii, Sov. Phys. Solid State, 13*;1790, 1972.

98. D.V. Lang, L.C. Kimerling, and S.Y. Leung, *J. Appl. Phys., 47*;3587, 1976.

99. D. Pons, A. Mircea, and J. Bourgoin, *J. Appl. Phys., 51*;4150, 1980.

100. D. Pons and J. Bourgoin, *Phys. Rev. Lett., 47*;1293, 1981.

101. W.W. Williams and H.B. Bebb in *Semiconductors and Semimetals (eds., R.K. Willardson, and A. C. Beer), vol. 8, New York;Academic, p. 321, 1972.*

102. *V. Swaminathan, Bull. Mater. Sci* (India) *4*;403, 1982.

103. R.J. Wagner, J.J. Krebs, G.H. Stauss, and A.M. White, *Solid State Comm., 36*;15, 1981.

104. N.D. Wilsey and T.A. Kennedy in "Mciroscopic Identification of Electronic Defects in Semiconductors" (eds., N.M. Johnson, S.G. Bishop, and G. Watkins), Materials Research Society Sumposia Proceedings, vol. 46, p. 309, 1985.

105. A. Bharatti, K.P. Gopinathan, C.S. Sundar, and B. Viswanathan, *Pramana*, (India), *13*;625, 1979.

106. S. Dannefaer, B. Hogg, and D. Kerr, in Proceedings of the 13th International Conference on Defects in Semiconductors, (eds., L.C. Kimerling, and T.M. Parsey, Jr.), The Metallurgical Society of AIME, Pennsylvania p. 1029, 1985.

107. S. Dannefaer and D. Kerr, *J. Appl. Phys. 60*;591, 1986; see also S. Dannefaer, B. Hogg and D. Kerr, *Phys. Rev., 30B*;3355, 1984.

108. G. Dlubek, O. Brummer, F. Plazaola and P. Hautojärvi, *J. Phys. C 19*;331, 1986.

109. F. Sette, S.J. Pearton, J.M. Poate, J.E. Rowe, and J. Stohr, *Phys. Rev. Lett., 56*;2637, 1986; see also T.N. Morgan, *Phys. Rev. Lett. 58*;1280, 1987.

110. D.V. Lang, R.A. Logan, and M. Jaros, *Phys. Rev. B19*;1015, 1979.

111. D.V. Lang, R.A. Logan, and L.C. Kimmerling, Proc. 13th Conf. on Physics of Semiconductors, (ed., F.G. Fumi) North Holland, Amsterdam, p. 615, 1976.

112. V. Swaminathan, N.E. Schumaker, J.L. Zilko, W.R. Wagner, and C.A. Parsons, *J. Appl. Phys. 52*;412, 1981.

113. V. Swaminathan and W.T. Tsang, *Appl. Phys. Lett. 38*;347, 1981.

114. W.R. Wagner, *J. Appl. Phys. 49*;173, 1978.

115. V. Narayanamurti, R.A. Logan, and M.A. Chin, *Phys. Rev. Lett. 43*;1536, 1979.

116. V.N. Brudnyi, *Appl. Phys. A, 29*;219, 1982.

117. H. Temkin and V.V. Dutt, in ''Defects in Semiconductors II'' (eds., S. Mahajan and J.W. Corbett) Materials Research Society Symposia Proceedings, vol. 14; p. 253, 1983.

118. M. Levinson, J.L. Benton, and L.C. Kimerling, *Phys. Rev. B 27*;6216, 1983.

119. M. Levinson and M. Stavola, In Proceedings of the 13th International Conference on Defects in Semiconductors, (eds. L.C. Kimerling and J.M. Parsey, Jr.), The Metallurgical Society of AIME, Pennsylvania, p. 1133, 1985.

120. J.L. Pelloie, G. Guillot, A. Nouailhat, and A.G. Antolini, *J. Appl. Phys. 59*;1536, 1986.

121. O. Ueda, S. Komiya, and S. Isozumi, *Japan J. Appl. Phys. 23*;L394, 1984.

122. M.B. Panish, in ''Phase diagrams,'' vol. 3, New York;Academic, p. 53, 1970.

123. M.B. Panish, *J. Phys. Chem. Solids, 27*;291, 1966; see also M.B. Panish, *J. Electrochem. Soc. 113*;224, 1966.

124. M.B. Panish and H.C. Casey, Jr., *J. Phys. Chem. Solids, 28*;1673, 1967.

125. M.B. Panish, *J. Less Common Metals, 10*;416, 1966.

126. M.B. Panish, *J. Appl. Phys., 44*;2659, 1973.

127. M.B. Panish, ibid, *41*;3195, 1970.

128. M.B. Panish, *J. Electrochem. Soc., 113*;1226, 1966.

129. M.B. Panish, ibid, *44*;2676, 1973.

130. M.B. Panish, *J. Appl. Phys., 44*;2667, 1973.

131. H.C. Casey, Jr., M.B. Panish, and K.B. Wolfstirn, *J. Phys. Chem. Solids, 32*;571, 1971.

132. U. Merten and A.P. Hatcher, *J. Phys. Chem. Solids, 23*;533, 1962.

133. K.H. Zschauer and A. Vogel, in *Gallium Arsenide and Related Compounds*. Inst. Phys. Conf. Ser. *9*;100, 1970.

134. F.E. Rosztoczy and K.B. Wolfstirn, *J. Appl. Phys., 42*;426, 1971.

135. L.R. Dawson, Extended Abstracts, Electrochemical Society Meeting, Boston, *J. Electrochem. Soc., 126*;127C, 1979.

136. C.L. Reynolds, S.F. Nygren, and C.A. Gaw, *Materials Lett., 4*;439, 1986.

137. E. Kuphal, *J. Cryst. Growth, 54*;117, 1981.

138. F. Fiedler, H.H. Wehmann, and A. Schlachetzki, *J. Cryst. Growth, 74*;27, 1986.

139. M.G. Astles, F.G.H. Smith, and E.W. Williams, *J. Electrochem. Soc., 120*;1750, 1973.

140. I. Umebu and P.N. Robson, *J. Cryst. Growth, 53*;292, 1981.

141. N. Tamari, *J. Electron. Mater., 11*;611, 1982.

142. A. Perronnet, J. Magnabal, D. Sigogne, D. Huet, and J. Benoit, *J. de Physique*, (Paris) *43*;C5-73, 1982.

143. T.P. Pearsall and J.P. Hirtz, *J. Cryst. Growth, 54*;127, 1981.

144. M.M. Tashima, L.W. Cook, and G.E. Stillman, *Appl. Phys. Lett., 39*;960, 1981.

145. Y. Takeda, M. Kuzuhara, and A. Sasaki, *Japan J. Appl. Phys., 19*;899, 1980.

146. K. Kuphal and D. Fritzche, *J. Electron. Mater., 12*;743, 1983.

147. D.T.J. Hurle, *J. Phys. Chem. Solids, 40*;627, 1979.

148. D.T.J. Hurle, ibid., *40*;639, 1979.

149. D.T.J. Hurle, ibid. *40*;647, 1979.

150. U. Konig, U. Langmann, K. Heime, L.J. Balk, and E. Kubalek, *J. Cryst. Growth, 36*;165, 1976.

151. D.R. Ketchow, *J. Electrochem. Soc. 121*;1237, 1974.

152. I. Teramoto, *J. Phys. Chem. Solids, 33*;2089, 1972.

153. W.G. Rado, W.J. Johnson, and R.L. Crawley, *J. Appl. Phys., 43*;2763, 1972.

154. J.V. DiLorenzo and G.E. Moore, Jr., *J. Electrochem. Soc., 118*;1823, 1971.

155. D.J. Ashen, P.J. Dean, D.T.J. Hurle, J.B. Mullin, A.M. White, and P.D. Greene, *J. Phys. Chem. Solids, 36*;1041, 1975.

156. M. Heyen, H. Bruch, K.H. Bachem, and P. Balk. *J. Cryst. Growth, 42*;127, 1977.

157. R. Sankaran, *J. Cryst. Growth, 50*;859, 1980.

158. P. Kupper, H. Bruch, M. Heyen, and P. Balk, *J. Electron. Mater., 5*;455, 1976.

159. M. Maier, B. Hanel, and P. Balk, *J. Appl. Phys., 52*;342, 1981.

160. J. Hallais, J.P. Andre, A. Mircea-Roussel, M. Mahieu, J. Varon, M.C. Boissy, and A.-T. Vink, *J. Electron. Mat., 10*;665, 1981.

161. J.P. Andre, M. Boulou, and A. Mircea-Roussel, *J. Crystal. Growth, 55*;192, 1981.

162. G. Keil, M. Le Metayer, A. Cuquel, and D. LePollotec, *Rev. Phys. Appl., 17*;405, 1982.

163. V. Swaminathan, J.L. Zilko, and S.F. Nygren, *Mat. Lett., 2*;308, 1984.

164. S.J. Bass and P.E. Oliver, in *Gallium Arsenide and Related Compounds*, Inst. of Phys. Conf. Series *33b*;1, 1977.

165. S.J. Bass, *J. Cryst. Growth, 47*;613, 1979.

166. C.R. Lewis, W.T. Dietze, and M.J. Ludowise, *Electron Letters, 18*;569, 1982.

167. K. Tamamura, T. Ohkata, H. Kawai, and C. Kajima, *J. Appl. Phys., 59*;3549, 1986.

168. J.D. Parsons and F.G. Krojenbrink, *J. Electrochem. Soc., 130*;1782, 1982.

169. N. Bottka, R.S. Sillman, and W.F. Tseng, *J. Cryst. Growth, 68*;54, 1984.

170. J. Chevrier, E. Horache, L. Goldstein, and N.T. Linh, *J. Appl. Phys., 53*;3247, 1982.

171. J. Chevrier, A. Huber, and N.J. Linh, *J. Appl. Phys., 51*;815, 1980.

172. M. Oishi, S. Nojima, and H. Asaki, *Japan J. Appl. Phys., 24*;L380, 1985.

173. J.P. Hirtz, M. Razeghi, M. Bonnet, and J.P. Duchemin, in "GaInAsP Alloy Semiconductors," (ed. T. P. Pearsall), New York;Wiley, p. 61, 1982.

174. A. Mircea, *J. Electron. Mater., 13*;603, 1984.

175. A.W. Nelson and L.D. Westbrook, *J. Appl. Phys. 55*;3103, 1984.

176. J.J. Yang, R.P. Ruth, and H.M. Manasevit, *J. Appl. Phys., 52*;6729, 1981.

177. A.W. Nelson and L.D. Westbrook, *J. Cryst. Growth, 68*;102, 1984.

178. E. Veuhoff, M. Maier, K.H. Bachem, and P. Balk, *J. Cryst. Growth, 53*;598, 1981.

179. L. Hollan and C. Schiller, *J. Cryst. Growth, 13/14*;319, 1972.

180. J.B. Mullin, *J. Cryst. Growth, 42*;77, 1977.

181. K. Jacobs, *J. Cryst. Growth, 56*;362, 1982.

182. A.Y. Cho, *J. Appl. Phys., 46*;1722, 1975.

183. T. Ishibashi, S. Tarucha, and H. Okamoto, *Japan J. Appl. Phys., 21*;L476, 1982.

184. K.Y. Cheng, A.Y. Cho, and W.R. Wagner, *Appl. Phys. Lett., 39*;607, 1981.

185. M.B. Panish and S. Sumski, *J. Appl. Phys. 55*;3571, 1984.

186. M. Ilegems, *J. Appl. Phys., 48*;1278, 1977.

187. K.Y. Cheng, A.Y. Cho, and W.A. Bonner, *J. Appl. Phys. 52*;4672, 1981.

188. A.Y. Cho and M.B. Panish, *J. Appl. Phys., 43*;5118, 1972.

189. M. Ilegems, R. Dingle, and L.W. Rupp, Jr., *J. Appl. Phys., 46*;3059, 1975.

190. A.Y. Cho and I. Hayashi, *J. Appl. Phys., 42*;4422, 1971.

191. R. Heckingbottom, C.J. Todd, and G.I. Davies, *J. Electrochem. Soc., 127*;444, 1980.

192. R. Heckingbottom, G.J. Davies, and K.A. Prior, *Surface Sci., 132*;375, 1983.

193. A. Iliadis, K.A. Prior, C.R. Stanley, T. Martin, and G.J. Davies, *J. Appl. Phys., 60*;213, 1986.

194. C.E.C. Wood and B.A. Joyce, *J. Appl. Phys., 49*;4854, 1978.

195. A.Y. Cho, *Thin Solid Films, 100*;291, 1983.

196. C.E.C. Wood, J. Woodcock, and J.J. Harris, in *GaAs and Related Compounds,* Inst. Phys. Conf. Ser. *45*;28, 1979.

197. H. Kunzel, A. Fischer, and K. Ploog, *Appl. Phys., 22*;23, 1980.

198. Y.G. Chai, R. Chow, and C.E.C. Wood, *Appl. Phys. Lett., 39*;800, 1980.

199. A. Munoz-Yague and S. Baceiredo, *J. Electrochem. Soc., 129*;2108, 1982.

200. J.M. Ballingal and C.E.C. Wood, *Appl. Phys. Lett., 41*;947, 1982.

201. W.I. Wang, E.E. Mendez, T.S. Kuan, and L. Esahi, *Appl. Phys. Lett., 47*;826, 1983.

202. J.D. Grange, *Vacuum, 32*;477, 1982.

203. R. Dingle, C. Weisbuch, H.L. Stormer, H. Morkoc, and A.Y. Cho, *Appl. Phys. Lett., 40*;507, 1982.

204. H. Temkin and J.C.M. Hwang, *Appl. Phys. Lett., 42*;178, 1983.

205. M. Heiblum, E.E. Mendez, and L. Osterling, *J. Appl. Phys., 54*;6982, 1983.

206. J.B. Clegg, C.T. Foxon, and G. Weimann, *J. Appl. Phys., 53*;4518, 1982.

207. P.D. Dapkus, H.M. Manasevit, K.L. Hess, T.S. Low, and G.E. Stillman, *J. Cryst. Growth, 55*;10, 1981.

208. K.H. Goetz, D. Bimberg, H. Jurgensen, J. Selders, A.V. Solomonov, G.F. Glinskii, and M. Razeghi, *J. Appl. Phys., 54*;4543, 1983.

209. C.P. Kuo, R.M. Cohen, K.L. Fry, and G.B. Stringfellow, *J. Electron. Materials, 14;231, 1985.*

*210. K.L. Hess, P.D. Dapkus, H.M. Manasevit, T.S. Low, B.J. Skromme, and G.E. Stillman, J. Electron. Materials, 11*;1115, 1982.

211. C.S. Kang and P.E. Greene, in *GaAs and Related Compounds*, Inst. of Phys. Soc. *7*;18, 1969.

212. P.J. Anthony, J.L. Zilko, V. Swaminathan, N.E. Schumaker, W.R. Wagner, and J.C. Norberg, *Appl. Phys. Lett., 38*;434, 1981.

213. A. Chandra and L.F. Eastman, *J. Electrochem. Soc., 127*;211, 1980.

214. J. Hornstra, *J. Phys. Chem. Solids. 5*;129, 1958.

215. D.B. Holt, *J. Phys. Chem. Solids. 23*;1353, 1962.

216. W.T. Read, *Phil. Mag. 45*;775, 1954.

217. P. Haasen, *Acta Metall., 5*;598, 1957.

218. J.P. Hirth and J. Lothe, ''Theory of Dislocations,'' 2nd ed., New York;McGraw Hill, p. 373-382, 1982.

219. For explanation of the notation see the foreword to Conference of Dislocations in Tetrahedrally Coordinated Semiconductors in Hunfeld, 1978, in *J. de Physique 40*;Colloque C6, 1979.

220. D.J.H. Cokayne and A. Hons, *J. de Physique 40*;Colloque C6, 11, 1979.

221. A. Gomez and P.B. Hirsch, *Phil. Mag. A38*;733, 1978.

222. H. Gottschalk, G. Patzer, and H. Alexander, *Phys. Stat. Solidi 45a*;207, 1978.

223. C.B. Carter, J.S. Roberts, and C.E.C. Wood, *Appl. Phys. Lett. 38*;805, 1981.

224. S. Mader and A.E. Blakeslee, *Appl. Phys. Lett. 25*;365, 1974.

225. H. Alexander, *J. de Physique 40*;Collogque C6, 1, 1979.

226. P.B. Hirsch, *J. de Physique 40*;Colloque C6, 27, 1979.

227. The A(g) and B(g) notations instead of α and β are used to denote the dislocations introduced by bending since it is now widely accepted that the mobile dislocations are of the glide set configurations. However, it should be mentioned that experimentally the core structure of the dislocation that is the carrier of plastic deformation it not unambiguously known.

228. A. Olsen and J.C.H. Spence, *Phil. Mag., 43*;945, 1981.

229. G.R. Anstis, P.B. Hirsch, C.J. Humphreys, J.L. Hutchinson, and A. Ourmazd, *Inst. Phys. Conf. 60*;15, 1981.

230. A. Bourret, J. Thibault - Desseaux, and C. D'Anterroches, *Inst. Phys. Conf. 60*;9, 1981.

231. A. Bourret, J. Thibault - Desseaux, C. D'Anterroches, J.M. Berisson, and A. DeCrecy, *J. Microscopy, 129*;337, 1983.

232. P.B. Hirsch, in "Defects in Semiconductors" (eds., J. Narayan and T.Y. Tan), *Mat. Res.Soc. Proc., 2*;257, 1981.

233. E. Weber and H. Alexander, *J. de Physique 40*;Colloque C6, p. 101, 1979.

234. S. Marklund, *Phys. Stat. Solidi 92b*;83, 1979.

235. R. Jones, *Journ. de Physique, 40*;Colloque C6, 33, 1979.

236. S. Marklund, *J. de Physique 44*;Colloque C4, 25, 1983.

237. W. Shockley, *Phys. Rev. 91*;228, 1953.

238. R. Labusch and W. Schröter, *Inst. Phys. Conf.* Ser. no. 23, p. 56, 1975; See also in *Dislocation in Solids*, (ed. F.R.N. Nabarro), vol. 5, Chap. 20, North Holland, Amsterdam, 1980.

239. H.R. Zeller, in *Festkorperprobleme*, (ed. O. Madelung), vol. XIII. Braunschweig;Springer-Verlag, 1973.

240. A. Ourmazd, *Contemp. Phys. 25*;251, 1984.

241. R. Jones, S. Oberg, and S. Marklund, *Phil. Mag. 43B*;839, 1981.

242. A.L. Lin and R.H. Bube, *J. Appl. Phys. 46*;5302, 1975.

243. A.L. Esquivel, S. Sen, and W.N. Lin, *J. Appl. Phys. 47*;2588, 1976.

244. D. Gwinner and R. Labusch, *J. de Physique 40*;Colloq. C6, 75, 1979.

245. Y. Kadota and K. Chino, *Japan J. Appl. Phys. 22*;1563, 1983.

246. H. Nakata and T. Ninomiya, *J. Phys. Soc. (Japan) 42*;552, 1977.

247. J.D. Venables and R.M. Boudy, *J. Appl. Phys. 29*;1025, 1958.

248. H. Nakata and T. Ninomiya, *J. Phys. Soc. Japan, 47*;1912, 1979.

249. I. Ishida, K. Maeda, and S. Takeuchi, *Appl. Phys. 21*;257, 1980.

250. D. Laister and G.M. Jenkins, *J. Mat. Science, 8*;1218, 1973.

251. V. Swaminathan and S.M. Copley, *J. Am. Ceram. Soc. 58*;482, 1975.

252. G.T. Brown, B. Cockayne, and W.R. MacEwan, *J. Mat. Sci. 15*;1469, 1980.

253. D. Brasen and W.A. Bonner, *J. Mat. Sci. 61*;167, 1983.

254. G.T. Brown, B. Cockayne, W.R. MacEwan, and D.J. Ashen, *J. Mat. Sci. Lett. 2*;667, 1983.

255. G. Muller, R. Rupp, J. Volkl, H. Wolf, and W. Blum, *J. Cryst. Growth, 71*;771, 1985.

256. H. Siethoff and W. Schroter, *Z. Metallkde. 75*;475, 1984.

257. W. Schroter and H. Siethoff, *Z. Metallkde, 75*;482, 1984.

258. V. Swaminathan, Mechanical behavior of GaAs single crystals. Ph.D Thesis (University of Southern California, unpublished) 1975.

259. E. Schmid and W. Boas "Kristallplastizitalt." (1935), English edition, "Plasticity of Crystals." London;F. A. Hughes and Company, Ltd., 1950.

260. S. Guruswamy, R.S. Rai, K.T. Faber, and J.P. Hirth, *J. Appl. Phys. 62*;4130, 1987.

261. H. Alexander and P. Haasen, in *Solid State Phys.*, (eds. F. Seitz and D. Twinbull), vol. 22, New York;Academic, p. 27, 1968.

262. H. Gottschalk, G. Patzer, and H. Alexander, *Phys. Stat. Sol. 45a*;207, 1978.

263. M.S. Abrahams and L. Ekstrom, *Acta Met. 8*;654, 1960.

264. J.R. Patel and A.R. Chaudhuri, *Phys. Rev., 143;601, 1966.*

265. *N.P. Sazhin, M.G. Milvidskii, V.B. Osvenskii, and O.G. Stolyarov, Sov. Phys. Solid State 8*;1223, 1966.

266. V.B. Osvenskii, G.G. Stolyarov, and M.G. Milvidskii, *Sov. Phys. Solid Stat 10*;2540, 1969.

267. T. Ninomiya, *J. de Physique, 40*;Colloq. C6;143, 1979.

268. V. Swaminathan, W.R. Wagner, and P.J. Anthony, *J. Electrochem. Soc. 130*;2468, 1983.

269. G.T. Brown, B. Cockayne, and W.R. MacEwan, *J. Cryst. Growth, 51*;369, 1981.

270. S.K. Choi, M. Mihara, and T. Ninomiya, *Japan J. Appl. Phys. 16*;737, 1977.

271. H. Nagai, *Japan J. Appl. Phys. 20*;793, 1981.

272. V. Celli, M. Kabler, T. Ninomiya, and R. Thomson, *Phys. Rev. 131*;58, 1963.

273. T. Ninomiya, R. Thomson, and F. Garcia-Moliner, *J. App. Phys. 35*;3607, 1964.

274. H.J. Muller, *Acta Met. 26*;963, 1978).

275. P.B. Hirsch, *J. de Physique 40*;Colloque C6, 117, 1979.

276. R. Jones, *J. de Physique 44*;Colloque C4;61, 1983.

277. H.L. Frisch and J.R. Patel, *Phys. Rev. Lett. 18*;784, 1967.

278. P. Haasen, *Phys. Stat. Solidi 28a*;145, 1975.

279. Ref. 218, p. 532-545.

280. R. Jones, *Phil. Mag. B42*;213, 1980.

281. V. Swaminathan and S.M. Copley, *J. Appl. Phys., 47*;4405, 1976.

282. G.Y. Chin, L.G. Van Uitert, M.L. Green, G.J. Zydzik, and T.Y. Kometani, *J. Am. Ceram. Soc. 56*;369, 1973.

283. J.J. Gilman, *J. Appl. Phys. 45*;508, 1974.

284. R.T. Chen and W.G. Spitzer, *J. Electron. Mat. 10*;1085, 1981.

285. S. Mahajan, W.A. Bonner, A.K. Chin, and D.C. Miller, *Appl. Phys. Lett. 35*;165, 1979.

286. D.Y. Watts and A.F.W. Willoughby, *Mat. Lett. 2*;355, 1984.

287. D. Brasen, *J. Mater. Sci. 11*;791, 1976.

288. D.Y. Watts and A.F.W. Willoughby, *J. Appl. Phys. 56*;1869, 1984.

289. D. Brasen, *J. Mater. Sci. 13*;1776, 1978.

290. M.S. Abrahams, J. Blanc, and C.J. Biocchi, *Appl. Phys. Lett. 21*;185, 1972.

291. G.H. Olsen, M.S. Abrahams, and T.J. Zamerowski, *J. Electrochem. Soc. 121*;1650, 1974.

292. P.B. Hirsch, P. Pirouz, S.G. Roberts, and P.D. Warren, *Phil. Mag. B52*;759, 1985.

293. J. Rabier, H. Garem, J.L. Demenet, and P. Veyssiere, *Phil. Mag. A51*;L67, 1985.

294. A. Lefebvre, P. Francois, and J. Di Persio, *J. Physique Lett. 46*;1023, 1985.

295. Y. Androussi, P. Francois, J. Di Persio, G. Vanderschaeve, and A. Lefebvre, 14th International Conference on Defects in Semiconductors, Paris Aug. 1986.

296. B. Mutaftschiev, in *Dislocation in Solids*, (ed. F. R. N. Nabarro), vol. 5, (North-Holland, Amsterdam, p. 59, 1980.

297. J.C. Brice and G.D. King, *Nature 209*;1346, 1966.

298. J.C. Brice, *J. Cryst. Growth 7*;9, 1970.

299. D.E. Holmes, R.T. Chen, K.R. Elliot, and C.G. Kirkpatrick, *Appl. Phys. Lett. 40*;46, 1982.

300. L.B. Ta, H.M. Hobgood, A. Rohatgi, and R.N. Thomas, *J. Appl. Phys. 53*;5771, 1982.

301. J.M. Parsey, Jr., Y. Nanishi, J. Lagowski, and H.C. Gatos, *J. Electrochem. Soc. 128*;936, 1981; *129*;388, 1982.

302. J. Lagowski, H.C. Gatos, T. Aoyama, and D.C. Lin, *Appl. Phys. Lett. 45*;680, 1984.

303. A.S. Jordan and J.M. Parsey, Jr., *J. Cryst. Growth, 79*;280, 1986.

304. A.S. Jordan, A.R. VonNeida, and R. Caruso, *J. Cryst. Growth 79*;243, 1986.

305. M.G. Milvidskii and E.P. Bochkarev, *J. Cryst. Growth 44*;61, 1978.

306. A.S. Jordan, A.R. VonNeida, and R. Caruso, *J. Cryst. Growth 70*;555, 1984.

307. D.C. Bennett and B. Sawyer, *Bell Syst. Tech. J. 35*;637, 1956.

308. A.S. Jordan, R. Caruso, and A.R. VonNeida, *Bell Syst. Tech. J. 59*;593, 1980.

309. A.S. Jordan, G.T. Brown, B. Cockayne, D. Brasen, and W.A. Bonner, *J. Appl. Phys. 58*;4383, 1985.

310. N. Kobayashi and T. Iwaki, *J. Cryst. Growth 73*;96, 1985.

311. S. Shinoyama, C. Uemura, A. Yamamoto, and S. Tohno, *Japan J. Appl. Phys. 19*;331, 1980.

312. M.S.S. Young, A.S. Jordan, A.R. VonNeida, and R. Carusso, unpublished.

313. M. Duseaux, *J. Cryst. Growth 61*;576, 1983.

314. G. Jacob, in ''Semi-insulating III-V Materials,'' (eds. S. Makram-Ebeid and B. Tuck), UK;Shiva, Nantwich p. 2, 1982.

315. G. Jacob, *J. Cryst. Growth 58*;455, 1982.

316. J. Osaka and K. Hoshikawa, in ''Semi-insulating III-V Materials,'' (eds. D.C. Look and J.S. Blakemore), UK;Shiva, Nantwich, p. 126, 1984.

317. Y. Seki, H. Watanabe, and J. Matsui, *J. Appl. Phys. 49*;822, 1978.

318. K. Katagiri, S. Yamazaki, A. Takagi, O. Oda, M. Araki, and I. Tsuboya, in ''GaAs and Related Compounds,'' Inst. Phys. Conf. Ser. *79*, p. 67, 1986.

319. J. Matsui, H. Watanabe, and Y. Seki, *J. Cryst. Growth 46*;563, 1979.

320. G. Jacob, J.P. Farges, C. Schemali, M. Duseaux, J. Hallais, W.J. Bartels, and P.J. Roksnoer, *J. Cryst. Growth 57*;245, 1982.

321. S. McGuigan, R.N. Thomas, D.L. Barrett, H.M. Hobgood, and B.W. Swanson, *Appl. Phys. Lett. 48*;1377, 1986.

322. S. Guruswamy, J.P. Hirth, and K.T. Faber, *Mat. Res. Soc. Proc., 53*;329, 1986.

323. A. Djemel and J. Castaing, *Euro Phys. Lett. 2*;611, 1986.

324. M.G. Tabache, E.D. Bourret, and A.G. Elliot, *Appl. Phys. Lett. 49*;289, 1986.

325. H.M. Hobgood, S. McGuigan, J.A. Spitznagel, and R.N. Thomas, *Appl. Phys. Lett. 48*;1654, 1986.

326. H. Ehrenreich and J.P. Hirth, *Appl. Phys. Lett. 46*;668, 1985.

327. I. Yonenaga, K. Sumino, and K. Yamada, *Appl. Phys. Lett. 48*;326, 1986.

# CHAPTER 7

# DEFECTS AND DEVICE PROPERTIES

## 7.1 INTRODUCTION

The presence of native defects and dislocations has a profound influence in determining the performance of semiconductor devices. However, the correlation between device characteristics and defects has remained, for most part, elusive. Only in certain instances, such as the dark line defects and the degradation of GaAs-AlGaAs lasers, the effects of defects are clear cut. The reason for this is that it is often difficult to isolate pure dislocation effects when possibility of dislocation-point defect interaction exists. In the case of native defects, it has been even more difficult to correlate device parameters with any specific defect. The implication of such defects is often based on circumstantial evidences.

Defects can be present in the device for a variety of reasons. They can be introduced during epitaxial growth. Dislocations in the substrate can propagate into the epitaxial layers during growth and end up in the active medium. Similarly, point defects can be introduced during growth. The different epitaxial growth conditions favor either excess group III or group V conditions and accordingly native defects characteristic of the specific nonstoichiometry are introduced into the layers. Substrates are heated at high temperatures before commencement of epitaxial growth. Since these temperatures are sometimes higher than the congruent evaporation temperature of the substrates, preferential evaporation of the group V atom occurs, leaving behind a group III rich surface. Although measures are taken to suppress the evaporation of the group V atom, some loss occurs and the resulting nonstoichiometric defects on the surface have never been fully understood nor their effects on device performance have been established.

Besides grown-in defects, defects are also introduced during device fabrication steps, such as implantation, stripe delineation using dielectrics, metallization and so on. Also with dielectrics and metallizations, differences in the thermal expansion coefficients between them and the semiconductor introduces stresses in the device which add an extra dimension to the complexity of the defect interactions in the device by affecting defect motion. In light emitting devices such as lasers and LEDs, nonradiative recombination enhanced defect motion is an important mode of device degradation.

If reliability assurance of a device has to be made based on accelerated life-test results, it is imperative that degradation mechanisms are clearly understood. In the development of $Al_xGa_{1-x}As$ lasers, one witnessed a steady increase in life times to $10^5$ hours from a few seconds when the first continuous operation at room temperature was demonstrated, once the degradation in lasers caused by dark-line defects became understood. Because of the reaped benefits from the study of defects in as far as their effects on device performance are concerned, it has been a field of active research in recent years. In this chapter, we discuss the effects of defects in both electronic and photonic devices.

## 7.2 INTERFACE EFFECTS

In both electronic and photonic devices, the physical and electronic properties of interfaces, such as formed between the semiconductor and other materials such as metals and dielectrics or between different semiconductors, play a crucial role both in determining the suitable process technology for fabricating the device and the stability and reliability of the device once a particular technology is selected. For example, in fabricating a metal-semiconductor field effect transistor (MESFET) for integrated circuits, the height of the Schottky barrier formed between the metal and the semiconductor is an important parameter to be considered. In the case of GaAs, the Schottky barrier height is nearly independent of the metal used to form the contact and is of the order 0.8 to 0.9 eV for n-type GaAs and 0.5 to 0.6 eV for p-type GaAs. The large barrier height for n-type material keeps the carriers away from the surface and thus makes the depletion type MESFET technology a viable one for GaAs integrated circuits. On the other hand, in InP, the Schottky barrier height is only a few tenths of an electron volt which makes it unsuitable for MESFET technology. In contrast to this, the choice of the material system is reversed when one considers the metal-insulator-semiconductor FET (MISFET) technology. In the case of GaAs, there exists a high density of interface states at the semiconductor-insulator interface which pin the Fermi level near midgap. This rules out inversion mode MISFET GaAs devices. By the same token, in InP-insulator interfaces, Fermi level is pinned near the conduction band which allows MISFET technology.

The importance of interfaces has, therefore, made the study of interface properties an area of intense research. Although understanding of the physical and electronic properties of the interfaces has been largely on an empirical level, the application of modern surface analytical techniques holds the promise for more basic understanding and hence better and more reliable electronic and photonic devices. In this section, we review briefly some of the issues concerning interface effects in GaAs and InP.

### 7.2.1 Metal-Semiconductor Interface

Since almost all semiconductor devices require electrical contacts, the metal-semiconductor interface plays a crucial role in determining the performance of the device. The first step in understanding the metal-semiconductor interface is the understanding of atomically clean surfaces of the semiconductor. The barrier height between the metal and the semiconductor can be rather sensitive to the methods of preparation of the interface and hence the properties of a freshly cleaved surface or a surface cleaned in ultra-high vacuum can be very different from that of the surfaces under normal processing conditions.

In the absence of surface states in the energy gap, the formation of Schottky barrier at the metal-semiconductor interface is illustrated in Fig. 7.1. For a n-type semiconductor shown in Fig. 7.1 when the work function of the metal, $\phi_m$, is greater than the work function of the semiconductor, $\phi_s$, electrons are transferred from the semiconductor into the metal leaving behind

**Figure 7.1** Schematic view of the band diagrams for contact between a metal and an n-type semiconductor: (a) $\phi_m > \phi_s$ and there are no interface states (b) same as (a) but $\phi_m < \phi_s$ (c) same as (a) but with a high density of interface states [6].

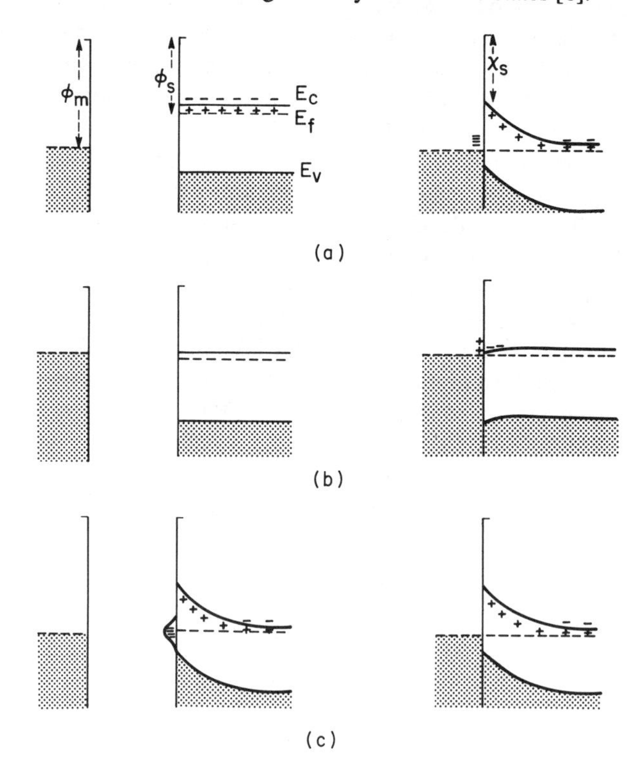

positively charged donor ions in the space charge layer of the semiconductor. If the surface dipoles that contribute to $\phi_m$ and $\chi_s$ are unchanged, then the Schottky barrier height is equal to $\phi_m - \chi_s$ where $\chi_s$ is the electron affinity of the semiconductor [1]. Naturally, the barrier height will vary when different metals form the interface with the semiconductor depending upon the work function of the metal. If $\phi_m < \phi_s$, then electrons are injected into the semiconductor and the Schottky barrier height is zero (Fig. 7.1(b)). That is, an "ohmic" contact is formed. The situations shown in Figs. 7.1(a) and 7.1(b) are reversed for a p-type semiconductor; that is, a barrier is formed for $\phi_m < \phi_s$ and a ohmic contact for $\phi_m > \phi_s$.

A more realistic picture of the Schottky barrier formation would be one where surface states in the energy gap are assumed, as shown in Fig. 7.1(c). If acceptor-like surface states exist, a space charge layer is formed even in the absence of the metal. For a high density of surface states, Schottky barrier height is independent of the metal since charge transfer occurs mainly via these states [2]. In the presence of surface states of density $D_s$ ($cm^{-2}\ eV^{-1}$) uniformly distributed in the energy gap, the Schottky barrier height, $\phi_b$ is given by [3, 4]

$$\phi_b = \gamma(\phi_m - \chi_s) + (1-\gamma)(E_g - \phi_0) \tag{7.1}$$

where $E_g$ is the energy band gap, $\phi_0$ is the energy up to which the surface states are occupied and empty above it, and $\gamma$ is related to $D_s$

$$\gamma = \frac{\varepsilon_1 \varepsilon_0}{\varepsilon_1 \varepsilon_0 + e\delta D_s} \tag{7.2}$$

where $\delta$ is the width of an insulating layer of permittivity, $\varepsilon_1$ separating the metal and the semiconductor, $\varepsilon_0$ is free space permittivity and e is the electronic charge. Inspection of Eqs. (7.1) and (7.2) would show that for $D_s = 0$, $\phi_b = \phi_m - \chi_s$ and for large $D_s$, $\phi_b = E_g - \phi_0$, corresponding to the two cases of Schottky barrier formation with and without the surface states as discussed. The two barrier heights are referred to as the Schottky limit and Bardeen limit, respectively. Interface states of density $10^{12} - 10^{13}$ $cm^{-2}\ eV^{-1}$ are enough to affect Schottky barrier height. The terminology, pinning of the Fermi level $E_F$, applies to the second situation when $\phi_b$ is independent of the metal. It should be noted that $\phi_b$ can be lower than that given by Eq. (7.1) because of the electric field at the metal-semiconductor interface and by image forces. The contribution of the first term is, however, only a few meV. Other things being equal, image force lowering effect would tend to yield a lower value of the barrier height measured by current-voltage technique compared to capacitance voltage technique.

The Schottky barrier formation described by Eq. (7.1) while it explains the pinning of the Fermi level as encountered in III-V semiconductors such as GaAs and InP, is nevertheless too simple. Heine [5] noted that since the metal on the surface of the semiconductor can change the form of the potential and of the charge associated with the "dangling" bonds that are responsible for the surface states, the pinning states signify the microscopic nature of the interfacial region between the metal and the semiconductor. In this sense, the pinning states are really interface states. The microscopic nature of the interfacial region is determined by several parameters such as whether the metal when deposited on the semiconductor is stationary or can diffuse along the surface, how abrupt the interface is, whether the metal atoms are incorporated into the semiconductor, and if so, whether they occupy substitutional or interstitial sites, and whether they are electrically active, whether in the process of occupying sites in the semiconductor the metal atoms displace matrix constituents and create native defects at the interface, and so on [6]. The formation of the native defects at the interface is the underlying theme of the defect models of Fermi level pinning in III-V semiconductors.

Before a detailed picture of the origin of the interface states that pin, the Fermi level at the semiconductor-metal or semiconductor-insulator interface can be attempted, the understanding of the atomically clean free surface of the semiconductor is imperative. In this connection, the (110) cleaved surface of GaAs is the most extensively studied. The ideal atomic arrangement on (001), (110), and (111) surfaces of a zinc blende type semiconductor is illustrated in Fig. 7.2. The polar (100) and (111) surfaces are differentiated from the (110) surface in that they consist of all group III or group V atoms unlike the (110) surface which consists of equal numbers of both atoms. In general, clean free surfaces of semiconductors do not retain their ideal structure but undergo surface reconstruction. This is particularly true for the polar (100) and (111) surfaces. The nonpolar (110) surface undergoes no reconstruction but the surface atoms assume a relaxed configuration shown in Fig. 7.2. The As atoms move outwards and the Ga atoms move inwards, consistent with the pentavalent As and trivalent Ga left behind after bonds are broken from the surface, minimizing their bond energies by assuming $s^2p^3$ and $sp^2$ configuration rather than the $sp^3$ configuration of the bulk bonds [7,8]. As a result of the relaxation of the surface atoms, the filled As surface states and the empty Ga surface states move out of the energy gap leaving no states in the gap as illustrated in Fig. 7.3 [9 - 11]. Several experiments have confirmed the absence of surface states in GaAs [12 - 15].

**Figure 7.2** The structure of (001), (110) and (111) surfaces of GaAs [6].

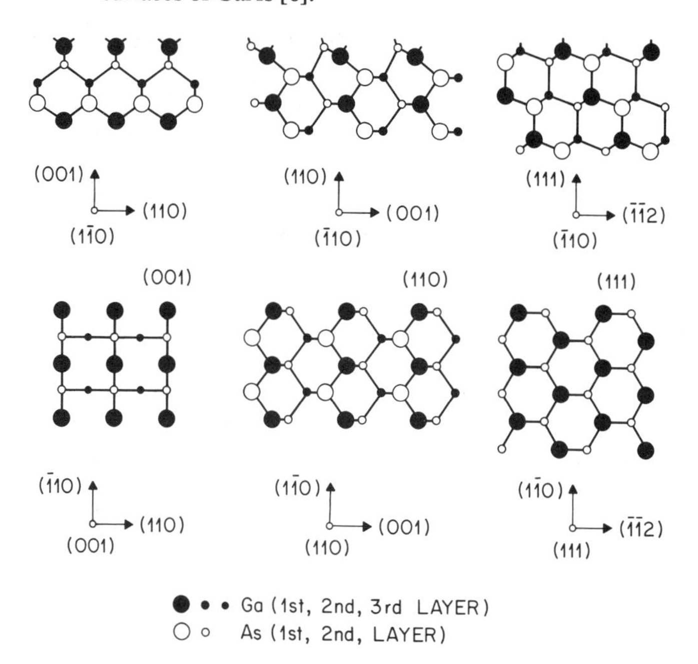

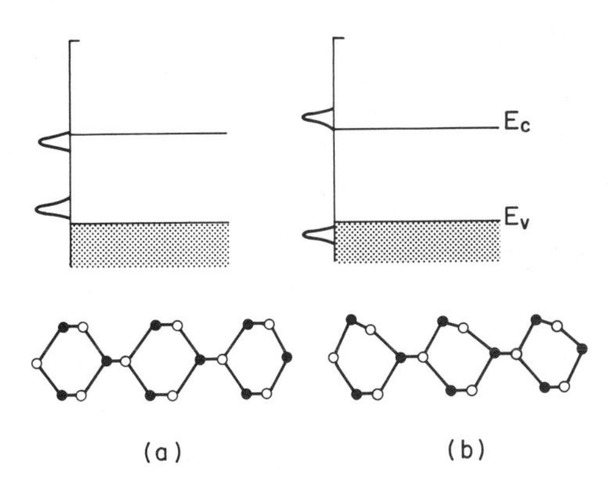


**Figure 7.3** (a) Surface states in the band gap for an unrelaxed surface (b) The surface states are driven from the gap by the relaxation of surface atoms [6].

The absence of surface states in the gap for the clean high quality (110) surfaces means that there will be no pinning of the Fermi level. The situation can be drastically different if cleaving process leaves steps and other imperfections. Then gap states may be expected. The chemical composition of the surface may be far from stoichiometric which can also lead to gap states and Fermi level pinning. Street et al., [16, 17] have observed a new photoluminescence band in vacuum cleaved InP containing steps and they have associated the band with defects at the steps. The presence of defects associated with the cleaving damage and/or that associated with nonstoichiometry at the surface, their interactions with the metals and adsorbed gases on the semiconductor surface and the intermixing of the semiconductor and the metal would all play an important role in determining Fermi level pinning.

**Metal on atomically clean semiconductor surfaces**

Although the interfaces formed in real electronic devices are very different from those formed between atomically clean semiconductors (e.g., a vacuum cleaved (110) surface) and metals or insulators, in order to understand the physical principles of Schottky barrier formation on semiconductors, it is important to measure barrier heights when the interfaces are formed under well controlled and identifiable conditions. Schottky barriers are generally measured in one of four ways [18]: (a) measuring the current-voltage characteristics and fitting them to the appropriate theories of carrier transport. The Schottky barrier measured this way can be lower than the true value because of electric field in the semiconductor and image-force lowering [4]. (b) measuring the capacitance of a diode as a function of applied voltage. The barrier height measured this way does not include the contribution from image-force lowering since no carriers are transported across the interface during the capacitance measurement. The barrier height can be overestimated in the presence of an interfacial layer such as an oxide layer [19, 20]. (c) measuring the spectral response of the diode photocurrent. Errors of the order of 50 meV can be introduced in the determination of barrier height caused by thermal excitation and quantum-mechanical tunneling effects [21]. (d) measuring the shift of core electron emission peaks in the semiconductor by photoelectron spectroscopy during the deposition of the metal for metal coverages of the order of monolayers or even fraction of a monolayer [22]. For the reasons noted, there can be as much as 0.1 eV or more variation in the Schottky barrier heights determined by the different methods for a given metal.

The method of photoelectron spectroscopy is the one most ideally suited for studying Schottky barrier formation *in situ* as metals are deposited on vacuum cleaved (110) surfaces.

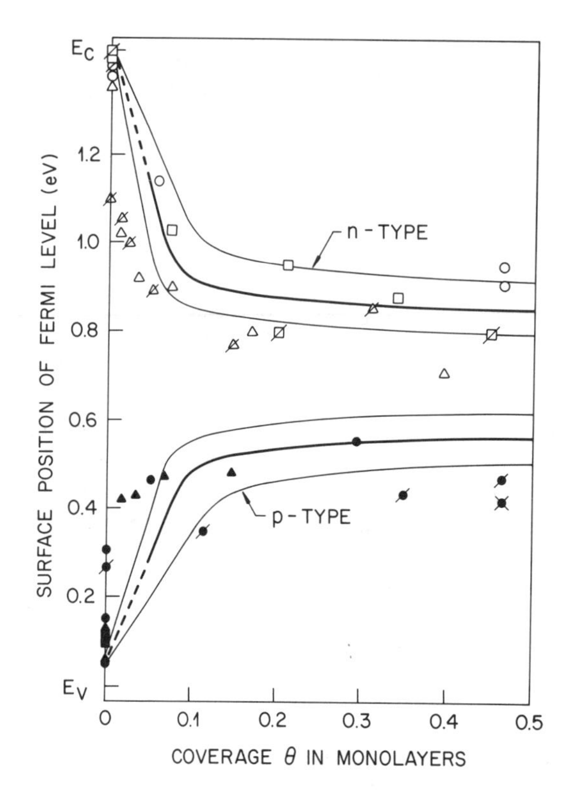


**Figure 7.4** Surface position of the Fermi level versus adatom coverage on the GaAs (110) surface. Squares represent Al; circles represent Ga; and triangles represent In. Data for Al, Ga and In overlayers are from photoemission measurements [33]. Solid lines are for Ge overlayers obtained by contact potential difference measurements [24].

Especially in the last few years, the availability of tunable monochromatic radiation, ranging in photon energies from the ultraviolet to the x-ray region, from a synchrotron radiation source has provided tremendous insights into the metal-semiconductor interaction. By following the changes in the Fermi-level position at the surface as a function of submonolayer metal coverage, new energy levels created in that process can be inferred. Information about the chemistry of the metal-semiconductor interaction can be obtained by the energy shifts in the core levels.

Figure 7.4 shows Fermi level position at the surface as a function of Al, Ga, and In metal coverage on n- and p-type (110) GaAs [23]. Several points can be made from the data shown in Fig. 7.4. First, the Fermi level position reaches a constant value for n- or p-type GaAs after only about 0.1 monolayer coverage of metal. Second, the Fermi level position in either n- or p-type material is independent of the metal. In n-type GaAs, the Fermi level is pinned at ~0.8 eV above the valence band and in p-type material it is pinned at ~0.5 eV above the valence band. Third, similar pinning positions were obtained with different measurement techniques. Figure 7.4 also shows the Fermi level position obtained as a function of Ge overlayers using the Kelvin probe (contact potential) measurements [24]. Since the Schottky barrier heights were in close agreement with the pinning positions, it is believed that both pinning and barrier height formation are controlled by the same mechanisms. Similar behavior was exhibited by InP also in that constant pinning positions irrespective of the type of metal coverage were obtained for n- and p-type semiconductor. In the case of GaAs, the condition of the surface that is, vacuum cleaved, sputtered or annealed, did not change the pinning energies very much.

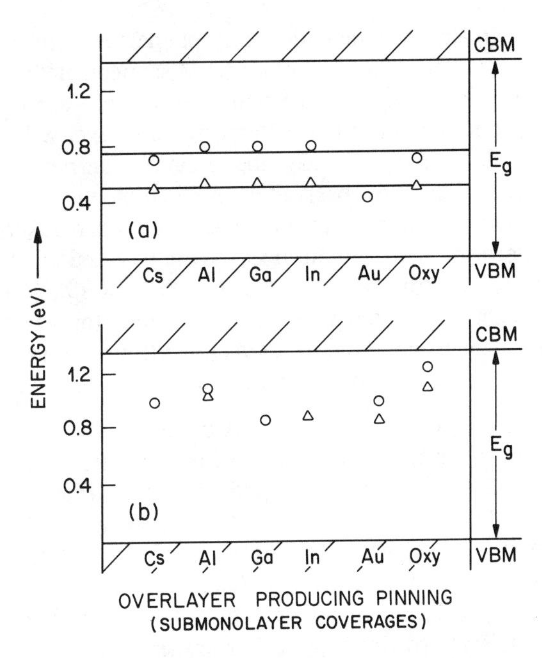


**Figure 7.5** Stable Fermi level positions for a wide range of overlayers on a) GaAs and b) InP (110) surfaces. Circles represent n-type and triangles represent p-type material [23].

Figure 7.5 shows the Fermi level positions measured by photoelectron spectroscopy for a range of metals including very electropositive Cs and also the very electronegative non metal O [23]. Note that in InP, the pinning energies are closer to the conduction band for both n- and p-type material unlike in GaAs where they are near the middle of the energy gap. More detailed measurements by photo emission, C-V and I-V techniques in InP indicate that the pinning positions fall into two regions; one 0.1 to 0.3 eV below the conduction band and the other near the middle of the gap as shown in Fig. 7.6 [25, 26]. The surface condition of the InP substrate and exposure of the surface to gases affected the pinning position. For metal coverage on etched surfaces the Fermi level was pinned closer to mid gap. For Au on InP, the barrier height decreased when the surface was exposed to $10^9$L of water [27].

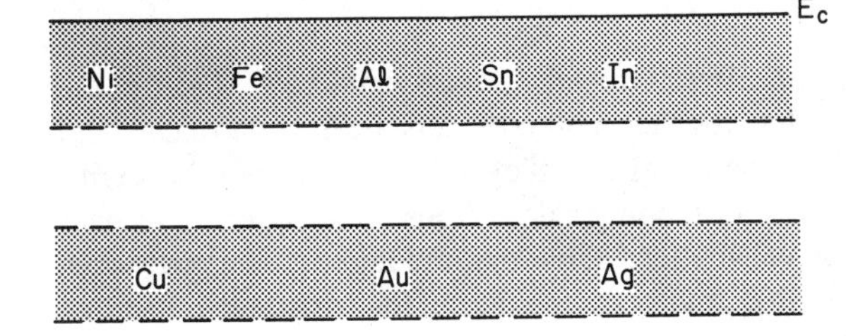


**Figure 7.6** Fermi level positions for various metals on cleaved n-type InP crystals [25,26].

Several models have been proposed to explain the rather intriguing phenomenon of Fermi level pinning and Schottky barrier formation in GaAs, InP and other III-V semiconductors and their alloys. Equation (7.1) shows that depending on the interface states density, one can obtain for $\phi_b$ the value corresponding either to the Schottky limit or to the Bardeen limit. Even a density of $10^{12}-10^{13}$ $cm^{-2}$ $eV^{-1}$ interface states is enough to make the Schottky barrier height independent of the metal work function. Since pinning is observed even on an atomically clean vacuum cleaved (110) surfaces of the semiconductor which does not contain any inherent surface states, it is necessary to postulate that interface states are formed by the interaction of the deposited metal and the relaxed (110) surface. Kurtin et al., [28] showed that for strong ionic semiconductors, the Schottky barrier height is a function of $\phi_m$ while for covalent semiconductors, it is independent of $\phi_m$. But reanalysis of the data which led to this proposition makes this distinction between the covalent and ionic semiconductors less certain [29].

Mead and Spitzer [30] developed an empirical model for the Schottky barrier height according to which $\phi_b$ is approximately 2/3 $E_g$. This rule gives $\phi_b$ (GaAs) ~ 0.9 eV in rough agreement with the value of 0.8 eV at least in n-type GaAs. Similarly this rule is applicable for GaP and AlAs. However, for InP, InAs, and GaSb this rule does not give the correct barrier heights. Further extension of the 2/3 $E_g$ rule by McCaldin et al., [31] who used Au as a reference metal to a variety of p-type semiconductors gave the "common anion rule." According to this rule, $\phi_b$ varied inversely with the anion electronegativity and semiconductors which have the same anion, for example, GaAs and InAs, should have the same barrier height. For ternary $Ga_xIn_{1-x}As$ and $Al_xGa_{1-x}As$, the "common anion rule" would imply that the Schottky barrier height of p-type material is independent of x. Such a behavior is observed for $Ga_xIn_{1-x}As$ but not for $Al_xGa_{1-x}As$.

Since the atomic relaxation of atoms on clean cleaved (110) surfaces of GaAs and InP drives the intrinsic surface states out of the energy gap, it is conceivable that the adsorption of metal atoms reduces the surface relaxation and keeps the surface states in the gap. But a wide variety of metals giving the same pinning energy raises doubts about such a situation. On the other hand, the presence of the metal may give rise to extrinsic metal-induced gap states which may pin the Fermi level. For Al-GaAs contact, calculations treating the metal as a structureless "jellium," show that the metal wave functions penetrate into the semiconductor [32]. These calculations have not been done for all metals and do not explain the variation in $\phi_b$ between different semiconductors.

To account for the lack of variation in $\phi_b$ for different metals on cleaved GaAs, Spicer et al., [13] proposed the "defect model." This model states that native defects such as vacancies and antisite defects are created by the deposition of the metal and they subsequently pin the Fermi level. Support for this model is provided by the tight-binding calculations of defect energies near the surface of (110) GaAs and InP [33]. These calculations show that the highest occupied level of cation or anion vacancies, for example, $V_{Ga}$, $V_{In}$, and $V_{As}$, and $V_p$, in the surface layer and in the second atomic layer moves towards lower energy compared to their positions in the bulk. In particular, the energy level of $V_p$ is closer to the conduction band edge while that of $V_{As}$ is about 0.5 eV above the valence band in agreement with the surface Fermi level positions in InP and GaAs, respectively. The applicability of these tight binding calculations was also illustrated for $Al_xGa_{1-x}As$ and $Ga_xIn_{1-x}As$. Figure 7.7 shows the calculated energy levels of $V_{As}$ near surface and the measured Fermi level pinning energies in $Al_xGa_{1-x}As$ and $Ga_xIn_{1-x}As$ [34]. The

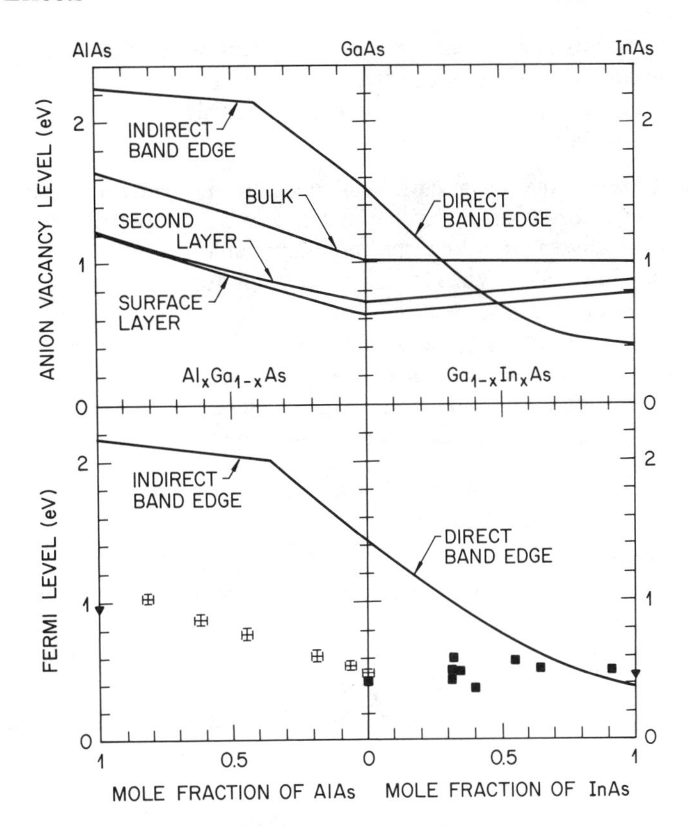


**Figure 7.7** The defect model for Schottky carrier formation. The predicted pinning energies based on pinning by anion vacancies are shown in (a) for (AlGa)As and (GaIn)As. The measured pinning levels are shown in (b) [34].

correlation between the $V_{As}$ energy levels and the pinning positions strongly suggests that $V_{As}$ can easily lead to the pinning behavior. Allen and Dow [35] considered pinning by antisite defects and found trends similar to that seen for $V_{As}$ in Fig. 7.7. The defect model has certain natural appeal in the case of surfaces prepared by sputtering or annealing where they can be expected to be nonstoichiometric. In the case of vacuum cleaved surfaces, it has to be assumed that the deposition of even a fraction of a monolayer of a metal generated native defects at the surface of the order $10^{12}$ cm$^{-2}$.

Since the chemical reactivity of the adsorbed metal and the semiconductor is an important parameter in affecting the Schottky barrier formation, Brillson [36] proposed that the heat of the interface reaction, defined as the difference in semiconductor heat of formation and the heat of formation of the metal-anion complex, is related to the Schottky barrier height. For large negative heat of interface reaction, giving the most abrupt interfaces [37], there is a large interfacial dipole which reduces the barrier height because of charge transfer between the adsorbed metal atoms and the semiconductor surface atoms. The effect of the chemical reaction at the interface on $\phi_b$ is illustrated in Fig. 7.8 [38] for barriers on atomically clean n-type InP. For negative heat of reaction, meaning stable metal-phosphorus compound, (e.g., Ni, Fe, Al) the barrier heights are low compared to positive heat of reaction, meaning less stable metal-P compound (e.g., Cu, Ag, Au).

The interface between the metal and semiconductor in many instances is not always atomically abrupt. Even for deposition of the metals at room temperature there is considerable interfacial reaction for coverages below one monolayer and the interfaces are disordered, for example, Al on (110) GaAs [39]. When interfaces are imperfect and disordered, one other model besides the "defect model" for the Schottky barrier heights is the effective-work-function (EWF) model [40, 41]. This model states that the barrier height is not related to the work function of the metal but is related to the weighted average of the work functions of the interface phases present either prior to the deposition of the metal on chemically etched "real" surfaces or formed during metallization. Thus $\phi_b$ in a n-type semiconductor is given by

$$\phi_b = \phi_{eff} - \chi_s \tag{7.3}$$

where $\phi_{eff}$ is the weighted average of the work functions of the interface phases. In III-V semiconductors such as GaAs and InP, it is known that metallization of Au leads to excess Ga or In in the metal leaving behind the group V atoms at the interface. In this case $\phi_{eff}$ equals the work function of the group V atom. In the case of GaAs with excess As at the metal-semiconductor interface $\phi_b \sim \phi_{As} - \chi_s$ which gives $\phi_b \sim 0.9$ eV. This is in agreement with the value of $\sim 0.8$ eV frequently observed for many metallizations. Table 7.1 lists the values of the barriers for Au on several III-V semiconductors. It can be seen that $\phi_{eff} \approx \phi_V$, the work function of the group V atom.

The EWF model is particularly attractive in explaining the significant variations in $\phi_b$ obtained on the same metal-semiconductor system by several workers and also the variations

**Figure 7.8** Schottky barrier height (eV) on atomically clean n-type InP as a function of the heat of reaction for the most stable phosphorous compound which can form between the contact metal and InP [38].

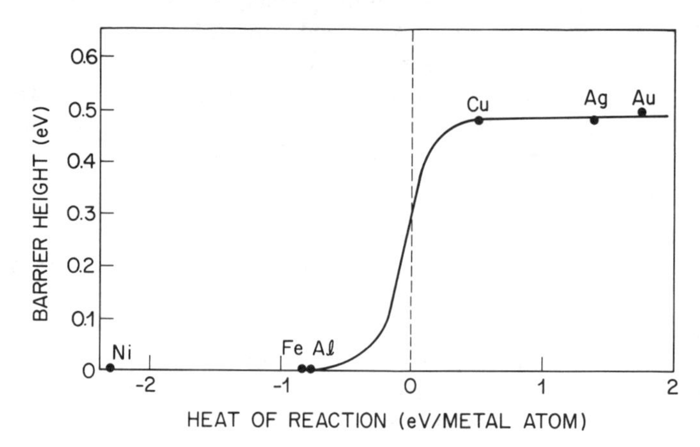

**Table 7.1**
Au Schottky barriers on III-V semiconductors [40]

| Anion | III-V | $\phi_{bp}$ | $\chi_s + \phi_{bn} = \phi_{eff}$ | $\phi_V$ |
|---|---|---|---|---|
| P | GaP | 0.96 | 4.9 | 5.0 |
| | InP | 0.85 | 4.9 | 5.0 |
| As | AlAs | 0.9 | 4.7-5.1 | 5.0 (4.8) |
| | GaAs | 0.5 | 5.0 | 5.0 |
| | InAs | 0.3-0.5 | 4.8-5.0 | 5.0 |
| | GaInAs | 0.7-0.5 | 4.8-4.9 | 5.0 |
| Sb | AlSb | 0.54 | 4.7 | 4.8 (4.7) |
| | GaSb | 0.1 | 4.7 | 4.8 |
| | InSb | ~0.1 | 4.8 (77K) | 4.8 |

$\phi_V$ values in parenthesis are measured values. $\phi_{bp}$ and $\phi_{bn}$ are, respectively, Schottky barrier heights on p- and n-type semiconductors.

depending upon the measurement technique, whether C-V or I-V. Since the metallurgical reactions and the final reaction phases would be determined by kinetic factors, differences in the starting surfaces and metallization conditions can easily be the source of variations in $\phi_b$. Further, the C-V technique measures an average value of $\phi_b$ associated with all the interfacial phases while I-V measures the value associated with the lowest barrier phase. In the case of the formation of the barrier for very low metal coverages on vacuum cleaved (110) surfaces where the assumption of group V rich interfacial phases may not be appropriate, the EWF model assumes that the impinging metal atom knocks out the group III and group V atoms pair wise from the lattice onto the surface. $\phi_{eff}$ is then taken to be an appropriate weighted average of the work function of the atoms or microclusters on the surface.

In summary, though several models have been proposed to explain the Schottky barrier formation in III-V semiconductors, no one model explains all the observed results. In the case of chemically prepared (100) surfaces as encountered in device fabrication, the situation is even more complex compared to that in atomically clean (110) surfaces. This makes it difficult to develop processes that will produce reproducible properties of the metal-semiconductor junctions in various device applications.

**Ohmic contacts**

In the fabrication of GaAs and InP based electronic and opto-electronic devices, obtaining ohmic contacts with low resistance is an important consideration for optimum device operation and reliability. In describing the Schottky barrier formation, we saw that for $\phi_m < \phi_s$ (Fig. 7.1(b) there is no barrier for injection of electrons into the semiconductor. In other words, the metal-semiconductor interface is ohmic. A parameter which describes ohmic behavior is the contact resistance, $r_c$, which should be as low as possible for a good ohmic contact. It depends on such factors as the barrier height, doping level, and effective mass of majority carriers. The functional dependence of $r_c$ on these factors is determined by the mechanism of carrier transport across the barrier that is operative. For typical doping concentrations encountered in various devices, the transport of charge carriers across the interface, and the associated space charge region is via thermionic emission or thermionic-field emission. For n-type GaAs at room temperature the transition from thermionic emission to thermionic field emission occurs for doping concentrations in the range $10^{17} - 10^{18}$ $cm^{-3}$. In thermionic emission, the transport of carriers across the top of the barrier gives a current-voltage characteristic which is diode like [42]

$$J = J_0 \exp \left( \frac{qV}{nkT} - 1 \right) \tag{7.4}$$

where n is the ideality factor. Equation (7.4) gives for the contact resistivity

$$r_c = \left( \frac{dJ}{dV} \right)^{-1}_{v=0} = \frac{nkT}{qJ_0} \tag{7.5}$$

where $J_0 = A^* T^2 \exp(-q\phi_b / kT)$. $A^*$ is called the Richardson's constant and is given by $4\pi qm^* k^2 / h^3$, where $m^*$ is the effective mass of the majority carriers, k is the Boltzman's constant and h is Planck's constant. Note that $J_0$ is independent of doping but is a function of $\phi_b$. The current-voltage characteristic for thermionic-field emission is somewhat more complicated than the simple diode expression given in Eq. (7.4). In this case, $r_c$ is dependent on the doping level, N, besides $\phi_b$ and $m^*$.

For very high doping levels, $>10^{18}$ $cm^{-3}$, as the width of the space charge region decreases, field emission takes over. Quantum mechanical tunneling of electrons through the barrier at energies near the Fermi level occurs. This mode of current transport, which is essentially temperature independent, can give rise to ohmic contacts in III-V semiconductors. The current-voltage behavior is given by a expression similar to that in Eq. (7.4) [43].

$$J = J_0' \exp \left( \frac{qV}{E_0} \right) \tag{7.6}$$

where $E_0$ is a parameter having dimensions of energy and is proportional to $(N / m^*)^{1/2}$ and $J_0'$ is similar to $J_0$ besides having other constants such as $E_0$ and Fermi energy in the semiconductor. Equation (7.5) then gives

$$r_c \propto \exp \left[ \phi_B \left( \frac{m}{N} \right)^{1/2} \right] \tag{7.7}$$

When current transport occurs entirely by thermionic emission, to obtain low $r_c$, $\phi_b$ has to be reduced. On the other hand, increasing the doping level, which is relatively easier than reducing $\phi_b$, leads to quantum mechanical tunneling and $r_c$ is inversely proportional to $N^{1/2}$. This approach of increasing the dopant concentration in the semiconductor just below the metal surface is widely used to obtain low resistance ohmic contacts in semiconductors. The thickness of the heavily doped region should be larger than the depletion width of the metal-semiconductor junction. Figure 7.9 illustrates the band diagrams for ohmic contacts in n- and p-type semiconductors with $n^+$ and $p^+$ regions [18]. For the contact potential of the $n^+ / n$ or $p^+ / p$

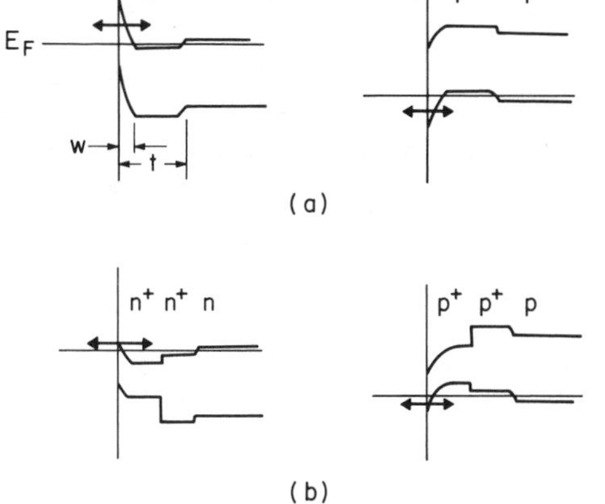


**Figure 7.9** Schematic band diagrams for ohmic contacts at zero bias. Current flow by majority carriers is shown by the arrows. (a) Homojunction $n^+$-n or $p^+$-p-contacts. (b) Heterojunction $n^+$-$n^+$-n or $p^+$-$p^+$-p contacts [18].

junction less than or equal to kT/q, current flow across the barrier is governed by field emission. One way to combine low $\phi_b$ and high N to reduce $r_c$ is to form a heterojunction contact by introducing a heavily doped narrow band gap semiconductor as shown in Fig. 7.9(b).

The heavily doped $n^+$ or $p^+$ regions can be formed by diffusion, ion implantation, epitaxy and alloying. While all these four techniques have been used, alloying is the most widely used one. Diffusion can be accomplished on wide area or selectively in areas defined by some dielectric mask on the wafer. It is done typically at temperatures $\geq 0.5\ T_m$, where $T_m$ is the melting point in degrees Kelvin either in a closed ampoule or in a semiopen tube furnace using either the pure element, or a suitable alloy or compound of the dopant species. Since the diffusion temperatures are above the congruent evaporation temperatures of GaAs and InP, some provision must be made to prevent the preferential loss of the group V element. This is accomplished by providing excess As or P vapor pressure in the diffusion ampoule. When diffusion is done selectively using a mask, lateral diffusion under the mask should be considered in designing the time, temperature of diffusion, and the mask width. Stress at the interface of the dielectric mask and the semiconductor can assist lateral diffusion complicating matter further [44]. In some cases, laser assisted diffusion has been performed as for example, in forming $n^+$ regions in GaAs using Sn from a spin-on film of $SnO_2$ / $SiO_2$ [45] and $p^+$ regions in InP using Cd from the photolysis of diethyl Cd [46].

The maximum doping achieved by diffusion is determined by the solid solubility of the dopant at the diffusion temperature. At about 800°C the solid solubility limit of the commonly used dopants is of the order $10^{19}$ $cm^{-3}$. Doping levels in excess of the solid solubility limit can be achieved by ion implantation followed by a high temperature furnace annealing or rapid thermal annealing to activate the dopant. Once again loss of the group V constituent must be suppressed. Instead of furnace annealing, laser or electron-beam assisted annealing has also been employed. Ohmic contacts with $r_c \sim 2 \times 10^{-5}\Omega$ $cm^2$ have been reported in Te-implanted and laser annealed GaAs [47]. Electron beam annealing has been used to produce contacts with $r_c < 6 \times 10^{-6}\Omega$ $cm^2$ in Se-implanted GaAs [48,49]. Similarly, ohmic contacts of $r_c \sim (0.5\text{–}2) \times 10^{-4}\Omega$ $cm^2$ have been reported in Zn and Cd implanted InP where laser annealing was used to activate the dopants [50].

Thin layers of heavily doped regions have also been grown epitaxially especially by MBE or VPE where dopant incorporation may be governed by kinetics considerations and thus dopant concentrations greater than the solubility limit may be achieved. Barnes and Cho [51] obtained $r_c \sim 2 \times 10^{-6} \Omega\ cm^2$ using Sn as the dopant in MBE grown GaAs. Tsang [52] reported low resistance contacts in n- and p-type GaAs using, respectively, Sn and Be as the dopants. Since no alloying or annealing is required in the contact formation these epitaxially doped regions can give not only low $r_c$ but also can produce contacts with smooth morphology. Further, *in situ* metallization under high vacuum MBE conditions can lead to absence of oxide contamination at the interface.

Alloying is the most widely used method for forming ohmic contacts in GaAs and InP. A thin film of a metal containing the dopant atom is deposited on the semiconductor by some means such as evaporation, sputtering, or electroplating. The wafer is then heated to a temperature above the melting temperature of the metallic film. The molten film dissolves a thin layer of the semiconductor and on cooling the dopant in the film gets incorporated into the semiconductor. Generally, an eutectic alloy of the metal-dopant is chosen in order to keep the alloying temperatures low. Uniform wetting of the surface of the semiconductor wafer by the molten alloy is important to obtain smooth surface morphology. Towards achieving this, a nonreactive cap layer may be deposited over the metallic film. In addition, careful preparation of the semiconductor surface, alloying temperature and time, the thickness of the metallic film and the order of metal depositions if multilayer metallization is used are all important. Preparation of the wafer surface involves the usual degreasing step using solvents. Removal of the surface oxide layer, if necessary, may be achieved by thermal desorption [53] or sputter cleaning prior to metal deposition [54]. However, the surface damage introduced by the sputtering may prove to be more harmful than the presence of the oxide layer. Each laboratory may develop its own surface preparation conditions giving an unique but undetermined starting surface and also different temperature-time cycle for alloying. These differences make comparison of the results of different workers difficult. In this sense, ohmic contact technology is more an art than science.

In the case of GaAs and InP alloying is usually done at temperatures $\leq 400°C$ for times not exceeding a few minutes. Usually alloying is done in a open tube furnace in a flowing $H_2$ or $N_2$ gas or preferably forming gas (15% $H_2$ 85% $N_2$). Measures to suppress the preferential evaporation of the volatile group V component may not be needed but if necessary, excess group V source should be provided. Another solution to suppress the loss of group V element is to employ pulsed electron beam [55] or laser annealing [56, 57] or rapid thermal annealing [58, 59] to heat the surfaces rapidly. Alternately, low temperature solid phase epitaxy, that is, sintering, can be used to form ohmic contacts [60].

Table 7.2 lists the commonly used ohmic contacts to GaAs, InP, GaInAs, AlGaAs, and GaInAsP semiconductors. Generally, Au or Ag based alloys containing Zn or Be for contacting p-type material and Ge or Sn for contacting n-type material are used. The Au-Ge contact which is widely used for n-type material was first developed for GaAs Gunn diodes [63]. The composition is that of the eutectic composition (88% Au, 12% Ge by weight). A layer of Ni deposited on top of the Au-Ge film improves the wetting and increases the solubility of GaAs. Typically, the thicknesses of the layers are 1000 to 3000 Å for Au-Ge and 100 to 500 Å for Ni and they are deposited by electron beam evaporation. The contact is alloyed at a temperature slightly above

**Table 7.2**
Alloyed ohmic contacts to GaAs, InP, AlGaAs, GaInAs and GaInAsP [18]

| Semiconductor | Type | Contact Material | Minimum $r_c$ ($\Omega cm^2$) | Majority Carrier Concentration ($cm^{-3}$) | Ref. |
|---|---|---|---|---|---|
| GaAs | p | Au-Zn | $r_c \approx (1.8\times10^{18})/p^{1.3}$ | $10^{17}-10^{19}$ | 61 |
| | p | Ag-Zn | $2\times10^{-5}$ | $2\times10^{17}$ | 62 |
| | n | Au-Ge-Ni | $r_c \approx (1.8\times10^{12})/n$ | $10^{15}-10^{19}$ | 63-66 |
| $Al_{0.4}Ga_{0.6}As$ | p | Al | $2\times10^{-5}$ | $2\times10^{19}$ | 65 |
| | p | Au-Zn | $8\times10^{-6}$ | $2\times10^{19}$ | 65 |
| | n | Au-Ge-Ni | $2\times10^{-4}$ | $1\times10^{18}$ | 65 |
| InP | p | Au-Be | $r_c \approx (1\times10^{14})/p$ | $10^{16}-10^{19}$ | 67 |
| | p | Au-Zn | $5\times10^{-3}$-$1.1\times10^{-4}$ | $10^{16}-10^{18}$ | 68, 69 |
| | n | Au-Ge-Ni | $3\times10^{-5}$-$8\times10^{-7}$ | $3\times10^{16}$-$8\times10^{17}$ | 70, 71 |
| | n | Au-Sn | $1.8\times10^{-6}$ | $3\times10^{18}$ | 72 |
| $Ga_{0.47}In_{0.53}As$ | p | Au-Zn | $2\times10^{-5}$ | $5\times10^{18}$ | 73 |
| | n | Au-Ge-Ni | $5\times10^{-7}$ | $1\times10^{17}$ | 74 |
| | n | Au-Sn | | | 75 |
| GaInAsP | p | Au-Zn | $10^{-4}-10^{-5}$ | $2\times10^{18}$-$5\times10^{18}$ | 73 |
| | n | Au-Ge-Ni | $5.8\times10^{-6}$ | $1\times10^{17}$ | 74 |

Au-Ge eutectic temperature (360°C). Time and temperature of the alloying cycle plus the heating and cooling rates vary from laboratory to laboratory. But use of rapid heating and cooling rates has been reported to give better results [58, 59].

For p-type material, Au-Zn and Au-Be alloy contacts are most often used. The Zn or Be concentration varies between 5 to 20 percent. In the case of Au-Zn, contact alloying takes place at 400 to 500°C for a few minutes. An additional layer of Au under the Au-Zn may help to improve its adhesion on the semiconductor [61]. Thicker Au may also be deposited on top of the p-contact to reduce the sheet resistance of the contact metal and/or to facilitate bonding of the device to any heat sink. Since excess Au can lead to excessive out diffusion of the group III metal, barrier layers are introduced between the Au-Zn contact and the thicker Au layer. For example, the p-metallization for GaInAsP injection lasers consisted of 50Å-150Å-800Å thick e-beam evaporated Au-Zn-Au layer which was alloyed at 430°C for 20 seconds in forming gas. After alloying, a 1000Å thick Ti and 1500Å thick Pt layers were sputter deposited on top. The Ti-Pt layer acted as a barrier between the Au-Zn p-contact and thicker Au top layer [76].

In the case of InP, it is more difficult to form low-resistance ohmic contacts to p-type material than to n-type material (Table 7.2). This follows from the fact that $\phi_b$ (p-type) > $\phi_b$ (n-type) and $m_h^* > m_e^*$. Both the conditions make tunneling of holes through the barrier more difficult than tunneling of electrons. Thus according to Eq. (7.7) for comparable doping levels, $r_c$ (p-type) > $r_c$ (n-type). For GaAs, $\phi_b$ (n-type) > $\phi_b$ (p-type) and Eq. (7.7) would give $r_c$ (n-type) > $r_c$ (p-type). But experimentally, the differences between $r_c$ for n- and p-type GaAs are not significant (Table 7.2).

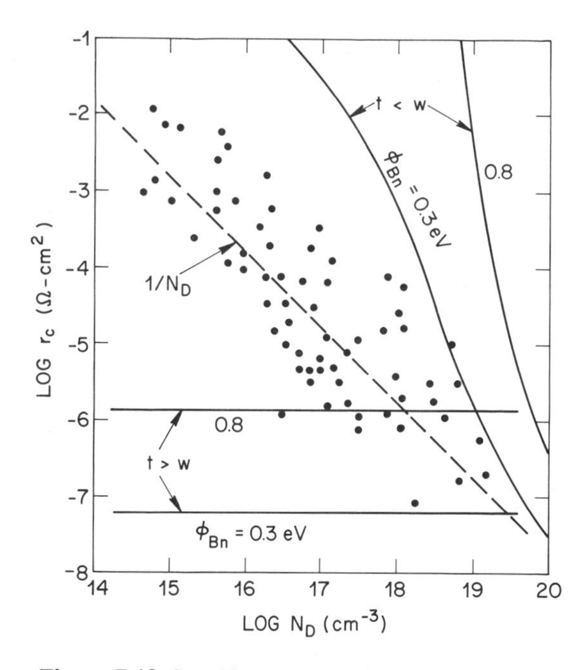


**Figure 7.10** Specific contact resistance $r_c$ versus substrate doping $N_D$ for Au-Ge-Ni alloyed contacts to n-type GaAs. The data points follow a $N_D^{-1}$ dependence. Ge is presumed to be incorporated as a donor to form a $n^+$ layer of thickness t and w is the width of the space charge region at the metal-semiconductor interface. Assuming uniform contact area, two theoretical cases (solid lines) are shown: for $t < w$, $r_c$ is controlled by the substrate doping and varies with $N_D$ and for $t > w$, $r_c$ is controlled by the concentration of the Ge (taken to be $6 \times 10^{19}$ cmref–3) and independent of $N_D$ [64].

Equation (7.7) also predicts that $r_c \propto 1/N^{\frac{1}{2}}$, where N is the concentration in the $n^+$ or $p^+$ layer (Fig. 7.9) whose thickness is greater than the width of the space-charge region at the metal-semiconductor interface. Thus, when Au-Ge or Au-Zn contacts are used, Ge or Zn would be incorporated in the semiconductor. Since Ge acts as a donor and Zn acts as an acceptor, $n^+$ or $p^+$ regions are expected to form. However, data indicates that $r_c$ varies inversely with substrate carrier concentration, $N_D$ as shown in Fig. 7.10 for Au-Ge contact on GaAs [64].

Braslau [64] proposed a model to explain the inverse dependence of $r_c$ on $N_D$ for the Au-Ge contact. Since the surface morphology of the alloyed contact indicated nonplanar and nonuniform alloying, Braslau postulated that current flows through small localized regions of Ge-rich islands as shown in Fig. 7.11. The contact resistivity is reduced at the localized regions because of field enhancement at the sharp protrusions. In the regions adjacent to the Ge-rich protrusions,

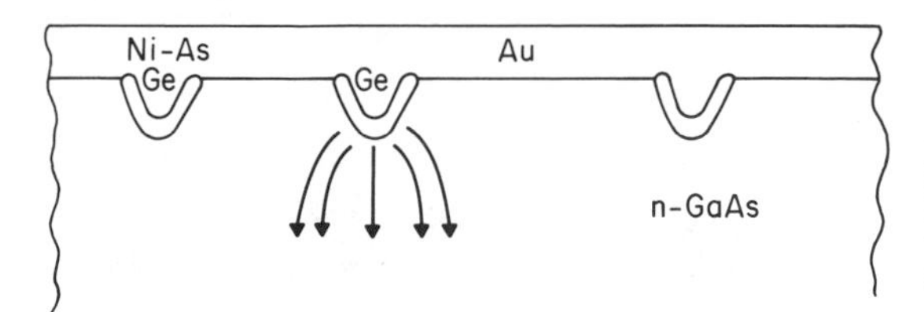


**Figure 7.11** Structural model to explain the $N_D^{-1}$ dependence of $r_c$ in Au-Ge-Ni alloyed contacts as shown in Fig. 7.10 [64].

resistivity is determined by the underlying substrate doping according to Eq. (7.7). The measured $r_c$ is a combination of these two factors and is given by

$$(r_c)_{meas} = A_{av}^2 \left[ \frac{\rho}{\pi R_{Av}} + \frac{r_c}{2f\,R_{Av}^2} \right] \tag{7.8}$$

where $A_{av}$ and $R_{av}$ are, respectively, the average separation between protrusions and the radius of the protrusion, f is the field enhancement factor caused by the protrusions and $\rho$ is the substrate resistivity which is proportional to $1/N_D$. Braslau [64] shows that for $\rho > 10^{-3}\,\Omega$ cm, the second term in Eq. (7.8) can be neglected which then gives the observed $r_c \propto 1/N_D$ dependence. The straight line fit to the data points in Fig. 7.10 was done using Eq. (7.8) from the measured values of $A_{av}$ and $R_{av}$. The inverse dependence of $r_c$ on substrate doping concentration has also been observed for alloyed ohmic contacts to p-type GaAs [61] as well as for Au-Zn and Au-Be contacts to p-type InP [18].

In the case of alloy semiconductors Au-Ge for n-type and Au-Zn for p-type material are often used as ohmic contacts. Nakano et al., [73] found that for Au-Zn contacts to p-type GaInAsP, $r_c$ decreased with decreasing band gap, that is, going from InP to GaInAs by nearly an order of magnitude. Because of this reason, in lasers and LEDs made of GaInAsP-InP system, p-contacts are generally made to a heavily doped GaInAsP or GaInAs layer rather than to InP. The smaller barrier $\phi_b$ in the ternary and the quaternary plus the possibility of increasing p-doping in them compared to the binary are responsible for the lower $r_c$. Nakano et al., [73] demonstrated the dependence of $r_c$ on $\phi_b$ and N according to Eq. (7.7) illustrating quantum mechanical tunneling as the mechanism of carrier transport across the barrier.

### Heterojunction contacts

Equation (7.7) shows that for quantum mechanical tunneling through the barrier, $r_c$ can be lowered by lowering $\phi_b$. One way to decrease $\phi_b$ is to introduce a semiconductor with a lower $\phi_b$ between the semiconductor to be contacted and the metal. An example of the heterojunction contact is the Ge/GaAs contact in n-type material [77, 78]. The conduction band edge for this system is shown in Fig. 7.12 [77]. $\phi_b \sim 0.5$ eV for Ge compared to 0.8 eV for GaAs. Using a Ge layer which was deposited at low temperature by MBE and doped with As to give a doping level of $10^{20}$ cm$^{-3}$, Stall et al., [78] obtained $r_c \sim 5 \times 10^{-8}\,\Omega$ cm$^2$ for $N_D$ (GaAs) $= 1.5 \times 10^{18}$ cm$^{-3}$ and $r_c \sim 1.5 \times 10^{-7}\,\Omega$ cm$^2$ for $N_D$ (GaAs) $= 1.0 \times 10^{17}$ cm$^{-3}$. Woodall et al., [79] made a low resistance heterojunction contact to GaAs using InAs-GaAs heterojunction. Instead of using an abrupt InAs/GaAs junction which produces a large barrier between InAs and GaAs because of the conduction band discontinuity, they introduced a layer of graded $Ga_xIn_{1-x}As$ between the two semiconductors with $x = 0$ at the GaAs interface and $x = 1$ at the InAs interface. It is also

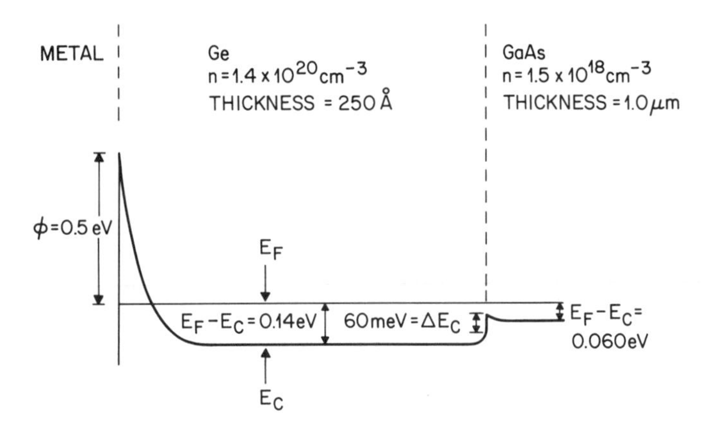


**Figure 7.12** Conduction band edge of a $n^+$ Ge/$n^+$ GaAs heterojunction contact grown by MBE [77].

possible to omit the InAs layer and deposit metal directly on the $Ga_xIn_{1-x}As$ ($x \gtrsim 0.8$) layer since the barrier height of the ternary is $\leq 0$ for $0.8 \leq x \leq 1.0$ [80]. Using Ag contact to a 0.25 μm thick graded $Ga_xIn_{1-x}As$ layer, with a doping level of $3 \times 10^{18}$ $cm^{-3}$ in the first 0.1 μm, grown on top of a 0.5 μm n-GaAs ($N_D = 2 \times 10^{17}$ $cm^{-3}$) layer, Woodall et al., [79] measured $r_c$ in the range $10^{-6} - 5 \times 10^{-7} \Omega\, cm^2$.

### 7.2.2 Insulator-Semiconductor Interface

There has been a considerable interest over the years to develop the metal-insulator-semiconductor FET (MISFET) structures using III-V compounds similar to the MIS structures in Si, in order to reap the benefits of the large low-field peak electron velocities in III-V semiconductors compared to Si. In the Si MIS structures, the insulator is thermally oxidized silicon. The thermally grown oxide on Si has all the properties required for a gate insulator. It is highly resistive, mechanically strong, electronically stable, and produces an insulator-semiconductor interface with low interface state density. In contrast, grown native oxides (homomorphic) on III-V semiconductors cannot match the superior qualities of the thermally grown oxide on Si.

The homomorphic dielectrics on III-V semiconductors such as GaAs and InP suffer from several disadvantages. They are compositionally inhomogeneous. They contain variable quantities of the group III or group V atoms. There is also considerable variation in their crystalline phase, order, and morphology as a function of the method used to grow them. They do not have high resistivity. Arsenic oxides are thermodynamically unstable in the presence of GaAs, InAs or GaInAs and oxide-substrate reaction releases elemental As at the interface. The anodically grown oxides on GaAs or InP are soft and hygroscopic and do not lend themselves readily to conventional photolithographic processing.

Since the grown oxide does not meet the requirements of a gate insulator, attention has been focussed on developing a suitable heteromorphic or synthetic dielectric insulator for the MIS structures in III-V semiconductors. Besides their use as a gate insulator, dielectrics are also needed as (a) a passivation layer to protect the surface of the device from external ambient and to ensure long term stability, (b) an encapsulant to protect the surfaces during heat treatment of wafers as, for example, after ion-implantation, (c) an insulating layer for isolating interconnection lines, and (d) a protective coating and/or as a coating to give different reflectivity mirror facets in semiconductor light emitting devices.

**Effects of bad dielectric-semiconductor interface**

Localized states can be generated at the interface between the semiconductor and the dielectric as well as in the bulk of the dielectric layers. They can be generated as a result of nonstoichiometry either in the semiconductor or in the dielectric [81]. Nonstoichiometric defects can be introduced in the semiconductor during deposition of the dielectric at elevated temperatures due to preferential evaporation of the group V element. The native oxide which is present on the surface of the semiconductor prior to dielectric deposition can act as a source of traps [82]. Buildup of excess As at the interface during dielectric growth has been supposed to be the origin of interface defects in the case of GaAs [83]. The presence of interface states can give rise to trapping, scattering, and nonradiative recombination all of which are deleterious to device performance.

Figure 7.13 shows schematically the presence of surface and interface states which can trap charge carriers in a MIS device [84]. The trapping of charge can alter the "quiescent" or zero bias condition of the semiconductor surface which can give rise to unwanted surface enhancement or inversion. This can lead to interdevice surface current leakage paths and degrade integrated circuit operation. Further modulation of the surface potential by means of externally applied fields will be affected if the traps can respond to the frequency of the external signal. This is particularly a problem for low frequency operation. In a FET, traps can lead to reduction of current for a given voltage as well as degradation in transconductance. Also traps are responsible for long term drift in device characteristics. Lile [85] and Shinoda and Kobayashi [86] have shown that the trap concentration should be $\leq 10^{11}$ $cm^{-2}$ $eV^{-1}$ in the region of surface potential accessed by the device for acceptable device performance.

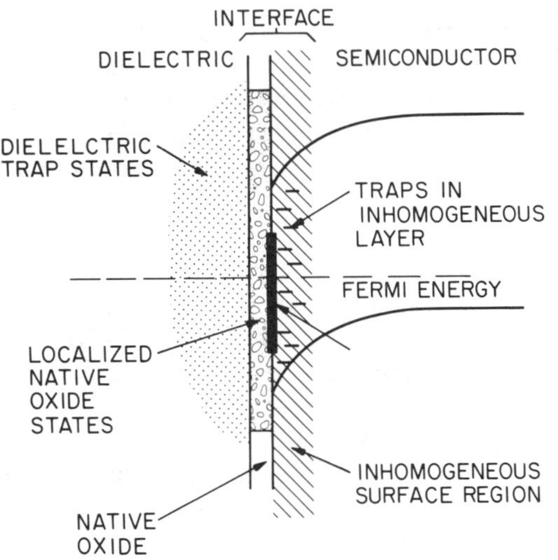


**Figure 7.13** Schematic band diagram illustrating the potential sources of surface and interfacial trapping to be expected in an MIS device [84].

In devices such as enhancement-mode MISFET and the surface channel charge coupled device (CCD) where charge carriers are moving in a narrow inversion layer close to the semiconductor-dielectric interface, drift mobility of carriers is reduced caused by scattering in the surface channel. The scattering is determined by surface roughness, trap density, and their occupation [87]. Generally, surface mobility is not greater than two-thirds of the bulk mobility value [84]. However, in the case of InP surface mobility close to the bulk value has been reported [88, 89].

If carriers are generated within a diffusion length from the surface, recombination can occur via a surface/interface defect level. Such a recombination is generally nonradiative and it reduces the minority carrier lifetime. Traps which act as nonradiative centers in the presence of excess carriers can also contribute to generation current when a deficiency of carriers occurs. Such nonradiative recombination paths or generation-recombination current are undesired in light-emitting devices, CCD senor arrays, APDs, and bipolar junction transistors. In the absence of any interface defects, a dielectric passivation layer can reduce the surface recombination velocity by reducing the density of intrinsic surface states. But the reduction in the surface recombination velocity will very much depend on the type of the dielectric, the method of deposition, and the surface preparation prior to deposition. For example, in InP, a surface recombination velocity of $2 \times 10^3$ cm/sec was obtained using $Al_2O_3$ [90] compared to a value of $10^4$ cm/sec obtained with $SiO_2$ [91].

In the case of heterostructures consisting of layers of different composition as in the case of optoelectronic devices, interface recombination is one of the important nonradiative channels [92]. In double heterostructure lasers, two-dimensional electron gas structures and the like, the optical and electrical properties of the active layer can be improved by introducing single or superlattice buffer layers before the first interface of the active layer. This improvement occurs as a result of reduction in the interface recombination velocity [93, 94] caused by interface smoothing and/or impurity gettering.

From the preceding arguments it is clear that the quality of the dielectric-semiconductor interface has a profound influence on device performance. Accordingly, technological control and understanding of the dielectric-semiconductor interfaces have received considerable attention over the years. The control of the interfaces involves combining the desired qualities of the dielectric such as mechanical stability, electrical resistivity, pin hole free morphology, breakdown strength, and last but not the least, low density of bulk and interfacial traps and selecting a suitable technique for depositing the dielectric taking into consideration the various pre- and post-deposition conditions [95].

**Interfacial constraints in devices**

The presence of localized states in the interfacial region imposes serious limitations on the performance of electronic and optoelectronic devices. In the case of GaAs MESFETs, shown schematically in Fig. 7.14(a), besides the metal gate-semiconductor interfaces, the interfacial region between the conducting channel of the device located between the source and drain electrodes and the semi-insulating (SI) substrate affects profoundly device performance. The operation of an FET consists of the electrostatic modulation of a current flowing between the source and drain electrodes. The modulation of the space charge region under the gate located between the source and drain electrodes controls the conductance between them. The presence of

(a)

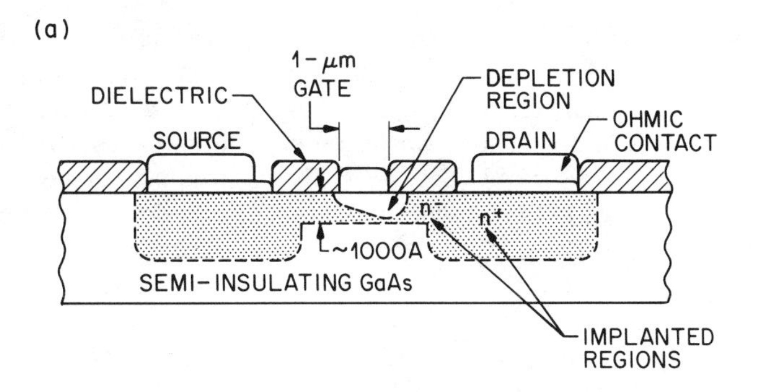


(b)

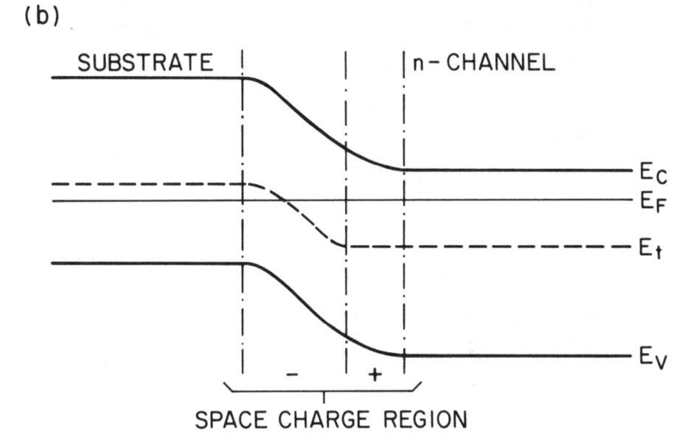


**Figure 7.14** (a) Cross section of a conventional ion implanted GaAs MESFET (b) Energy band diagram of the channel-substrate interface of a MESFET with a deep trapping level $E_t$ near midgap [97].

defects and deep level impurities at the conducting channel ($n^-$ region in Fig. 7.14(a))-SI substrate interface has been implicated in the poor performance of the MESFET as characterized by hysteresis in the drain current and source-drain voltage behavior, long term drift and decay of the drain current if a negative bias is applied to the substrate relative to the source, and changes in the rf parameters. Using the DLTS method to characterize the interfacial traps in the MESFETs, Adlerstein [96] identified Cr and O-related deep levels as well as the electron trap EL2 in the substrate-channel interfacial region.

The space charge region present between the n-channel and SI substrate interface (Fig. 7.14(b)) can affect channel thickness and in turn MESFET parameters, a phenomenon known as back gating [97]. The interaction between adjacent MESFET and current flow in the SI substrate is termed as side gating. Interfacial defects also affect side gating which can be of great significance in VLSI circuits. Lee et al., [98] found that the onset of side gating, that is, the voltage between the side gate and MESFET that begins to affect the performance of the device, coincides with the threshold for space-charge-limited current flow in the presence of trapping [99]. This suggests that a high density of traps in the SI substrate in fact increases the threshold for side gating. But the presence of such traps would degrade the FET performance in other ways. Kocot and Stolte [100] found that side gating is reduced in FETs made on undoped buffer layers grown on Cr-doped SI substrates compared to those made on undoped SI substrates with or without a buffer layer. It is known that out diffusion of Cr from the substrate occurs when the substrate is heated to high temperatures (Sec. 7.3). The reduced side gating in the case of Cr-doped substrates may be related to the presence of Cr related deep levels at the channel-substrate interface identified by DLTS [101].

**GaAs-dielectric interface**

In order to realize the potential advantages of the MISFET structures, extensive studies have been made concerning both homomorphic and heteromorphic dielectric layers, particularly on GaAs. Several methods of depositing the homomorphic dielectric layers such as anodic oxidation, microwave plasma oxidation, rf gaseous oxidation, and magnetically confined plasma beam oxidation have been tried [97]. Similarly, in the case of heteromorphic layers, pyrolysis of silane, chemical vapor deposition of silicon nitride and silicon oxynitride layers, and pyrolysis of aluminum isopropylate to deposit aluminum oxide have been tried [97]. Notwithstanding the type and method of deposition of the dielectric layer, the GaAs-dielectric interface is characterized by a high density of interface states ($\sim 10^{13}$ $cm^{-2}$ $eV^{-1}$) which pin the Fermi-level near the middle of the energy gap. In both n- and p-type MIS structures, the Fermi-level is pinned 0.8-0.9 eV below the conduction band and 0.6-0.7 eV above the valence band, respectively [101]. These values are similar to those obtained on atomically clean GaAs surfaces. The Fermi-level pinning near the middle of the gap and the limited displacement of the surface potential reduce the dynamic range of GaAs MISFETs. Besides, the interfacial traps have a time constant of 0.1 seconds which limit the low frequency operation of the device.

To circumvent the problem of the high density of defects associated with the dielectric- GaAs interface, alternative schemes have been tried to produce an insulating gate for MIS structures. Pruniaux et al., [102] used a semi-insulating GaAs layer as a gate insulator. Proton bombardment at 25 keV and $10^{14}/cm^2$ dose was used to convert about 0.2 $\mu m$ of a 1.3 $\mu m$ thick n-type GaAs epitaxial layer with a carrier concentration of $8.8 \times 10^{15}/cm^3$ to a semi-insulating gate region. Similarly, Macksey et al., [103] used 30 keV $Ar^+$ ion bombardment to produce semi-insulating gate. However, ion bombardment also generates defects which can pin the Fermi level near the middle of the gap [104] and hence the same problems as in dielectric gate MIS structures prevail.

Casey et al., [105, 106] used $Al_{0.5}Ga_{0.5}As$ grown by MBE as the gate insulating layer. They found that the C-V curves of MIS structures made with this layer did not exhibit hysteresis or any anomalous frequency dispersion of the capacitance. Further DLTS measurements indicated trap density not greater than $10^{11}$ $cm^{-2}$ $eV^{-1}$. Similar MIS structures with undoped, Cr or O doped $Al_{0.5}Ga_{0.5}As$ layers have been reported [107].

**InP-dielectric Interface**

The larger peak electron velocity, higher thermal conductivity and more suitable dielectric interfacial properties of InP compared to GaAs have motivated a substantial research effort in developing InP based MISFET structures. The superior quality of the dielectric-substrate interface stems from two factors. First, the dielectric-InP interface is characterized by a low density ($10^{11} - 10^{12}$ $cm^{-2}$ $eV^{-1}$) of long time constant traps [108]. Second, the Fermi level is

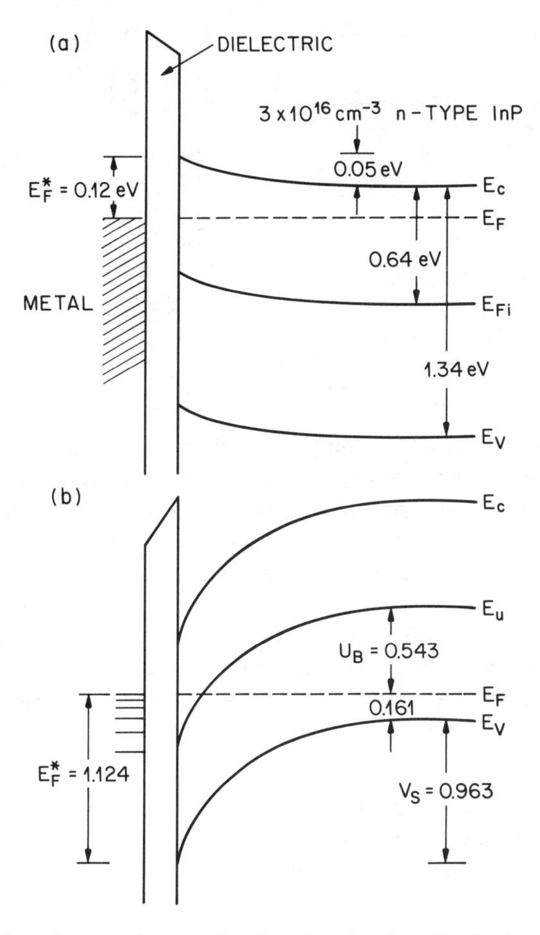


**Figure 7.15** Equilibrium energy band diagram of (a) an n-type and (b) p-type InP MIS capacitor with a $SiO_2$ dielectric layer and Al gate [97].

pinned near the conduction band edge for both n-type and p-type material. Meiners [108] determined in InP MIS capacitor with a $SiO_2$ dielectric layer, the Fermi level to be within 0.1 to 0.2 eV below the conduction band in both n- and p-type substrates, in very good agreement with the XPS measurements made on cleaved (110) surfaces as a function of oxygen exposure [109], as well as those obtained on (100) surfaces covered by a native oxide [110]. Figure 7.15 shows the equilibrium band diagram of a (100) n-type InP and a p-type InP MIS capacitor with a $SiO_2$ dielectric layer, illustrating the Fermi level near the conduction band in both cases. The pinning of the Fermi level near the conduction band in p-type material allows the displacement of the Fermi level over a large portion of the band gap and also inversion of the surface to make possible enhancement-mode-inversion layer MISFET transistor. The low-interface density in the case of the InP-dielectric interface also gives rise to superior low-frequency performance as illustrated in Fig. 7.16 [111]. The GaAs MISFET shown in Fig. 7.16 was made with an electrochemically anodized gate insulator while the InP MISFET consisted of a $SiO_2$ gate dielectric layer. The frequency-independent gain of the InP devices is as a result of the absence of slow traps which would otherwise respond to the low-frequency (<100 Hz) excitation as in the case of the GaAs devices.

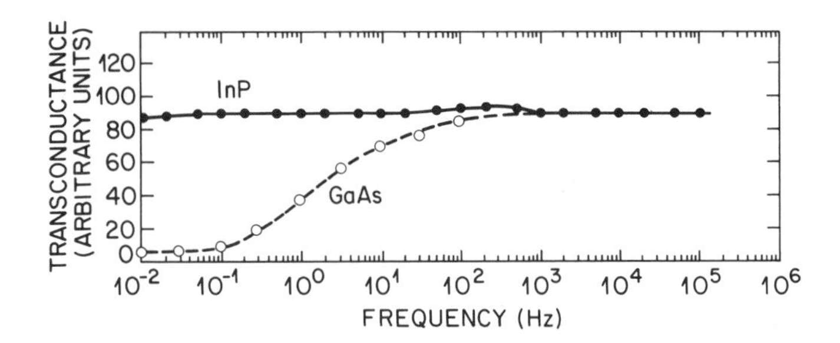


**Figure 7.16** Low frequency dispersion in transconductance for representative InP and GaAs MISFETs [111].

Despite the superior characteristics of the InP MISFETs, long term drift in drain current is a limiting factor in their performance. This phenomenon, which is far more serious in GaAs MIS structures, results from the injection of electrons from the InP channel into localized interface states located in the dielectric or in the region of the semiconductor-dielectric interface. The electron exchange into trap centers gives rise to hysteresis in C-V measurements and drift of FET characteristics. In this connection, surface treatment prior to dielectric deposition seems to have a beneficial effect. Figure 7.17 shows the drift of normalized drain current for an enhancement-mode FET on InP made with and without a severe HCl vapor etch prior to deposition of the $Al_2O_3$ gate insulator [112,113]. The reduced drift achieved with the HCl etch is believed to occur because of the removal of surface native oxide and hence the traps present in it by the etch.

Sawada and Hasegawa [114] proposed that a disordered or contaminated metamorphic layer exists near the interfacial region of the semiconductor-dielectric interface and that it traps carriers. Use of etchants to remove surface damage and/or low temperature dielectric deposition conditions would reduce the extent of the disordered region and hence minimize device drift. In using

**Figure 7.17** Normalized drain current vs time characteristics for an enhancement-mode FET on InP following application of a +2V change in input gate voltage [112,113].

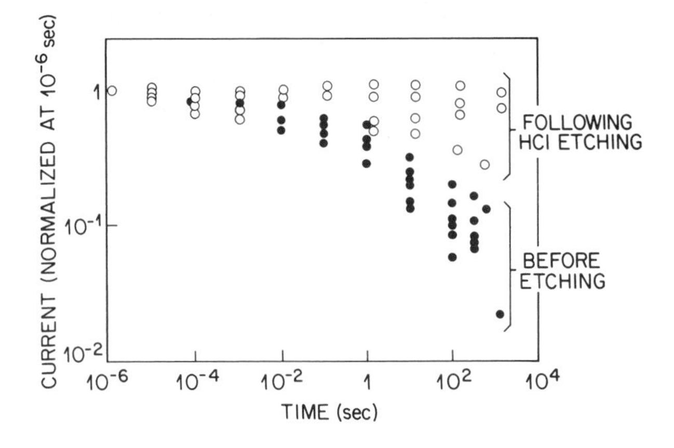

surface pre-etch treatments it should be borne in mind that excessive surface treatment, while beneficial for device stability may affect electron surface transport as measured by mobility, transconductance and $I_{DSS}$.

**Ternary and quaternary alloys**

The barrier heights of GaInAsP alloys grown lattice matched to InP are not large enough (0.2 eV) [80, 115] for MESFET applications but may be suitable for MISFET devices. $Ga_{0.47}In_{0.53}As$ lattice matched to InP is of particular interest since it has a peak electron velocity which is larger than InP and GaAs as shown in Fig. 7.18 [116]. Since this material system is used for making detectors in fiber optical communication systems operating at 1.3 to 1.5 μm wavelength range, the potential of integration of photodetector and FET circuitry is tremendous. Several types of MISFET structures have been reported in GaInAs and GaInAsP material systems with reasonable device characteristics. A summary of these results can be found in Refs. [84] and [97]. The increase in mobility as a function of composition is also duplicated in the increase of field-effect mobility measured on a series of inversion-mode MISFETs made on p-type $Ga_xIn_{1-x}As_yP_{1-y}$ grown lattice matched to InP ($y = 2.2x$) with various compositions [117, 118]. These devices employed $Al_2O_3$ gate dielectric. Figure 7.19 shows the field-effect mobility along with bulk mobility as a function of As mole fraction. The surface state distributions deduced from a Terman analysis of 1-MHz C-V results from MIS diodes made at the same time as the MISFETs are also shown in Fig. 7.19. The increase in mobility with composition can be attributed to

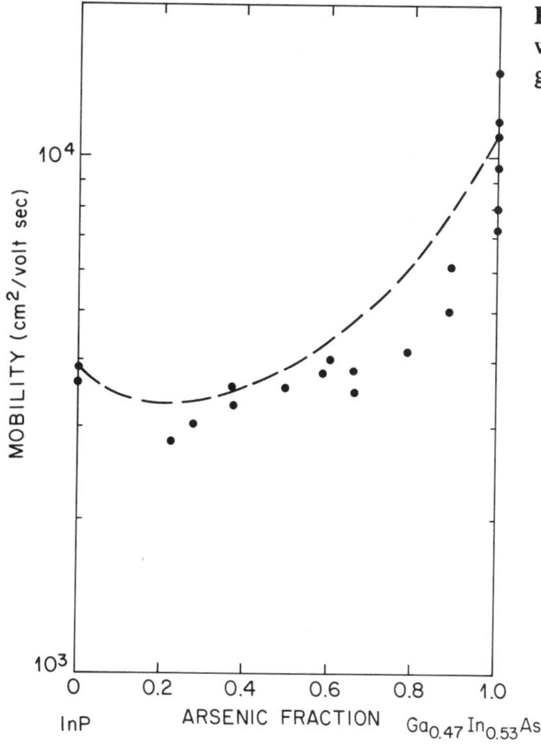


**Figure 7.18** Variation of electron mobility with composition for $Ga_xIn_{1-x}As_yP_{1-y}$ grown lattice matched to InP [116].

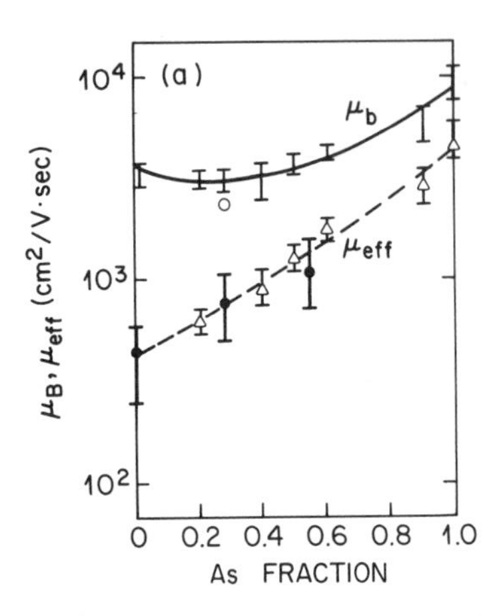


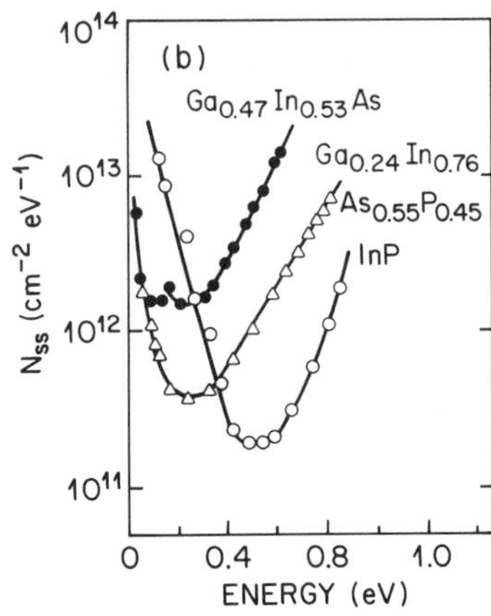


**Figure 7.19** Compositional dependence of $Ga_xIn_{1-x}As_yP_{1-y}$ MISFET; (a) Bulk mobility $\mu_B$ and effective mobility as a function of As mole fraction (b) Energy dependence of the interface state density for InP, $Ga_{0.24}In_{0.76}As_{0.55}P_{0.45}$ and $Ga_{0.47}In_{0.53}As$ [117,118].

decreasing surface state concentrations for the ternary and quaternary alloys over the surface potential range of interest near the conduction band edge.

## 7.3 OUT-DIFFUSION OF IMPURITIES AND THERMAL CONVERSION

In the last section we saw that the presence of Cr, O-related deep traps at the channel-substrate interface can severely affect the performance of MESFETs. Further, the threshold voltage for the onset of side gating is higher in FET made on Cr-doped substrates. Similarly, deep energy levels were found to be present at the interface of Cr-Au Schottky barrier gates and the semiconductor. The presence of deep traps at these interfaces is linked with the fast outdiffusion of Cr during high temperature processing steps. In the case of Cr-compensated GaAs SI substrates, the fast outdiffusion of Cr to the surface depletes Cr in the near surface region. Since the semi-insulating behavior is achieved by the compensation of the background donors by the deep Cr acceptors, depletion of Cr also leads to conversion from semi-insulating behavior to n-type conductivity. The out-diffusion of impurities and the associated changes in the electrical properties of the substrates are presented in this section.

Since type conversion is intimately related to diffusion of impurities either into a semiconductor or out of it, it is worthwhile to review briefly diffusion equations. Diffusion occurs when concentration gradients exist so as to make the system homogeneous. In one dimension, if the concentration gradient exists along the x axis then the flux of matter along the concentration gradient is given by Fick's first law of diffusion [119]

$$J = -D \left[ \frac{\partial c}{\partial x} \right] \tag{7.9}$$

where c is the concentration of the diffusing species and D is the diffusion coefficient. If a steady state does not exist, that is, if the concentration is changing with time, Eq. (7.9) has to be modified taking into account the conservation of matter. Under such circumstances, Fick's second law of diffusion is obtained

$$\frac{\partial c}{\partial t} = -\nabla J = \frac{\partial}{\partial x}\left[D\,\frac{\partial c}{\partial x}\right] \tag{7.10}$$

Equation (7.10) is called the continuity equation. If D does not depend on x, then Eq. (7.10) gives

$$\frac{\partial c}{\partial t} = D\,\frac{\partial^2 c}{\partial x^2} \tag{7.11}$$

If the surface concentration of a specimen, which is initially free of a solute, is kept at a value $c_o$ of the solute from time $t = 0$, the solute diffuses into the specimen. For example, the solid can be kept into contact with a vapor containing the solute at some appropriately high temperature. The boundary conditions for this situation are

$$\begin{aligned} c &= c_o \quad t > 0,\ x = 0 \\ c &= o \quad t = 0,\ x > 0 \end{aligned} \tag{7.12}$$

and the solution to Eq. (7.11) with these boundary conditions is

$$c(x,t) = c_o\left[1 - \operatorname{erf}\left(\frac{x}{2\sqrt{Dt}}\right)\right] \tag{7.13}$$

Instead of diffusion into the solid from the surface as depicted by Eq. (7.13), if a homogeneous sample which initially contains a solute of concentration $c_1$ is heated to a high temperature, the solute diffuses out. If the solute that collects at the surface is removed such that the surface concentration is held at $c = o$ for all $t > 0$, then the boundary conditions are

$$\begin{aligned} c &= c_1 \quad t = 0,\ x > 0 \\ c &= 0 \quad t > 0,\ x = 0 \end{aligned} \tag{7.14}$$

and the solution to the diffusion equation becomes

$$c(x,t) = c_1 \operatorname{erf}\left(\frac{x}{2\sqrt{Dt}}\right) \tag{7.15}$$

Equations (7.13) and (7.15) describe the cases of indiffusion and outdiffusion, respectively, and they are plotted in Fig. 7.20 [120].

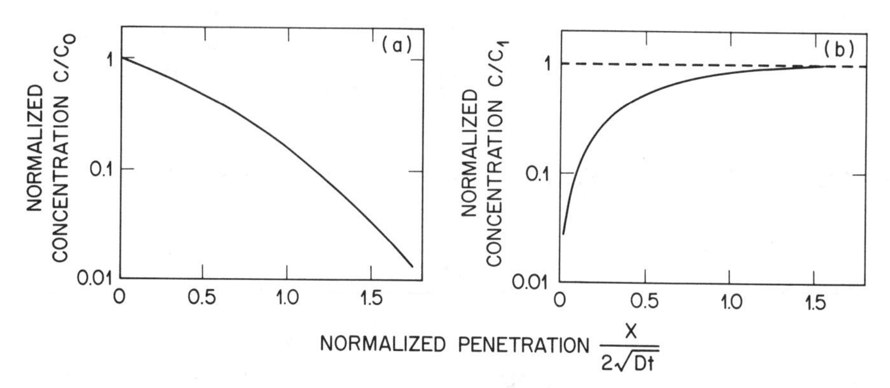


**Figure 7.20** Normalized concentration versus normalized diffusion depth for (a) indiffusion and (b) outdiffusion [120].

A special case of Eq. (7.15) is diffusion from a homogeneously doped substrate into a growing epitaxial layer. In this case, if no dopant is lost from the surface such that $c = c_1/2$ at $x = 0$ for all $t > 0$, the solution becomes [120]

$$c = \frac{c_1}{2}\left[1 + \mathrm{erf}\left(\frac{x - vt}{2\sqrt{Dt}}\right)\right] \tag{7.16}$$

where v is the growth velocity or the rate of movement of the epitaxial layer-substrate interface from the origin $x = 0$ taken at the growing surface. Let us now examine out diffusion and redistribution of dopants in GaAs and InP.

**GaAs substrates**

Direct ion implantation into semi-insulating GaAs substrates is considered to be the most viable low-cost manufacturing technology for the fabrication of large scale integrated circuits. The availability of SI GaAs substrates has given a large impetus to the development of this technology. In this respect, the quality of SI substrates and its effect on IC performance has been studied rather extensively over the last several years [121]. When ion implantation is used, defects resulting from the damage production are stable at room temperature and compensate the electrical activity of the implanted ions. So high temperature heat treatments are required to anneal out the damage and activate the implanted ions. Typically, annealing temperatures are in the range 800 to 900°C. This temperature range is much above the congruent evaporation temperature of ~650°C [122]. Therefore measures to suppress the preferential evaporation of As from the substrate must be taken. Generally, this is achieved by using thin dielectric films such as $SiO_2$, $Si_3N_4$, phosphorus or arsenic doped glass or combinations of these to encapsulate the substrate. The method of protecting the surface against thermal degradation using these films has been reviewed by Oberstar and Streetman [123]. Instead of the thin film encapsulation technique, capless annealing in an As rich ambient [124] or encapsulated annealing in an As ambient over pressure [125] have also been used to suppress As loss.

Apart from the preferential loss of As from the surface during the high temperature heat treatments, there is also the possibility of redistribution of impurities and/or defects present in the SI substrate. This redistribution gives rise to a conductive surface layer which has adverse effects on device performance. Hence, thermal stability of the SI substrates is essential for direct ion implantation device fabrication. In SI GaAs substrates which contain Cr as the deep acceptor compensating the shallow donors, during high temperature heat treatment Cr outdiffusion occurs leaving behind uncompensated donors. As a result, the surface becomes n-type to a depth of even up to several microns [126, 127]. Type conversion of the surface to p-type conductivity is the one more commonly observed [120]. This occurs because of rapid outdiffusion of another transition element, namely Mn which is invariably present as a contaminant in Cr-doped substrates. Mn is relatively a shallow acceptor in GaAs and it occupies the Cr-depleted sites and thus gives rise to the p-type surface layer. It has also been suggested that Mn gets incorporated from the ambient during heat treatment [128].

## Redistribution of Cr, Fe, and Mn

As we stated earlier, the type conversion of Cr-doped SI substrates results from the outdiffusion of Cr. Figure 7.21 shows the atomic concentration of Cr as a function of depth from the surface for two different Cr-doped SI horizontal Bridgman grown GaAs substrates after a high temperature heat treatment [129]. The atomic concentration profiles are obtained by secondary

**Figure 7.21** Depth profile of Cr after anneal for two different horizontal Bridgman grown GaAs substrates: closed circles represent material that did not type convert and closed triangles represent material that did convert. The dashed line indicates shallow donor concentration [129].

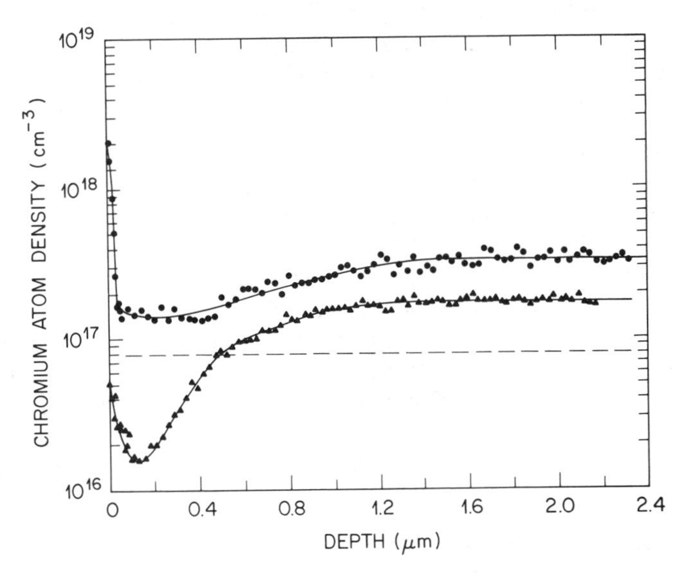

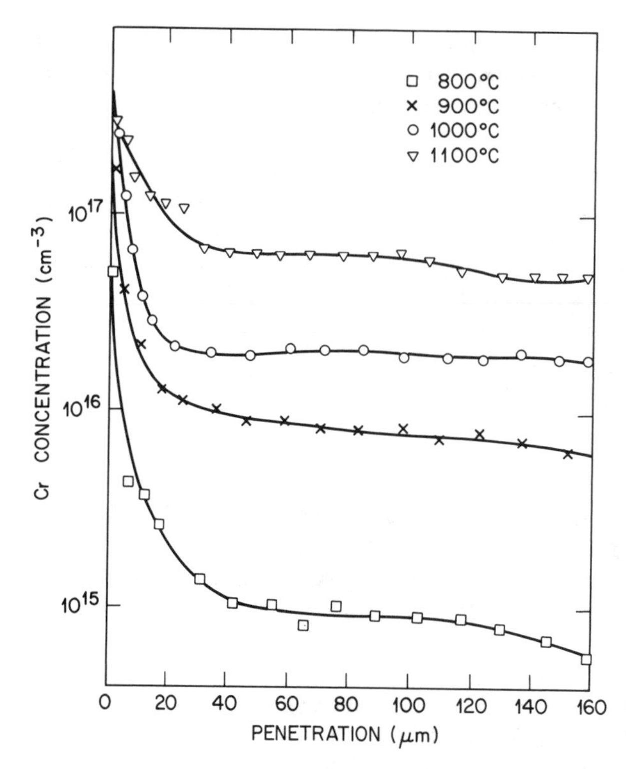


**Figure 7.22** Diffusion profiles for Cr in GaAs for diffusion time of 4h at different temperatures [130].

ion mass spectrometry (SIMS) which is one of the techniques used to study outdiffusion phenomenon. The shallow donor concentration as indicated by the dashed line is $8 \times 10^{16}$ $cm^{-3}$. Clearly, in one of the substrates rapid outdiffusion of Cr has occurred. The Cr concentration falls below the shallow donor concentration near the surface and hence this substrate showed type conversion. In the substrate which did not undergo type conversion Cr outdiffusion is not severe and Cr concentration stays above the shallow donor concentration at all depths.

Figure 7.21 illustrates two points. First, Cr outdiffusion is not reproducible as shown by the different results on the two substrates. Second, the outdiffusion profile does not resemble the profile shown in Fig. 7.20. This simply indicates that outdiffusion of Cr is not a simple process which can be described by Eq. (7.15). Of course, even indiffusion of Cr does not follow the profile in Fig. 7.20. Tuck and Adegboyega [130] measured the indiffusion and outdiffusion of Cr in GaAs using radiotracer Cr as the diffusing species. Figure 7.22 shows the diffusion profiles for 4h diffusion times at four different temperatures. Each diffusion profile is marked by a high surface concentration and a nearly constant level in the interior of the crystal. Cr has penetrated the whole sample even at the lowest diffusion temperature indicating rapid diffusion of Cr. Note once again that the indiffusion of Cr does not follow the profile shown in Fig. 7.20. As for the electrical or optical behavior of the Cr centers those introduced by diffusion behaved in the same way as those introduced during bulk crystal growth.

Tuck and Adegboyega [130] also studied the outdiffusion of radiotracer Cr. For this experiment, they first diffused Cr into a GaAs substrate at 1100°C which gave a profile similar to

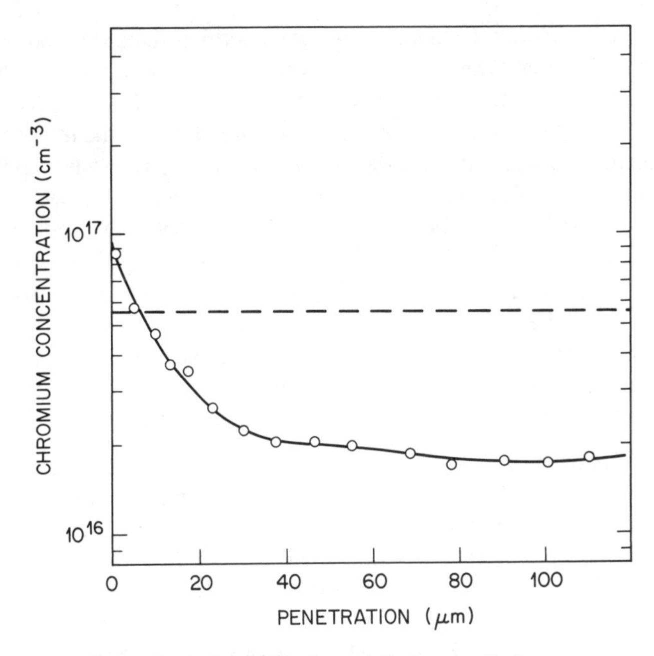


**Figure 7.23** Out-diffusion of Cr from a GaAs sample which has been homogeneously doped with radiotracer Cr after annealing at 750°C. The dashed line shows the original Cr concentration [130].

that shown in Fig. 7.22. The highly doped surface regions were etched off, leaving a sample homogeneously doped with radioactive Cr to a level of $5 \times 10^{16}$ $cm^{-3}$. Outdiffusion experiments were then carried out at 750°C for 1h. The resulting profile is shown in Fig. 7.23. Once again, the outdiffusion profile does not resemble the expected profile shown in Fig. 7.20 but qualitatively resembles the one in Fig. 7.21 obtained for normal Cr outdiffusion. The Cr level in the interior of the crystal has dropped and a surface peak has developed with a concentration greater than the original doping level. The radiotracer diffusion establishes that the development of the surface peak is caused by the outdiffusion of Cr rather than by the introduction of Cr into the crystal from the ambient during the heat treatment. The rapid indiffusion or outdiffusion of Cr into GaAs suggest that Cr diffuses through the lattice interstitially [130].

Interesting effects occur regarding the redistribution of Cr in GaAs after high temperature annealing in two other situations when Cr is implanted into GaAs and when other elements are implanted into Cr-doped substrates. Since high temperature heat treatment is invariably carried out following implantation to activate the dopants, redistribution of Cr occurs. The concentration profiles of Cr for short annealing times are not what would be expected from simple diffusion theory. Usually a peak of Cr occurs in the interior of the crystal, approximately near the peak of the implantation induced damage. This pile up of Cr at the damage regions has been explained as defect gettering effects which will be discussed further in Sec. 7.4. For longer annealing times, the interior peak disappears and concentration profiles resembling the diffusion profiles shown in Fig. 7.22 develop.

Similar to Cr, other transition elements which show similar redistribution on annealing are Fe and Mn. Both these elements are generally present as contaminants in Cr-doped SI GaAs substrates [129]. Using low temperature photoluminescence and SIMS, Yu [131] studied the accumulation of Fe in the near surface region after heat treatment in the temperature range 400 to 900°C. The photoluminescence spectrum showed the characteristic bands near 0.37 eV caused by intracenter $Fe^{2+}$ transitions. Accumulation of Fe near the surface was seen in the SIMS depth profiles as shown for example in Fig. 7.24. The Fe redistribution after a 900°C annealing treatment resembles that of Cr in Fig. 7.21. For heat treatments at 600 to 650°C, a broad pile up of Fe in the interior of the crystal at a depth of 1 μm was also seen in addition to the accumulation near the surface. Yu [131] explained these features in terms of a slow substitutional and a fast interstitial Fe out-diffusion. In the heat-treated samples an acceptor level caused by Fe with an ionization energy of ~0.5 eV [132] was also identified. This acceptor will compensate shallow donors and hence can give rise to moderate p-type conversion after heat treatment [131, 132].

The p-type conversion of the surface region after heat treatment is generally believed to be caused by the redistribution of Mn present as a contaminant in both SI and n-type GaAs. Mn on the Ga site is an acceptor with an ionization energy of ~0.1 eV and evidence for Mn in the type converted region has been obtained by temperature dependent Hall measurement [132] low temperature photoluminescence [133] and secondary ion mass spectrometry [133]. In some

**Figure 7.24** Depth profile of Fe in a GaAs sample heat treated in $H_2$ at 900°C without encapsulation [131].

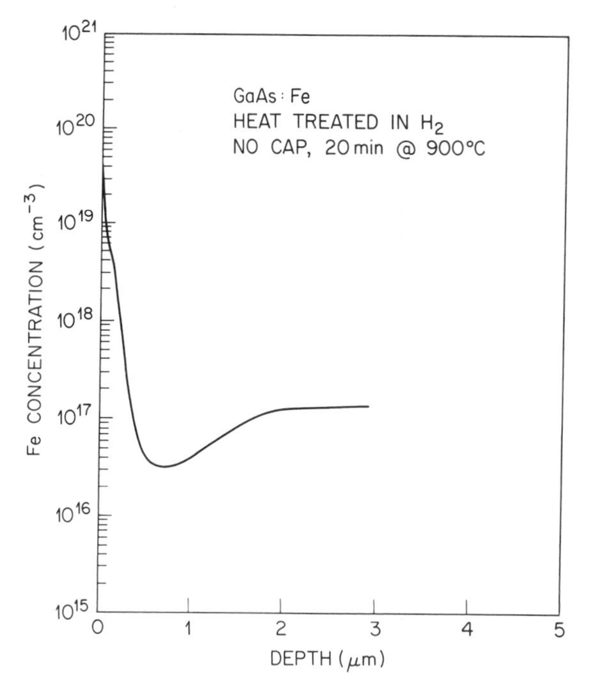

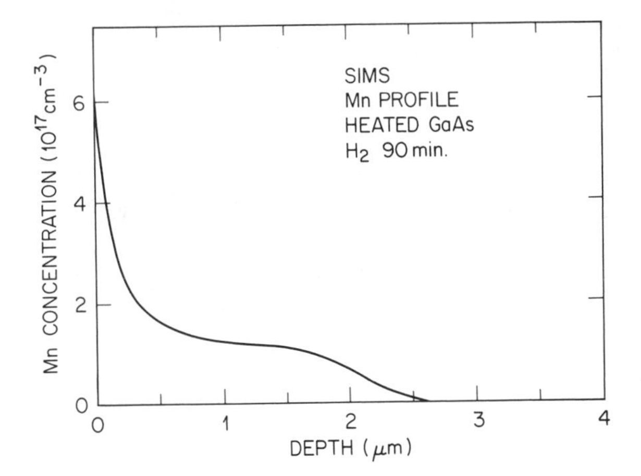


**Figure 7.25** Depth profile of Mn in a GaAs sample annealed at 740°C in $H_2$ [133].

cases, the density of acceptors in the type converted region was greater than the total Mn concentration requiring additional acceptor centers to account for the p-type conversion.

Similar to Fe and Cr, the outdiffusion profile of Mn can not be explained by the simple diffusion models (Fig. 7.20). On the other hand, the indiffusion of Mn from an external source follows the expected complementary error function profile except in a few instances where the diffusion profiles suggested two overlapping profiles corresponding to a slow and a fast diffusion component [134]. The Mn concentration profile obtained after a heat treatment in $H_2$ ambient at 740°C for 90 minutes is shown in Fig. 7.25 [133]. A concentration of $\gtrsim 10^{17}$ $cm^{-3}$ in about 1 to 3 μm region from the surface is observed even though the concentration in the bulk of the sample before heat treatment is only of the order of $10^{15}$ $cm^{-3}$. A series of successive heat treatments performed after etching away the Mn-rich surface layer each time, showed that the surface accumulation of Mn decreased in each step. This suggested that the source of the accumulation at the surface is not external to the sample but that the surface is indeed gettering Mn from the interior of the crystal.

The presence of Mn in the surface region was also confirmed by low-temperature photoluminescence which indicated the characteristic 1.41 eV emission band seen in deliberately Mn doped sample. This is shown in Fig. 7.26 [133]. The depth profile of the 1.41 eV band was found to be in good agreement with the SIMS profile lending credence to the association of the band with Mn.

Apart from the anomalous out diffusion profile, the Mn surface concentration and consequently p-type conversion exhibited a dependence on the ambient during the heat treatment. The Mn concentration at the surface and p-type conversion decreased drastically during annealing at 740°C when the ambient was changed in the sequence vacuum, hydrogen and argon [133]. However, if the annealing temperature was above ~ 850°C, p-type conversion was observed even with argon ambient [135]. Jordan and Nikolakopoulou [136] have suggested that the effect of the

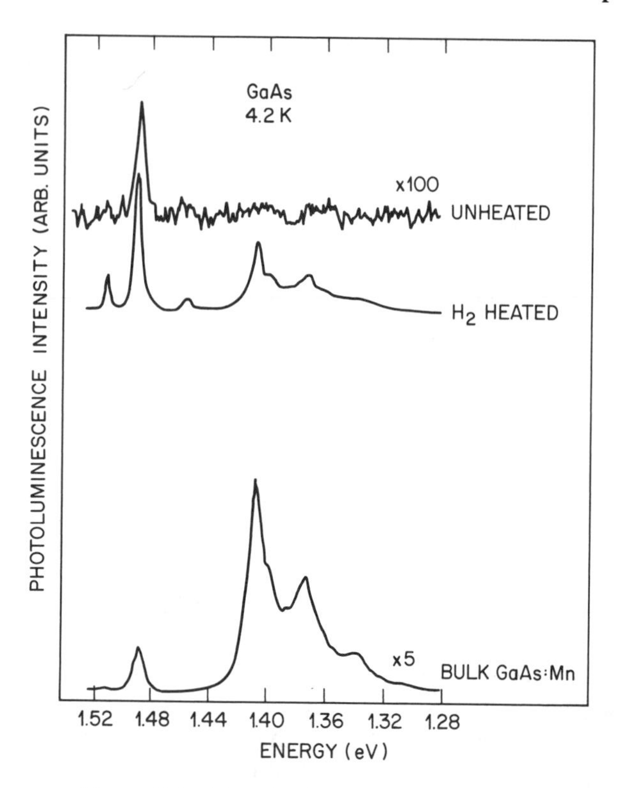


**Figure 7.26** Photoluminescence spectra at 4.2K of unheated, heated in $H_2$ ambient and bulk Mn doped GaAs samples. The band at 1.41 eV is a zero-phonon transition associated with Mn acceptors. Note the presence of this band in the sample heated in $H_2$ ambient [133].

ambient on Mn outdiffusion comes from its role in establishing the equilibrium Ga and As vacancy concentrations at the surface. They conjuctured that for annealing in vacuum the equilibrium concentrations are established more rapidly than for annealing in a gaseous ambient. They also suggested that a light $H_2$ molecule would not efficiently suppress As evaporation from the surface as would a large Ar atom. Hence equilibrium vacancy concentration at the surface would be more quickly established in a $H_2$ ambient than in a Ar ambient.

Jordan and Nikolakopoulou [136] also modelled the Mn outdiffusion by assuming a substitutional-interstitial diffusion mechanism as has been done for Cr [130]. It is assumed that Mn exists as substitutional $Mn_{Ga}$ and interstitial $Mn_i$, the concentration of the former being much greater than the latter. Diffusion occurs by the movement of $Mn_i$ which has a larger diffusion coefficient than $Mn_{Ga}$. The capture of a moving $Mn_i$ by a gallium vacancy, $V_{Ga}$, results in $Mn_{Ga}$ and impurity diffusion has taken place. At any lattice location the interchange between $Mn_i$ and $Mn_{Ga}$ can be described by the following quasi chemical reaction.

$$V_{Ga} + Mn_i^+ \rightleftharpoons Mn_{Ga}^- + 2e^+ \qquad (7.17)$$

Here $V_{Ga}$ is assumed to be neutral. The outdiffusion of Mn can be qualitatively described on the basis of Eq. (7.17), by making the following assumptions. First, the total amount of Mn in the sample is conserved during the diffusion process. Second, before annealing the Mn atoms are present as $Mn_{Ga}$ and that the concentrations of $V_{Ga}$ and $Mn_i$ are negligible. Third, at the annealing temperature the equilibrium vacancy concentration at the external surface is established instantaneously. When the sample is heated to a high temperature the low concentration of $V_{Ga}$ in the interior of crystal forces reaction Eq. (7.17) to the left generating $V_{Ga}$ and $Mn_i$. The interstitials move rapidly towards the surface where they are captured by the indiffusing $V_{Ga}$ generated at the surface and reaction Eq. (7.17) proceeds to the right. Thus accumulation of $Mn_{Ga}$ at the surface occurs.

By choosing the values for the diffusion constants of $V_{Ga}$ and $Mn_i$, the vacancy concentration and the reaction constants for the forward and reverse reaction in Eq. (7.17) that give the best fit to the SIMS data of Klein et al., [133] for Mn concentration versus depth and also by assuming that the excess $V_{Ga}$ in the interior of the crystal is removed by donor-$V_{Ga}$ pair formation so as to keep the supply of $Mn_i$ towards the surface, Jordan and Nikolakopoulou [136] generated Mn concentration versus depth profiles as a function of annealing time as shown in Fig. 7.27. The experimental data of Klein et al. [133] is also shown. The theoretical curves show the gradual increase with time in the Mn concentration from the original background level of $2 \times 10^{15}$ cm$^{-3}$.

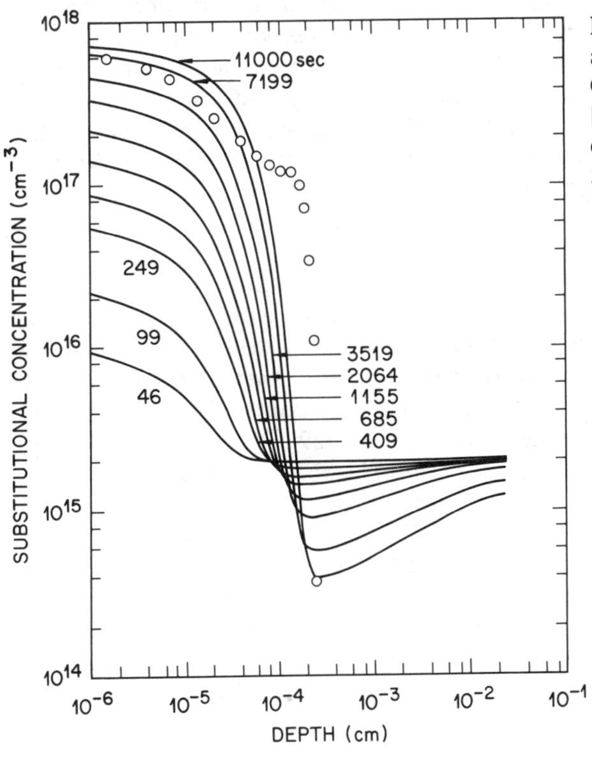


**Figure 7.27** Substitutional concentration as a function of depth from the surface for a GaAs wafer for various annealing times [136]. The data points are for Mn out-diffusion in a wafer annealed at 740°C for 5400 sec [133].

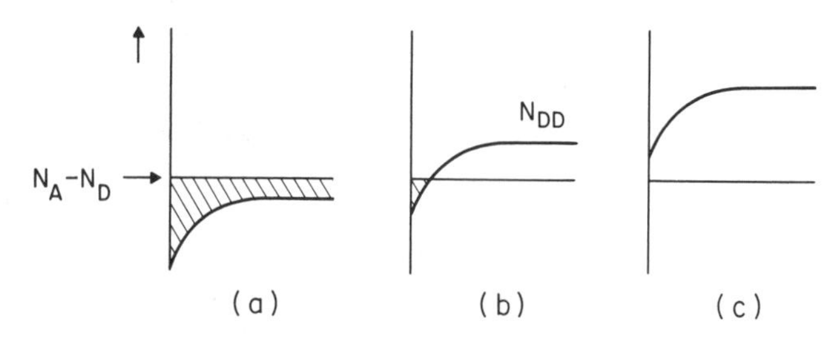


**Figure 7.28** Schematic error function outdiffusion profiles of the deep donor EL2 ($N_{DD}$) from a semi-insulating GaAs with $N_A > N_D$. The cross hatched region represents p-type region: (a) $N_A - N_D > N_{DD}$, the material is p-type (b) $N_{DD} > N_A - N_D$ except at surface where the material is type the converted (c) $N_{DD} > N_A - N_D$ throughout and the material is n-type and semi-insulating [140].

The lines converge as the bulk is depleted with increasing time. The theoretical curve corresponding to the annealing time of 5359 seconds agrees well with the experimental data up to a depth of 0.5 μm. Beyond this depth, the inflection and the rapid fall of the experimental diffusion profile cannot be explained by the model.

The agreement between the model and experiment would perhaps improve if effect of built-in electric field on diffusion is also included. For example, the inflection in the experimental diffusion profile can be caused by the built-in field at the p-n junction which is formed near the surface between the originally weakly n-type sample and the p-type converted region. Since the interstitials are expected to be donors, diffusing $Mn_i$ will accumulate near the p-side of the junction because of the electric field of the junction retarding motion of positively charged species. That inflection in the diffusion profile occurs when electric field effect is taken into account has been demonstrated by Shaw and Wells [137] and Anthony [138].

**Redistribution of EL2**

It is well known that the deep donor defect named EL2 is responsible for the semi-insulating behavior in undoped GaAs. The deep donor compensates net $N_A - N_D$ shallow acceptors where $N_A$ and $N_D$ are total concentrations of shallow acceptors and donors to give high resistivity [139]. Similar to the transition elements, outdiffusion of EL2 has been reported to be the cause of thermal conversion of surface conductivity in undoped SI GaAs [139]. Near the surface p-type conductivity can result if the concentration of EL2 falls below $N_A - N_D$ as illustrated in Fig. 7.28 [140]. $N_{DD}$ is shown to diffuse according to an error function profile in Fig. 7.28. When the concentration of EL2 ($N_{DD}$) stays above $N_A - N_D$, even at the surface no type conversion occurs as shown in Fig. 7.28(c). The outdiffusion of EL2 was found to be very much dependent on the annealing ambient. Annealing in a $H_2$ ambient caused significant outdiffusion while annealing in a $N_2$ ambient did not affect the concentration of EL2 [141]. Samples annealed in excess As pressure showed no or reduced outdiffusion [142, 143] indicating that the preferential loss of As from the surface during annealing has an influence on EL2 outdiffusion. The conversion to p-type conductivity as indicated in Fig. 7.28(a), has been correlated with residual carbon concentration suggesting that the uncompensated acceptor is residual carbon in the bulk GaAs

[143, 144]. The outdiffusion profiles of EL2 also indicated a very rapid diffusion process characterized by a diffusion constant of $10^{-10} - 10^{-12}$ $cm^2 sec^{-1}$ at 800 to 850°C [142, 143] which is considerably greater than self-diffusion coefficients in GaAs [145].

It has been suggested that the decrease in EL2 for annealing in a $H_2$ ambient may be related to the passivation of the EL2 centers [141]. Lagowski et al., [146] have reported that annealing in a hydrogen plasma at 300°C for 2h resulted in a decrease in the EL2 concentration in Bridgman grown crystals. Similarly, Pearton and Tavendale [147] have shown a decrease in electron traps in LPE GaAs layers after a hydrogen plasma anneal. Such effects are very much consistent with hydrogen passivation of shallow and deep centers in GaAs [148]. If the decrease in EL2 concentration is the result of hydrogen passivation of the defects responsible for EL2, then the unusually large outdiffusion coefficient could be explained as caused by indiffusion of atomic hydrogen interstitials [141]. At 800 to 850°, some dissociation of molecular hydrogen may occur to give atomic hydrogen which then rapidly diffuses into GaAs and passivate EL2 centers.

Oxygen implantation has been shown to reduce the outdiffusion of EL2 annealed with a $Si_3N_4$ cap as illustrated in Fig. 7.29 [149]. In the unimplanted sample the EL2 concentration has

**Figure 7.29** EL2 depth profiles obtained from DLTS measurements in GaAs samples before and after implantation with Se (fluence $3.5 \times 10^{12}$ cmref–2) and O (fluence $10^{13}$ and $10^{14}$ cmref–2) [149].

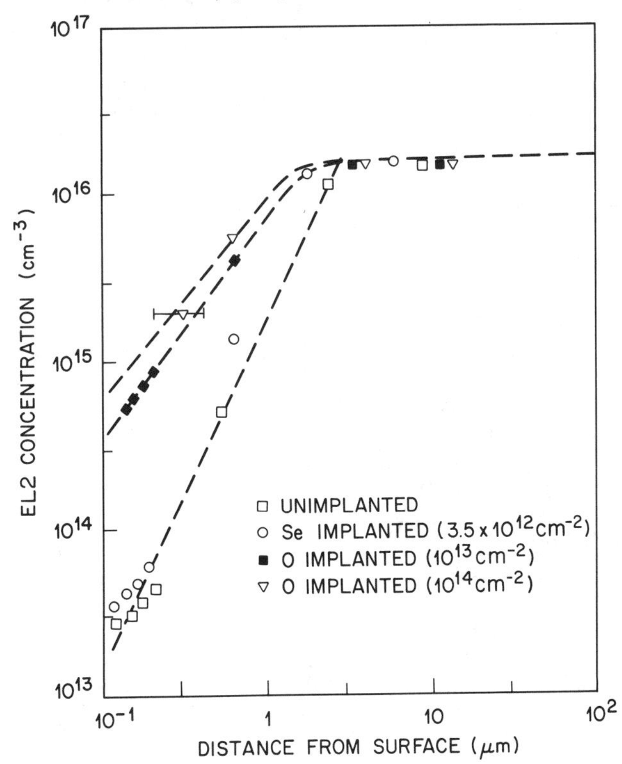

fallen by three orders of magnitude at a depth of ~3000 Å from the surface after annealing at 870°C for 20 minutes. However, when the sample is implanted with oxygen at 100 keV to a dose of $10^{13}$–$10^{14}$ $cm^{-2}$ and annealed in the same way, the reduction in EL2 is considerably less. As can be seen from Fig. 7.29, the reduction in the outdiffusion is only weakly dependent on the oxygen implant dose. Further, the effect is caused by oxygen itself and not caused by the implant damage since an implantation with Se with a similar dose and a similar projected range has no effect on EL2 outdiffusion. This effect of oxygen raises the query about the possible connection between oxygen and EL2 centers.

**Type conversion in n-type substrates**

Type conversion of the near surface region on annealing can also be a problem in n-type GaAs substrates. The n-type substrates are used for the epitaxial growth of (Al,Ga)As and GaAs layers which are used for fabricating photonic devices. The substrates are kept at 600 to 800°C for times ranging from a few minutes to a few hours prior to epitaxial growth. The effect of the heat treatment can be either to reduce the surface electron concentration and give rise to a high resistivity n-layer [150 - 152] or to cause complete type conversion and produce a p-type layer on the surface [151, 153, 154]. Both situations will have significant effect on the performance of such devices.

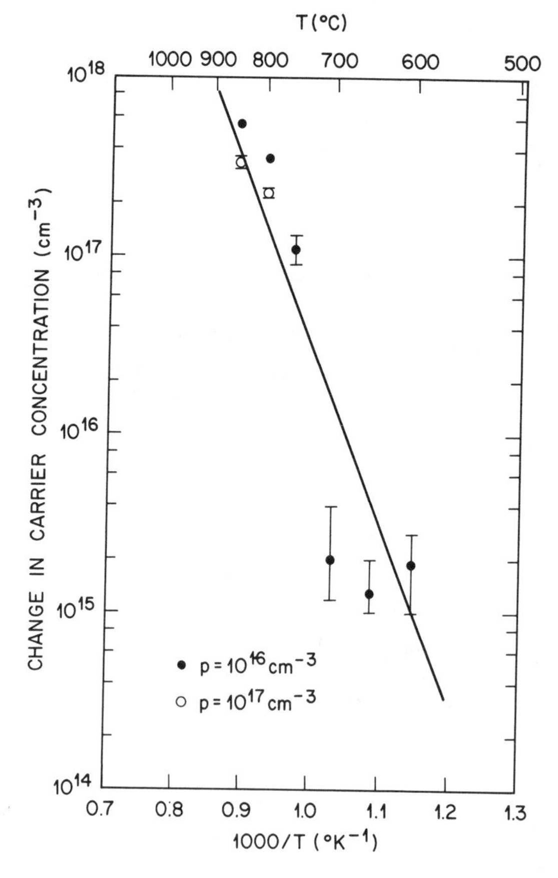


**Figure 7.30** Increase in surface hole concentration in p-type LPE GaAs as a function of annealing temperature in flowing hydrogen for 15 min [153]. Fitting an Arrehenius function to the data gives an activation energy of 2.0 eV which is interpreted as the enthalpy of formation of the acceptors created by the heat-treatment.

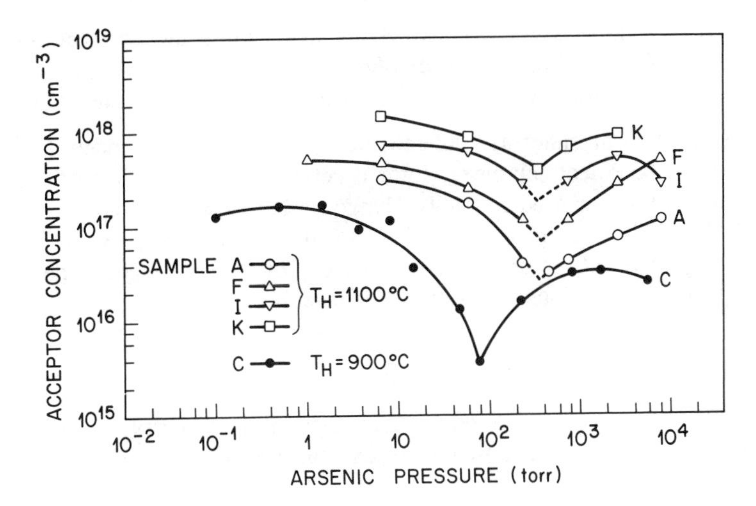


**Figure 7.31** Acceptor concentration in heat-treated n-type GaAs at 1100°C and 900°C for 67 hrs as a function of arsenic vapor pressure. The initial electron concentration of the samples are A) $8–9 \times 10^{16}$ cmref–3 F) $5.4 \times 10^{17}$ cmref–3 I) $9.5 \times 10^{17}$ cmref–3 K) $2.8 \times 10^{18}$ cmref–3 C) $1.8 \times 10^{17}$ cmref–3 [156].

When complete type conversion occurs, the surface p-concentration can be in the range $10^{14}$–$10^{16}$ cm$^{-3}$ [153, 154]. Typically, a region ranging in thickness from submicron to a few microns only will be type converted and after this region is removed the initial carrier concentration is recovered [153]. The surface hole concentration was also found to increase when a p-type GaAs substrate was heat-treated suggesting that acceptors are created during the annealing [153] for both n- and p-type substrates. This is illustrated in Fig. 7.30 which shows increase in surface hole concentration in p-type GaAs liquid phase epitaxial layers as a function of heat treatment temperature in flowing hydrogen [153]. The change in carrier concentration is higher for higher heat-treatment temperature and fitting an Arrehenius function to the data points gives an activation energy of 2.0 eV which is interpreted to be the enthalpy of formation of the acceptor defect. Incidentally, a similar value has also been reported by Potts and Pearson [155] who measured the change in lattice parameter of bulk GaAs crystals quenched from high temperatures. They associated the measured enthalpy of formation with arsenic vacancy, $V_{As}$. This means that $V_{As}$ is also involved in the thermal conversion process.

To investigate the nature of acceptors created under high temperature heat treatments, Munoz et al., [153] and Nishizawa et al., [156] performed experiments as a function of As over pressure. Results obtained by Nishizawa et al., [156] in n-type Te-doped GaAs crystals are shown in Fig. 7.31. It can be seen that acceptors are created both at low- and at high-arsenic pressures. Further, the concentration of acceptors formed is proportional to the initial donor concentration. Based on these results, it was suggested that the acceptors are defects associated with the donor impurity and native defects formed under excess or deficient As conditions, the defects being, respectively, gallium vacancy, $V_{Ga}$, arsenic interstitial, $As_i$ arsenic antisite defect, $As_{Ga}$, and $V_{As}$,

$Ga_i$ and $Ga_{As}$. Of these defects, the ones which are likely to behave as acceptors are $V_{Ga}$ and $Ga_{As}$ [see Sec. 6.4]. If no impurities were involved in the type conversion, these native defects alone would explain the acceptor formation at high- and low-arsenic pressures, respectively.

If impurities are present either as contaminant or as intentionally added dopants, then the acceptors formed can be defect complexes formed between the native defects and the impurity atoms. For example, $Ga_{As}$ can be a doubly charged acceptor which forms a complex with a singly charged $Te_{As}$ donor to give a singly charged acceptor complex ($Te^{+}_{As}\ G^{--}_{As}$). If an amphoteric impurity such as Si is present, then transfer from the metal site to non-metal site can occur under As deficient conditions according to the reaction

$$Si^{+}_{Ga} + V_{As} + e \rightarrow Si^{-}_{As} + V_{Ga} + h \tag{7.18}$$

On the gallium site Si is a donor while on the arsenic site it is an acceptor. Thus site-transfer of Si which is usually present as contaminant can explain the formation of acceptors in undoped GaAs crystals.

That reaction Eq. (7.18) is a distinct possibility during high temperature treatment of GaAs crystals is borne out in many photoluminescence investigations. The near band edge photoluminescence spectrum from a heat-treated n-type GaAs shows clearly the emission band which has been associated with the $Si_{As}$ acceptors [157]. In addition to the near band edge emission, a band near 1.4 eV has also been associated with a ($Si_{As}$–$V_{As}$) complex [157 - 160]. In addition to Si, the contamination of heat-treated crystals by Cu, which gives an acceptor level at 0.156 eV, has been reported by many authors. The source of Cu is believed to be the quartz containers used in the heat-treatment experiments. As evidence of Cu contamination, the photoluminescence spectrum from heat-treated GaAs crystals invariably shows the emission band near 1.35 eV which has been associated with the Cu acceptor [161 - 163].

If the formation of acceptors during heat treatment is predominantly controlled by impurities (e.g., Si, Cu), then the participation of native defects may be completely overshadowed. Under experimental conditions which were supposed to have strictly excluded impurity effects, Chiang and Pearson [164] showed that annealing under high- As pressures leads to the formation of acceptors ($V_{Ga}$) and under a medium- to low-arsenic pressure leads to the formation of donors ($V_{As}$). One of their experimental results for the formation of donors in an undoped GaAs crystal ($n \sim 3 \times 10^{16}\ cm^{-3}$) on heat treatment at 800°C and low arsenic pressure is shown in Fig. 7.32. Chiang and Pearson have fitted the experimental results with theoretical concentration profiles calculated assuming thermodynamic equilibrium between the surface of the crystal and external ambient and complementary error function diffusion profiles for $V_{Ga}$ and $V_{As}$, as shown in Fig. 7.32. The values of the diffusion coefficients of $V_{Ga}$ and $V_{As}$ obtained by these authors are given

$$\begin{aligned} D(V_{Ga}) &= 2.1 \times 10^{-3} \exp\left(\frac{-2.1eV}{kT}\right) \\ D(V_{As}) &= 7.9 \times 10^{3} \exp\left(\frac{-4.0eV}{kT}\right) \end{aligned} \tag{7.19}$$

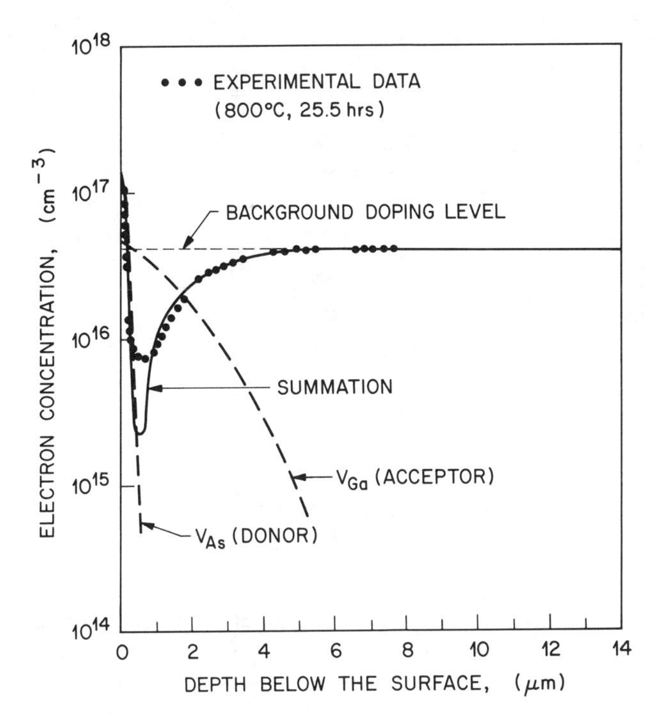


**Figure 7.32** Comparison of experimental results with theoretical calculations assuming complementary error function profiles for gallium and arsenic vacancies. Sample was originally undoped with $n \sim 3 \times 10^{16}$ cmref–3 [164].

Equation (7.19) shows that the diffusivity of $V_{Ga}$ is larger than that of $V_{As}$, meaning that $V_{Ga}$ will be the dominant species in mass transport processes at the heat-treatment temperatures even if the concentration of $V_{As}$ is greater than that of $V_{Ga}$. It should, however, be noted that these values for the vacancy diffusivities are only rough estimates. Combining these diffusivities with the concentration of vacancies estimated by Chiang and Pearson [164] in the same work, are inconsistent with the self-diffusion coefficients in GaAs. Palfrey et al., [145, 165] have measured the self-diffusion coefficients in the As and Ga sublattice as

$$D_{As} = (5.5 \pm 2.4) \times 10^{-4} \exp\left(\frac{-3.0 \pm 0.04\text{eV}}{kT}\right)$$
$$D_{Ga} = (4.0 \pm 16.0) \times 10^{-5} \exp\left(\frac{-2.6 \pm 0.5\text{eV}}{kT}\right) \tag{7.20}$$

If diffusion in the sublattice occurs via a vacancy mechanism, then the self-diffusion coefficients can be expressed as [166]

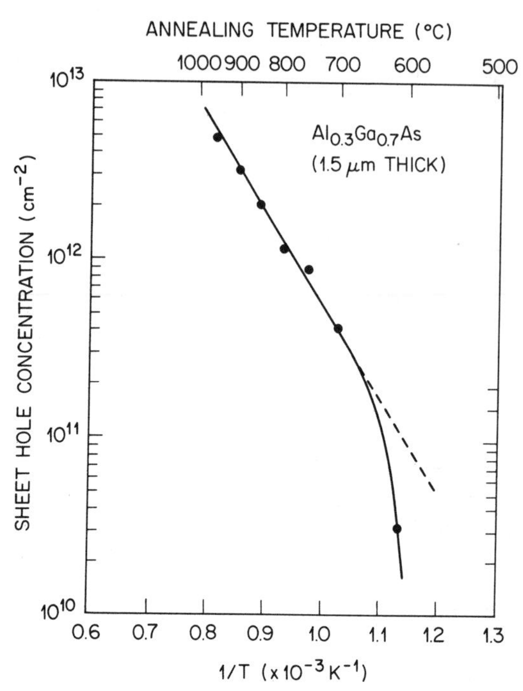


**Figure 7.33** Room temperature sheet hole concentration as a function of annealing temperature for an $Al_{0.3}Ga_{0.7}As$ [169].

$$D = f[V]\, D_V \tag{7.21}$$

where f is the correlation factor and [V] and $D_V$ denote, respectively, vacancy concentration and diffusivity. Combining Eqs. (7.19) and (7.20) gives vacancy concentrations which are very much different from the estimates of Chiang and Pearson [164].

Earlier we discussed the outdiffusion of EL2 in undoped semi-insulating GaAs and the possibility of forming slightly p-type surface skin caused by the presence of uncompensated shallow acceptors. In contrast to this behavior, Lagowski et al., [167] have reported an "inverted thermal conversion" phenomenon in GaAs crystals by which conversion from conducting to semi-insulating behavior occurs on annealing at 850°C. The inverted thermal conversion is obtained by cooling the crystals rapidly from very high temperature (about 40°C below the melting point) such that migration of point defects does not occur and that the formation of EL2 is prevented. Subsequent annealing in the 700 to 900°C range induces point defect mobility and EL2 is formed which then compensates residual acceptors giving rise to semi-insulating behavior. In this process of inverted thermal conversion it appears possible to control the EL2 density and achieve resistivities one order of magnitude greater (~ $10^9$ Ωcm at 300K) than those attained in standard semi-insulating crystals (e.g., $10^7$–$10^8$ Ωcm) [168].

Thermal conversion on annealing has also been reported in the alloy $Al_xGa_{1-x}As$ grown by MBE [169]. For x in the range 0.05 to 0.3, undoped high resistivity epilayers were converted to low resistivity p-type after annealing above the growth temperature of 650°C. Figure 7.33 shows the room temperature sheet hole concentration and sheet Hall mobility as a function of annealing temperature for an $Al_xGa_{1-x}As$ epilayer (x ~ 0.3, 1.5 µm thick). Fitting the data to the Arrehenius law gives an activation energy of 1.06 eV for the thermal introduction of acceptors into the layer. The acceptors are believed to be C being driven into solution on annealing from certain C-rich phases (e.g., Al-C compounds) present in the as-grown layer.

**InP substrates**

The redistribution of impurities during high temperature annealing also occurs in InP but only a few elements have been studied. One such element is Fe which is used routinely as a dopant to obtain semi-insulating InP. Fe is a deep acceptor in InP and it compensates the residual shallow donors to give high resistivity. Similar to GaAs, diffusion of Cr and Fe into InP does not follow the expected error function diffusion profiles. Figure 7.34 shows the diffusion of Fe into InP for 15 min at 900°C. Although for this shorter diffusion time the bulk section of the profile is not horizontal as in Fig. 7.22 for Cr diffusion into GaAs, there is still considerable penetration of Fe into the specimen. In the diffused specimens no measurable acceptor activity of Fe was found [170, 171], which led to the conjucture that when Fe and Cr are incorporated into InP, a significant fraction of the dopants are on nonsubstitutional sites. This hypothesis also lends support to the generally accepted view that the rapid diffusion of Cr and Fe in GaAs and InP occurs via an interstitial mechanism.

Outdiffusion of Fe from the bulk to the surface has been reported to occur both in Fe-doped semi-insulating InP substrates as well as in other InP substrates where Fe can be present as a

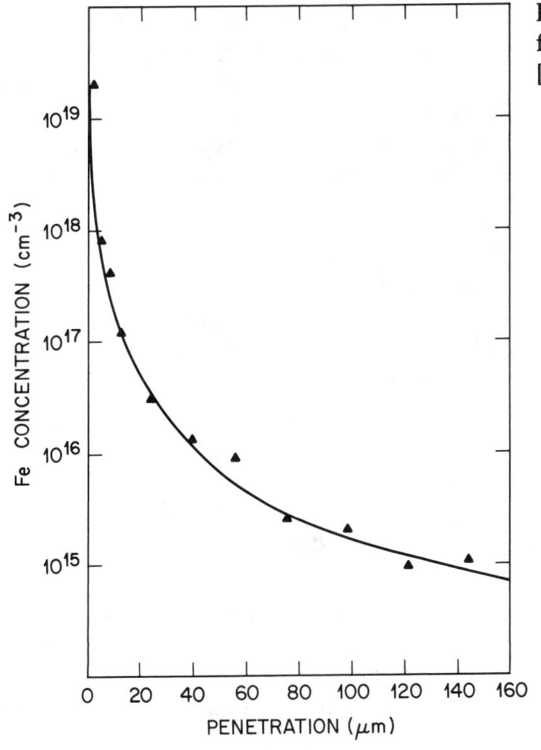


**Figure 7.34** Diffusion profile of Fe in InP for diffusion time of 15 min at 900°C [170,171].

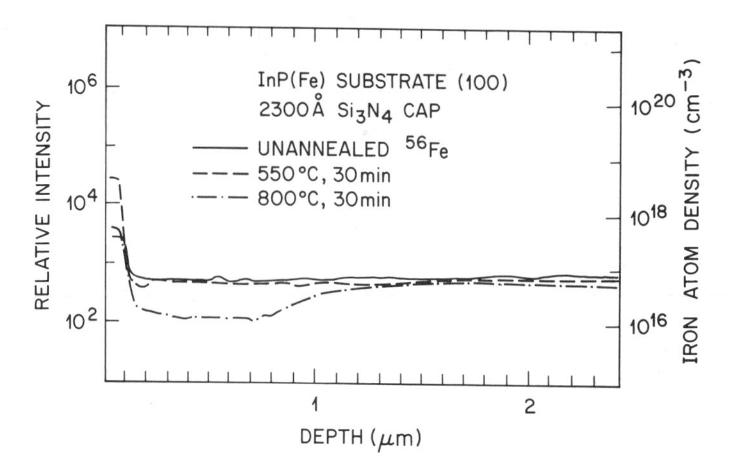


**Figure 7.35** Depth profiles of Fe in a semi-insulating InP substrate after 30 min anneals at 550 and 800°C with a $Si_3N_4$ encapsulant [173].

contaminant [172 - 175]. Figure 7.35 shows concentration of Fe as a function of depth obtained by secondary ion mass spectrometry from a Fe-doped semi-insulating InP substrate following annealing at 550°C and 800°C for 30 minutes with a $Si_3N_4$ encapsulant [173]. While the Fe depth profiles of the unannealed sample and sample annealed at 550°C are very similar, the profile from the sample after the 800°C annealing treatment shows clearly a depletion region for about 1 μm after an accumulation region extending to a depth of ~1000 Å from the surface. Since Fe is a deep acceptor compensating shallow donors in InP, depletion of Fe near the surface would give rise to n-type conduction in the surface region. Such n-type surface layers after annealing Fe-doped semi-insulating InP crystals in the temperature range 680 to 750°C have been reported [176 - 177].

Evidence for Fe accumulation near the surface upon annealing has also been provided by low temperature photoluminescence spectroscopy. The emission band at 0.35 eV associated with an internal transition of $Fe^{2+}$ is readily seen in heat-treated InP crystals as illustrated in Fig. 7.36 [175]. By calibrating the strength of the 0.35 eV transition with Fe concentration measured by some analytical means, the unknown Fe concentration in a crystal can be easily measured using photoluminescence. Using this method, it is estimated that nominally undoped crystals can have $\sim 10^{14}$ $cm^{-3}$ Fe [174].

Besides Fe, other impurities which migrate towards the surface upon annealing are Cr and Mn [173, 174]. A major problem with such migration and accumulation of impurities at the surface of substrates is that when epitaxial layers are grown on them, the impurities diffuse into the layers thereby seriously undermining the purity [172, 178 - 184]. Figure 7.37 shows Fe profiles in layers grown by vapor phase epitaxy on a Fe-doped, Sn-doped, and Cr-doped InP substrates [178]. The striking feature in the depth profiles is the accumulation of Fe at the substrate-epitaxial layer interface. Further, the epitaxial layers contain Fe in the range $10^{15}$–$10^{16}$ $cm^{-3}$. At this level the electrical properties of the layers can be affected. For example, if all the Fe are electrically active, a high resistive n-type layer may be obtained instead of an intended n-layer of

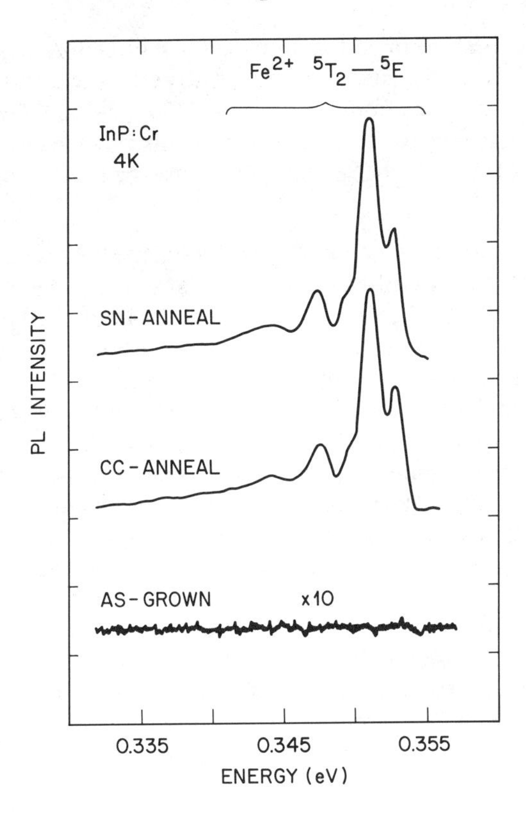


**Figure 7.36** Photoluminescence spectra at 4.2K of annealed Cr-doped InP samples showing the $Fe^{2+}$ intracenter transition. Samples were annealed with a $Si_3N_4$ encapsulant (SN anneal) or a InP cover wafer (CC anneal) [175].

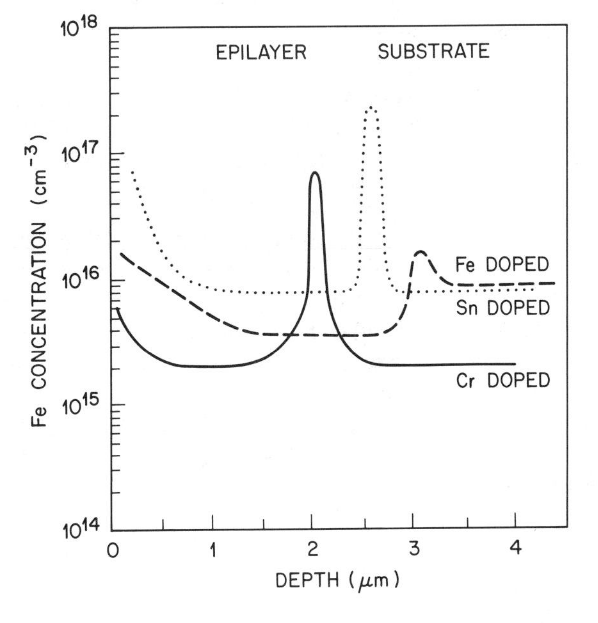


**Figure 7.37** Fe depth profiles in epitaxial InP layers grown on Fe-doped, Sn-doped and Cr-doped substrates [178]. Note the accumulation at the epilayer-substrate interface.

carrier concentration $10^{15}$–$10^{16}$ $cm^{-3}$. It has been reported that in the growth of GaInAs layers on InP substrates by MBE, as much as 2/3 of the carrier concentration and 40 percent reduction in mobility can be attributed to outdiffusion of impurities from the substrate [179]. Since the type converted material is confined to the near surface region it can be removed either by mechanical polishing [179 - 181] or by *in situ* etching of the substrate in the growth reactor itself prior to epitaxial growth [172].

Redistribution on annealing is not confined to transition metal impurities alone but occurs for the commonly present shallow impurities as well. A well-known example is the redistribution of Si in GaAs from one site to another as described by Eq. (7.18). As discussed earlier, the p-type thermal conversion in GaAs can be explained by this site exchange reaction. Chin et al., [185] have reported extremely rapid out-diffusion of n-type impurities such as S, Sn, and Se in InP upon heat treatment in evacuated ampoule at 550°C. They measured diffusivities of the order $10^{-8}$ $cm^2$ $sec^{-1}$ which are two orders of magnitude greater than that of rapidly diffusing Zn impurity. Such a rapid outdiffusion of the n-type impurities is surprising especially in view of the fact that they do not diffuse rapidly into InP [186]. The mechanism for this rapid outdiffusion is not understood although pipe diffusion along dislocations has been ruled out as a possible one [185, 187]. In connection with outdiffusion of n-type impurities, the annealing ambient was found to have a profound effect [179, 188]. The outdiffusion showed a drastic increase as the ambient changed in the order $PH_3 + H_2$ mixture, to inert gas (He or Ar) to forming gas to $H_2$ to vacuum. Annealing in air also suppressed outdiffusion [188]. This effect of the ambient is very similar to that found for Mn outdiffusion in GaAs as discussed before and may be related to evaporation of P from the surface and to how rapidly the equilibrium vacancy concentrations are established at the surface. Loss of P is presumably reduced by the P overpressure established by the $PH_3$ ambient or by the formation of an impervious oxide as in the case of annealing in air [188].

In the case of Zn-doped p-type InP heat treatment at temperatures between 300 and 500°C gave rise to an increase in the hole concentration extending as much as 60 μm into the bulk at 500°C [189, 190]. Figure 7.38 shows the depth profiles of hole concentration in InP:Zn substrates with initial hole density $1 \times 10^{16}$ $cm^{-3}$, after heat treatment at 500°C for 30 minutes in various ambients [190]. The increase was greatest in hydrogen ambient and was completely suppressed when the surface of the substrate is encapsulated with $SiO_2$ during the heat treatment (not shown in the figure).

Zn is a fast diffusing impurity in InP and the rapid outdiffusion of Zn interstitials to the surface upon heat treatment and their conversion to substitutional Zn via Eq. (7.17) would give rise to the increase in hole concentration. An analysis similar to the Mn outdiffusion in GaAs would be applicable in this case also. However, Wong and Bube [190] were able to fit the depth profiles in Fig. 7.38 as out-diffusion of $Zn_i$ following Fick's second law by simply assuming that the maximum increase of hole density after the 500°C heat treatment is equal to the Zn interstitial concentration at equilibrium and by assuming a certain evaporation rate of the dopant from the surface. They obtain a value of $3 \times 10^{-9}$ $cm^2$ $sec^{-1}$ for the outdiffusion coefficient at 500°C which is nearly three orders of magnitude greater than the diffusion coefficient of Zn into n-type InP substrate [191 -1 93].

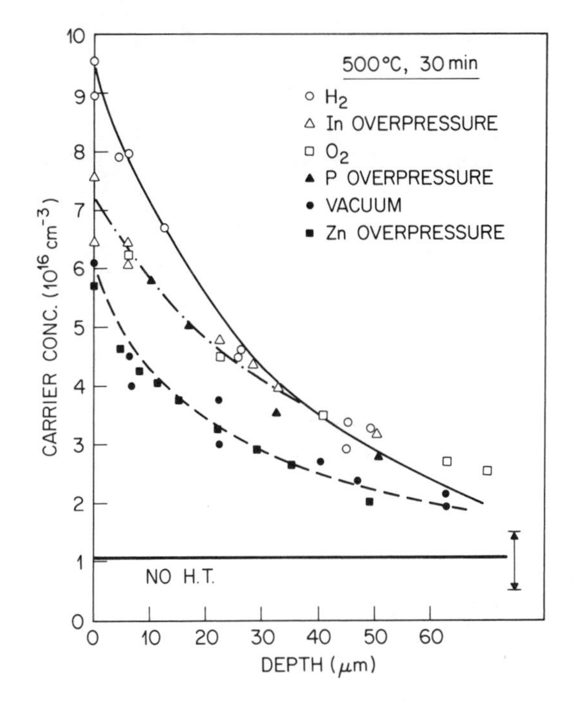


**Figure 7.38** Hole concentration versus depth in p-type InP:Zn crystals with initial hole concentration of $10^{16}$ cmref–3, after heat-treatment at 500°C for 30 min in various ambients [190].

## 7.4 DEFECT GETTERING

### 7.4.1 Gettering at Implantation Damage

When a dopant is introduced by ion-implantation, it is necessary to perform an annealing cycle at fairly high temperatures in order to activate the implanted atoms and to remove the damage. During this heat treatment impurity redistribution would naturally occur but because of the simultaneous annealing of the damage it would be a great deal more complex than the situations discussed in Sec. 7.3. The damage region can getter impurities and give rise to preferential segregation. An example of this is illustrated by the Cr redistribution in Cr-doped GaAs following S implantation and annealing at 840°C as shown in Fig. 7.39. The Cr atoms preferentially move to the region of implantation damage as evidenced by the concentration peak in that region [194]. Since Cr is a deep acceptor, its redistribution during the implant activation annealing would affect the electrical profile of the implanted donor species.

The pile-up of Cr usually occurs under the peak of implant profile at a depth approximately equal to the range of the implanted species, $R_p$. Sometimes two Cr peaks can be observed one below and the other above $R_p$ [194 - 196]. Measurement of Cr concentration profiles by SIMS together with transmission electron microscopic study of the damage regions further confirmed that Cr peaks correspond to the damaged regions [197]. Therefore, gettering of Cr atoms by the damage region during the post-implantation annealing cycle is generally believed to be the cause of the peaks in the concentration-depth profiles (Fig. 7.39).

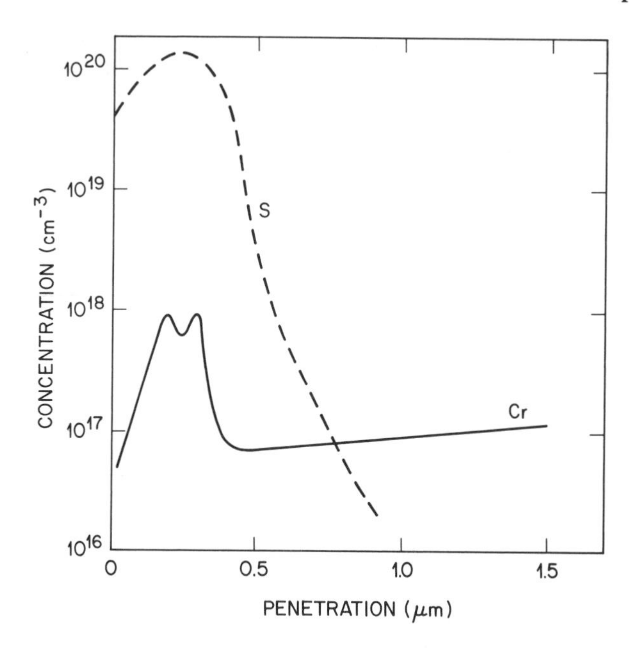


**Figure 7.39** Cr redistribution in GaAs after S implantation at 300 keV, $4 \times 10^{15}$ cmref–2 and annealing at 840°C for 20 min. The original S profile is also shown (dashed line) [194].

The build-up of Cr at damage regions upon annealing also occurs when Cr itself is the implant atom [198, 199]. Figure 7.40 shows Cr concentration profile in InP after implantation and after annealing at 750°C for 15 minutes [199]. Besides Cr accumulation near the surface, two other pile-ups can be seen, one below and another above $R_p$. Note that the Cr concentration at the latter location is even higher than the initial maximum concentration, presumably caused by the formation of metal rich phosphides. The peaks which develop for short annealing times disappear for longer annealing times and the Cr distribution then looks the same as a normal indiffusion profile (Fig. 7.22). Besides Cr, pile-up of dopants in the implant region has also been observed for Mn [200], Zn [201] in GaAs, and Fe [173] in InP.

Implantation of lighter atoms such as Be, because of the relatively small extent of the implant damage, does not give rise to build-up of Cr atoms [200]. Instead, the Cr concentration is depleted in the implanted region and the concentration profile simply resembles the profile caused by the outdiffusion of Cr during heat treatment (Fig. 7.21).

The redistribution of impurities in implanted crystals depends on such factors as implantation fluence and damage, stoichiometry in the implanted layer and thermal stress, especially if an encapsulant is used to protect the surface from loss of group V atom during post-implantation heat treatment. A larger fluence would be expected to give rise to a more pronounced build-up of atoms in the damage region [200]. As observed for Cr (Fig. 7.40) pile-up can occur both below and above $R_p$. The pile-up of atoms near $R_p$ or slightly below it is expected since the maximum

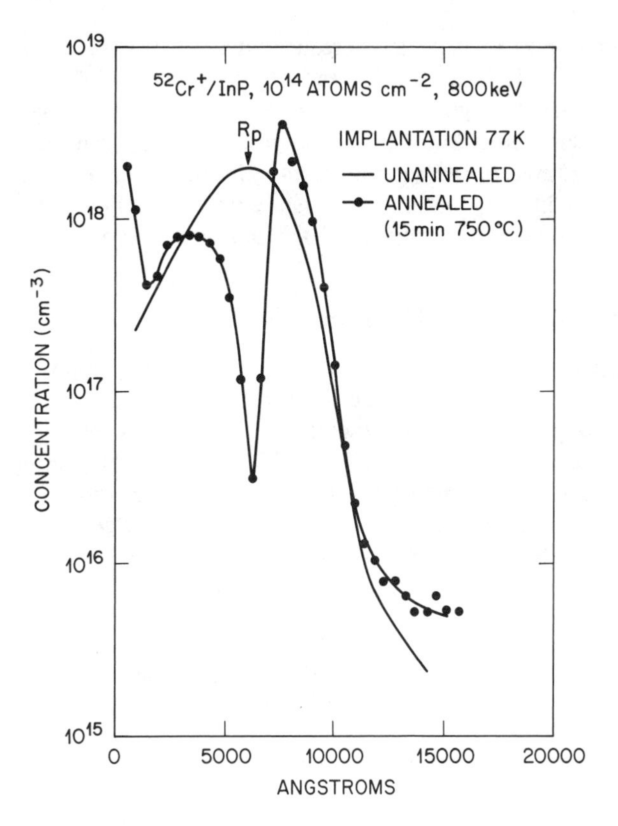


**Figure 7.40** Cr depth profiles in InP for a 800 keV $10^{14}$ cmref–2 Cr implant at 77K: unannealed (solid line); annealed at 750°C for 15 min (solid line with closed circles) [199].

of damage usually occurs around there. In order to explain the pile-up beyond $R_p$, Magee et al., [202] proposed, based on the calculations of stoichiometric imbalances caused by implantation [203], that damage is produced in the form of native interstitials beyond $R_p$ by the recoiled group III and group V atoms. Gettering of Cr atoms in the interstitial rich regions then explains the peak.

### 7.4.2 Strain Induced Gettering

In addition to the pile-up in the implant damage region, pile-up of impurity atoms is also seen near the interface of the encapsulant and the substrate after the heat treatment. This may have been caused by strain induced gettering, the origin of strain being the thermal expansion mismatch between the encapsulant and the substrate [204]. Strain induced gettering can also play a very important role if rapid thermal annealing (RTA) treatment is used to remove ion implantation damage and electrically activate implanted dopants. RTA is beginning to be more widely used than conventional furnace annealing techniques because of its advantages resulting from short annealing times, typically 1 to 20 seconds, compared to 20 to 60 minutes for furnace annealing. The shorter times are expected to reduce diffusion of implant dopants and background impurities. However, very high thermal stresses leading to plastic deformation in the crystal can develop caused by temperature gradients during the heating and cooling portions of the RTA cycles [205, 206]. Associated with the plastic deformation is the generation of dislocations and point defects which can give rise to significant impurity redistribution at considerable distances beyond $R_p$ [207]. This implies that the temperature-time annealing cycles by RTA should be carefully controlled to minimize thermal gradients and the deleterious effects of strain induced gettering.

### 7.4.3 Heterostructure Gettering at Heterostructure Interfaces

In multilayer structures consisting of layers of different compositions, interfaces between the layers may not always be perfect and atomically smooth. A certain amount of interfacial disorder may be present, resulting in a high concentration of defects. Such defect laden interfacial regions would naturally act as gettering sites for impurities and dopants during growth and device processing steps. In addition, the interfacial region may also be strained because of the thermal expansion mismatch between the layers leading to strain induced gettering as well. An example of interfacial impurity gettering is shown for the case of Zn at GaInAs/InP interface in Fig. 7.41 [208]. When Zn is diffused into a n-type GaInAs layer grown by LPE or MBE directly on an n-

**Figure 7.41** SIMS depth profile of a LPE grown $Ga_{0.47}In_{0.53}As$/InP wafer after Zn diffusion from a solid source in a tube furnace: (a) Diffusion at 575°C for 10 min. $10^4$ counts/s correspond to Zn concentration of $7 \times 10^{19}$ cmref–3. (b) Diffusion at 625°C for 15 min. $10^4$ counts/s correspond to a Zn concentration of $9 \times 10^{19}$ cmref–3 [208].

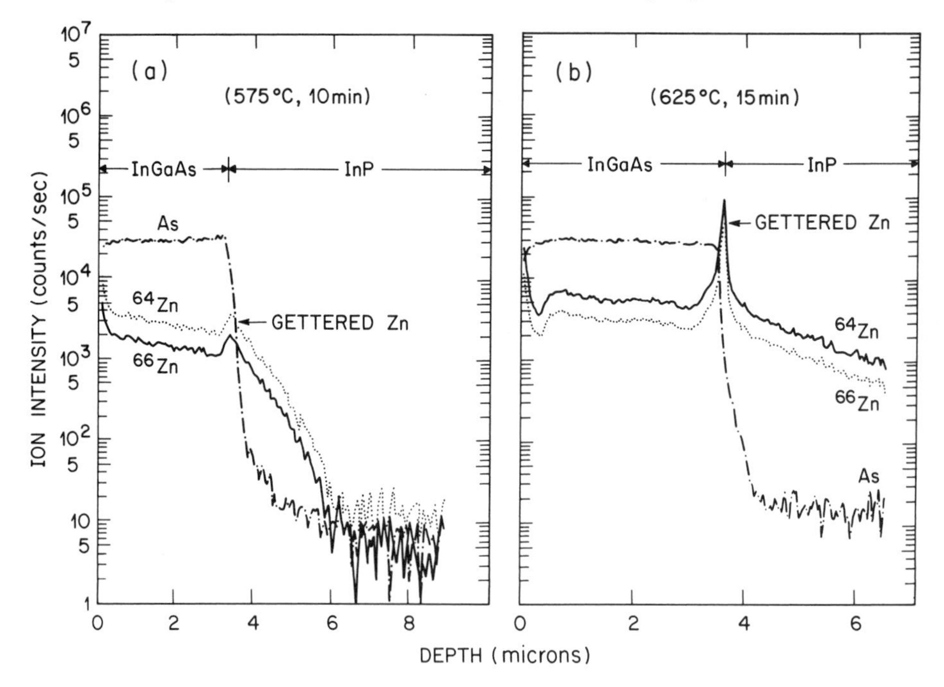

type InP buffer layer, it accumulates at the GaInAs/InP interface as shown in Fig. 7.41. The amount of accumulation is a function of the temperature and duration of the diffusion. Note the significant increase in the Zn concentration at the interface when the temperature increases from 575°C to 625°C. The Zn concentration at the interface for the higher temperature is about $10^{21}$ $cm^{-3}$ which is considerably higher than the solid solubility of Zn in bulk GaInAs ($1.2 \times 10^{19}$ $cm^{-3}$) and InP ($\sim 3 \times 10^{18}$ $cm^{-3}$). Such a high concentration of impurities at heterointerfaces can alter the electrical properties near the interfacial regions and further can act as a source of defects in active regions of the device which adjoin these interfaces. One can expect that the interfacial gettering seen for Zn in GaInAs/InP applies to other mobile impurities such as Li, Cu, Cd, Au, Fe, and so on, as well as to other heterointerfaces.

### 7.4.4 Gettering at Dislocations

Dislocations in semiconductors, the same as in metals, can be efficient gettering sites for impurities and native defects. The gettering at dislocations is a result of the elastic interaction between the stress field of the dislocation and the strain field of the point defect [209]. The point defects can be impurity or dopant atoms either on substitutional or interstitial sites in the lattice. They also include native defects like vacancies and self-interstitials. In both cases, the point defects are centers of expansion or contraction and thus they interact only with the hydrostatic component of the stress field of the dislocation. For example, there will be an increase in the concentration of interstitials which act as centers of expansion in the tensile region of an edge dislocation but a decrease in the concentration of vacancies which act as centers of contraction. A similar picture holds good for impurity atoms which are centers of expansion or contraction.

This effect of point defect gettering at dislocations plays an important role in many situations in GaAs and InP materials and devices. One such situation pertains to the liquid encapsulated Czochralski (LEC) grown semi-insulating GaAs substrates used for fabricating integrated circuits by direct ion-implantation technology. The semi-insulating LEC GaAs crystals typically exhibit dislocation densities ranging from $10^4$ to $10^5$ $cm^{-2}$. These dislocations are potential gettering centers for impurities and defects in the crystal. Gettering at dislocations is revealed by the variation in the luminescence efficiency in the crystal between regions of different dislocation densities.

Generally, luminescence efficiency is found to be high in regions of high dislocation density and low in regions of low dislocation density [210 - 212]. This surprising result, since dislocations are generally expected to behave as strong nonradiative centers, is believed to be caused by the effect of gettering of nonradiative centers at dislocations. Dislocations attract nonradiative impurities and nonradiative defects leaving a region around them denuded of these centers. These denuded regions will naturally have a higher luminescence efficiency than that in areas far away from dislocations. Therefore, regions of high dislocation density will have a higher luminescence than regions of low dislocation density. This correspondence between dislocation density and luminescence efficiency is illustrated in Fig. 7.42 which shows a one-dimensional photoluminescence scan along the <110> direction of a LEC GaAs wafer and the dislocation distribution along the same line [212]. The luminescence output matches very closely the dislocation distribution. Further, the dislocation distribution across a (100) LEC wafer generally exhibits a W-shaped pattern (i.e., high dislocation densities at the middle and edges,

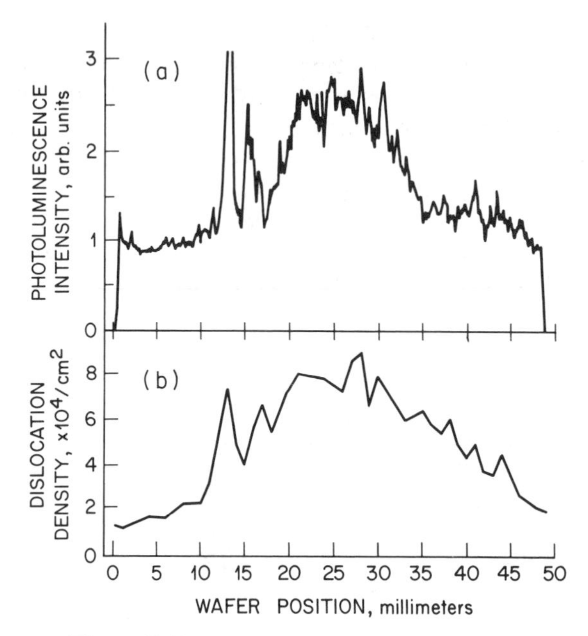


**Figure 7.42** (a) One dimensional photoluminescence line scan along <110> direction of a LEC GaAs wafer (b) Dislocation distribution along the same line [212]. Copyright, 1985, IEEE.

low in between as shown in Fig. 2.15). This shape is very closely reproduced by the luminescence efficiency profile [212, 213].

The unique similarity between dislocation distribution and radiative efficiency is also exhibited by spatially resolved luminescence maps obtained from LEC GaAs wafers [214 - 216]. The dislocation structure in the crystal manifests as a cellular structure, with a cell size of a few hundred micrometers. The interior of the cells contains fewer dislocations compared to the cell boundaries. The luminescence map also reveals the cellular structure of dislocations with the cell regions being dark and the cell boundaries being bright corresponding respectively, to low- and high-radiative efficiency. The maximum intensity variation between the cell interior and cell boundaries is a factor of 2 to 3. Only at the core of individual dislocations, which appear as arrays of dark spots along the cell boundaries [214, 217], the efficiency is lower consistent with their nonradiative character.

This relationship between dislocations and radiative efficiency is also observed in low dislocation density LEC grown In doped GaAs, though, between regions far from dislocations and regions surrounding dislocations the luminescence intensity differed by two to three orders of magnitude [218]. This large variation in luminescence intensity indicated that the carrier lifetimes are significantly altered near dislocations caused by the gettering phenomenon.

### 7.4.5 Gettering at Back Surface Damage

The increase in the radiative efficiency described is a result of an intrinsic gettering process at dislocations. As opposed to this, extrinsic gettering has also been used to achieve improvement in the quality of substrates and epitaxial layers grown on them. Dastidar et al., [219] reported improvement in mobility by a factor of two to three achieved by impurity gettering in GaAs using a borosilicate glass layer. Bozler et al., [220] used ion implantation damage followed by a 750°C anneal to getter impurities in semi-insulating GaAs substrates. After removal of the damaged material, epitaxial layers were grown on the substrates. The carrier concentration profiles for the epitaxial layers grown on gettered substrates did not exhibit the carrier depletion near the substrate-epitaxial layer interface that was seen for those grown on ungettered substrates. These results indicated that the concentration of out-diffusing compensating impurities or defects from the substrate into the epitaxial layer during growth was greatly reduced by the ion-implantation damage gettering.

Mechanical balk surface damage gettering was used to getter Cr and Au in GaAs [221, 222]. Macroscopic mechanical damage was introduced at the balk surface of substrates by a rotary abrasive unit while the front polished surface was protected by a $SiO_2$ film. Even at annealing temperatures as low as 300°C back surface damage was found to getter impurities [223]. After the damage process, the substrate was annealed in the temperature range 750 to 900°C using $Si_3N_4$ as the encapsulant to protect the front surface. Using transmission/scanning electron microscopy and SIMS, it was found that during this annealing treatment, dislocation lines introduced at the back surface by the mechanical damage are effective gettering sites for the impurities.

The back surface damage gettering is illustrated for the case of Cr in GaAs in Fig. 7.43 which shows the measured dislocation density as a function of depth from the back surface as well as the SIMS Cr concentration profile measured before and after annealing at 750°C [224]. The Cr profile after annealing clearly shows the gettering effect in the damage regions. The gettered Cr is essentially concentrated in the 1.5 μm damage region. In the case of Au, a similar result was found except that the gettered Au precipitated along dislocation lines [222] while Cr showed no evidence of precipitation. The beneficial effect of the mechanical damage gettering was realized when pre-gettered Cr-doped GaAs substrates were used for epitaxial growth. The defect density in epitaxial layers grown on gettered substrates was significantly less compared to those grown on regular substrates. In addition, the outdiffusion of Cr into the epitaxial layer was reduced by a factor of two to three in the gettered substrates [224].

Mechanical back surface damage was also found to be effective in improving the quality of undoped semi-insulating GaAs substrates [225]. The damage was introduced by bead blasting and the gettering heat treatment was carried out at ~500°C. When discrete MESFETs were fabricated on the gettered substrates by direct implantation, an increase in rf gain and about 40 percent increase in dc transconductance were observed while the gate capacitance at zero bias was unchanged. These improvements in device characteristics, without any significant change in other material parameters such as dislocation density and distribution in and immediately under the active layer, are supposed to have occurred because of the gettering of some fast diffusing

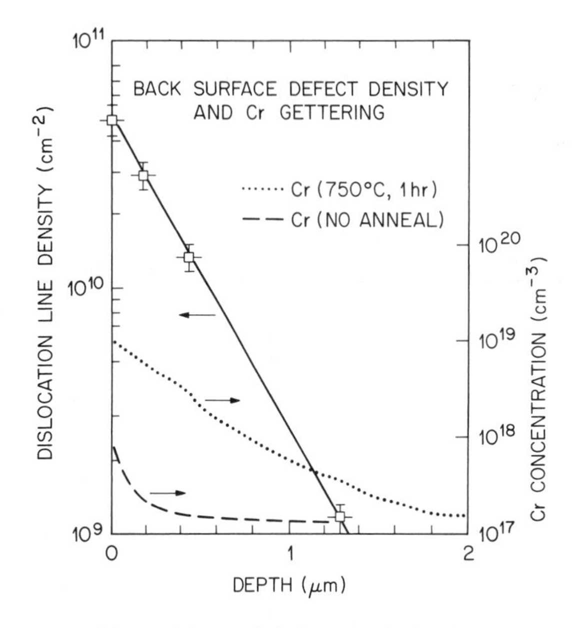


**Figure 7.43** Dislocation density and Cr concentration profiles of back surface damaged and annealed GaAs [224].

unwanted impurities and defects at the back surface damage. The nature of the diffusing species which are gettered is not identified, though spatially resolved cathodoluminescence mapping of gettered substrates reveal a reversal of luminescence contrast of the cellular dislocation structure. Instead of the commonly observed (Sec. 7.4.4) bright cell boundaries and dark cell interiors, the cell interiors had higher luminescence efficiency than the cell boundaries. A similar reversal of contrast was also observed in the case of undamaged LEC GaAs substrates annealed at ~800°C [226]. In both situations, the removal of nonradiative centers from the cell interiors gave the increased luminescence efficiency.

In spite of the improvement in the device properties back surface damage gettering is not a viable technological process because of the difficulty in achieving reproducibility and uniformity [225]. On the other hand, annealing treatments of either bulk ingots [213] or wafers [227, 228] without back surface damage appear to be reproducible and improve uniformity in comparison to as-grown crystals. Perhaps a judicious combination of these two techniques will provide the improvement in uniformity of device characteristics as well as an increase in absolute device performance.

## 7.5 PHOTONIC DEVICES

### 7.5.1 Recombination Enhanced Defect Motion

When electron-hole recombination occurs, the energy may be released as photons (luminescence), used to excite another carrier to higher energy levels (Auger process) or transferred to the lattice as heat (phonon production). In the last situation, the excess energy may be deposited locally as vibrational energy at the recombination center which then increases rate of defect reactions such as diffusion, dissociation, and annihilation. Particularly, the recombination enhanced diffusion (RED) of defects plays a principal role in degradation of light emitting devices such as lasers and LEDs.

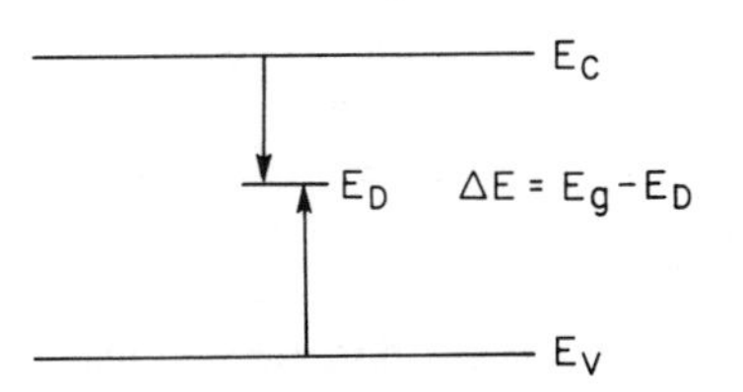


**Figure 7.44** Schematic band diagram showing a deep energy level at $E_D$ in the band gap. $\Delta E$ is the energy released in a recombination process involving $E_D$ and this is transferred as vibrational energy to the defect.

The first suggestion of recombination enhanced defect reaction was provided by the work of Gold and Weisberg on GaAs tunnel diodes [229]. The degradation rate of these diodes was found to be proportional to the nonradiative recombination component of the junction current. Gold and Weisberg suggested a ''phonon-kick'' mechanism by which the energy released by the recombination event (the maximum being equal to $E_g$) generates Frenkel defects.

If all the electronic energy released in recombination process is transferred as vibrational energy to the defect, the barrier for motion is reduced by an amount $\Delta E$ given by (see Fig. 7.44)

$$\Delta E = E_g - E_D \tag{7.22}$$

When defect motion is purely thermally activated, for a first-order reaction one can write the reaction rate as [230]

$$R_T = \nu \frac{\exp \frac{\Delta S}{k}}{N_j} \exp \frac{-E_T}{kT} \tag{7.23}$$

where $\nu$ is the jump frequency ($k\,\theta_D / h \sim 10^{13}$ sec$^{-1}$), $\Delta S$ is the change in entropy associated with the jump, $N_j$ is the number of diffusion jumps required to reach the final state and $E_T$ is barrier energy to be overcome by the defect. When RED occurs, the reaction rate is given in two limiting cases as [230]

$$R_{RED} \simeq \eta \frac{\exp \frac{\Delta S}{k}}{N_j} R_R \exp \left[ - \frac{E_T - \Delta E}{kT} \right] \tag{7.24}$$

when $\Delta E < E_T$ and

$$R_{RED} \simeq \eta \frac{\exp \frac{\Delta S}{k}}{N_j} R_R \tag{7.25}$$

when $\Delta E > E_T$. Higher is $R_R$ the recombination rate, higher is $R_{RED}$ (Fig. 7.45) and $\eta$ is an efficiency factor related to the fraction of recombination events resulting in a successful reaction.

The concept of RED and its description according to Eqs. (7.24 and 7.25) are neatly demonstrated by the annealing of defects in 1 MeV electron irradiated GaAs [231, 232]. These defects annealed at a faster rate under minority carrier injection as illustrated in Fig. 7.46. The RED rate saturates at high injection levels (Fig. 7.46(c)) and the saturated RED rate increases linearly with majority carrier concentration (Fig. 7.46(b)) indicating that for this defect $\sigma_p > \sigma_n$ where $\sigma_p$ ($\sigma_n$) is the capture cross-section for holes (electrons). The reduction in the activation energy by an amount equal to the hole transition energy is clearly illustrated in Fig. 7.46(a). It should be noted that the observed activation energy for the RED process ($E_{sat}$ in Fig. 7.46(a)) involves not only $E_T - \Delta E$ but also an additional energy representing the Arrehenius temperature dependent capture cross-section process.

**Figure 7.45** Schematic diagram of the temperature dependence of recombination enhanced kinetics. The observed activation energy (slope) is lowered by the energy deposited by the electronic transition (Fig. 7.44) and the absolute rate is proportional to the recombination rate [230].

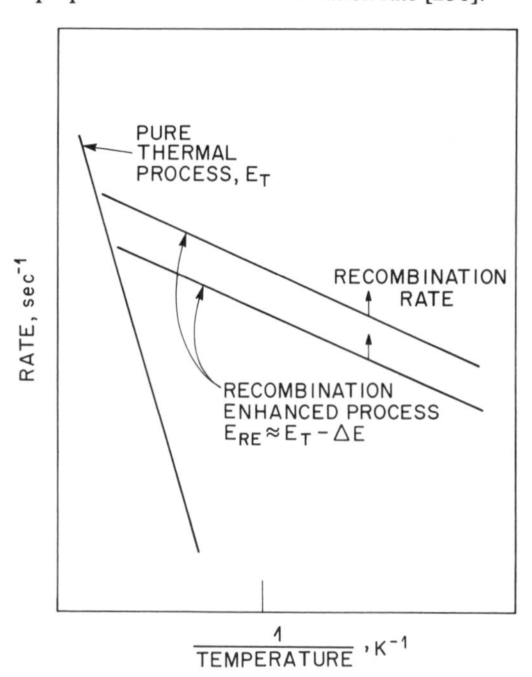

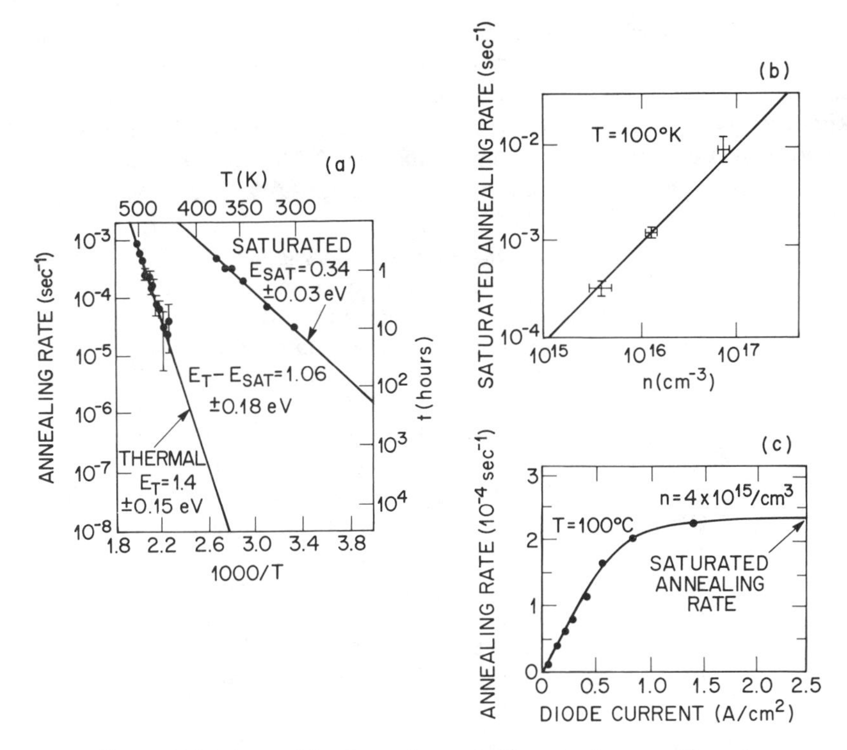


**Figure 7.46** Recombination enhanced effects in the annealing of 1-MeV electron bombardment damage in GaAs. (a) Reduced activation energy observed for injection anneal. (b) Dependence of annealing rate on majority carrier (n) concentration. (b) Dependence of annealing rate on injection level [230]. The reduction in the activation energy is equal to the hole transition energy.

### 7.5.2 Degradation in Lasers and Light Emitting Diodes

Dislocations have been closely linked with the degradation of light emitting devices. This follows from the fact that dislocations cause nonradiative recombination and decrease luminescence efficiency. Ettenberg [233] showed that dislocations reduce the minority carrier diffusion length and when the spacing between them is comparable to the diffusion length then luminescence efficiency decreases. By doing spatially resolved photoluminescence in GaAs and InP at a spatial resolution of 3 μm, Bohm and Fischer [234] found that the half-width of the photoluminescence reduction around dislocations is larger than the diffusion length. They suggested that the quenching of luminescence near dislocations is caused by enhanced bulk nonradiative recombination.

That dislocations act as nonradiative centers is clearly shown in the cathodoluminescence micrograph in Fig. 7.47 from a $Al_xGa_{1-x}As_{1-y}P_y$ epitaxial film on a GaAs substrate [235]. cathodoluminescence micrographs showing the recombination characteristics of the misfit dislocations parallel to [110] and $[1\bar{1}0]$ directions. Figure 7.47(c) is the bright field electron micrograph from the framed area in Figs. 7.47(a) and 7.47(b). Dark contrast in the EBIC indicates reduced carrier collection efficiency and dark contrast in the cathodoluminescence micrograph indicates reduced luminescence caused by nonradiative recombination in the vicinity of the misfit dislocations. The dislocations labelled $D_2$, $D_3$, and $D_4$ which show dark contrast are dissociated 60 degrees dislocations. The dislocation labeled $D_1$ is an edge sessile dislocation and does not cause nonradiative recombination as evidenced from the absence of contrast in

**Figure 7.47** Misfit dislocation network in $Al_xGa_{1-x}As_yP_{1-y}$ epitaxial film grown on (100) GaAs substrate by LPE. (a) EBIC micrograph (b) Monochromatic ($\lambda = 790$ nm) cathodoluminescence (CL) micrograph of the same area as in (a). (c) Bright field electron transmission micrograph of the same area as in (a). The operating reflections are (220) and $(2\bar{2}0)$ and the incident electron beam is parallel to (001). Dislocations labelled $D_2$, $D_3$ and $D_4$ are misfit dislocations parallel to [110] and $[1\bar{1}0]$ direction and they are of the 60° type. The dislocation labelled $D_1$ shows no contrast for the $(2\bar{2}0)$ reflection in (c) and is a sessile edge dislocation with Burgers vector parallel to the [110] in the plane of the interface. Note the absence of contrast for $D_1$ in (b) [235].

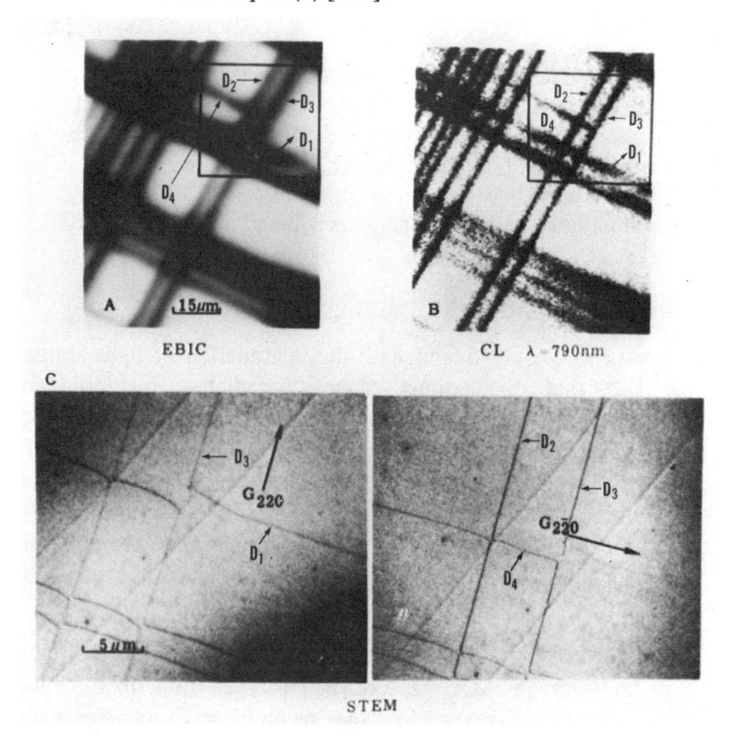

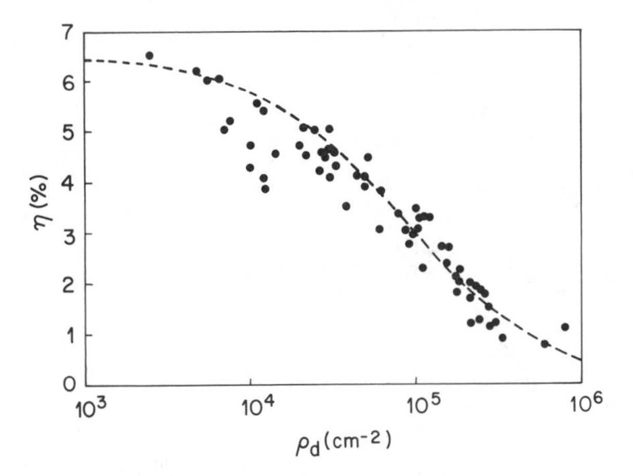


**Figure 7.48** Efficiency versus dislocation density for 45 individual $Al_xGa_{1-x}As$:Si LEDs. The dashed line is a calculated curve [237]. Reprinted with permission by the Electrochemical Society, Inc.

Figs. 7.47(a) and 7.47(b). The absence of recombination at $D_1$ is taken to imply that perhaps the core of the dislocation is reconstructed leaving no dangling bonds [235].

Sometimes it is difficult to separate pure dislocation effects from effects produced by dislocation-point defect interactions. It is well known that impurities segregate at dislocations because of the stress fields associated with the dislocations. For example, a fast diffusing impurity such as Cu in GaAs segregates near dislocations and quenches band edge luminescence. In such cases the quenching of luminescence may be mistaken to be caused by nonradiative recombination near dislocations. Heinke [236] noted a large effect of luminescence quenching in GaAs when fresh dislocations are introduced by bending compared to as-grown dislocations. Bohm and Fischer [234] suggested that this difference was in fact due to segregation of Cu at the deformation induced dislocations.

The relation between dislocation and luminescence efficiency is rather well illustrated in Fig. 7.48 which shows the external quantum efficiency of graded band gap Si-doped $Al_xGa_{1-x}As$ LEDs as a function of dislocation density [237]. The dislocations in the epitaxial layer have propagated from the substrates implying that low dislocation density substrates would produce a fewer dislocations in the epitaxial layer and hence more efficient LEDs.

Dislocations are also known to be the cause of the rapid degradation of $Al_xGa_{1-x}As$-GaAs lasers and LEDs. The degradation of lasers can be divided into three categories; (a) a rapid degradation which occurs after relatively short duration of operation, (b) a slow degradation occurring during long term operation, and (c) a catastrophic degradation process which occurs at high optical power densities caused by mirror facet damage. In the case of LEDs degradation proceeds via process (a) and (b). The rapid degradation of lasers and LEDs has been linked to linear defects known as dark line defects (DLDs) [238]. The DLDs develop mostly along <100> direction but sometimes along <110> direction as well. The nature and origin of the DLDs have

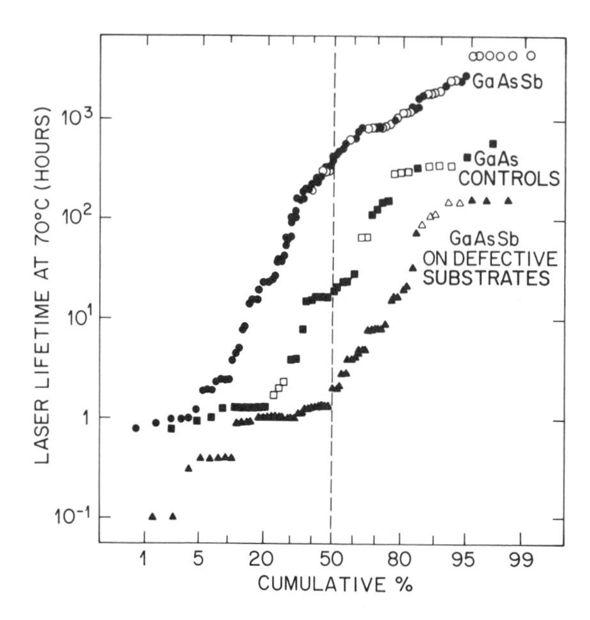


**Figure 7.49** Normal probability plot of the time to failure (solid symbols) or the accumulated testing time (open symbols) for Ga(As,Sb) lasers operated at 3 mW at 70°C. The Ga(As,Sb) devices have GaSb concentrations of between 0.4 and 2 mole % while the GaAs controls have GaSb concentrations of zero or less than 0.1 mole %. These were grown on GaAs substrates having dislocation densities $<3\times10^3$ cmref–2. The Ga(As,Sb) devices were grown on substrates that had dislocation densities of $\sim4\times10^4$ cmref–2 [244].

been identified by transmission electron microscopy as three dimensional dislocation networks consisting of long dislocation dipoles and small dislocation loops [239]. The DLDs originate from precursor defects in the active layer which are classified as dark spot defects (DSDs) since they appear as dark regions in luminescent images from the active layer [240 - 243]. The DSDs themselves originate from threading dislocations in the active layer which in turn propagate from the substrate.

Since dislocations propagate from the substrate and act as sources for DLDs, the use of low dislocation density substrates for epitaxial growth should yield highly reliable devices. Figure 7.49 compares the lifetime at 70°C of two classes of $Al_xGa_{1-x}As$ injection lasers having Ga(As,Sb) active layers. The devices which have poor 70°C median lifetimes were made from LPE wafers that were grown on substrates having a dislocation density of $\sim4\times10^4$ cm$^{-2}$ as compared to the good devices for which the substrates had dislocation densities $<3\times10^3$ cm$^{-2}$ [244].

From the very beginning of the discovery of DLDs, it was realized that the growth of DLDs and subsequent degradation of the device was aided by recombination enhanced motion of defects. Kimerling et al., [245] showed in *in situ* TEM studies that electron beam enhanced dislocation climb motion occurred in $Al_xGa_{1-x}As_{1-y}P_y$ which contained point defects introduced by a prior 1 MeV electron bombardment. The enhancement occurred not only under electrical injection conditions as in lasers or LEDs but as well as under optical injection conditions [246 - 248]. In fact, the correspondence between the mode of degradation under the two injection conditions made the optical pumping technique a convenient tool to study degradation in the devices or even in the double heterostructure wafer prior to processing.

The growth rate of DLDs is typically $\sim10^{-6}$ cm/sec and the growth rate is dependent on the injection current of the laser [249]. In some cases, <110> oriented DLDs have much higher growth rates of $10^{-3} - 10^{-2}$ cm/sec [242, 250]. These DLDs have been shown to be caused by recombination enhanced dislocation glide [251] consistent with their high growth rates and consistent with the fact that the glide direction in the zinc blende structure is the <110> direction. It has also been observed that the <110> DLDs can be sources of <100> DLDs [252]. Since process induced stresses can induce the <110> DLDs, they should be kept as low as possible to avoid the rapid degradation of the devices.

The <100> oriented DLDs which consist predominantly of dislocation dipoles and dislocation loops have been suggested to arise because of a dislocation climb process [239, 251]. The dipoles are interstitial in character lying in {110} planes [253, 254]. Although the <100> dipoles are predominant, in some instances dipoles along <110> direction have also been observed [255]. The small dislocation loops inside the main dipole have been identified to be of vacancy type [253, 254]. Based on the similarity between the dislocation configuration in degraded lasers and LEDs and that found in fcc metals caused by climb of dissociated dislocations, Petroff [256] suggested that the dislocations forming the dipole may indeed be dissociated into partials. This is consistent with the general observation that dislocations in zinc blende semiconductors are dissociated (Sec. 6.7). However, in the climb models that have been proposed to explain the <100> DLD structure, dissociation has not been considered.

The vacancy loops inside the main dipole of the <100> DLD structure are supposed to be the by-product of the climbing dipole. Generally, the absorption or emission of point defects at jogs in the dislocation core would produce dislocation climb motion. Since in a zinc blende semiconductor two fcc lattices are involved, climb process requires addition or removal of two kinds of atoms of the structure. The interstitial character of the dipoles would imply dislocation climb either by the absorption of interstitials or emission of vacancies of both atoms.

Since a supersaturation of point defects associated with each sublattice is not likely, Petroff and Kimerling [257] proposed a new point defect model for dislocation climb in zinc blende semiconductors which requires abundance of point defects in one sublattice only. Their model is illustrated in Fig. 7.50. In Fig. 7.50(b) an As dislocation in GaAs in the shuffle set is shown. $Ga_i$ is chosen to be the excess point defect. $Ga_i$ migrates to the climbing dislocation and attaches itself at the core, creating an arsenic vacancy at the core $V_{As}^{C}$ in that process. An As atom fills this vacancy and $V_{As}$ in the bulk is created. This vacancy moves around to relieve the tensile stress at the dislocation and the dislocation has completed the climb process. The excess $V_{As}$ left behind the climbing dislocation along with $V_{Ga}$ created by the reaction $V_{As} + Ga_{Ga} \rightarrow Ga_{As} + V_{Ga}$, cluster and collapse to form the vacancy like prismatic dislocation loops, which are observed inside the <100> DLD dipoles. In the case of lasers or LEDs grown by LPE, group III interstitials are likely to be the excess point defects since growth occurs from a group III metal solution. In the case of structures grown by VPE, group V atom is presumed to be in excess yielding $V_{Ga}$ as the defect left behind the climbing dislocation.

In this model, the diffusion of the $Ga_i$ towards the dislocation core would determine the rate of dislocation climb. Under carrier injection conditions, the migration of $Ga_i$ would be assisted by recombination enhanced motion. A further outcome of the model is the formation of antisite defects in the As-sublattice. A point defect concentration of $\sim 10^{-3}$ atom fraction is suggested by the observed dislocation structure. Such a high concentration of defects is supposed to be generated at the hetero-interfaces of the epitaxial layers.

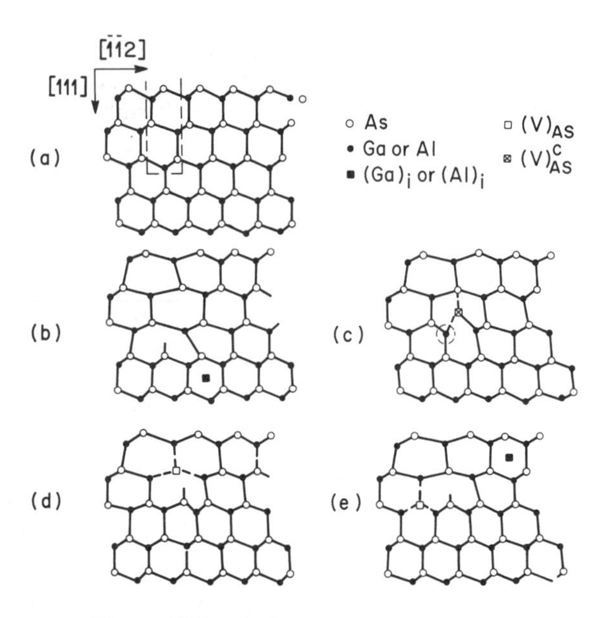


**Figure 7.50** (a) Zinc blende lattice projected normal to the (1$\bar{1}$0) plane. The dotted line indicates material removed to produce the dislocation (b) 60° dislocation with b=a/2 [110]. (c-e) Schematic of the point defect diffusion required for the dislocation climb (see text) [257].

In view of the uncertainty of the existence of a high concentration of native interstitials required for the climb process in solution, O'Hara, et al., [258] proposed an intrinsic defect generation process. According to this process the energy released by the electron-hole recombination process at the dislocation allows a host atom on a substitutional site near the dislocation to move onto the next dislocation and creating a vacancy. This *intrinsic* model requires emission of both group III and group V vacancies. To substantiate that the defects needed for the dislocation climb come from the energy released by the electron-hole recombination and are not present at thermal equilibrium, Hutchinson et al., [259] investigated dislocation dipole structure in Te-doped n-type GaAs under optical pumping. The sample showed small interstitial dislocation loops after an annealing treatment at 880°C. The concentration of interstitials in these loops is estimated to be of the order of $10^{17}$ cm$^{-3}$. The presence of the interstitial loops suggested that no excess interstitials are present in solution. However, under optical pumping the sample developed the dislocation dipole structure characteristic of degraded lasers, indicating that dislocation climb has occurred in spite of the low interstitial concentration. This experiment according to Hutchinson et al., [259] confirmed the vacancy emission model for the climb process. Only some of the vacancies emitted by this process are observed as vacancy type prismatic loops inside the dipoles and the majority of them are believed to form submicroscopic clusters not observable in the transmission electron microscope.

The *intrinsic* point defect model would predict that there should be no saturation of the climb process. On the other hand, in the *extrinsic* process, once the available point defects are consumed by the climbing dislocation the formation of dislocation loops and the growth of the dislocation dipole network would stop. Petroff and Kimerling [257] observed saturation of the dipole growth under electron beam injection conditions. Also the dislocation networks which are fully developed failed to show any further growth under carrier injection. Further, in broad area optically pumped laser devices, new defect structures appeared only on the freshly formed DLD network showing that the point defects enabling the climb process are localized. Thus some experimental results are not explained by either the extrinsic or the intrinsic model-the saturation effects by the intrinsic model or the growth of DLDs in samples containing no interstitials by the extrinsic model. While each model has some deficiencies, in the intrinsic model it is difficult to envisage the creation of point defects entirely by the recombination energy, which can atmost only be equal to the band gap, viz. 1.4 eV for GaAs, when the defect formation energies are likely to be greater than this value.

In another experiment, it was shown that the point defect needed for the climb process may be provided by an existing defect. By a scanning DLTS technique, Lang et al., [260] found that near the dislocation climb network, the concentration of the DX center, a common defect in n-type $Al_xGa_{1-x}As$ ($x \geq 0.30$) cladding layer of the laser structure at $10^{17}$ $cm^{-3}$ level, decreased by ~40% (Fig. 7.51). This suggested that there is a relationship between the DX center and the dislocation network. Although the change in the concentration of the DX center could have occurred either as a result of supplying the point defects directly for dislocation climb or as a result of an indirect interaction with point defects generated by the climb process, Lang et al., [260] conjectured that the former mechanism is the most reasonable one. They proposed that the

**Figure 7.51** EBIC and scanning DLTS intensities as a function of the beam position for a beam scanning across a <100> dark line defect [260]. Near the DLD the DLTS signal decreased by ~40%.

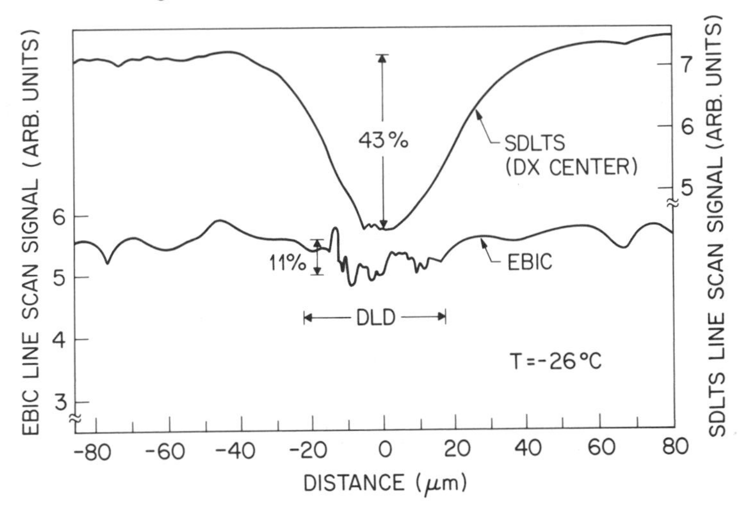

decay of the DX center under electron-hole recombination proceeds by the emission of $Ga_i$ needed for the dislocation climb in the extrinsic model.

Consistent with the picture of two types of dislocations (α or β dislocations either of the shuffle set or of the glide set, see Sec. 6.7.1) and their different velocities, Hutchinston and Dobson [261] observed anisotropic <100> DLD growth; that is, the climb of one type of dislocation is greater than that of the other. Similarly, Imai et al., [262] observed a fast and a slow growth component in elongation of <100> DLDs under optical pumping, reflecting probably the asymmetry in climb of α and β dislocations.

### <110> DLD and recombination enhanced glide

The dark line defects which lie along <110> directions in degraded lasers, the <110> DLDs, have been identified by transmission electron microscopic study as relatively straight dislocations lying on the glide plane [263]. The growth rate of <110> DLDs is also faster than that of <100> DLDs [242, 250]. The <110> DLDs which are formed by dislocation glide as opposed to dislocation climb are also assisted by recombination enhanced motion. Recombination enhanced dislocation glide motion has been observed in GaAs-AlGaAs laser diodes under current injection [252, 263], optical pumping [250, 264, 265] and electron beam injection [252, 266, 267] with [252, 264, 265] or without [266, 267] an externally applied stress.

Maeda et al., [268] measured the velocity of α and β dislocations under electron beam injection in a scanning electron microscope containing a bending apparatus as a function of temperature and stress. Their results are shown in Fig. 7.52. The velocity of both α and β dislocations follow an Arrhenius relation in dark as well as under carrier injection conditions but

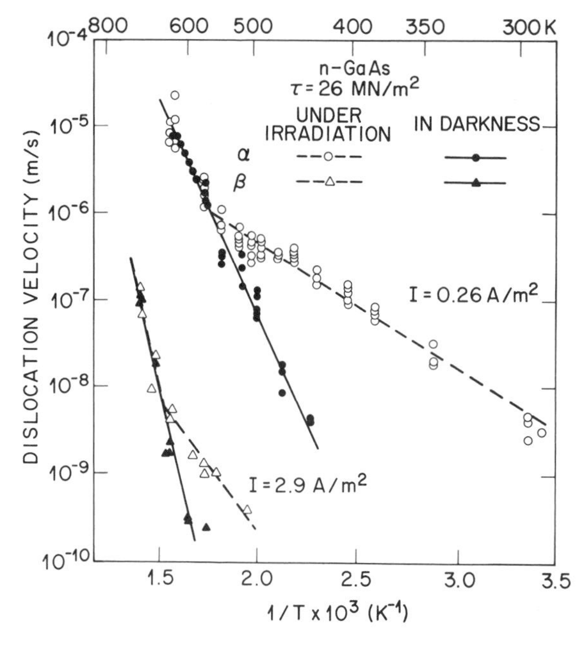


**Figure 7.52** Temperature dependence of dislocation velocities of α and β dislocations in GaAs under 30 keV electron beam irradiation and in darkness [268].

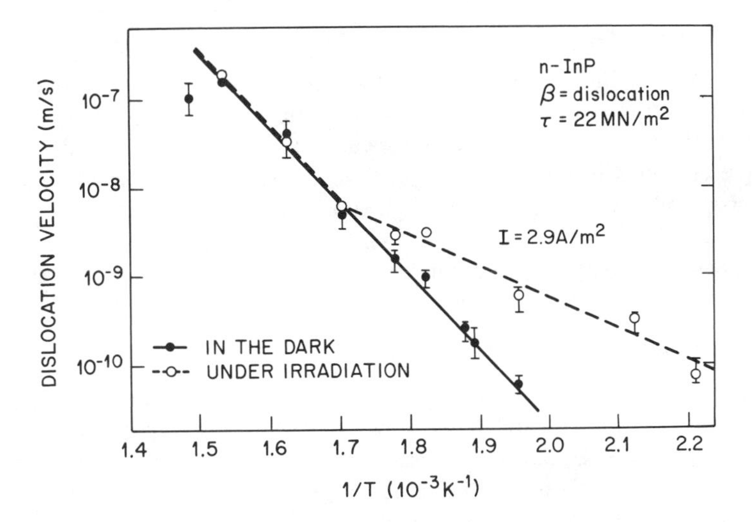


**Figure 7.53** Temperature dependence of β dislocation velocity in InP under 30 keV electron beam irradiation and in darkness [270].

the activation energies are reduced by ~0.7 eV for α dislocations and 1.1 eV for β dislocations in the latter case. These results clearly illustrate recombination enhanced dislocation glide motion. The enhancement was observed only below a certain critical temperature, $T_c$. From measurements of irradiation intensity I and stress τ dependences, Maeda et al. [268] expressed the dislocation velocity under carrier injection as

$$V = V_d^{\circ} \exp \frac{-E_d(\tau)}{kT} +$$

$$V_i^{*} \left(\frac{I}{I_o}\right)^{0.84} \exp \left\{ \frac{-(E_d(\tau) - \Delta E)}{kT} \right\} \tag{7.26}$$

where ΔE, the reduction in the activation energy caused by the recombination enhancement is independent of τ and I. Maeda et al., proposed that the observed ΔE values (see Fig. 7.52) are the energies released during the recombination events at the dislocations. Maeda and Takeuchi [269] found that ΔE is different for the same dislocation depending on the conductivity type of the sample which suggested a possible difference in the minority carrier capture cross section of the dislocation levels that differs according to the dislocation charge state in thermal equilibrium.

Recombination enhanced glide of dislocation has also been observed in InP [270]. Fig. 7.53 shows the temperature dependence of velocity of β dislocation under dark and under irradiation of 30 kV electron beam [270]. The value of ΔE is ~0.9 eV which is comparable to the value for the β dislocation in GaAs.

**Rapid degradation of LEDs**

Just as the lasers, AlGaAs LEDs also show a rapid and gradual mode of degradation [271]. Transmission electron microscopic investigations of rapidly degraded LEDs indicated that the defect structure produced is very much similar to that in degraded lasers. Both <100> and <110> DLDs were present though the former were predominant [272, 273]. The <100> DLDs consisted to develop rather readily in material with a high density of defects and their propagation required minority injection [274]. Therefore, the use of low dislocation density GaAs substrates and general cleanliness during epitaxy to minimize defect density in the epitaxial layers have served to reduce the incidence of DLDs and thus to improve reliability [275].

**Gradual degradation in lasers and LEDs**

Even after elimination of the rapid degradation mode by employing high quality substrates combined with careful growth and processing procedures, there is still a *gradual* mode of degradation exemplified by a gradual increase in drive current for lasers or a gradual decrease in output power for LEDs. This mode of degradation is shown schematically in Fig. 7.54 [276]. Table 7.3 summarizes the characteristics of the gradual degradation in AlGaAs and GaInAsP devices [277, 278]. The devices which have degraded by accelerated aging at elevated temperatures show dark areas containing extrinsic (interstitial) Frank type dislocation loops with Burgers vectors a/3 <111>. These loops are supposed to have been formed by the condensation of point defects whose migration is assisted by the recombination enhanced motion.

Chu et al., [278] have studied in detail the defect mechanisms in degradation of 1.3 μm GaInAsP/InP wavelength channeled substrate buried heterostructure lasers. They studied the defect structure in devices which showed gradual and rapid degradation under accelerated aging. In devices which showed gradual degradation the defect mechanism was associated with the nucleation of extrinsic dislocation loops along the V-groove {111} sidewall interfaces between the Cd diffused p-InP and LPE grown n-InP buffer inside the groove. These loops subsequently grow out of the interfaces into the buffer layer assisted by recombination enhanced defect motion. Some of the loops which entered the active region eventually became dark line defects. The extrinsic nature of the loops implied that the {111} sidewall interfaces as well as the GaInAsP active region contained a high density of interstitials.

Besides the presence of dislocation loops in degraded devices, the creation of point defects during gradual degradation is also indicated by the increase in the concentration of deep levels as

**Figure 7.54** Schematic diagrams of various degradation modes in lasers and LEDs [276].

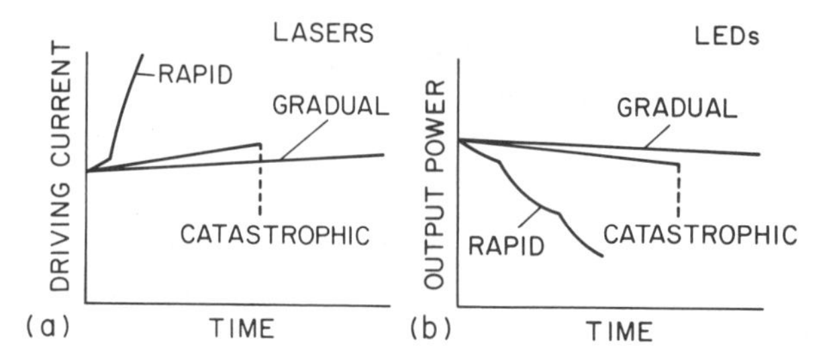

**Table 7.3**
Features of Gradual Degradation in AlGaAs/GaAs and GaInAsP/InP Devices [276 - 278]

| Material | Appearance | Defect Structure | Mechanism |
|---|---|---|---|
| AlGaAs/GaAs | Uniform darkening or DSD formation<br><br>increase in deep level defects | Dislocation loop, stacking fault | Formation of point defect clusters or loops by climb |
| GaInAsP/InP | Uniform darkening or DSD formation | Precipitates Multiple dislocations | |
| | Dark line defects | Extrinsic dislocation loops | Formation of point defects at epilayer-substrate interface and their subsequent growth in the active layer due to recombination assisted diffusion |

measured by the DLTS technique. In proton-bombarded stripe geometry AlGaAs lasers, Lang et al. [279] found that an electron trap at 0.89 eV, which is introduced by the proton damage, increased in concentration by more than an order of magnitude in the first 100h of aging at 70°C, and more slowly thereafter, in moderately long-lived (~ $10^3$ h at 70°C) lasers. However, this trap has been found to decrease in lasers with very long life times (extrapolated room temperature life time in excess of 50 yr). Uji et al., [280] found in AlGaAs lasers the concentration of the hole trap at 0.24 eV to increase during accelerated aging at junction temperatures in the range 120 to 320°C. Their results are shown in Fig. 7.55 as normalized DLTS signal of the 0.24 eV trap and threshold increase as a function of aging time at 320°C for two lasers, A and B. Laser B which showed less degradation than laser A, also showed less increase in the 0.24 eV signal. The lack of increase in 0.24 eV signal in laser C which is kept at the operating temperature without current injection clearly shows that the growth of 0.24 eV defect requires carrier recombination. Uji et al., further observed that the activation energy for the rate of increase of 0.24 eV defect as the temperature is increased is nearly the same as that for the rate of increase in threshold current.

In spite of these strong correlations between the 0.24 eV trap and laser degradation, the estimated increase in threshold caused by the trap is significantly smaller than the observed value suggesting that other effects have to be invoked. One such effect is caused by stress in the device introduced by various steps during its fabrication. This is discussed in more detail in Sec. 7.5.3. Interaction between the contact metals, especially Au, and the semiconductor has also been shown to be the cause of gradual degradation. In Burrus type GaAs LEDs, gold bearing precipitates were found in the contact area near the junction in degraded devices [281]. In Ti/Au Schottky barrier restricted AlGaAs LEDs, the electromigration of Au and interaction with Ga

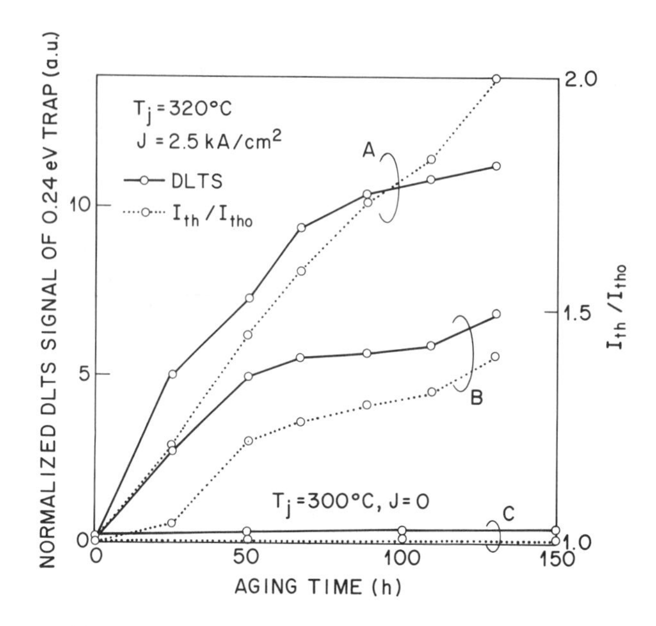


**Figure 7.55** Normalized DLTS signal of 0.24 eV trap and pulsed threshold current $I_{th}/I_{th0}$ as a function of aging time for lasers A and B. $I_{th0}$ is the initial pulsed threshold current. Junction temperature is 320°C. Laser C was kept at 300°C without current [280].

formed AuGa alloy from which dark lines initiated which led to degradation [282]. The use of 1000 Å Pt or Pd as a barrier layer prevented Au migration and improved device reliability.

Gradual degradation in lasers also occurs due to facet related effects. In AlGaAs lasers mirror erosion caused by oxidation of unprotected mirror surfaces has been found to occur both in the early stages and after long hours in long term life testing experiments [283]. The oxidation has been enhanced by optical and electrical excitation. The oxide decreases the mirror reflectivity and leads to increased thresholds. In the aging experiments which are done under constant current conditions, the increased thresholds result in the degradation of the optical power output.

With regard to facet erosion and degradation in AlGaAs lasers, Nash and co-workers [284 - 285] showed that the improvement in the long-term reliability of lasers aged at room temperature in a dry $N_2$ ambient [283, 287 - 290] is caused by the suppression of an initially occurring and temporally saturable mode of degradation (tens of hours at 70°C and hundreds of hours at 25°C and outputs ~3 mW/facet). The saturable mode of degradation is caused by the increase in the nonradiative recombination caused by defects formed at the oxide semiconductor interface [291]. Besides the initial saturable mode of degradation, facet erosion also appears to induce gradual erosion beyond ~10,000 h of aging caused by reflectivity decrease especially at 70°C and 85 percent RH ambient [286].

To suppress facet erosion and improve reliability both in terms of eliminating the initial saturable mode of degradation and improving the long term aging, protection of the mirror facets with half-wavelength dielectric films has been adopted in AlGaAs lasers. Although $Al_2O_3$ is the preferred coating material [288, 292, 293], other materials such as $Si_3N_4$ [294], C [289], and $SiO_2$ [290] have also been used. However, in order to fully gain the advantage of the coatings as mirror facet protections, it was found to be necessary to apply the coatings soon after cleaving the mirrors and minimizing exposure to laboratory ambient [286].

An important distinction with regard to long-term reliability of AlGaAs lasers containing GaAs active layers and AlGaAs active layers is that for uncoated lasers, the GaAs active layer lasers generally had inferior reliability than AlGaAs active layer lasers [295 - 297]. On the one hand, the poor reliability of GaAs active layer lasers may have been related to the increased susceptibility of GaAs mirror facets to oxidation as suggested by the greater degradation in photoluminescence intensity under optical excitation in GaAs compared to AlGaAs [291]. On the other hand, the work of Hayakawa et al. [298] showed that degradation of lasers with Al composition <8 percent is primarily caused by DLDs formed in the immediate vicinity of the mirror surfaces. The role of Al is believed to be one of making the material more resistant to DLD formation. The mechanical damage produced by the cleaving operation served as the generation sites for DLDs. In fact, with or without facet coatings, DLDs and dark spots have been found near the mirror surfaces [277,286] suggesting that facet oxidation is not the cause of the defects.

Hayagawa et al., also found that oxidation increased with increasing Al in the range 8 to 17 percent, contrary to the result of Suzuki and Ogawa [291] in their photoluminescence study, and was responsible for degradation in the AlGaAs active layer lasers for these compositions. The enhanced oxidation of higher Al percentage layers has been suggested to be the result of enhanced oxidation in the beginning stages of oxidation. $Al_2O_3$ facet coatings suppressed the oxidation and the associated degradation in these lasers. Another support for the mechanical cleaving damage induced degradation of GaAs active layer lasers was provided by the work of Wolf et al., [299] who made GaAs active layer lasers which had very low degradation rates at 100°C better than even AlGaAs active layer lasers. These lasers had sputtered $Al_2O_3$ facet coatings and presumably the cleaving induced mechanical damage, the source of DLDs, was removed by ion-milling or back-sputtering just prior to $Al_2O_3$ deposition.

Besides suppressing facet erosion, $Al_2O_3$ facet coatings when applied immediately following cleaving of the mirrors, also helped to increase catastrophic damage thresholds [287, 289, 292, 300], suppress the growth of self-pulsations during cw aging in a 70°C dry $N_2$ ambient [301], reduce the transport of degradation promoting Cu to the facets [284], suppress light-jump development [302], and increase the power levels at which "kinks" in the light-current characteristics develop [302].

### Catastrophic degradation

Catastrophic degradation is marked by sudden failure of the device during operation (Fig. 7.54). In the case of lasers catastrophic degradation occurs because of damage or erosion caused by the high optical power density at mirror facets. Devices failed this way show several <110>

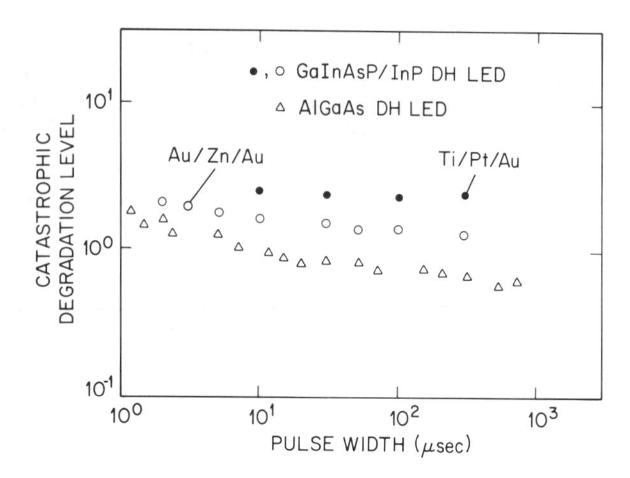


**Figure 7.56** Dependence of pulse width on the degradation level for catastrophically degraded AlGaAs LEDs and GaInAsP/InP LEDs with Au/Zn/Au and Ti/Pt/Au p-electrodes [276].

DLDs to have generated from the mirror surface [276]. This degradation process is known as catastrophic optical damage or COD and occurs at power levels of typically several $MW/cm^2$. Local melting and rapid cooling and a subsequent thermal runaway situation generate defects near the mirror which lead to rapid degradation [278].

In LEDs catastrophic degradation occurs for reasons other than COD since the mirror facets are absent. In AlGaAs devices recombination enhanced glide of <110> DLDs is observed under the application of high currents in pulsed operation [303]. Figure 7.56 shows the dependence of pulse width on the catastrophic degradation level in GaInAsP/InP and AlGaAs/GaAs LEDs [276]. It can be seen that the degradation level is higher for the GaInAsP LEDs compared to the AlGaAs LEDs due to the absence of recombination enhanced glide in the former. Further, Ti/Pt/Au p-contact is better than Au/Zn/Au contact presumably because of less interaction between the Au and the semiconductor.

As stated earlier, coating of the mirror facets with dielectric films helps to increase the threshold of COD [287, 289, 292, 300]. Yonezu et al., [304] made a window stripe laser wherein the central regions of the active region were heavily doped leaving the regions near the mirrors lightly doped. This way the emission wavelength was decreased so as to make the mirror regions more transparent than was the case with uniform doping. Besides these approaches, it is important to prepare the facets as flawless as possible so as to avoid mechanical damage which seeds DLD growth. It is also important to reduct internal/external stresses so that recombination enhanced glide of <110> DLD would not occur.

### 7.5.3 Process Related Effects

Many of the device process steps such as, diffusion, dielectric stripe delineation, ion bombardment for stripe delineation, contact metallization, cleaving of mirror facet, bonding, and so on, can introduce material defects which can affect device reliability.

**Contact metallization**

In the case of GaInAsP/InP devices rapid degradation caused by dislocation climb as in AlGaAs/GaAs devices has not been observed during operation at room temperature [305 - 309]. The dark defects observed including <110> type DLDs were found to be introduced during crystal growth and device fabrication. Some of the DSDs were found to be caused by the reaction between the contact metals and the semiconductor.

During gradual degradation also many DSDs were observed in the light emitting region of GaInAsP/InP LEDs aged at 200°C for 100h (Fig. 7.57). The DSDs were identified as <100> or <110> oriented bar shaped precipitates 0.5 – 1.0 μm in length. The precipitates were In rich and were thought to be formed at unidentified nucleation sites [310]. On the other hand, in GaInAsP/InP lasers <100> DLDs and DSDs were observed [311]. The DSDs were identified as plate like precipitates in {111} planes [312]. They were found to be rich in As and Ga and to penetrate from the p-cladding layer through the active layer to the n-cladding layer [313].

Chin et al., [314] and Mahajan et al., [315] have suggested that electromigration of Au from the p-contact as the origin of the DSDs. The failure to see Au in the precipitates by energy dispersive X-ray analysis was attributed to the poor sensitivity of the technique. Camlibel et al., [316] found the interaction between Au and the semiconductor to be very nonuniform. In localized regions the alloyed contact was found to have penetrated into the p-InP confining layer and under heat treatment Au-rich particles were formed in the active layer. The gradual degradation mode caused by DSD formation was completely eliminated when non alloyed Pt contact was used thereby further suggesting the involvement of Au in [317] DSD formation.

In the catastrophic degradation of GaInAsP/InP and AlGaAs/GaAs devices also contact metallizations seem to play a role. From Fig. 7.56 it can be seen that when Ti/Pt/Au p-contacts were used the catastrophic damage level is higher compared to when Au/Zn/Au contacts were used. This is once again conjuctured to be caused by reduced metal-semiconductor interaction with the Ti/Pt/Au contact.

**Figure 7.57** Formation of DSDs in GaInAsP/InP LEDs aged at 200°C for 100 h. (a) EL image showing the DSD in the light emitting region (b) TEM is micrograph of bar shaped precipitates corresponding to the DSD in (a) [276].

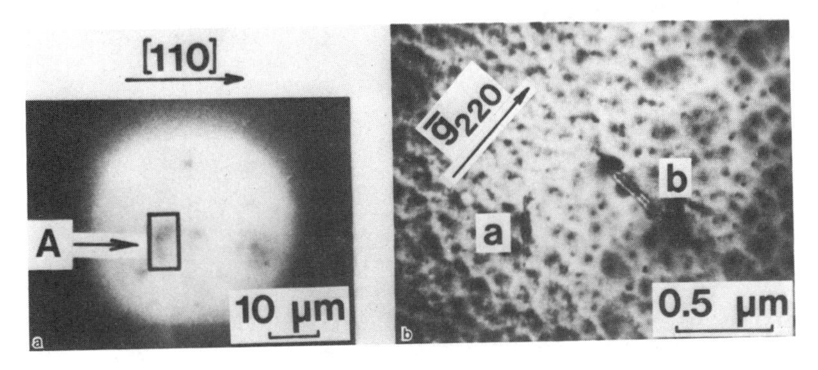

### Device stress and degradation

Strain in the device has long been considered to be one of the factors limiting its reliability. Strain can be introduced into the device by heteroepitaxial stresses [318], dielectric masks [319, 320], contact metallizations [321], bonding [322, 323] and scribing, and cleaving [324]. To minimize heteroepitaxial stresses, it is necessary to grow the epitaxial layers at near lattice matched compositions. Generally, perfect lattice match is aimed near the growth temperature. Because of the difference in the thermal expansion coefficients of the individual layers, stresses are introduced at room temperature. The average layer stress in AlGaAs layers grown on GaAs substrate has been determined from the radius of curvature of the substrate measured by the x-ray automatic Bragg angle control method and the results are shown in Fig. 7.58 [325].

Rozgonyi et al., [326, 327] investigated the stress compensation in the AlGaAs layers by the addition of small amounts of phosphorus. Since P is an isovalent dopant in AlGaAs, small additions of it does not affect significantly the electrical or optical properties. Afromowitz and

**Figure 7.58** Average layer stress as a function of composition for $Al_xGa_{1-x}As$. The circles and squares represent data from two different LPE reactors [325].

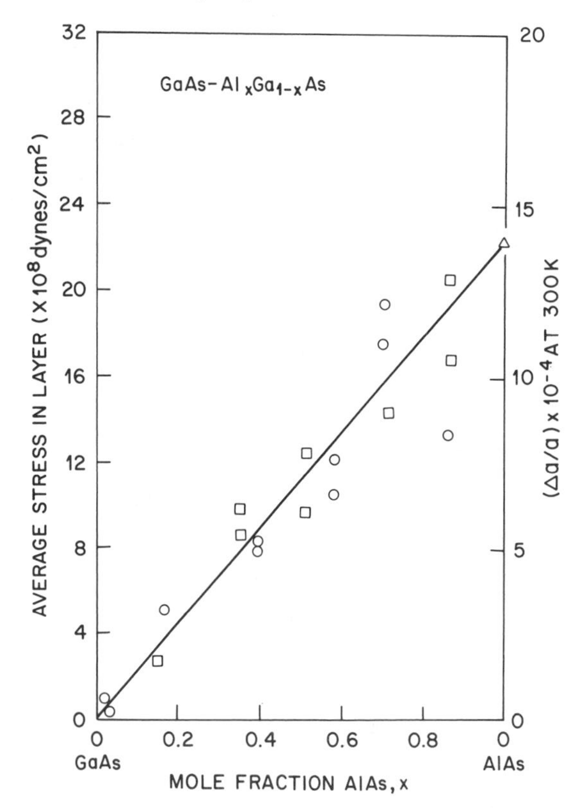

Rode [328] investigated the limitations on stress compensation by P in the heterostructure growth of AlGaAs/GaAs laser structures. They found that a factor of 10 reduction in the stress may be obtained for 2 μm thick quarternary layers. It should be borne in mind that the addition of P to achieve lattice matched condition at room temperature produces lattice mismatch at the growth temperature. If this mismatch is large, then unidirectional array of edge dislocations and cross hatched misfit dislocations are formed. In some instances, the layer thickness and composition can be such that substrate threading dislocations bend over at the epilayer-substrate interface caused by the misfit strain and thus dislocation free epitaxial layer can be obtained.

In the case of AlGaAs/GaAs double heterostructures, the GaAs active layer has a tensile stress since its unstrained lattice parameter is less than that of AlGaAs. However, by adding small amounts of Al to the active layer, the active layer stress changes to compression with the crossover occurring at an Al composition of 5 percent [318]. This means that a active layer with zero stress can be obtained. The improved reliability of AlGaAs lasers when small amounts of Al is added to the active layer (Sec. 7.5.2) may in fact have its origin either in this reduction in stress or in the compressive state of stress in the active layer.

In the GaInAsP/InP laser structures, the thermal expansion coefficient of InP is smaller than that of GaInAsP which for lattice matched condition at the growth temperature gives rise to a tensile stress of the order of $\sim 10^8$ dynes $cm^{-2}$ in the active layer at room temperature. At this level of tensile stress in the active layer of lasers, the TM mode (electrical vector perpendicular to the junction) gain is higher than that of the TE mode (electrical vector parallel to the junction) gain. The TE to TM mode change and the stability of the TM mode depend on the magnitude of stress and operating current [329]. The undersirable consequences of the polarization change of the lasing mode are the presence of nonlinearities in the light-current characteristics and spectral changes which severely limit the usefulness of the lasers for high speed communication systems. To achieve near zero stress in the active layer at room temperature, Liu and Chen [330] proposed inserting an $Ga_{0.47}In_{0.53}As$ buffer layer in the GaInAsP/InP double heterostructure.

The problem of lattice mismatch and differential thermal expansion between the layers and the resultant stress in the active layer becomes more complicated in the case of buried heterostructure lasers. Such structures are desired for achieving single lateral mode operation. One such structure, the channeled substrate buried heterostructure, for making visible AlGaAs/GaAs laser is shown in Fig. 7.59 [331]. Because of the nonplanar geometry of the structure stress gradients exist near the edges of the channeled stripe as shown in Fig. 7.60. The stress distribution in Fig. 7.60 gives rise to maximum volume dilation near the edge of the stripe towards which interstitial atoms move and aggregate relieving the stress and the total energy of the system. The interstitial atoms condense to form dislocation loops. They then form dipoles which extend into the stripe by climb eventually causing degradation of the device.

Strain induced degradation of the device can also be caused by the dielectric masks used to define the current confirming stripe in lasers. Goodwin et al., [332] studied the degradation in oxide stripe defined AlGaAs lasers as a function of the shape of the oxide profile and its thickness. They found that the initial degradation rate was dependent on the magnitude of the stress which in turn was determined by the shape of the oxide profile and its thickness. In the case of an oxide window tapered at its edges to form a distinct chamber, the peak shear stress was lower by a factor of ~2 compared to the normal sharply defined edges. Similarly, reducing the oxide thickness resulted in lower active layer stress. Whenever the active layer stress was lower, there was improvement in the initial degradation rate.

Wakefield [319] and Robertson, et al., [320] studied the degradation behavior of the oxide stripe geometry AlGaAs lasers by photoluminescence, cathodoluminescence, and electron beam

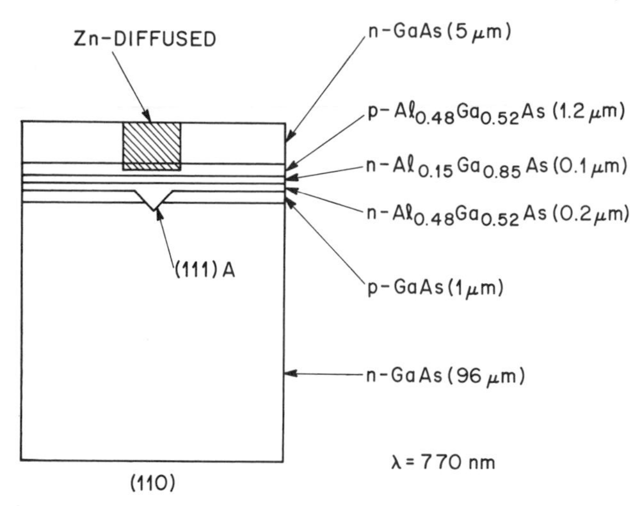


**Figure 7.59** Schematic view of AlGaAs/GaAs channel substrate planar laser [331].

**Figure 7.60** Calculated stress distributions in the active layer of a AlGaAs/GaAs channel substrate planar laser emitting at 770 nm plotted against X, the distance from the center of the stripe. On both edges of the strip (X = ±2.4 μm) steep stress gradients exist [331].

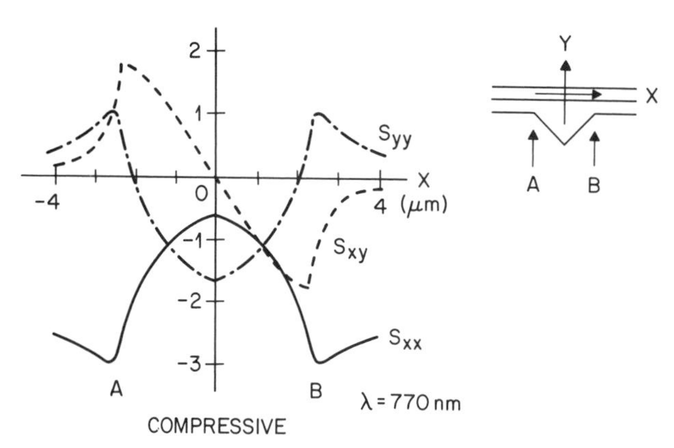

induced current techniques. The principal finding of their study was the presence of dark and bright bands running parallel to the stripe, together with brightened mottled regions extending well outside. The dark and bright bands were associated with regions of maximum compression and between regions of maximum compression and maximum tension in the active layer, respectively. In addition, broad dark parallel bands associated with maximum tension at the substrate-n/AlGaAs cladding layer interface were also seen. To explain the dark features along the active stripe these authors postulated strain and recombination enhanced motion of point defects to the strain regions where the defects act as nonradiative centers. Vacancy type defects are expected to move towards region of compression and interstitials toward regions of tension. In support of such a model the appearance of small prismatic dislocation loops at the edges of the strips was cited. These loops were supposed to have been formed by the aggregation of the native defects moving in the strain field. In GaInAsP/InP double heterostructure wafers Chin et al., [333] interpreted the dark regions concentric with the p-contacts in cathodoluminescence images as due to stress induced migration of phosphorous vacancies.

The thickness of the different metallic layers used for n- and p-metallizations should be judiciously chosen so as to minimize stresses in the device. The stresses are caused by the differences in the thermal expansion coefficients of the adjoining layers. For Ti-Pt bilayer metallizations on GaAs substrates and annealed at 450°C, Henein and Wagner [334] showed by radius of curvature measurements that for the appropriate choice of the thickness ratio of Ti/Pt layers, near zero stress in the substrate can be achieved. This is because increasing the thickness of Ti increases the tensile stress while the opposite is true for Pt, as shown in Fig. 7.61. For

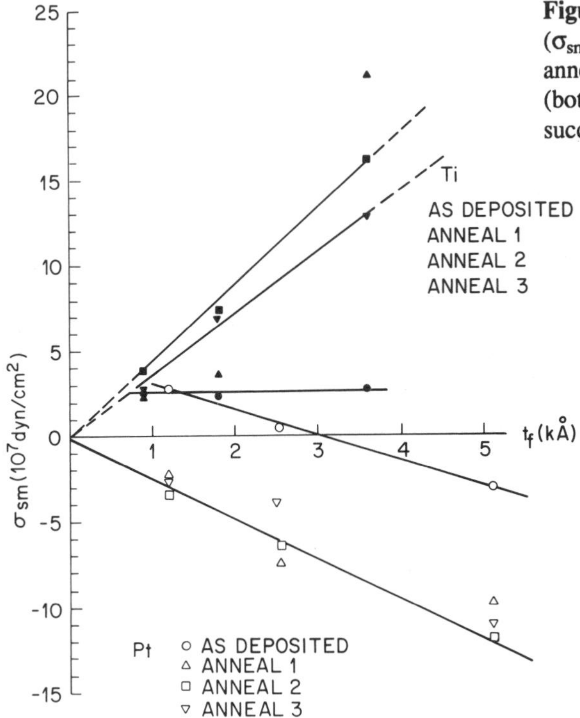


**Figure 7.61** Maximum stress in the substrate ($\sigma_{sm}$) versus film thickness for as deposited and annealed GaAs samples with Ti (top) and Pt (bottom). The stresses obtained after three successive annealing at 450°C are shown [334].

thicknesses of ~1000 Å Ti and ~1500 Å Pt, typical values used in AlGaAs/GaAs lasers, the stress in the layer underneath the metals is $\leq 1 \times 10^7$ dyne $cm^{-2}$, a value low enough not to cause any concern regarding laser reliability.

To determine the metallization induced stresses in GaInAsP/InP heterostructures, Swaminathan, et al., [321] measured the radius of curvature of double heterostructure wafers after epitaxial growth and after broad area p- and n-metallizations. The p-side metallization consisted of an e-beam evaporated Au-Zn-Au layer (50 Å – 150 Å – 800 Å) which was alloyed at 430°C for 20 seconds in forming gas (15% $H_2$: 85% $N_2$). As part of the p-contact there was also a ~2500 Å thick Ti/Pt layer which acts as a barrier between the Au-Zn contact and the Au bonding pads. This layer was annealed at 250°C for 5 minutes. The n-side metallization which was done before the Ti/Pt layers were deposited consisted of depositing ~2900 Å thick Au-Ge layer and alloying at 350°C. Just before the n-metallization the substrate was also chemo-mechanically polished down to a thickness of ~90 μm from an initial thickness of ~250 μm. From the measured radius of curvature after each step the stresses in the different layers were calculated using the formalism developed by Olsen and Ettenberg [318] which is based on the well known Timoshenko's "bimetal thermostat" solutions [335]. Table 7.4 shows the stresses in the different layers of a double heterostructure wafer grown by LPE consisting of a n-type InP substrate, 0.2 μm thick n-InP buffer layer, 0.4 μm thick GaInAsP active layer, 0.7 μm thick p-InP cladding layer and a 0.6 μm thick GaInAsP capping layer. The active layer composition was such to give a lasing wavelength of 1.3 μm. It can be seen from the table that the stress in the active layer is tensile at all times and is minimum after n-metallization and maximum after annealing the Ti/Pt layer. The minimum and maximum values of stress are estimated to be $9 \times 10^7$ dynes $cm^{-2}$ and $1.4 \times 10^8$ dynes $cm^{-2}$, respectively. These values are less by an order of magnitude than the fracture limit of GaAsInP [336]. The change in the active layer stress caused by the metallizations is only ~$2 \times 10^7$ dynes $cm^{-2}$. At these stress levels no strain related degradation would be expected.

Very early in the development of AlGaAs/GaAs lasers it was recognized that bonding strain is a controlling factor in the rapid degradation of devices [322]. The strain fields in the lasers were observed by the strain induced birefringence. The technique involves viewing the transmitted light through the laser facets under crossed polarizers. GaAs exhibits strong birefringence and strain in the crystal induces rotation of the polarization of the light, the rate of rotation determined by the nature and magnitude of the strain field. When viewed under crossed polarizers, strained regions transmit some light as can be seen in Fig. 7.62(b) [337]. The angle $\beta$ in Fig. 7.62 is the polarizer angle and the variation of the transmitted intensity for the different angles indicates whether the strain is normal (tensile or compressive) or shear strain [338]. The maximum in intensity for $\beta = 45°$ indicates normal stress in the laser diode.

Koyama et al., [339] used the stress birefringence effect for observing the strain fields in AlGaAs/GaAs laser diodes bonded to Si submounts. In conjunction with finite element calculations of the bonding stress they arrived at an optimum thickness of 0.1 mm for the Si submount to minimize bonding stress in lasers.

The thickness of the Au bonding pad on AlGaAs/GaAs lasers was also found to be a critical parameter determining device reliability [323]. It was found that for bonding pad thickness in the range 10 to 12 μm, stresses approaching the fracture stress ($>5 \times 10^8$ dynes $cm^{-2}$) can exist in the

**Table 7.4**

Stress in the various layers of a GaInAsP/InP heterostructure wafer grown by LPE as a function of metallization steps employed in the fabrication of lasers. Stresses are given in units of $10^6$ dynes $cm^{-2}$

| Metallization Step | Stress in | | | | | | | |
|---|---|---|---|---|---|---|---|---|
| | n-metals | InP† Substrate | n-InP Buffer | GaInAsP Active layer | p-InP Cladding layer | p-GaInAsP Capping layer | p-metals | Ti/Pt Barrier layer |
| As-grown | | 1.0<br>−2.0 | −2 | 128 | −2 | 128 | | |
| After p-metallization | | 8.4<br>−16.7 | −16 | 117 | −13 | 117 | 9383 | |
| After n-metallization | 1648 | 2.4<br>−36.5 | −36 | 93 | −37 | 93 | 9367 | |
| After annealing Ti/Pt Barrier layer | 1612 | −57.5<br>15 | 15 | 146 | 16 | 146 | 9417 | −6400 |

† The two values correspond to bottom and top of the substrate.

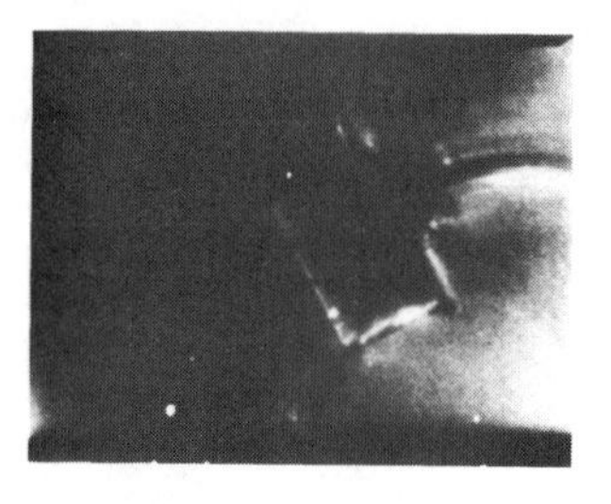


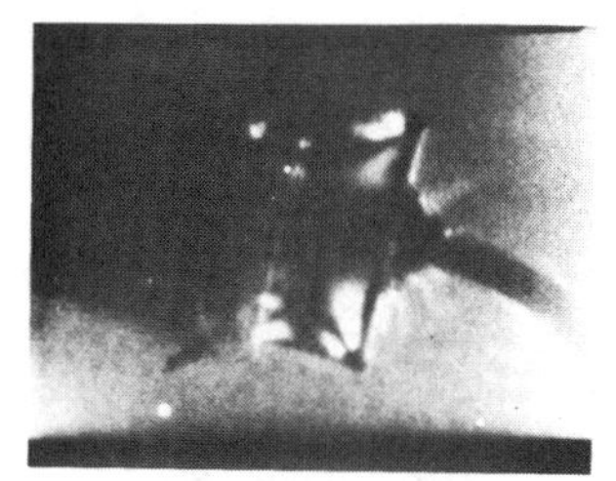


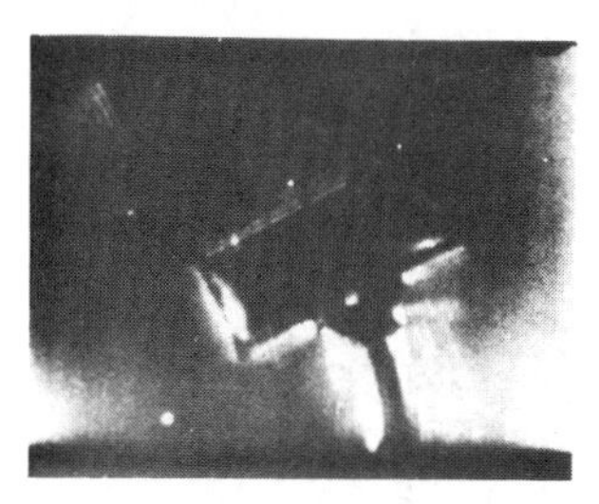


**Figure 7.62** Stress birefringence in AlGaAs/GaAs lasers. $\beta$ the angle between the polarizer axis and the [100] direction, is (a) 0° (b) 45° (c) 90°C. With the transmitted intensities maximum at 45° and minimum at 0 and 90°. The bonding wire is also seen in the figures. The magnification is 158X [338].

devices. The high stress caused aggregates of microcracks along <100> directions in the epitaxial layers. The regions in the stripe where the cracks intersected were nonradiative, causing failure of the device. The remedial action to prevent the Au bonding pad induced degradation was obviously to reduce the thickness of the pad and devices with 0.1 to 0.2μm thick pad did not show any microcracks. Evidence for bonding stress induced degradation in homojunction graded band gap Si-doped AlGaAs LEDs, via <110> DLD glide, has also been cited by Chin et al., [340].

### Other processing effects

Processing operations such as dicing, scribing, and so on, introduce mechanical damage in the semiconductor. The defects created by such damage can then move towards the active regions of the device under the combined action of process induced and/or built-in stresses and temperature and ultimately lead to device failure. Laister and Jenkins [341] showed that dislocation loops can be produced by scribing GaAs surface with a sapphire stylus which penetrated to a depth of at least 50 μm after annealing at 1040°C. In the earlier days of AlGaAs/GaAs lasers when the cavity was formed by mechanical sawing operation rapid degradation caused by dislocation networks occurred inevitably. Diffusion is another source of defects in devices. Rapid degradation in AlGaAs/GaAs lasers where the stripe was defined by Zn diffusion has been shown to be linked primarily with defects introduced during the diffusion process [342].

### 7.5.4 Degradation Modes in Photodetectors

The first generation light wave communication systems which employed AlGaAs/GaAs lasers and LEDs as sources utilized silicon detectors as receivers. As the wavelength of minimum loss and dispersion in the fibers moved to longer wavelengths viz., 1.30 to 1.55 μm, the sources were replaced by GaInAsP/InP lasers and LEDs. Similarly, the receivers were replaced by germanium avalanche photodiodes (APDs). However Ge APDs suffer from high dark current and excess noise compared to Si APDs. Figure 7.63 shows the spectral response of semiconductor detectors useful in the optical communication systems [343]. Besides Ge, GaInAs/InP and AlGaSb/GaSb photodiodes are suitable in the 1.30 to 1.55 μm wavelength range. The GaSb photodiodes have a relatively large dark current and hence GaInAs/InP photodiodes are the preferred choice for this wavelength range. They have certain advantages over Ge APDs. The ternary compound is a direct band gap semiconductor with a lower dielectric constant than Ge. This results in high efficiency and high speed of response for a lower depletion layer width. The absorbing region can be covered by a higher band gap InP window layer to eliminate surface recombination effects. Further the unequal electron and hole ionization coefficients gives rise to reduced noise.

There are two schemes for fabricating the GaInAs photodiodes: the APD and the p-i-n photodiodes. In the early stages the diodes had mesa geometry. The various schemes of fabricating mesa geometry photodiodes are illustrated in Fig. 7.64. The simplicity of fabrication

**Figure 7.63** Spectral response of Si and GaInAs photodiodes. Also shown is the loss spectrum of the silica fibre [343].

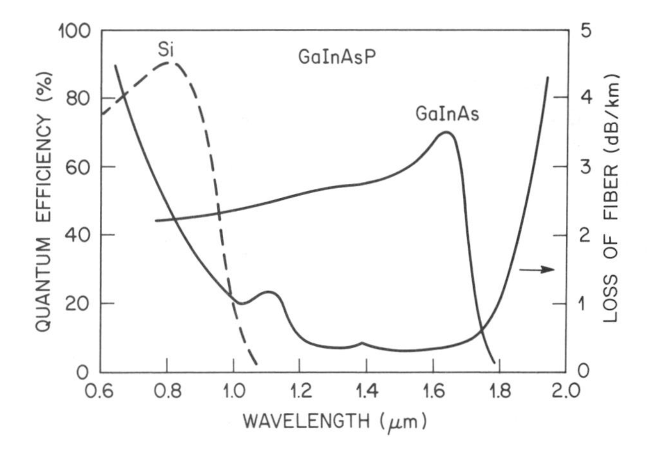

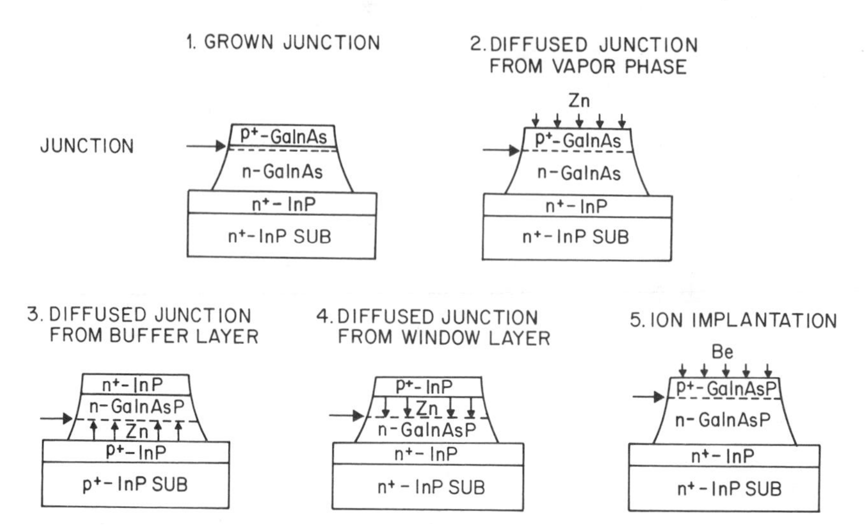


**Figure 7.64** Schematic views of various mesa type photodiodes in which the p-n junction is formed by different methods [343].

of these diodes is outweighed by the expected poor long-term reliability because of the exposed p-n function perimeter. In this respect, planar-type diodes such as shown in Fig. 7.65 would be expected to face better compared to mesa-type diodes since the surfaces can be passivated to protect the p-n junction.

Table 7.5 lists the potential failure modes in photodiodes and the associated device parameters. Two degradation modes can be realized, a catastrophic degradation mode and a gradual degradation mode. The former occurs as a result of a break in the electrical continuity due to die- or wire-bond failures or fracture in the chip, and so on. Gradual degradation in the detector performance can occur because of changes with time in its quantum efficiency, dark current, capacitance and response time.

**Figure 7.65** Schematic views of planar photodiodes with guard ring structures made from (a) $Ga_{0.47}In_{0.53}As$ and (b) InP [343].

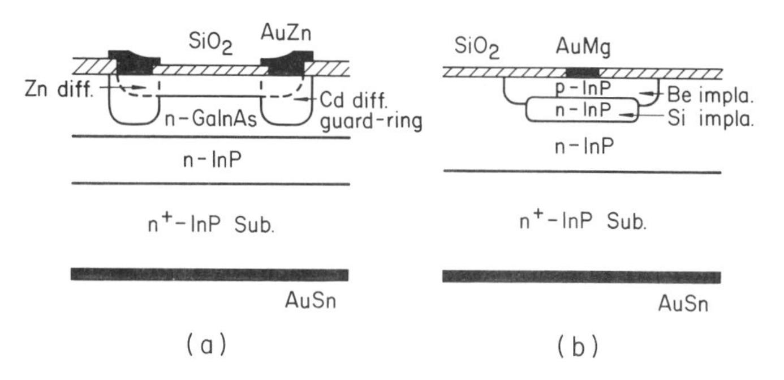

**Table 7.5**

The failure mechanisms and the affected device parameters in GaInAs/InP photodetectors [344]

| Device Parameter | Failure Mechanism | Conditions to promote failure mode in accelerated aging tests |
|---|---|---|
| Dark current | Bulk leakage (junction degradation, local breakdown)<br>Surface leakage | High-temperature bias over voltage<br>High humidity |
| Capacitance | Diffusion, doping variations | High temperature |
| Quantum | Dark spots or dark line defects<br>Degradation of AR coating | High temperature |
| Rise and fall times | Formation of traps | High-temperature bias |
| Electrical continuity | Open wire or chip bond<br>Fractured chip | Thermal cycling<br>Centrifuge<br>Vibration<br>Shock |

By means of an extensive aging studies on planar GaInAs p-i-n photodiodes done to assure their reliability for undersea optical communication systems and to screen flowed or weak devices, Saul et al., [344] concluded that among the failure mechanisms listed in Table 7.5, the increase in dark current is the primary failure mode. Any change in device capacitance caused by thermally induced changes in the diffusion profile is not expected to occur since device operating temperature is very much less than the temperature at which diffusion is done. Similarly, change in the quantum efficiency as a result of formation of dark-line defects or dark spots as in the case of lasers and LEDs is not likely since operating current densities are several orders of magnitude lower in the photodiodes compared to the light emitting devices.

In terms of minimizing detector degradation because of dark current change, the sealing of the junction from the ambient by surface passivation is an important step for both mesa- and planar-type photodiodes. The increase in dark current can be caused by the field-assisted migration of impurities from the surface into the detector's active region resulting in increased generation-recombination current. In mesa diodes contamination along the mesa walls can result in surface leakage. It is important, however, to choose the right passivating material and the technique for depositing it onto the semiconductor because poor passivation can be as bad as no passivation. For example, sputtered $SiO_2$ film on GaInAs was supposed to increase dark current [345], while chemical vapor deposited (CVD) $SiO_2$ on both GaInAs and GaInAsP surfaces did not increase dark current [346]. Diadiuk et al., [347] found that for devices consisting entirely of InP, passivation was achieved with plasma deposited $Si_3N_4$ and for those with a GaInAsP layer but with the p-n junction in InP, passivation was achieved with polyimide. However, when the p-n junction was in the quaternary layer, neither of these films provided passivation. In that case, a photoresist film sprayed with $SF_6$ as the propellant gas provided passivation and yielded devices with low leakage current and sharp breakdown. These effects were supposed to have been produced by the $SF_6$ ambient which reduced excess surface charge and fringing fields at the exposed junction owing to the high electron affinity and dielectric constant of $SF_6$. The GaInAs planar p-i-n diodes to be used for undersea communication applications that Saul et al., [344] described were passivated with $Si_3N_4$ which was also used as the diffusion mask.

Gradual degradation caused by problems associated with metallizations such as electromigration is generally not expected in photodiodes due to the low current density and low electric field in these devices. However, Tashiro et al., [348] attributed a nonrecoverable high temperature (>200°C) degradation in planar heterojunction GaInAs photodiodes to thermal deterioration in the p-side AuZn contacts. They suggested that migration of Au, AuIn precipitates and/or native defects into the depletion region increased the generation-recombination currents. Similar degradation phenomenon has also been suggested in mesa geometry devices [349, 350]. On the other hand, Chin et al., [351] found in their planar diodes no evidence of contact metal penetration when CrAu instead of AuZn p-contact was used. Cr served as a barrier for Au migration in this case.

Degradation owing to stress effects caused by metallizations can be greatly minimized by restricting the metal contact to the photodiode [344]. The metallization procedure also appears to be important in realizing high reliability. Lichtman et al., [352] found that for accelerated aging under a bias in high humidity conditions a fivefold increase in mean times to failure was achieved when e-beam evaporated p-contacts were used instead of plated p-contacts, presumably as a result of reduced contamination and the absence of pores in the metallization.

Tashiro et al., [348] found a recoverable surface degradation mode that occurs below 200°C in planar heterojunction GaInAs p-i-n diodes. These diodes which exhibited 10 fold increase in the room temperature dark current after an accelerated aging test consisting of an increase in ambient temperature in 50°C steps starting from 150°C under a 100 μA reverse current bias, showed a recovery of the dark current to almost to the initial value after annealing at 150°C for 168h without any bias voltage. They suggested that the degradation is caused by hole trapping at the interface between $Si_3N_4$ and the semiconductor or hole injection into the $Si_3N_4$ passivation film. The high kinetic energy of the holes under breakdown would have exceeded the potential barrier between the $Si_3N_4$ film and the quaternary cap and have caused hole injection into the dielectric. Thus positively charged traps in the film induce electron accumulation which becomes a leakage source. At elevated temperatures, the trapped holes are neutralized by recombination with electrons causing the recovery of the dark current.

In a detailed failure mode analysis of planar Zn-diffused GaInAs p-i-n diodes that failed during long term aging ($> 10^3$ h) as well as those that failed during short term aging ($< 10^2$ h) at high reverse bias, Chin et al., [351] found that a majority of the diodes failed as a result of a single localized leakage source located at the perimeter of the p-n junction. They found three types of leakage sources: (a) a microplasma, (b) a microplasma associated with a region of high recombination rate, and (c) a microplasma associated with a thermally damaged region. Further, by analysis of a large number of devices before and after aging they found that leakage paths are developed from microplasmas initially present in the device. Pinholes in the $SiN_x$ diffusion mask at distances of <5 µm from the edge of the p-n junction were identified as the major source of microplasmas. Unintentional Zn diffusion through the pinholes produced shallow p-n junctions which locally increased the electric field and the leakage current.

Figure 7.66 shows a series of EBIC images illustrating the existence of microplasma in a photodiode before aging and its development after aging [351]. In Fig. 7.66(a), an irregularity consisting of a bright circular region of 2 µm in diameter intersecting the perimeter of the p-n

**Figure 7.66** High magnification EBIC images from a planar $Ga_{0.47}In_{0.53}As$ p-i-n photodiode showing the development of microplasma at an irregularity with aging: (a) without bias (b) with −20V bias, a microplasma is observed at the perimeter of the irregularity (c) dark region of high recombination is observed at the irregularity after the device failed in aging (d) with −10V bias, the location of the microplasma appears at the location of the dark region in (c) [351].

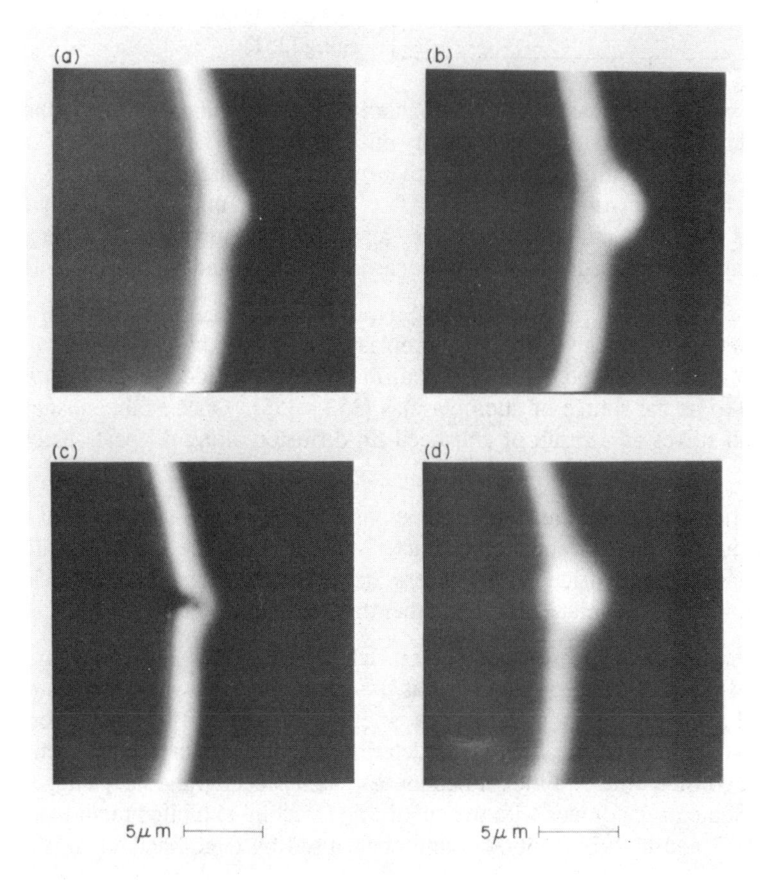

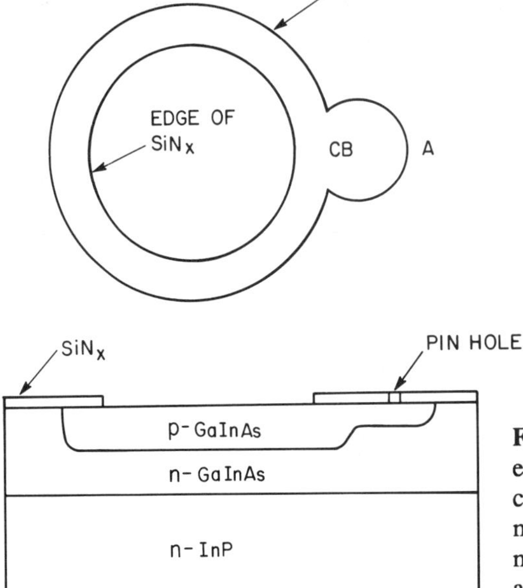


**Figure 7.67** Schematic representation of the effect of a pin hole in the $SiN_2$ mask, centered at B on the p-n junction. A microplasma is originally located at A and moves to C after the device fails during aging [351].

junction is shown. The formation of this irregularity by unintentional diffusion through a pinhole in the $SiN_x$ film is illustrated schematically in Fig. 7.67. When a –20V bias is applied, a microplasma is observed at the perimeter of the irregularity, point A in Fig. 7.67. This is the site where junction breakdown is expected. After aging, the microplasma moves to point C (Fig. 7.67) and the EBIC image shows a dark region of high recombination (Fig. 7.66(c)). The enhanced leakage current leads to a thermal runaway condition and eventually to device failure.

Even in the absence of pinholes in the $SiN_x$ diffusion mask, enhanced Zn diffusion along threading dislocations can give rise to microplasma sites and high leakage current in unaged devices [351]. Both in InP and GaInAsP/InP diffused junction diodes threading dislocations have been observed to be the source of microplasmas [353 - 355]. Once again enhanced electric field at the diffusion spikes as a result of enhanced Zn diffusion along dislocations caused localized breakdown.

The deleterious effect of threading dislocations is not expected to be present in devices with grown in p-n junctions. Poor junction characteristics in avalanche GaInAs diodes with InP p-n junctions produced by hydride VPE process have been attributed to crystal imperfections, presumably stoichiometric related defects, rather than to dislocations [356].

In a bias-temperature life tests of planar GaInAs/InP APDs fabricated from VPE grown wafers, Matsushima, et al., [357] found that diodes fabricated without a guard ring structure exhibited a slight increase in dark current after 10,000 h at 180°C and low bias (~10V) conditions. They attributed this increase to deterioration at the interface between the passivating plasma deposited $SiN_x$ film and the semiconductor, as has been previously suggested by Tashiro et al., [348]. Some of the diodes with the guard ring structure exhibited rapid increase in the dark current at 180°C and high bias (50V) conditions caused by edge-breakdown phenomena in the periphery of the guard ring. In a similar InP/GaInAsP/GaInAs VPE grown APDs with guard rings, Sudo et al., [358] found that in bias-temperature tests, in failed APDs only the dark current increased by nearly 250 times from 50 nA. From wavelength selective light induced current (equivalent to EBIC) measurements they identified the source of degradation to the surface of guard ring junction periphery formed in the InP cap layer. The dark current increase is supposed to result from local avalanche multiplication and degradation induced surface defects.

## 7.6 MATERIAL ASPECTS OF FIELD EFFECT TRANSISTORS

### 7.6.1 Introduction

In the last decade integrated circuit technologies based on GaAs have emerged. These technologies have been developed because transistor (switching speed)×(power consumption) products lower than those for Si based transistors are attainable. This is true not only because intrinsic mobilities and electron velocities in the III-V alloys can exceed those attainable in Si, but also because band gap engineering and selective doping can be used to achieve still higher mobilities.

In this section we will first discuss heterostructure field effect transistors (FETs) followed by a discussion of the DX center which is primarily blamed for the deleterious persistent photoconductivity (PPC) of selectively doped heterostructure transistors (SDHTs) of AlGaAs/GaAs. We will also discuss the collapse of drain current-drain voltage characteristic which occurs in the absence of any illumination.

We will, furthermore, discuss the older GaAs metal-semiconductor FET (MESFET) technology followed by a discussion of two major materials related issues in GaAs MESFETs. These are the effects of the deep level center EL2 and of threading dislocations on FET properties.

### 7.6.2 Heterostructure Field Effect Transistors

#### AlGaAs/GaAs SDHTs

A number of reviews on SDHTs have been published in recent years (see Refs. [359] and [360] and references therein). The basic SDHT device structure is shown in Fig. 7.68 and contains four epitaxial layers grown on a semi-insulating GaAs substrate. An undoped GaAs buffer layer (layer A) which is usually called the channel layer is first grown on the substrate. Next come the undoped spacer layer (layer B), the donor layer (layer C), and the capping layer

**Figure 7.68** Basic device structure of an SDHT [359].

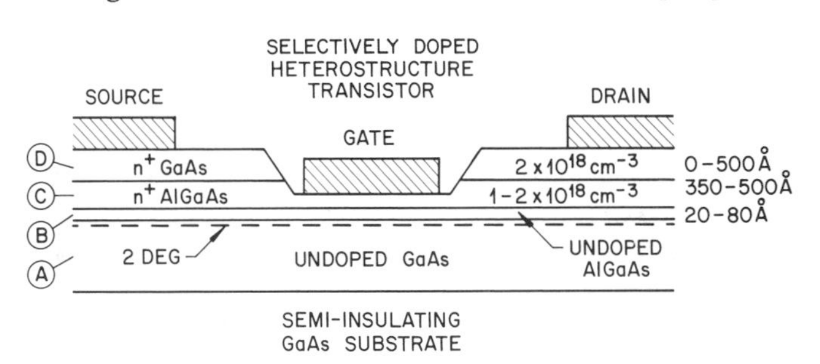

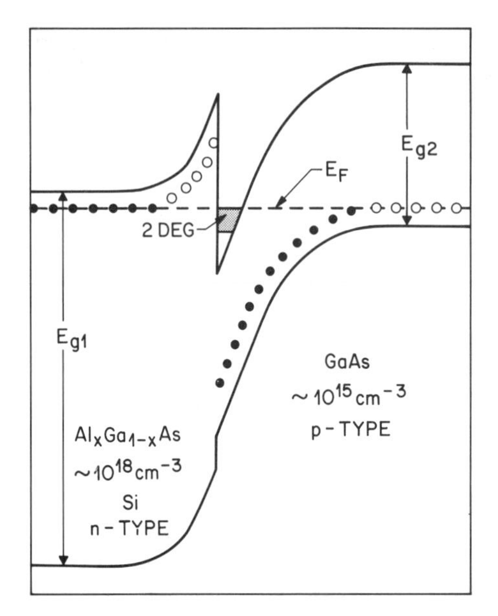


**Figure 7.69** Heterostructure selectively doped to create a two dimensional electron gas [359].

(layer D). Typical layer thicknesses and doping concentrations are indicated in Fig. 7.68. The central concept of selective heterostructure doping which leads to very high mobilities is illustrated for an n-AlGaAs/p-GaAs interface in the band diagram in Fig. 7.69. As a result of the offset in the conduction bands, electrons released by the donors in the AlGaAs spill into the GaAs and form a quasi two-dimensional sheet of charge which is analogous to the accumulation layer near the semiconductor surface under the gate of a Si MOSFET. This sheet of charge is usually referred to as a two-dimensional electron gas (2DEG). If the GaAs is not doped and is highly pure then the mobility of the electrons can be very high because ionized impurity scattering is very small. This central concept of charge separation is an extension of original ideas and experiments of Dingle et al., [361] and of Stormer et al., [362] on multiquantum well structures incorporating alternating layers of GaAs and AlGaAs. Of the two layers of each superlattice period only the AlGaAs barrier layers were doped to form a modulation doped structure. This resulted in an accumulation of electrons in the GaAs well where they were removed from the ionized donor charge in the AlGaAs barriers. It was quickly realized, however, that only a single heterointerface was needed to achieve the necessary charge separation and to confine electrons to the GaAs side of an AlGaAs/GaAs interface [362]. Since these original discoveries the essential concept of charge separation has been utilized in FETs by many workers. The transistors have been given several different names. The main ones are listed in Table 7.6 which is from Dingle et al., [359].

There are three major differences between SDHTs and GaAs MESFETs. The enhanced mobility is beneficial for those properties sensitive to the low-field behavior of the transistor. A low-parasitic resistance and a low "knee" voltage in the drain current-drain voltage characteristic are the result. The lower "knee" voltage leads to lower operating points determined by circuit load lines and to lower supply voltages. The second difference is the confinement of the carriers

**Table 7.6** [359]

| Acronym | Name | Origin |
|---|---|---|
| HEMT | *H*igh *E*lectron *M*obility *T*ransistor | Fujitsu |
| MODFET | *Mod*ulation doped FET, similar to MOSFET | University Illinois, Cornell, Honeywell |
| TEGFET | *T*wo-dimensional *E*lectron *G*as FET | Thomson-CSF |
| SDHT | *S*electively *D*oped *H*eterostructure *T*ransistor | AT&T Bell Labs. |

in a 2DEG very close to the gate which leads to high transconductance (the derivative of the source-drain current with respect to the gate voltage) and to high output resistance. The transconductance depends critically on the gate-2DEG separation which can be as small as ~250 Å. This leads to the highest transconductances known in FETs [359]. The third major way in which the two types of FETs are different is that the gate is formed on $Al_xGa_{1-x}As$ in SDHTs and on GaAs in MESFETs. Schottky barrier heights greater than one volt have been reported for $Al_xGa_{1-x}As$ [363, 364] which is a considerable enhancement over barrier heights on GaAs (0.75-0.80 eV) [359].

Enhanced mobilities are achieved because of the undoped $Al_xGa_{1-x}As$ spacer layer shown in Fig. 7.68. Mobilities in excess of 550,000 $cm^2/V$-s at 4K can be achieved for spacer layer thicknesses in excess of 150 Å [359]. However, increased spacer layer thicknesses result in decreased transconductances [365] because the distance between the gate and the 2DEG is increased. This trade-off between mobility and transconductance results in an optimum spacer layer thickness of between 20 and 75 Å.

There is also an optimum range of x, the Al mole fraction. For x less than 0.2 the conduction band discontinuity is sufficiently small that the quantum mechanical wave function of the 2DEG is not confined to the undoped GaAs channel layer but lies partially within the $Al_xGa_{1-x}As$ layer above it. This results in a decreased mobility. For high Al mole fractions contact resistances are higher because of the larger band gap, and lattice mismatch is higher. The mismatch of AlAs and GaAs is 0.127 percent and for values of x less than unity the mismatch scales proportionately (Vegard's law, see Sec. 1.5). Alloy layers with high values of x are, therefore, more likely to be problematical because of the built-in stress. Most device researchers have chosen x values between 0.25 and 0.33 [359], and it is in this range that mobilities are the highest [366].

Furthermore, Al mole fractions of 0.25 or less are found to reduce the problems of persistent photoconductivity and drain I-V collapse. When illuminated at temperatures below 140K SDHTs are conductive from source to drain long after the illumination is removed. The conductivity can persist for as long as a day or more because of the photogenerated sheet carrier concentration. When samples are warmed the sheet carrier density decays quickly. The source of the photogenerated carriers is most likely a defect-donor complex known as the DX center which occurs in doped $Al_xGa_{1-x}As$ [367], although other explanations have been put forward [359, 360]. The DX center is discussed in detail in Sec. 7.6.3. This defect is also believed to be responsible [368] for the collapse problem which occurs at 77K. As shown in Fig. 7.70, above a certain drain bias the drain current collapses at low drain voltages [360, 369]. At room temperature an apparently related phenomenon occurs. This is a hysteresis in the source to drain current versus source to drain voltage characteristic. Both this hysteresis and the collapse which occurs at 77K can be eliminated by shining light on the device. However, this is impractical for most applications.

Workers in the field have instead pursued means to reduce the DX center concentration or to render it less effective. One approach is to replace the $Al_xGa_{1-x}As$ layer which is directly on the gate (see Fig. 7.68) with a AlAs/GaAs superlattice having a short period (~40 Å) [370]. A band diagram for such a structure is shown in Fig. 7.71. In this structure only the GaAs wells are doped (with Si usually). Undoped $Al_xGa_{1-x}As$ layers may also be used for the barriers. With thin wells the electron ground state lies at a considerably higher energy than the conduction band edge of GaAs, and the wave function is spread over the entire superlattice. The electrons spill

**Figure 7.70** Current-voltage characteristics of an SDHT in the dark and under illumination with visible light. The characteristic in the dark is "collapsed" [360].

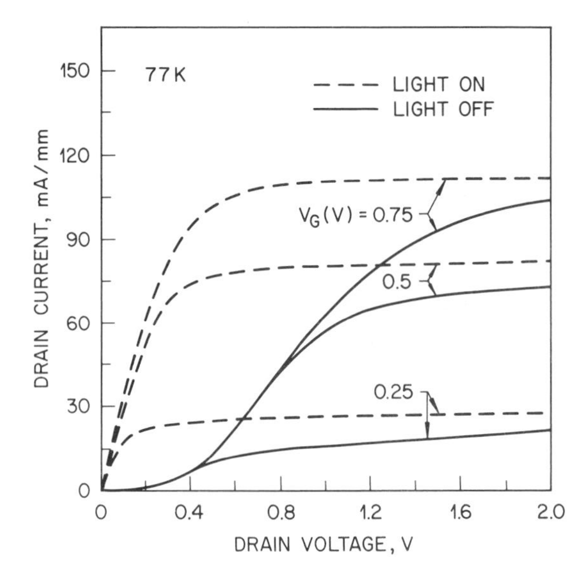

over into the GaAs channel layer to form a 2DEG just as in a conventional SDHT structure. This approach has been shown to be effective in eliminating light sensitivity and I-V collapse as shown in Figs. 7.72 and 7.73. These figures show the I-V characteristic and the transconductance, respectively, for a one micron gate device in which a $Al_{0.6}Ga_{0.4}As$/n-GaAs superlattice having a period of 30 Å was incorporated [371]. A second approach involves passivation of the DX center by hydrogenation [372].

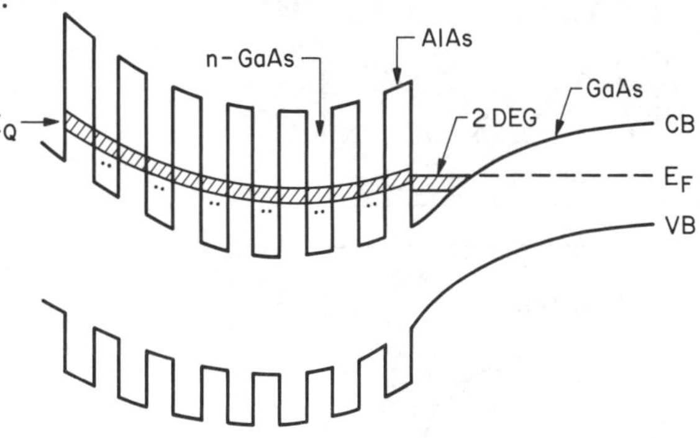


**Figure 7.71** Band bending diagram of a superlattice-SDHT. The AlGaAs in the conventional SDHT structure has been replaced with a GaAs/AlAs superlattice [360].

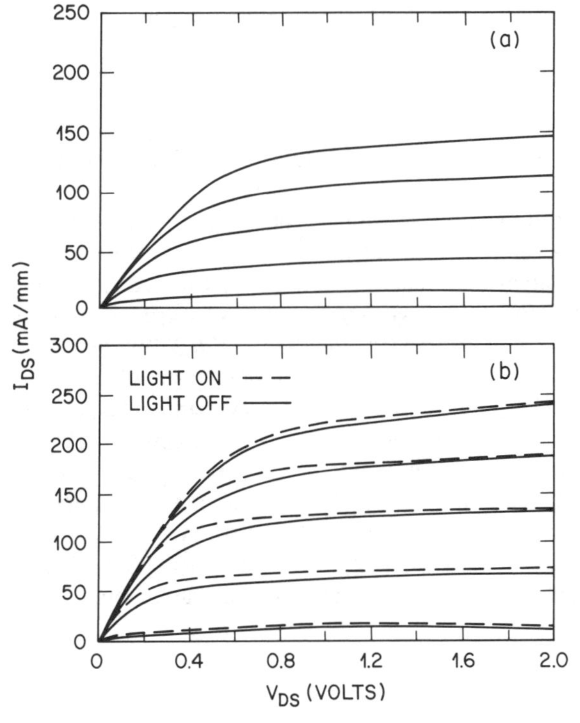


**Figure 7.72** Current-voltage characteristics of a superlattice-SDHT at (a) 300K and (b) 77K. The 77K characteristics in the dark and under illumination are quite similar and show that the "collapse" problem is largely eliminated [371]. Copyright, 1986, IEEE.

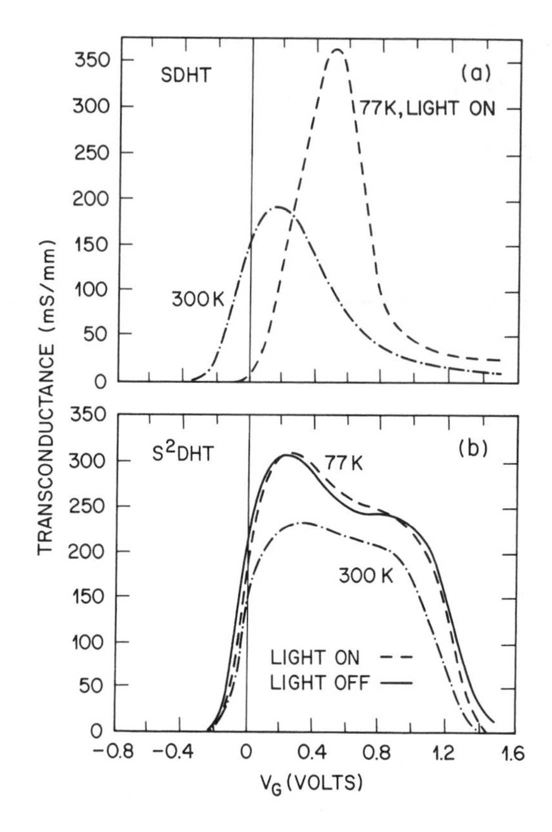


**Figure 7.73** Transductance as a function of gate bias for an enhancement-mode SDHT in (a) and for a superlattice-SDHT in (b). The transconductance of superlattice SDHT exhibits only a weak sensitivity to light and is not as temperature dependent [371]. Copyright, 1986, IEEE.

### SDHTs made with other III-V alloys

A third alternative is to make SDHTs with other heterostructures [359, 360]. Attractive alternatives are $Al_{0.48}In_{0.52}As/Ga_{0.47}In_{0.53}As$ and $Ga_{0.47}In_{0.53}As/InP$ heteroepitaxial combinations. Both of these are lattice matched to an InP substrate. The choice of $Ga_{0.47}In_{0.53}As$ for the channel material is a good one since an effective saturated drift velocity of $2\times10^7$ cm/sec at 300K has been reported [373], which is larger than that of GaAs ($\sim1.2\times10^7$ cm/sec). The presence of a 2DEG has been demonstrated for $Al_{0.48}In_{0.52}As/Ga_{0.47}In_{0.53}As$ grown by MBE [374] and for $Ga_{0.47}In_{0.53}As/InP$ grown by MOCVD [375] and VPE [376,377]. For these alternate heterostructures the two key properties of mobility and sheet carrier density of the 2DEG are reported to be attractive for SDHTs [359,360]. Because Schottky barrier heights for $Ga_{0.47}In_{0.53}As$ and for n-InP are low, gates are difficult to form for this heterostructure. Schottky barrier heights for $Al_{0.48}In_{0.52}As$, on the other hand, are reported to be sufficiently high ($\sim0.8$ eV) that good gates are readily formed [378]. In addition, device isolation in an integrated circuit is also available since undoped $Al_{0.48}In_{0.52}As$ can be grown semi-insulating by MBE [379]. With the advent of two inch diameter (and larger) InP substrates, integrated circuits using SDHTs with $Ga_{0.47}In_{0.53}As$ channel layers appear poised to play an important role in future high speed electronic circuit applications.

### 7.6.3 DX Centers

**Defect properties relevant to deep level effects observed for SDHTs**

Both persistent photoconductivity (PPC) and source-drain I-V collapse are generally acknowledged to arise from deep levels which exist in $Al_xGa_{1-x}As$ alloys. These deep levels are commonly referred to as DX centers, and these centers have been the object of a large number of studies by solid state physicists interested in point defect physics.

A group of such defects with similar properties have been discovered. They all involve a donor and are thought to be responsible for PPC. This group includes $Al_xGa_{1-x}As$ doped with Se, Te, Si, and Sn and $GaAs_yP_{1-y}$ doped with S and Te [380].

A consensus seems to have been reached in recent years that the dramatic optical and electrical properties of DX centers are best explained by invoking large localized lattice relaxations. A convenient vehicle with which to describe these relaxations is the configuration coordinate diagram shown in Fig. 7.74 (Sec. 5.2.9). In this figure the abscissa is an unspecified configuration coordinate (dimension of length) of the defect. The energy E is the barrier to electron emission as observed by DLTS (0.33 eV for $Al_xGa_{1-x}As$:Te [380]), and $E_B$ is the barrier to the capture of an electron from a delocalized state in the conduction band. PPC is explained by the capture barrier, $E_B$, since at low temperatures electrons would remain in the conduction band. Source to drain I-V collapse in the dark is explained by the emission barrier, E, since electrons remain localized at the DX center and are unavailable for conduction until a bias voltage sufficient to excite them over the barrier is applied. The energy $E_n$ is the photoexcitation threshold which is known to exhibit a large Stokes shift relative to E. The fact that source to

**Figure 7.74** Typical configuration coordinate diagram used for DX centers [380].

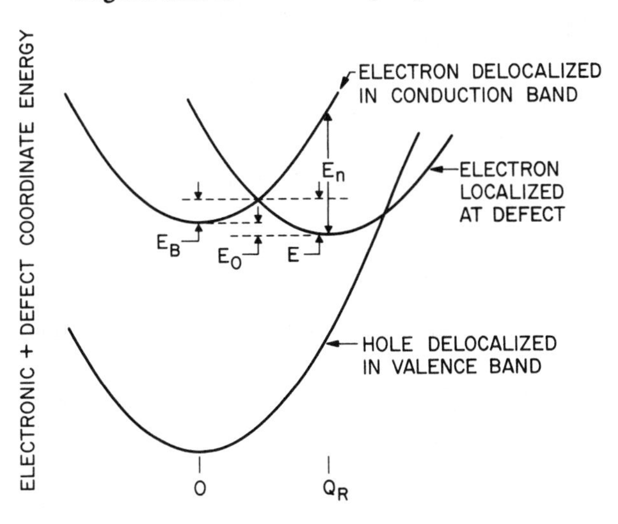

**Table 7.7**

Summary of DX Centers in AlGaAs and GaAsP

| Donor Impurity | E (eV) | $E_B$ (eV) | $E_0$ (eV) | $E_n$ (eV) | Reference |
|---|---|---|---|---|---|
| AlGaAs: | | | | | |
| Se | 0.28±0.03 | 0.18±0.02 | 0.10±0.05 | 0.85±0.1 | 382 |
| Te | 0.28±0.03 | 0.18±0.02 | 0.10±0.05 | 0.83±0.1 | 382, 383 |
| Si | 0.43±0.05 | 0.33±0.05 | 0.10±0.05 | 1.25±0.1 | 382 |
| Sn | 0.19±0.02 | <0.1 | 0.10±0.05 | 1.1±0.1 | 382 |
| GaAsP: | | | | | |
| S | 0.35 | 0.15±0.03 | 0.20±0.03 | 1.53 | 383 |
| Te | 0.19±0.02 | 0.12±0.03 | 0.07 | 0.65±0.05 | 384 |

drain I-V collapse is not observed if an $Al_xGa_{1-x}As$:Si/GaAs SDHT is illuminated is explained by the photoexcitation of carriers and subsequent relaxation into the conduction band. The energy $E_0$ is the thermal activation energy of the resistivity observed under thermal equilibrium conditions (e.g., in Hall measurements made at several temperatures). A summary of the various energies for DX centers is given in Table 7.7.

Interpretation of DLTS measurements for DX centers is severely complicated by the large ratio of DX center concentration to the carrier concentration which usually occurs. The name "DX center" was originally coined [381, 382] to describe the near proportionality of the DX concentration to the shallow donor concentration (the D in DX) and to describe the lack of effective-mass-like behavior (the X in DX). Large values for $\Delta C/C$ in DLTS measurements (here C is the capacitance) usually lead to nonexponential capacitance transients. In spite of the difficulties, DLTS results can be interpreted in a physically consistent manner [382]. The DLTS signal strength caused by DX centers in Te doped $Al_xGa_{1-x}As$ is shown in Fig. 7.75. We note that the reduced signal strength for $x \lesssim 0.35$ coupled with observations that PPC and I-V collapse in SDHTs are reduced for low values of x suggest that the DLTS signal strength is indicative of PPC and I-V collapse (or lack thereof) in SDHTs. The decreased DLTS signal strength for $x \lesssim 0.35$ apparently is not caused by a decreased concentration of DX centers, however. The apparent DX concentration is reduced because the DX level lies increasingly above the $\Gamma$ conduction band minimum as the Al mole fraction is decreased. At equilibrium the DX center will in these cases be above the Fermi level and will be only partially occupied because of the decreasing Boltzmann tail of the Fermi function [380]. Since DLTS signal strength is a measure of the occupancy of a deep level, DX signal strength is reduced at low Al mole fractions.

The DX center does not appear to be a single clearly identifiable defect with the same configuration in all $Al_xGa_{1-x}As$ samples. This is inferred from the DLTS spectra shown in

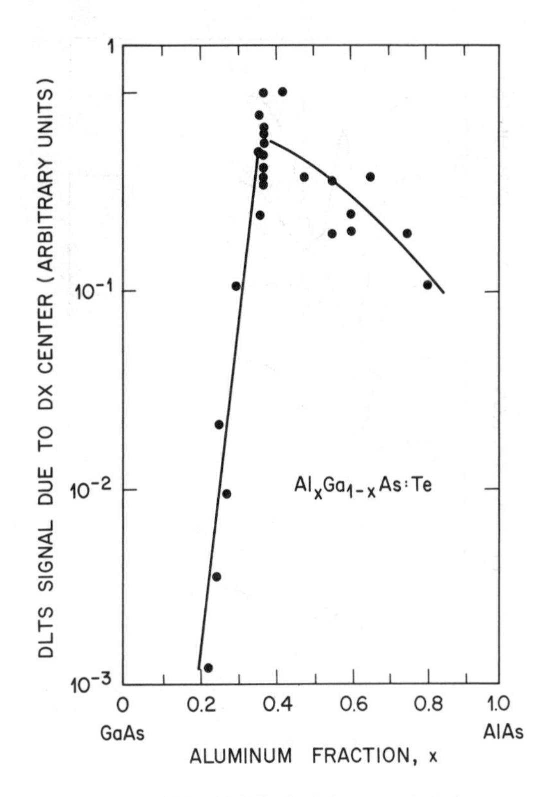


**Figure 7.75** DLTS signal strength due to DX centers as a function of aluminum fraction [380].

Fig. 7.76 [380]. A great variation in peak shape and width is seen, and in some samples a double peak occurs. Peak shape variations are not found to correlate with sample stoichiometry but occur more or less at random [380]. Lang has speculated that these variations are as a result of particular arrangements of Ga and Al around the defect which vary almost randomly [380]. However, DX centers can apparently also exist in binary GaAs [385], and this implies that although alloy fluctuations in the near neighbor environment can influence the emission behavior, an alloy is not essential for a DX center to occur.

### Microscopic models

A consensus on the microscopic structure of DX centers has not yet been reached [380]. However, structural information is available from the ballistic phonon absorption measurements of Narayanamurti et al., [386]. A model for DX center is shown in Fig. 7.77 [380]. This model is consistent with the phonon absorption measurements for both the cation site donors (group IV-Sn) and for the anion site donors (group VI-Te). However, alternate models involving only an off-center donor and no anion vacancy are being developed by Baraff and Schluter [387].

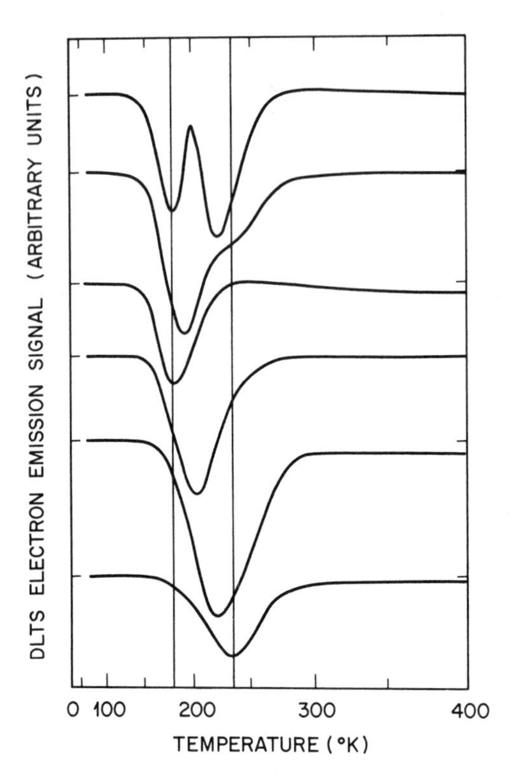


**Figure 7.76** DLTS spectra of DX centers in Te-doped $Al_xGa_{1-x}As$ samples showing the range of peak positions and shapes observed [380].

**Figure 7.77** Ball and stick model of point defects proposed for the DX center [380].

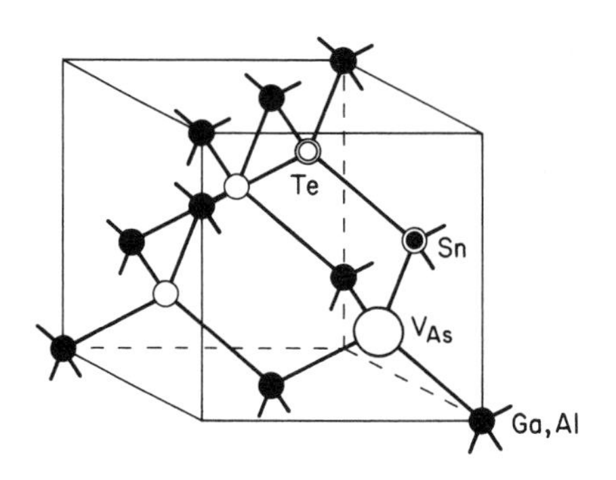

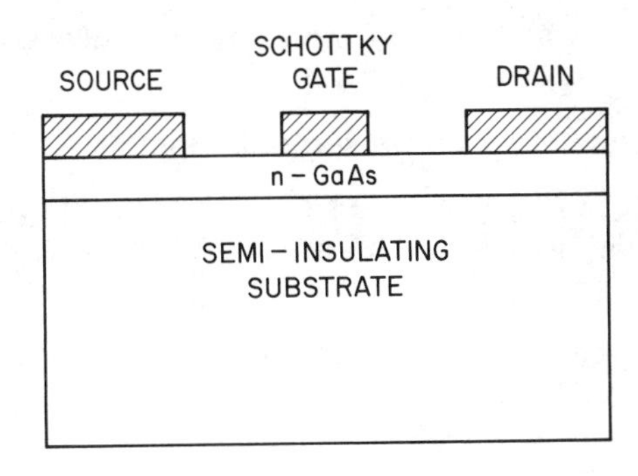


**Figure 7.78** Schematic illustration of a basic GaAs FET.

### 7.6.4 GaAs FETs

Microwave frequency transistors can be made by ion implantation into semi-insulating GaAs substrates [388, 389]. A schematic cross section of such an FET is shown in Fig. 7.78. The conducting channel can also be formed by growing a doped epitaxial layer on a S [388, 389]. The technology surrounding the manufacture of GaAs FETs is more mature than that of SDHTs. Integrated circuits on GaAs wafers are now commercially available. The cost of these circuits is ultimately related to the manufacturing yield which seems still to be significantly influenced by defects. Dislocations and/or EL2 deep level defects can seriously affect the essential device parameters. The basic D.C. characteristic of a GaAs FET is shown in Fig. 7.79. At a given gate to source voltage, $V_{GS}$, the drain to source current, $I_{DS}$, saturates as the drain to source voltage is raised because the velocity of the electrons under the gate saturates. The saturated current at $V_{GS} = 0$ is called $I_{DSS}$. Furthermore, the channel can be pinched off by the depletion region under the gate. This occurs if the Schottky gate is sufficiently reverse biased. Of course, the depth of depletion will depend on the carrier concentration in the channel under the gate, and if there are doping nonuniformities across a wafer, the minimum value of $V_{GS}$ needed for conduction between the source and the drain will vary across the wafer. This voltage is known as the threshold voltage, $V_{th}$. Uniformity in $V_{th}$ is an obviously desirable characteristic of the gates of an integrated circuit. In what follows we discuss reports made on dislocations in GaAs

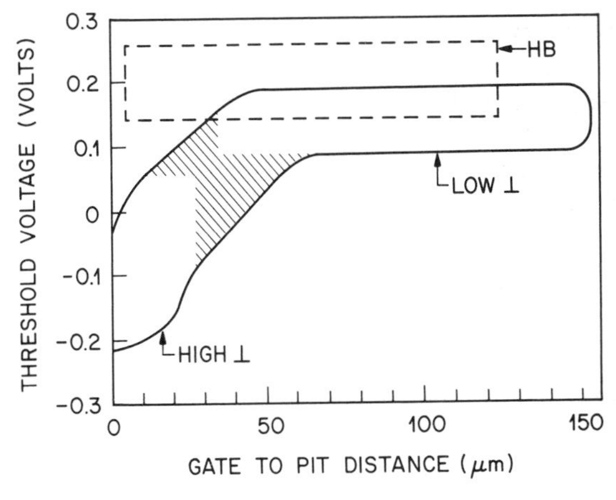


**Figure 7.79** Range of $V_{th}$ as a function of gate-to-etch pit distance. The high and low dislocation density areas are separated by a hatched region where the proximity effects starts to break down. The upper boxed region represents data for GaAs grown by the horizontal Bridgman technique [392].

substrates and on EL2 defects and on correlations of the distribution of these defects to the distribution of $V_{th}$.

**Correlation of dislocation distribution to threshold voltage distribution**

A correlation of the distribution of dislocations to FET performance has been reported for both undoped [390] and Cr-doped semi-insulating (SI) GaAs grown by the liquid encapsulated Czochralski (LEC) technique.

For the undoped case, two adjacent (100)-oriented slices were used to establish the correlation. One of them was etched in molten KOH to reveal the variation in etch pit density (EPD). The EPD distribution agrees quite well with the predictions of Jordan et al., [391, 392]. In particular, a "W" shaped variation of the EPD along diagonals through the center of the slice is exhibited (Sec. 2.1.3). FETs were fabricated on the other slice. An active layer was first formed in the SI wafer by ion implantation of Si at 60 keV. The fluence was $2.0 \times 10^{12}$ $cm^{-2}$, and the wafer was annealed at 800°C for 20 minutes to activate the implanted Si. Subsequently 404 FETs were processed with a $1.5 \times 1.0$ mm repetition lattice. Ohmic contacts were obtained by alloying Au-Ge-Ni deposits while Schottky gates were obtained by Ti-Pt-Au evaporation. After the Schottky gates were formed, $V_{th}$ was measured for each device. The threshold voltage was defined as the gate voltage required to produce a value of 5 μA for $I_{DS}$ at a value of 1V for $V_{DS}$. More negative values of $V_{th}$ were found in those regions having a high dislocation density. The $V_{th}$ distribution showed a clear correlation with the EPD distribution. Nanishi et al., [393] also measured the distribution of $I_{DS}$ at a $V_{DS}$ of 2V. This was done prior to the formation of the Schottky gates. The $I_{DS}$ distribution mimicked the $V_{th}$ distribution. High $I_{DS}$ values were found for FETs at the periphery and near the center of the wafer. Similar results for the distribution of $I_{DS}$ were found for Cr-doped LEC GaAs [393].

These important results have been elucidated by spatially resolved cathodoluminescence studies and by a study of the influence of dislocations on the electrical activation of implanted Si ions. Chin et al., [394] reported that bands of dislocations in Cr-doped SI (100)-oriented GaAs wafers grown by LEC are surrounded by a ~50 μm wide region which exhibits a luminescence efficiency ~10% greater than that of a dislocation free region. The dislocations themselves corresponded to dark spots. These results for Cr-doped GaAs mimic those reported earlier for GaAs doped with Te, Se, and Si. In a report of a photoluminescence study Heinke et al. [395] furthermore concluded that a cylindrical volume exists around dislocations which is depleted of acceptors. That a Cottrell atmosphere of point defects exists around dislocations was confirmed by the work of Honda et al., [396]. These workers reported a clear correlation of the sheet resistance and sheet carrier concentration after Si ion implantation and annealing to the EPD distribution for both Cr-doped and undoped SI GaAs grown by the LEC technique. Carrier activation was found to be higher near dislocations. The distribution of the electron mobility was, however, not found to be correlated to the EPD distribution.

Further studies have revealed a dependence of $V_{th}$ for enhancement mode FETs on the distance from an FET gate. The results of Miyazawa et al., [397, 398] are encompassed in Fig. 7.79. Note that for gate to etch pit distances greater than 50 μm there is no effect on $V_{th}$ values. Note also that the range of the effect, 50 μm, is the same as the range of the bright cathodoluminescence region seen by Chin et al., [394].

There has been worldwide concern regarding these results because of their significance for large scale integration of GaAs FETs in integrated circuits [392]. In studies of depletion mode FETs made on In-alloyed substrates containing fewer than $3 \times 10^4$ $cm^{-2}$ dislocations Winston et al., [399] found no dependence of $V_{th}$ on the gate to etch pit distance. The resulting controversy prompted Miyazawa and Hyuga [400] to reexamine the proximity effect which Miyazawa and co-workers had first reported, and they confirmed earlier results. In these later studies conventional (i.e., not In-alloyed) LEC substrates were again used. We note that Fig. 7.79 is based specifically on the results of Miyazawa and Hyuga [399]. A possible explanation for the discrepant results of Miyazawa and co-workers and of Winston and co-workers was given by Miyazawa and Hyuga [400]. These authors suggested that differences might have come about because of the fact that enhancement mode device arrays are more sensitive to dislocations than are depletion mode arrays.

Dislocations in an LEC grown GaAs wafer are often not distributed uniformly but instead form a strongly networked cell structure, and the standard deviation of the local distribution of threshold voltages, $\sigma$ ($V_{th}$) can vary by as much as a factor of two as a result [401]. These results suggest that a high density of dislocations may actually be desirable if a small $\sigma$ ($V_{th}$) is desired. Typically, EPD values are in the range $10^4 - 10^5$ $cm^{-2}$ for LEC grown GaAs. We note that GaAs grown epitaxially on a Si substrate may be desirable for this reason since dislocation densities are typically at least $10^7$ $cm^{-2}$. Indeed, Fischer [402] has found that GaAs grown on Si can serve as useful substrate material for FETs and found no effect directly attributable to the dislocations [403]. Oddly, the high dislocation density may be a boon since a narrower distribution in $V_{th}$ is expected based on the work of Ishii et al., [401].

**EL2 Deep Level Point Defects**

The need for electrical isolation between the GaAs FETs of an integrated circuit requires that the FETs be formed on semi-insulating GaAs. GaAs substrates can be made SI by doping with Cr, but because the Cr can redistribute or outdiffuse during annealing (subsequent to ion implantation) [404] (Sec. 7.3) undoped SI GaAs appears to be a more attractive material for IC fabrication. The Fermi level of this material is pinned near the middle of the band gap by a deep donor center known widely as EL2. This center is now believed to be formed entirely of intrinsic point defect(s). In this section we discuss first the relevant basic properties as well as the nature of the EL2 center and then correlations of observed distributions of EL2 in boules and in wafers to observed distributions of dislocations (EPDs).

The existence of EL2 deep levels was first established from deep-level spectroscopic data obtained by thermally stimulated current measurements and by deep-level transient spectroscopy (DLTS) [405]. Martin et al., [406] first referred to the level as EL2, a label which is now very widely used. An example of a DLTS spectrum exhibiting a peak for EL2 is shown in Fig. 7.80 [407]. From the Arrhenius temperature dependence of the electron emission rate of EL2 an activation energy of 0.825 eV is obtained [405]. The temperature dependence of the electron capture rate has been measured and found to have an activation energy of 0.066 eV. [408]. From detailed balance we have that the EL2 level must lie 0.760 eV below the conduction band. These results demonstrate that EL2 is a midgap level.

DLTS measurements can only be made for EL2 in the presence of shallow level dopants, and this fact prevents the use of DLTS to monitor EL2 concentrations in SI GaAs wafers. To accomplish this the technique of infrared light transmittance mapping has been applied. Optical

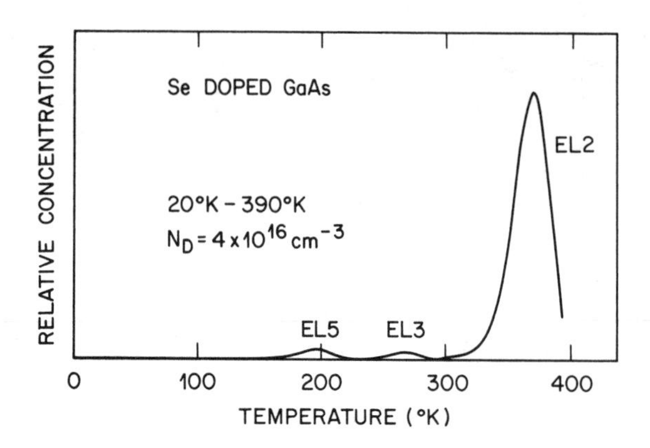


**Figure 7.80** DLTS spectrum of Se doped GaAs grown under As rich conditions [407].

absorption spectra for undoped SI GaAs are shown in Fig. 7.81 [409]. That the curve labeled 'a' represents absorption caused entirely by EL2 is demonstrated by the curves labeled 'b' and 'c' which were obtained after white light illumination for 1 and 10 minutes, respectively. Quenching of the absorption is caused by transfer of the EL2 defect from its normal state to its metastable

**Figure 7.81** Optical absorption spectra for the same undoped semi-insulating GaAs material. Curve a) was obtained after cooling in the dark, curves b) and c) after white light illumination for 1 and 10 min, respectively [409].

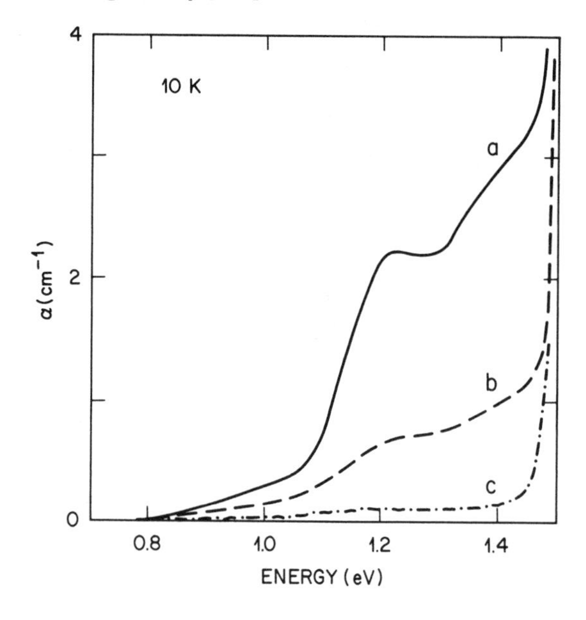

state [409]. Recovery to the normal state results when the material is heated to 140K for a few minutes. Since no other deep level exhibits such quenching, the absorption must be caused by EL2 alone [409]. Absorption coefficients were made quantitative measures of EL2 concentrations by accurate measurements of the concentration of EL2 using a capacitance technique. Calibration curves are shown in Fig. 7.82 [409]. This calibration curve is only valid when the defect is neutral, that is, when the Fermi level lies sufficiently above the EL2 level that it is completely filled with electrons. With this calibration the concentration of EL2 can easily be obtained for entire slices by scanning them at room temperature [409].

That EL2 is a donor was shown by Mircea et al. [410]. These workers measured the out diffusion of EL2 at 600-750°C from VPE material for which the electron concentration ($\sim 10^{15}$ $cm^{-3}$) was comparable with the EL2 concentration ($1-3\times10^{14}$ $cm^{-3}$). Depth profiles of the concentration of EL2 were obtained by DLTS, and space charge profiles were measured by making capacitance-voltage (C-V) measurements and etching. At low temperatures the space charge profile was flat and was not changed by the heat treatment. This is expected if the EL2 centers are neutral and are occupied by an electron. At high temperatures the space charge profile closely followed the outdiffusion profile measured by DLTS. This is expected if the EL2 centers are ionized. These results clearly showed that EL2 centers are neutral when they are occupied by an electron, and they must therefore be donors. In most instances the acceptor impurity compensated by the deep donor, EL2, in undoped SI GaAs is believed to be carbon. In principle, a variation in the concentration of either carbon or EL2 (or both) could result in a variation in resistivity.

**Figure 7.82** Optical absorption coefficient in undoped n-type GaAs as a function of the EL2 concentration measured by capacitance methods in the same materials. Open circles are data obtained at 5K and full circles at 300K [409].

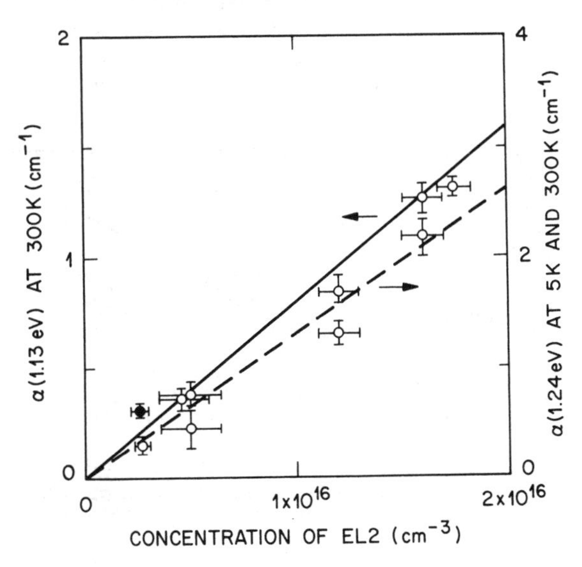

Oxygen is no longer believed to be directly involved in the compensation mechanism [405]. That high resistivity GaAs boules could be obtained by adding oxygen was reported by Hilsum and Rose-Innes in 1961 [411]. SI GaAs was later obtained with the horizontal Bridgman technique by introducing an oxygen partial pressure or by adding $Ga_2O_3$ to the melt [405]. These results are now believed to be caused in part to a reduced Si contamination since $Ga_2O$ gas prevents decomposition of quartz reactor walls [405]. Furthermore, a reduced concentration of Si contaminant was found in GaAs grown by vapor phase epitaxy (VPE) when oxygen was intentionally introduced. When oxygen is ion implanted into n-GaAs a strong increase in resistivity is obtained. The resistivity remains high even after the implanted GaAs has been annealed at a temperature of 900°C. However, the carrier removal which occurs because of the oxygen implantation is now believed to be caused by the formation of complexes between oxygen and donors rather than to the formation of compensating acceptors as first supposed [405].

EL2 defects occur in bulk crystals grown by the LEC and the Bridgman technique, and they occur in epitaxial layers grown by VPE and by metal-organic-chemical-vapor deposition (MOCVD). Studies of these materials have revealed that the concentration of EL2 and stoichiometry are correlated [412,413] (see Fig. 2.13). The concentration of EL2 was found to be larger for As rich melts than for Ga rich melts. In the Bridgman technique As losses during growth are reduced by tightly controlling the As overpressure, and a higher concentration ($\sim 1.5 \times 10^{16}$ cm$^{-3}$) of EL2 is typically observed. Results for the concentration of EL2 as a function of the As/Ga ratio in MOCVD grown GaAs are shown in Fig. 7.83 [405]. The data for GaAs grown by these four techniques (LEC, Bridgman, VPE, and MOCVD) indicate that the EL2 defect is more easily formed under As rich conditions.

EL2 is ordinarily absent in LPE [412, 414] and MBE grown [414, 415] epilayers. Martin and Makram-Ebeid have speculated that the absence of EL2 in MBE layers (grown under As-rich or Ga rich conditions) could be due either to the fact that the growth temperatures which have been

**Figure 7.83** EL2 concentration as a function of the As/Ga ratio with the pressure of trimethylgallium kept constant. The two symbols are for different substrates [405].

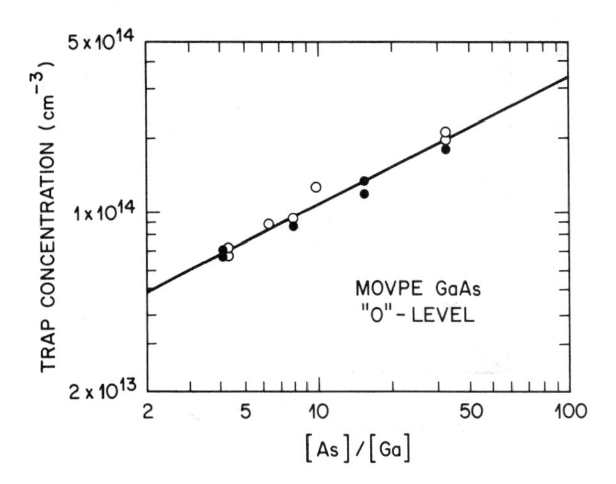

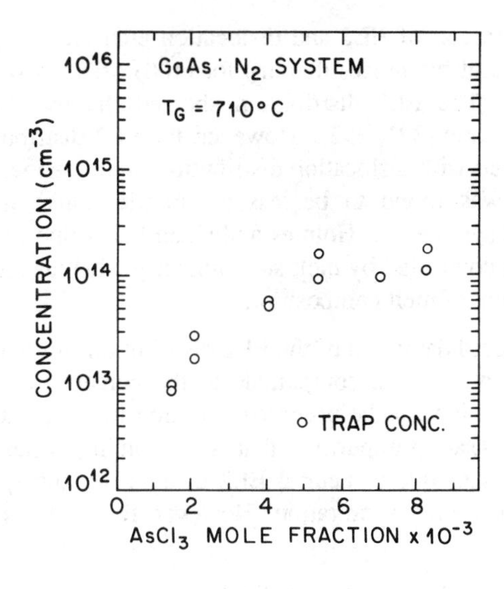


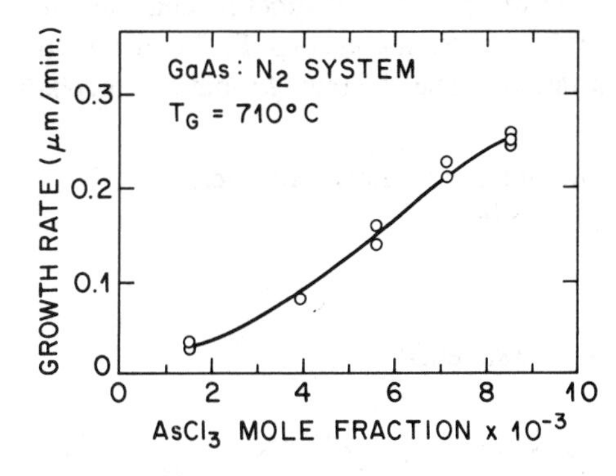


**Figure 7.84** EL2 concentration and growth rate as a function of the $AsCl_3$ mole fraction during trichloride vapor phase epitaxy [416].

used (~600°C) were too low for EL2 defects to be formed or to the slow growth rate inherent to MBE [404]. Note that because the growth rate is slow, equilibrium defect concentrations are more likely for MBE than for VPE, MOCVD, or LPE-grown GaAs. In the case of LPE, the growth rate is considerably larger than for MBE, but Ga rich conditions generally prevail. This could explain the lack of EL2 in LPE grown layers [405].

A strong correlation of the EL2 concentration on growth rate for trichloride-VPE-grown layers was reported by Ozeki et al., [416]. Both the trap concentration and growth rate were found to be dependent on the $AsCl_3$ mole fraction as shown in Figs. 7.84(a) and 7.84(b). If EL2 were a simple substitutional impurity, then its concentration would be lower at higher growth rates. Since the opposite was found, it appears that EL2 is not a simple substitutional impurity.

Microstructural information for EL2 defects is available from electron paramagnetic resonance (EPR) experiments (Sec. 5.4.4). EPR spectra for as-grown undoped SI GaAs crystals exhibiting four lines attributable to an $As_{Ga}$ antisite defect were first reported by Wagner et al., [417]. The antisite defect observed by EPR is generally identified as $As_{Ga}^{4+}$. However, the EPR experiments cannot rule out complexes involving $As_{Ga}^{4+}$ bound to near neighbor defects [405].

A correlation between the concentration of EL2 and dislocation etch pit density (EPD) has been established (see reference [418] and references therein) for (100) oriented wafers cut from the seed end of (100) LEC boules. The EL2 distribution showed the four-fold symmetry characteristic of the dislocation distribution [391, 392]. However, the EL2 distribution in wafers from the tail end of the boules correlated with dislocation distribution in only about 50 percent of the crystals. The EL2 distribution was found to be less symmetric unlike the dislocation distribution which remained four fold symmetric. Holmes and Chen [418] find that the average EL2 concentration along the crystal is controlled by melt stoichiometry whereas radial variations in the EL2 concentration are independent of melt composition.

For the (100) wafers a typical standard deviation of the EL2 distribution (measured by optical absorption) was found to be 10 percent, a value comparable to the percent ionization of EL2 defects [418]. Another possible explanation for the observed variation in optical absorption is a variation in the concentration of compensating impurities, that is, carbon impurities. Because the infrared absorption technique is only sensitive to neutral EL2 centers, an absorption variation could also be caused by a variation in carbon concentration. However, EL2 profiles are found not be correlated to carbon profiles [418].

Holmes and Chen [418] conclude that there are two distinct mechanisms for the formation of EL2. One is stoichiometry related and determines the average EL2 concentration and the second causes the observed intra-wafer variation. The second formation mechanism is likely a

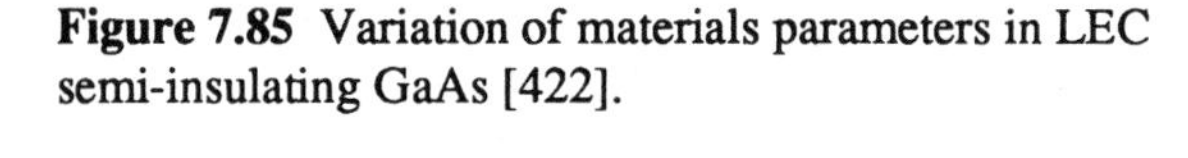

**Figure 7.85** Variation of materials parameters in LEC semi-insulating GaAs [422].

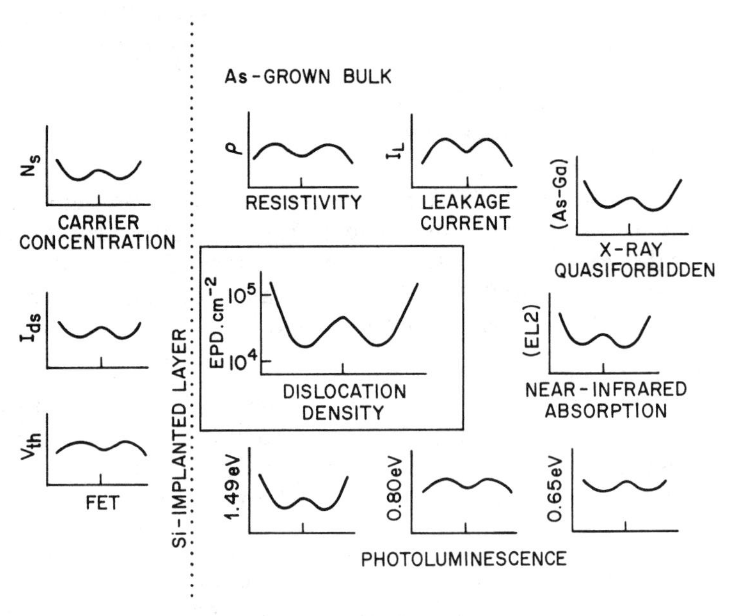

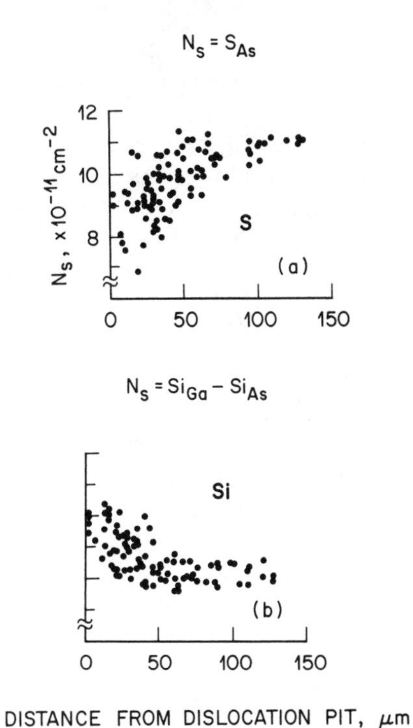


**Figure 7.86** Dislocation proximity effect in a) S-implanted and b) Si-implanted GaAs [422].

dislocation related mechanism involving dislocation climb and/or dislocation gettering [418]. That dislocation climb during plastic deformation results in the formation of EL2 has been proposed by Weber et al., [419] and is supported by the DLTS results of Ishida et al. [420] who showed that the concentration of EL2 was increased after plastic deformation. Gettering of EL2 to dislocation cores has also been considered to explain the intricate cells and bands observed by high spatial resolution infrared imaging [421]. Presumably the second formation mechanism is superimposed on a uniform background determined by the stoichiometry [418].

Correlations between FET threshold voltages, EPD distribution, and EL2 distribution have recently been reviewed by Miyazawa et al., [422]. As illustrated schematically in Fig. 7.85 many different materials properties have been found to exhibit either "W" or "M" shaped patterns when this variation across a wafer diameter is mapped [422]. Miyazawa et al., have arrived at a defect model which takes into consideration all these observations. In this model As-rich fine precipitates or aggregates at dislocation lines and an As depleted environment within roughly 50 μm of a dislocation is proposed. Support for this was found by making sheet carrier concentration, $N_s$, measurements using the van der Pauw method (Hall effect) on $40 \times 40\,\mu m^2$ chips made at 400 μm intervals [423]. The measured $N_s$ values are plotted versus the distance between the nearest etch pit and the center of the chip in Fig. 7.86. A proximity effect reminiscent of that found for FET threshold voltage, $V_{th}$, is clearly seen for both S and Si ion implants. (Such a proximity effect is expected since $N_s$ is inversely proportional to $V_{th}$.) Since Si is amphoteric, the measured $N_s$ for the Si case corresponds to the difference between the number of Si atoms on Ga sites and the number on As sites. One is led to the conclusion that either there are more vacant Ga sites available for occupation by a Si atom near a dislocation or there are fewer vacant As sites. The results shown in Fig. 7.86 for S corroborate this conclusion. Since S donors reside on As sites a decreased number of empty As sites near a dislocation is implied.

In summary, Miyazawa et al., [422] propose the model illustrated schematically in Fig. 7.87. In this model dislocation cores are decorated with As and are surrounded by a ~ 50 μm wide zone depleted of As vacancies and rich in As interstitials. This model is consistent with the experimental observations that (a) semi-insulating undoped GaAs is As rich, (b) As-rich precipitates or aggregates exist on dislocation lines, (c) the concentration of EL2 is higher around dislocations, (d) the concentration of EL2 increases with excess As fraction in crystals.

As was discussed earlier, a leading candidate for the EL2 defect is an $As_{Ga}$ antisite defect associated with another intrinsic point defect. The model shown in Fig. 7.87 is consistent with higher $As_{Ga}$ concentrations since the following two reactions are expected:

$$As_i + V_{As} \rightleftarrows As_{As}$$

$$As_i + V_{Ga} \rightleftarrows As_{Ga}$$

As a result, these reactions lead to a decrease in the concentration of As vacancies (as in Fig. 7.87) and to an increase in the concentration of $As_{Ga}$ antisites (EL2) around dislocations.

**Figure 7.87** Schematic illustration of the defect model around dislocations in semi-insulating GaAs [422].

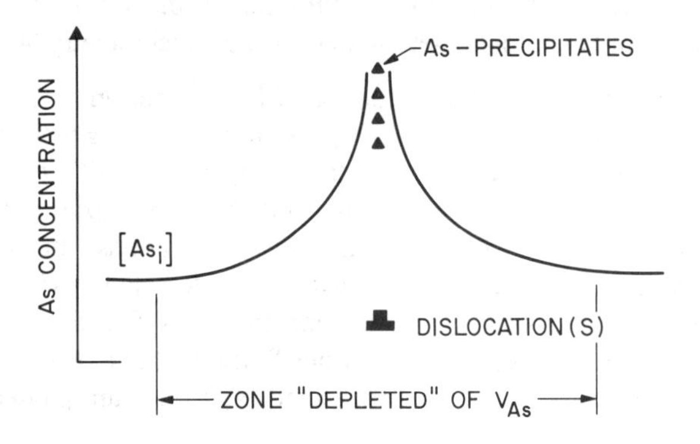

## REFERENCES

1. W. Schottky, *Z. Phys. 113*;367, 1939.

2. J. Bardeen, *Phys. Rev. 71*;717, 1947.

3. A.M. Cowley and S.M. Sze, *J. Appl. Phys. 36*;3212, 1965.

4. E.H. Rhoderick, *Metal-Semiconductor Contacts* Oxford;Clarendon Press, 1978.

5. V. Heine, *Phys. Rev. A138*;1689, 1965.

6. R.H. Williams, in *Physics and Chemistry of III-V Compound Semiconductor Interfaces*, (ed. C.W. Wilmsen) New York;Plenum, Chap. 1, 1985.

7. S.Y. Tong, A.R Lubinsky, B.J. Mrstik, and M.A. Van Hove, *Phys. Rev. B17*;3303, 1978.

8. R.J. Mayer, C.B. Duke, A. Paton, A. Kahn, E. So, J.L. Yeh, and P. Mark, *Phys. Rev. B19*;5194, 1979.

9. D.J. Chadi, *Phys. Rev. B19*;2074, 1979.

10. J.R. Chelikowsky and M.L. Cohen, *Phys. Rev. B20*;4150, 1979.

11. C.M. Bertoni, O. Bisi, C. Calandra, and F. Mangi, *Inst. Phys. Conf. Ser. 43*;191, 1978.

12. P. Chiaradia, G. Chiarotti, F. Ciccacci, R. Momeo, S. Nannarone, P. Sasaroli, and S. Selci, *Surf. Sci. 99*;70, 1980.

13. W.E. Spicer, I. Lindau, P. Skeath, and C.Y. Su, *J. Vac. Sci. Technol. 17*;1019, 1980.

14. A. Huijser and J. Van Laar, *Surf. Sci. 52*;202, 1975.

15. A. Huijser, J. Van Laar, and T.L. Van Rooy, *Surf. Sci. 62*;472, 1977.

16. R.A. Street, R.H. Williams, and R.S. Bauer, *J. Vac. Sci. Technol. 17*;1001, 1980.

17. R.A. Street and R.H. Williams, *J. Appl. Phys. 52*;402, 1981.

18. G.Y. Robinson, in *Physics and Chemistry of III-V Compound Semiconductor Interfaces*, (ed. C.W. Wilmsen) New York;Plenum, Chap. 2, 1985.

19. A.M. Cowley, *J. Appl. Phys. 37*;3024, 1966.

20. B.R. Pruniaux and A.C. Adams, *J. Appl. Phys. 43*;1980, 1972.

21. C.L. Anderson, C.R. Crowell, and T.W. Kao, *Solid State Electron. 18*;705, 1975.

22. R.H. Williams, G.P. Srivastava, and I.T. McGovern, *Rept. Progr. Phys. 43*;1357, 1980.

23. W.E. Spicer and S.J. Eglash, in *VLSI Electronics-Microstructure Science*, vol. 10 Surface and Interface Effects in VLSI (eds. N.G. Einspruch and R.S. Bauer) New York;Academic, p. 79, 1985.

24. W. Monch and H. Gant, *Phys. Rev. Lett. 48*;512, 1982.

25. R.H. Williams, *J. Vac. Sci. Technol. 18*;929, 1981.

26. E. Hokelek and G.Y. Robinson, *Appl. Phys. Lett. 40*;426, 1982.

27. V. Montgomery and R.H. Williams, *J. Phys. C;15*;5887, 1982.

28. S.G. Kurtin, T.C. McGill, and C.A. Mead, *Phys. Rev. Lett. 22*;1433, 1969.

29. M. Schluter, *Phys. Rev. B;17*;5044, 1978.

30. C.A. Mead and W.G. Spitzer, *Phys. Rev. 34*;A713, 1964.

31. J.O. McCaldin, T.C. McGill, and C.A. Mead, *Phys. Rev. Lett. 36*;56, 1976.

32. S.G. Louie, J.R. Chelikowsky, and M.L. Cohen, *Phys. Rev. B15*;2154, 1977.

33. M.S. Daw and D.L. Smith, *Appl. Phys. Lett. 36*;690, 1980.

34. M.S. Daw and D.L. Smith, *Solid State Comm. 37*;205, 1981.

35. R.E. Allen and J.D. Dow, *Phys. Rev. B* 25;1423, 1982.

36. L.J. Brillson, *Phys. Rev. Lett. 40*;260, 1978.

37. L.J. Brillson, C.F. Brucker, N.G. Stoffel, A.D. Katnani, and G. Margaritondo, *Phys. Rev. Lett. 46*;838, 1981.

38. P. Skeath, C.Y. Su, I. Lindau, and W.E. Spicer, *J. Vac. Sci. Technol. 17*;874, 1980.

39. J.L. Freeouf and J.M. Woodall, *Appl. Phys. Lett. 39*;727, 1981.

40. J.M. Woodall, N. Braslau, and J.L. Freeouf, in *Phys. of Thin Films* (eds. M.H. Francombe and J.L. Vossen) New York;Academic, vol. 13, p. 199, 1981.

41. R.H. Williams, V. Montgomery, and R.R. Varma, *J. Phys. C11*;L735, 1978.

42. S.M. Sze, *Physics of Semiconductor Devices,* New York;Wiley, Chap. 8, 1969.

43. F.A. Padovani and R. Stratton, *Solid State Electron 9*;695, 1966.

44. S. Singh, unpublished.

45. Y.I. Nissim, J.F. Gibbons, and R.B. Gold, *IEEE Electron Devices ED28*;607, 1981.

46. T.F. Deutsch, D.J. Ehrlich, R.M. Osgood, Jr., and Z.L. Liau, *Appl. Phys. Lett. 36*;847, 1980.

47. P.A. Barnes, H.J. Leamy, J.M. Poate, and S.D. Ferris, *Appl. Phys. Lett. 33*;965, 1978.

48. P.A. Pianetta, C.A. Stolte, and J.L. Hansen, *Appl. Phys. Lett. 36*;597, 1980.

49. R.L. Mozzi, W. Fabian, and F.J. Pierkarski, *Appl. Phys. Lett. 35*;337, 1979.

50. Z.L. Liau, N.L. DeMeo, J.P. Donnelly, D.E. Mull, R. Bradbury, and J.P. Lorenzo, Paper presented at Materials Research Society Meeting, Cambridge, MA, Nov. 1979.

51. P.A. Barnes and A.Y. Cho, *Appl. Phys. Lett. 33*;651, 1978.

52. W.T. Tsang, *Appl. Phys. Lett. 33*;1022, 1978.

53. W.T. Anderson, A. Christou, and J.E. Davey, *IEEE J. Solid State Circuits SC-13*;430, 1978.

54. K. Heime, U. Konig, E. Kohn, and A. Wortmann, *Solid State Electron 17*;835, 1974.

55. C.P. Lee, J.L. Tandom, and P.J. Stocker, *Electron. Lett. 16*;849, 1980.

56. S. Margalit, D. Fekete, D.M. Pepper, C.P. Lee, and A. Yariv, *Appl. Phys. Lett. 33*;346, 1978.

57. W.T. Anderson, A. Christou, and J.F. Giuliani, *IEEE Electron Dev. Lett. EDL-2*;115, 1981.

58. N. Yokoyama, S. Ohkawa, and H. Ishikawa, *Jap. J. Appl. Phys. 14*;1071, 1975.

59. R.P. Gupta and J. Freyer, *Int. J. Electron. 47*;459, 1979.

60. A.K. Sinha, T.E. Smith, and H.J. Levinstein, *IEEE Trans. Electron. Devices ED-22*;218, 1975.

61. T. Saanda and O. Wada, *Jap. J. Appl. Phys. 19*;L491, 1980.

62. O. Ishihara, K. Nishitani, H. Sawano, and S. Mitsui, *Jap. J. Appl. Phys. 15*;1411, 1976.

63. N. Braslau, J.B. Gunn, and J.L. Staples, *Solid State Electron. 10*;381, 1967.

64. N. Braslau, *J. Vac. Sci. Technol. 19*;803, 1981.

65. N. Braslau, *Thin Solid Films 104*;391, 1983.

66. K.K. Shih and J.M. Blum, *Solid State Electron. 15*;1177, 1972.

67. A.J. Valois and G.Y. Robinson, *Solid State Electron. 25*;973, 1982.

68. E. Kuphal, *Solid State Electron. 24*;69, 1981.

69. W. Tseng, A. Christou, H. Day, J. Davey, and B. Wilkins, *J. Vac. Sci. Technol. 19*;623, 1981.

70. G. Weimann and W. Schlapp, *Phys. Stat. Solidi A50*;K219, 1978.

71. H.T. Mills and H.L. Hartnagel, *Int. J. Electron. 46*;65, 1979.

72. P.A. Barnes and R.S. Williams, *Solid State Electron. 24*;907, 1981.

73. Y. Nakano, S. Takahashi, and Y. Toyoshima, *Jap. J. Appl. Phys. 19*;L495, 1980.

74. H. Morkoc, T.J. Drummond, and C.M. Stanchek, *IEEE Trans. Electron. Devices ED-28*;1, 1981.

75. H.H. Wieder, A.R. Clawson, D.I. Elder, and D.A. Collins, *IEEE Electron Dev. Lett. EDL-2*;73, 1981.

76. V. Swaminathan, J. Lopata, and J.W. Lee, *Mat. Res. Soc. Symp. Proc. Vol. 77*;p. 779, 1987.

77. R.A. Stall, C.E.C. Wood, and L.F. Eastman, *Electron. Lett. 15*;800, 1979.

78. R.A. Stall, C.E.C. Wood, K. Board, N. Dandekar, L.F. Eastman, and J. Devlin, *J. Appl. Phys. 52*;4062, 1981.

79. J.M. Woodall, J.L. Freeouf, G.D. Pettit, T. Jackson, and P. Kirchner, *J. Vac. Sci. Technol. 19*;626, 1981.

80. K. Kajiyama, Y. Mizushima, and S. Sakata, *Appl. Phys. Lett. 23*;458, 1973.

81. T. Sawada and H. Hasegawa, *Thin Solid Films 56*;183, 1979.

82. M. Okamura and T. Kobayashi, *Jap. J. Appl. Phys. 19*;2143, 1980.

83. R.P.H. Chang, T.T. Sheng, C.C. Chang, and J.J. Coleman, *Appl. Phys. Lett. 33*;341, 1978.

84. D. L. Lile, in *Physics and Chemistry of III-V Compound Semiconductor Interfaces,* (ed. C.W. Wilmsen) New York;Plenum, Chap. 6, 1985

85. D.L. Lile, *Solid State Electron. 21*;1199, 1978.

86. Y. Shinoda and T. Kobayashi, *J. Appl. Phys. 52*;6386, 1981.

87. D.A. Bagles, D.H. Laughlin, B.T. Moore, B.L. Eastep, D.K. Ferry, and C.W. Wilmsen, *Inst. Phys. Conf. Ser. 56*;259, 1980.

88. M.J. Taylor, D.L. Lile, and A.K. Nedoluha, *J. Vac. Sci. Technol. B2*;522, 1984.

89. K.P. Pande and D. Gutierrez, *Appl. Phys. Lett. 46*;416, 1985.

90. P.N. Favennec, M. LeContellec, H. L'Harridon, G.P. Pelous, and J. Richard, *Appl. Phys. Lett. 34*;807, 1979.

91. J. Stannard and R.L. Henry, *Appl. Phys. Lett. 35*;86, 1979.

92. D. Bimberg, in *Solid State Devices* (eds. P. Balk and O.G. Folberth) The Netherlands;Elsvier, p. 101, 1986.

93. P. Dawson and K. Woodbridge, *Appl. Phys. Lett. 45*;1227, 1984.

94. B. Sermage, F. Alexandre, J. L. Lievin, R. Azoulay, M.El. Kaim, H. LePherson, and J.A. Marzin, *Inst. Phys. Conf. Ser. 74*;345, 1985.

95. J.F. Wager and C.W. Wilmsen, in *Physics and Chemistry of III-V Compound Semiconductor Interfaces,* ed. C.W. Wilmsen New York;Plenum, Chap. 3, 1985.

96. M.G. Adlerstein, *Electron. Lett. 12*;297, 1976.

97. H.H. Wieder, in *VLSI electronics microstructure science* (ed. N.G. Einspurch and R.S. Bauer) vol. 10 "Surface and Interface Effects in VLSI" New York;Academic, Chap. 5, 1985.

98. C.P. Lee, S.J. Lee, and B.M. Welch, *IEEE Electron. Dev. Lett. EDL-3*;97, 1982.

99. M.A. Lampert and P. Mark, ''Current Injection in Solids,'' New York;Academic, 1970.

100. C. Kocot and C.A. Stolte, *IEEE Trans. Electron. Devices ED-29*;1059, 1982.

101. L.G. Meiners, *Thin Solid Films 56*;201, 1979.

102. B.R. Pruniaux, J.C. North, and A.V. Payer, *IEEE Trans. Electron Devices ED-19,*;672, 1972.

103. H.M. Macksey, D.W. Shaw, and W.R. Wisserman, *Electron. Lett. 12*;192, 1976.

104. L.G. Meiners, *J. Appl. Phys. 50*;1154, 1979.

105. H.C. Casey,Jr., A.Y. Cho, and E.H. Nicollian, *Appl. Phys. Lett. 32*;679, 1978.

106. H.C. Casey,Jr., A.Y. Cho, D.V. Lang, E.H. Nicollian, and P.W. Foy, *J. Appl. Phys. 50*;3484, 1979.

107. J.P. Andre, C. Schiller, A. Mitonneau, A. Briere, and J.Y. Supied, *Inst. Phys. Conf. Ser. 65*;117, 1983.

108. L.G. Meiners, *J. Vac. Sci. Technol. 19*;373, 1981.

109. W.E. Spicer, P.W. Chye, P.R. Skeath, C.Y. Su, and I. Lindau, *J. Vac. Sci. Technol. 16*;1422, 1979.

110. J.R. Waldrop, S.P. Kowalizyk, and R.W. Grant, *Appl. Phys. Lett. 42*;454, 1983.

111. D.L. Lile and D.A. Collins, *Thin Solid Films, 56*;225, 1979.

112. M. Okamura and T. Kobayashi, *Jap. J. Appl. Phys. 19*;2143, 1980.

113. M. Okamura and T. Kobayashi, *Japan J. Appl. Phys. 19*;2151, 1980.

114. T. Sawada and H. Hasegawa, *Thin Solid Films 56*;183, 1979.

115. H.H. Wieder, *Appl. Phys. Lett. 38*;170, 1981.

116. R.F. Leheny, R.E. Nahory, M.A. Pollack, A.A. Ballman, E.D. Beebe, J.C. DeWinter, and R.J. Martin, *IEEE Trans. Electron. Dev. Lett. EDL-1*;110, 1980.

117. Y. Shinoda and T. Kobayashi, *J. Appl. Phys. 52*;6386, 1981.

118. Y. Shinoda and T. Kobayashi, Proc. of Ninth International Symposium on GaAs and Related Compounds, Japan, 1981.

119. P.G. Shewmon, *Diffusion in Solids,* New York;McGraw Hill, 1963.

120. B. Tuck, in *Semi-insulating III-V Materials,* ed. D.C. Look and J.S. Blakemore, Shiva Publishing Ltd, England;Nantwich, p. 2, 1984.

121. J.V. Dilorenzo, A.S. Jordan, A.R. VonNeida, and P. O'Connor, in *Semi-insulating III-V Materials,* eds. D.C. Look and J.S. Blakemore Shiva Publishing Ltd., England;Nantwich, p. 308, 1984.

122. C.T. Foxon, J.A. Harvey, and B.A. Joyce, *J. Phys. Chem. Solids 34*;1693, 1973.

123. J.D. Oberstar and B.G. Streetman, *Thin Solid Films, 103*;17, 1983.

124. J. Kasahara, M. Arai, and N. Watanabe, *J. Appl. Phys. 50*;541, 1979.

125. P.M. Campbell, O. Aina, and B.J. Baliga, *Appl. Phys. Lett. 45*;95, 1984.

126. P.K. Vasudei, R.G. Wilson, C.A. Evans, and V.R. Deline, *Solid State Electron 26*;565, 1983.

127. M. Oshima, K. Watanabe, and S. Miyazawa, *J. Electrochem. Soc. 131*;130, 1984.

128. J.B. Clegg, G.B. Scott, J. Hallais, and A. Mircea-Roussel, *J. Appl. Phys. 52*;1110, 1981.

129. P.T. Greiling and C.F. Krumm, in *VLSI Electronics,* vol. 11, "GaAs Microelectronics," (eds. N.G. Einspruch and W.R. Wisseman), New York;Academic, p. 133, 1985.

130. B. Tuck and G.A. Adegboyega, *J. Phys. D: Appl. Phys. 12*;1895, 1979.

131. P.W. Yu, *J. Appl. Phys. 52*;5786, 1981.

132. D.C. Look, P.W. Yu, J.E. Ehret, Y.K. Yeo, and R. Kwor, in *Semi-insulating III-V Materials,* (eds. S. Makram-Ebeid and B. Tuck) Nantwich, England;Shiva Publishing Ltd, p. 372, 1982.

133. P.B. Klein, P.E.R. Nordquist, and P.G. Siebenmann, *J. Appl. Phys. 51*;4861, 1980.

134. M.S. Seltzer, *J. Phys. Chem. Solids, 26*;243, 1915.

135. A. Mircea-Roussel, G. Jacob, and J.P. Hallais, in *Semi-insulating III-V Materials,* (ed. G.J. Rees Nantwich, England;Shiva Publishing, p. 133, 1980.

136. A.S. Jordan and G.A. Nikolakopoulou, *J. Appl. Phys. 55*;4194, 1984.

137. D. Shaw and A.L.J. Wells, *Brit. J. Appl. Phys. 17*;999, 1966.

138. P.J. Anthony, *Solid State Electronics 25*;1003, 1982.

139. S. Makram-Ebeid, P. Langlade, and G.M. Martin, in *Semi-insulating III-V Materials,* (eds. D.C. Look and J.S. Blakemore), Nantwich, England;Shiva Publishing Ltd, p. 181, 1984.

140. W.M. Duncan and G.H. Westphal, in *VLSI Electronics – GaAs Microelectronics,* (eds. N.G. Einspruch and W.R. Wisseman), New York;Academic, vol. 11, p. 41, 1985.

141. B. Hughes and C. Li, *Inst. Phys. Conf. Ser. 65*;57, 1982.

142. S. Makram-Ebeid, D. Gautard, P. Devillard, and G.M. Martin, *Appl. Phys. Lett. 40*;161, 1982.

143. M. Matsui and T. Kazuno, *Appl. Phys. Lett. 51*;658, 1987.

144. T. Obokata, H. Okada, T. Katsumata, and T. Fukuda, *Jap. J. Appl. Phys. 25*;L179, 1986.

145. H.D. Palfrey, M. Brown, and A.F.W. Willoughby, *J. Electrochem. Soc. 128*;2224, 1981.

146. J. Lagowski, M. Kaminska, J.M. Parsey, Jr., H.C. Gatos, and M. Lichtensteiger, *Appl. Phys. Lett. 41*;1078, 1982.

147. S.J. Pearton and A.J. Tavendale, *Electron. Lett. 18*;715, 1982.

148. W.C. Dautremont-Smith, *Mat. Res. Soc. Symp. Proc. 104*;313, 1988.

149. G.M. Martin, S. Makram-Ebeid, N.T. Phuoc, M. Berth, and C. Venger, in *Semi-insulating III-V Materials*,, (eds. S. Makram-Ebeid and B. Tuck), Nantwich, England;Shiva Publishing Ltd, p. 275, 1982.

150. F. Hasegawa and T. Saito, *Jap. J. Appl. Phys. 7*;1125, 1540, 1968.

151. J.S. Harris, Y. Nannichi, G.L. Pearson, and G.F. Day, *J. Appl. Phys. 40*;4575, 1969.

152. F. Hasegawa, *J. Electrochem. Soc. 119*;930, 1972.

153. E. Munoz, W.L. Snyder, and J.L. Moll, *Appl. Phys. Lett. 16*;262, 1970.

154. T.G. Blocker, R.H. Cox, and T.E. Hasty, *Solid State Commun. 8*;1317, 1970.

155. H.R. Potts and G.L. Pearson, *J. Appl. Phys. 37*;2098, 1966.

156. J. Nishizawa, H. Otsuka, S. Yamakoshi, and K. Ishida, *Jap. J. Appl. Phys. 13*;46, 1974.

157. W.Y. Lum and H.H. Wieder, *J. Appl. Phys. 49*;6187, 1978.

158. E.V.K. Rao and N. Duhamel, *J. Appl. Phys. 49*;3457, 1978.

159. T. Itoh and M. Takeuchi, *Jap. J. Appl. Phys. 16*;227, 1977.

160. H. Birey and J. Sites, *J. Appl. Phys. 51*;619, 1980.

161. S. Nojima, *J. Appl. Phys. 53*;7602, 1982; see also other references therein.

162. H.J. Guislain, L. DeWolf, and P. Clauus, *J. Electron. Mater. 7*;83, 1978.

163. Z.G. Wang, H.P. Gislason, and B. Monemar, *J. Appl. Phys. 58*;230, 1985.

164. S.Y. Chiang and G.L. Pearson, *J. Appl. Phys. 46*;2986, 1975.

165. H.D. Palfrey, M. Brown, and A.F.W. Willoughby, *J. Electron. Mat. 12*;863, 1983.

166. F.A. Kroger, *The Chemistry of Imperfect Crystals* Amsterdam;North Holland, vol. III, 1973.

167. J. Lagowski, H.C. Gatos, C.H. Kang, M. Skowronski, K.Y. Ko, and D.G. Lin, *Appl. Phys. Lett. 49*;892, 1986.

168. C.H. Kang, J. Lagowski, and H.C. Gatos, *J. Appl. Phys. 62*;3482, 1987.

169. S. Adachi and S. Yamahata, *Appl. Phys. Lett. 51*;1266, 1987.

170. E.J. Foulkes, PhD Thesis, Nottingham University, (1983) loc. cit Ref. 118.

171. M.R. Brozel, E.J. Foulkes, and B. Tuck, *Phys. Stat. Solidi 72a*;K159, 1982.

172. D.E. Holmes, R.G. Wilson, and P.W. Yu, *J. Appl. Phys. 52*;3396, 1981.

173. J.D. Oberstar, B.G. Streetman, J.E. Baker, and P. Williams, *J. Electrochem. Soc. 128*;1814, 1981.

174. L. Eaves, A.W. Smith, M.S. Skolnick, and B. Cockayne, *J. Appl. Phys. 53*;4955, 1982.

175. B.V. Shanabrook, P.B. Klein, P.G. Siebenmann, H.B. Dietrich, R.L. Henry, and S.G. Bishop, in *Semi-insulating III-V Materials,* (eds. S. Makram-Ebeid and B. Tuck), Nantwich, England;Shiva Publishing Ltd, p. 310, 1982.

176. J.P. Donnelly and C.E. Hurwitz, *Appl. Phys. Lett. 31*;418, 1977.

177. D.E. Davies, J.P. Lorenzo, and T.G. Ryan, *Solid State Electron. 21*;981, 1978.

178. J. Chevrier, M. Armand, A.M. Huber, and N.T. Linh, *J. Electron. Mat. 9*;745, 1980.

179. A.S. Brown, S.C. Palmateer, G.W. Wicks, L.F. Eastman, A.R. Calawa, and C. Hitzman, in *Semi-insulating III-V Materials,* (eds. D.C. Look and J.S. Blakemore), Nantwich, England;Shiva Publishing Ltd, p. 36.

180. S.L. Palmateer, W.J. Schaff, A. Galuska, J.D. Berry, and L.F. Eastman, *Appl. Phys. Lett. 42*;183, 1983.

181. S.C. Palmateer, P.A. Maki, M.A. Hollis, L.F. Eastman, C. Hitzman, and I. Ward, *Inst. Phys. Conf. Ser. 65*;149, 1982.

182. H. Morkoc, T. Andrews, and S.B. Hyder, *Electron. Lett. 14*;715, 1978.

183. L. Jastrzebski, J. Lagowski, and H. C. Gatos, *J. Electrochem. Soc. 126*;2231, 1979.

184. P.Trung Dung and M. Laznicka, *Phys. Stat. Sol. 92a*;K113, 1985.

185. A.K. Chin, I. Camliber, B.V. Dutt, V. Swaminathan, W.A. Bonner, and A.A. Ballman, *Appl. Phys. Lett. 42*;901, 1983.

186. H.C. Casey, in *Atomic diffusion in semiconductors,* (ed. D. Shaw) New York;Plenum, Chap. 6, 1973.

187. A.K. Chin, I. Camlibel, T.T. Sheng, and W.A. Bonner, *Appl. Phys. Lett. 43*;495, 1983.

188. B.V. Dutt, A.K. Chin, I. Camlibel, and W.A. Bonner, *J. Appl. Phys. 56*;1630, 1984.

189. K. Tsubaki and K. Sugiyama, *Jap. J. Appl. Phys. 19*;1789, 1980.

190. C.C. Daniel Wong and R.H. Bube, *J. Appl. Phys. 55*;3804, 1984.

191. N. Chand and P.A. Houston, *J. Electron. Mat. 11*;37, 1988.

192. G.J. Van Gurp, P.R. Boudewign, M.N.C. Kempenars, and D.L.A. Tjaden, *J. Appl. Phys. 61*;1846, 1987.

193. H.B. Serreze and H.S. Marek, *Appl. Phys. Lett. 51*;2031, 1987.

194. C.A. Evans, C.G. Hopkins, J.C. Norberg, V.R. Deline, R.J. Blattner, R.G. Wilson, D.M. Jamba, and Y.S. Park, in *Semi-insulating III-V Materials,* (eds. G.J. Rees) Nantwich, England;Shiva Publishing Ltd, p. 138, 1980.

195. P.K. Vasudev, R.G. Wilson, and C.A. Evans, *Appl. Phys. Lett. 37*;308, 1980.

196. C.A. Evans, V.R. Deline, T.W. Sigmon, and A. Lidow, *Appl. Phys. Lett. 35*;291, 1979.

197. D.K. Sadana, J. Washburn, and T. Zee, *Inst. Phys. Conf. Ser. 63*;359, 1983.

198. F. Simondet, C. Venger, and G.M. Martin, *Appl. Phys. 23*;21, 1980.

199. M. Gauneau, H. L'Haridon, A. Rupert, and M. Salvi, *J. Appl. Phys. 53*;6823, 1982.

200. H. Kanber, M. Feng, and J.M. Whelan, in *Advanced Semiconductor Processing and Characterization of Electronic and Optical Materials*, SPIE vol. 463, p. 69, 1984.

201. J. Kasahara and N. Watanabe in *Semi-insulating III-V Materials,* (eds. S. Makram-Ebeid and B. Tuck), Nantwich, England;Shiva Publishing Ltd., p. 238, 1982.

202. T.J. Magee, H. Kawayoshi, R.D. Ormond, L.A. Christel, J.F. Gibbons, C.G. Hopkins, C.A. Evans, Jr., and D.S. Day, *Appl. Phys. Lett. 39*;906, 1981.

203. L.A. Christel and J.F. Gibbons, *J. Appl. Phys. 52*;5050, 1981.

204. A. Lidow, J.F. Gibbons, T. Magee, and J. Peng, *J. Appl. Phys. 49*;5213, 1978.

205. R.T. Blunt, M.S.M. Lamb, and R. Szweda, *Appl. Phys. Lett. 47*;304, 1985.

206. G. Bentini, L. Correra, and C. Donolato, *J. Appl. Phys. 56*;2922, 1984.

207. H. Kanber and J.M. Whelan, *J. Electrochem. Soc. 134*;2596, 1987.

208. M. Geva and T.E. Seidel, *J. Appl. Phys. 59*;2408, 1986.

209. J.P. Hirth and J. Lothe, *Theory of Dislocations,* 2nd ed. New York;Wiley, Chap. 14, p. 497, p. 506, 1982.

210. K. Kitahara, K. Nagai, and S. Shibatomi, *J. Electrochem. Soc. 129*;880, 1982.

211. K. Kitahara, M. Ozeki, and A. Shibatomi, *Appl. Phys. Lett. 42*;188, 1983.

212. H.J. Hovel, *IEEE Trans. Electron. Dev. ED-32*;233, 1985.

213. M. Yokogawa, S. Nishine, M. Sasaki, K. Matsumoto, K. Fujita, and S. Akai, *Jap. J. Appl. Phys. 23*;L339, 1984.

214. A.K. Chin, A.R. VonNeida, and R. Caruso, *J. Electrochem. Soc. 129*;2386, 1982.

215. S. Miyazawa, Y. Ishii, S. Ishida, and Y. Nanishi, *Appl. Phys. Lett. 43*;853, 1983.

216. B. Wakefield, P.A. Leigh, M.H. Lyons, and C.R. Elliot, *Appl. Phys. Lett. 45*, 66 1984.

217. T. Kamejima, F. Shimura, Y. Matsumoto, H. Watanabe, and J. Mitsui, *J. Appl. Phys. 21*;L721, 1982.

218. A.T. Hunter, *Appl. Phys. Lett. 47*;715, 1985.

219. P.R. Dastidar, K. Ravindran, and H.C. Pant, *Electron. Lett. 5*;553, 1969.

220. C.O. Bozler, J.P. Donnelly, W.T. Lindley, and R.A. Reynolds, *Appl. Phys. Lett. 29*;698, 1976.

221. T.J. Magee, J. Peng, J.D. Hong, C.A. Evans, and V.R. Deline, *Phys. Stat. Sol. A 55*;169, 1979.

222. T.J. Magee, J. Peng, J.D. Hong, W. Katz, and C.A. Evans, *Phys. Stat. Sol. A55*;161, 1979.

223. T.J. Magee, J. Hung, V.R. Deline, and C.A. Evans, *Appl. Phys. Lett. 37*;53, 1980.

224. T.J. Magee, J. Peng, J.D. Hong, C.A. Evans, V.R. Deline, and R.M. Malbon, *Appl. Phys. Lett. 35*;277, 1979.

225. F.C. Wang and M. Bujatti, *IEEE Trans. Electron Dev. ED-32*;2839, 1985.

226. A.K. Chin, I. Camlibel, R. Caruso, M.S.S. Young, and A.R. VonNeida, *J. Appl. Phys. 57*;2203, 1985.

227. S. Miyazawa, T. Honda, Y. Ishii, and S. Ishida, *Appl. Phys. Lett. 44*;410, 1984.

228. K. Kitahara, M. Ozeki, and A. Shibatomi, *Jap. J. Appl. Phys. 23*;207, 1984.

229. R.D. Gold and L.R. Weisberg, *Solid State Electron. 7*;811, 1964.

230. L.C. Kimerling, *Solid State Electron. 21*;1391, 1978.

231. D.V. Lang and L.C. Kimerling, *Phys. Rev. Lett. 33*;489, 1974.

232. L.C. Kimerling and D.V. Lang, *Lattice Defects in Semiconductors, Inst. Phys. Conf. Ser. 23*;p. 589, 1975.

233. M. Ettenberg, *J. Appl. Phys. 45*;901, 1974.

234. K. Bohm and B. Fischer, *J. Appl. Phys. 50*;5453, 1979.

235. P.M. Petroff, R.A. Logan, and A. Savage, *Phys. Rev. Lett. 44*;289, 1980.

236. W. Heinke, *Inst. Phys. Conf. Ser. 23*;380, 1975; see also W. Heinke and H.J. Queisser, *Phys. Rev. Lett. 33*;1082, 1974.

237. R.J. Roedel, A.R. Von Neida, R. Caruso, and L.R. Dawson, *J. Electrochem. Soc. 126*;637, 1979.

238. B.C. DeLoach, B.W. Hakki, R.L. Hartman, and L.A. D'Asaro, *Proc. IEEE 61*;1042, 1973.

239. P.M. Petroff and R.L. Hartman, *Appl. Phys. Lett. 23*;469, 1973.

240. H. Yonezu, T. Kamejima, M. Ueno, and I. Sakuma, *Jap. J. Appl. Phys. 13*;1679, 1974.

241. R. Ito, H. Nakajima, and O. Nakada, *Jap. J. Appl. Phys. 13*;1321, 1974.

242. R. Ito, H. Nakashima, S. Kishino, and O. Nakada, *IEEE J. Quant. Electron. QE-11*;551, 1975.

243. K. Ishida and T. Kamejima, *J. Electron. Mat. 8*;57, 1979.

244. P.J. Anthony, R.L. Hartman, N.E. Schumaker, and W.R. Wagner, *J. Appl. Phys. 53*;756, 1982.

245. L.C. Kimerling, P. Petroff, and H.J. Leamy, *Appl. Phys. Lett. 28*;297, 1976.

246. W.D. Johnston, Jr. and B.I. Miller, *Appl. Phys. Lett. 23*;192, 1973.

247. P.M. Petroff, W.D. Johnston, Jr., and R.L. Hartman, *Appl. Phys. Lett. 25*;226, 1974.

248. P.M. Petroff, L.C. Kimerling, and W.D. Johnston, Jr., *Inst. Phys. Conf. Ser. 31*;362, 1977.

249. P.M. Petroff and R.L. Hartman, *J. Appl. Phys. 45*;3899, 1974.

250. B.A. Monemar and G.R. Woolhouse, *Inst. Phys. Conf. Ser. 33a*;400, 1977.

251. P.W. Hutchinson and P.S. Dobson, *Phil. Mag. 32*;745, 1975.

252. T. Kamejima, K. Ishida, and J. Matshi, *Jap. J. Appl. Phys. 16*;233, 1977.

253. P.W. Hutchinson, P.S. Dobson, S. O'Hara, and D.H. Newman, *Appl. Phys. Lett. 26*;250, 1975.

254. P.M. Petroff, O.G. Lorimor, and J.M. Ralston, *J. Appl. Phys. 47*;1583, 1976.

255. P.S. Dobson, P.W. Hutchinson, S. O'Hara, and D.H. Newman, *Inst. Phys. Conf. Ser. 33a*;419, 1977.

256. P.M. Petroff, *J. de Physique,* 40, C6 201, 1979.

257. P.M. Petroff and L.C. Kimerling, *Appl. Phys. Lett. 29*;461, 1976.

258. S. O'Hara, P.W. Hutchinson, and P.S. Dobson, *Appl. Phys. Lett. 30*;368, 1977.

259. P.W. Hutchinson, P.S. Dobson, B. Wakefield, and S. O'Hara, *Solid State Electron. 21*;1413, 1978.

260. D.V. Lang, P.M. Petroff, R.A. Logan, and W.D. Johnston, Jr., *Phys. Rev. Lett. 42*;1353, 1979.

261. P.W. Hutchinson and P.S. Dobson, *Phil. Mag. A41*;601, 1980.

262. H. Imai, T. Fujiwara, K. Segi, M. Takusagawa, and H. Takanashi, *Jap. J. Appl. Phys. 18*;589, 1979.

263. K. Ishida, T. Kamejima, and J. Matsui, *Appl. Phys. Lett. 31*;397, 1977.

264. S. Kishino, N. Chinone, H. Nakashima, and R. Ito, *Appl. Phys. Lett. 29*;488, 1976.

265. H. Nakashima, S. Kishino, N. Chinone, and R. Ito, *J. Appl. Phys. 48*;2771, 1977.

266. A.K. Chin, V.G. Keramidas, W.D. Johnston, Jr., S. Mahajan, and D.D. Roccasecca, *J. Appl. Phys. 51*;978, 1980.

267. K. Maeda and S. Takeuchi, *Jap. J. Appl. Phys. 20*;L165, 1981.

268. K. Maeda, M. Sato, A. Kubo, and S. Takeuchi, *J. Appl. Phys. 54*;161, 1983.

269. K. Maeda and S. Takeuchi, in *Dislocation in Solids,* Univ. Tokyo, Tokyo, p. 433, 1985.

270. K. Maeda and S. Takeuchi, *Appl. Phys. Lett. 42*;664, 1983.

271. S. Yamakoshi, O. Hasegawa, H. Hamaguchi, M. Abe, and T. Yamaoka, *Appl. Phys. Lett. 31*;627, 1977.

272. O. Ueda, S. Isozumi, T. Kotani, and T. Yamaoka, *J. Appl. Phys. 48*;3950, 1977.

273. O. Ueda, S. Isozumi, S. Yamakoshi, and T. Kotani, *J. Appl. Phys. 50*;765, 1979.

274. A.K. Chin, V.G. Keramidas, W.D. Johnston, Jr., S. Mahajan, and D.D. Roccasecca, *J. Appl. Phys. 51*;978, 1980.

275. C.L. Zipfel, *Semiconductors and Semimetals, 22*;Part C, p. 239, 1985.

276. O. Ueda, *J. Electrochem. Soc. 135*;11c, 1988.

277. J. Matsui, in *Defects in Semiconductors II*, (eds. S. Mahajan and J. W. Corbett), *Mat. Res. Soc. Symp. Proc. 14*;477, 1983.

278. S.N.G. Chu, S. Nakahara, M.E. Twigg, L.A. Koszi, E.F. Flynn, A.K. Chin, B.P. Segner, and W.D. Johnston, Jr., *J. Appl. Phys. 63*;611, 1988.

279. D.V. Lang, R.L. Hartman, and N.E. Schumaker, *J. Appl. Phys. 47*;4986, 1976.

280. T. Uji, T. Suzuki, and T. Kamejima, *Appl. Phys. Lett. 36*;655, 1980.

281. S.D. Hersee and D.J. Stirland, *Inst. Phys. Conf. Ser. 33a*;370, 1977.

282. A.K. Chin, C.L. Zipfel, and B.V. Dutt, *Jpn. J. Appl. Phys. 21*;1308, 1982.

283. T. Yuasa, M. Ogawa, K. Endo, and H. Yonezu, *Appl. Phys. Lett. 32*;119, 1978.

284. F.R. Nash, R.L. Hartman, N.M. Denkin, and R.W. Dixon, *J. Appl. Phys. 50*;3122, 1979.

285. F.R. Nash and R.L. Hartman, *J. Appl. Phys. 50*;3133, 1979.

286. F.R. Nash and R.L. Hartman, *IEEE J. Quant. Electron. QE-16*;1022, 1980.

287. Y. Shima, N. Chinone, and R. Ito, *Appl. Phys. Lett. 31*;625, 1977.

288. I. Ladany, M. Ettenberg, H.F. Lockwood, and H. Kressel, *Appl. Phys. Lett. 30*;87, 1977.

289. T. Furuse, T. Suzuki, S. Matsumoto, K. Nishida, and Y. Nannichi, *Appl. Phys. Lett. 33*;317, 1978.

290. T. Yuasa, K. Endo, T. Torikai, and H. Yonezu, *Appl. Phys. Lett. 34*;685, 1979.

291. T. Suzuki and M. Ogawa, *Appl. Phys. Lett. 31*;473, 1977.

292. H. Imai, M. Morimota, H. Sudo, T. Fujimara, and M. Takusagawa, *Appl. Phys. Lett. 33*;1011, 1978.

293. M. Ettenberg and H. Kressel, *IEEE J. Quant. Electron. QE-16*;186, 1980.

294. H. Namizaki, S. Takamiya, M. Ishii, and W. Susaki, *J. Appl. Phys. 50*;3743, 1979.

295. A. Thompson, *IEEE J. Quant. Electron. QE-15*;11, 1979.

296. J.A.F. Peek, *IEEE J. Quant. Electron. QE-17*;781, 1981.

297. R.L. Hartman and F.R. Nash (unpublished).

298. T. Hayakawa, S. Yamamoto, T. Sakurai, and T. Hijikata, *J. Appl. Phys. 52*;6068, 1981.

299. H.D. Wolf, K. Mettler, and K.H. Zschauer, *Jap. J. Appl. Phys. 20*;L693, 1981.

300. M. Ettenberg, H.S. Sommers, H. Kressel, and H.F. Lockwood, *Appl. Phys. Lett. 18*;571, 1971.

301. F.R. Nash, R.L. Hartman, T.L. Paoli, and R.W. Dixon, *Appl. Phys. Lett.* *35*;905, 1979.

302. F.R. Nash, T.L. Paoli, and R.L. Hartman, *J. Appl. Phys.* *52*;48, 1981.

303. O. Ueda, S. Yamakoshi, T. Sanada, I. Umebu, T. Kotani, and O. Hasegawa, *J. Appl. Phys.* *53*;9170, 1982.

304. H. Yonezu, I. Sakuma, T. Kamejima, M. Jeno, K. Iwamoto, I. Hino, and I. Hayashi, *Appl. Phys. Lett.* *34*;637, 1979.

305. S. Yamakoshi, M. Abe, O. Wada, S. Komiya, and T. Sakurai, *IEEE J. Quant. Electron.* *QE-17*;167, 1981.

306. T. Yamamoto, K. Sukai, S. Akiba, and Y. Suematsu, *IEEE J. Quant. Electron.* *QE-14*;95, 1978.

307. T. Yamamoto, K. Sakai, and S. Akiba, *IEEE J. Quant. Electron.* *QE-15*;684, 1979.

308. O. Ueda, S. Yamakoshi, and T. Kotani, *J. Appl. Phys.* *53*;2991, 1982.

309. O. Ueda, S. Yamakoshi, and T. Yainaoka, *Jap. J. Appl. Phys.* *19*;L251, 1980.

310. O. Ueda, S. Komiya, S. Yamakoshi, and T. Kotani, *Jap. J. Appl. Phys.* *20*;1201, 1981.

311. M. Fukuda, K. Wakita, and G. Iwane, *Jap. J. Appl. Phys.* *20*;L87, 1981.

312. K. Wakita, H. Takaoka, M. Seki, and M. Fukuda, *Appl. Phys. Lett.* *40*;525, 1982.

313. M. Seki, M. Fukuda, and K. Wakita, *Appl. Phys. Lett.* *40*;115, 1982.

314. A.K. Chin, C.L. Zipfel, S. Mahajan, F. Ermanis, and M.A. DiGiuseppe, *Appl. Phys. Lett.* *41*;555, 1982.

315. S. Mahajan, A.K. Chin, C.L. Zipfel, D. Brasen, B.H. Chin, R.T. Tung, and S. Nakahara, *Mater. Lett.* *2*;184, 1984.

316. I. Camlibel, A.K. Chin, F. Ermanis, M.A. DiGiuseppe, J.A. Lourenco, and W.A. Bonner, *J. Electrochem. Soc.* *129*;2585, 1982.

317. A.K. Chin, C.L. Zipfel, M. Geva, I. Camlibel, P. Skeath, and B.H. Chin, *Appl. Phys. Lett.* *45*;37, 1984.

318. G.H. Olsen and M. Ettenberg, *J. Appl. Phys.* *48*;2543, 1977.

319. B. Wakefield, *J. Appl. Phys.* *50*;7914, 1979.

320. M.J. Robertson, B. Wakefield, and P. Hutchinson, *J. Appl. Phys.* *52*;4462, 1981.

321. V. Swaminathan, J. Lopata, and J.W. Lee, *Mat. Res. Soc. Symp. Proc.* *77*;779, 1987.

322. R.L. Hartman and A.R. Hartman, *Appl. Phys. Lett.* *23*;147, 1973.

323. V. Swaminathan, W.R. Wagner, P.J. Anthony, G. Henein, and L.A. Koszi, *J. Appl. Phys.* *54*;3763, 1983.

324. C.R. Elliot, J.C. Regnault, and B. Wakefield, *Inst. Phys. Conf. Ser.* *65*;553, 1983.

325. G.A. Rozgonyi, P.M. Petroff, and M.B. Panish, *J. Cryst. Growth* *27*;106, 1974.

326. G.A. Rozgonyi and M.B. Panish, *Appl. Phys. Lett. 23*;533, 1973.

327. G.A. Rozgonyi, P.M. Petroff, and M.B. Panish, *Appl. Phys. Lett. 24*;251, 1974.

328. M.A. Afromowitz and D.L. Rode, *J. Appl. Phys. 45*;4738, 1974.

329. V. Swaminathan, P. Parayanthal, and R.L. Hartman, *Appl. Phys. Lett. 52*;1461, 1988.

330. J.M. Liu and Y.C. Chen, Materials Research Society Meeting, Boston, 1983; see also Y.C. Chen and J.M. Liu, *Appl. Phys. Lett. 45*;731, 1984.

331. M. Ikeda, O. Ueda, S. Komiya, and I. Umebu, *J. Appl. Phys. 58*;2448, 1985.

332. A.R. Goodwin, P.A. Kirby, I.G.A. Davies, and R.S. Baulcomb, *Appl. Phys. Lett. 34*;647, 1979.

333. A.K. Chin, M.A. DiGiuseppe, and W.A. Bonner, *Mat. Lett. 1*;19, 1982.

334. G.E. Henein and W.R. Wagner, *J. Appl. Phys. 54*;6395, 1983.

335. S. Timoshenko, *J. Opt. Soc. Am., 11*;23, 1925.

336. V. Swaminathan, unpublished.

337. C.A. Green, unpublished.

338. H. Booyens and J.H. Basson, *J. Appl. Phys. 51*;4368, 1980.

339. H. Koyama, T. Nishioka, K. Isshiki, H. Namizaki, and S. Kawazu, *Appl. Phys. Lett. 43*;733, 1983.

340. A.K. Chin, W.C. King, T.J. Leonard, R.J. Roedel, C.L. Zipfel, V.G. Keramidas, and F. Ermanis, *J. Electrochem. Soc. 128*;661, 1981.

341. D. Laister and G.M. Jenkins, *Solid State Electron. 13*;1200, 1970.

342. Y. Shinoda and T. Kwaakami, *Jap. J. Appl. Phys. 16*;1271, 1977.

343. Y. Matsushima and K. Sakai, in ''GaInAsP alloy semiconductors,'' (ed. T.P. Pearsall) New York;Wiley, p. 413, 1982.

344. R.H. Saul, F.S. Chen, and P.W. Schumate, Jr., *AT&T Technical Journal 64*;861, 1985.

345. N. Susa, Y. Yamauchi, and H. Kanbe, *IEEE J. Quant. Electron. QE-16*;542, 1980.

346. K. Taguchi, Y. Matsumoto, and K. Nishida, *Electron. Lett. 15*;453, 1979.

347. V. Diadiuk, C.A. Armiento, S.H. Groves, and C.E. Hurwitz, *IEEE Electron Dev. Lett. EDL-1*;177, 1980.

348. Y. Tashiro, K. Taguchi, Y. Sugimoto, T. Torikai, and K. Nishida, *J. Lightwave Tech. LT-1*;269, 1983.

349. G.H. Olsen, N.J. DiGiuseppe, P.P. Webb, T.J. Zamerowski, J.R. Appert, and M.G. Harvey, Device Research Conf. Burlington, VT, 1983.

350. P.P. Webb and G.H. Olsen, *IEEE Trans. Elect. Dev. ED-30*;395, 1983.

351. A.K. Chin, F.S. Chen, and F. Ermanis, *J. Appl. Phys. 55*;1596, 1984.

352. L.S. Lichtmann, P.A. Kohl, and R.H. Burton, in *Tech. Digest IEEE Specialist* Conf. Light Emitting Diodes and Photodetectors, Ottawa, Canada p. 150, 1982.

353. F. Capasso, P.M. Petroff, W.A. Bonner, and S. Sumski, *IEEE Electron. Dev. Lett. 1*;27, 1980.

354. T.P. Lee and C.A. Burrus, *Appl. Phys. Lett. 36*;587, 1980.

355. T.P. Lee, C.A. Burrus, and A.G. Dentai, I.E.D.M. digest, Washington, D. C. 1978.

356. N. Susa, Y. Yamauchi, and H. Ando, *J. Appl. Phys. 53*;7044, 1982.

357. Y. Matsushima, Y. Noda, and Y. Kushiro, *IEEE J. Quant. Electron. QE-21*;1257, 1985.

358. H. Sudo, M. Suzuki, and N. Miyahara, *IEEE Electron Dev. Lett.* EDL-8;386, 1987.

359. R. Dingle, M.D. Feuer, C.W. Tu, ''Selectively doped heterostructure transistors: materials, devices, and circuits,'' Chap. 6 in *VLSI Electronics, Microstructure Science*, vol. 11, (ed. Einspruch and Wisseman), New York;Academic Press, 1985

360. C.W. Tu, R.H. Hendel, and R. Dingle, ''Molecular Beam Epitaxy and the Technology of Selectively Doped Heterostructure Transistors,'' Chap. 4 in *Gallium Arsenide Technology*, (ed. D.K. Ferry), Indianapolis;Macmillan, Inc., 1985.

361. R. Dingle, W. Wiegmann, and C.H. Henry, *Phys. Rev. Lett. 33*;827, 1974; R. Dingle and W. Wiegmann, *IEEE J. of Quantum Electron. 10*;79, 1974.

362. H.L. Stormer, R. Dingle, A.C. Gossard, W. Wiegmann, and M.D. Sturge, *Sol. State Commun. 29*;705, 1979.

363. D. Delagebeaudeuf, M. Laviron, P. Delecluse, P.N. Tung, J. Chaplart, and N.T. Linh, *Electronics Lett. 18*;103, 1982.

364. M.D. Feuer, R.H. Hendel, R.A. Kiehl, J.C.M. Hwang, V.G. Keramidas, C.L. Allyn, and R. Dingle, *IEEE Electron Device Lett.* EDL-4;306, 1983.

365. T.J. Drummond, R. Fischer, S.L. Su, W.G. Lyons, H. Morkoc, K. Lee, and M.S. Shur, *Appl. Phys. Lett. 42*;262, 1983.

366. R. Dingle, W. Wiegmann, and C.H. Henry, *Phys. Rev. Lett. 33*;827, 1974.

367. D.V. Lang, R.A. Logan, and M. Jaros, *Phys. Rev. B* 19;1015, 1979.

368. A.J. Valois, G.Y. Robinson, K. Lee, and M.S. Shur, *J. Vac. Science Technol. B1*;190, 1983.

369. T.J. Drummond, R.J. Fischer, W.F. Kop, H. Morkoc, K. Lee, and M.S. Shur, *IEEE Trans. on Elec. Devices ED-30*;1806, 1983.

370. T. Baba, T. Mizutani, and M. Ogawa, *Jap. J. of Applied Phys.* 22;L627, 1983.

371. C.W. Tu, W.L. Jones, R.F. Kopf, L.D. Urbanek, and S.S. Pei, *IEEE Electron Device Lett.* EDL-7;552, 1986.

372. J.C. Nabity, M. Stavola, J. Lopata, W.C. Dautremont-Smith, C.W. Tu, and S.J. Pearton, *Appl. Phys. Lett. 50*;921, 1987.

373. P. O'Connor, T.P. Pearsall, K.Y. Cheng, A.Y. Cho, J.C.M. Hwang, and K. Alavi, *IEEE Electron Device Lett. EDL-3*;64, 1982.

374. A. Kastalsky, R. Dingle, K.Y. Cheng, and A.Y. Cho, *Appl. Phys. Lett. 41*;274, 1982.

375. M. Razeghi and J.P. Duchemin, *J. Vac. Science Technology B1*;262, 1983.

376. J. Komeno, M. Takikawa, and M. Ozeki, *Electronic Lett. 19*;473, 1983.

377. H.M. Cox, *J. Cryst. Growth 69*;641, 1984.

378. T.P. Pearsall, R.H. Hendel, P. O'Connor, K. Alavi, and A.Y. Cho, *IEEE Electron Device Lett. EDL-4*;5, 1983.

379. A.T. Macrander, J. Hsieh, J. Ren, and J. Patel, *J. Cryst. Growth, 92*;83, 1983.

380. D.V. Lang, "DX Centers in III-V Alloys," Chap. 7 in *Deep Centers in Semiconductors*, (ed. S.T. Pantelides), New York;Gordon and Breach, 1986.

381. D.V. Lang and R.A. Logan, in *Physics of Semiconductors 1978* Inst. Phys. Conf. Ser. No. 43, p. 433, 1979.

382. D.V. Lang, R.A. Logan, and M. Jaros, *Phys. Rev. B19*,;1015, 1979.

383. R.A. Craven, *J. Appl. Phys. 50*;6334, 1979.

384. I.D. Henning and H. Thomas, *Solid-State Electron 25*;325, 1982.

385. T.N. Theis, P.M. Mooney, and S.L. Wright, *Bulletin of the American Physical Society 32*;554, 1987.

386. V. Narayanamurti, R.A. Logan, and M.A. Chin, *Phys. Rev. Lett. 43*;1536, 1979.

387. G.A. Baraff and M. Schluter, private communication.

388. J.V. DiLorenzo and D.D. Khandelwal (eds), *GaAs FET Principles and Technology*, Dedham, MA, Artech House, 1982.

389. M.J. Howes and D.V. Morgan (eds), *Gallium Arsenide Materials, Devices, and Circuits*, New York;Wiley, Ch. 10, 1985.

390. Y. Nanishi, S. Ishida, T. Honda, H. Yamazaki, and S. Miyazawa, *Jap. J. Appl. Phys. 21*;L335, 1982.

391. A.S. Jordan, R. Caruso, and A.R. Von Neida, *Bell Syst. Tech. J. 59*;593, 1980.

392. A.S. Jordan, A.R. Von Neida, and R. Caruso, *J. Cryst. Growth 76*;243, 1986.

393. Y. Nanishi, S. Ishida, and S. Miyazawa, *Jap. J. Appl. Phys. 22*;L54, 1983.

394. A.K. Chin, A.R. Von Neida, and R. Caruso, *J. Electrochem. Soc. 129*;2386, 1982.

395. W. Heinke and H.J. Queisser, *Phys. Rev. Lett. 33*;1082, 1974.

396. T. Honda, Y. Ishii, S. Miyazawa, H. Yamazaki, and Y. Nanishi, *Jap. J. Appl. Phys. 22*;L270, 1983.

397. S. Miyazawa, Y. Ishii, S. Ishida, and Y. Nanishi, *2Appl. Phys. Lett. 43*;853, 1983.

398. S. Miyazawa, T. Honda, Y. Ishii, and S. Ishida, *Appl. Phys. Lett. 44*;410, 1984.

399. H.V. Winston, A.T. Hunter, H.M. Olsen, R.P. Bryan, and R.E. Lee, in *Semi-Insulating III-V Materials, Kah-nee-ta*, 1984, (eds. D.C. Look and J.S. Blakemore)1 Nantwich, England;Shiva, p. 402, 1984;*Appl. Phys. Lett. 45*;447, 1984.

400. S. Miyazawa and F. Hyuga, *IEEE Transactions on Electron Devices, ED-33*;227, 1986.

401. Y. Ishii, S. Miyazawa, and S. Ishida, *IEEE Trans. on Electron Devices ED-31*;800, 1984.

402. R. Fischer, "An Investigation of Gallium Arsenide Materials and Devices Grown on Silicon Substrates," Ph.D. Thesis, University of Illinois, Urbana, Illinois, 1986.

403. R. Fischer, private communication.

404. B. Tuck, in *Semi-Insulating III-V Materials, Kah-nee-ta*, 1984, (eds. D.C. Look and J.S. Blakemore) Shiva, Nantwich, p. 2, 1984.

405. G.M. Martin and S. Makram-Ebeid, "The Mid-Gap Donor Level EL2 in GaAs," Ch. 6 in *Deep Centers in Semiconductors*, (ed. S.T. Pantelides) New York;Gordon and Breach, 1986.

406. G.M. Martin, A. Mitonneau, and A. Mircea, *Electron. Lett. 13*;191, 1977.

407. K.R. Elliott, R.T. Chen, S.G. Greenbaum, and R.J. Wagner, in *Semi-Insulating III-V Materials, Kah-nee-ta*, 1984, (eds. D.C. Look and J.S. Blakemore) Shiva, Nantwich, p. 239, 1984.

408. A. Mitonneau, A. Mircea, G.M. Martin, and D. Pons, *Rev. Phys. Appl. 14*;853, 1979.

409. G.M. Martin, *Appl. Phys. Lett. 39*;747, 1981.

410. A. Mircea, A. Mitonneau, L. Hollan, A. Briére, *Appl. Phys. 11*;153, 1976.

411. C. Hilsum and A.C. Rose-Innes, *Semiconducting III-V Compounds*, New York;Pergamon Press, 1961.

412. D.E. Holmes, R.T. Chen, K.R. Elliot, C.G. Kirkpatrick, P.W. Yu, *IEEE Trans. on Elect. Dev. ED-29*;1045, 1982.

413. D.V. Lang and R. A. Logan, *J. Electron. Mat. 4*;1053, 1975.

414. G.M. Martin, A. Mitonneau, A. Mircea, *Electron. Lett. 13*;191, 1977.

415. D.V. Lang, A.Y. Cho, A.C. Gossard, M. Ilegems, W. Wiegmann, *J. Appl. Phys. 47*;2558, 1976.

416. M. Ozeki, J. Komeno, A. Shibatomi, and S. Ohkawa, *J. Appl. Phys. 50*;4808, 1979.

417. J.R. Wagner, J.J. Krebs, G.H. Strauss, A.M. White, *Solid State Commun. 36*;15, 1980.

418. D.E. Holmes and R.T. Chen, *J. Appl. Phys. 55*;3588, 1984.

419. E.R. Weber, H. Ennen, U. Kaufmann, J. Windschief, and J. Schneider, *J. Appl. Phys. 53*;6140, 1982.

420. T. Ishida, K. Maeda, and S. Takeuki, *Appl. Phys. 21*;257, 1980.

421. M.R. Brozel, I. Grant, R.M. Ware, and D.J. Stirland, *Appl. Phys. Lett. 42*;610, 1983.

422. S. Miyazawa, K. Watanabe, J. Osaka, and K. Ikuta, *Revue Phys. Appl. 23*;727, 1988.

423. F. Hyuga, *Jpn. J. Appl. Phys. 24*;L160, 1985.

# INDEX

A

**B**



**C**

**E**

**K**



**L**

M



N

O



P

### Q



### R

### W



### X



### Y



### Z